KYNURENINE AND SEROTONIN PATHWAYS

Progress in Tryptophan Research

ADVANCES IN EXPERIMENTAL MEDICINE AND BIOLOGY

A Continuation Order Plan is available for this series. A continuation order will bring delivery of each new volume immediately upon publication. Volumes are billed only upon actual shipment. For further information please contact the publisher.

KYNURENINE AND SEROTONIN PATHWAYS
Progress in Tryptophan Research

Edited by

Robert Schwarcz

Maryland Psychiatric Research Center
University of Maryland School of Medicine
Baltimore, Maryland

Simon N. Young

McGill University
Montreal, Quebec, Canada

and

Raymond R. Brown

University of Wisconsin
Madison, Wisconsin

PLENUM PRESS • **NEW YORK AND LONDON**

Library of Congress Cataloging-in-Publication Data

International Study Group for Tryptophan Research. Meeting (6th :
  1989 : Baltimore, Md.)
    Kynurenine and serotonin pathways : progress in tryptophan
  research / edited by Robert Schwarcz, Simon N. Young, and Raymond R.
  Brown.
        p.   cm. -- (Advances in experimental medicine and biology ; v.
  294)
    "Proceedings of the International Study Group for Tryptophan
  Research, Sixth International Meeting, held May 9-12, 1989, in
  Baltimore, Maryland"--T.p. verso.
    Includes bibliographical references and index.
    ISBN 0-306-43929-8
    1. Tryptophan--Metabolism--Congresses. 2. Tryptophan-
  -Pathophysiology--Congresses. 3. Kynurenine--Metabolism-
  -Congresses. 4. Serotonin--Metabolism--Congresses.   I. Schwarcz,
  Robert. II. Young, Simon N. III. Brown, Raymond R. IV. Title.
  V. Series.
    [DNLM: 1. Kynurenine--congresses. 2. Serotonin--congresses.   W1
  AD559 v. 294 / QU 60 I566k 1989]
  QP562.T7I58 1989
  599'.019245--dc20
  DNLM/DLC
  for Library of Congress   91-21251
                                                                 CIP

Proceedings of the International Study Group for
Tryptophan Research, Sixth International Meeting,
held May 9-12, 1989, in Baltimore, Maryland

ISBN 978-1-4684-5954-8   ISBN 978-1-4684-5952-4 (eBook)
DOI 10.1007/978-1-4684-5952-4

© 1991 Plenum Press, New York
Softcover reprint of the hardcover 1st edition 1991

A Division of Plenum Publishing Corporation
233 Spring Street, New York, N.Y. 10013

SIR FREDERICK GOWLAND HOPKINS
1861 - 1947

(Courtesy: National Library of Medicine)

PREFACE

   The 6th triennial meeting of the International Study Group for Tryptophan Research (ISTRY) was held May 9-12, 1989 in Baltimore, Maryland (USA).
From the wide variety of topics and disciplines represented, as documented in
this volume, it is clear that tryptophan research and ISTRY are alive and
well.

   ISTRY traces its origins to at a tryptophan symposium organized in 1971
by H. Schievelbein at Hohenried near Munich (Germany). Up to that time there
had been occasional international tryptophan conferences at irregular intervals. A number of participants at the Hohenried meeting felt that an international tryptophan organization should be formed to organize regular meetings and to foster collaboration and information exchange on tryptophan-re-
lated topics. Thanks mainly to the founding work of H. Schievelbein and W.
Kochen, an executive committee was elected and ISTRY was born. The inaugural
meeting in 1974 was held in Padova (Italy) to honor L. Musajo, one of the
foremost pioneers in tryptophan studies. This first ISTRY meeting was successfully organized by L. Musajo, G. Allegri, A. De Antoni, and C. Costa, and
was critical in assuring the viability of the new organization. Subsequent
meetings were held in 1977 in Madison, Wisconsin (USA), organized by R.R.
Brown, D.P. Rose, and W.E. Knox, honoring C.P. Berg; 1980 in Kyoto (Japan),
organized by O. Hayaishi, R. Kido, Y. Ishimura, T. Deguchi, T. Hino, T.
Kawaichi, T. Segawa, Y. Shibata, and K. Soda, honoring Y. Kotake; 1983 in
Martinsried (Germany), organized by W. Kochen, H. Schlossberger, B. Linzen,
and H. Steinhart, honoring A. Butenandt; and 1986 in Cardiff, Wales (United
Kingdom), organized by A.A.-B. Badawy, D.A. Bender, K.J. Collard, G. Curzon,
M.H. Joseph, and C.I. Pogson, honoring the discoverer of tryptophan, F.
Gowland Hopkins. The meeting in Baltimore was organized by R. Schwarcz, C.
Speciale, and E. Okuno, honoring B. Witkop.

   During the formative years of ISTRY, W. Kochen served as president from
1971 to 1980, R.R. Brown from 1980 to 1983, G. Allegri from 1983 to 1986, and
R. Kido from 1987 to 1990. At the 1983 meeting in Martinsried, a Constitu-
tion and By-laws were adopted by the membership, defining the organizational
structure and procedures of conducting the business of ISTRY.

   It is of interest to note the changes in the emphasis of tryptophan
studies over many years. Early meetings dealt mainly with the delineation of
metabolic pathways in various animals and plants, the development of analytical methods, the identification and synthesis of metabolites, and the examination of biological activity. Many studies dealt with tryptophan-niacin
relationships. With the advent of improved analytical methods and the discovery of 5-hydroxytryptophan and related neurochemicals, interests have expanded to include neurochemical, psychiatric and cardiovascular studies.
More recently, immunobiology and neuroimmunobiology has been added with the
finding that interferon-related immune stimulation induces marked increases
in tryptophan metabolism. This suggests an increasing importance of tryp-

tophan studies in a wide variety of human diseases ranging from autoimmune conditions to AIDS, from organ transplantation to tissue inflammatory processes. With the reports in late 1989 of an eosinophilia myalgia syndrome (EMS) associated with tryptophan dietary supplementation, there will be still newer problems for future study.

We wish to thank all of the participants of this meeting for lively, open discussions, and our speakers and poster presenters for submission of manuscripts. We also thank the University of Maryland for providing excellent conference facilities, and the local organizing committee (particularly C. Speciale and E. Okuno) for all the special efforts which are necessary for a successful meeting. We also wish to express our sincere gratitude to our editorial assistant, Mrs. Joyce Burgess. Her determination, endurance and quality-consciousness were essential for the publication of the Proceedings volume.

Several pharmaceutical companies made financial contributions, which were indispensable for the organization of the Baltimore meeting. The support of American Cyanamid Company (USA), ASTRA Research Center (Sweden), Burroughs Wellcome Company (USA), Degussa AG (Germany), Fresenius AG (Germany), Glaxo Inc. (USA), ICI Pharmaceuticals Group (USA), Merrell Dow Pharmaceutical Inc. (USA), Nova Pharmaceutical Corporation (USA), NutraSweet Company (USA), Polifarma S.p.a. (Italy) and Servier (France) is therefore gratefully acknowledged.

The next meeting of ISTRY will be organized by I. Ishiguro and will be held near Nagoya (Japan) in 1992. At the early meeting in Hohenried, H. Schievelbein presented each participant with a large heart-shaped cake decorated with the words "Tryptophan, du bist meine Freude" (tryptophan, you are my joy). W. Kochen later added "but also my suffering". On behalf of the Executive Committee of ISTRY, we hope that all members and friends of ISTRY will plan to "enjoy and suffer" with us in Nagoya in 1992.

Baltimore, Montreal and Madison   R. Schwarcz
Spring, 1991   S.N. Young
                                                       R.R. Brown

CONTENTS

SESSION III.   TRYPTOPHAN METABOLISM: KYNURENINES

SESSION IV.   BIOLOGICAL EFFECTS: SEROTONIN AND
               MELATONIN

SESSION I

TRYPTOPHAN AND ITS METABOLITES: CHEMICAL AND ANALYTICAL ISSUES

RETRO-, INTRO- AND PERSPECTIVES OF TRYPTO-FUN

B. Witkop

Institute Scholar
National Institutes of Health
Bethesda, Maryland 20892
USA

FREDERICK GOWLAND HOPKINS (1861-1947): GODFATHER OF THE ISTRY SYMPOSIA

"I have a feeling, Dale, that if we can find the meaning of a color reaction like that [the reaction of Adamkiewicz], it may open up a way to new
knowledge of the structure of the old protein molecule itself". Shortly after Hopkins had said these words to his student Henry Dale (1948), he identified glyoxylic acid as the contaminant needed to give the blue color with
commercial acetic acid and strong sulfuric acid, known now as the Hopkins-
Cole-Adamkiewicz reaction, with most proteins. In 1901, he isolated the
chromogenic protein constituent and named it (-)-tryptophan (Hopkins and
Cole, 1901), after R. Neumeister who in 1890 observed it in impure form on
digestion of protein with trypsin.

Hopkins, a naturalist, as early as 1889, just 100 years ago, took an interest in the yellow pigment of butterflies (Hopkins, 1889), a subject that
was continued by my teacher Heinrich Wieland (1925) and laid the groundstone
for folic acid, a cofactor for tryptophan 5-hydroxylase.

When Hopkins was knighted in 1925 for his contributions to nutrition and
"vitamins", his department journal, "Brighter Biochemistry", devised a coat-
of-arms for him with the inscription "Gluta thy own!" In 1929, he shared the
Nobel Prize with Christian Eijkman who as early as 1890 had discovered the
connection between beri-beri and decorticated rice.

OXYTRYPTOPHAN REVISITED

Exactly 50 years ago in the Laboratory of Heinrich Wieland I isolated 1-
(+)-α-tryptophan or oxindolyl-L-alanine by hydrolysis of the poisonous cy-
clopeptide phalloidin from the death-cap Amanita Phalloides (Wieland and Witkop, 1940). Paul Walden in his highly readable History of Chemistry (1941)
lists oxytryptophan as a new amino acid. But Theodor Wieland showed that in
phalloidin as well as amanitin, tryptophan is modified by an S-cysteine substituent in the α-indole position (Witkop, 1983), the so-called tryptathio-
nine. Although there are oxidole alkaloids, such as gelsemine (Witkop,
1948), metabolic conversion of an indole to an oxidole was first observed by
Axelrod, Freter and Witkop in the inactivation of lysergic acid diethylamide,
the hallucinogen LSD, by microsomes.

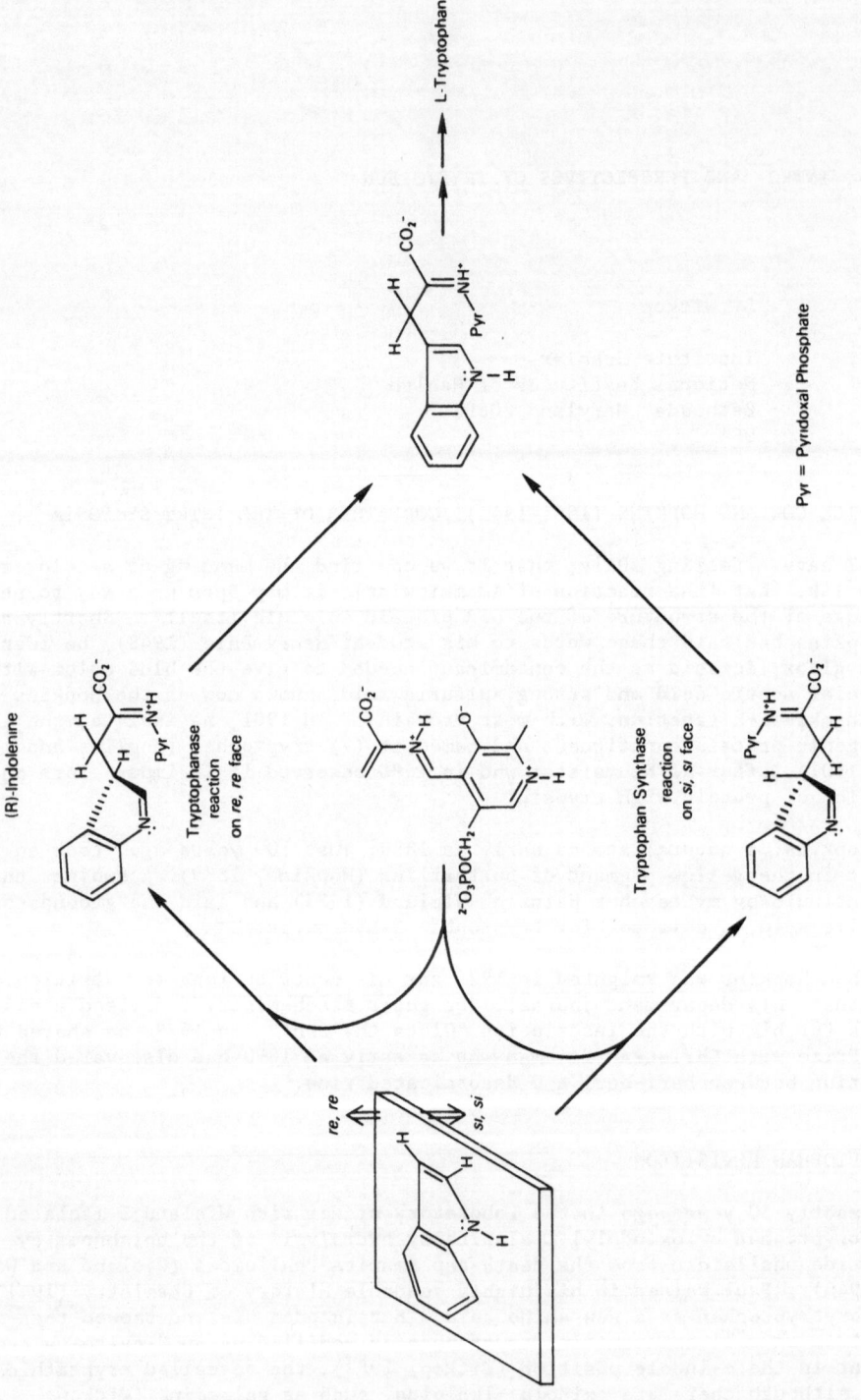

Fig. 1. Dihydrotryptophan and its progeny.

4

Very recently the non-physiological oxidolyl-L-alanine, because of its tetrahedtral geometry at C-3, became a substrate analog and inhibitor in the tryptophan synthase (EC 4.1.2.20) system which catalyzes the replacement of the β-hydroxyl group of L-serine with indole to yield L-tryptophan and water (Roy et al., 1988).

Both photoreduction (Yonemitsu et al., 1966) as well as reduction in the ground state (Daly et al., 1967) provided a diastereoisomeric mixture of 2,3-dihydro-L-tryptophans which Phillips, Miles and Cohen separated by high performance liquid chromatography (1985). The absolute stereochemistry of one of these diastereoisomers was established by conversion of the dibenzoyl derivative to a lysergic acid derivative of known absolute stereochemistry (Rebek et al., 1984).

Linus Pauling in 1946 reflected on the importance of analogs for transition states or of analogs as reaction intermediates. The two newly separated diastereoisomers of dihydrotryptophan make a valuable contribution to this concept. Diastereosiomer A, or 2,3-dihydro-αS,3R-tryptophan, is a potent inhibitor of tryptophanase. Diastereoisomer B, or 2,3-dihydro-αS,3S-tryptophan, inhibits tryptophan synthase in kinetic assays and is a good ligand in spectrophotometric binding studies. Phillips, Miles and Cohen suggest the preceding Scheme (Rebek et al., 1984) for the stereochemical course of the reaction of indole in the active sites of tryptophan synthase and tryptophanase.

5-HYDROXYTRYPTOPHAN AND THE NIH SHIFT

5-Hydroxytryptophan was synthesized in 1953 (Ek and Witkop) and, unlike its earlier isomer, α-hydroxytryptophan, became a pillar in physiology and neurochemistry. It was not until 1966 that, in collaboration with S. Udenfriend, the discovery of the "NIH-Shift" was made: when the aromatic hydroxyl group was enzymatically introduced into 5-^{3}H-L-tryptophan, the tritium atom was not replaced but migrated from position 5 to position 4 (Elk and Witkop, 1953. The presumable arene oxide intermediates open a new chapter in labile metabolites, nature or drug-derived, and directed attention to their importance for the etiology of metabolic diseases or long-term drug or cancer therapy.

GRAMICIDIN A: CHANNELS AND PORES

My unforgotten collaborator, Erhard Gross (1928-1981), spent many years on the tedious purification and separation of the tryptophan-rich gramicidins A, B and C (Gross and Witkop, 1965), some of the first N-formylated natural peptides that had perplexed other experts such as Richard L. M. Synge (NP 1952).

The molecule is a pentadecapeptide with the following sequence:

Formyl-L-Val-Gly-L-Ala-D-Leu-L-Ala-D-Val-L-Val-D-Val-L-Trp-D-Leu-L-Trp-D-Leu-L-Trp-D-Leu-L-Trp-ethanolamine

which was proven by synthesis (Sarges and Witkop, 1965). What is interesting for our symposium is the heavy involvement of 4 tryptophan residues in the architecture of channels and pores. Gramicidin A recently became the cover story of "Science" because B. A. ("Bonnie") Wallace accomplished an X-ray analysis of the crystalline complex with cesium. In her own words:

"Alternating residues in the sequence have opposite chirality. The active form of the molecule is a dimer. This relatively small polypeptide is capable of adopting several different conformations, depending on its environment. Its two major conformers have been designated the "channel" and the "pore" structures. The channel corresponds to the predominant conducting form in membranes and appears to be an amino terminal-to-amino terminal helical dimer, similar to the type of structure originally proposed by Urry et al. The orientation and general folding motif of the channel have been established by nuclear magnetic resonance, fluorescence, and conductance measurements, among others. The pore apparently corresponds to the minor conducting form detected in black lipid membrane preparations, which has a very long mean channel lifetime. The pore is also the predominant form found in organic solutions, and is one of the family of antiparallel intertwined double helices first proposed for the gramicidin structure by Veatch et al. The channel and pore forms are readily distinguished by their distinctly different circular dichroism spectra and by their differential responses to the binding of ions" (Wallace and Ravikumar, 1988).

The basic question is still: How are charged ions transported across biological membranes? The structural polymorphism in transmembrane channels was summed up as follows:

"In summary, gramicidin A appears to be a structure whose polymorphism may underlie a dynamic mechanism of ion transfer. Although much more remains to be learned about gramicidin function, particularly regarding details of mechanism and the origins of ion transport specificity, the crystallographic studies finally provide a structural basis for definitive spectroscopic studies and computation simulations" (Salemme, 1988).

NEW CHIRAL ASPECTS OF ALKALOIDS

The following chapter by Masako Nakagawa's topic will be entitled "Tryptophan and Related Alkaloids". It would be presumptuous of me to steal any of her new and impressive thunder. It is safe to say that her contributions to sophisticated indole chemistry are second to none (Hino and Nakagawa, 1988).

Allow me here to reminisce on some of the early chemistry of indole alkaloids, such as yohimbine (Witkop, 1943), an adrenergic blocking agent, sempervirine (Woodward and Witkop, 1949) and other anhydronium bases (Witkop, 1953). Since 1894 we have known that harmaline produces high-frequency tremor in mammals (Neuner and Tappeiner, 1894), but only recently have we learned that - in neurophysiological terms - harmaline "acts on the inferior olive and that burst activation of Purkinje cells was generated by large all-or-none depolarizations similar to climbing fiber excitatory postsynaptic potentials" (Llinas and Volkind, 1973).

M. Ozaki helped to develop tetrahydroharmanes as inhibitors of monamine oxidase (Ozaki et al., 1960), but only recently has the tricyclic condensation product of L-tryptophan ethyl ester with formaldehyde attracted attention as a - presumably exogenous - inhibitor of benzodiazepine receptors (Braestrup et al., 1980).

Natural (-)-physostigmine has been the classical inhibitor of acetylcholine esterase; according to Arnold Brossi, its unnatural isomer (+)-physostigmine (Yu and Brossi, 1988) is less toxic, still binds to the enzyme and is an ideal prophylactic and antidote in case of organo-phosphorous poisoning (Albuquerque et al., 1988; Kawabuchi et al., 1988). Arnold Brossi has developed an extensive program on the effect of chirality in general on pharmacological action (Brossi et al., 1988).

6

FROM VOMIPYRIN TO PQQ

Pyrroloquinoline in the laboratory of Heinrich Wieland was a well-known tricyclic degradation product, i.e., vomipyrin, derived from vomicine, an indole alkaloid from strychnos nux vomica (Wieland and Horner, 1937).

Originally Wieland and Horner formulated vomipyrin as a 5,6-pyrroquin-oline (Wieland and Horner, 1988) instead of the correct 7,8-pyrroquinoline synthesized by Robinson (Robinson and Stephen, 1948). This 5,6-pyrroquin-oline was synthesized by Wieland and Horner and is the basis for pyrrolo-quinoline quinone (PQQ). Johannis A. Duine has discovered a novel redox co-factor of several important diverse enzymes (Duine and Jongejan, 1989). Its structure is derived from the 5,6-pyrroquinoline first synthesized in Wie-land's laboratory. As Table 1 shows, PQQ occurs in oxidases, dioxygenases, decarboxylases and probably in choline dehydrogenase (Duine, 1989).

PQQ containing bacterial dehydrogenases, the so-called quinoproteins have been found in the periplasm of gram-negative bacteria. As Duine sums this up: "They are involved in the primary oxidation step of substrates like alcohols (besides methanol dehydrogenase, different quino-protein dehydro-genases exist for oxidation of ethanol, glycerol, quinate, polyvinyl alcohol, and polyethylene glycol), amines (methylamine dehydrogenase), and aldose sugars (glucose dehydrogenase). Depending on the nature of the dehydro-genase, electron transfer occurs to a special cytochrome c, cytochrome b, copper protein (amicyanin) or the bacterial ubiquinones. Certain quino-protein dehydrogenases are involved in incomplete microbial oxidations (e.g., vinegar and gluconic acid production), while others function in a special respiratory chain, providing the organism with an auxilliary energy system. In other bacterial species, possessing a set of dehydrogenases catalyzing the same reaction but with different cofactors, the role of quinoproteins is still unclear. Not all bacteria are able to provide their quinoprotein apo-enzyme with PQQ, but nevertheless they produce the apoenzyme constitutively. Therefore, depending on the growth substrate and the occurrence of other metabolic pathways in the organism, supplementation of the medium with PQQ is stimulatory or indispensable for growth in these cases. Excretion of PQQ is not restricted to bacteria or to dehydrogenases. Several copper-containing oxidases from a variety of organisms appear to be quinoproteins: bovine plas-ma amine oxidase; hog kidney diamine oxidase; human placental lysyl oxidase; fungal galactose oxidase; bacterial methylamine oxidase; pea seedling amine oxidase. Examples are also found in other subclasses of enzymes and in the group of iron-containing enzymes: bovine adrenal medulla dopamine β-hydroxy-lase; soybean lipoxygenase-1; bacterial nitrile hydratase. Moreover, strong arguments exist to propose a role for PQQ as cofactor in several pyridoxal

Vomicine   VOMIPYRINE

PQQ

Table 1. Plant and mammalian quinoproteins

Enzyme	Organism	Enzyme	Organism
OXIDASES		DECARBOXYLASES	
Amine oxidases (EC 1.4.3.6)		Dopa decarboxylase (EC 4.1.1.28)	Pig kidney
Plasma amine oxidase	Bovine serum	OTHERS (?)	
Diamine oxidase	Pig kidney	Choline dehydrogenases	Dog liver
Lysyl oxidase	Human placenta and arteria, chicken cartillage	Lipoxygenases	Mammals
MONO-OXYGENASES		Peptidyl glycine mono-oxygenase	Mammals
Dopamine β-hydroxylase (EC 1.14.17.1)	Bovine medulla	Atypical pyridoxo-protein amino acid decarboxylase and transaminases	Plants and mammals
DEOXYGENASES			
Lipoxygenase-1 (EC 1.13.11.12)	Soybean		

phosphate (PLP)-dependent enzymes, e.g. in certain ω-aminotransferases and in decarboxylases. In fact, the presence of PQQ has recently been established in dopa decarboxylase. From the published spectroscopic and mechanistic data of some of these enzymes, a function of PQQ as cofactor can be deduced. This means that the prevailing view on the mechanism of these enzymes, once thought to be well understood, has to be reconsidered."

"The current list of quinoproteins shows that many are involved in the degradation or biosynthesis of mammalian bioregulators. Therefore, design of inhibitors blocking the action or the biosynthesis of PQQ can be expected to shed light on several physiological phenomena" (Duine, 1989).

TRYPTOPHAN 5-HYDROXYLASE and 5-HYDROXYTRYPTOPHAN DECARBOXYLASE

About 1% of the body's tryptophan is converted to 5-hydroxytryptophan (5-OHT) by tryptophan hydroxylase [oxygen oxidoreductase (5-hydroxylating) EC 1.14.16.4] which requires tetrahydropteridine as a cofactor. In the biosynthesis of 5-OHT, hydroxylation is the rate-limiting step (Bender, 1982). Udenfriend's conversion of synthetic D,L-5-hydroxytryptophan (Elk and Witkop, 1953) to serotonin by aromatic amino decarboxylase has become a classical experiment (Udenfriend et al., 1953). We shall hear more about 5-hydroxytryptophan decarboxylase from M. Ebadi in the course of this Symposium, but here are Johannis Duine's ideas about this important enzyme:

"Dihydroxyphenylalanine (DOPA) decarboxylase (DDC) or aromatic amino acid decarboxylase (EC 4.1.1.28) from pig kidney has been intensively studied. It contains PLP (pyridoxal phosphate), and the absorption spectrum shows maxima at 335 and 420 nm. After removal of PLP, a shoulder at 335 nm is still observed in the spectrum of the apoenzyme. Although it has been suggested that this belongs to a vitamin B6-like compound, the structure has never been elucidated. Since the properties of this compound were reminiscent of covalently bound PQQ in enzymes previously though to contain PLP, we attempted its identification. The hydrazine method failed in this case, but using the hexanol extraction procedure, the presence of one PLP and one PQQ per enzyme molecule could be established (Groen et al., 1988)."

These ideas are further elaborated in a letter to me dated January 24, 1989, in which Professor J.A. Duine emphasizes the historical connections with indole chemistry:

"Please find enclosed a manuscript on glutamate decarboxylase from E. coli, for submission to FEBS Letters (van der Meer et al., 1989). Besides the surprise of the existence of a second PLP/PQQ-enzyme, it appears now that E. coli is able to produce a quinoprotein, but (as we know already) not free PQQ. An interesting development which will interest you is the discovery of an enzyme with pro-PQQ (mentioned in the Discussion section). As far as we can see now, pro-PQQ has the indole ring but the attached glutamic acid still requires cyclisation and of course oxidation (paper submitted to Nature), reactions which obviously occur during application of our hydrazine method. The hot topic in my laboratory is now, what is the real structure of the cofactor in all the mammalian quinoproteins (is it PQQ or pro-PQQ and what is the exact structure of the latter)? If so, it seems that an interesting evolutionary relationship of amino acid cofactors can be indicated: enzymes with a tyrosyl free radical (several of them have been reported, e.g. ribonucleotide reductase); enzymes with pro-PQQ (in fact a cyclized tyrosine); enzymes with genuine PQQ (occurring in a number of bacterial dehydrogenases).

I was very delighted with your last letter and its enclosures. When I received it last summer, I read with interest the historical information unknown to me (there are several surprising coincidences: the cofactor of methanol dehydrogenase was once thought to be a pteridine instead of PQQ (Wieland); phenylhydrazine was indispensable in our search for mammalian quinoproteins (Fischer); ^{13}C-labelled PQQ was crucial in our study on the biosynthesis (Kamen)). Vomipyrine was so far unknown to me. Unfortunately, this year I wrote already a review on PQQ for the series Vitamins and Hormones so that I have to wait for another opportunity before I can indicate the relationship with PQQ (although the ring-linkage is different!).

Now after rereading your letter and reprints at this moment, I realize that the information given in it on the indole industry is highly relevant in the light of the discovery of pro-PQQ. One of the approaches we will follow is an attempt to derivatize the hydroxyl groups in pro-PQQ in the enzyme (as we expect that this derivative ill be stable enough for isolation). Hopefully, we can announce the structure soon to you."

Table 2. Physiological functions of serotonin (5HT)

Vigilance & sleep	Sleep-wakefulness control system
	Slow wave sleep-rem sleep inducer
Learning & memory	Punitive system Memory inhibition
Pain	Threshold elevation
Movement	
Mood	
Miscellaneous	Agressive behavior Sexual behavior
	Sensory reactivity

SEROTONIN AND ITS RECEPTORS

Over the last 30 years the role of the enigmatic hormone of Irvine Page, serotonin, as a novel neurotransmitter, has grown both in its physiological functions as well as in the etiology of nervous disorders.

That there must be special receptors for serotonin was first proposed by Gaddum (Gaddum and Picarelli, 1957). Snyder's radioligand binding technology expanded this finding to $5-HT_1$ and $5-HT_2$ recognition sites (van der Meer et al., 1989).

Most neurotransmitters activate multiple receptor sites: dopamine qualifies for at least two receptors, histamine moves up to three, acetylcholine so far four and noradrenaline five. At the moment serotonin seems to be the winner with at least seven receptor subtypes on the basis of pharmacological criteria. These receptors are mostly single subunit proteins and members of the G protein receptor superfamily (Hartig, 1989). Genomic clones have been reported for at least three 5-HT receptors. Their transmembrane segments of the receptor clones show 78% sequence homology. In the end, what is involved is direct activation of ion channels. The debate on receptor classification (Hartig, 1989) and models (Kilpatrick et al., 1987) for the structure of G protein-coupled 5-HT receptors is more active than ever. The "enigmatic hormone" of Irvine Page has produced an analog of Edward Elgar's Enigma Variations (Opus 36) in which over the whole set a large theme "goes". The identity of this mysterious, unheard theme has provoked many ingenious guesses but no final explanation (Fozard, 1987).

ASKING THE RIGHT QUESTIONS

The questions asked in the Abstracts of this lecture have, in part, been answered. Some are somewhat trivial, such as the serotonin derivative in your morning coffee: it is the 5-hydroxytryptamide of arachidic acid (C_{20}). Many answers to the other questions will be found in David A. Bender's excellent review (1982). Some of the topics of the upcoming lectures could be

Table 3. Role of serotonin (5HT) in nervous disorders

Normal aging	↓5HTP-DC ↑MAO A
Senile dementia Alzheimer type	↓5HT ↓5HIAA
Parkinson's disease	↓5HT ↓5HIAA
Epilepsy	↑5HT(?)
Depression	↓5HIAA
Schizophrenia	
Alcoholism	↓5HT ↓5HIAA

phrased as questions. The interesting inhibition of molting of crustaceans by xanthurenic acid to be reported by Dr. Yoko Naya raises the question of similar antioxidant properties in other organisms. We are always looking for the causes of cancer and aging that may be due to oxidative damage of DNA. We are aware of the potential danger of reactive oxygen species, such as the superoxide radical (O_2-) hydrogen peroxide (H_2O_2) the hydroxyl radical ($\cdot$OH) and singlet oxygen which are all present in normal metabolism. The question is, are there tryptophan metabolites that may promote or repair oxidative damage of DNA? The idea that in mitochondria oxidative DNA damage is high, repair mechanisms almost absent and integration into nuclear genomes possible (Richter, 1988), deserves mention here.

REFERENCES

Albuquerque, E.X., Alkondon, M., Deshpande, S.S., Cintra, W.M., Aracava, Y., and Brossi, A., 1988, The role of carbamates and oximes in reversing toxicity of organophosphorus compounds: a perspective into mechanisms, in: "Elsevier Science Publications", Vol. 26, Elsevier, Amsterdam.

Bender, D.A., 1982, "Biochemistry of Tryptophan in Health and Disease", Molecular Aspects of Medicine, Pergamon Press, pp. 103-197.

Bradley, P., 1987, 5-HT_2 receptors in the brain, Nature, 330.

Braestrup, E., Nielson, M., and Olsen, C.E., 1980, Proc. Nat. Acad. Sci. USA, 77:2288.

Brossi, A., Schönenberger, B., Clark, O.E., and Ray, R., 1988, Inhibition of acetylcholinesterase from electric eel by (-)- and (+)-physostigmine and related compounds, FEBS Lett., 201:190.

Dale, H., 1948, Frederick Gowland Hopkins, Obituary Notices, Proc. Royal Soc., A: 114.

Daly, J.W., Jerina, D.M., and B. Witkop, 1972, Arene oxides and the NIH shift: The metabolism, toxicity and carcinogenicity of aromatic compound, Experientia, 28:1129.

Daly, J.W., Mauger, A.B., Yonemitsu, O., Antonov, V.K., Takase, K., and B. Witkop, 1967, Biochemistry, 6:648.

Duine, J.A., and Jongejan, J.A., 1989, Ann. Rev. Biochem., 58:403-426.

Duine, J.A., 1989, Pyrroloquinoline quinone (PQQ): a novel redox cofactor, in: "Vitamins and Hormones", in press.

Elk, A., and B. Witkop, 1953, J. Amer. Chem. Soc., 75:500.

Fozard, J.R., 1987, 5-HT: The enigma variations, Trends Pharmacol. Sci., 8:501.

Freter, K., Axelrod, J., and B. Witkop, 1957, J. Amer. Chem. Soc., 79:3191.

Gaddum, J.H., and Picarelli, Z.P., 1957, Brit. J. Pharmacol., 12:323.

Groen, B.W., van der Meer, R.A., and Duine, J.A., 1988, FEBS Lett., 237:98.

Gross, E., and Witkop, B., 1965, Gramicidin. IX. The preparation of gramicidine A, B and C, Biochemistry 4:2495.

Hartig, P.R., 1989, Molecular biology of 5-HT receptors, Trends Pharmacol. Sci., 10:64.

Hino, T., and Nakagawa, M., 1988, "The Alkaloids", Academic Press, New York.

Hopkins, F. G., and S. W. Cole, 1901, 1903, J. Physiol., 27:418; 29:451.

Hopkins, F. G., 1889, Proc. Chem. Soc., 5:117.

Kawabuchi, M., Boyne, A.F., Deshpande, S.S., Cintra, W.M., Brossi, A., and Albuquerque, E.X., 1988, Enantiomer (+)-physostigmine prevents organophosphate-induced subjunctional damage at the neuromuscular synapse by a mechanims not related to cholinesterase carbamylation, Synapse, 2:139.

Kilpatrick, G.J., Jones, B.J., and Tyers, M.B., 1987, Identification and distribution of $5HT_3$-receptors in rat brain using radioligand binding, Nature, 330:746.

Llinas, R., and Volkind, R.A., 1973, Exp. Brain Res., 18:69.

Neuner, A., and Tappeiner, H., 1894, Arch. exp. Pathol. Pharmacol., 35,I:69.

Ozaki, M., Weissbach, H., Azaki, A., Witkop, B., and Udenfriend, S., 1960, J. Med. Pharm. Chem., 2:591.

Phillips, R.S., Miles, E.W., and L. A. Cohen, 1985, Differential inhibitions of tryptophan synthase and of tryptophanase by the two diastereoisomers of 2,3-dihyro-L-tryptophan: implications for the stereochemistry of the reaction intermediates, J. Biol. Chem., 260:14665.

Rebek, Jr., J., Tai, D.F., and Y. K. Shue, 1984, J. Amer. Chem. Soc., 106:1813.

Renson, J., Daly, J.W., Witkop, B., and Udenfriend, S., 1966, Biochem. Biophys. Res. Comm., 5:504.

Richter, C., 1988, Do mitochondrial DNA fragments promote cancer and aging?, FEBS Lett., 241:1.

Roy, M., Miles, E.W., Phillips, R.S., and M. F. Dunn, 1988, Biochemistry, 27:8661.

Robinson, R., and Stephen, S., 1948, Nature, 162:177.

Salemme, F.R., 1988, Structural polymorphism in transmembrane channels, Science, 241:145 & 230.

Sarges, R., and Witkop, B., 1965, Gramicidin, VI. The synthesis of valine- and isoleucine-gramicidin A. J. Amer. Chem. Soc., 87:2020.

Udenfriend, S., Clark, C.T., and Titus, E., 1953, 5-Hydroxytryptophan decarboxylase: a new route of metabolism of tryptophan, J. Amer. Chem. Soc., 75:501.

van der Meer, R.A., Groen, B.W., and Duine, J.A., 1989, On the biosynthesis of free and covalently bound PQQ glutamic acid decarboxylase from Escherichia coli is a pyridoxo-quinoprotein, FEBS Lett., 246:109-122.

Walden, P., 1941, "Geschichte der Organischen Chemie seit 1880", Julius Springer, Berlin.

Wallace, B.A., and Ravikumar, K., 1988, The Gramicidin pore: crystal structure of a cesium complex, Science, 241:182.

Wieland, H., and Horner, L., 1937, Liebigs Ann., 528:75.

Wieland, H., and Horner, L., 1938, Synthese des 5,6-Pyrro-chinolins, Liebigs Ann., 536:89.

Wieland, H., and Schopf, C., 1925, Ber., 58:2178.

Wieland, H., and Witkop, B., 1940, Liebigs Ann., 543:171.

Witkop, B., 1943, Liebigs Ann., 554:83.

Witkop, B., 1948, The final structure came much later, J. Amer. Chem. Soc., 70: 1424; see J.E. Saxton, 1960, The alkaloids of gelsemium species, The Alkaloids 7:153.

Witkop, B., 1953, Studies on anhydronium bases, J. Amer. Chem. Soc., 75:3361.

Witkop, B., 1983, Tradition und Thematik in der Naturstoffchemie, Naturwiss. Rundschau 36:261; cf. T. Wieland, 1986, "Peptides of Poisonous Amanita Mushrooms", Springer, New York, Berlin, Heidelberg.

Woodward, R.B., and Witkop, B., 1949, The structure of sempervirine, J. Amer. Chem. Soc., 71:379.

Yonemitsu, O., Cerutti, P., and B. Witkop, 1966, J. Amer. Chem. Soc., 88: 3941.

Yu, Q.S., and Brossi, A., 1988, Practical synthesis of unnatural (+)-physo-stigmine and carbamate analogues, Heterocycles 27:745.

TRYPTOPHAN AND RELATED ALKALOIDS

M. Nakagawa and T. Hino

Faculty of Pharmaceutical Sciences
Chiba University
Japan

INTRODUCTION

Tryptophan is metabolized to a number of biologically important substances (Hayaishi, 1962, 1974). It is now well established that tryptophan
is metabolized to NAD via formylkynurenine which was formed by the oxidative
cleavage of the 2,3-bond of the indole ring catalyzed by tryptophan-2,3-di-
oxygenase. On the other hand, the hydroxylation of tryptophan at 5-position
accompanied by the NIH shift is another important reaction catalyzed by
tryptophan 5-hydroxylase (Daly et al., 1972; Jerina and Daly, 1974; Boyd and
Berchtold, 1979). Furthermore, tryptophan also serves as an important precursor for the biosynthesis of indole alkaloids (Scheme 1).

It is obvious from the structures of some representative indole alka-

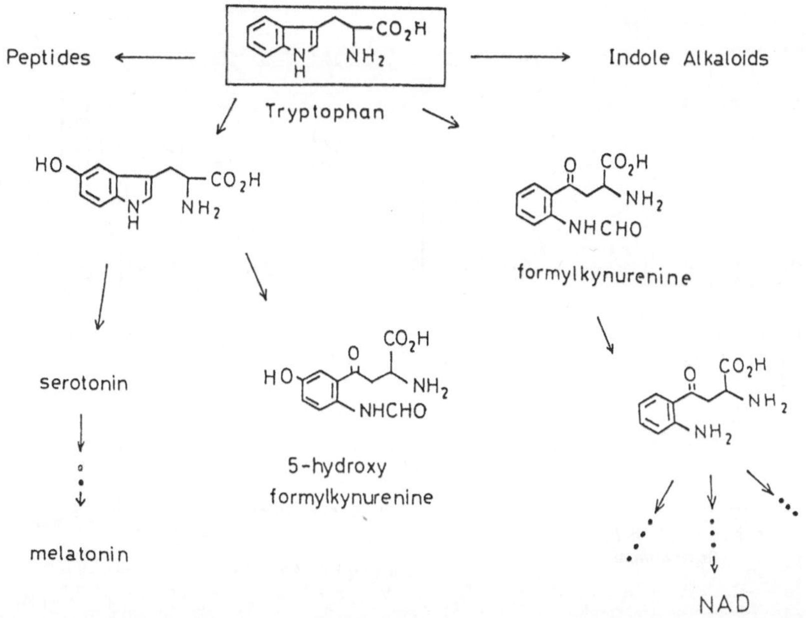

Scheme 1

Kynurenine and Serotonin Pathways
Edited by R. Schwarcz *et al.*, Plenum Press, New York, 1991

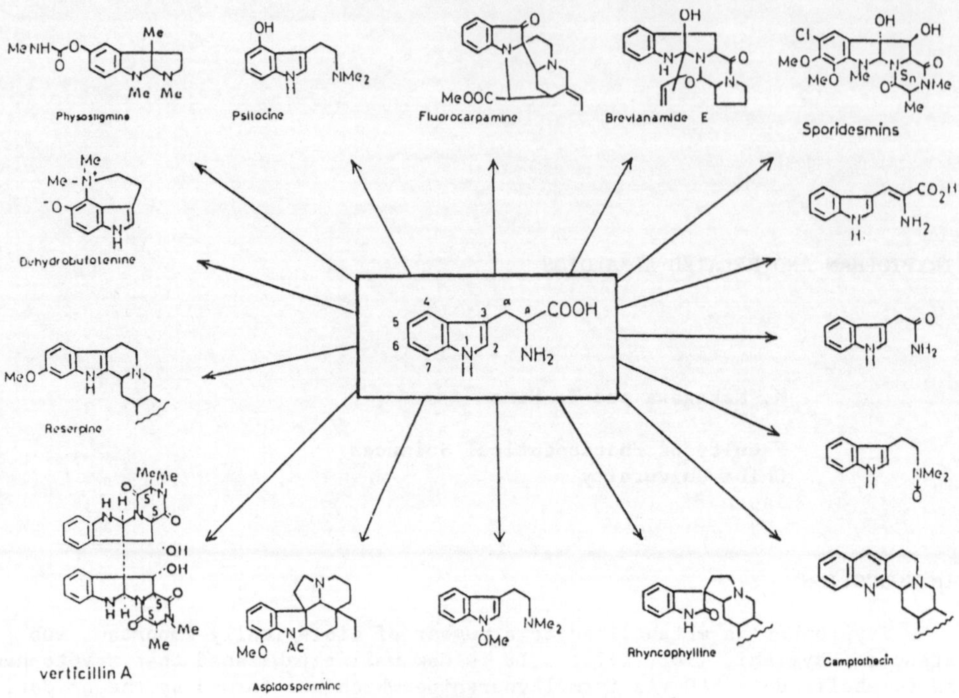

Scheme 2

loids, secondary metabolites of tryptophan, shown in Scheme 2, that the biological oxidation occurs not only at 2, 3- or 5-positions but also occurs at all the positions of the tryptophan moiety including the side chain. In addition, the oxidative coupling reaction at the 3-position must be involved in the formation of the dimeric indole alkaloids like folicanthine and verticillin A.

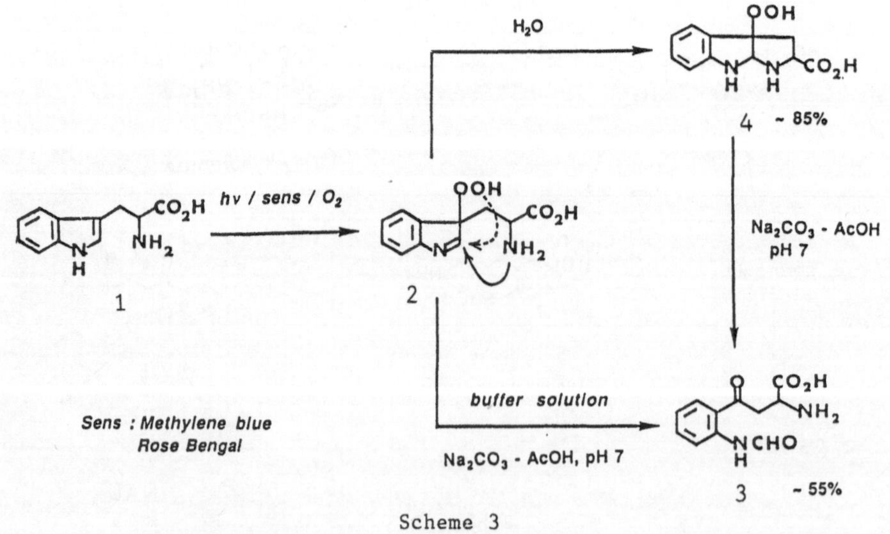

Scheme 3

16

We have reported (Nakagawa, 1980) that tryptophan 1 reacted with singlet oxygen produced by the dye-sensitized photooxidation in aqueous solution to give the tricyclic hydroperoxide 4, whereas the similar oxygenation of 1 in a buffer solution of Na_2CO_3-AcOH (pH 7) provided formylkynurenine 3 (Scheme 3).

This report deals with further results on the dye-sensitized photooxygenation of tryptophan in various conditions in an effort to understand further the mechanism of the biological oxidation of tryptophan and the total synthesis of biologically active indole alkaloids.

I. Dye-sensitized photooxygenation of L-tryptophan at various pH (Nakagawa et al. 1984c; Nakagawa et al., 1985b)

Despite intensive work on the photooxygenation of tryptophan in the presence of a dye as a function of pH, no report of the nature of the oxidation products has been given, but O_2 uptake has been measured. Therefore, we reexamined the oxgenation of 1 over the pH range 1-9.

We first investigated the relative rate of photooxidation of tryptophan in acetate buffers in the range pH 3.6-6.2 and in phosphate buffers in the range pH 5.9-8.4 using methylene blue as the sensitizer at all pHs.

Table 1. Photo-oxygenation of Tryptophan at various pH

Buffer	pH	Reaction time (hr)	5 (%)		6 (%)
			UV	HPLC	
NaOAc–HOAc	3.6	3.5	58	66	—
	4.0	3.5	54	63	—
	4.6	4.0	62	69	—
	5.3	1.5	68	69	—
	6.2	1.0	47	46	—
Na_2HPO_4–KH_2PO_4	5.9	2.0	56	74	—
	7.1	1.0	58	59	Trace
	7.7	1.0	18	15	21
	8.4	1.0	14	15	17

The reaction mixture was reduced with Me_2S and left overnight.

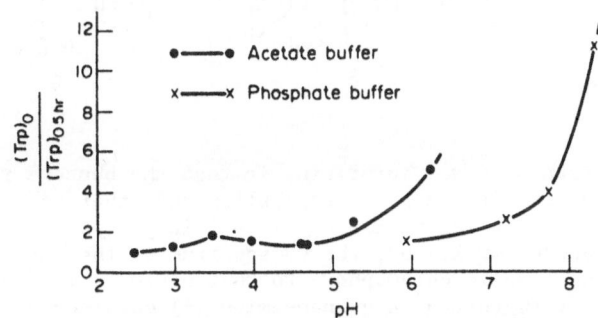

Fig. 1. Photooxidation of tryptophan: pH dependency.

The results are shown in Fig. 1. The pH of the reaction mixture had a profound influence on the reaction rate. In accord with previous reports, tryptophan was oxidized more rapidly with increasing pH.

We next carried out a product analysis of the reaction mixture over the

17

range pH 3.6-8.4. The oxygenation of a buffer solution of L-tryptophan was carried out by irradiation (hv > 550 nm) at 0.5°C in the presence of methylene blue and with Me_2S and was left overnight prior to ion exchange column chromatography (Amberlite CG-50). Lyophilization of the elution with water provided the products summarized in Table 1. In acidic or neutral conditions, the hydroxide, 5 (a 1:1 mixture of cis and trans isomers) was the sole product isolated. In contrast, an alkaline phosphate buffer solution (pH 7.7) 5 became the minor product and a new product, 5-hydroxyformylkynurenine (6) was obtained. A similar result was obtained at pH 8.4. The structure of 6 was assigned from its spectral and chemical properties. Hydrolysis of 6 with trifluoroacetic acid, basification and acylation with methyl chloroformate and then esterification with diazomethane followed by acetylation provided the crystalline derivative 7.

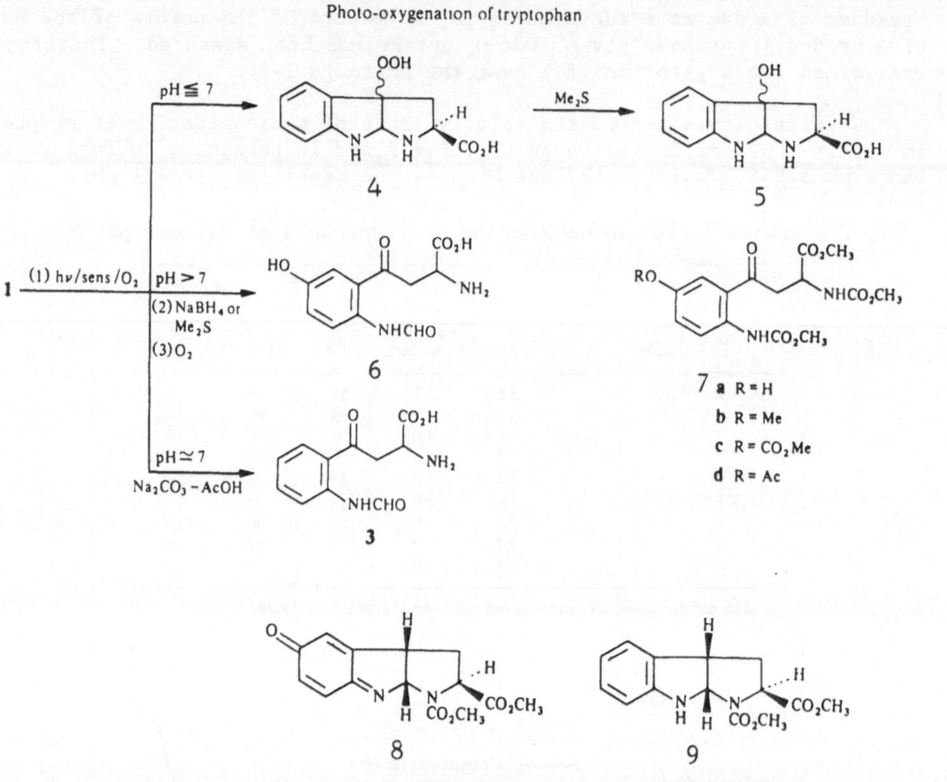

Photo-oxygenation of tryptophan

Scheme 4

The formation of 6 was surprising in that the benzene ring was hydroxylated by the dye-sensitized photooxygenation and, therefore, the mechanism for the formation of 6 was examined. When 1 in phosphate buffer, pH 7.8, was irradiated as above for 1.5 hr, the UV spectrum of the reaction mixture changed from the indolic chromophore to that having a maximum of 269 nm, which was similar to that of a quinoneimine (8) obtained by the $Ph(OAc)_4$ or Fremy's salt oxidation of 9. The reaction was also followed by high performance liquid chromatography (HPLC), showing the exclusive transformation of 1 (t_R ≈ 24.5 min) to the more polar substance (t_R ≈ 5.8 min), probably 11. Accordingly, when the reaction mixture was reduced with $NaBH_4$ immediately under N_2 followed by immediate neutralization with dilute HCl and work-up, 3a,5-dihydroxypyrroloindole (12) was obtained in 95% yield as a mixture of cis and trans isomers and 6 was not obtained. In contrast to 5, 12 was unstable under basic conditions and the surprisingly facile autoxidation of 12 to 6 occurred.

Consequently, the sequence to obtain **6** from **1** was best performed in 47% yield, without isolating **12**, by treating the NaBH$_4$ reduction mixture with oxygen for 3 hr at room temperature. Under similar conditions **3** did not convert to **6** and was recovered unchanged, implying that **3** is not an intermediate for formation to **6**. On the other hand, **4** and **5** were converted to **12** in 42% and 16% yields, respectively, by methylene blue sensitized photooxygenation under these conditions whereas, in the absence of methylene blue, **4** and **5** were not oxidized to **6**. Furthermore, the similar photooxygenation of **5** followed by immediate reduction with NaBH$_4$ gave **12** in 53% yield, but **9** was recovered unchanged when treated with O$_2$ in alkaline phosphate buffer. These results demonstrated the intermediacy of **12** for **6**.

The yield of **6** was increased when the reaction mixture was reduced with NaBH$_4$ followed by immediate oxidation with O$_2$.

A possible rationale for the oxidation leading from tryptophan (**1**) to 5-hydroxyformylkynurenine **6** is outlined in Scheme 3. We concluded that the initial hydroperoxidation at the *para*-position of primary product, **4**,

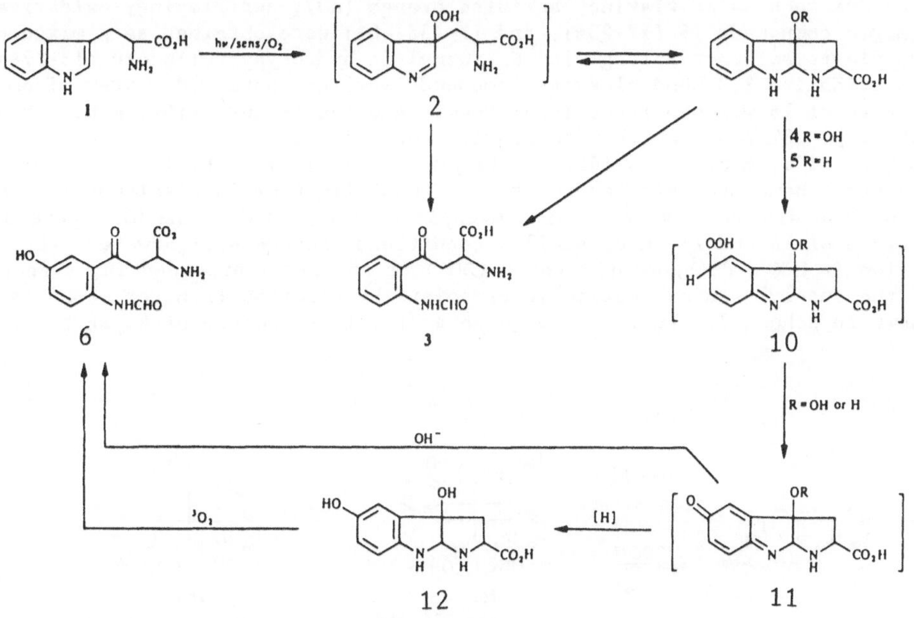

Scheme 5

probably by singlet oxygen in alkaline phosphate buffers, gave the quinone-imine **11** *via* **10**, which was converted to **12** on treatment with NaBH$_4$. When Me$_2$S was used, we conclude that disproportionation between intermediates must have occurred to form **12** in accord with the observation for the low yield of **6**. Supporting evidence for the transformation of **5** to **12** *via* a quinoneimine was demonstrated by the Fremy's salt oxidation of **5** to **12** in a phosphate buffer followed by NaBH$_4$ reduction. However, the mechanism for oxidation of **12** to **10** is not yet clear but may well involve the initial oxidation of the phenolate anion of **12** with triplet oxygen, since **12** is stable in neutral media in the presence of O$_2$ as well as in alkaline phosphate buffer in Ar.

II. **Dye-sensitized photooxygenation of L-tryptophan in formic acid** (Hino et al., 1978; Nakagawa et al., 1981)

Proflavine-sensitized photooxygenation of tryptophan and N$_b$-benzyloxycarboxyltryptophan in formic acid has been reported to give kynurenine-type compounds in good yields. As the tricyclic hydroperoxide 13 rearranges to 1,4-benzoxazine derivative 5 in methanol-HCl at room temperature (Nakagawa, 1980) we reexamined proflavine-sensitized photooxygenation of tryptophan and N_b-benzyloxycarbonyltryptophan in formic acid, but neither kynurenine-type compound nor 1,4-benzoxazine derivative was obtained, and polymeric compounds were isolated. Proflavine-sensitized photooxygenation of N_b-methoxycarbonyl-tryptamine (15) in formic acid, however, was found to give dimeric products which have not previously been obtained in the sensitized photooxygenation of indole derivatives and were converted to folicanthine and chimonanthine.

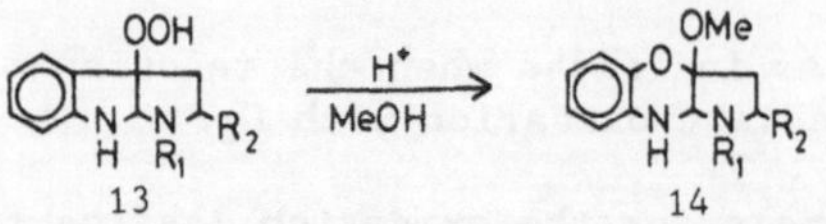

When N_b-methoxycarbonyltryptamine 15 in thoroughly O_2-saturated formic acid was irradiated with a halogen lamp for 1 hr in the presence of an acridine dye such as proflavine, acridine orange (AO), acriflavine, oxidative dimeric compounds 16 (17-23%), and 17 (38-40%) were obtained as a mixture of two diastereoisomers along with N_a-formyl-3a-hydroxypyrrolindole (13-27%), and oxidative 2,3-bond cleavage compounds were not obtained. Stereoisomeric mixture of 16 was separated into racemi- and mesoisomers (16a, m.p. 255-256°; 16b, m.p. 282-284°) which were readily converted to racemi- and meso- folicanthine (18a, m.p. 167.5-168.5°; 18b, m.p. 174-175°) by LiAlH₄ reduction. On the other hand, the similar reaction without light or in a solvent like MeOH or CF_3CO_2H did not give 2. Using methylene blue or toluidine blue gave only a trace of 16 and 17 under similar conditions (Scheme 6). However, the reaction in HCO₂H-dicyanoanthrathene (DCA) or chloranil provided 16, suggestive of the indolyl radical cation intermediate by electron transfer mechanism, as shown in Scheme 7. In fact, 15 quenches the fluorescence of AO at 575 nm in

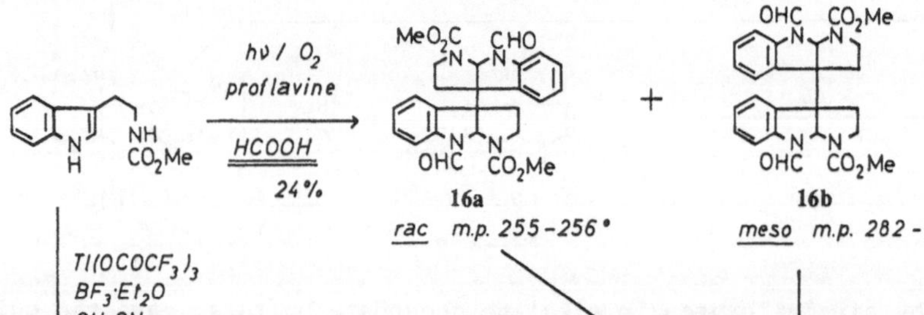

Scheme 6

20

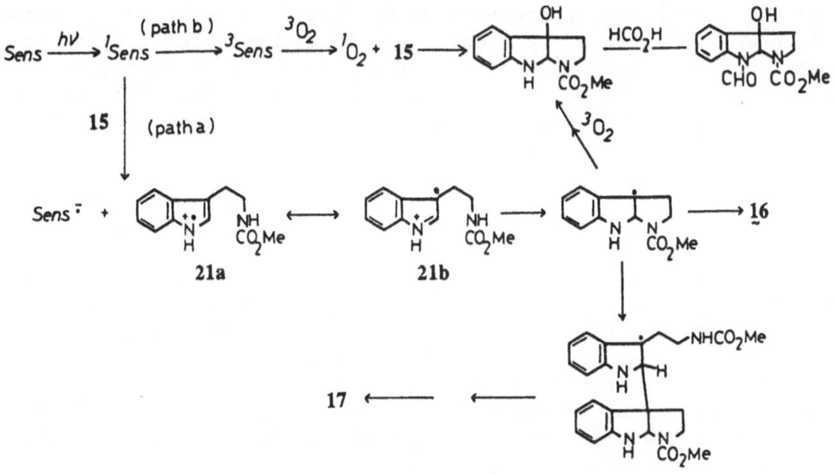

$$\text{Sens} \xrightarrow{h\nu} {}^1\text{Sens} \xrightarrow{(\text{path b})} {}^3\text{Sens} \xrightarrow{{}^3O_2} {}^1O_2 + 15 \longrightarrow$$

(structures)

OH / CO₂Me / H / N

$\xrightarrow{HCO_2H}$ OH / CHO CO₂Me

$\nwarrow {}^3O_2$

$15 \Big| (\text{path a})$

Sens⁻ + (structure 21a) ⟷ (structure 21b) → (structure) ⟶ **16**

21a   **21b**

(structure with NHCO₂Me)

17 ⟵ ⟵ (structure)

Scheme 7

HCO₂H. Furthermore, the oxidation of 15 with thallium (III) trifluoracetate (TTFA) in acetonitrile gave the deformylated 16, which was converted to chimonanthine (rac, 18b, m.p. 184-186°; meso, 18c, m.p. 198-203°). In light of these results, we intended to synthesize ditryptophenaline 20 by analogous oxidative coupling of cyclo-L-N-methylphenyl-alanyl-L-tryptophanyl 19. Irradiation of 19 in HCO₂H with proflavine, chloranil, DCA under a variety of conditions did not give the corresponding dimeric compounds. However, an alternate reagent, TTFA was employed successfully to produce the desired dimer, ditryptophenaline 20 in 5% yield, which was readily crystallized from CH₂Cl₂-MeOH to give m.p. 196-203°(Lit. m.p. 204-205°) and was identical through spectral (IR, UV, NMR, $[\alpha]_D{}^{33}$ - 318.1°, high resolution mass) and chromatographic comparison with an authentic sample of the natural material. Since relative configuration of 20 has been reported, our total synthesis established the absolute configuration of 20 as shown.

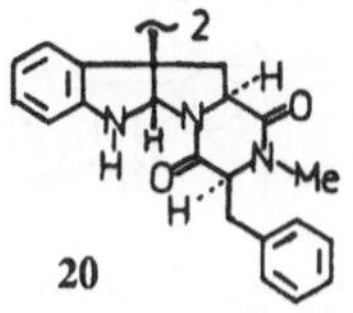

(structure 20)

20

III. Oxidative transformation of tryptophan to 3-(2-aminophenyl)-2-pyrro-lidone and kynurenine (Nakagawa et al., 1980; Nakagawa et al., 1985a)

Oxytryptamine, 3-(2-aminoethyl)-2-indolinone, 22, has been known to be unstable as a free base. Attempts to synthesize 22 in free form met with no success and gave a complicated mixture of unidentified products. Thereby, 22 has been prepared as its salt form such as hydrochloride. The instability of 22 has been ascribed to its transformation to 3-(o-aminophenyl)-2-pyrroli-done 25 by intramolecular acyl migration via 23 or 24. However, while the conversion of oxytryptamine derivatives to the corresponding isomeric 2-pyr-rolidones is precedented, the isolating of 4 itself has not been demonstra-ted. In 1958, Witkop has shown that by refluxing 1 in 2 N sodium hydroxide under nitrogen, the oxidole ring was opened to give α-(0-amino-phenyl)-γ-aminobutyric acid 23, which was isolated as the bisbenzyloxycarbonyl de-rivative.

When oxytryptamine hydrochloride was basified with excess sodium hydroxide in open air followed by treatment with methyl chlorformate, we unexpectedly obtained kynurenamine derivatives 26 and 27 beside 28b. But pyrrolidone 25 was not found in the reaction mixture. The yield of kynurenine derivatives (26 and 27) increased up to 57% when oxytryptamine was treated with a base like EtONa in an oxygen atmosphere.

Compounds 26, 27, and 28 must, therefore, be formed by atmospheric oxidation, suggesting the instability of 1 might be associated with its susceptibility to triplet oxygen under alkaline conditions. A possible pathway to account for the unexpected formation of 26, 27 and 28b is shown below, in which all the products are derived from the key intermediate 29. Either cyclization of 29 to a dioxetane 30 which collapsed to 31 followed by subsequent decarboxylation or oxidative decarboxylation of 32 which arises from the hydrolytic ring opening of 29 might result in the formation of 26. The formation of 27 can be envisioned to occur from 31 by intramolecular cyclization. This rationalization is supported by the conversion of 28c to 20, 27, and 28b.

On the other hand, oxytryptamine (22) underwent N,N'-transacylation to give the isomeric product, the desired 3-(o-aminophenyl)-2-pyrrolidone (4, m.p. 120.5-121.5°C) in 30% yield, accompanied with 28a (28%) and 26a (7%) in argon atmosphere.

Likewise, oxytryptophans 33, which are readily obtained by dye-sensitized photooxygenation of tryptophan followed by acid treatment, underwent a facile N,N'-transacylation to give the 3-(2-aminophenyl)-2-pyrrolidones 34 in the absence of oxygen, which were isolated as N-acyl derivatives (35-37), whereas in the presence of oxygen 33 were oxidized to kynurenine.

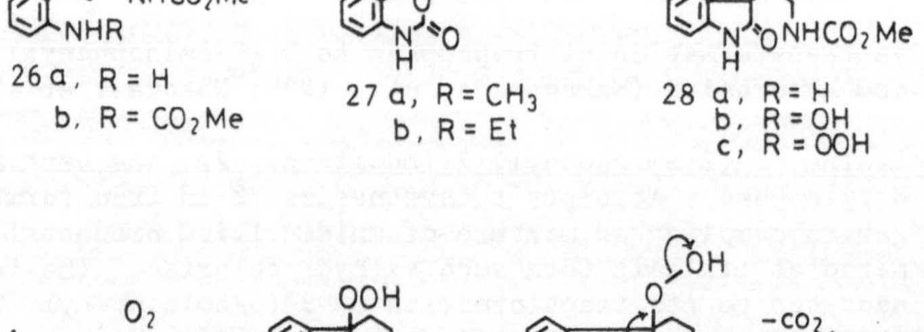

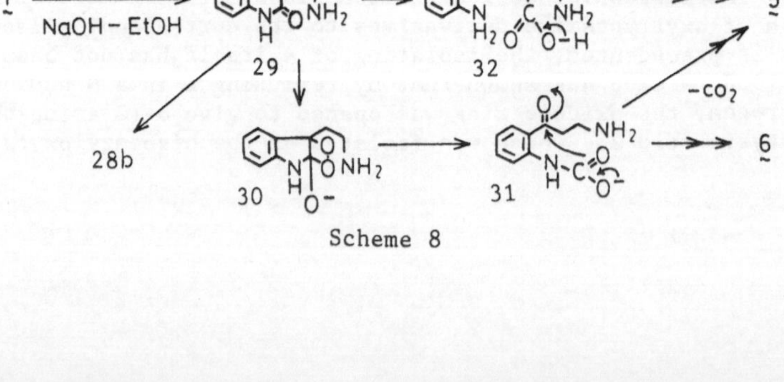

Scheme 8

22

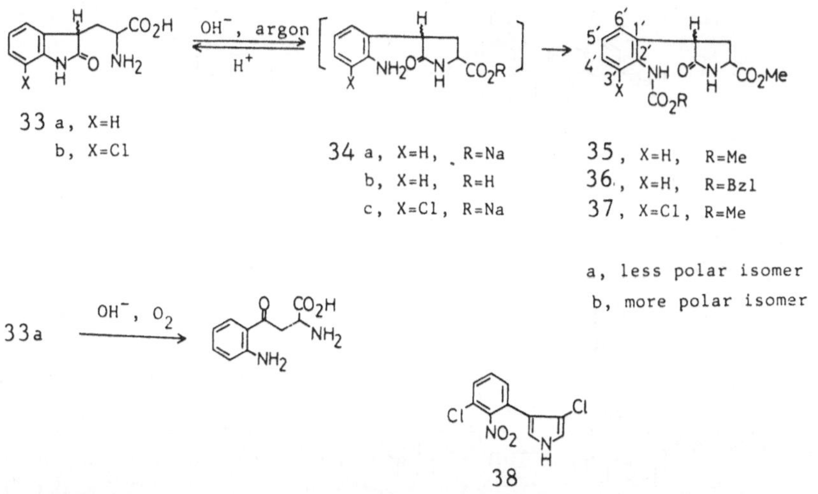

33 a, X=H
b, X=Cl

34 a, X=H, R=Na
b, X=H, R=H
c, X=Cl, R=Na

35, X=H, R=Me
36, X=H, R=Bzl
37, X=Cl, R=Me

a, less polar isomer
b, more polar isomer

33a $\xrightarrow{\text{OH}^-,\ O_2}$

38

Scheme 9

Recently, L- and D-tryptophans have been shown to be efficient precursors for the biosynthesis of pyrrolnitrin 38, and plausible biosynthetic pathways have been proposed (Gorman and Lively, 1967; Martin et al., 1972; Chang et al., 1976). But our present result implicates an alternative pathway for the biosynthesis of pyrrolnitrin.

It is interesting to note that oxytryptophan 33a was once postulated as a biological intermediate between tryptophan and kynurenine, but this pathway was excluded due to the fact 3a was not metabolized to kynurenine. However, the present results show that chemically, 3a readily converts to kynurenine.

IV. Total synthesis of biologically active indole alkaloids

1. **Total synthesis of fumitremorgin B (Nakagawa et al., 1986a; Nakagawa et al., 1986b; Kodato et al., 1988)**

Fumitremorgin B (FTB) (39) is a structually unique and potentially biologically important family of mycotoxins. There are currently six known natural products from a variety of *Aspergillus* and *Penicillium* species in this family, all of which are characterized by the presence of a 2,5-piperazinedione ring formed from 6-methoxy-L-tryptophan and L-proline. They cause severe tremorgenic reactions in mice on either oral or intraperitoneal administration.

Our original plan to 39 required the pentacycle 40 as the key intermediate, which had called for the oxidative cyclization of prenylated cyclo-(6-methoxy-L-tryptophyl-L-prolyl) 41, derived from the methoxylation of the cyclic tautomer (Hino et al., 1981). We have obtained the tetrahydro-β-carboline 42 by the oxidative cyclization of 1,2-diisopentyltryptamine derivative as a preliminary study for a total synthesis of 39 (Nakagawa et al., 1983a). However, a similar oxidation of the tetrahydroderivative of 41 to the corresponding pentacyclic compound was unsuccessful.

Our new approach to FTB (39) was based on the synthesis of the common ring system, optically active pentacycle 40 from 6-methoxy-L-tryptophan and L-proline *via* the Pictet-Spengler reaction and subsequent hydroxylation as shown in Scheme 10. The introduction of the 6-methoxy group was carried out *via* the cyclic tautomer of N_b-methoxycarbonyl-tryptophan ester, a method

23

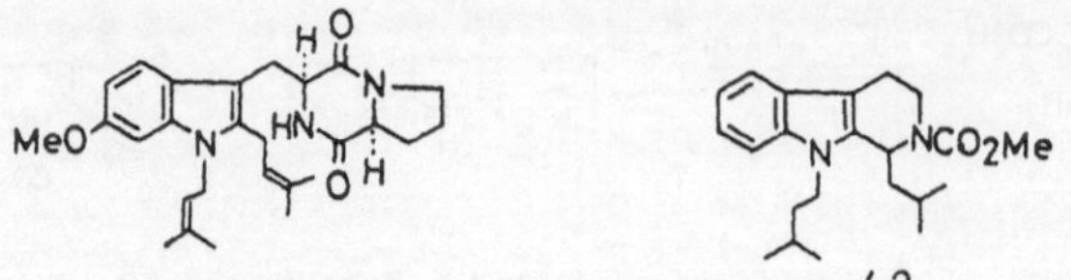

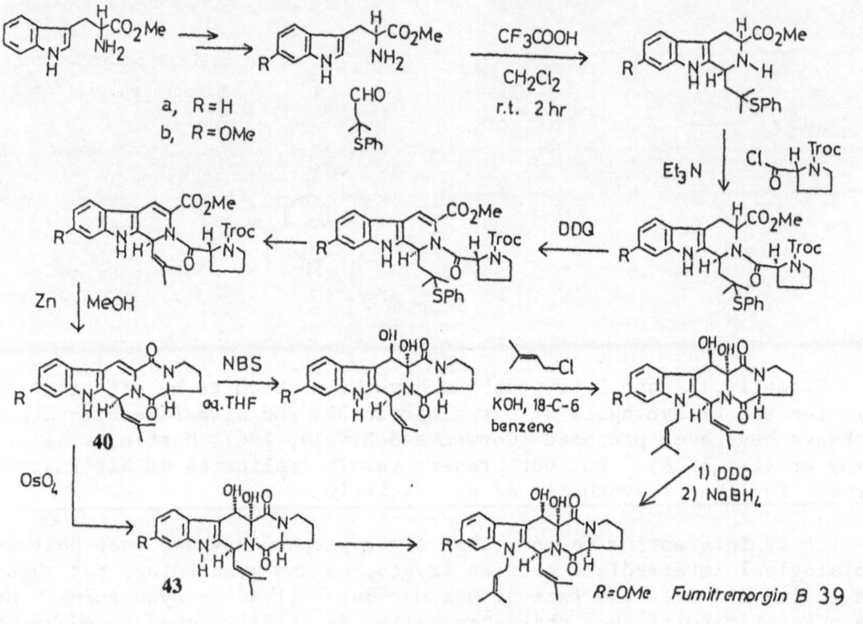

Scheme 10

which we had previously reported to be an efficient means of introducing a methoxy group into the 6-position of tryptophan (Taniguchi et al., 1984).

The direct α-cis hydroxylation of 40 was achieved by OsO₄-N-methyl-morpholine N-oxide, and pyridine to give 43. Subsequent prenylation of 43 gave fumitremorgin B which was identical in all respects (mp, mmp, IR, UV, IH-NMR, and CD spectra, chromatographic mobility) with natural fumitremorgin B by direct comparison.

2. **Total synthesis of tryptoquivaline (Nakagawa et al., 1984a) and tryptoquivaline G (Nakagawa et al., 1983b; Nakagawa et al., 1984b)**

(+)-Tryptoquivaline 44a, a tremorgenic mycotoxin, is the major metabolite among 14 tryptoquivalines isolated from *Aspergillus clavatus* and *fumigatus*. It has a highly sterically hindered substituent at the 2 and 3 positions of quinazolinone ring, whereas tryptoquivaline G 44b bears no substituent at the 2-position.

The lack of a standard method for the synthesis of sterically hindered 2,3-dialkylquinazolinone in the literature has prevented the total synthesis of tryptoquivaline itself.

We have achieved an abbreviated facile biogenetic type total synthesis of (+)-tryptoquivaline 44a utilizing the newly employed oxidative double cyclization of N-acyltryptophan precursor 48, which allowed an efficient formation of the unique ring system of 44 in one step. (Scheme 11) (+)-Trypoquivaline G 44b and its antipode were also synthesized by a similar method.

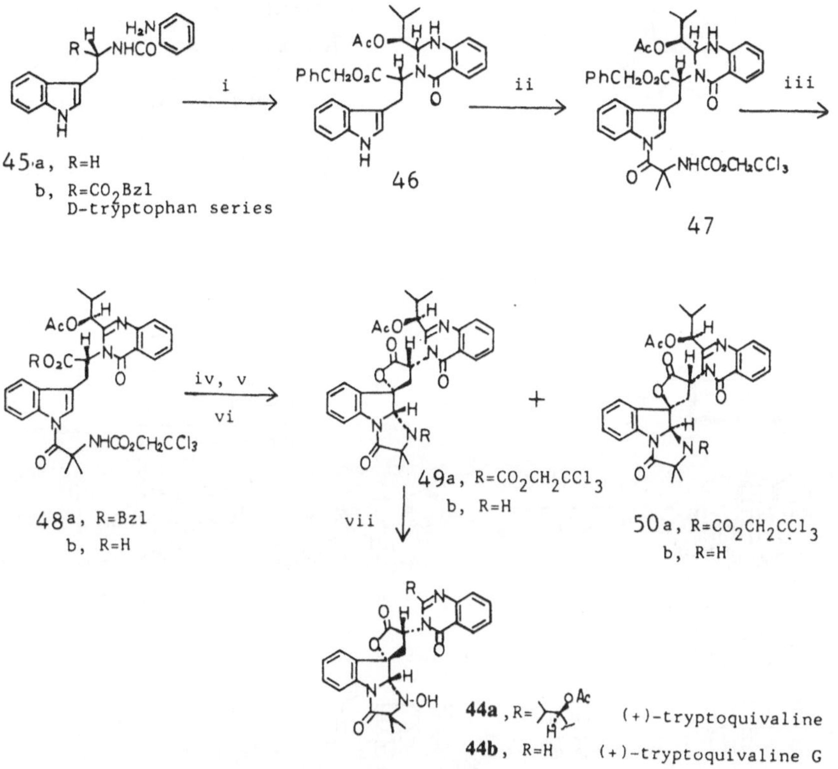

Scheme 11

i, 7, molecular sieves 4A, TsOH, CH_2Cl_2, r.t.; ii, $CCl_3CH_2O_2CNHCMe_2CO_2C_6H_4$-p-
NO_2, 10, KF, MeCN, 18-crown-6, EtN(i-Pr)$_2$, 35°C, 4 h; iii, DDQ, CHCl$_3$, 30°C,
3 h; iv, H_2, Pd/C; v, N-iodosuccinimide(3 equiv), CF_3CO_2H, reflux; vi, Zn,
AcOH; vii, m-ClC$_6$H$_4$CO$_3$H.

3. Total synthesis of (-)-eudistomin L and (-)-debromoeudistomin L (Nakagawa et al., 1986c; Nakagawa et al., 1988; Nakagawa et al., 1989)

In 1984, Rinehart and Kobayashi reported the isolation of the first naturally occurring tetrahydro-ß-carbolines incorporating an oxthiazepine ring,
the eudistomins 51a-d from the colonial tunicate *Eudistoma olivaceum*. More
recently, the sulfoxide of eudistomin K and the unsubstituted eudistomin 51e
were isolated from *Ritterella sigillinoids*. These compounds display potent
activity against *Herpes simplex* virus, type I (HSV-I). This fact, coupled
with the unusual structural features, has attracted interest in 51 as a synthetic target and several groups have reported preliminary results.

We have recently completed the first total synthesis of (-)-eudistomin L
(51a) and (-)-debromoeudistomin L (51e) in an optically pure form possessing
the natural configuration as shown in Scheme 12.

The synthesis also provides direct evidence for the absolute configuration of eudistomins.

ACKNOWLEDGEMENT

We are grateful for support of this research by a Grant-in Aid for

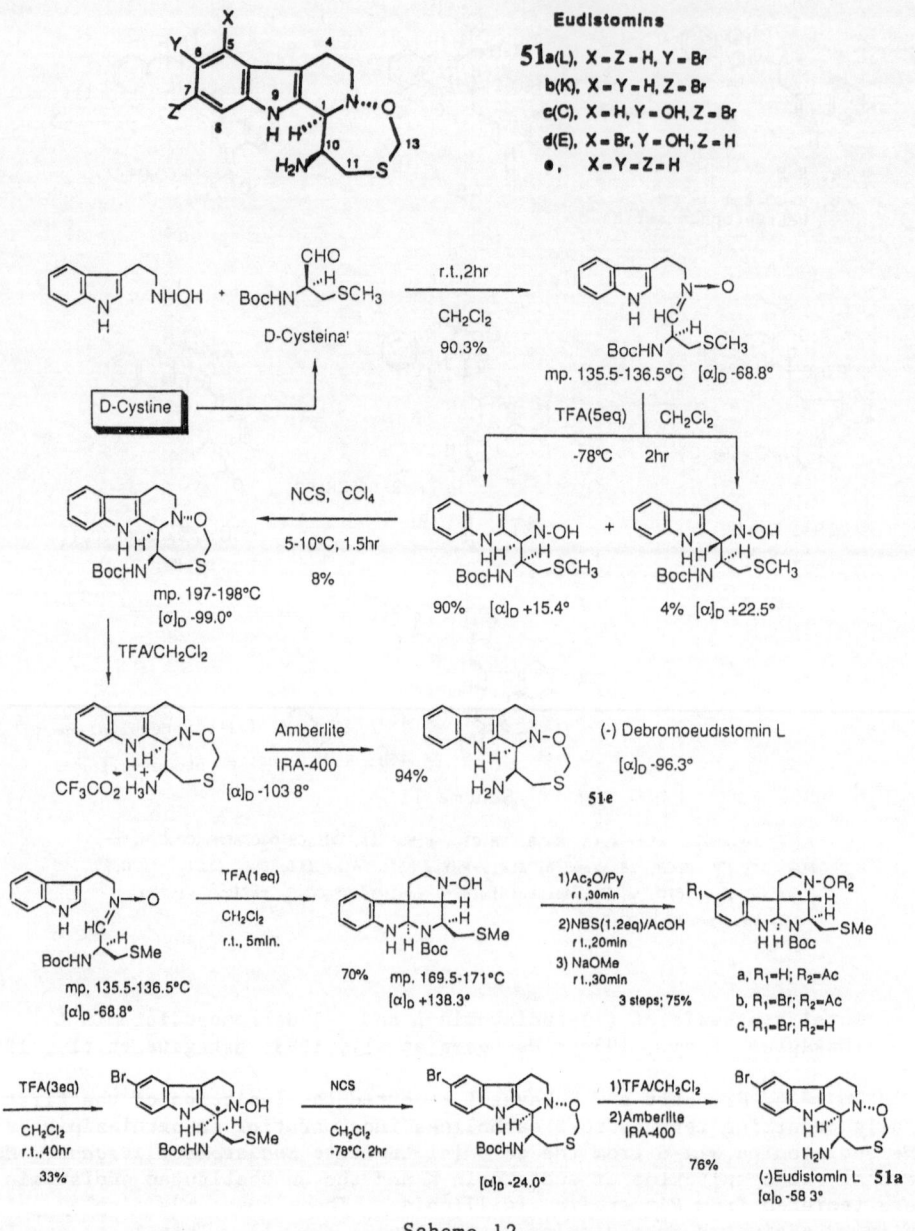

Eudistomins

51a(L), X = Z = H, Y = Br
b(K), X = Y = H, Z = Br
c(C), X = H, Y = OH, Z = Br
d(E), X = Br, Y = OH, Z = H
e, X = Y = Z = H

Scheme 12

Scientific Research (62470134 and 63105005) from the Ministry of Education, Science, and Culture, Japan and Uehara Memorial Foundation.

REFERENCES

Boyd, D.R., and Berchtold, G.A., 1979, Aromatization of arene 1,2-oxides. 1-Carboxy-and 1-carboalkoxybenzene oxides, J. Am. Chem. Soc., 101:2470-2474, and references cited therein.

Chang, C.-J., Floss, H.G., Hurley, L.H., and Zmijewski, M., 1976, Application of long-range spin-spin couplings in biosynthetic studies, J. **Org. Chem.**, 41:2932-2934.

Daly, J.W., Jerina, D.M., and Witkop, B., 1972, Arene oxides and the NIH shift: the metabolism, toxicity and carcinogenicity of aromatic compounds, **Experientia**, 28:1129-1264.

Gorman, M., and Lively, D.H.,1967, Pyrrolnitrin: a new mode of tryptophan metabolism, in: **"Antibiotics"** Vol. II, D. Gottlieb, and P.D. Shaw, eds., Springer-Verlag, New York, pp. 433-438.

Hayaishi, O., ed., 1962, **"Oxygenases"**, Academic Press, New York.

Hayaishi, O., ed., 1974, **"Molecular Mechanisms of Oxygen Activation"**, Chapter 1, History and Scope, Academic Press, New York, pp. 1-29.

Hino, T., Kodato, S., Takahashi, T., Yamaguchi, H., and Nakagawa, M., 1978, Oxidative dimerization of N^b-methoxycarbonyltryptamines by dye-sensitized photooxygenation in formic acid. Synthesis of (±)-folicanthine and (±)-chimonanthine, **Tetrah. Lett.**, 19:4913-4916.

Hino, T., Taniguchi, M., Yamamoto, I., Yamaguchi, K., and Nakagawa, M., 1981, Cyclic tautomers of tryptamines and tryptophans. V. Formation and reactions of cyclic tautomers of cyclo-L-tryptophanyl-L-proline, **Tetrah. Lett.**, 22:2565-2568.

Jerina, D.M., and Daly, J.W., 1974, Arene oxides: a new aspect of drug metabolism, **Science**, 185:573-582.

Kodato, S., Nakagawa, M., Hongu, M., Kawate, T., and Hino, T., 1988, Total synthesis of (+)-fumitremorgin B, its epimeric isomers, and demethoxy derivatives, **Tetrahedron**, 44:359-377.

Martin, L.L., Chang, C.-J., Floss, H.G., Mabe, J.A., Hagaman, E.W., and Wenkert, E., 1972, A ^{13}C nuclear magnetic resonance study on the biosynthesis of pyrrolnitrin from tryptophan by pseudomonas, J. **Am. Chem. Soc.**, 94:8942-8944.

Nakagawa, M., 1980, Oxygenation of tryptophan to formylkynurenine-dye-sensitized photooxygenation, in: **"Biochemical and Medical Aspects of Tryptophan Metabolism"**, Hayaishi, O., Ishimura, U., and Kido, R., eds., Elsevier/North Holland, pp. 49-58.

Nakagawa, M., Fukushima, H., Kawate, T., Hongu, M., Kodato, S., Une, T., Taniguchi, M., and Hino, T., 1986a, Synthetic approach to the total synthesis of fumitremorgins II synthesis of optically active pentacyclic intermediates and their dehydrogenation, **Tetrah. Lett.**, 27:3235-3238.

Nakagawa, M., Ito, M., Hasegawa, Y., Akashi, S., and Hino, T., 1984a, Total synthesis (+)-tryptoquivaline, **Tetrah. Lett.**, 25:3865-3868.

Nakagawa, M., Kato, S., Fukazawa, H., Hasegawa, Y., Miyazawa, J., and Hino, T., 1985a, Oxidative transformation of tryptophan to 2-(2-aminophenyl)-2-pyrrolidone and kynurenine, **Tetrah. Lett.**, 26:5871-5874.

Nakagawa, M., Kodato, S., Hongu, M., Kawate, T., and Hino, T., 1986b, Total synthesis of fumitremorgen B, **Tetrah. Lett.**, 27:6217-6220.

Nakagawa, M., Liu, J.-J., and Hino, T., 1989, Total synthesis of (-)-eudistomin L and (-)-debromoeudistomin L, J. **Am. Chem. Soc.**, 111:2721-2722.

Nakagawa, M., Liu, J.-J, Ogata, K., and Hino, T., 1986c, Synthetic approaches to eudistomins. Part 1. synthesis of 1-amino-3-thiaindolo[2,3-a] quinolizidine, **Tetrah. Lett.**, 27:6087-6090.

Nakagawa, M., Liu, J.-J., Ogata, K., and Hino, T., 1988, New evidence for the presence of a spiroindolenine intermediate in Pictet-Spengler reaction of N^b-hydroxytryptamine, **J.C.S. Chem. Comm.**, 463-464.

Nakagawa, M., Maruyama, T., Hirakoso, K., and Hino, T., 1980, Reactivity of oxytryptamine conversion to 3-(o-aminophenyl)-2-pyrrolidone and kynurenamine, **Tetrah. Lett.**, 21:4839-4842.

Nakagawa, M., Matsuki, K., and Hino, T., 1983a, A new synthesis of betacarboline, **Tetrah. Lett.**, 34:2171-2174.

Nakagawa, M., Sodeoka, M., Yamaguchi, K., and Hino, T., 1984b, Synthesis of the imidazo[1,2-α]indole-spirolactone ring system by oxidative double cyclization. A synthetic approach to tryptoquivalines, **Chem. Pharmacol. Bull.**, 32:1373-1384.

Nakagawa, M., Sugumi, H., Kodato, S., and Hino, T., 1981, Oxidative dimerization of N^b-acyltryptophans; total synthesis and absolute configuration
of ditryptophenaline, Tetrah. Lett., 22:5323-5326.

Nakagawa, M., Taniguchi, M., Sodeoka, M., Ito, M., Yamaguchi, K., and Hino,
T., 1983b, Total synthesis of (+)- and (-)-tryptoquivaline G by biomimetic
double cyclization, J. Am. Chem. Soc., 105:3709-3710.

Nakagawa, M., Yokoyama, Y., Kato, S., and Hino, T., 1984c, Oxidative transformation to 5-hydroxy-N-formylkynurenine, Heterocycles, 22:59-62.

Nakagawa, M., Yokoyama, Y., Kato, S., and Hino, T., 1985b, Dye-sensitized
photo-oxygenation of tryptophan, Tetrahedron, 41:2125-2132.

Sakan, T., and Hayaishi, O., 1950, α-hydroxytryptophan, not an intermediate
between tryptophan and kynurenine, J. Biol. Chem., 186:177-180.

Taniguchi, M., Anjiki, T., Nakagawa, M., and Hino, T., 1984, Formation and
reactions of the cyclic tautomers of tryptophans and tryptamines. VII.
Hydroxylation of tryptophans and tryptamines, Chem. Pharm. Bull., 32:2544-
2554.

STABILITY OF TRYPTOPHAN IN PEPTIDES AGAINST OXIDATION AND IRRADIATION

H. Steinhart

Institute of Biochemistry and Food Chemistry
University of Hamburg
2000 Hamburg
Germany

INTRODUCTION

Tryptophan (trp) is an essential amino acid for humans and most animals. It is, however, not the first and second limiting essential amino acid, but about 30% of the plant proteins used for human nutrition show a deficiency of trp. Besides cysteine, methionine and lysine, it belongs to the most reactive amino acids, which are endangered especially during food manufacturing processes. Friedman and Cuq (1988) recently reviewed the transformations of trp in foods. They report a lack of information of data concerning the stability of trp against industrial or home processing. Processing techniques, however, can cause losses in the trp content of food. Oxidations are of high significance. The most important reason for the instability of trp, especially against oxidizing agents and irradiation, is the bulky nonpolar aromatic side chain.

In order to understand the reactions of trp with oxidizing agents and against irradiation in oligopeptides, it is first necessary to study the reactions of the free trp, then to use model peptides. The results of the studies with free tryptophan are published elsewhere (Kell, 1988; Troeder, 1988).

The parameters temperature (T) and pH-value were changed in the experiments in which oxidation of trp in peptides was studied. The dipeptides ala-trp and phe-trp were oxidized with H_2O_2. The irradiation experiments were conducted by using a Co-60 bomb as irradiation source. The dipeptides trp-gly, gly-trp, trp-ala, ala-trp, trp-leu, leu-trp, trp-phe, phe-trp, met-trp, and the tripeptides gly-trp-gly, ala-trp-ala, leu-trp-leu were under investigation. The concentrations of trp containing peptides and the energy of irradiation were changed. In order to eliminate oxidation reactions, the trp solutions were treated with an inert gas (Ar) before starting the irradiation procedure. In other experiments, however, N_2O was used to study the combined effect of irradiation and of oxidation. In all cases the solvent was water. The reaction products of both experimental series which were determined by analytical means were water soluble and not volatile. New GC and HPLC techniques were used in order to determine the reaction products.

OXIDATION OF PEPTIDE BOUND TRYPTOPHAN WITH H₂O₂

Experimental

Oxidation of dipeptides at pH 7 and 8: 612.5 μmoles of the above mentioned dipeptides were dissolved in 100 ml phosphate buffer of pH 7 or 8 at 50°C in an ultrasonic bath. Either 2 ml 1 N H_2O_2 or 2 ml H_2O (control system) were added to 8 ml of these solutions. The solutions were stirred under reflux for 1 hour at 25, 60 or 100°C. The reaction mixture was then immediately cooled in an ice bath and lyophilized. After the addition of 10 ml methanol, the solutions were centrifuged and filtered. An aliquot of 200 μl was pipetted into amber vials with teflon lined screw caps and concentrated to dryness with N_2. The residues were either hydrolyzed or derivatized.

Derivatization of the dipeptides: Introducing a modified method for the derivatization, trp and most of the degradation products can be determined very easily by gas chromatography (gc). The esterification is conducted by using isopropanol, and acylating is performed with trifluoroacetic acid anhydride (TFAA). It is important to choose proper reaction conditions (higher concentration of TFAA, reaction time 15 hours).

One μmole of the dipeptides under investigation was weighed into 4 ml amber vials with teflon lined screw caps and 1 ml 1.5 N HCl/isopropanol was added. The vials were closed with teflon tape before being screwed up. They were treated in an ultrasonic bath for 1 min and heated at 100°C for 1 hour. The solutions were concentrated to dryness with N_2. 200 ml TFAA/methylene chloride (100 μl TFAA diluted with methylene chloride to 10 ml = 0.0714 mmole TFFA/ml, freshly prepared), and 200 ml methylene chloride were added. The solutions were treated in an ultrasonic bath for 5 min and heated at 100°C for 45 min, then concentrated to dryness and filled up with 200 ml internal standard solution. These conditions yield derivatives with one trifluoroacyl group on the amino nitrogen, confirmed by ms. The derivatization is linear between 0.2 and 5 μmoles/ml. Kynurenine (kyn) and N-formyl-kynurenine cannot be distinguished because the same derivative is formed for both compounds. The reason is that the formyl group is replaced by the trifluoroacyl group.

Hydrolysis of dipeptides: An acid hydrolysis method proposed by Yokote et al. (1986) was modified. This method is rapid and the reagents are easy to remove as compared to often used alkaline hydrolysis procedures. Also adsorptions on precipitating salts will not occur. Thioglycolic acid, however, which was used by Yokote et al. (1986) in order to protect trp against oxidation during the hydrolysis, disturbs the gc determination. The reducing group of the molecule is the thiol group. Therefore other compounds with thiol groups may cause the same reducing effect. A volatile mercaptan, butylmercaptan, was used because it is easy to remove from the reaction mixture. Absence of O_2 was obtained by gassing with Ar.

One μl of the respective dipeptide was weighed into 4 ml amber vials with teflon lined screw caps and carefully gassed with Ar for 2 min. 200 μl HCl conc.; 100 μl trifluoroacetic acid and 10 μl butylmercaptan were added and the vials were closed. The solutions were heated at 160-165°C for 25 min, and concentrated to dryness with N_2 at 80°C. The residues were derivatized for gc determination. Recovery of trp was 100% (ala-trp) and 84% (phe-trp). Recovery of ala was 100% and of phe 94%.

Apparatus: Gc determinations of trp degradation products were carried out with a Hewlett Packard gc 5700 A equipped with a capillar-gc addition (Fa. Gerstel) using a 60 m DB 5 fused silica-column, 0.32 mm i.d., 0.25 μm film thickness. Myristic acid methylester was used as internal standard for quantitative determinations. Identities were confirmed by gc-mass spectro-

Table 1. Oxidation of ala-trp

| | | | Recoveries | | | | |
| | | | Before hydrolysis | | | After hydrolysis | |
pH	°C	Control	ala-trp	ala	trp	ala	trp
7.0	25	76.9	2.9	1.8	-	84.8	2.2
	60	86.7	4.1	2.7	-	78.0	1.9
	100	81.0	3.0	2.7	-	67.4	2.0
8.0	25	87.8	4.8	1.3	0.4	94.4	2.5
	60	93.7	4.0	4.2	0.2	88.3	3.3
	100	95.1	2.6	3.1	0.2	86.6	0.4

Recoveries are expressed in %; - : below detection limit (0.1 mole/100 moles).

metry (ms) (Finnigan MAT 311 A ms) using the same column as described above and by comparing with standard compounds. Operating conditions included: electron energy 70 eV, accelerator voltage 3000 kV, electron impact (CI) and chemical ionization (EI) measurements.

RESULTS

Oxidation of dipeptides: Results of the treatment of ala-trp and phe-trp with and without addition of H_2O_2 at pH 7 and 8 are summarized in tables 1 and 2. The values are the averages of 2 determinations.

In general, the amounts of free amino acids before hydrolysis do not exceed 1% with the exception of ala (2.7%). The recoveries of the dipeptides range between 76.9% and 96.9% with an average of about 90%. These values indicate that the oxidation conditions do not cause hydrolysis of the dipeptides in higher amounts. The chemical modifications of the dipeptides occur almost completely during the oxidation with H_2O_2. The recoveries of the dipeptides are only 3%. But new compounds cannot be determined by gc in

Table 2. Oxidation of phe-trp

| | | | Recoveries | | | | |
| | | | Before hydrolysis | | | After hydrolysis | |
pH	°C	Control	phe-trp	phe	trp	phe	trp
7.0	25	83.9	3.1	0.3	0.7	79.5	15.0
	60	96.3	3.0	0.3	-	73.8	7.7
	100	87.2	2.5	0.7	-	78.3	2.6
8.0	25	96.9	4.4	0.5	-	66.1	5.2
	60	94.9	3.9	0.6	-	62.2	8.4
	100	88.7	3.0	0.7	-	76.1	0.4

Recoveries are expressed in %; - : below detection limit (0.1 mole/100 moles).

Table 3. Identified degradation products formed by oxidation of ala-trp and phe-trp

pH	°C	Dipeptide	ser	ind	kyn	5-OHT	oia
7.0	25	ala-trp	-	-	-	-	13.0
	60		-	-	-	-	7.0
	100		-	-	1.5	-	-
8.0	25		-	-	3.3	9.2	14.2
	60		-	-	1.4	9.5	18.4
	100		-	-	4.7	5.2	2.1
7.0	25	phe-trp	-	-	-	-	-
	60				2.6	4.1	32.1
	100		-	-	2.2	-	21.5
8.0	25		-	-	0.9	-	17.0
	60		6.9	2.6	1.3	-	11.9
	100		7.5	5.0	3.5	-	29.5

Degradation products are expressed in %; for abbreviations see text;
- : below detection limit (0.1 mole/100 moles)

amounts which may explain the high losses. The recoveries of ala and phe after hydrolysis of the oxidized dipeptides range between 62.2% and 94.4%. The recoveries of ala are mostly higher compared to phe. In contrast to the high recoveries of ala and phe, the rates for trp are very low. In treated ala-trp, recovery of trp ranges between 0.4% and 3.3%, in treated phe-trp between 0.4% and 15.0%. The experiments show that trp is destroyed in dipeptides under oxidizing conditions very fast, even at room temperature and neutral pH value.

Degradation products: The identified oxidation products by gc are listed in Table 3. The values are averages of 2 determinations.

Oxindolylalanine (oia) is formed most effectively at almost all reaction conditions. The highest amounts are formed in the case of ala-trp at pH 8 and at lower temperatures, in the case of phe-trp at higher temperatures. Smaller amounts of ser, indoleacetic acid (ind), kynurenine + N-formylkynurenine (kyn) and 5-hydroxy-tryptophan (5-OHT) are formed at different reaction conditions. The amounts normally correspond with those found with free trp. In the case of phe-trp, however, higher amounts of ind, ser and oia are detected as compared to free trp.

DISCUSSION

The determined degradation products of trp in the dipeptides cannot wholly explain the loss of trp at each reaction condition. Only lower quantities of unidentified volatile minor components appear in the gc after derivatization. A higher amount of degradation products, not volatile even after derivatization, seems to arise, which cannot be determined with gc analysis. A large part of these unidentified degradation products may consist of polymeric compounds. This is also indicated by yellow to brown colored products and partly brown precipitates in the reaction solutions. Polymerisations possibly occur at alkaline as well as at acidic reaction conditions. The identity of the emerging products has to be clarified in further studies, especially because some of these products may be toxic.

The attack on the pyrrole ring is the main step of the oxidation of trp. The resulting products of the attack are oia, kyn and 3-OH-kynurenine which, however, could only be detected after oxidation of free trp. At the same time, oxidative cleavages of C-C bonds happen. The resulting products are ser, ind, and in the case of free trp also gly, ala, asp, and aminobenzoic acid.

The influence of the neighboring amino acids in peptides containing trp on the degradation of trp was investigated with the dipeptides ala-trp and phe-trp. The reactions were carried out under similar reaction conditions in order to ascertain whether similar reaction products can be detected. The reactions were, however, only conducted in neutral and weak alkaline media because differences in reaction products seemed to be evident between these pH conditions. Losses of dipeptides during oxidative reactions result mainly from degradation of trp in the dipeptides, because recoveries of ala and phe after hydrolysis are of the same amount as recoveries in the control systems. Compounds which could explain the noticed trp losses could not be detected by gc. This result can be explained by a modification of trp in the dipeptide. Ala and phe were set free by hydrolysis, but trp was no longer in a form detectable by gc. The formation of polymers seems to be possible as well.

Losses of trp by oxidation of ala-trp are of the same amount as by oxidation under similar conditions of free trp. Loss of trp after oxidation of phe-trp under similar conditions are much lower. An electrophilic attack on trp in the dipeptide phe-trp is more difficult because the phenyl ring of phe exerts a -I effect. This effect cannot occur at ala-trp, and the attack of trp is therefore not influenced. Trp is not more stable against oxidative attacks.

Hardly any differences in the formation of degradation products of ala-trp and phe-trp can be noticed at relatively mild conditions at pH 7.0. The formation of the main oxidation product oia is of the same quantity as for oxidation of free trp under similar conditions (Kell, 1988). Differences in the formation of degradation products of the di-peptides are obvious at pH 8.0. The reason is the increase of radical attacks.

These results indicate that it is not appropriate to compare results of reactions of free amino acids with results obtained with the same amino acids bound in peptides. The expected influences of adjoining amino acids on the reactivity of the amino acid under investigation may increase for peptides with more than two amino acids and for proteins. The influence of further adjoining amino acids on the oxidation behavior of trp should be studied in more detail in order to understand such complex systems.

IRRADIATION OF PEPTIDE BOUND TRYPTOPHAN WITH GAMMA RAYS

Experimental

Irradiation: Aqueous solutions of different trp containing di- and tri-peptides within a concentration range of 10 to 360 mg peptides/100 ml water were used for the irradiation experiments. 10 ml of a trp base solution were pipetted into a 20 ml sample vessel (polyethylene without softener). The trp solutions were either gassed for 30 min with Ar or with N_2O in order to get them free of oxygen. The prepared samples were irradiated in an irradiation source, working with Co-60. The intensity of irradiation of the samples differed from 5 to 60 kGy.

Enzymatic hydrolysis of the peptides: Hydrolysis of the trp containing peptides after irradiation was necessary in order to determine the trp losses during the irradiation of the peptides. Procedures in alkaline media are

33

Table 4. Optimal hydrolysis times with Pronase for different
          oligopeptides

Peptide	Time (hours)	Recovery of amino acids (in %)
trp-gly	8	trp 85.2; gly 82.0
gly-trp	24	trp 67.0; gly 67.8
trp-leu	6	trp 97.9; leu 85.1
leu-trp	15	trp 100 ; leu 65.3
trp-phe	2	trp 92.6; phe 78.4
phe-trp	2	trp 92.8; phe 76.8
gly-trp-gly	4	trp 65.1; gly 55.9
ala-trp-ala	4	trp 89.8; ala 84.5
leu-trp-leu	6	trp 80.0; leu 62.3

normally used for this purpose (Vangala and Menden, 1970; Lucas and Sotelo, 1980). The drawbacks of these methods are the occurrence of racemisation of the trp and the loss of trp during the removal of the cations (Steinhart, 1978). The available enzymatic methods are mild procedures without those drawbacks. The disadvantages are, however, that there are not suitable enzymes which may split all possible peptide bonds in proteins and peptides, and that the hydrolysis takes a very long time. But here the aim was the splitting of oligopeptides. It was therefore proposed to use a mild enzymatic procedure. Kanner and Fennema (1987) used Pronase E for this purpose. This enzyme mixture splits unspecifically, and normally only short hydrolysis times are needed. The described method was optimized for these experiments. The enzyme had to be removed from the reaction mixture after hydrolysis because it disturbed the derivatization of the amino acids for the following gc determination. This was achieved by gel filtration.

One ml of the sample solution is pipetted into amber vials with screw caps, and the pH value is brought to 7.4 with 0.01 N NaOH, if necessary. One ml of Pronase solution (0.25 mg Pronase E/ml) is added. Activity of Pronase E was 8 DMC-U/mg according to the proposal of Lin et al. (1969). Pronase E isolated from Streptomyces griseus (EC 3.4.24.4) was purchased from Fa. Serva, Heidelberg. The incubation temperature was 40°C. The optimal incubation time varied in dependence on the investigated peptide. Table 4 shows the dependence of the optimal hydrolysis time on the splitting of the respective peptides. The separation of the enzyme from the reaction mixture after finishing the hydrolysis is carried out by gel filtration with the gel column PD-10 (Fa. Pharmacia) and the gel Sephadex G-25 M, volume 9, 1 ml. The elution is carried through with water, the first 6 ml are discarded, 30 ml are collected. The eluate is ready for derivatization after lyophilization.

Gas chromatography: The gc conditions are the same as described above.

RESULTS

Irradiation of dipeptides: The amount of decomposition of trp in the oligopeptides is not dependent on the amounts of irradiated oligopeptides in the chosen ranges. Fig. 1 shows the losses of trp in the peptides trp-gly and gly-trp. The losses increase with the irradiation dose. The lowest losses of trp occur by irradiating with 10 kGy. They increase up to 100% by irradiating with 60 kGy. They are generally lower when the reaction mixture is gassed with Ar compared to N_2O. There is also an influence of the position of the trp in the peptide. The losses are smaller when the amino group of the trp is free.

34

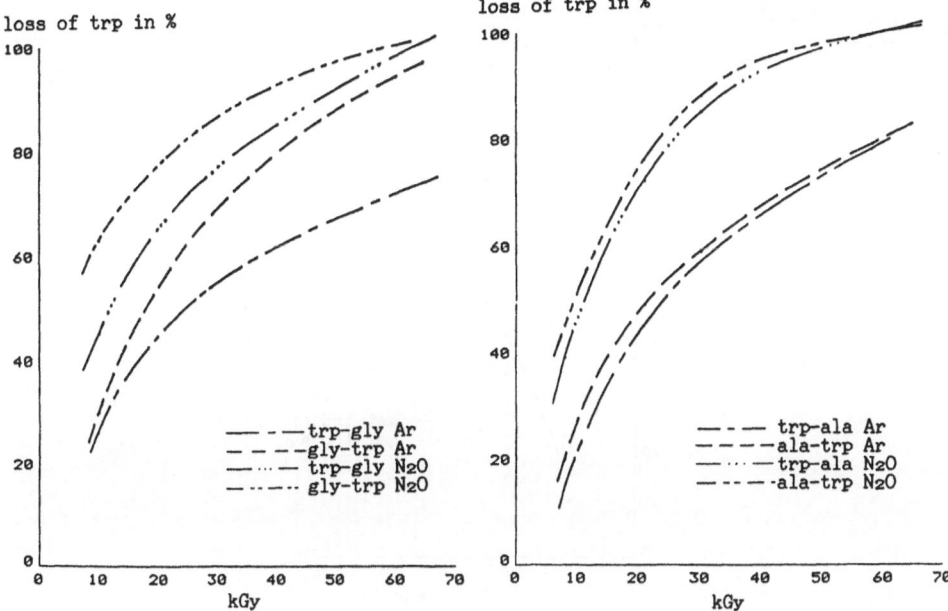

Fig. 1. Loss of trp after irradiation of trp-gly and gly-trp in Ar and in N_2O atmosphere.

Fig. 2. Loss of trp after irradiation of trp-ala and ala-trp in Ar and in N_2O atmosphere.

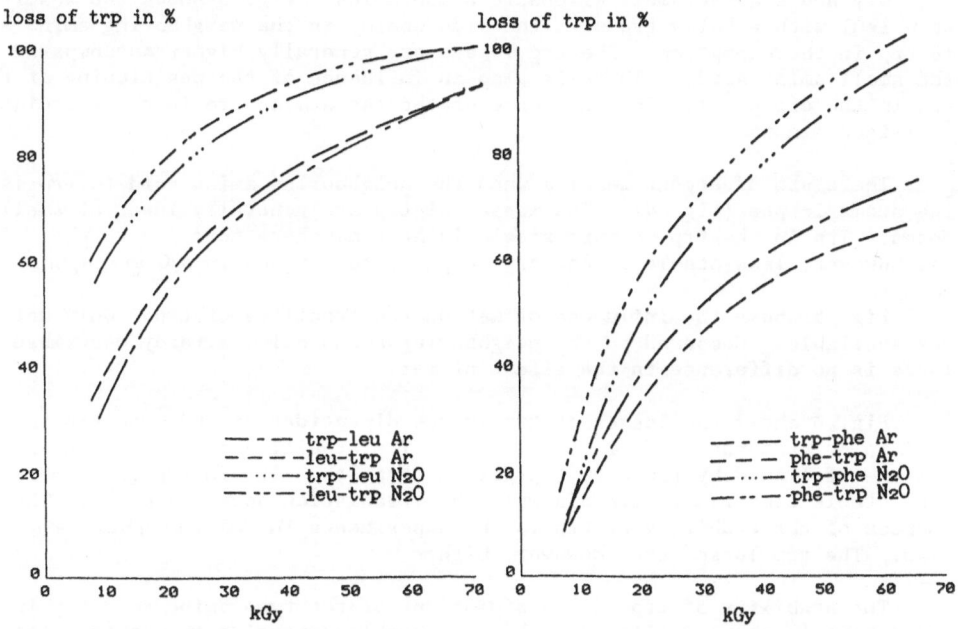

Fig. 3. Loss of trp after irradiation of trp-leu and leu-trp in Ar and in N_2O atmosphere.

Fig. 4. Loss of trp after irradiation of trp-phe and phe-trp in Ar and in N_2O atmosphere.

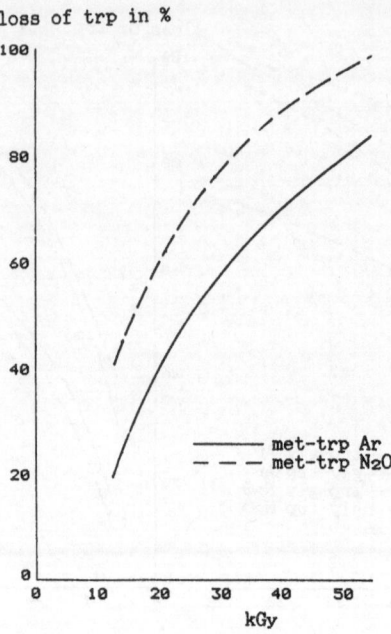

loss of trp in %

Fig. 5. Loss of trp after irradiation of met-trp in Ar and in N_2O atmos-
         phere.

        Similar results are obtained by irradiating trp-ala and ala-trp (Fig.
2). But the influence of the trp position in this dipeptide is smaller.

        Gly and ala are small aliphatic amino acids. Fig. 3 shows the results
when leu, with a bulky group in the side chain, is the neighboring amino acid
to trp in the dipeptide. The trp losses are generally higher as compared to
the small amino acids. There is also an influence of the positioning of the
trp in the dipeptide. The influence of the gas atmosphere in the irradiation
vial is dominant.

        There are divergent results when the neighboring amino acid to trp is
the aromatic phe (Fig. 4). The losses of trp are generally lower at small
doses. Trp in phe-trp is more stable in Ar atmosphere than in trp-phe. Trp
is, however, less stable in phe-trp compared to trp-phe in N_2O atmosphere.

        Fig. 5 shows the influence of met on the stability of trp. Only met-trp
was available. Compared to the neighboring amino acids already described,
there is no difference in the effect of met.

        Fig. 6 shows the losses of trp in the dipeptides of the type trp-amino
acid in Ar atmosphere after irradiation. The stability of trp is highest in
trp-phe, followed by trp-ala, trp-gly and trp-leu. Trp in trp-gly remains
more stable compared to trp-ala when the irradiation dose increases. The sequences of the stability of trp in the experiments in N_2O atmosphere are similar. The trp losses are, however, higher.

        The stability of trp in the dipeptides of the type amino acid-trp in Ar
atmosphere is shown in Fig. 7. The most stable dipeptide is phe-trp, followed by ala-trp, and similar losses of trp are found in gly-trp and leu-trp.
The sequences in the experiments in N_2O atmosphere are comparable to those in
Ar atmosphere, but the losses are higher.

        Irradiation of tripeptides: The tripeptides gly-trp-gly, ala-trp-ala

36

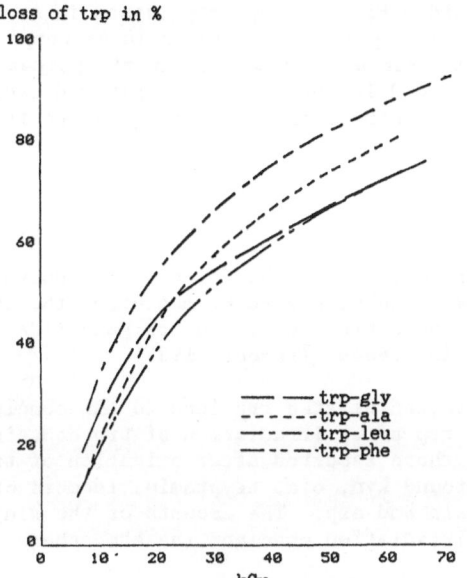

Fig. 6. Loss of trp after
irradiation of dipeptides
of the type trp-amino acid
Ar atmosphere.

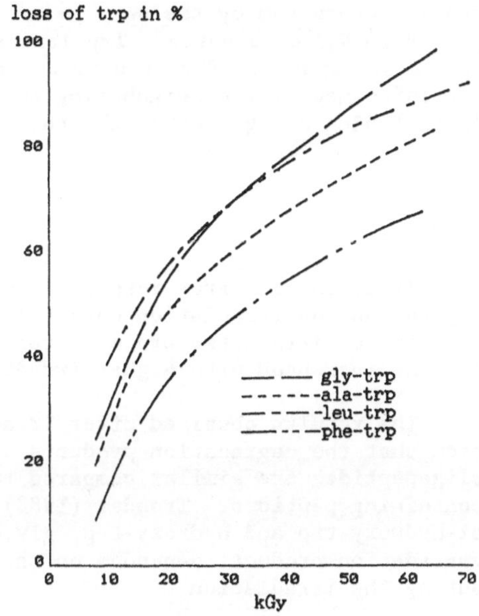

Fig. 7. Loss of trp after
irradiation of dipeptides
of the type amino acid-trp
in Ar atmosphere.

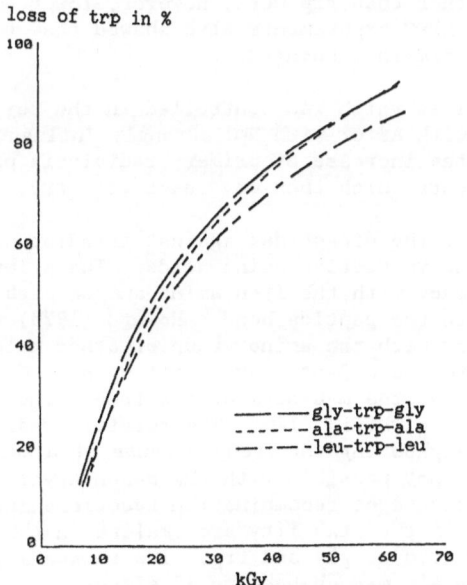

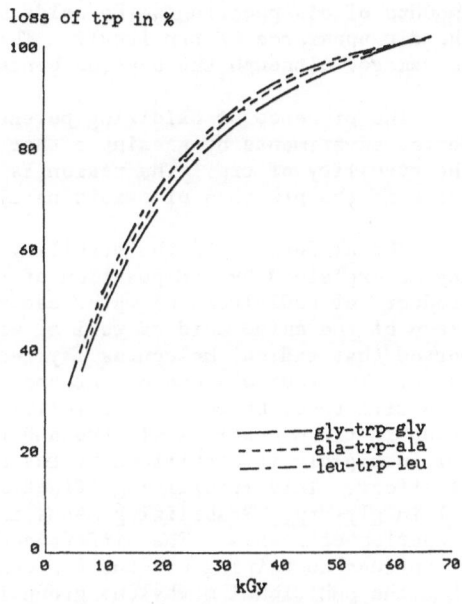

Fig. 8. Loss of trp after irradiation of tripeptides in
Ar atmosphere.

Fig. 9. Loss of trp after irradiation of tripeptides in
N₂O atmosphere.

and leu-trp-leu were irradiated. The amino and the carboxyl groups of the trp are protected by the same amino acid. Fig. 8 shows trp losses in Ar and Fig. 9 in N_2O atmosphere. Trp losses are significantly lower in Ar compared to N_2O atmosphere. The results are comparable to those of the dipeptides. The influence of the neighboring amino acid is, however, strongly reduced. Establishing a sequence of the three adjoining amino acids in question is of use now.

DISCUSSION

It is obvious from results with free trp that the formation of polymers depends on the irradiation dose. This could be proved by measuring the extinction of irradiated trp mixtures. The extinction in the oligopeptide solutions increased with higher irradiation doses (Troeder, 1988).

The results obtained after irradiation of free trp lead to the conclusion that the degradation products of trp after irradiation of trp containing oligopeptides are similar compared to those reported after oxidation of trp containing peptides. Troeder (1988) found kyn, oia, tryptamin, isomers of di-hydroxy-trp and hydroxy-trp, gly, ala and asp. The amounts of the single degradation products depended on the irradiation dose and the atmosphere during the irradiation.

The most important parameters for the stability of the trp containing dipeptides are the position of the trp, the neighboring amino acid and the oxidation/reduction status of the reaction mixtures. This status was controlled in our experiments by gassing either with Ar or with N_2O. The destruction of the dipeptides increases within the irradiation range of 10 - 70 kGy according to a logarithmic function. The increase of the losses in the lower energy range is higher than in the higher energy range. Additional experiments with HPLC separation of the dipeptides showed that not only was trp damaged during the irradiation but also the neighboring amino acids. The amounts of disappearing amino acids other than trp were, however, lower than the disappearance of trp itself. The HPLC experiments also showed that trp is damaged although the peptide bonds remain unchanged.

The presence of oxidizing potentials which was controlled in the reported experiments by gassing either with Ar or with N_2O strongly influenced the stability of trp. The reason is the increase of primary radiolysis products in the presence of oxidizing agents which then may react with trp.

The difference in the stability of the dipeptides against irradiation may be explained by the position of the respective amino acids. The primary products of radiolysis of water can react with the free amino and/or carboxyl group of the amino acid as well as with the peptide bond. Howard (1973) reported that radical molecules may react with the amino group of other molecules. The radicals convert at once to the α-C-atom by splitting H_2. There is a difference between the reactivity of the α-C-atom of the free amino group in the dipeptides gly-trp and trp-gly (Fig. 10). The tertiary radical in trp-gly is stabilized by the neighboring CH_2 group because of a small +I effect. This stabilizing effect is not possible with the secondary radical in gly-trp. Stabilizing means increase of recombination reactions and reduction of losses. The differences in the stability are smaller in the dipeptides containing ala and trp compared to gly and trp. The reason may be that the additional methylene group in ala may change the +I effect.

A possible explanation for the relatively high protecting effect of the phe may be the -I effect of the phenyl ring which makes the electrophilic attack of the OH radical more difficult. Simic (1978) found a range for the reaction velocity of aliphatic amino acids with OH radicals: leu > gly > ala.

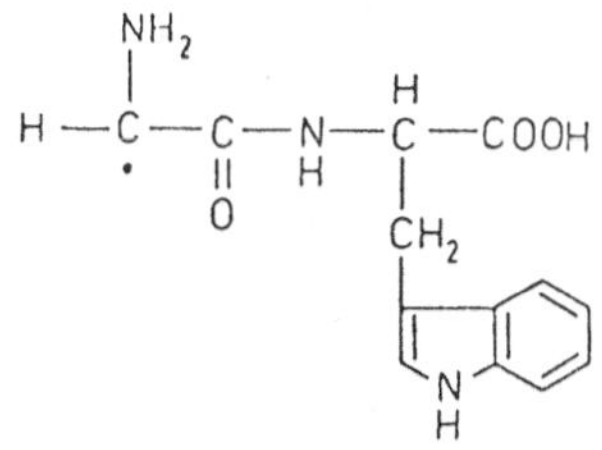

Fig. 10. Structure of the molecule radicals trp-gly and gly-trp.

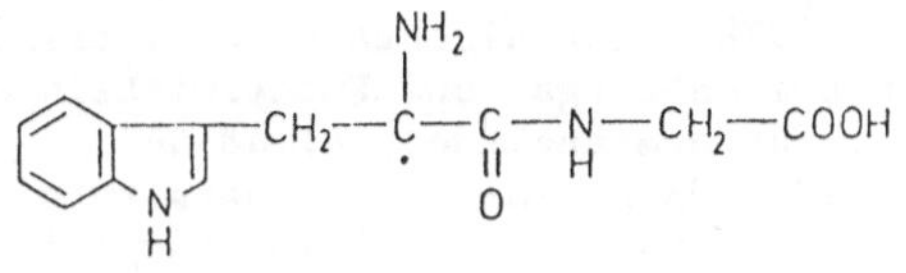

The recovery rates of trp in the dipeptides in these experiments correspond to this range. The lower the reaction velocity of the OH radical with amino acids, the lower are the losses of trp. Therefore the stability of the trp in dipeptides against irradiation depends at least on the I effect and on the reaction velocity of the OH radical on the neighboring amino acid.

The trp in the tripeptides is more stable as compared to the dipeptides. The differences between the amino acids neighboring trp are also small. Only the influences of oxidation potentials are similar to the investigations with dipeptides. This may be explained by steric effects and also by changes in the chemical stability of possible intermediary products. It is difficult to infer the stability of trp in peptides and proteins from these investigations. However the results of the experiments also show that there are manifold effects of neighboring amino acids on the stability of trp.

The importance of the reported results (oxidizing and irradiation experiments) lies in changes of the trp content in food during processing. There are oxidizing potentials in all steps of food processing. The influence of the irradiation on trp containing peptides is important in connection with the irradiation of food in order to impede the growth of microorganisms.

REFERENCES

Friedman, M., and Cuq, J. C., 1988, Chemistry, analysis, nutritional value, and toxicology of tryptophan in food. A review, J. Agric. Food Chem., 36:1079-1093.

Howard, J.A. 1973, Homogeneous liquid-phase autoxidations, in: "Free Radicals, Part II", Kochi, J.K., ed., Wiley, New York, pp. 3-62.

Kanner, J. D., and Fennema, O., 1987, Photooxidation of tryptophan in the presence of riboflavin, J. Agric. Food Chem., 35:71-76.

Kell, G., 1988, Ph.D. Thesis, University of Hamburg.

Lin, Y, Means, G. E., and Feeney, R.E., 1969, The action of proteolytic enzymes on N, N-dimethyl proteins, J. Biol. Chem., 244:789-793.

Lucas, B., and Sotelo, A., 1980, Effects of different alkalies, temperature and hydrolysis times on tryptophan determination of pure proteins and of food, Anal. Biochem., 109:192-197.

Simic, M.G., 1978, Radiation chemistry of amino acids and peptides in aqueous solutions, J. Agric. Food Chem., 26:6-14.

Steinhart, H., 1978, Eine direkte fluorimetrische Bestimmungsmethode für Tryptophan aus Nahrungs- und Futtermittelhydrolysaten, **Z. Tierphysiol. Tierernähr. Futtermittelkde.**, 41:48-56.

Troeder, U., 1988, Ph.D. Thesis, University of Hamburg.

Vangala, R. R., and Menden, E., 1970, Vergleich verschiedener Methoden zur Bestimmung von Tryptophan in Proteinen, **Z. Lebensmitt. Forsch.**, 142:195-204.

Yokote, Y., Arai, K. M., and Akahane, K., 1986, Recovery of tryptophan from 25 minute acid hydrolysates of protein, **Anal. Biochem.**, 152:245-249.

MASS SPECTROMETRIC DETERMINATIONS OF TRYPTOPHAN AND ITS METABOLITES

S.P. Markey, R.L. Boni, J.A. Yergey[1], and M.P. Heyes

Section on Analytical Biochemistry
Laboratory of Clinical Science
National Institute of Mental Health

[1]Laboratory of Clinical Studies
NIAAA
Bethesda, Maryland 20892
USA

INTRODUCTION

The unique character of a chemical compound is determined by its molecular architecture. That architecture is a sum not only of all of the atoms in the compound and their masses, but of their particular relationship in space, i.e., the molecular bonds, sub-structural components, and their stereochemistry. Consequently, a physical tool which can measure both the summed masses of the elements in a compound and reflect the subtleties of their arrangement is very useful in both quantitative and qualitative investigations in biochemistry.

Mass spectrometers can measure the mass of small quantities of most organic compounds accurately, and because of the techniques associated with the mass measurement process, these instruments generally reflect very subtle differences in molecular structure. There is now a large variety of mass spectrometers employed in biochemistry, but there are several principles which they share. First, a neutral organic molecule cannot be directed in space, but once ionized, it can be accelerated, steered, focussed, or otherwise directed by either electric or magnetic fields. Second, an ionized molecule can only be directed by external fields if its movement is unrestricted by the action of other molecules. Consequently, all mass spectrometers require high vacuum in the region of the mass analyzer. In fact, most of the cost and physical size of mass spectrometers is determined by this requirement. Third, the quantity of ionized compound required for modern detectors is in the range of hundreds or thousands of ions (10^{-21} or 10^{-20} moles), so that the instrumentation is inherently sensitive. However, due to the low efficiency of most ionization processes, current detection limits are 10^{-15} to 10^{-18} moles, and those limits are for the most favorable cases. Nevertheless, mass spectrometry remains among the most sensitive and specific of the physical chemical measurement methods.

What these general characteristics mean to biochemists studying tryptophan metabolism is that mass spectrometry is an appropriate tool for high sensitivity measurements requiring a high degree of structural differentiation, but that there will be many alternative specific approaches to any

measurement to be made. Two major components of compound differentiation
available to the mass spectroscopist are the choice of how a compound is to
be presented to the instrument (gas, solid, liquid, chromatographic effluent,
etc.) and the method of its subsequent ionization (electron impact, fast atom
bombardment, thermospray, etc.). In most of the examples which follow, on-
line gas chromatography has been the method of choice for sample introduction
into the mass spectrometer. This is because most biological fluids and their
extracts contain very complex mixtures of materials. By coupling a separation technique with mass analysis, an additional element of selectivity is
introduced. Further, online and automated gas or liquid chromatography provides a convenient way to admit samples sequentially to analytical instrumentation. A chromatographic column can absorb most of the chemical insult
which would otherwise be directed into an ionization chamber. Interposing a
removable and disposable guard element between the analyst and the instrumentation minimizes maintenance and maximizes usable instrument time. The result of improvements in all of the elements of chromatographic-mass spectrometric systems is that this instrumentation is highly reliable, faster, less
expensive, easier to use, and consequently, a preferred analytical method for
many applications. Gas chromatography requires that the compounds of interest be stable in the vapor phase, a requirement not met by tryptophan metabolites unless chemically derivatized for that purpose. Much of the analytical literature describing measurements of metabolites by gas chromatography-mass spectrometry (GC/MS) is largely devoted to derivatization chemistry.

While sensitivity and selectivity are attainable with mass spectrometric instrumentation, there are many other sensitive and selective detectors
for gas and liquid chromatography, as well as highly sensitive immunoassay
methods. Many of these alternatives are both less costly and simpler to use.
However, there is another property of organic compounds which defines the
utility of mass spectrometry in biochemistry, and that is the occurrence of
stable isotopes of carbon (^{13}C), nitrogen (^{15}N), oxygen (^{18}O), and hydrogen
(^{2}H or D). Physically and chemically, all of the isotopes of an element are
nearly indistinguishable. In nature, the low abundance of stable, non-radioactive, isotopes of carbon, nitrogen, oxygen, and hydrogen makes their relative contribution to a complex molecule small. That is, while ^{13}C, the most
common of these, occurs at an abundance of 1.1% of ^{12}C, the chance of having
more than one ^{13}C in any given molecule is $n(1.1\%)^y$, where n is the number of
carbon atoms in a molecule, and y equals the number of multiple ^{13}C atoms.
Consequently, molecules intentionally synthesized containing a high abundance
of one or more stable isotopes will be unique in nature, while retaining the
chemical properties of the natural material. Stable isotope labeled variants
of compounds are known as "isotopomers". Isotopomers will generally be indistinguishable from the natural material with regard to chemical reactivity
and physical properties, with the exception of molecules highly enriched with
^{2}H. That is, the mass of the rarer isotopes of carbon, nitrogen, and oxygen
differs by approximately ten percent from each of the respective more abundant isotopes, whereas the hydrogen-deuterium mass difference (1 vs. 2 daltons, or 100%) confers significant differences in bond strength and polarity.
Only mass selective detection will differentiate most isotopomers, and thus
they are ideal species for tracing metabolic pathways, measuring kinetics, or
as internal standards for quantitative analysis.

After reviewing the tryptophan literature, we have chosen several examples of qualitative and quantitative measurements which illustrate mass
spectrometry as a tool in tryptophan metabolism studies.

QUALITATIVE APPLICATIONS OF MASS SPECTROMETRY

Characterization of an unknown metabolite: The unambiguous demonstration of the presence of a known compound, and the determination of the structure of a previously unknown compound, are common qualitative applications of mass spectrometry. Most tryptophan metabolite structures were determined by classical chemical structural methods before the availability of organic mass spectrometers. However, detailed examination of the metabolism of well-known compounds can still reveal unsuspected structures. One such example is the recent report of a new cysteine adduct of the aldehyde produced by enzymatic oxidation of tryptamine in rat brain homogenates. Susilo et al. (1988) traced the metabolism of ^{14}C-tryptamine in homogenates, and in addition to the expected indole-3-acetaldehyde and indole-3-acetic acid metabolites, they reported the detection of an unknown compound by radio-thin layer chromatography. The new metabolite was extractable into polar organic solvents; was formed more readily by brain than by liver homogenates; was formed more avidly in the presence of non-ionic detergents; was not formed when monoamine oxidase inhibitors were added; and became the major metabolite of tryptamine when low substrate concentrations (μM) were used. The unknown was isolated by high performance liquid chromatography, and characterized by fast atom bombardment mass spectrometry. In this technique, the sample, solubilized in glycerol and placed inside the spectrometer, is subjected to a beam of high energy atoms. The atom beam causes the sample on the glycerol surface to vaporize and ionize (frequently by proton attachment) by a sputtering mechanism. The tryptamine metabolite mass spectrum contained a prominent ion at m/z 263[*], arising from the protonation of a neutral molecule of molecular weight 262. An ion at m/z 130 indicated that the indole nucleus was unsubstituted, as an analogous fragment is seen in the spectra of tryptamine and tryptophan. The non-enzymatic condensation of cysteine with indole-3-acetaldehyde to produce a thiazolidine derivative was postulated and then proven by comparison with a synthetic standard of (4R)-2-(3-indolylmethyl)-1,3-thiazolidine-4-carboxylic acid.

Tryptamine — Indole-3-acetaldehyde
(4R)-2-(3-indolylmethyl)-1,3-thiazolidine-4-carboxylic acid — L-cysteine

Confirmation of the presence of a known compound: Other examples of the qualitative uses of mass spectrometry are in the identifications of known compounds as detected in previously undocumented sites. For example, 6-hydroxymelatonin was first reported in 1961 (Kopin et al.) from the urine of rats fed radiolabeled melatonin. However, it was not detectable in normal human urine until Sisak et al. (1979) found a suitable derivative for GC/MS, and published a complete mass spectrum for material isolated from a urine extract in comparison with authentic material. Similarly, another putative melatonin metabolite, 5-methoxy-indole-3-acetic acid (5-MIAA), had only been detected in rat urine when exogenous methoxyindoles were administered (see review, Higa and Markey, 1985). This compound was qualitatively identified

---

[*]The notation m/z (mass/charge) used in mass spectrometry derives from the fact that spectrometers do not measure mass directly, but indicate a value relative to the charge (positive or negative) on each ion.

in normal human urine using GC/MS (ibid.) Mass spectra produced when derivatized urine extracts were analyzed were compared with those produced by authentic compound. The presence of 5-MIAA in human urine and its metabolic origin remain somewhat puzzling because quantification by gc-ms indicated that its temporal pattern of excretion was not correlated to that of melatonin or 6-hydroxymelatonin. The only known synthesis of 5-methoxyindoles occurs in the pineal gland and is circadian, implying that 5-MIAA in human urine may be of bacterial or dietary origin.

QUANTIFICATION OF TRYPTOPHAN METABOLITES

The use of mass spectrometry for quantitative measurements has a long history in biochemistry, dating to the work of Rittenberg and colleagues beginning in 1940 (see review by Caprioli, 1972). The attractiveness for quantification by mass spectrometry derives from the principle of isotope dilution - that is, isotopomers are inseparable by normal laboratory manipulations, and by adding a fixed quantity of a uniquely labeled isotopomer to a sample, the measurement of the ratio of the natural isotopomer to the unique one provides a direct measure of the unknown quantity. While the principle is simple to state, there are significant technical details which limit the scope and practicality of isotope dilution mass spectrometry. An example of the details of one assay can be used as the basis for generalizing with regard to all tryptophan metabolites.

A prototype assay: Bertilsson et al. (1972) published the first GC/MS assay for 5-hydroxyindole-3-acetic acid (5-HIAA), based upon the work of Hammar et al. (1968) who demonstrated that GC/MS provided the requisite separation to take isotope dilution experiments practical for complex mixtures. First, Bertilsson et al. selected suitable derivatization procedures to transform 5-HIAA into a volatile, readily chromatographable substance. They chose the methyl ester, di-hexafluoroacetyl derivative, a product formed readily and with simple reagents. Next, they synthesized an isotopomer of 5-HIAA, one containing two deuterium atoms on the 2-carbon atom. The isotopic purity of the deuteriated isotopomer was > 99%, so that every molecule would be 2 daltons heavier than native 5-HIAA. The molecular weight of derivatized 5-HIAA was 597, and the d2 isotopomer, 599. Mass spectra of the pure 5-HIAA contained prominent ions at m/z 597 (the molecular or parent ion) and 538 for a fragment representing loss of the carboxymethyl sidechain. These structurally specific ions were shifted by 2 daltons to m/z 599 and 540 for d2-5-HIAA. By choosing mass spectrometric conditions to record continuously the relative signals at each of these four selected ions, a quantitative signal was produced when the derivatized isotopomers co-eluted from the chromatographic column into the mass spectrometer. The assay was applied to cerebrospinal fluid to which a fixed quantity of d2-5-HIAA was added prior to extraction, derivatization and gas chromatography. Thus, after the addition of the internal standard, ion ratios proportional to the original amount of 5-HIAA would result, regardless of sample spills, non-quantitative transfers, etc. The magnitude of the mass spectrometric signal is dependent upon an analyst's skill and instrument performance, but the ratio is only dependent upon the accuracy of the addition of internal standard isotopomer.

This example of 5-HIAA quantification by GC/MS illustrates several generalizations which are applicable to all mass spectrometric assays to the present day.

(1) Derivatization for favorable GC and MS characteristics is critical to the success of the assay. The chemical reactions should be simple, nearly quantitative and result in derivatives which are stable for storage, free of excess reagents, and which exhibit structure specific ions mass spectrometrically (especially, molecular ions).

Table I. GC/MS Studies of Tryptophan and Metabolites

COMPOUND	DERIVATIVE	INTERNAL STANDARD	SAMPLE	IONIZ.	REFERENCE
TRYPTOPHAN					
	TFA/Me ester	d7-Tryptophan	plasma	-C.I.	Hayashi et al.(1986)
	DMF/Et ester	3,3-d2-tryptophan	bacterial prep.	+C.I.	Vicchio et al. (1987)
	PFP/Me ester	3,3-d2-tryptophan	brain	E.I.	Artigas & Gelpi (1979)
	TMS/TFA		standards	E.I.	Donike, et al. (1977)
	2,4 DNP		standards	-C.I.	Williams et al. (1986)
	t-BDMS		protein hydrol.	E.I.	Mawhinney et al. (1986)
	TMS-ß-Carboline		standards	E.I.	Middleditch (1975)
	PFP/nBu ester		standards	+C.I.,-C.I.	Low & Duffield (1984)
5-HYDROXYINDOLE-3-ACETIC ACID					
	HFB/Me ester	2,2-d2-5HIAA	csf	E.I.	Bertilsson et al. (1972)
	PFP/Me ester	2,2-d2-5HIAA	plasma,csf,urine	E.I.	Davis et al. (1986)
	PFP/Me ester	2,2-d2-5HIAA	brain	E.I.	Artigas & Gelpi (1979)
INDOLE-3-ACETIC ACID					
	PFP/Me ester	2,2-d2-IAA	brain	E.I.	Artigas & Gelpi (1979)
	PFB ester		plants	-C.I.	Netting& Milborrow (1988)
	PFP/Me ester	2,2-d2-IAA	plasma,urine	E.I.	Davis & Durden (1987)
SEROTONIN (5HT)					
	PFP	d4-5HT	brain	E.I.	Artigas & Gelpi (1979)
	Ac,Pr-PFP	d4-5HT	tiss.,csf,plat.	-C.I.	Markey et al. (1981)
	PFP	d4-5HT	urine	E.I.	Curtius et al. (1980)
TRYPTAMINE					
	PFP	d4-tryptamine	brain	E.I.	Artigas & Gelpi (1979)
	PFP	d4-tryptamine	brain	-C.I.	Artigas & Gelpi (1979)
	PFP	d4-tryptamine	brain	-C.I.	Durden & Bolton (1988)
N-METHYLTRYPTAMINE					
	TMS	d3,C13,N15-NMT	urine	+C.I.	Walker et al. (1984)
MELATONIN					
	PFP	d4-melatonin	plasma	-C.I.	Lewy & Markey (1978)
	PFP	d4-melatonin	plasma	E.I.	Lee & Esnaud (1988a,b)
	PFP	d4-melatonin	pineal	E.I.	Beck & Pevet (1984)
6-HYDROXYMELATONIN					
	t-BDMS/PFP	d4-6-hydroxymelatonin	urine	-C.I.	Tetsuo et al. (1981)
	PFP	d3-6-hydroxymelatonin	urine	E.I.	Fellenberg et al. (1980)
	PFP	d3-6-hydroxymelatonin	urine	E.I.	Francis et al. (1987)
5-METHOXYTRYPTOPHOL					
	PFP	d4-5-MTOL	pineal	E.I.	Beck & Pevet (1984)
5-METHOXYINDOLE ACETIC ACID					
	PFP/TFE ester	d2-5MIAA	pineal	E.I.	Beck & Pevet (1984)
	PFP/TFE ester	d3-5MIAA	urine	-C.I.	Higa & Markey (1985)
N-ACETYLSEROTONIN					
	PFP	d4-NAS	csf	-C.I.	Taylor et al. (1985)
	PFP	d2-NAS	urine	E.I.	Young et al. (1985)
QUINOLINIC ACID					
	HFIP ester	O-18 QUIN	brain,plasma	-C.I.	Heyes & Markey (1988)
	HFIP ester	2,4-Pyr-dicarboxylate	brain	E.I.	Wolfensberger et al.(1983)
	HFIP ester	2,4-Pyr-dicarboxylate	brain	E.I.	Moroni et al.(1984)
KYNURENIC ACID					
	TFA/Me ester	3-Hydroxy-2-napthoic	brain	E.I.	Carla et al. (1988)

(2) Deuteriated internal standards remain the most common isotopomers encountered in the mass spectrometric literature. Many are commercially available or are readily synthesized by exchange labeling with deuterium oxide or from commercial precursors or reagents.

(3) The monitoring of more than a single ion for each isotopomer adds selectivity to the assay. The presence of chemical interference with the same m/z and elution characteristics is predictably common, especially when higher sensitivity instrumentation is utilized for the analysis of biological samples. Monitoring two or more ions for each isotopomer permits comparison of two or more ratios for consistency, and the detection of probable error if discrepant values occur.

(4) The use of chemical analogues as internal standards rather than isotopomers may be expedient, but is not satisfactory for quantitative measures. Bertilsson et al. tested 5-hydroxyindole-3-propionic acid, and found the resulting ratios irreproducible, an observation which since has been repeated for other GC/MS assays.

Tryptophan metabolite GC/MS assays: A literature search of mass spectrometric assays of all tryptophan metabolites indicates that not all of the more than twenty kynurenine and indoleamine metabolites have been studied using mass spectrometry. Table I is a survey of GC/MS assays reported for tryptophan and 11 of its metabolites.

It is interesting to note that GC/MS quantitative assay development has concentrated on the serotonin-indoleamine branch of tryptophan metabolism, and not on the major kynurenine pathway metabolites. This is not surprising because mass spectrometric methods are usually applied first to the areas requiring greatest specificity and sensitivity to resolve biochemical questions, and to compounds for which there are not other chromatographic or enzymatic assay methods. Further, it is apparent from Table I that new assays are continuously being developed and refined for the same set of compounds. The use of quantitative GC/MS to resolve issues in biochemistry, and the rationale for new assay proliferation with technical advantages are illustrated in the following examples.

Are there tissues in mammals beside the pineal gland which produce melatonin? This question had been addressed by numerous biochemists in the pineal field, with a general positive consensus of opinion because the melatonin biosynthetic enzymes could be measured in several specialized tissue regions, and radioimmunoassays measured circulating melatonin in rodents following pineal gland removal. However, with the development of mass spectrometric assays for melatonin and its major urinary metabolite 6-hydroxymelatonin, this question could be answered by direct measures. Following pinealectomy, both plasma melatonin (Lewy et al., 1980) and urinary 6-hydroxymelatonin (Markey and Buell, 1982) diminished by > 97% to levels below detectability in the fluids from male rats. The question of extra-pineal melatonin synthesis having been resolved, analysts using immunoassay techniques have been able to refine their procedures so that fluids from pinealectomized animals provide a positive control for non-specific cross-reactivity. Both GC/MS and immunoassays are now used routinely for quantification of melatonin to address questions regarding its possible hormonal function in man.

Do schizophrenics synthesize methylated tryptamines from tryptamine by an aberrant metabolic pathway? The indoleamine-methylation hypothesis of schizophrenia has been an attractive biochemical notion which has been easier to state than to resolve with analytical data. That is, because methylated tryptamines induce schizophrenic behavior when administered in pharmacological doses to man, perhaps methylated tryptamines are causative agents in individuals with enhanced capacity to form them from dietary tryptamine substrates. Measurement of tryptamine metabolites in urine has not resolved this hypothesis because trace quantities are present in many foods, making urine blanks impractical. The hypothesis was unambiguously tested by Walker et al. (1984), who administered a unique isotopomer, [^{13}C,^{15}N]tryptamine, to patients. They quantified the amount of N-methylated product formed from this specific precursor by schizophrenics. The GC/MS assay sensitivity and selectivity were such that methylation of 0.0001% of the administered 75 mg dose could have been detected. They showed that only amounts less than that could have been formed. Thus, dietary tryptamine is not converted to N-methyltryptamine in man, and the formation of N,N-dimethyltryptamine is precluded.

Can abnormal brain concentrations of the excitotoxic tryptophan metabolite quinolinic acid explain various neuropathies of unknown origin? This question, like the preceding two questions, has been addressed by designing a sensitive GC/MS assay (Heyes and Markey, 1988a). The answers are the subject of several other chapters in this volume, but discussion of the development of a suitable quantitative assay for quinolinic acid (QUIN) is illustrative of the current state-of-the-art in GC/MS. The general principles derived from the consideration of the 5-HIAA assay of Bertilsson et al. apply to the QUIN assay development, but at each stage refinements in techniques have resulted in higher sensitivity limits than in the past.

(1) Chemical derivatization of QUIN with hexafluoroisopropanol (HFIP) to form the di-HFIP ester was known to provide a volatile derivative with excellent gas chromatographic and electron ionization mass spectrometric properties (Wolfensberger et al., 1983). Use of negative chemical ionization enhanced the ionization process by several orders of magnitude, with the result that sub-picogram amounts of standards could be detected. Additionally, the major ions in the spectra are structurally specific, being the molecular anion (m/z 467) and m/z 316 for the loss of C_3HF_6.

(2) An ^{18}[O]$_4$ isotopomer internal standard was synthesized conveniently by exchange labeling QUIN with H_2^{18}[O] with HCl catalysis (Heyes and Markey, 1988b). This isotopomer has the advantages of simple preparation and multiple isotope incorporation, but the disadvantage that the isotope can be lost during derivatization. If QUIN esterification is performed in the usual manner utilizing mixed anhydride intermediates or acid catalysis, the ^{18}O labels will be lost due to back-exchange. However, with an acyl imidazole catalyst to form a mixed anhydride without traces of acid, an [^{18}O]$_2$QUIN-HFIP diester is formed quantitatively. Thus, the second criterion of establishing a highly sensitive and selective GC/MS assay was met with the availability of a QUIN isotopomer which could be added at the beginning of each sample assay.

(3) During the past ten years, major advances in high resolution capillary GC techniques and materials have made high speed and high resolution sample separations routine. Incorporating capillary GC procedures into the QUIN assay increased sensitivity of detection as well as ease of sample introduction. Present procedures use retention gap pre-columns which permit automated high temperature on-column injections at the rate of one every 3 minutes. The resulting chromatographic peaks are 3-5 seconds wide, or approximately one-tenth the peak width with packed column chromatography. An added advantage of improved chromatography and mass spectrometric sensitivity is the fact that extracts from smaller brain regions or fluid volumes can be used, reducing the amount of material injected into the GC/MS, and significantly extending system performance (Fig. 1).

(4) Commercial data systems which control spectrometers, record data, and report area and peak higher ratios for multiple ions have enhanced the usability of GC/MS systems, and reduced operator related measurement errors.

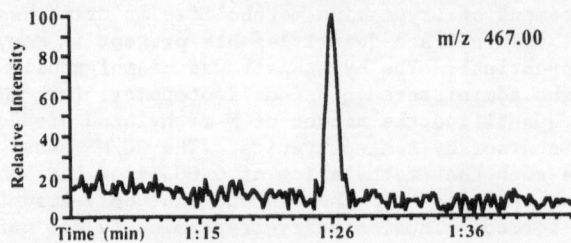

Fig. 1.   A solution containing 1.6 fg (10^6 molecules) of the dihexafluoro-
          propanol ester of quinolinic acid injected on a 15m, 0.25 mm id,
          DB-5 fused silica capillary column was monitored at m/z 467 using
          a negative channel ionization mass spectrometer. Rapid cycle
          time of repeat injections was possible due to isothermal elution
          at 110°C with on-column injection into a retention gap.

Together, the above four factors have improved the sensitivity of the
QUIN assay to the low femtogram (attomole) level, well below that required to
measure QUIN in cerebrospinal fluid. The same techniques could be applied to
other tryptophan and kynurenine pathway metabolites, although it would be
easier technically to use ^{13}C-multiple labeled internal standards, simplify-
ing derivatization chemistry requirements.

FUTURE MASS SPECTROMETRIC INVESTIGATIONS IN TRYPTOPHAN RESEARCH

Most of the research developments in mass spectrometry of the past five
years have not, as yet, been applied to problems in tryptophan metabolism.
Consequently, there remain several powerful tools for the biochemistry to add
to the ones discussed in this review. For example, Johnson et al. (1984)
have demonstrated tandem mass spectrometry for the trace determination of
tryptolines in crude brain extracts, and Artigas and Gelpi (1987) have surveyed the potential of thermospray liquid chromatography-mass spectrometry in
neurochemistry. However, neither of these techniques, nor the emerging techniques of continuous flow-liquid secondary ion mass spectrometry (Wang et
al., 1989) or reaction interface mass spectrometry (Markey and Abramson,
1982) have been applied to this field. The future mass spectrometric instrumentation clearly offers required tools to the tryptophan biochemist.

REFERENCES

Artigas, F., and Gelpi, E., 1979, A new mass fragmentographic method for the
    simultaneous analysis of tryptophan, tryptamine, indole-3-acetic acid,
    serotonin, and 5-hydroxyindole-3-acetic acid in the same sample of rat
    brain, Anal. Biochem., 92:233-42.
Artigas, F., and Gelpi, E., 1987, Evaluation of the potential of thermospray
    liquid chromatography-mass spectrometry in neurochemistry. J. Chromatogr.,
    394:123-134.
Beck, O., and Pevet, P., 1984, Analysis of melatonin, 5-methoxytryptophol and
    5-methoxyindoleacetic acid in the pineal gland and retina of hamster by
    capillary column gas chromatography-mass spectrometry. J. Chromatogr.,
    311:1-8.
Bertilsson, L., Atkinson, A.J., Jr., Althaus, J.R., Härfast, A., Lindgren,
    J.E., and Holmstedt, B., 1972, Quantitative determination of 5-hydroxy-
    indole-3-acetic acid in cerebrospinal fluid by gas chromatography-mass
    spectrometry, Anal. Chem., 44:1434-1438.

Caprioli, R.M., 1972, Use of stable isotopes, in: "Biochemical Applications of Mass Spectrometry", G.R. Waller, ed., Wiley, New York.

Carlá, V., Lombardi, G., Beni, M., Russi, P., Moneti, G., and Moroni, F., 1988, Identification and measurement of kynurenic acid in the rat brain and other organs, Anal. Biochem., 169:89-94.

Curtius, H.-Ch., Farner-Wegmann, H., Niederwieser, A., and Rey, F., 1980, in vivo measurement of tryptophan-5-hydroxylase activity using stable isotopes and GC/MS, in: "Biochemical and Medical Aspects of Tryptophan Metabolism", Hayaishi, O., Ishimura, Y., and Kido, R., eds., Elsevier/North-Holland, Amsterdam, pp. 281-290.

Davis, B. A., and Durden, D.A., 1987, A comparison of the gas chromatographic and mass spectrometric properties of the pentafluoropropionyl and heptafluorobutyryl derivatives of the methyl, trifluoroethyl, pentafluoropropyl and hexafluoroisopropyl esters of twelve acidic metabolites of biogenic amines, Biomed. Environ. Mass. Spectrom., 14:197-206.

Davis, B.A., Durden, D.A., and Boulton, A.A., 1986, Simultaneous analysis of twelve biogenic amine metabolites in plasma, cerebrospinal fluid and urine by capillary column gas chromatography-high-resolution mass spectrometry with selected-ion monitoring. J. Chromatogr., 374:227-238.

Donike, M., Gola R., and Jaenicke, L., 1977, Determination of indolalkylamines after selective derivatisation. J. Chromatogr., 134:385-395.

Durden, D.A., and Boulton, A.A., 1988, Analysis of tryptamine at the femtomole level in tissue using negative ion chemical ionization gas chromatography-mass spectrometry, J. Chromatogr., 440:253-259.

Fellenberg, A.J., Phillipou, G., and Seamark, R.F., 1980, Specific quantitation of urinary 6-hydroxymelatonin sulphate by gas chromatography mass spectrometry, Biomed. Mass. Spectrom., 7:84-87.

Francis, P.L., Leone, A.M., Young, I.M., Stovell, P., and Silman, R.E., 1987, Gas chromatographic-mass spectrometric assay for 6-hydroxymelatonin sulfate and 6-hydroxymelatonin glucuronide in urine, Clin. Chem., 33:453-457.

Hammar, C.G., Holmstedt, B., and Ryhage, R., 1968, Identification of chlorpromazine and its metabolites in human blood by a new method, Anal. Biochem., 25:532-548.

Hayashi, T., Shimamura, M., Matsudea, F., Minatogawa, Y., Naruse, H., and Iida, Y., 1986, Sensitive determination of deuterated and non-capillary gas chromatography and negative ion chemical ionization mass spectrometry, J. Chromatogr., 383:259-269.

Heyes, M.P., and Markey, S.P., 1988a, Quantification of quinolinic acid in rat brain, whole blood, and plasma by gas chromatography and negative chemical ionization mass spectrometry: effects of systemic L-tryptophan administration on brain and blood quinolinic acid concentrations, Anal. Biochem., 174:349-359.

Heyes, M.P., and Markey, S.P., 1988b, (^{18}O) quinolinic acid: its esterification without back exchange for use as internal standard in the quantification of brain and CSF quinolinic acid, Biomed. Environ. Mass. Spectrom., 15:291-293.

Higa, S., and Markey S.P., 1985, Identification and quantification of 5-methyoxyindole-3-acetic acid in human urine, Anal. Biochem., 144:86-93.

Johnson, J.V., Yost, R.A., and Faull, K.F., 1984, Tandem mass spectrometry for the trace determination of tryptolines in crude brain extracts, Anal. Chem., 56:1655-1661.

Kopin, I.J., Pare, C.M.B., Axelrod, J., and Weissbach, H., 1961, The fate of melatonin in animals, J. Biol. Chem., 236:3072-3075.

Lee, C.R., and Esnaud, H., 1988a, Determination of melatonin in blood plasma, using capillary gas chromatography and electron impact medium-resolution mass spectrometry, Biomed. Environ. Mass. Spectrom., 15:249-252.

Lee, C.R., and Esnaud, H., 1988b, Determination of melatonin by GC-MS: problems with solid phase extraction (SPE) columns, Biomed. Environ. Mass. Spectrom., 15:677-679.

Lewy, A.J., and Markey, S.P., 1978, Analysis of melatonin in human plasma by gas chromatography negative chemical ionization mass spectrometry, Science, 201:741-743.

Lewy, A.J., Tetsuo, M., Markey, S.P., Goodwin, F.K., and Kopin, I.J., 1980, Pinealectomy abolishes plasma melatonin in the rat, J. Clin. Endocrinol. Metab., 50:204-205.

Low, G.K.-C., and Duffield, A.M., 1984, Positive and negative ion chemical ionization mass spectra of amino acid carboxy-n-butyl ester N-pentafluoropropionate derivatives, Biomed. Mass Spectrom., 11:223-229.

Markey, S.P., and Abramson, F.P., 1982, Capillary gas chromatography/mass spectrometry with a microwave discharge interface for determination of radioactive-carbon-containing compounds, Anal. Chem., 54:2375-2376.

Markey, S.P., and Buell, P.E., 1982, Pinealectomy abolishes 6-hydroxymelatonin excretion by male rats, Endocrinol., 111:425-426.

Markey, S.P., Colburn, R.W., and Johannessen, J.N., 1981, Efficient extraction and mass spectrometric assay of serotonin in biological fluids, Biomed. Mass. Spectrom., 8:301-304.

Mawhinney, T.P., Robinett, R.S. Atalay, A., and Madson, M.A., 1986, Analysis of amino acids as their tert.-butyldimethylsilyl derivatives by gasliquid chromatography and mass spectrometry, J. Chromatogr., 358:231-242.

Middleditch, B. S., 1975, A novel derivatization procedure for the vaporphase analysis of tryptophan, Analyt. Lett., 8:397-401.

Moroni, F., Russi, P., Lombardi, G., Beni, M., and Carlá, V., 1988, Presence of kynurenic acid in the mammalian brain, J. Neurochem., 51:177-180.

Netting, A.G., and Milborrow, B.V., 1988, Methane chemical ionization mass spectrometry of the pentafluorobenzyl derivatives of abscisic acid its metabolites and other plant growth regulators, Biomed. Environ. Mass Spectrom., 17:281-286.

Sisak, M.E., Markey, S.P., Colburn, R.W., Zavadil, A.P., and Kopin, I.J., 1979, Identification of 6-hydroxymelatonin in normal human urine by gas chromatography-mass spectrometry, Life Sci., 25:803-806.

Susilo, R., Damm, H., and Rommelspacher, H., 1988, Formation of a new biogenic aldehyde adduct by incubation of tryptamine with rat brain tissue, J. Neurochem., 50:1817-1824.

Taylor, P.A., Garrick, N.A., Tamarkin, L., Murphy, D.L., and Markey, S.P., 1985, Diurnal rhythms of N-acetylserotonin and serotonin in cerebrospinal fluid in monkeys, Science, 228:900.

Tetsuo, M., Markey, S.P., Colburn, R.W., and Kopin, I.J., 1981, Quantitative analysis of 6-hydroxymelatonin in human urine by gas chromatography-negative chemical ionization mass spectrometry, Anal. Biochem., 110:208-215.

Vicchio, D., Speedie, M.K., and Callery, P.S., 1987, Gas chromatographic-mass spectrometric determination of tryptophan transaminase-catalyzed deuterium exchange, J. Chromatogr., 415:104-109.

Walker, R.W., Mandel, L.R., Delisi, L., Wyatt, R.J., and Vandenheuvel, W.J., 1984, Capillary column gas-liquid chromatography selected ion monitoring assay for [^{13}C,^{15}N]tryptamine in schizophrenia patients, J. Chromatogr., 289:223-229.

Wang, T.L., Shih, M., Markey, S.P., and Duncan, M.W., 1989, Quantitative analysis of low molecular weight polar compounds by continuous flow liquid secondary ion tandem mass spectrometry, Anal. Chem., 61:1013-1016.

Williams, T.D., Vachon, L., and Anderegg, R.J., 1986, Negative ion chemical ionization mass spectrometry of 2,4-dinitrophenyl amino acid esters, Anal. Biochem., 153:372-379.

Wolfensberger, M., Amsler, U., Cuénod, M., Foster, A.C., Whetsell, W.O., Jr., and Schwarcz, R., 1983, Identification of quinolinic acid in rat and human brain tissue, Neurosci. Lett., 41:247-252.

Young, I.M., Leone, R.M., Francis, P., Stovell, P., and Silman, R.E., 1985, Melatonin is metabolized to N-acetyl serotonin and 6-hydroxymelatonin in man, J. Clin. Endocrinol. Metab., 60:114.

SIGNAL-TO-NOISE OPTIMIZATION OF HPLC-FLUOROMETRIC SYSTEMS AND THEIR

APPLICATION TO THE ANALYSIS OF INDOLES

G.M. Anderson

Child Study Center and Department of Laboratory Medicine
Yale University School of Medicine
New Haven, Connecticut 06510
USA

ABSTRACT

The signal-to-noise optimization of high performance liquid chromato-graphic (HPLC) flow-cell fluorometric systems is described and the possibili-ties for further improving limits of detection for indoles is discussed. Ap-plication of HPLC-fluorometry to analyses in rat cerebrospinal fluid (CSF), human lumbar CSF, and human blood is presented. Finally, the intriguing clinical finding of hyperserotonemia in autism is discussed.

SIGNAL-TO-NOISE OPTIMIZATION

Introduction: Tryptophan, like most indoles, has an absorptivity ap-roaching 10^4 with absorption maximum near 220 and 280 nm. Emission of indol-ic fluorescence usually occurs with a maximum near 350 nm and with quantum efficiencies of 15-40%. The relatively specific nature of the emission spec-trum, coupled with the resolving power of HPLC, makes HPLC-fluorometry an ex-tremely selective way to determine tryptophan and other indoles, such as serotonin. The sensitivity or, more properly, the detectability, of such systems is defined by the amount or concentration producing a signal twice the peak-to-peak noise. In order to decrease limits, one must increase the signal-to-noise ratio either by increasing the signal from a given amount of analyte or by decreasing the system noise. The theory and practice of sig-nal-to-noise optimization of fluorometric systems has been developed and re-viewed by Winefordner and colleagues (1966).

Results and Discussion: In our initial efforts (Anderson, 1978; Ander-son et al., 1979) at improving signal-to-noise ratios in HPLC-fluorometry systems the signal (in amps) was found to agree closely with that predicted by a relatively simple signal equation (Equation 1).

$$S(amps) = (2.3 \ \varepsilon bc I_{sx} A T_\lambda)(\Phi_f \Lambda)(t_\lambda \Omega P)(\gamma G) \qquad \text{Equation 1}$$

The bracketed terms describe, in order, the energy absorbed, energy emitted, energy collected, and the photomultiplier (PMT) amplification. The terms are defined as follows: absorptivity; b, path length; c, concentration; I_{sx}, source intensity; A, cell cross-section; T_λ, excitation filter or monochromator transmittance;, Φ_ℓ, quantum efficiency; Λ, energy loss term; t_λ, emission filter or monochromator transmittance; Ω, geometric efficiency; P, reflective loss term; γ, photocathode responsivity; G, photomultiplier gain. Of note is that the signal is directly proportional to the source intensity (I_{sx}) as well as the collection efficiency (the third term). More difficult is establishing which noise type predominates in a particular system.

In the Aminco fluorometer examined earlier (Anderson, 1978; Anderson et al., 1979) and in the more recently characterized Shimadzu RF 350 fluorometer, the limiting noise was found to be photon shot noise. This type of noise is described by Equation 2 - it should be noted that the noise increases as the square root of the background light level (i_B).

$$\text{Noise (amps)} - (2\, Q_e\, \Delta f\, G\, i_B)^{1/2} \qquad \text{Equation 2}$$

By increasing the source intensity and increasing the collection efficiency of emitted radiation, the Aminco system achieved detection limits of, for example, 18 pg for tryptophan, 17 pg for 5-HT, and 6 pg for indoleacetic acid. Retention times were 4-7 minutes with peak volumes of approximately 500 μl (Anderson, 1978; Anderson et al., 1979).

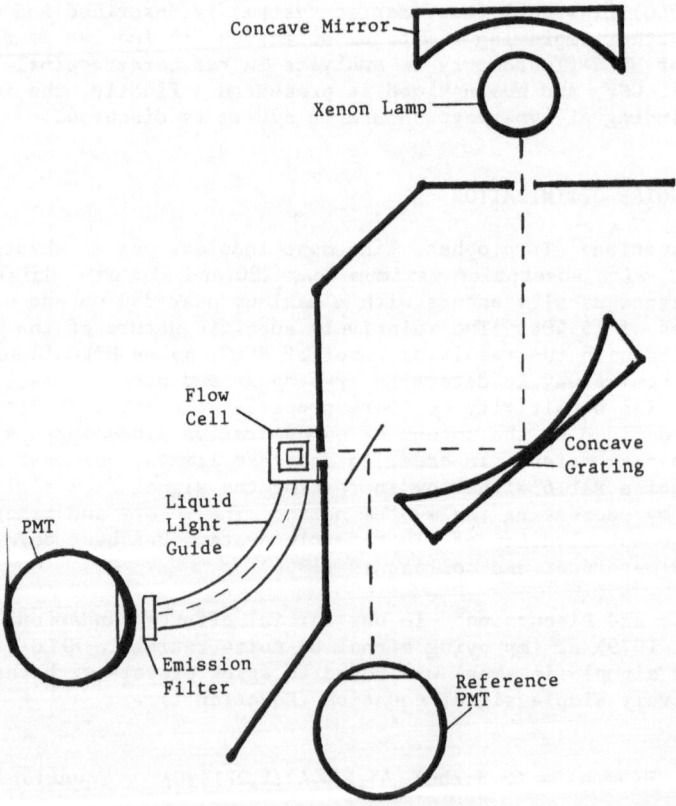

Fig. 1. Schematic diagram of modified Schimadzu RF 350 fluorometer. Top view, not drawn to scale. See text for details.

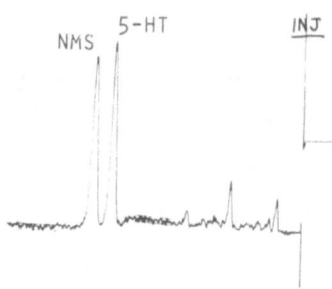

Fig. 2.   Ten picogram standards of serotonin (5-HT) and N-methylserotonin
          (NMS) separated on 2 x 250 mm reverse-phase column and detected
          using a modified Shimadzu fluorometer. Detection limit (S/N = 2)
          of 700 femtograms was obtained; retention time: 8 min.

More recently, the Shimadzu RF350 flurometer has been modified by doub-
ling the excitation monochrometer entrance slit and by replacing the emission
monochrometer with a liquid light guide (250 x 5 mm, Oriel Corp., Stratford,
CT) and Corning 7-51 350 nm glass filter (see Fig. 1). The modifications
have increased the source intensity 4-fold and improved the collection efficiency 4-fold. As expected, the signal was increased approximately 16-
fold. Although background light levels also increased 16-fold, noise increased by the square root of 16, resulting in a 4-fold improvement in signal-to-noise ratio for the indoles of interest. When this optimized system
was coupled with 2 mm I.D. narrow-bore HPLC separation, a further 3-fold improvement in detectability was obtained due to 3-fold lower peak volumes.

This has enabled us to achieve sub-picogram detection limits for serotonin and other indoles at analytically useful retention times. Shown in
Fig. 2 is the detection of 10 pg of serotonin and N-methylserotonin after
reverse phase separation on a 2 x 250 mm C_{18} column. A detection limit of
700 femtograms can be calculated for 5-HT. When 100 μl of sample is injected, a concentration detection limit of 5×10^{-11} M or 50 picomolar is
obtained. Lower detection limits can be obtained for other indoles and at
shorter retention times. Further improvements in chromatographic-fluorometric systems appear possible; however, for the type of system described, a
10-fold increase in source intensity and a 10-fold decrease in background
light will result only in a 10-fold improvement in detectability.

More promising at this time is the use of isotachoelectrophoresis coupled with a laser source sheath-cell fluorometric detector. A recently described capillary zone electrophoresis (or CZE) system has a detection limit
for a fluorescein isothiocyanate derivative of arginine of 5×10^{-21} mole
(approximatelyl 2 attograms, or 0.002 fg) (Cheng and Dovichi, 1988). However, the 1 nl injection volume resulted in a concentration detection limit
of 5×10^{-12} M - only 10-fold better than that obtained for 5-HT with the
modified Shimadzu. If one allows approximately an order of magnitude for
differences in absorptivity and in U.V. versus visible source intensity, the
systems are roughly equivalent in terms of concentration detection limits.
However, if isotachoelectrophoresis is used instead of CZE, much greater injection volumes can be employed; in this case the practical advantage of the
laser source sheath-cell can be realized. For instance, if a 10 μl injection was employed, a concentration detection limit for 5-HT of 5×10^{-13} M
(0.5 pmolar) might be obtained, 100-fold lower than that presently obtained
with the modified Shimadzu fluorometer.

APPLICATIONS

   Rat cerebrospinal fluid (CSF):   Initial applications of the optimized
Aminco filter fluorometer were to the determination of tryptophan metabolites
in rat cisternal CSF (Anderson and Purdy, 1979; Young et al., 1980a,b).   More
recently we used a modified Shimadzu fluorometer to measure serotonin, along
with tryptophan and 5-HIAA, in rat CSF (Anderson et al., 1987).   As shown in
Fig. 3, 5-HT, 5-HIAA, tryptophan (TRP), and the internal standard, N-methylserotonin (NMS) were determined in rat cisternal CSF by directly injecting
25-50 µl of CSF on an HPLC-fluorometric system.   Serotonin was determined
florometrically with an absolute detection limit of 5 pg, equivalent to a
concentration detection unit (S/N - 2) of 100 pg/ml when 50 µl of CSF was
injected.   Samples were determined with within-day coefficients of variation
of 2-7%.

   When one examines the changes in CSF 5-HT, CSF 5-HIAA, brain 5-HT, and
brain 5-HIAA seen after drug treatment, it is apparent that, in nearly every
case, CSF 5-HT is superior to the other measures in reflecting changes in
functionally active 5-HT (Anderson et al., 1987).   After treatment with fenfluramine, CSF 5-HIAA increased slightly while brain 5-HIAA was unchanged,
and brain 5-HT actually declined by over 50%.   However, the large increases
in CSF 5-HT clearly reflected the potent serotonergic quality of fenfluramine.   When amitriptyline was administered, CSF and brain 5-HIAA, as well as
brain 5-HT, were unchanged.   As increase in CSF 5-HT (2.8-fold) was the only
the only indication that amitriptyline acts to increase serotonergic functioning.   As expected, brain 5-HIAA decreased and brain 5-HT increased after
pargyline or pargyline plus TRP: the effects of 5-HIAA are directly opposite
to the treatments' effect on serotonin function and the changes in brain 5-HT
are not proportional to the observed changes in functional activity (Anderson
et al., 1987).

   It would appear, from the data presented, that CSF 5-HT is superior to
CSF 5-HIAA, brain 5-HIAA, or brain 5-HT in assessing changes in functionally

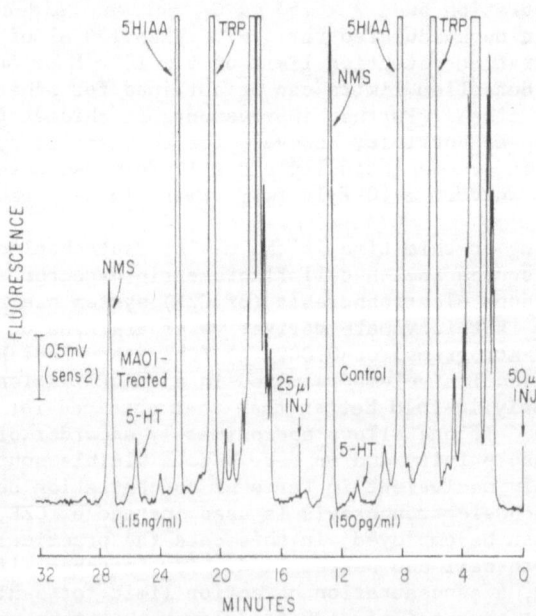

Fig. 3.   HPLC-fluorometric determination of TRP, 5-HIAA, 5-HT and NMS
          (internal standard) in directly-injected rat CSF.   See text for
          chromatographic conditions.

active 5-HT. This is partly a consequence of the fact that the latter measures do not always reflect changes in the compartmentalization of 5-HT. It should also be pointed out that the absolute levels of 5-HT observed in ratcisternal CSF are 400-500 pg/ml. This concentration is similar to the 300 pg/ml estimated by Kalén and coworkers (1988) to be present in ECF of the caudate using in vivo dialysis.

In summarizing the rat CSF work, one can suggest that cisternal CSF 5-HT appears to be a good index of functionally active 5-HT and that CSF 5-HT levels are similar to ECF 5-HT levels estimated in periventricular brain areas. Measurement of 5-HT in rat cisternal CSF offers an alternative to in vivo dialysis when anatomical resolution is not critical. The measure offers advantages in terms of cost and simplicity as well as allowing long-term or chronic studies.

Human CSF: Further improvement in the 5-HT detection limit to the 700 fg level allowed us to examine the issue of 5-HT in human lumbar CSF. We and others have thought that the measurement of 5-HT in human lumbar CSF might prove useful in assessing serotinergic functioning in depression, suicidality, Tourette's syndrome, obsessive-compulsive disorder, autism, and impulsive behavior.

A review of the literature reveals at least 28 reports concerning human lumbar CSF 5-HT. Early studies utilized either bioassays or fluorometric assays with detection limits in the low ng/ml range. In 1979, Ternaux et al. reported a mean level of 69 pg/ml using a radioenzymatic method. Beginning in 1983, 11 reports of CSF 5-HT determinations using HPLC have appeared. In all but two instances, HPLC separation was followed by electrochemical detection. The mean values reported using the HPLC methods have ranged from 60 pg/ml to 12.7 ng/ml. We have carried out an interlaboratory comparison study using both HPLC-fluorometry and HPLC-electrochemical detection, in order to ascertain which, if any, of the previously reported means accurately reflected human CSF 5-HT levels.

A typical HPLC-F chromatogram is shown in Fig. 4. Separation was effected using a 2 mm x 250 mm column of 5 μm Altex ultrasphere C_{18} packing. As mentioned, the Shimadzu RF-350 fluorometer had been modified by increasing the width of the excitation monochrometer entrance slit and by coupling a liquid light guide directly to the flow cell. The light guide was routed to the window of the photomultiplier. An excitation wavelength of 285 nm was

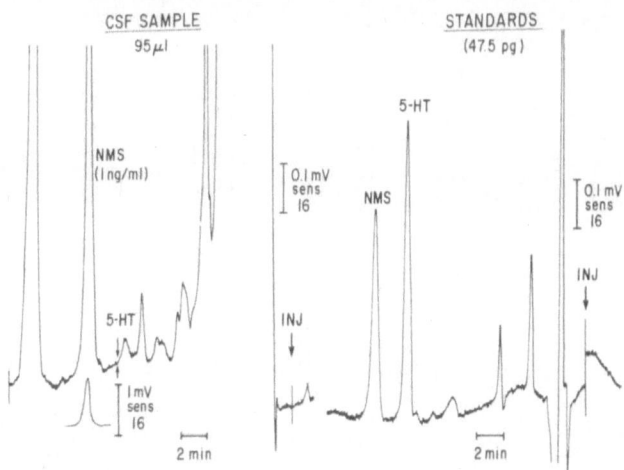

Fig. 4. HPLC-fluorometric determination of 5-HT and NMS in human lumbar CSF. Less than 11 pg/ml of 5-HT was observed in the sample shown.

selected; the approximate excitation bandpass was 30 nm. A detection limit for 5-HT of approximately 700 fg was obtained, giving concentration detection limits of 4-15 pg/ml (approximately 25-80 pmolar) when 65-100 μl of CSF was injected. From our study of 19 normal control subjects, it appears that extremely low levels of 5-HT - less than 10 pg/ml - are normally present in human lumbar CSF (Anderson et al., 1990). From these data, one can conclude that all previously reported means for lumbar CSF 5-HT are in error. The extremely low levels presumed actually to be present make it unlikely that the measure will ever be of clinical utility. For example, it would be difficult to rule out minute platelet contamination.

Serotonin in blood: Another area where the improved detectibility has been useful is in studying blood 5-HT. The vast majority of whole blood 5-HT (typically 150 ng/ml) is contained within the platelet (about 600 ng/10^9 platelets). Whole blood serotonin can be easily determined along with indoleacetic acid, indolepropionic acid and N-methylserotonin, as shown in Fig. 5 using dodecyl (C_{12}) sulfate ion-pairing mobile phase and a 4.6 mm x 250 mm C_{18} reverse phase column (Anderson et al., 1987). Alternatively, serotonin can be determined, along with tryptophan and the internal standard 5-hydroxytryptophan, by reverse phase HPLC without using an ion-pair agent (Anderson et al., 1981; Anderson et al., 1987).

More difficult has been the measurement of the low levels of free 5-HT found in platelet-poor plasma (PPP). Previous reports of mean PPP levels of 5-HT have ranged from 1.4-28 ng/ml. We have determined 5-HT in both PPP and in PPP ultrafiltrate. Determinations of PPP 5-HT were made using a C_{12} ion-pairing mobile phase and a 2 mm x 250 mm, C_{18} reverse phase column. Fifty μl of deproteinized plasma was injected directly on the chromatographic system (Fig. 6). An identical chromatographic system was used to determine 5-HT in plasma ultrafiltrate (Anderson et al., 1987).

Steady-state model for plasma free 5-HT: The availability of methods for determining plasma and ultrafiltrate 5-HT has enabled us to test a model developed to describe quantitatively the factors controlling plasma free 5-HT

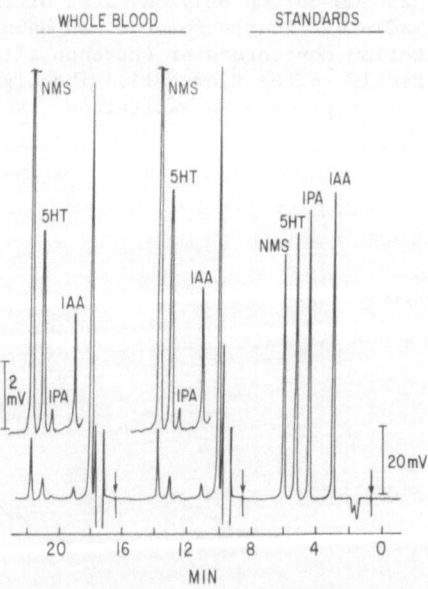

Fig. 5.   Indoleacetic acid and (IAA) indolepropionic acid (IPA), 5-HT, and NMS determined by HPLC-fluorometry in two different human whole blood samples.

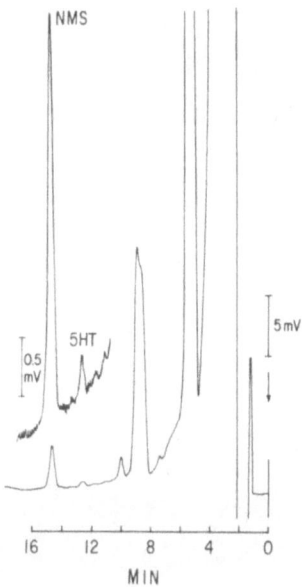

Fig. 6. HPLC-fluorometric determination of 5-HT and the internal standard,
NMS, in 75 µl of deproteinized human platelet-poor plasma.

and platelet 5-HT levels (Anderson et al., 1987). The steady-state model
diagrammed in Fig. 7 is similar to physiological flow, or perfusion models
developed for describing drug concentrations after intravenous infusion.
Critical present variables are the rate of gut production of 5-HT, and the
lung, liver, kidney, and capillary bed extraction efficiencies. The variables to be calculated are the proximal venous, the arterial, and the distal
venous concentration of blood free 5-HT. Protein binding and platelet storage of 5-HT should not affect blood free 5-HT levels at steady state. Blood
flow rates (in ml/min) through the various organs are assumed to be constant
and are those reported for a "standard" man.

When best estimates of the pre-set variables are substituted in a model-
derived equation, a predicted value of 5-HT in distal venous plasma ultrafiltrate of 304 pg/ml is obtained. This compares well with the mean value
observed in seven normal control subjects of 387 ± 84 pg/ml. Levels observed
in PPP were somewhat higher, averaging 578 ± 277 pg/ml, apparently due to the
presence of protein-bound 5-HT (approximately 30% of the total). The plasma
levels of 5-HT observed are 2- to 3-fold lower than the lowest previous reports; although analytical interferences probably contributed to some of the
error, care must also be taken to avoid platelet contamination and 5-HT release. The model should permit a more systematic approach to examining the
etiology of observed group differences in platelet 5-HT levels. In addition,
reports of an effect of 5-HT on leukocytes, intestinal motility, platelets,
vascular tissue, and smooth muscle cell mitogenesis make the model potentially relevant to understanding the role of plasma free 5-HT in a number of
processes.

Hyperserotonemia of autism: We have fruitfully applied blood, plasma,
and urine 5-HT measures, and the model, to the issue of hyperserotonemia and
autism. As shown in Fig. 8, the group mean whole blood level of 5-HT in

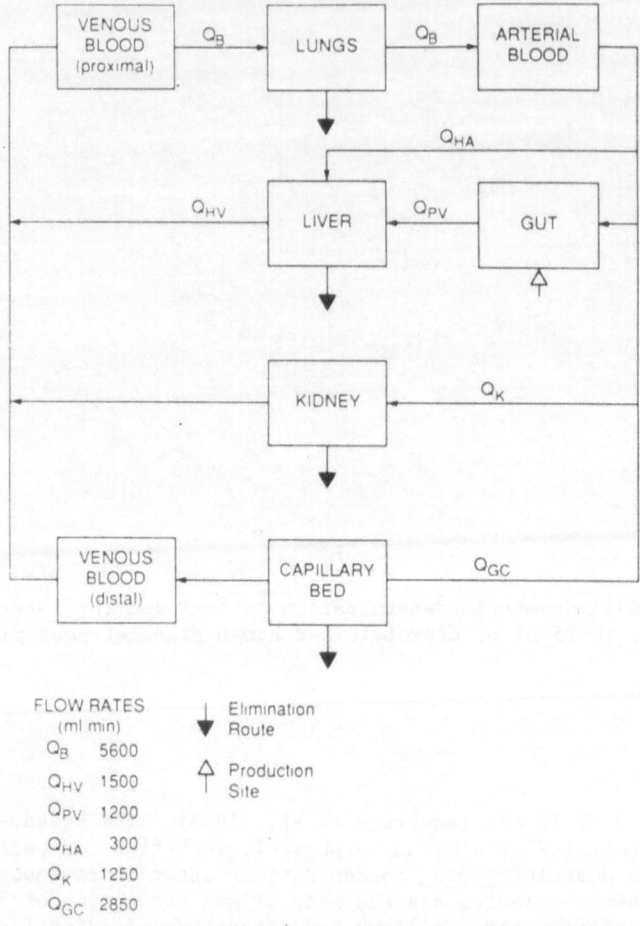

FLOW RATES
(ml min)

Q_B	5600
Q_{HV}	1500
Q_{PV}	1200
Q_{HA}	300
Q_K	1250
Q_{GC}	2850

↓ Elimination Route

△ Production Site

Fig. 7. Diagram of model used to estimate free blood levels of 5-HT based on the gut production rates, organ extraction efficiencies, and blood flow rates.

autistic subjects is 50% higher than in normal control subjects. Possible causes for this well-replicated and robust finding might be: 1. decreased catabolism of 5-HT, 2. increased synthesis of 5-HT, or 3. altered platelet uptake/storage of 5-HT. More simply stated, the question is whether or not the platelet is exposed to higher levels of serotonin.

The data summarized in Table 1 suggest that 5-HT synthesis and catabolism are not altered in autism (Anderson, 1987; Anderson et al., 1987). The plasma (Cooke et al., 1988; Anderson, unpublished data) and urine 5-HT measurements (Anderson et al., 1989) especially, strongly suggest that the platelets of autistic subjects are not exposed to increased concentrations of serotonin.

One can conclude tentatively that some aspect of platelet 5-HT uptake, storage, or release is altered. We are presently examining these possibilities in autistic subjects by carrying out studies of platelet function and stimulus-response coupling, as well as performing platelet protein inventor-

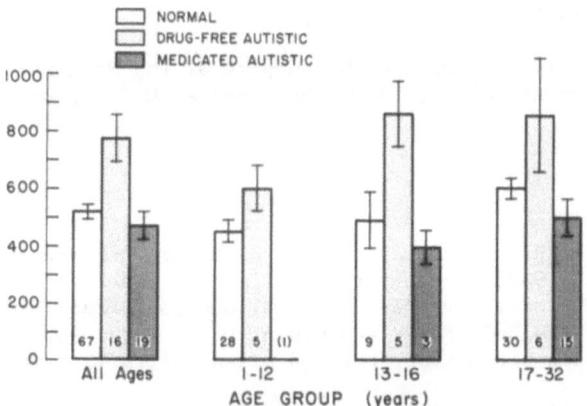

Fig. 8. Whole blood serotonin levels in normal and autistic subjects.
Groups means of normal subjects and unmedicated autistic subjects
were significantly different (P < 0.01).

ies using giant-gel two-dimensional gel electrophoresis of total platelet
proteins and of dense granule-associated proteins.

SUMMARY

The signal-to-noise optimization of HPLC-fluorometric systems has led to
subpicogram detection limits for serotonin and related indoles. The improved
detectability has permitted the development of useful CSF and plasma analyses. Further improvements in indole detection limits will be difficult using HPLC-fluorometry; however, ITE coupled with laser-source sheath flow-cell
detection is a promising technique in this regard.

Table 1. Findings relevant to the hyperserotonemia of autism

Reference	Sample	Analyte/ Measure	Findings in autistic subjects
Minderaa et al., 1987	Urine	5-HIAA	Normal
Anderson et al., 1989	Urine	5-HT	Normal
Anderson et al., 1989	Urine	5-HT/5HIAA	Normal
Anderson et al., 1987	Blood	TRP	Normal
Young et al., 1982	Platelet	MAO-B	Normal
Geller et al., 1988	Platelet	Count and volume	Normal
Cooke et al., 1988 Anderson, unpublished	Plasma	5-HT	Normal
Anderson et al., 1984	Platelet	Imipramine binding	Normal
Yuweiler et al., 1975 Rothman et al., 1980 Katsui et al., 1986	Platelet	5-HT uptake	Normal/Inc.
Yuweiler et al., 1975	Platelet	5-HT efflux	Normal
McBride et al., 1989	Platelet	5-HT$_2$ B$_{max}$	Decreased
McBride et al., 1989	Platelet	5-HT augmented aggregation	Decreased

ACKNOWLEDGEMENTS

We gratefully acknowledge the support of the National Institutes of Health (MHCRC grant 30929), the Gettner Research Fund, and the Marks Foundation.

REFERENCES

Anderson, G.M., 1978, Doctoral thesis, Dept. of Chemistry, McGill University.

Anderson, G.M., 1987, Monoamines in autism: an update of neurochemical research of a pervasive developmental disorder, Med. Biol., 65:67-74.

Anderson, G.M., Feibel, F.C., and Cohen, D.J., 1987, Determination of serotonin in whole blood, PRP, PPP and plasma ultrafiltrate, Life Sci., 40: 1063-1070.

Anderson, G.M., Freedman, D.X., Cohen, D.J., Hoder, E.L., Volkmar, F.R., Paul, R., McPhedran, P., Minderaa, R.B., Young, J.G., and Hansen, C.R., Jr., 1987, Whole blood serotonin in autistic and normal subjects, J. Child Psych. Psychol., 28:885-900.

Anderson, G.M., Mefford, I.M., Tolliver, T.A., Riddle, M.A., Olame, D., Lechman, J.F., and Cohen, D.J., 1990, Serotonin in human lumbar cerebrospinal fluid: a reassessment, Life Sci., 46:247-255.

Anderson, G.M., Minderaa, R.B., Choe, S.C., Volkmar, F.R., and Cohen, D.J., 1989, The issue of hyperserotonemia and platelet serotonin exposure: preliminary study, J. Aut. Dev. Dis., 19:349-351.

Anderson, G.M., Minderaa, R.B., Van Bentham, P.P.G., Volkmar, F.R., and Cohen, D.J., 1984, Platelet imipramine binding in autistic subjects, Psych. Res., 11:133-141.

Anderson, G.M., and Purdy, W.C., 1979, Liquid chromatographic-fluorometric system for the determination of indoles in physiological samples, Anal. Chem., 51:283-287.

Anderson, G.M., Purdy, W.C., and Young, S.N., 1979, The determination of neurologically important tryptophan metabolites in brain and of cerebrospinal fluid, in: "Trace Organic Analysis: a New Frontier in Analytical Chemistry", National Bureau of Stands Publication, U.S., Spec. #519, pp. 411-418.

Anderson, G.M., Stevenson, J.M., and Cohen, D.J., 1987, Steady-state model for plasma free and platelet serotonin in man, Life Sci., 41:1777-1785.

Anderson, G.M., Teff, K.L., and Young, S.N., 1987, Serotonin in cisternal cerebrospinal fluid of the rat: measurement and use as an index of functionally active serotonin, Life Sci., 40:2253-2260.

Anderson, G.M., Young, J.G., Cohen, D.G., Schlicht, K.R., and Patel, N., 1981, Liquid-chromatographic determination of serotonin and tryptophan in whole blood and plasma, Clin. Chem., 27:775-776.

Cheng, Y.F., and Dovichi, N.J., 1988, Ultramicro amino acid analysis, Science, 242:562-563.

Cooke, E.H., Leventhal, B.L., and Freeman, D.X., 1988, Free serotonin in plasma, autistic children and their first-degree relatives, Biol. Psych., 24:488-491.

Geller, E., Yuwiler, A., Freeman, B.J., and Ritvo, E., 1988, Platelet size, number, and serotonin content in blood of autistic, childhood schizophrenic and normal children, J. Aut. Dev. Dis., 18:119-126.

Kalén, P., Strecker, R.E., Rosengren, E., and Björklund, A., 1988, Endogenous release of neuronal serotonin and 5-hydroxyindoleactic acid in the caudate-putamen of the rat, J. Neurochem., 51:1422-1435.

Katsui, T., Okuda, M., Usuda, S., and Koizumi, T., 1986, Kinetics of ^{3}H-serotonin uptake by platelets in infantile autism and developmental language disorder (including five pairs of twins), J. Aut. Dev. Dis., 16:69-76.

McBride, A.P., Anderson, G.M., Hertzig, M.E., Sweeney, J.A., Kream, J., Cohen, D.J., and Mann, J.J., 1989, Serotonergic responsivity in male young adults with autistic disorder, Arch. Gen. Psych., 46:213-221.

Minderaa, R.B., Anderson, G.M., Volkmar, F.R., Akkerhuis, G.W., and Cohen, D.J., 1987, Urinary 5-HIAA and whole blood 5HT and tryptophan in autistic and normal subjects, **Biol. Psych.**, 22:933-940.

Rotman, A., Caplan, R., and Szekeley, G.A., 1980, Platelet uptake of serotonin in psychotic children, **Psychopharmacology**, 67:245-248.

St. John, P.A., McCorthy, W.J., and Winefordner, J.D., 1966, Applications of signal-to-noise theory in molecular luminescense spectrometry, **Anal. Chem.**, 38:1828-1835.

Ternaux, J.P., Mattei, J.F., Faudon, M., Berrit, M.C., Ardisson, J.P., and Giraud, F., Peripheral and central 5-hydroxytryptamine in trisomy 21, **Life Sci.**, 25:2017-2022.

Young, J.G., Kavanaugh, M.E., Anderson, G.M., Shaywitz, B.A., and Cohen, D.J., 1982, Clinical neurochemistry of autism and associated disorders, **J. Aut. Dev. Dis.**, 12:147-165.

Young, S.N., Anderson, G.M., Gauthier, S.G., and Purdy, W.C., 1980a, The origin of indoleacetic acid and indolepropionic acid in rat and human cerebrospinal fluid, **J. Neurochem.**, 34:1087-1092.

Young, S.N., Anderson, G.M., and Purdy, W.C., 1980b, Indoleamine metabolism in rat brain studied through measurements of tryptophan, 5-hydroxy-indoleacetic and incoleacetic acids in cerebrospinal fluid, **J. Neurochem.**, 34:309-315.

Yuwiler, A., Ritvo, E., Geller, E., Glousman, R., Schneiderman, G., and Matsumo, D., 1975, Uptake and efflux of serotonin from platelets of autistic and normal children, **J. Aut. Childhd. Schiz.**, 5:83-98.

BRAIN INDOLE METABOLISM ASSESSED USING IN VIVO DIALYSIS

G. Sarna

Institute of Neurology
London WC1N 1PG
UK

INTRODUCTION

The relationships between tryptophan (Trp) and its major CNS metabolites serotonin (5-HT) and 5-hydroxyindoleacetic acid (5-HIAA) may be studied by analysis of post-mortem tissue or by analyzing concentrations in the extra-cellular fluid. Major advances in our understanding of the physiological im-portance of Trp supply to the brain and subsequent 5-HT synthesis and metabo-lism was derived, in part, from regional brain tissue analysis following e.g. Trp loading, stress, fasting (Fernstrom and Wurtman, 1971; Knott and Curzon, 1974; Curzon and Marsden, 1975). However, tissue analysis does not differ-entiate between functionally active substrate concentrations available to receptors in the extracellular compartment and intracellular concentrations. Furthermore, as only a single value is obtained from each animal, large num-bers of animals are required to obtain temporal profiles of drug action. The primary advantage of in vivo monitoring of extracellular substrate concentra-tions is that repeated measurements may be made in the same animal over time. When applied to the conscious, freely moving animal, this allows associations between neurochemical changes and their roles in behaviors to be determined.

A variety of in vivo methods had been described (Sarna et al., 1983; Westerink et al., 1987; Knott, 1988). These include microdialysis, sampling of cerebrospinal fluid (CSF), push-pull, cortical-cup, ventricular-cisternal perfusion and voltammetry. The present chapter will focus specifically on our studies with cerebral microdialysis in investigations of the roles of 5-HT following physiological, pharmacological and neurological interventions.

PRINCIPLES OF MICRODIALYSIS

The use of cerebral dialysis was first described by Bito et al. (1966). A small dialysis bag containing artificial CSF was implanted into the brain and the contents allowed to equilibrate with the extracellular fluid com-partment (Fig. 1A). After 10 weeks the bag was removed and the dialysate analysed for different amino acids. Delgado et al. (1972) and Ungerstedt and Pycock (1974) subsequently reported a perfusion-dialysis system whereby a microdialysis bag is continuously perfused with artificial CSF, the dialysate collected over timed periods (e.g. 20 min) and analysed (Fig. 1B). As well as sampled extracellular substrates, drugs may be delivered to the surround-ing fluid. The recent widespread use of microdialysis may be attributed to

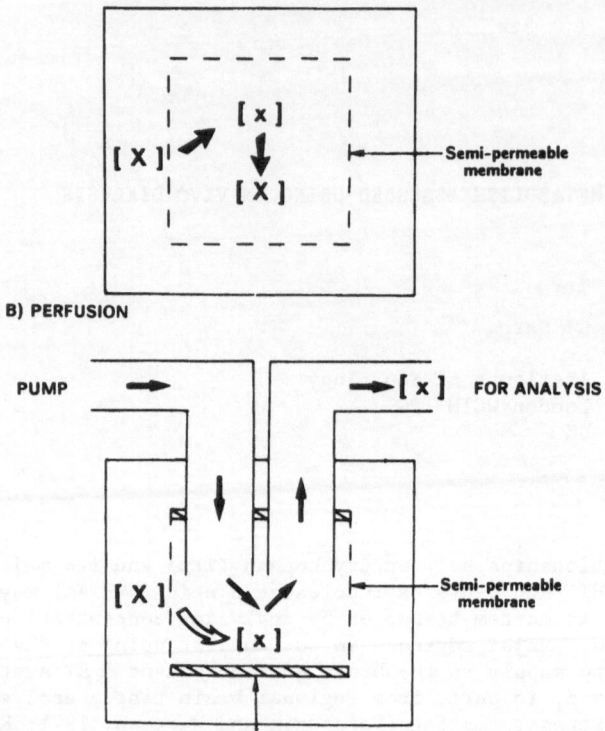

A) NO PERFUSION

[X]   [x]   [X]   Semi-permeable membrane

B) PERFUSION

PUMP   [x] FOR ANALYSIS

[X]   [x]   Semi-permeable membrane

SEALED

Fig. 1.   Schematic illustration of substrate (X) transfer across a
dialysis membrane.

the major improvements in analytical methods, especially high-performance
chromatography (HPLC) with electrochemical and fluorometric detection (Hutson
et al., 1985; Church and Justice, 1986). It is now possible to separate all
biogenic amines and their metabolites within a single chromatographic run in
less than 15 min from injected sample volumes of only a few microlitres (Fig.
2).

Dialysis probe

There now many different designs of dialysis probes (Knott, 1988) and we
currently use the concentric probe shown in Fig. 3. It is similar in construction to a push-pull cannula but the presence of the dialysis membrane
reduces potential tissue damage due to flowing perfusion fluid being in direct contact with brain tissue.

Factors influencing recovery

A number of factors influence the passage of substrates across the
dialysis membrane including membrane composition, chemical interactions with
membrane, molecular weight, molecular shape, perfusion rate, surface area,

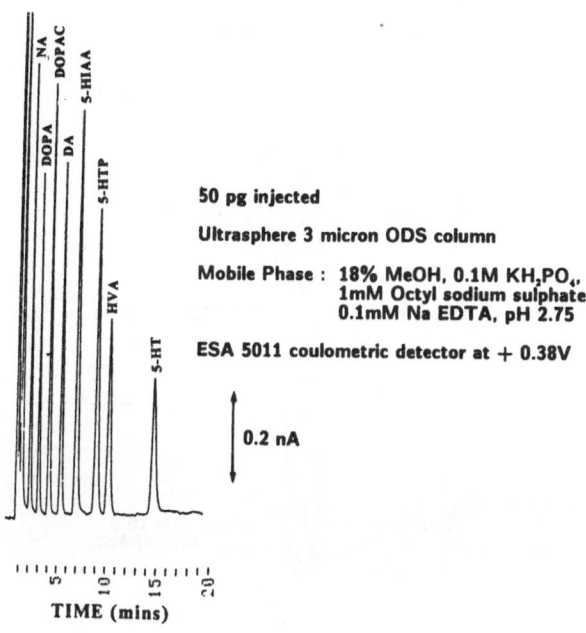

Fig. 2. HPLC with electrochemical detection, trace of 50 pg standards.

temperature and osmotic pressure (Church and Justice, 1986). The most commonly used membranes are cellulose based (e.g. Cuprophan, EnkaGlanzstoff AG) and have a cut off of 5,000-10,000 dalton molecular weight. Diaflo hollow firbres (Amicon) are available with a cut off up to 100,000 dalton molecular weight. In theory, any substance which can diffuse freely across the membrane without significant binding problems (either to the membraneor collection apparatus) could be recovered. A feature of microdialysis in the determination of indoleamines (MW approx. 200) is that the membrane restricts the entry of protein, minimizing enzymatic degradation, and allows direct injection of dialysates onto HPLC columns without purification (Church and Justice, 1986). Fig. 4 shows the effect of increasing flow rate on the relative percentage recovery of substances obtained from a probe immersed in a solution containing 100 ng/ml standards. The relative recovery approaches 100% as flow is reduced to zero. Absolute recovery will increase with flow rate until pressure in the dialysis bag causes outflow of fluid or rupture of the membrane. In vitro recoveries should not be directly equated with in vivo recoveries as the rate of diffusion in brain tissue will be lower than that in a liquid and the effective volume of extracellular fluid being dialysed must be compensated for. Methods for obtaining brain absolute extracellular concentrations from dialysed concentrations have been reported (Lerma et al., 1986; Alexander et al., 1988). The temporal resolution of dialysis is dependent on the sensitivity of the assay employed and we routinely use 20 mins for HPLC analysis of 5-HT although the much greater extracellular 5-HIAA concentrations would allow 1-2 minute sampling.

Implantation of the dialysis probes is an invasive procedure and may cause trauma e.g. breakdown of the blood-brain barrier especially with probes larger than the one shown in Fig. 3 (Westerink and De Vries, 1988). Recent histological, autoradiographic and release studies (Benveniste et al., 1984; Benveniste and Diemer, 1987; Benveniste et al., 1987) have shown that the cerebral dialysis technique may best be used within a specific temporal window following implantation of the probes. The blood-brain barrier appears to

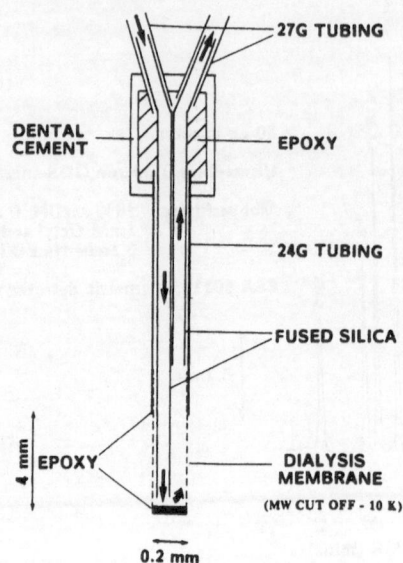

ART. CSF
INFLOW 0.25-1.0 ul/min   OUTFLOW

27G TUBING

DENTAL
CEMENT   EPOXY

24G TUBING

FUSED SILICA

4 mm EPOXY   DIALYSIS
                          MEMBRANE
                          (MW CUT OFF - 10 K)

0.2 mm

Fig. 3.   Details of a concentric dialysis probe.

be intact to $Na^{99m}TcO_4$ (Tossmann and Ungerstedt, 1986) and AIB (Benveniste et al., 1984) following the first hour or so of implantation but subsequent changes have not been detailed. Varying baseline concentrations and altered local glucose phosphorylation and cerebral blood flow during the first few hours following implantation (Benveniste et al., 1987; Kalén et al., 1988), suggests that perfusions should not be carried out immediately. However, perfusion following some days of implantation is complicated by cellular and macrophage infiltration (Benveniste et al., 1987). The marked rise in striatal dopamine after influsion of 60 mmol KCl almost disappears 48 hours after implantation probably because a glial barrier prevents K^+ reaching dopaminergic terminals (Westerink et al., 1987). Dialysis membranes with more biocompatibility with brain are required.

METHODS USED IN OUR LABORATORY

Animals and surgery: Individually housed male Sprague-Dawley rats (Charles River, U.K.) weighing 250-350 g were maintained on a 12 h light-dark cycle (lights on 06.00 h) and allowed food (RRF diet, Charles River U.K.) and water and libitum. Procedures employed for probe implantation have been previously detailed, (Hutson et al., 1985; Adell et al., 1989). Following anesthesia (Sagatal BDH, 60mg/kg i.p.) dialysis probes [as described previously (Hutson et al., 1985) or as shown in Fig. 3] were implanted into specific brain regions. Twenty-four hours following recovery from anesthesia, dialysis probes were perfused with artificial CSF (composition in mM: NaCl, 125; KCl, 2.5; $MgCl_2$, 1.18; $CaCl_2$, 1.26) at 0.25-1.2 μl/min. Successive 20 min samples were collected into tubes containing 10 μl 0.01 M perchloric acid and 0.2% cysteine hydrochloride and stored in a freezer at -70°C until required for analysis. In a separate study, samples of dialysate from freely moving

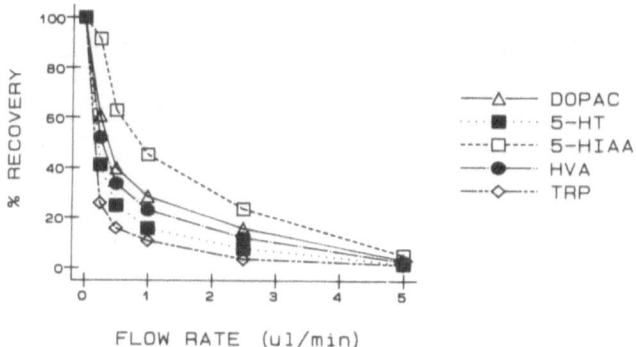

Fig. 4. Effect of increasing perfusion flow rate on the relative percentage recovery of substances obtained from a probe immersed in a solution containing standards.

animals were injected directly ('on-line') into the HPLC. Fig. 5 shows a schematic illustration of concurrent analysis of dialysates and behaviour. A computer was used to control injections, record activity, record food/water intake and analyze the HPLC data. More complex behaviors (e.g. 5-HT syndrome) were recorded on video for subsequent scoring.

Biochemical analysis: Dialysate samples were assayed for Trp, 5-HT and 5-HIAA by HPLC with electrochemical detection. The system used an Altex ultrasphere 3 micron column (4.6 mm x 7.5 cm, Beckman Ltd., U.K.). The mobile phase (composition shown in Fig. 2) was filtered through a 0.2 micron cellulose nitrate filter (Millipore Ltd, London) and degassed with helium before use. The electrochemical detector for 5-HT and 5-HIAA was an ESA Coulochem model 5100A (Severn Analytical LTD, U.K.) with a dual analytical cell (model 5011). Electrode 1 was set at - 0.04 V and electrode 2 at + 0.38 V with respect to palladium reference electrodes. Trp was determined by a BAS electrochemical detector (model 4a, Anachem, U.K.) set at 0.97 V.

STUDIES USING INTRACEREBRAL MICRODIALYSIS

Physiological

Evidence suggests that serotonergic systems are involved in feeding (see Curzon, this volume). Drugs that enhance serotonergic neurotransmission reduce feeding, (Blundell, 1984) whereas decreased transmission leads to increased food intake (Dourish et al., 1985). Also, intake of meals of varying carbohydrate/protein composition may alter tryptophan availability to the brain and influence 5-HT synthesis (Blundell, 1984). Using different feeding paradigms two recent dialysis studies have shown increases in extraneuronal 5-HIAA (Hutson et al., 1986) and more interestingly in 5-HT (Sarna et al., 1984). We are currently investigating whether there are changes in indoleamine metabolism in the freely feeding rat using the on-line procedure described above. Fig. 6 shows changes in extraneuronal Trp and 5-HIAA in relation to feeding episodes over a 24 hour period in a single rat. The complex patterns shown have not been fully resolved but illustrate how dialysis may be used to attack fundamental questions in relation to feeding: e.g. to what extent do extracellular concentrations reflect anticipatory events, circadian variation, altered neuronal firing and altered availability of tryptophan to the brain?

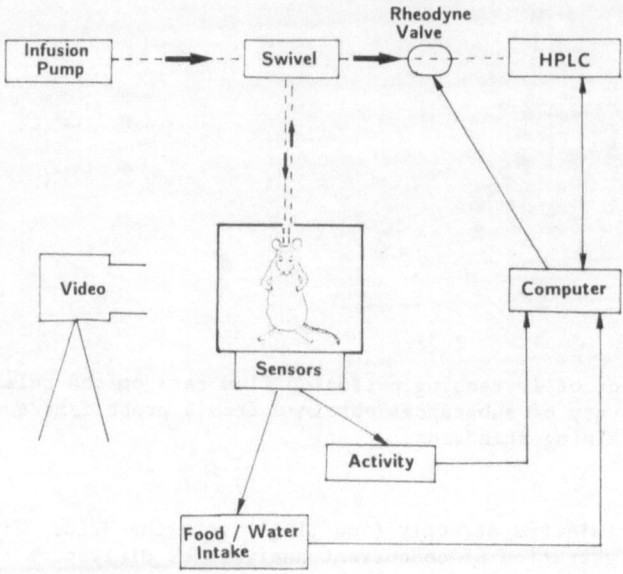

Fig. 5.   Schematic illustration of on-line dialysis with concurrent analysis
of animal behavior. A computer is used to inject dialysate into
the HPLC, record motor activity and food and water intake. More
complex behaviors are recorded on videotape.

PHARMACOLOGICAL APPLICATIONS OF MICRODIALYSIS

Effect of tryptophan loads on 5-HT metabolism?

We have obtained detailed profiles of changes in concentrations of Trp
and 5-HIAA in different CNS compartments including whole brain/striatal tissue, extraneuronal space (striatal dialysis) and CSF following a systemic injection of Trp (50 mg/kg) (Hutson et al., 1985). Of particular interest was
the finding of parallel and proportionate changes for both Trp and 5-HIAA in
each of these compartments. However, De Simoni et al. (1987) who were unable
to demonstrate an increase in extraneuronal caudate 5-HIAA on giving Trp using in vivo voltammetry although whole brain 5-HIAA increased. The differences between tissue and extracellular measures was attributed to compartmen-
tation of 5-HIAA and a contribution of endothelial cell 5-HT metabolism to
tissue 5-HIAA. This study raises important questions about the validity of
measuring tissue 5-HIAA as an index of neuronal 5-HT metabolism. Therefore,
we compared and contrasted the effect of exogenous L-Trp on extraneuronal Trp
and 5-HIAA in two regions of brain differing in 5-HT innervation, the striatum (rich innervation) and cerebellum (poor innervation) (Chan-Palay, 1976;
Azmitia and Segal, 1978; Takeuchi et al., 1982; Steinbusch, 1984; Bonanno et
al., 1986) in order to determine the potential sources of raised 5-HIAA concentrations.

Dialysis probes were implanted into the cerebellum and striatum of individual rats. Following injection of Trp (50 mg/kg, i.p.), its concentration rose approximately five fold in both regions and fell to approximately
pre-injection values within 3 hours. Fig. 7 shows the data from an individual rat. The pharmacokinetic profiles of dialysate Trp (peak concentration,
time to peak concentration, area under curves and half-times for decline)
were essentially similar in both regions. The concentrations of 5-HIAA in
striatal dialysate rose significantly more slowly than those of TRP. Cerebellar dialysate 5-HIAA concentrations did not increase above detectable

68

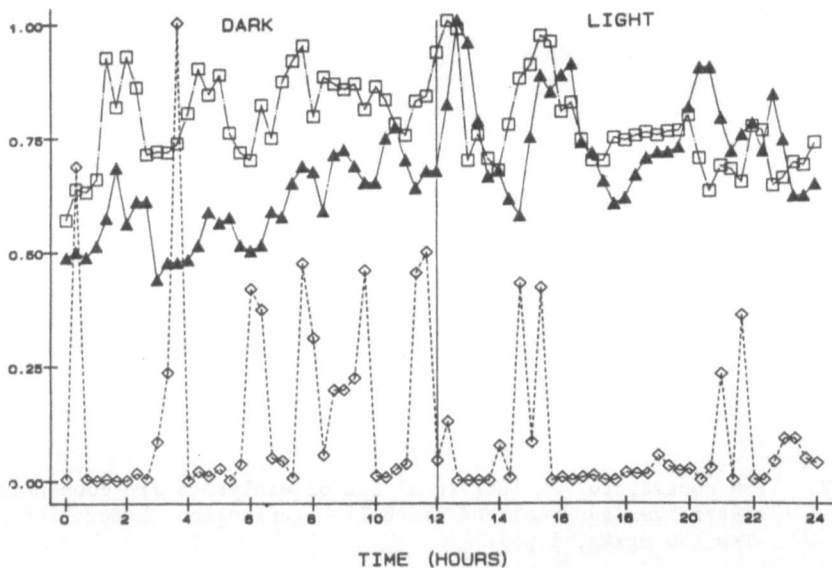

Fig. 6. Dialysate variations of Trp and 5-HIAA (frontal cortex) in relation
        to food intake, measured over 20 min periods. Trp (triangles),
        5-HIAA (squares), food intake (deltoids). Values shown are from a
        single rat and are expressed as a proportion of the maximum value
        obtained over the 24 hour period.

levels (Fig. 8) despite equivalent increases in extracellular Trp in both
regions of brain.

    In the present study the dialysis probes were implanted for 24 h prior
to perfusion in order to overcome some of the methodological problems associated with acute implantation (see above). At this time local glucose phosphorylation and cerebral blood flow are restored to normal and gliosis is not
apparent (Benveniste and Diemer, 1987; Benveniste et al., 1987). The basal
concentrations of striatal dialysate Trp and 5-HIAA and half-times for Trp
clearance were approximately one third of those found previously (Hutson et
al., 1985) when samples were taken six hours after implantation. The time to
peak concentration and area under curves for both Trp and 5-HIAA were comparable with the previous findings. In both studies 5-HIAA increased approximately two-fold in agreement with tryptophan hydroxylase, the rate limiting
enzyme in 5-HT synthesis being normally half saturated (Neckers et al.,
1977).

    The transport capacity of the carrier for tryptophan at the blood-brain
barrier is one of the variables influencing the supply of tryptophan to the
brain (Gessa and Tagliamonte, 1974; Curzon and Marsden, 1975; Hardebo et al.,
1980; Sarna et al., 1982; Tracqui et al., 1983). Dialysate tryptophan concentrations reflect net transport changes across the barrier and similar kinetic profiles were obtained in both striatum and cerebellum suggesting similar transport characteristics in agreement with unidirectional transport
rates as determined autoradiographically (Lookingland et al., 1986).

    The source of the rise in striatal dialysate 5-HIAA, which peaks approx-
imately one hour after the peak tryptophan response, demands discussion as it
is conceivable that the implantation of the probe caused a breakdown of the
blood-brain barrier so that systemic 5-HIAA or platelet 5-HIT entered the
brain extracellular fluid. Another possible non-neuronal source are endothe-

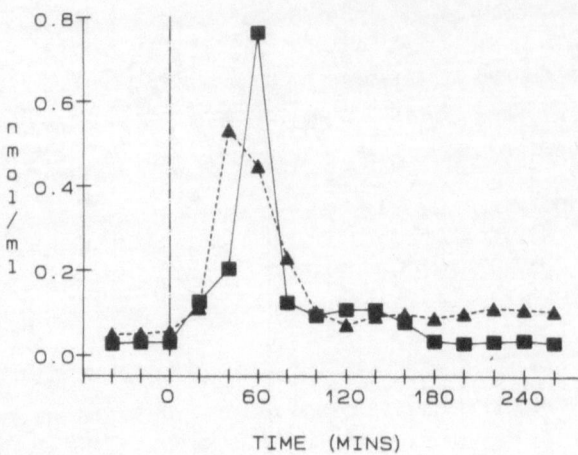

Fig. 7.   Time courses for an individual rat of dialysate Trp concentrations
          in striatum (squares) and cerebellum (triangles) before and after
          L-Trp (50 mg/kg, i.p.).

lial cells which can synthesize and metabolize 5-HT (Edvinsson et al., 1984;
Manikij et al., 1984; Defeudis, 1986; Kalaria and Harik, 1986). However, in
the above circumstances, cerebellar dialysate 5-HIAA would have increased
substantially. Therefore, extracellular 5-HIAA is likely to be largely neuronally derived. Also, evidence does not point to a diffusion barrier restricting transport of 5-HIAA between neurones and the extracellular space
(Burns et al., 1976; Tracqui et al., 1983), or to its active reuptake into
neurones. Thus extracellular and tissue concentrations of 5-HIAA parallel
each other (Sarna et al., 1983; Hutson et al., 1985).

A considerable controversy still surrounds the degree to which tryptophan availability determines 5-HT function and whether 5-HIAA concentrations
(intra- or extra-cellular) reflect 5-HT release (Baumann and Waldmeier, 1984;
Auerbach and Lipton, 1985; Commissiong, 1985; Wolf et al., 1985; Curzon,
1986; Anderson et al., 1987). Our findings indicate that changes of extracellular 5-HIAA reflect changes of 5-HT metabolism, at least in the absence
of altered 5-HIAA egress from the brain.

Do tryptophan loads influence extracellular 5-HT concentrations?

There are conflicting reports on whether precursor loading directly influences 5-HT release (Holman and Vogt, 1972; Ternaux et al., 1977; Elks et
al., 1979; Hery and Ternaux, 1981; Suter and Conard, 1983; Lookingland et
al., 1986; Sleight et al., 1988). However, we have so far been unable to
show significant changes in extracellular 5HT in the frontal cortex of rats
injected with Trp (50 mg/kg, i.p.; Fig. 9). The animals had been adapted to
a reverse light-dark for two weeks and Trp injections made 1 hour following
lights off, when rats are normally active. A recent dialysis study has shown
that Trp loads cause a massive increase of extracellular quinolinic acid (Trp
metabolite from the kynurenine pathway) (During et al., 1989). It is conceivable this can influence extracellular 5-HT concentrations.

Lack of parallelism between dialysate 5-HIAA and 5-HT concentrations

Numerous dialysis experiments reveal a lack of parallelism between extraneuronal 5-HT and 5-HIAA concentrations, e.g. (a) Trp loading increased

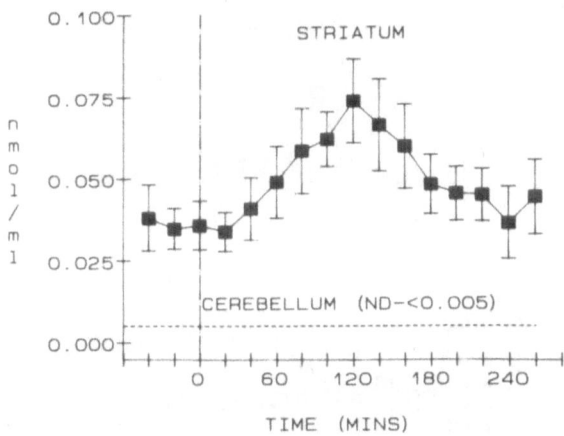

Fig. 8. Mean time courses of dialysate 5-HIAA concentrations in striatum
and cerebellum (not detected, < 0.005 nmol/ml) following L-Trp
(50 mg/kg, i.p.). Values are means ± SEM.

5-HIAA without influencing 5-HT (Fig. 9); (b) Local depolarization by application of a bolus of 100 mM KCl through the probe decreased dialysate 5-HIAA
but increased 5-HT (Fig. 10); (c) Infusion of the 5-HT$_{1A}$ agonist 8-hydroxy-2-
di(n-propylamino)tetralin (8-OH-DPAT) decreased extracellular 5-HT dose de-
pendently but the decrease of 5-HIAA was <u>inversely</u> related to dose (Hutson et
al., 1989) (Fig. 11). Changes of either tissue or extracellular 5-HIAA
(including CSF) should therefore be interpreted with caution.

Do extracellular 5-HT concentrations reflect neuronal activity?

The following evidence supports the view that dialysate 5-HT concentrations reflect serotonergic activity (see also Kalén et al., 1988): (a) Higher levels of 5-HT were found in awake freely moving animals than when anesthetised; (b) 5-HT levels showed a marked increase during depolarization of
the terminals with KCl (Fig. 10); (c) Decreased 5-HT levels were seen on removing calcium from the perfusion medium (Kalén et al., 1988); (d) Blockade
of neuronal impulse flow with tetrodotoxin decreased 5-HT concentrations
(Kalén et al., 1988); (e) Baseline concentrations of 5-HT are increased by
reuptake inhibitors (indalpine) (Kalén et al., 1988) citalopram (Sharp et
al., 1989); (f) Electrical stimulation of cell bodies (raphe neurons) can induce release of 5-HT in hippocampal dialysates; (g) Inhibition of neuronal
firing by application of 8-OH-DPAT onto the dorsal raphe decreased basal 5-HT
concentrations at a distal site (hippocampus) (Hutson et al., 1989). Changes
of extracellular 5-HT in a terminal region following electrical stimulation
or application of serotonergic agonists at a <u>distal</u> cell body further implies
that a considerable fraction of dialysate 5-HT originates from neurones and
not from non-neuronal sources such as platelets.

Dialysis together with behavioral studies

An important application of <u>in vivo</u> monitoring methods is in the study
of the neurochemical mediation of the behavioral effects of drugs. Thus,
intracerebral microdialysis has been used to determine the relation between
changes in extraneuronal 5-HT induced by p-chloroamphetamine (pCA) and of
components of the 5-HT syndrome and other aspects of motor behaviour in the
same animals (Hutson and Curzon, 1989). pCA markedly and transiently increased extracellular 5-HT in the frontal cortex (similar to control values

71

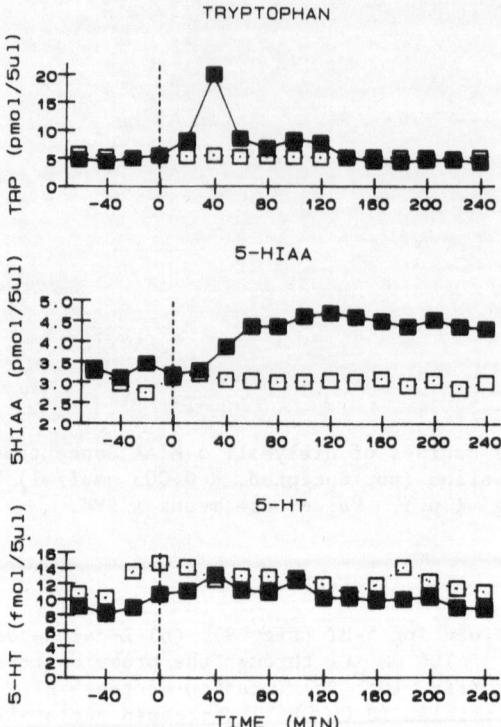

Fig. 9. Mean time courses of frontal cortex dialysate Trp, 5-HIAA and 5-HT
concentrations after i.p. injection of L-Trp (50 mg/kg).

shown in Fig. 12) with extracellular 5-HIAA values decreasing moderately.
The time course of 5-HT changes correlated positively ($P < 0.001$) with three
components of the 5-HT syndrome (reciprocal forepaw treading, headweaving,
wet dog shakes) but did not correlate significantly with locomotion and correlated negatively ($P < 0.01$) with grooming.

Drugs which release neuronal 5-HT (e.g. pCA) or are agonists at 5-HT
receptors (e.g. 5-methoxy-NN-dimethyltryptamine, 5-MeODMT) cause the 5-HT
syndrome not only in normal rats but also in rats given reserpine which de-
pletes brain 5-HT by disrupting vesicular storage. Consequently, it has been
suggested that pCA induces the syndrome by releasing 5-HT from a non-vesicular pool (Kuhn et al., 1985). However, as it seemed conceivable that reserpine both reduces the release of 5-HT by pCA and increases the response of
receptors to it, so that the behavioral response is unaltered, we used intracerebral dialysis to monitor the change of extracellular 5-HT after pCA in
normal and reserpine treated rats (Hutson and Curzon, 1989).

Rats were given either reserpine (2.5 mg/kg, i.p.) or vehicle 24 hours
before pCA (5.0 mg/kg, i.p.) or 5-MeODMT (5 mg/kg, i.p.). They were placed
in individual perspex cages and behavior was recorded on videotape for 1
hour. In a separate experiment rats were implanted with dialysis probes in
the frontal cortex and given either reserpine (2.5 mg/kg, i.p.) or vehicle
and allowed to recover overnight prior to perfusion. The effect of reserpine (2.5 mg/kg, i.p.) on brain amine concentrations was determined in a
separate group of animals.

HIPPOCAMPAL 5-HT

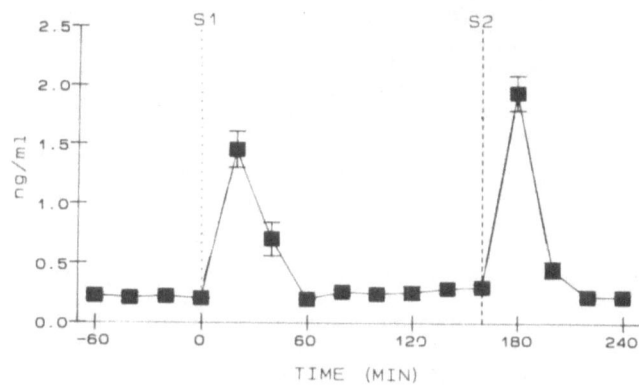

HIPPOCAMPAL 5-HIAA

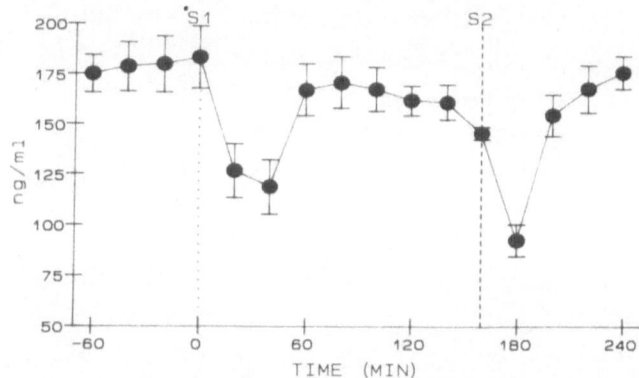

Fig. 10. Effects of the addition of KCl (100 mM) to the perfusion fluid for 20 minute periods on hippocampal 5-HT and 5-HIAA concentrations. The concentration of NaCl was reduced to maintain isoosmolarity. Values are means ± SEM (N - 5).

Reserpine decreased rat brain 5-HT by 86% 24 hours later but most components of the 5-HT syndrome induced by pCA or 5-MeODMT were unaffected. Before injection of pCA, extraneuronal 5-HT was measurable in dialysates from normal rats (0.006 ± 0.0001 pmol/5 µl) but not from reserpinised rats. However, following pCA, dialysate 5-HT concentration rose markedly in both vehicle and reserpine treated rats, peaking at 20-40 min after administration (Fig. 12). There are no significant difference between the groups in the net increase of 5-HT when expressed as the area under the curves over the 1 hour period after pCA. Dialysate 5-HIAA values were essentially unchanged by reserpine administration.

These results indicate that pCA causes the 5-HT syndrome by releasing 5-HT from a non-vesicular pool. Also, as reserpine without pCA markedly reduced dialysate 5-HT, basal release of 5-HT probably occurs from reserpine-sensitive vesicular stores.

Application of dialysis to neurological models

In vivo cerebral dialysis has recently been applied to study experimental hepatic encephalopathy (Hamberger and Nyström, 1984), status epilepticus

A : DIALYSATE 5-HT

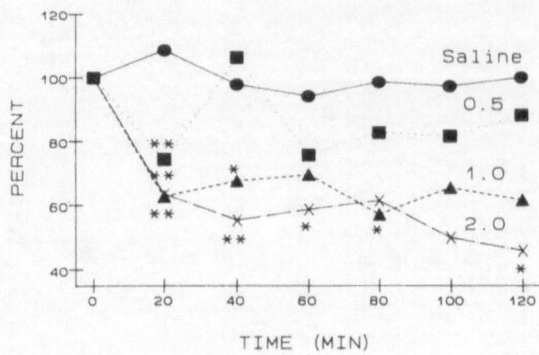

TIME (MIN)

B : DIALYSATE 5-HIAA

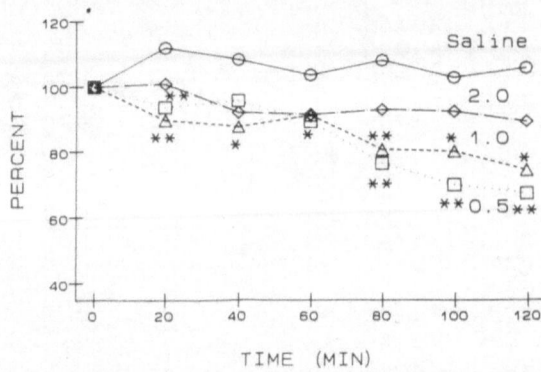

TIME (MIN)

Fig. 11.   Percentage changes in hippocampal dialysate 5-HT (A) and 5-HIAA (B)
following infusion of 0.9% saline or 8-OH-DPAT (0.5-2.0 µg) into
the dorsal raphe.   Values are means ± SEM (N = 5 per group).
*P < 0.01.

(Lehmann et al., 1985), efficacy of transplanted grafts (Zetterström et al.,
1986) and cerebral ischaemia (Hagberg et al., 1985).   Cerebrovascular diseases are one of the most frequent causes of death and disability, with crucial social implications.   It is imperative to improve our understanding of
abnormalities of brain changes during and after episodes of reduced blood
supply in order to develop drugs which can protect the brain against ischemia-induced damage.   In vivo procedures which monitor central changes during
experimental ischemia are powerful tools in such studies.   For example, in
vivo dialysis revealed a massive increase in extracellular glutamate (Benveniste et al., 1984; Hagberg et al., 1985) which could well have neurotoxic effects since glutamate antagonists ameliorate effects of ischemia in some animal models (Foster et al., 1987; Ozyurt et al., 1988).   Similar DA changes in
the striatum may also have a pathological role as 6-hydroxydopamine (6-OHDA)
lesions of the DAergic nigrostriatal pathway can confer protection (Globus et
al., 1988).   We find a differential effect of cerebral ischemia on extracellular concentrations of 5-HT and dopamine (DA) (Sarna et al., 1989).

These experiments were performed in rats anesthetized throughout with
halothane (1-2%) in $O_2:N_2O$ (1:1).   Concentric dialysis probes were implanted
in the left striatum and hippocampus a day prior to the experiment.   Follow-

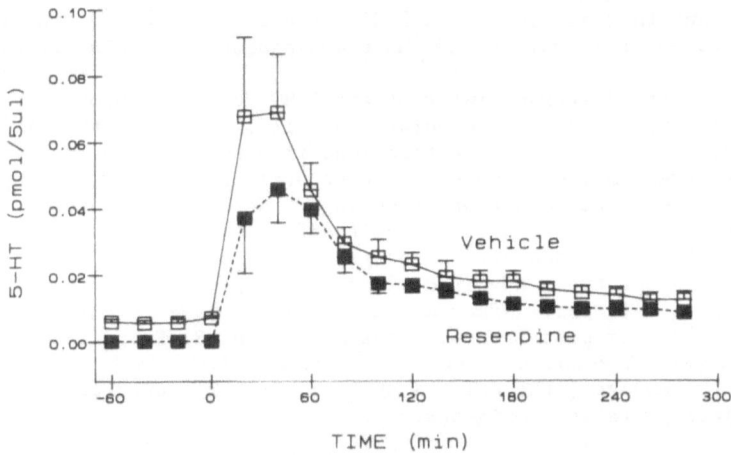

Fig. 12. Effects of reserpine (2.5 mg/kg, i.p.) on dialysate 5-HT concentration in frontal cortex before and after pCA (5 mg/kg, i.p. at time 0).

ing a stabilization period, 20 min cerebral ischemia was produced by 4-vessel occlusion (Pulsinelli et al., 1982) and followed by 80 min reperfusion and cardiac arrest. Measurements were continued for a further 60 mins.

    In 6 our of 9 experiments there are rapid and significant increases (P < 0.01) in striatal DA and in both striatal and hippocampal 5-HT concentrations (x 8-10) during ischemia. Striatal homovanillic acid and 5-HIAA were significantly decreased by 50% and 60% respectively (P < 0.01). All baseline values were restored on 20 min reperfusion. Following cardiac arrest, striatal DA rose again approximately as before (x 300-400) but striatal 5-HT rose more markedly than during ischemia (x 100 and x 36 respectively, P < 0.01). This differential ischemia-induced response of DA and 5-HT is shown in (Fig. 13).

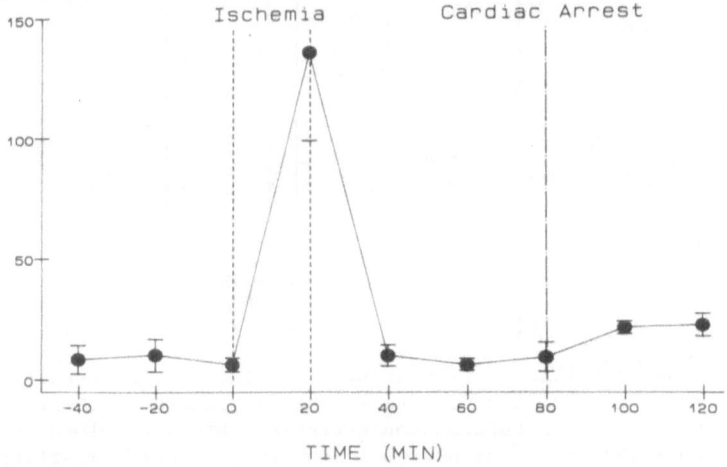

Fig. 13. Effect of experimental cerebral ischemia and cardiac arrest on striatal dialysate DA/5-HT ratio.

A significant increase in the DA/5-HT ratio was apparent during ischemia as compared to baseline values and those subsequent to cardiac arrest.

The increased extracellular DA and 5-HT during ischemia may simply reflect a massive efflux from depolarized cells. However, the differential effects suggest that selective alterations in disposition or metabolism of the two transmitters may be involved. These preliminary findings suggest that a number of fundamental questions need to be addressed in future studies e.g. how do degree and duration of ischemia affect extracellular 5-HT, DA and GLU? do they influence subsequent neurological and behavioral deficits? are specific 5-HT receptor types involved? if so, do appropriate antagonists protect against these deficits? Nakayama et al. (1988) have recently reported that both pre- and 1-hr post-ischemic treatment with (S)-Emopamil (said to be a 5-HT$_2$ antagonist) reduced infarct size after middle cerebral artery occlusion in the rat. However, the calcium channel blocker properties of this drug renders interpretation ambiguous.

CONCLUSIONS

In *vivo* cerebral microdialysis coupled with recent advances in HPLC analytical procedures provide powerful tools in the study of tryptophan and

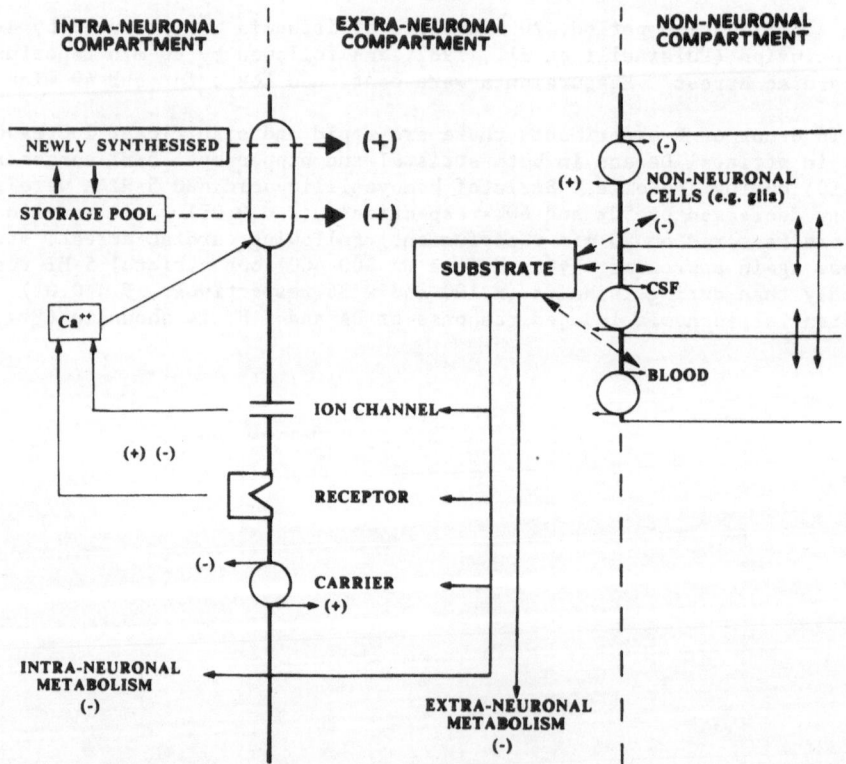

Fig. 14.   Schematic illustration of possible factors influencing the net concentration of a substrate in the extraneuronal compartment of brain. Extraneuronal concentrations are determined by possible contributions from neuronal and non-neuronal compartments and by different mechanisms e.g. release, diffusion, active transport, autoreceptor activation. Factors increasing and decreasing extraneuronal concentrations are indicated by (+) and (-), respectively.

5-HT metabolism. The use of microdialysis in the conscious, freely moving animal enables the study of the behavioral effects of drugs, and its use in neurological models could lead to potential therapeutic measures. Current evidence suggest that both dialysate 5-HIAA and 5-HT are neuronally derived. The dialysis approach may also prove valuable in characterising the effects of physiological, pharmacological and neurological interventions on the various mechanisms (Fig. 14) controlling the disposition of central transmitters, their precursors and metabolites.

REFERENCES

Adell, A., Sarna, G.S., Hutson, P.H., and Curzon, G., 1989, An in vivo dialysis and behavioral study of the release of 5-HT by p-chloro-amphetamine in reserpinized rats, Brit. J. Pharmacol., 97:206-212.
Alexander, G.M., Grothusen, J.R., and Schwartzman, R.J., 1988, Flow dependent changes in the effective surface area of microdialysis probes, Life Sci., 43:595-601.
Anderson, G.M., Teff, K.L., and Young, S.N., 1987, Serotonin in cisternal cerebrospinal fluid of the rat: measurement and use as an index of functionally active serotonin, Life Sci., 40:2253-2260.
Auerbach, S., and Lipton, P., 1985, Regulation of serotonin release from the in vitro rat hippocampus: effects of alterations in levels of depolarization and rates of serotonin metabolism, J. Neurochem., 44:1116-1130.
Azmitia, E.C., and Segal, M., 1978, An autoradiographic analysis of the differential ascending projections of the dorsal and median raphe nuclei in the rat, J. Comp. Neurol., 179:641-669.
Baumann, P.A., and Waldmeier, P.C., 1984, Negative feedback control of serotonin release in vivo: comparison of 5-hydroxy-indoleacetic acid levels measured by voltammetry in conscious rats and by biochemical techniques, Neuroscience, 11:195-204.
Benveniste, H., and Diemer, N.H., 1987, Cellular reactions to implantation of a microdialysis tube in the rat hippocampus, Acta Neuropathol., 74:234-238.
Benveniste, H., Drejer, J., Schousboe, A., and Diemer, N.H., 1984, Elevation of the extracellular concentrations of glutamate and aspartate in rat hippocampus during transient cerebral ischemia monitored by intracerebral dialysis, J. Neurochem., 43:1369-1374.
Benveniste, H., Drejer, J., Schousboe, A., and Diemer, N.H., 1987, Regional cerebral glucose phosphorylation and blood flow after insertion of a microdialysis fiber through the dorsal hippocampus in the rat, J. Neurochem., 49:729-734.
Bito, L., Davson, H., Levin, E., Murray, M., and Sider, N., 1966, The concentrations of free amino acids and other electrolytes in cerebrospinal fluid, in vivo dialysate of brain, and blood plasma of the dog, J. Neurochem., 13:1057-1067.
Blundell, J.E., 1984, Serotonin and appetite, Neuropharmacology, 23:1537-1551.
Bonanno, G., Maura, G., and Raiteri, M., 1986, Pharmacological characterization of release regulating serotonin autoreceptors in rat cerebellum, Eur. J. Pharmacol., 126:317.
Burns, D., London, J., Brunswick, D.J., Pring, M., Garfinkel, D., Rabinowitz, J.L., and Mendels, J., 1976, A kinetic analysis of 5-hydroxyindoleacetic acid excretion from rat brain and CSF, Biol. Psych., 11:125-157.
Chan-Palay, V., 1976, Serotonin afferents from raphe nuclei to the cerebellum in mammals, Exp. Brain Res., Suppl., 1:20-25.
Church, W.H., and Justice, J.B., 1986, Sampling considerations for on-line microbore liquid chromatography of brain dialysate, 58:1649-1656.
Commissiong, J.W., 1985, Monoamine metabolites: their relationship and lack of relationship to monoaminergic neuronal activity, Biochem. Pharmacol., 34:289-290.

Cudennec, A., Duverger, D., Serrano, A., Scatton, B., and Mackenzie, E.T., 1988, Influence of ascending serotonergic pathways on glucose use in the conscious rat brain. II. Effects of electrical stimulation of the rostral raphe nuclei, **Brain Res.**, 444:227-246.

Curzon, G., 1986, Critique: serotonin neurochemistry revisited: a new look at some old axioms, Neurochem. Int., 2:155-159.

Curzon, G., and Marsden, C.A., 1975, Metabolism of a tryptophan load in the hypothalamus and other brain regions, J. Neurochem., 25:251-256.

Defeudis, F.V., 1986, New studies on cerebrovascular endothelium - possible reference to the interpretation of 'precursor-loading' experiments, **Trends Pharmacol. Sci.**, 7:51-52.

Delgado, J.M.R., Defeudis, F.V., Roth, R.H., Ryugo, D.K., and Mitruka, B.M., 1972, Dialtrode for long term intracerebral perfusion in the awake cat brain, J. Arch. Int. Pharmacodyn., 198:9-21.

De Simoni, M.G., Sokola, A., Fodritto, F., Dal Toso, G., and Algeri, S., 1987, Functional meaning of tryptophan-induced increase of 5-HT metabolism as clarified by in vivo voltammetry, **Brain Res.**, 411:89-94.

Dourish, C.T., Hutson, P.H., and Curzon, G., 1985, Characteristics of feeding induced by the serotonin agonist 8-hydroxy-2-(di-n-propylamino)tetralin (8-OH-DPAT), **Brain Res. Bull.**, 15:377-384.

During, M.J., Heyes, M.P., Freese, A., Markey, S.P., Martin, J.B., and Roth, R.H., 1989, Quinolinic acid concentrations in striatal extracellular fluid reach potentially neurotoxic levels following systemic L-tryptophan loading, **Brain Res.**, 476:384-387.

Edvinsson, L., Birath, E., Uddman, R., Lee, T.J.-F., Duerger, D., Mackenzie, E.T., and Scatton, B., 1984, Indoleaminergic mechanisms in brain vessels; localization, concentration, uptake and in vitro responses of 5-hydroxy-tryptamine, **Acta Physiol. Scand.**, 121:291-299.

Elks, M.L., Youngblood, W.W., and Kizer, J.S., 1979, Serotonin synthesis and release in brain slices: independence of tryptophan, **Brain Res.**, 172: 471-486.

Fernstrom, J.D., and Wurtman, R.J., 1971, Brain serotonin content: physiological dependence on plasma tryptophan levels, **Science**, 173:149-151.

Foster, A.C., Gill, R., Kemp, J.A., and Woodruff, G.N., 1987, Systemic administration of MK-801 prevents N-methyl-D-aspartate-induced neuronal degeneration in rat brain. Neurosci. Lett., 76:307-311.

Gessa, G.L., and Tagliamonte, A., 1974, Serum free tryptophan: control of brain concentrations of tryptophan and of synthesis of 5-hydroxytryptamine, in: "Ciba Foundation Symposium 22: Aromatic Acids in the Brain", Excerpta Medica, Amsterdam, pp. 207-216.

Globus, M.P.-T, Busto, R., Dietrich, W.D., Martinez, E., Valdes, I., and Ginsberg, M.D., 1988, Intra-ischemic extracellular release of dopamine and glutamate is associated with striatal vulnerability to ischemia, **Neurosci. Lett.**, 91:36-40.

Hagberg, H., Lehmann, A., Sandberg, M., Nystrom, B., Jacobson, I., and Hamberger, A., 1985, Ischemia-induced shift of inhibitory and excitatory amino acids from intra- to extracellular compartments, J. **Cereb. Blood Flow Metab.**, 5:413-419.

Hamberger, A., and Nyström, B., 1984, Extra- and intracellular amino acids in the hippocampus during development of hepatic encephalopathy, **Neurochem. Res.**, 9:1182-1192.

Hardebo, J.E., Emson, P.C., Falck, B., Owman, C., and Rosengren, E., 1980, Enzymes reacted to monoamine transmitter metabolism in brain microvessels, J. Neurochem., 35:1388-1393.

Hery, F., and Ternaux, J.P., 1981, Regulation of release processes in central serotonergic neurons, J. **Physiol. (Paris)**, 77:287-301.

Holman, R.B., and Vogt, M., 1972, Release of 5-hydroxytryptamine from caudate nucleus and septum, J. **Physiol.**, 223:243-254.

Hutson, P.H., and Curzon, G., 1989, Concurrent determination of effects of p-chloroamphetamine on central extracellular 5-hydroxytryptamine concentration and behavior, **Br. J. Pharmacol.**, 96:801-806.

Hutson, P.H., Sarna, G.S., and Curzon, G., 1986, Neuropharmacokinetic applications of in vivo monitoring techniques, Ann. N.Y. Acad. Sci., 473:549-
552.

Hutson, P.H., Sarna, G.S., O'Connell, M.T. and Curzon, G., 1989, Hippocampal
5-HT synthesis and release in vivo is decreased by infusion of 8-OH-DPAT
into the nucleus raphe dorsalis, Neurosci. Lett., in press.

Hutson, P.H., Sarna, G.S., Kantamaneni, B.D., and Curzon, G., 1985, Monitoring the effect of a tryptophan load on brain indole metabolism in freely
moving rats by simultaneous cerebrospinal fluid monitoring and brain
dialysis, J. Neurochem., 44:1266-1273.

Hutson, P.H., Sarna, G.S., Sahakian, B.J., Dourish, C.T., and Curzon, G.,
1986, Monitoring 5-HT metabolism in the brain of the freely moving rat,
Ann. N.Y. Acad. Sci., 473:321-335.

Kalaria, R.N., and Harik, S.I., 1987, Blood-brain barrier monoamine oxidase:
enzyme characterization in cerebral microvessels and other tissues from
six mammalian species, including human, J. Neurochem., 49:856-864.

Kalén, P., Strecker, R.E., Rosengren, E., and Björklund, A., 1988, Endogenous
release of neuronal serotonin and 5-hydroxyindoleacetic acid in the caudate-putamen of the rat as revealed by intracerebral dialysis coupled to
high-performance liquid chromatography with fluorimetric detection, J.
Neurochem., 51:1422-1435.

Knott, P.J., 1988, Modern methods for studying the release of serotonin, in:
"Neuronal Serotonin", Wiley, New York, pp. 93-127.

Knott, P.J., and Curzon, G., 1974, Effect of increased brain tryptophan on
5-hydroxytryptamine and 5-hydroxyindoleacetic acid in the hypothalamus and
other brain regions, J. Neurochem., 22:1065-1071.

Kuhn, D.M., Wolf, W.A., and Youdim, M.B.H., 1985, 5-Hydroxytryptamine release
in vivo from a cytoplasmic pool: studies on the behavioral syndrome in reserpinized rats, Br. J. Pharmacol., 84:121-129.

Lehmann, A., Hagberg, H., Jacobson, I., and Hamberger, A., 1985, Effects of
status epilepticus on extracellular amino-acids in the hippocampus, Brain
Res., 359:147-151.

Lerma, J., Herranz, A.S., Herras, O., Abraira, V., and Martin Del Rio, R.,
1986, In vivo determination of extracellular concentrations of amino acids
in the rat hippocampus, Brain Res., 383:145-155.

Lookingland, K.J., Shannon, N.J., Chapin, D.S., and Moore, K.E., 1986,
Exogenous tryptophan increases synthesis, storage and intraneuronal metabolism of 5-hydroxytryptamine in the rat hypothalamus, J. Neurochem., 47:
205-212.

Manikij, C., Spatz, M., Veki, Y., Nagatsu, I., and Bembry, J., 1984, Cerebrovascular endothelial cell culture: metabolism and synthesis of 5-hydroxy-
tryptamine, J. Neurochem., 43:316-319.

Mans, A.M., Biebuyck, J.F., Shelly, K., and Hawkins, J.P., 1982, Regional
blood-brain barrier permeability to amino acids after portacaval anastomosis, J. Neurochem., 38:705-717.

Nakayama, H., Ginsberg, M., and Dietrich, W.D., 1988, (S)-Emopamil, a novel
calcium channel blocker and serotonin S_2 antagonist markedly reduces infarct size following middle cerebral artery occlusion in the rat,
Neurology, 38:1667-1673.

Neckers, L.M., Biggio, G., Moja, E., and Meek, J.L., 1977, Modulation of
brain tryptophan hydroxylase activity by brain tryptophan content, J.
Pharmacol. Exp. Therap., 201:110-116.

Ozyurt, E., Graham, D.I., Woodruff, G.N., and McCulloch, J., 1988, Protective
effect of the glutamate antagonist MK-801 in focal cerebral ischemia in
the cat, J. Cereb. Blood Flow Metab., 8:138-143.

Pulsinelli, W.A., Brierley, J.B., and Plum, F., 1982, Temporal profile of
neuronal damage in a model of transient forebrain ischemia, Ann. Neurol.,
11:491-498.

Sarna, G.S., Hutson, P.H., and Curzon, G., 1984, Effect of alpha-methyl-
fluorodopa on dopamine metabolites: importance of conjugates and egress,
Europ. J. Pharmacol., 100:343-350.

Sarna, G.S., Hutson, P.H., Tricklebank, M.D., and Curzon, G.S., 1983, Determination of brain 5-hydroxytryptamine turnover in freely moving rats using repeated sampling of cerebrospinal fluid, J. Neurochem., 40:383-388.

Sarna, G.S., Obrenovitch, T.P., Matsumoto, T., Symon, L., and Curzon, G., 1989, Differential effect of cerebral ischaemia on dopamine and serotonin release as determined by in vivo brain dialysis, J. Cereb. Blood Flow Metab., in press.

Sarna, G.S., Tricklebank, M.D., Kantamaneni, B.D., Hunt, A., Patel, A.J., and Curzon, G., 1982, Effect of age on variables influencing the supply of tryptophan to the brain, J. Neurochem., 39:1283-1290.

Schwartz, D.H., McClane, S., Hernandez, L., and Hoebel, B.G., 1989, Feeding increases extracellular serotonin in the lateral hypothalamus of the rat as measured by microdialysis, Brain Res., 479:349-354.

Sharp, T., Bramwell, S.R., and Grahame-Smith, D., 1989, 5-HT$_1$ agonists reduce 5-hydroxytryptamine release in rat hippocampus in vivo as determined by brain microdialysis, Br. J. Pharmacol., 96:283-290.

Sleight, A.J., Marsden, C.A., Martin, K.F., and Palfreyman, M.G., 1988, Relationship between extracellular 5-hydroxytryptamine and behaviour following monoamine oxidase inhibition and L-tryptophan, Br. J. Pharmacol., 93:303-310.

Steinbusch, H.W.M., 1984, Serotonin-immunoreactive neurones and their projections in the CNS, in: "Handbook of Chemical Neuroanatomy, Vol. 3", Björklund, A., Hökfelt, T., and Kuhar, M.J., eds., Elsevier, Amsterdam, pp. 68-125.

Suter, H.A., and Collard, K.J., 1983, The regulation of 5-hydroxytryptamine release from superfused synaptosomes by 5-hydroxytryptamine and its immediate precursors, Neurochem. Res., 8:723-730.

Takeuchi, Y., Kimura, H., and Sano, Y., 1982, Immunohistochemical demonstration of serotonin-containing nerve fibers in the cerebellum, Cell Tiss. Res., 226:1-12.

Ternaux, J.P., Hery, F., Hamon, N., Bourgoin, S., and Glowinski, J., 1977, 5-HT release from ependymal surface of the caudate nucleus in 'encephale isolé' cats, Brain Res., 132:575-579.

Tossmann, V., and Ungerstedt, U., 1986, Microdialysis in the study of extracellular levels of amino acids in the rat brain, Acta Physiol. Scand., 128:9-14.

Tracqui, P., Morot-Gaudry, Y., Staub, J.F., Brezillon, P., Perault-Staub, A.M., Bourgoin, S., and Hamon, M., 1983, Model of brain serotonin metabolism, II. Physiological Interpretation, Am. J. Physiol., 244:R206-R215.

Ungerstedt, U., and Pycock, C.H., 1974, Functional correlates of dopamine neurotransmission, Bull. Schweiz. Akad. Med. Wiss., 30:44-55.

Westerink, B.H.C., Damsma, G., Rollema, H., De Vries, J.B., and Horn, A.S., 1987, Scope and limitations of in vivo brain dialysis: a comparison of its applications to various neurotransmitter systems, Life Sci., 41:1763-1776.

Westerink, B.H.C., and De Vries, J.B., 1988, Characterization of in vivo dopamine release determined by brain microdialysis after acute and subchronic implantations: methodological aspects, J. Neurochem., 51:683-687.

Wolf, W.A., Youdim, M.B.H., and Kuhn, D.M., 1985, Does brain 5-HIAA indicate serotonin release or monoamine oxidase activity?, Eur. J. Pharmacol., 109:381-387.

Zetterström, T., Brundin, P., Gage, F.H., Sharp, T., Isacson, O., Dunnett, S.B., Ungerstedt, U., and Björklund, A., 1986, Brain Res., 362: 344-349.

SUB-PICOGRAM DETERMINATION OF SEROTONIN USING HPLC WITH ELECTROCHEMICAL

DETECTION FOR MICRODIALYSIS STUDIES OF SEROTONIN RELEASE

C.W. Bradberry, J.S. Sprouse, G.K. Aghajanian, and R.H. Roth

Departments of Pharmacology and Psychiatry
Yale University School of Medicine
New Haven, Connecticut 06510
USA

INTRODUCTION

The technique of microdialysis offers a unique avenue for monitoring extracellular chemical dynamics in nervous tissue. In using this technique to
monitor neurotransmitters with high affinity uptake systems, the accurate determination of the extremely small amounts of material recovered represents
an analytical challenge. For determinations of serotonin (5-HT) in microdialysis perfusates, HPLC coupled with either fluorescence (Kalén et al.,
1988), or electrochemical (Brazell et al., 1985; Sharp et al., 1986) detection has been employed. Both means of detection offer high sensitivity and
selectivity. This manuscript presents a chromatographic system we have designed and constructed which offers extremely high sensitivity, suitable for
analysis of microdialysis perfusates. In addition, the cost of the components for this system are roughly an order of magnitude less than the cost of
a commercial system. The essential components are a pneumatic displacement
fluid pump, narrow bore HPLC column, and battery-powered potentiostat for the
electrochemical detection of 5-HT.

We have used this high sensitivity system in collaborative biochemical/
electrophysiological studies examining the effects of the novel psychedelic
agent, 3,4-methylenedioxymethamphetamine (MDMA). This compound has received
increasing attention because of its use a psychotherapeutic aid (Greer,
1983), recreational abuse (Peroutka, 1987), and emergency classification as a
Schedule I compound following reports of long-term neurotoxic effects
(Schmidt et al., 1986). In these studies we have measured the MDMA-induced
release of 5-HT from dorsal raphe brain slices using microdialysis probes
modified for use on the surface of a brain slice. Parallel extracellular
electrophysiological measures of the firing rate of typical 5-HT-containing
dorsal raphe neurons from slices in the same chamber have allowed us to correlate changes in 5-HT release with alterations in 5-HT neuronal firing.
This in vitro application represents a novel use of the microdialysis technique and one which we believe offers further potential applications.

METHODS

Chromatographic system

As modern HPLC methodology has developed, workers have been aware of the advantage of moving to higher efficiency (smaller particle size) columns, namely, higher resolution and increased sensitivity (Snyder and Kirkland, 1979). Further increases in sensitivity are predicted in theory as the diameter of the column is reduced (Guichon and Colin, 1984). Also, there is the economic advantage of reduced consumption of mobile phase and lesser amounts of packing material necessary with smaller columns. As column diameters reach true "microbore" dimensions (≤ 1 mm), it is often necessary to utilize alternative components compatible with small injection volumes and ultra-low volume detectors (≤ 0.2 μl). In reality, the theoretical increases in detection limits are diminished by the necessity of injecting smaller volumes, by diminished detector volumes in volume dependent spectroscopic detectors, and the diminished flow rates in flow rate dependent (Mefford, 1987) electrochemical detectors. In general, columns of intermediate size (so called "narrow bore" with inner diameters of approx. 2 mm) can be used with conventional injectors and detectors, yet offer reduced consumption of mobile phase and increased sensitivity. Slurry packing of the narrow bore columns is easier than packing microbore columns, and cheaper, in terms of packing material, than for 1/4" columns. Thus, as has been noted by others (Guichon and Colin, 1984), these columns represent a realistic compromise between the performance of the "old standard 1/4 inch columns" and the theoretical performance of microbore columns difficult to obtain in practice. We have used laboratory-packed narrow bore columns (2.1 mm I.D., 3 micron C-18 particles) and have found them to be extremely reliable and convenient.

For use with the narrow-bore columns, our chromatographic system utilizes a pneumatic displacement fluid pump (Bradberry and Roth, in preparation). This type of pump is one in which a pressurized fluid (usually a gas, though other fluids such as mercury have been used) displaces the mobile phase as it is pumped. Pneumatic displacement fluid pumps are an old idea (for review see Martin et al., 1978). However, for use with the increasingly popular narrow bore and microbore columns they represent an old idea worth reexamining. Pneumatic displacement fluid pumps are most compatible with situations requiring low volume flow rates as is the case with narrow- and microbore columns. The most significant benefit of using this type of pump is that they offer truly pulseless fluid flow. This is expensive to achieve at low flow rates, and absolutely necessary for high sensitivity measurements using electrochemical detectors because of the flow rate dependence of their output (Mefford, 1987).

A crucial problem with pneumatic displacement fluid pumps is the tendency of gases to dissolve into the mobile phase at the high pressures used to pump the fluid, emerging as bubbles in the detector at the low pressure end of the column. Many methods have been utilized to separate the gas and liquid including floating seals (Meek, 1976), or metal bellows (Bonnelycke, 1969). The use of a coil minimizes the surface interface where the gas can dissolve. Given the separation of the interface from the bulk of solution in the coil, only liquid without dissolved gas is pumped from the bottom of the coil for a length of time determined by the coil volume and set flow rate. The use of a coil pneumatic pump has previously been published (Hare, 1966; Meek, 1976), and was part of previously marketed systems from DuPont and others.

The specific pump we have used is a double coil pneumatic displacement fluid pump working off one regulator, but with separate drain/fill systems allowing one or both coils to be used. The two coils are housed in a rectangular 1/4" plexiglass container which maintains the necessary vertical orientation of the coils. Fig. 1 diagrams one of the coils. Type 316 stainless steel tubing (0.25" O.D. x 0.21" I.D.) was coiled by hand around an 11 cm diameter pipet washer to make a 17 element coil spanning 54 cm. The top and bottom elements of the coil extend to bulkhead fittings in the front panel of the plexiglass housing which provide the sole support of the coils; all other

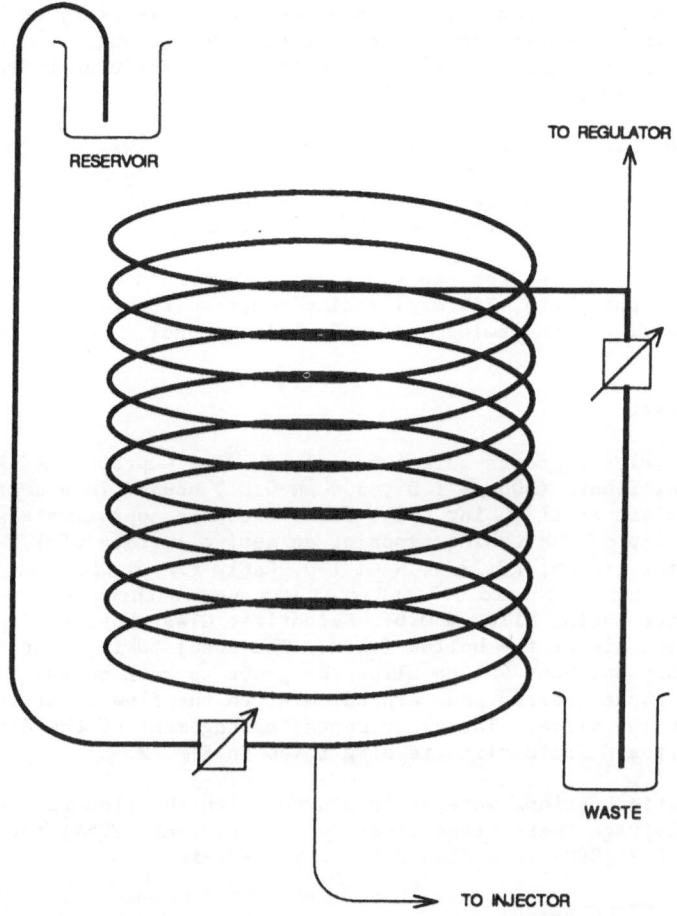

RESERVOIR

TO REGULATOR

WASTE

TO INJECTOR

Fig. 1.   Diagram of coil-based pneumatic displacement fluid pump used for
         HPLC mobile phase.

connections are made external to the plexiglass housing. High pressure nitrogen was used to pressurize the coil.

     To fill the coil with mobile phase, the two valves are opened allowing
mobile phase to syphon from the reservoir atop the plexiglass box through the
coil and out the waste tube. The 250 ml volume of mobile phase used to fill
the coil syphons through the coil (from the bottom up) in approximately one
minute. The volume of the coil itself is 150 ml. At a flow rate of 0.2 ml/
min (1800 psi), the volume of the coil is sufficient for six hours use without refilling.

     HPLC columns (10 cm x 2.1 mm I.D.) were packed with 3 micron C-18 mater-
ial (Shandon) using a Haskell 122X ratio pneumatic amplified slurry packer as
configured by Alltech Associates (Deerfield, IL; Bradberry and Roth, in preparation).

     The potentiostat used for the electrochemical detection was laboratory designed and constructed. It employs the simple circuitry required for applying a fixed potential to the three electrode assembly and converting the resulting current into a voltage for output to a strip chart recorder (Kissinger et al., 1973; Kissinger, 1984).

The thin layer amperometric electrode assembly was obtained from Bioanalytical Systems (West Lafayette, IN). In order to maximize sensitivity, Saran Wrap was used as the gasket material between the working and auxiliary electrodes as has been previously reported (Caliguri et al., 1985).

Mobile phase used for the experiments described herein was 0.05 M dibasic sodium phosphate, 350 mg/l sodium octanesulfonate, 0.1 mM disodium EDTA, 300 μl/l triethylamine, and 150 ml/l methanol. pH was unadjusted and was 5.6.

Dialysis measurements

Microdialysis probes were constructed using Cuprophan (Enka, West Germany) hollow fibers (300 μm I.D., 330 μm O.D.) housed in a section of 23 Gauge stainless steel tubing. The fiber extended approximately 2.0-2.5 mm beyond the tip of the tubing exposing an active surface of 1.5-2.0 mm. Perfusion buffer (in mM; KCl 2.4, NaCl 120, $CaCl_2$ 1.2, $MgCl_2$ 1.2, NaH_2PO_4 0.9, Na_2HPO_4 1.4, ascorbic acid 0.3, pH 7.4) was pumped through a section of vitreous silica tubing (170 μm O.D., Scientific Glass Engineering) which extended to the tip of the hollow fiber. The steel tubing containing the dialysis assembly was bent 90° to allow the probe to rest on the brain slice over the dorsal raphe nucleus and perpendicular to the flow of artificial CSF (ACSF) over the slice. The experimental arrangement of the dialysis probe and the perfused brain slice is diagrammed in Fig. 2.

Collection periods were at 10 minutes with the flow rate set at 2.0 μl/min. The average "percentage recovery" (Ungerstedt, 1984) for these probes was 9.7% ± 0.7 (SEM) at a flow rate of 2.0 μl/min.

Brain slice preparation

Details of the chamber, slice preparation, and extracellular electrophysiological recording procedures can be found in Sprouse et al. (1989). In brief, a gas-fluid interface slice chamber was used. Agents were administered in the ACSF flowing over the slice by means of a stopcock arrangement. Composition of the ACSF was (in mM): NaCl 125, KCl 5.0, $CaCl_2$ 2.0, $MgSO_4$ 2.0, $NaHCO_3$ 28, NaH_2PO_4 1.25, D-glucose 10. It was saturated with 95%/5% O_2/CO_2 and thermally equilibrated with a 33°C water bath immediately prior to flowing over the slice at a flow rate of 1-1.5 ml/min.

RESULTS

Chromatography

The routine limit of detection attained with our chromatographic system for 5-HT was 2 femtomoles (fmol) at a signal-to-noise ratio of 2. Fig. 3 illustrates chromatograms of a 20 fmol (3.5 pg) standard and a brain slice dialysate containing 1.5 fmol (0.26 pg) 5-HT. Without this high level of sensitivity, the experiments we describe below would not have been possible.

Effect of MDMA on 5-HT release

When MDMA was administered to the slice preparation under the same parameters as used for electrophysiological measurements, measurable quantities of 5-HT were detected in the dialysis perfusate. Fig. 4 presents results

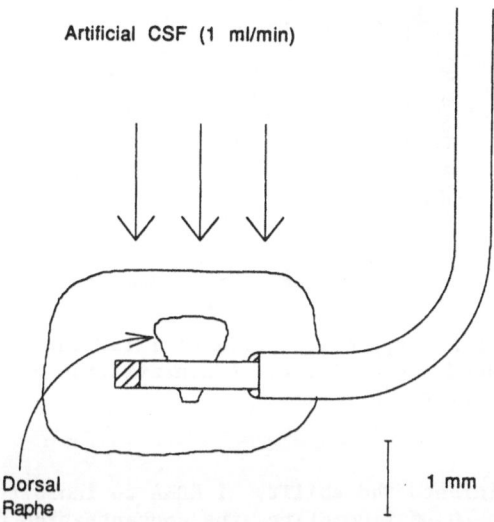

Artificial CSF (1 ml/min)

Dorsal
Raphe

1 mm

Fig. 2. Arrangement of dialysis probe on brain slice.

from a typical experiment in which MDMA at a concentration of 100 μM was applied to the slice for 3 min. As can be seen, the duration of the 5-HT release greatly exceeds the time of MDMA application, and agrees with the duration of action of MDMA on 5-HT dorsal raphe neurons (see Discussion).

Effects of tryptophan and fluoxetine on MDMA-induced 5-HT release

The effect of fluoxetine on MDMA release was tested because this com-

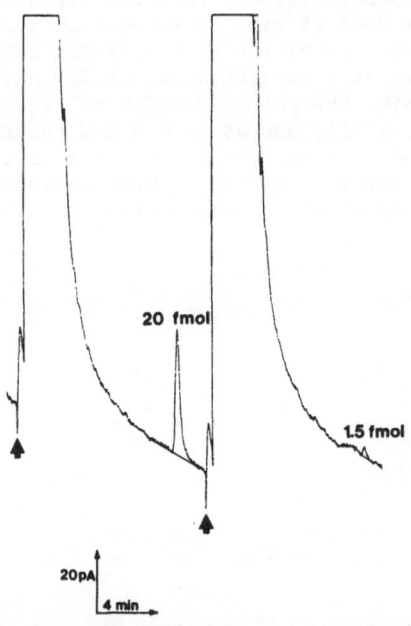

20 fmol

1.5 fmol

20pA

4 min

Fig. 3. Chromatograms of 20 fmol (3.5 pg) 5-HT standard followed by a brain slice dialysate containing 1.5 fmol (0.26 pg) 5-HT.

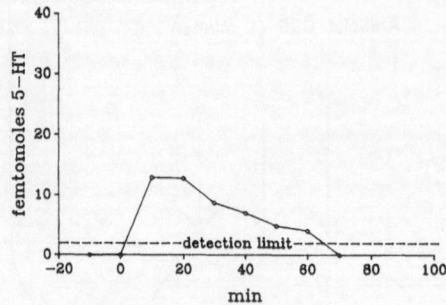

Fig. 4. Results from a typical experiment illustrating the time course of
         MDMA-induced 5-HT release. Administration of 100 μl MDMA for 3 min
         was begun at t = 0.

pound greatly diminished the ability of MDMA to inhibit the cell firing rate.
Pre-treatment with 20 μM fluoxetine, the concentration used in the firing
rate studies (Sprouse, et al., in press), significantly decreased the MDMA-
induced release of 5-HT. Tryptophan (Trp, 100 μM) had been found to potentiate the effects of MDMA on 5-HT neuronal firing. Pretreatment of the slice
with this same concentration of Trp significantly increased the amount of
5-HT released by MDMA, and also the duration of release. Fig. 5 illustrates
a typical experiment in which a slice had been pretreated with 100 μM Trp.
The effects of fluoxetine and Trp on MDMA-induced release of 5-HT for the
first ten minute sampling period (the one most relevant for comparison to the
firing rate measurements) are presented in Fig. 6.

DISCUSSION

        The electrophysiological effects of MDMA on 5-HT containing dorsal raphe
neuronal firing has been previously reported (Sprouse et al., 1989).   In
brief, application of MDMA (3 min) to the dorsal raphe brain slices caused a
concentration-dependent inhibition of the firing rate of identified 5-HT
neurons. When applied at a concentration of 100 μM, the firing rate was reduced to zero at 10 min, the point at which we record the highest release of
5-HT (Johnson et al., 1986), and at 60 min had returned on average to 40%　of
the baseline (pre-MDMA) firing rate. The (+) isomer was 2-3 fold more potent
than the (-) isomer, and the ability of MDMA to inhibit firing was greatly
diminished by pretreatment of the slice with the selective 5-HT reuptake in-

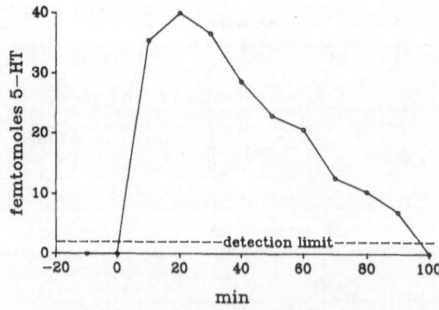

Fig. 5. Effect of pretreatment with 100 micromolar tryptophan on 5-HT
         release. Tryptophan administration began 40 min prior to MDMA
         administration (100 μM, 3 min) at t = 0.

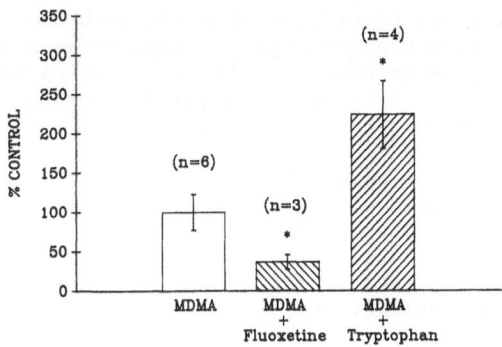

Fig. 6. Effects of 20 μM fluoxetine and 100 μM Trp pretreatment on MDMA-induced 5-HT release for first ten minute collection period following MDMA. 100% control: 11.9 ± 2.7 (SEM) fmol.

hibitor fluoxetine; the norepinephrine reuptake inhibitor desmethylimipramine had no effect. The efficacy of MDMA at inhibiting firing could be enhanced by prior administration of 100 μM Trp to the slice. The stereoselectivity of the action of MDMA, the ability of fluoxetine to prevent the MDMA-induced inhibition, and the enhancement of inhibition by Trp all indirectly suggested that MDMA was acting to inhibit 5-HT neuronal firing via indirectly released 5-HT (Johnson et al., 1986) acting on impulse-regulating somatodendritic autoreceptors. This conclusion is based on the following: 1) The (+) stereoisomer of MDMA has been shown to be more potent at releasing 5-HT from brain slices in vitro than the (-) isomer (Schmidt et al., 1987); 2) Fluoxetine blocks the ability of MDMA to release 5-HT in vitro (Schmidt et al., 1987); 3) Trp, by acting as a precursor to 5-HT has been shown to increase tissue content of 5-HT (Lookingland et al., 1986), and thus could be acting to make more 5-HT available for release. The possibility that MDMA could be acting primarily at a 5-HT somatodendritic receptor is weakened by the observation that the (-) isomer has a greater affinity for serotonin receptors than the (+) isomer, and the receptor for which it has the highest affinity is the $5-HT_2$ receptor (Lyon et al., 1986); the somatodendritic receptor has been shown to display characteristics of the $5-HT_{1A}$ subtype (Sprouse and Aghajanian, 1987).

The results of the effects of MDMA on 5-HT release we have presented agree completely with the indirect evidence that the mechanism of action through which MDMA inhibits the firing rate of 5-HT dorsal raphe neurons is one in which indirectly released 5-HT acts on impulse regulating somatodendritic autoreceptors. The time course of the firing rate reduction matches that of the release of 5-HT; fluoxetine blocks the effect on firing rate and the release of 5-HT; Trp enhances the effect on firing rate and also the release of 5-HT.

The use of microdialysis probes to monitor release from brain slices is a novel application, but one in the (admittedly short lived) tradition of using microdialysis probes as microsampling devices (Ungerstedt, 1984). In the present example, it is conceivable and practical also to collect the ACSF which has passed over the brain slice and by means of extracting the 5-HT from the large volume, obtain a sample concentrated enough for determination of 5-HT. However, the ease of using the microdialysis probe as an alternative made this approach unnecessary. There are other circumstances in which it would not be possible to collect a superfusate, for example a slice in a static (non-perfused) recording chamber. In such an instance, placing a dialysis probe next to an electrophysiological recording electrode would provide a means of correlating chemical dynamics with changes in firing rate. An additional use of such an arrangement would be as a means of administering small amounts of expensive compounds, without causing any mechanical disturbance which could disturb electrophysiological recording (as sometimes occurs by adding small drops onto the surface of a slice).

The consistently low noise characteristics of the chromatographic system we have developed, coupled with the high sensitivity of a narrow bore 3 micron particle size column were critical factors in allowing us to conduct these experiments. The low cost in combination with its high sensitivity should make this type of chromatographic system of interest to others making similar low level determinations.

ACKNOWLEDGEMENTS

Supported in part by USPHS grants MH 14092 and DA 05119, and by the Ribicoff Research Facilities, Connecticut Mental Health Center, State of Connecticut, Department of Mental Health.

REFERENCES

Bonnelycke, B.E., 1969, Pressure-elution pump for spacecraft chromatography, J. Chromat., 45:135-138.

Brazell, M.P., Marsden, C.A., Nisbet, A.P., and Routledge, C., 1985, The 5-HT$_1$ receptor agonist RU-24969 decreases 5-hydroxytryptamine (5-HT) release and metabolism in the rat frontal cortex in vitro and in vivo, Br. J. Pharmacol., 86:209-216.

Caliguri, E.J., Capella, P., Bottari, L., and Mefford, I.N., 1985, High-speed microbore liquid chromatography with electrochemical detection using 3 micrometer C-18 packing material, Anal. Chem., 57:2423-2425.

Greer, G., 1983, "MDMA: a New Psychotropic Compound and its Effects in Humans", George Greer, Santa Fe.

Guichon, G., and Colin, H., 1984, Narrow-bore and micro-bore columns in liquid chromatography, in: "Microcolumn High-Performance Liquid Chromatography", Journal of Chromatography Library, Vol. 28, Kucera, P., ed., Elsevier, Amsterdam, pp. 1-15.

Hare, E., 1966, Automatic multiple column amino acid analysis - the use of pressure elution in small bore ion-exchange columns, Fed. Proc., 25:709.

Johnson, M.P., Hoffman, A.J., and Nichols, D.E., 1986, Effects of the enantiomers of MDA, MDMA and related analogues on [^{3}H]serotonin and [^{3}H]dopamine release from superfused rat brain slices, Eur. J. Pharmacol., 132: 269-276.

Kalén, P., Strecker, R.E., Rosengren, E., and Björklund, A., 1988, Endogenous release of neuronal serotonin and 5-hydroxyindoleacetic acid in the caudate-putamen of the rat as revealed by intracerebral dialysis coupled to high-performance liquid chromatography with fluorimetric detection, J. Neurochem., 51:1422-1435.

Kissinger, P.T., Refshauge, C., Dreilling, R., and Adams, R.N., 1973, An electrochemical detector for liquid chromatography with picogram sensitivity, Anal. Lett., 6:465-477.

Kissinger, P.T., 1984, Introduction to analog instrumentation, in: "Laboratory Techniques in Electroanalytical Chemistry", Kissinger, P.T., and Heineman, W.R., eds., Marcel Dekker, New York, pp. 163-192.

Lookingland, K.J., Shannon, N.J., Chapin, D.S., and Moore, K.E., 1986, Exogenous tryptophan increases synthesis, storage, and intraneuronal metabolism of 5-hydroxytryptamine in the rat hypothalamus, J. Neurochem., 47:205-212.

Lyon, R.A., Glennon, R.A., and Titeler, M., 1986, 3,4-Methylenedioxymeth-
amphetamine (MDMA): stereoselective interactions at brain 5-HT$_1$ and 5-HT$_2$
receptors, Psychopharmacology, 88:525-526.

Martin, M., Guiochon, G., 1978, Pump systems, in: "Instrumentation for High-
Performance Liquid Chromatography", Huber, J.F.K., ed., Elsevier, Amsterdam, pp. 11-40.

Meek, J.L., 1976, Application of inexpensive equipment for high pressure liquid chromatography to assay for taurine, gamma-amino butyric acid, and
5-hydroxytryptophan, Anal. Chem., 48:375-379.

Mefford, I.N., 1987, Chromatographic approaches to signal and selectivity enhancement for determination of biogenic amines by liquid chromatography
with amperometric detection, Life Sci., 41:893-896.

Peroutka, S.J., 1987, Incidence of recreational use of 3,4-methylenedioxy-
methamphetamine (MDMA, "ecstasy") on an undergraduate campus, New Engl. J.
Med., 317:1542-1543.

Schmidt, C.J., Levin, J.A., and Lovenberg, W., 1987, In vitro and in vivo
neurochemical effects of methylenedioxymethamphetamine on striatal monoaminergic systems in the rat brain, Biochem. Pharmacol., 36:747-755.

Schmidt, C.J., Wu, L., and Lovenberg, W.A., 1986, Methylenedioxymethamphetamine: a potentially neurotoxic amphetamine analogue, Eur. J. Pharmacol.,
124:175-178.

Sharp, T., Zetterström, T., Christmanson, L., and Ungerstedt, U., 1986,
p-Chloroamphetamine releases both serotonin and dopamine into rat brain
dialysates in vivo, Neurosci. Lett., 72:320-324.

Snyder, L.R., and Kirkland, J.J., 1979, "Introduction to Modern Liquid
Chromatography", Wiley, New York.

Sprouse, J.S., and Aghajanian G.K., 1987, Electrophysiological responses of
serotoninergic dorsal raphe neurons to 5-HT$_{1A}$ and 5-HT$_{1B}$ agonists, Synapse,
1:3-9.

Sprouse, J.S., Bradberry, C.S., Roth, R.H., and Aghajanian, G.K., MDMA (3,4-
methylenedioxymethamphetamine) inhibits the firing of dorsal raphe neurons
in brain slices via release of serotonin, Eur. J. Pharmacol., 167:375-
383.

Ungerstedt, U., 1984, in: "Measurement of Neurotransmitter Release In
Vivo", Marsden, C.A., ed., Wiley, Chichester, England, pp. 81-105.

SESSION II

TRYPTOPHAN METABOLISM: SEROTONIN AND MELATONIN

AUTORADIOGRAPHIC MEASUREMENT OF THE RATE OF SEROTONIN SYNTHESIS IN THE RAT

BRAIN

M. Diksic[1] and T.L. Sourkes[2]

[1]Montreal Neurological Institute and Hospital
Department of Neurology and Neurosurgery

[2]Department of Psychiatry and Biochemistry
McGill University
Montreal, Quebec H3A 2B4
Canada

INTRODUCTION

The rate of serotonin synthesis has hitherto been measured by two different methodological approaches, steady- and nonsteady-state. In these methods, the rate of synthesis is estimated from the change in the concentration of a tryptophan metabolite after inhibition of an enzyme (tryptophan hydroxylase, aromatic amino acid decarboxylase (AAAD) or monoamine oxidase) or the inhibition of 5-hydroxyindoleacetic acid transport. These approaches can be challenged on the grounds that a feedback mechanism might have some effect on the measured synthesis rate. A major disadvantage of these methods is the necessity of separating different metabolites, a time-consuming process that generally requires tissue sampling, which is usually heterogeneous. Both pharmacological manipulation and tissue sampling have a profound effect on the estimation of the synthesis rates in all brain structures.

Conversion of L-tryptophan (Trp) into serotonin (5-HT) is accomplished in a two-step enzymatic process. First, Trp is hydroxylated by tryptophan hydroxylase (EC 1.14.16.4) to 5-hydroxy-L-tryptophan. This step requires molecular oxygen and co-factor tetrahydrobiopterin, and is generally accepted to be the rate-limiting step in the synthesis of serotonin in the brain (Moir and Eccleston, 1968). After synthesis, serotonin is stored in vesicles where it is protected from deamination by the action of monoamine oxidase. It is also generally accepted that the steady-state concentration of serotonin is closely controlled by synthesis and deamination.

Here we propose an autoradiographic method that uses α-[^{14}C]-methyl-L-tryptophan (α-MTrp), a tryptophan analog, which is the substrate for tryptophan hydroxylase and whose product, α-methyl-5-hydroxytryptophan, is decarboxylated by AAAD (Sourkes, 1971). Decarboxylation yields α-methylserotonin, (α-M5HT), a chemical and functional analog of serotonin (Missala and Sourkes, 1988; Montine and Sourkes, 1989; Nagahiro et al., 1990; Hamel, Diksic and Takada, to be published). α-M5HT is not a substrate for monoamine oxidase and as such stays in the brain for a very long time. This characteristic ensures that the total radioactivity present in brain tissue is related to the rate of serotonin synthesis.

Kynurenine and Serotonin Pathways
Edited by R. Schwarcz et al., Plenum Press, New York, 1991

THEORY

In devising a biological model to represent brain serotonin synthesis, we must make certain assumptions and accept a degree of approximation since the actual forces governing transfer of tracer and its metabolites from one compartment to another are disregarded. It is assumed that: 1) tracer kinetics are applicable; 2) tryptophan metabolism is in a steady-state and remains in a steady-state during the experiment, unchanged by the tracer injection; 3) all transfer coefficients are of the first order; 4) the latter permit us to describe the process with a set of ordinary differential equations with constant coefficients; 5) the mixing of the tracer and tracee is "instantaneous" in all compartments; 6) only the rate-limiting steps are considered; 7) the plasma is the only source of tryptophan for neurotransmitter synthesis (i.e., no significant amount derives from protein degradation); 8) incorporation of the tracer into brain proteins or alternative metabolic pathways is negligible; and 9) the loss of α-M5HT, the end metabolite of the tracer, from the metabolic pool is negligible during the experiment.

The biological model can be schematically represented as shown in Fig. 1. The first-order rate constants represent processes for transfer from plasma to brain (k_1; ml.g^{-1}.min^{-1} is the product of the k_1^*(min^{-1}) and the plasma volume in 1 g of brain) and back (k_2^*; min^{-1}). The constant for conversion of α-MTrp into α-M5HT or the transfer of α-MTrp into an irreversible compartment is k_3^*; (min^{-1}). The rate-limiting step for serotonin synthesis is hydroxylation of L-trp by tryptophan hydroxylase (Moir and Eccleston, 1968), an enzyme found only in serotonergic neurons. The same process is assumed to be the rate-limiting step in the conversion of α-MTrp into α-M5HT because very little α-methyl-5-hydroxy-L-tryptophan is found in brain tissue (unpublished results). However, in this model k_3^* (the rate-limiting step) does not have to represent the rate of hydroxylation of α-MTrp. It could indeed represent any step after which α-MTrp cannot return to the "precursor" pool which communicates with the plasma.

After a steady-state has been reached (Gjedde, 1982; Patlak et al.,

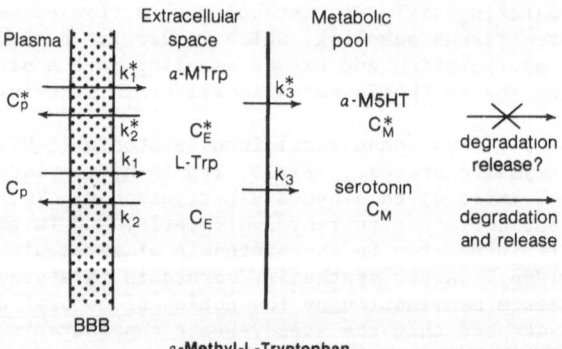

a-Methyl-L-Tryptophan

Fig. 1. A schematic representation of the biological model. The rate constants k_1, k_2, and k_3 (with asterisk are for tracer) are the constants for the transfer of tracer between different compartments. k_1 (used in equations) actually represents a true k_1^* (the first-order rate constant) multiplied by the plasma volume in a unit weight of brain. k_1^* is constant for a transfer from plasma to the brain. k_2 and k_3 represent, respectively, the rate constants for the transfer from brain to plasma and the transfer into an irreversible compartment from which the tracer cannot be moved back into the "precursor" pool. The constants without an asterisk are those for the tracee.

1983), the volume of distribution is a linear function of the exposure time [$\theta(T)$]. The steady-state is, of course, only apparent, for the model shown in Fig. 1 would reach a true steady-state only when the plasma concentration of the tracer is zero. Applying the principle of a net unidirectional transport of a tracer into the metabolic compartment, the tissue concentration of tracer can be described as (Gjedde, 1982; Patlak et al., 1983):

$$C_i^*(T) = K^* \cdot \int_o^T C_p^*(t) \cdot dt + V_{app} \cdot C_p^*(T) \tag{1}$$

where $C^*(T)$ is tissue tracer concentration at time T, K^* is a constant for unidirectional trapping (transfer into irreversible compartment), V_{app} is an apparent volume of the tracer distribution, $C_p^*(t)$ is plasma tracer concentration as a function of time, and $C_p^*(T)$ is the plasma tracer concentration at time T. Division of equation 1 by $C_p^*(T)$ and substitution of $\theta(T) = [\int_o^t C_p^*(t).dt]/C_p^*(T)$ (Gjedde, 1982; Patlak et al., 1983) yields an equation 2 where the tissue volume of distribution [$C_1^*(T)/C_p^*(T)$] is a linear function of the exposure time ($\theta(T)$) after the apparent steady-state is reached:

$$DV(T) = \frac{C_i^*(T)}{C_p^*(T)} = K^* \cdot \Theta(T) + V_{app} \tag{2}$$

In our work with rats, two discrete time points (60 and 150 min) were used. Even though rats were killed at the time of tracer injection, a set of different $\theta(T)$ was obtained because of a slight difference in the plasma clearance of the tracer in different animals. When animals with "large" differences in plasma tryptophan are used, equation 2 should be multiplied by the plasma free tryptophan (C_p; nmol/ml). [The reason for using free, instead of total, plasma tryptophan is discussed in detail elsewhere (Diksic et al., 1990)]. After multiplication by C_p, equation 3 is obtained:

$$V_o = C_p \cdot \frac{C_i^*(T)}{C_p^*(T)} = K^{C_p}\Theta(T) + V_{app} \cdot C_p \tag{3}$$

Since the objective of this model is the measurement of the rate of serotonin synthesis rather than the rate constant of tryptophan hydroxylase, variables K^* and K^{C_p} obtained from equation 2 or 3, respectively, are all we require. It can be shown (Phelps et al., 1979; Redies and Diksic, 1989) that the constant K^* is actually equal to $K_1^* k_3^*/(k_2^* + k_3^*)$ [where K_1^* to k_3^* are the transfer coefficients governing movement of the tracer between compartments (see Fig. 1), which can be described by the set of differential equations (Sokoloff et al., 1977)]. K^* is related to the rate of serotonin synthesis by equation 4 (Phelps et al., 1979; Redies and Diksic, 1989):

$$R = \frac{C_p}{LC} \cdot \frac{K_1^* k_3^*}{k_2^* + k_3^*} = \frac{C_p}{LC} \cdot K^* = \frac{K^{C_p}}{LC} \tag{4}$$

where $K^{C_p} = C_p K^*$ (K^* or K^{C_p} is estimated by the least-squares fit of experimental data to equations 2 or 3, respectively). The rate of synthesis is equal to the product of the plasma free-tryptophan concentrations (C_p) and K^*, divided by the lumped constant (LC), or K^{C_p}, divided by LC (Eq. 4).

The LC consists of several other constants, namely Michaelis-Menten constants for tracer (K_m^* and V_{max}^*) and tracee (K_m and V_{max}) and the ratio of the volume of distribution for tracer and tracee $LC = \frac{V_D^*}{V_D} \cdot \frac{K_m/V_{max}}{K_m^*/V_{max}^*}$.

Table 1. Physiological variables of rats used in
experiments

pH	7.41 ± 0.03
PaO_2 (mm Hg)	102 ± 14
$PaCO_2$ (mm Hg)	39.1 ± 3.2
Hematocrit (%)	43.7 ± 3.0
Mean blood pressure (mm Hg)	122 ± 8
Plasma total Trp (nmol/ml)	53.7 ± 12.2
Body weight (g)	201 ± 9

All values are the mean $\pm$ SD (N - 14).

V_D^* and V_D stand for the volume of distribution for the tracer and tracee, respectively. The Michaelis-Menten constants found in the literature, especially those for α-MTrp, differ greatly from laboratory to laboratory (see references in Table 2). Because of this, it should be noted that the absolute value of the synthesis rates presented in Table 3, and calculated with the LC estimated from ratios given in Table 1, are inversely proportional to

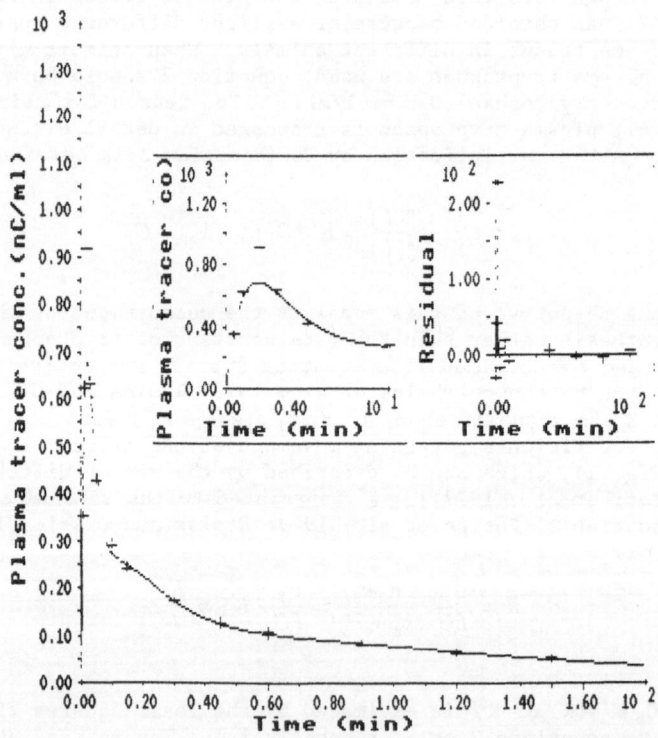

Fig. 2. A plasma input function obtained in a rat injected with 50 μCi of
α-MTrp. Experimental points (marked as +) were fitted to a function $y(t) - \Sigma^3_{i-1} A_i.t.e^{-\alpha i.t}$. A fitted curve is shown as a continuous line. One insert (left) represents the first fifteen
minutes of the plasma data points and fitted curve. Another insert (right) shows residuals plotted as a function of time to give
an impression of the goodness of fit.

the value of LC. An apparent LC can be estimated by equating the rate of unidirectional trapping of the tracer obtained in this work with the rates measured by other methods. However, the estimation of an apparent LC would also inherit errors associated with the determination of the rate of serotonin synthesis by those methods where pharmacological agents were used.

Because of binding by plasma protein, not all plasma tryptophan is available for transport to the brain (Bloxam and Curzon, 1978; Anderson, 1979). Tryptophan also competes with other branched-chain and aromatic amino acids for the same carrier (Pardridge, 1977; Yuwiler et al., 1977). Since our animals were deprived of food for about 18 hours, the total plasma amino acid concentrations were relatively constant (actual data not presented). We also assumed that the ratio between plasma free-tryptophan and other amino acids competing for the same carrier is stable and would not appreciably vary from animal to animal. (Confirmed in an independent measurement.). It has been suggested that this ratio is the most important single factor determining brain Trp concentration (Fernstrom and Wurtman, 1972; Bloxam and Curzon, 1978; Anderson, 1979) and probably a determining factor in the rate of serotonin synthesis. On this basis, the plasma free-tryptophan concentration was used in the calculation with equation 3 and 4. If there is a substantial variation in plasma Trp concentration, a "normalization" of data from different animals should be done by using equation 3 in the estimation of the constant $K^{Cp}(K^{Cp} = K^*.C_p)$. This procedure reduces the influence of animal-to-animal differences in plasma tryptophan on the slope and, hence, on the estimate of the rate of serotonin synthesis.

From our first set of experiments (Diksic et al., 1990), where the entire tissue time-radioactivity curve was measured, it was possible to estimate the half-life of the precursor pool as around 20 min (Diksic et al., 1990). On the basis of this, a two-time-point protocol was devised. In this protocol, rats are killed at 60 and 150 min after trace injection. The first point is selected to be about three to four half-lives of the precursor pool (reaching close to 90% of a transient equilibrium).

METHODS

Surgical procedure and brain slices

Details of animal preparation are given in Diksic et al. (1990) and Nagahiro et al. (1990) and will be outlined only briefly here. Under light halothane anesthesia (1% to 1.5%), rats were implanted with a PE50 polyethylene catheter in the femoral artery and vein. The lower part of the body was placed in a loosely fitting plaster cast (Sako et al., 1984) and animals were allowed to wake up. Physiological values (arterial pH, $PaCO_2$, PaO_2, blood pressure, hematocrit and rectal temperature) were measured before the beginning of the experiment. The body temperature was kept at about 37°C with a 100 W lamp. A summary of the physiological variables is given in Table 1. The animals were awake about two hours before being injected with about 50 μCi of tracer (α-[^{14}C]methyl-L-tryptophan; spec. activity: 55 mCi/mmol, New England Nuclear, Boston, MA) as a two min "bolus". Plasma samples, for determination of the input function, were taken at progressively increasing time intervals. (A representative plasma curve is shown in Fig. 2 along with a least-squares fit to a function given in the figure caption). At the end of the experiments, the rats were killed by a rodent guillotine and the brain quickly removed, usually within 1.5 min. All rats were killed between 13:00 and 15:00 h. The brain was mounted on a holder and cut into 30 μm slices, placed on a microscope glass slide, and co-exposed with standards, calibrated to a tissue equivalent, for about 3 weeks. After development, the concentration of tracer in selected structures was determined by comparing densities in those structures with a calibration curve (Sako et al., 1984).

Table 2. K_m/V_{max} ratios for hydroxylation of L-tryptophan
and α-methyltryptophan with tryptophan hydroxylase

Ratio[*] (mg h/ml)		Reference
L-Trp	α-MTrp	
7.76	63.6	Schirlin et al., 1988
10.99	36	Gál and Christiansen, 1975
10.44[1]		Hamon et al., 1977
7.36[2]		Hamon et al., 1977
15.55		Gál et al., 1975
10.3 ± 2.9	49.8	

[*]Ratios are calculated from the Michaelis-Menten constants
measured with 6-MPH$_4$ as cofactor. (6-MPH$_4$ is 2-amino-4-hydro-
xy-6-methyl-5,6,7,8-tetra-hydropterin). [1]L-Trp concentration
varied from 0.02 to 0.5 mM. The cofactor concentration was
constant at 0.32 mM. [2]Cofactor 6-MPH$_4$ concentration was
varied; see details in the reference cited.

Plasma total and free-tryptophan determination

Two 30 μl samples were deproteinized with 20 μl of 20% trichloroacetic
acid and used for the determination of plasma total tryptophan concentration
by HPLC (Krstulovic and Matzura, 1979; Missala and Sourkes, 1988). The plasma free fraction was determined in a separate set of animals, handled by an
equivalent procedure to remove any effects of handling, by filtering 50 μl of
plasma through an Ultrafree-MC filter (Millipore Canada Ltd., Cat. No. UF
3LGCOO) with a 10,000 MW cut-off point. Since the free-fraction was found to
be constant and agreed well with previous measurements, the value was not
determined in every animal.

Lumped constant estimation

The lumped constant (LC) is a conglomerate of other constants (Sokoloff
et al., 1977) and was estimated from the measurements of respective parameters (found in the literature). The LC is equal to $(V_m^*/K_m^*)/(V_m/K_m)\cdot\lambda$, where
V_m and K_m are Michaelis-Menten constants for tracee (L-Trp) and with asterisk
for tracer (α-MTrp). λ is the ratio of the volume of distribution in the
brain for tracer and tracee. The ratio of the volumes of distribution was
estimated from data of Missala and Sourkes (1988) where both distribution
volumes were measured in the same animal.

The average value of λ was estimated to be 2.21 ± 0.23 (mean ± SD) (Missala and Sourkes, 1988). The ratios of the Michaelis-Menten constants found
in the literature are given in Table 2. To minimize the impact of the measurements from different laboratories, we elected to use only those measurements where constants for both L-Trp and α-methyl-L-tryptophan were measured
(Table 2). By insertion of the value of λ and the ratios for the Michaelis-
Menten constants (Table 2) into the equation for LC, the lumped constant of
0.46 was estimated for the rat brain (Diksic et al., 1990). It was assumed
that the LC does not vary from one brain structure to another.

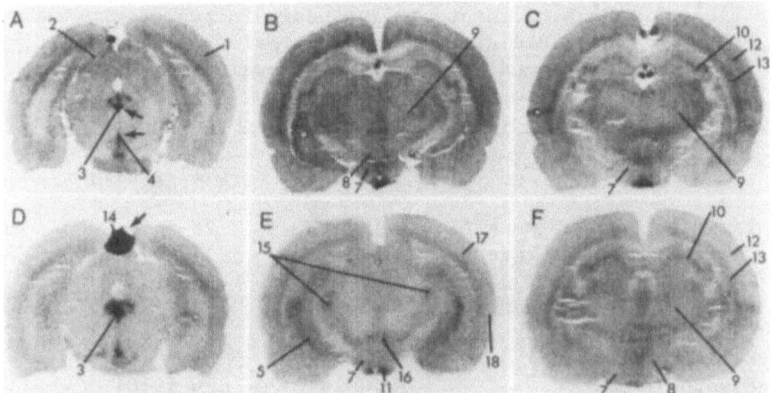

Fig. 3. A set of representative autoradiograms for 30 μm brain slices in
rats killed at 1 or 2.5 h after the beginning of the tracer injection (50 μCi of α-[^{14}C]MTrp). The sections are taken at the level
of the dorsal raphe nucleus (A,D), posterior hypothalamus (B,E), and
middle to anterior hypothalamus or anterior hippocampus (C,F). Some
of the structures clearly visualized are: 1) occipital cortex, 2)
superior colliculus, 3) dorsal raphe nucleus, 4) medial raphe nucleus, 5) hippocampus (ventral), 7) hypothalamus, 8) medial forebrain
bundle, 9) thalamus, 10) hippocampus (dorsal), 11) mammillary body,
12) parietal cortex, 13) VI layer of parietal cortex, 14) pineal
body, 15) medial geniculate nucleus, 16) ventral tegmental area, 17)
temporal cortex, and 18) perirhinal area.

RESULTS

Average values of the physiological variables for animals used in this
work are given in Table 1. Thus, the rats used were normal as far as these
variables are concerned. When the averages of these variables for animals
killed at 60 and 150 min were compared, there was no significant difference
between the two groups. The fraction of plasma free-tryptophan was also constant at 20 ± 2%. A representative autoradiogram for rats killed at 60 and
150 min after injection of about 50 μCi of ^{14}C-labelled α-MTrp is given in
Fig. 3. Noticeable is an accumulation of tracer in the areas rich in serotonergic cell bodies and reasonably well visualized projection areas (Descarries et al., 1982).

For the first time it is possible to show with high anatomical resolution inhomogeneity in the rate of serotonin synthesis in those structures.
Tracer distribution volumes as a function of the θ(T) [exposure time] are
given in Fig. 4. The straight lines represent the least-squares fit of experimental data according to equation 2. The slopes of these lines represent the rate of unidirectional trapping of the tracer. Uncertainties given
in Table 3 were calculated from the standard deviation of this slope (Bevington, 1969). When the slopes are converted into serotonin synthesis rates,
according to equation 4, the values given in Table 3 are obtained.

DISCUSSION

In general, two types of method have been used to date to estimate the
rate of serotonin synthesis in laboratory animals, steady- and nonsteady-
state. Both approaches have been used with grossly dissected and microdis-

Table 3. Rate of serotonin synthesis in representative
discrete structures of normal (untreated) rat
brain and in pineal body; evaluation of the
two-time-point method

	R (pmol/g/min)[*]
Parietal cortex; superficial layer	28 ± 11
Parietal cortex; Layer VI	37 ± 13
Thalamus	33 ± 13
Caudate (lateral part)	42 ± 12
Caudate (medial part)	58 ± 14
Nucleus accumbens	67 ± 19
Hypothalamus	26 ± 10
Medial geniculate body	23 ± 11
Hippocampus	47 ± 13
Dorsal raphe nucleus	166 ± 37
Medial raphe nucleus	95 ± 20
Raphe nuclei, average[1]	100 ± 23
Pineal body	251 ± 31

[*]Rates are given as an estimate ± SD. [Uncertainty was
calculated from the standard deviation of the slope (K^{Cp})
in equation 3 estimated from the least-squares fit (Bevington, 1969)]. [1]Average of dorsal, magnus, pontine, and
medial raphe nuclei.

sected brain. Certainly, micro-dissected brain permits a higher anatomical
resolution, but it is impossible to have a pure sample of any anatomical
structure with the exception of the pineal body, which lies outside the brain
and is therefore easily sampled. All these methods also require some pharmacological manipulation and time-consuming separation of L-tryptophan metabolites (e.g. Toppaz and Pujol, 1980; Long et al., 1982, 1983). The method
described here is a true steady-state method, does not require any separation
of metabolites, and has a high anatomical resolution (~100 μm). The anatomical resolution is strictly dictated by the energy of the ^{14}C-beta particle.
The method uses tracer kinetics and an analog of L-tryptophan, α-methyl-L-
typtophan, which is a substrate for tryptophan hydroxylase, the rate-limiting enzyme in the synthesis of serotonin.

The methods using an inhibitor (e.g., to block AAAD or MAO) have a draw-
back because the metabolite(s) are accumulated in the tissue which could, by
feed-back mechanisms, influence the rate of synthesis (activity or co-factor
of enzymes) or even the transport of L-Trp to the site of the enzyme. Mat-
ters could become even more complicated when an AAAD inhibitor is used because its effect on the concentration of catecholamines might also influence
their rate of synthesis. It has been reported that catecholamines influence
serotonin synthesis rate. From these interrelationships it is obvious that a
method requiring no pharmacological manipulation would probably be the most
accurate in estimating the serotonin synthesis rate. The method described
here has a high sensitivity because it uses a radioactive tracer in conjunction with sensitive X-ray film.

Data confirming in vivo conversion of α-MTrp into α-M5HT in the brain
are not presented here. The basic biochemical data and the time-dependent
conversion of ^{14}C-labelled α-MTrp into α-M5HT have permitted development of
this biological model (Diksic et al., 1990).

Basic data presented in other publications (Diksic et al., 1990; Nagahiro et al., 1990) show the feasibility of an autoradiographic method to measure the rate of serotonin synthesis in rat brain with high anatomical resolution ~100 μm). This method has allowed for the first time measurement of the serotonin synthesis rate in the living brain without the need for any pharmacological manipulation and with a high anatomical resolution. Unfortunately, these autoradiographic results cannot be directly compared to any other measurements where grossly dissected brain tissue is used and, as such, must stand on their own. The main reason for this is a discrete measurement in an anatomical structure in a 30 μm thick slice. Comparison with biochemical methods, even those where microdissection techniques were used (Tappaz and Pujol, 1980; Long et al., 1982), would only be possible if an integration of the synthesis rate in exactly the same volume of the brain as that measured in the tissue sampling method is done. From a practical point of view this is difficult, impractical, and probably impossible. For this reason, data presented here (Table 3) can only be compared to other biochemical measurements in a relative fashion. Even with this relative comparison one must bear in mind that any tissue sampling (even micro) will give a different admixture than that in autoradiography. To emphasize this, data from the biochemical literature are not presented in the same table. We have given (Table 4) a number of selected structures where the measurements of the synthesis rate were done on the micro-dissected rat brain. It must be accepted that the present autoradiographic method would set its own standards for the rate of serotonin synthesis in different brain structures, and these results could then be compared to other autoradiographic measurements made under different conditions (e.g., drugs, pathology).

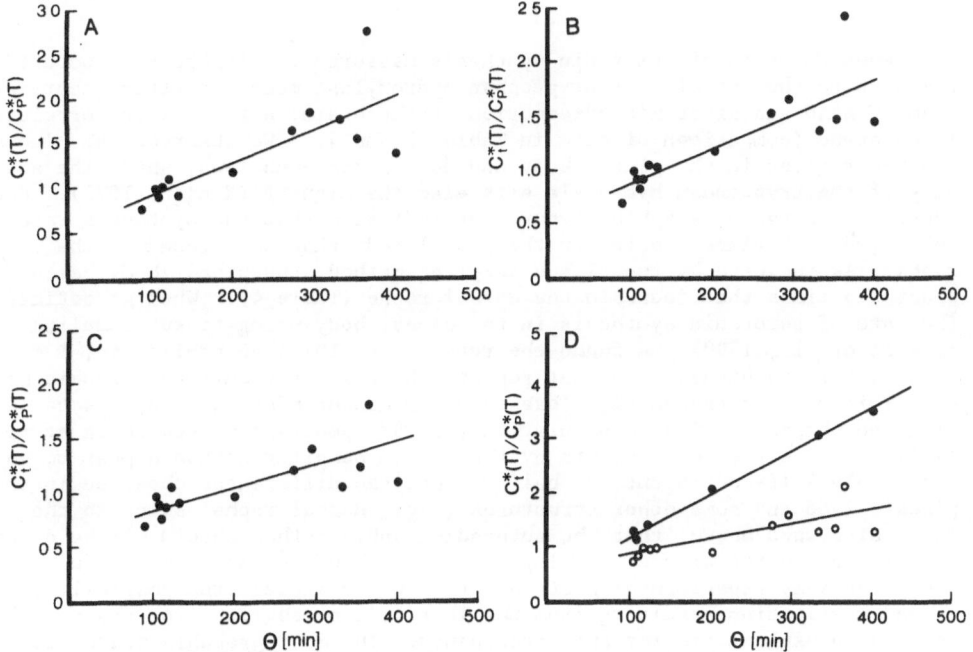

Fig. 4. A composite plot of the distribution volume for normal (untreated) rats killed at 60 or 150 min after injection as a function of θ (min). Data are taken at the level of nucleus accumbens (A), caudate nucleus (B), thalamus (C), and the dorsal raphe nucleus (solid dots) and hippocampus (open dots)(D). Note that the position of the points on the horizontal axis is not the same even if rats were killed at the same time after injection. The main reason for this is the difference in the plasma clearance of the tracer in different animals.

Table 4. Serotonin synthesis rate in certain discrete regions in the rat brain[*]

Structure	Rate (pmol/g/min)	
Dorsal raphe	359 ± 128 (155-570)	Long et al., 1982, 1983 Neckers and Meek, 1976 Tappaz and Pujol, 1980
Median raphe	217 ± 32 (194-262)	Neckers and Meek, 1976
Locus coeruleus	50 ± 14	Tappaz and Pujol, 1980
Hypothalamus	55 ± 8	Tappaz and Pujol, 1980
Anterior hypothalamus	51 ± 9	Long et al., 1982, 1983
Striatum	30 ± 5	Tappaz and Pujol, 1980
Hippocampus	24 ± 2	Tappaz and Pujol, 198
Cortex	16.4 ± 1.6	Tappaz and Pujol, 1980
N.amygdaloideus centralis	37 ± 2	Long et al., 1982
Pineal body	883	Lovenberg et al., 1967

[*]A simple average was calculated for the measurements in all structures. When more than one measurement was done, the range measurements is given in parentheses. Some rates were derived from measurements reported in the original publication per mg of protein by taking 100 mg of protein per g of brain tissue.

When the rate of serotonin synthesis measured in different structures is compared to the activity of tryptophan hydroxylase measured after inactivation of AAAD and after microdissection of the brain, a reasonable correlation is observed (comparison of data in Tables 3 and 4). We observed the highest synthesis rates in the pineal body and dorsal raphe nucleus, where the activity of the tryptophan hydroxylase is also the highest (Renson, 1973). However, this autoradiographic measurement indicates that the synthesis rate is only about 1.5 times greater in the pineal body than that found in the dorsal raphe. As measured by the tissue sampling method, the pineal body has a rate about 2.5 times that found in the dorsal raphe (Table 4). When we estimated the rate of serotonin synthesis in the pineal body using tissue sampling (Diksic et al., 1990), we found the rate to be 1107 ± 68 pmol/g/min, i.e. about 4.4 times higher than that reported here for the autoradiographic method. This rate is reasonably close to other measurements made by tissue sampling and pharmacological manipulation [cf. 883 pmol/g/min (Lovenberg et al., 1967)]. Since the measurements by the tissue sampling method appear to be relatively self-consistent, we believe that the differences observed in the pineal gland and some other structures (e.g., dorsal raphe) point to the fact, discussed above, that the autoradiographic method should not be directly compared to the tissue sampling. Since both of our measurements were done under the same experimental protocol, this would exclude the possibility of an underestimation resulting from the loss of a metabolite from the pineal body as an explanation for the discrepancy. The most probable reason for this large discrepancy is determination of the weight of this small structure. (The concentration per unit weight is determined, in the autoradiographic method, by comparison to standards calibrated as the tissue equivalent.).

A linearity observed in the pineal body distribution volume-exposure time plot (a correlation coefficient of 0.958 was obtained for the first 150 min data) also supports the notion that there is no loss of a metabolite from the pineal body (this linear correlation was obtained for the tissue sampling method). At this time, we do not have a satisfactory explanation for the much higher discrepancy between the tissue sampling and the autoradiographic methods.

Our data (Table 3) clearly show that the rate of serotonin synthesis is about 3 to 5 times greater in the dorsal raphe, the area rich in serotonin cell bodies, than in some projection areas (e.g., cortex and caudate). However, the data where microdissection (Table 4) was used indicate that this ratio is more than 20 (dorsal raphe-cortex). One explanation for this is certainly tissue sampling, because even when small samples are punched out from 1 mm of these slices, the "contamination" differs from that of an autoradiographic method where a region of interest is selected in an autoradiogram of a 30 μm tissue slice. This can be seen in Fig. 3 where some representative autoradiograms are given. A rough comparison of autoradiographic measurements (Table 3) and tissue sampling measurement (Table 4) shows some agreement in the rate in certain structures. As already mentioned, the greatest discrepancy is found in the pineal body (compare Tables 3 and 4).

In conclusion, the method we developed combines sensitivity and high anatomical resolution. By coupling this procedure with a microcomputer-based digitizer, one should be able to evaluate the changes in the serotonin synthesis rates in a large variety of structures and even in different parts of a structure (e.g., caudate nucleus, parietal cortex) and the differential influence of drugs on discrete regions of the brain [e.g., lithium treatment (Nagahiro et al., 1990)]. We would also like to emphasize once more that autoradiographic results should not be compared directly with any other biochemical measurements where tissue sampling was used. The use of powerful statistical methods would permit evaluation of the inter-structure relationship which could be especially important when the influence of a drug on the brain is tested.

ACKNOWLEDGEMENTS

The work described here was supported by grants from the MRC of Canada (No. MA 10232 and UI-0025), and from Merck-Frosst Canada Inc. Discussions with, and suggestions from Professor S. Zlobec, Department of Mathematics, McGill University relating to the optimization are greatly appreciated. We would also like to acknowledge the excellent secretarial help of Ms. C. Elliot.

REFERENCES

Anderson, G.H., 1979, Control of protein and energy intake: role of plasma amino acids and brain neurotransmitters, Can. J. Physiol. Pharmacol., 57:1043-1057.
Bevington, P.R., 1969, "Data Reduction and Error Analysis for the Physical Sciences", McGraw-Hill, New York, pp. 92-118.
Bloxam, D.L., and Curzon 1978, A study of proposed determinants of brain tryptophan concentration in rats after portocaval anastomosis or sham operation, J. Neurochem., 31:1255-1263.
Descarries, L., Watkins, K.C., Garcia, S., and Beaudet, A., 1982, The serotonin neurons in nucleus raphe dorsalis of adult rat: a light and electron microscope radioautographic study, J. Comp. Neurol., 207:239-254.
Diksic, M., Nagahiro, S., Sourkes, T.L., and Yamamoto, Y.L., 1990, A new method to measure brain serotonin synthesis in vivo: I. Theory and basic data for a biological model, J. Cereb. Blood Flow Metab., 10:1-12.
Fernstrom, J.D., and Wurtman, R.J., 1972, Brain serotonin content: physiological regulation by plasma neutral amino acids, Science, 178:414-416.
Gál, E.M., and Christiansen, P.A., 1975, Alpha-methyltryptophan: effects on cerebral monooxygenases in vitro and in vivo, J. Neurochem., 24:89-95.

Gjedde, A., 1982, Calculation of cerbral glucose phosphorylation from brain uptake of glucose analogs in vivo: a re-examination, **Brain Res. Rev.**, 4:237-274.

Hamon, M., Bourgoin, S., Artaud, F., and Héry, F., 1977, Rat brain stem tryptophan hydroxylase: mechanism of activation by calcium, **J. Neurochem.**, 28:811-818.

Krstulovic, A.M., and Matsura, C., 1979, Rapid analysis of tryptophan metabolites using reversed-phase high-performance chromatography with fluorometric detector, **J. Chromatogr.**, 163:72-76.

Long, J.B., Youngblood, W.Y., and Kizer, J.S., 1982, A microassay for simultaneous measurement of in vivo rates of tryptophan hydroxylation and levels of serotonin in discrete brain nuclei, **J. Neurosci. Meth.**, 6:45-48.

Long, J.B., Youngblood, W.Y., and Kizer, J.S., 1983, Regional differences in the response of serotonergic neurons in rat CNS to drugs, **Eur. J. Pharmacol.**, 88:89-97.

Lovenberg, W., Jequier, E., and Sjoerdsma, A., 1967, Tryptophan hydroxylation: measurement in pineal gland, brain stem and carcinoid tumor, **Science**, 155:217-219.

Missala, K., and Sourkes, T.L., 1988, Functional cerebral activity of an analogue of serotonin formed in situ, **Neurochem. Int.**, 12:209-214.

Moir, A.T.B., and Eccleston, D., 1968, The effects of precursor loading in the cerebral metabolism of 5-hydroxyindoles, **J. Neurochem.**, 15:1093-1108.

Montine, T.J., and Sourkes, T.L., 1989, Behaviour of alpha-methylserotonin in rat brain synaptosomes, **Neurochem. Int.**, 15:227-231.

Nagahiro, S., Takada, A., Diksic, M., Sourkes, T.L., Missala, K., and Yamamoto, Y.L., 1990, A new method to measure brain serotonin synthesis in vivo. II. A practical autoradiographic method tested in normal and lithium-treated rats, **J. Cereb. Blood Flow Metab.**, 10:13-21.

Neckers, L.M., and Meek, J.L., 1976, Measurement of 5HT turnover rate in discrete nuclei of rat brain, **Life Sci.**, 19:1579-1584.

Pardridge, W.M., 1977, Kinetics of competitive inhibition of neutral amino acid transport across the blood-brain barrier, **Neuroscience**, 2:103-108.

Patlak, S.C., Blasberg, R.G., and Fenstermacher, J.D., 1983, Graphic evaluation of blood-to-brain transfer constants from multiple time uptake data, **J. Cereb. Blood Flow Metab.**, 3:1-9.

Phelps, M.E., Huang, S.C., Hoffman, E.J., Selin M.S., Sokoloff, L., and Kuhl, D.E., 1979, Tomographic measurement of local cerebral glucose metabolic rate in humans with 2-[^{18}F]fluoro-2-deoxyglucose: validation of the method, **Ann. Neurol.**, 6:371-388.

Redies, C., and Diksic, M., 1989, The deoxyglucose method in the ferret brain. I. Methodological consideration, **J. Cereb. Blood Flow Metab.**, 9:35-42.

Renson J., 1973, Assays and properties of tryptophan 5-hydroxylase, in: "Serotonin and Behavior", Barchas J., and Usdin, E., eds., Academic Press, New York, pp. 19-32.

Sako, K., Diksic, M., Kato, A., Yamamoto, Y.L., and Feindel, W., 1984, Evaluation of [^{18}F]4-fluoroantipyrine as a new blood flow tracer for multi-radionuclide autoradiography, **J. Cereb. Blood Flow Metab.**, 4: 259-263.

Schirlin, D., Gerhart, F., Hornsperger, J.M., Hamon, M., Wagner, J., and Jung, M.J., 1988, Synthesis and biological properties of α-mono- and α-difluoromethyl derivatives of tryptophan and 5-hydroxytryptophan, **J. Med. Chem.**, 31:30-36.

Sokoloff, L., Reivich, M., Kennedy, C., Rosiers, H.D., Patlak, C.S., Pettigrew, K.D., Sakurada, O., and Shinohara, M., 1977, The [^{14}C] deoxyglucose method for the measurement of local cerebral glucose utilization: theory, procedure, and normal values in the conscious and anesthetized rat, **J. Neurochem.**, 28:897-910.

Sourkes, T.L., 1971, Alpha-methyltryptophan and its actions on tryptophan metabolism, **Fed. Proc.**, 30:897-903.

Tappaz, M.L., and Pujol, J-F., 1980, Estimation of the rate of tryptophan hydroxylation <u>in</u> <u>vivo</u>: a sensitive microassay in discrete rat brain nuclei, J. Neurochem., 34:933-940.

Yuwiler, A., Oldendorf, W.H., Geller, E., and Braun, L., 1977, Effect of albumin binding and amino acid competition of tryptophan uptake into brain, **J. Neurochem.**, 28:1015-1023.

AUTORADIOGRAPHIC STUDIES OF 5HT$_2$ RECEPTORS

J.M. Saavedra and A. Himeno

Section on Pharmacology
Laboratory of Clinical Science
NIMH
Bethesda, Maryland 20892
USA

INTRODUCTION

Serotonin has been implicated in normal brain activities, and alterations in serotonin metabolism or serotonin receptors have been linked to disorders such as depression and schizophrenia (Fuller, 1986).

In the brain, radioligand binding studies demonstrated the existence of distinct serotonin receptor subtypes. A specific increase in serotonin-2 (5-HT$_2$) receptors occurs in the frontal cortex of violent suicide victims (Mann et al., 1986). Antidepressant treatment downregulates 5-HT$_2$ receptors (Scott and Crews, 1986). This effect could be related to the therapeutic action of antidepressants on suicidal behavior and depression (Mann et al., 1986). In addition, some compounds with hallucinogenic properties, when administered in vivo, downregulate 5-HT$_2$ receptors (Buckholtz et al., 1985). The effects on 5-HT$_2$ receptors are common to apparently dissimilar classes of hallucinogens, appear to be critical for their psychotomimetic effects, and are apparently associated with 5-HT$_2$ agonist activity (Jacobs, 1987).

(±)DOI (2,5-dimethoxy-4-iodo-phenylisopropylamine) belongs to a class of hallucinogenic 5-substituted phenylalkylamines (Shulgin, 1982) that selectively binds to 5-HT$_2$ receptors (Glennon et al., 1986). The radioiodinated derivative of DOI ([^{125}I]-DOI) has been characterized as a 5-HT$_2$ agonist (Johnson et al., 1987; Glennon et al., 1988).

The availability of a selective, high specific activity, [^{125}I]-ligand for 5-HT$_2$ receptors made it possible to initiate quantitative autoradiographic studies on the distribution, characterization and regulation of 5-HT$_2$ receptors in selected areas of the rat brain. Using [^{125}I]-DOI, we have also been able to apply quantitative autoradiography to the study of 5-HT$_2$ receptors in human platelets.

MATERIALS AND METHODS

Animals and tissue preparation

Adult male Sprague-Dawley rats, 200 g body weight (Taconic Farms, Ger-

Kynurenine and Serotonin Pathways
Edited by R. Schwarcz et al., Plenum Press, New York, 1991

mantown, NY) were maintained under standard laboratory conditions with lights on from 6:00 a.m. to 6:00 p.m. for at least 1 week prior to sacrifice. Animals were killed by decapitation between 9:00 and 11:00 a.m., and their brains were immediately removed and frozen in isopentane at -30°C. Within 48 h, tissue sections (16 μm) were cut in a cryostat at -17°C and mounted on gelatin-coated glass slides, and placed under vacuum at 4°C overnight prior to incubation.

Autoradiography

Slides were preincubated for 30 min at room temperature in 50 mM Tris buffer (pH 7.4) containing 0.1% ascorbate, 0.1% bovine serum albumin and 4 mM $CaCl_2$. Following preincubation, slides containing adjacent brain sections were transferred to fresh buffer containing 200 pM of [^{125}I]-LSD (Amersham, Arlington Heights, IL) or [^{125}I]-R-DOI. Unlabeled LSD (obtained from the National Institute on Drug Abuse) was used to define non-specific binding for [^{125}I]-LSD, while unlabeled R-DOB (2,5-dimethoxy-4-bromo-phenylisopropyl-amine) was used to define non-specific binding for [^{125}I]-R-DOI. Non-specific binding was defined with adjacent sections incubated in the presence of 500 nM unlabeled drug. Under these conditions, non-specific binding was less than 10% of total binding for both ligands.

Slides were incubated for 60 min at room temperature in the presence of the radioligand, then washed 3 times (10 min) in ice-cold buffer, pH 7.4. Finally, slides were dipped in ice-cold distilled water (1 min) and dried under a cold stream of air. Slides were exposed to ^{3}H-sensitive Ultrofilm (LKB Industries, Rockville, MD) in X-ray cassettes at room temperature for 2 to 4 days. Films were developed in undiluted Kodak D-19 developer and optical densities were quantified by computerized densitometry as described elsewhere (Israel et al., 1984).

Drug treatment

Control and treated groups consisted of four rats each. Treated animals were given daily i.p. injections of ($\pm$)DOI (Research Biochemicals Inc., Natick, MA) in normal saline [1.0 mg/kg (2.8 μmol/kg)] between 9:00 and 11:00 a.m. for eight consecutive days.

Membrane binding

Animals were killed by decapitation 4 hours after the last ($\pm$)DOI injection, the brains removed, and the cortex dissected at 4°C. Tissues were homogenized in a glass homogenizer in 40 ml ice-cold 50 mM Tris buffer, pH 7.4, containing 0.1% ascorbate, 4 mM $CaCl_2$, and centrifuged at 35,000 x g for 20 minutes. The pellet was resuspended in 40 ml buffer. Following preincubation for 60 minutes at 37°C, the suspension was recentrifuged at 35,000 x g for 20 minutes, the supernatant was decanted, and the pellet was resuspended in a final volume of 40 ml and maintained at 4°C until used.

Binding in membrane homogenates was measured at the following single-point concentrations: [^{125}I]-R-DOI, 0.5 and 5.0 nM; [^{125}I]-LSD (Amersham, Arlington Heights, IL) 1.0 nM; and [^{3}H]-ketanserin (New England Nuclear, Wilmington, DE; specific activity: 61 Ci/mmol) 1.0 nM.

Assay solutions consisted of 50 μl buffer, 50 μl radioligand, and 400 μl membrane suspension (added last) in 12 x 75 mm disposable polystyrene tubes. Non-specific binding was defined at each concentration with 1 μM ritanserin (Janssen Pharmaceuticals, Beerse, Belgium). Total and non-specific binding at each concentration were determined in triplicate. Protein content of the membrane suspensions was determined using the method of Bradford (1976). Following addition of the membrane suspensions, tubes were incubated for 60 minutes at 37°C, vacuum filtered in a Brandel cell harvester over Schleicher and Schuell No. 32 glass fiber filters presoaked in 0.3% polyethylenimine, and washed three times with 5 ml ice-cold buffer. Following filtration, filters containing the [^{125}I] ligands were counted in a Beckman model 5500 gamma counter. Filters containing [^{3}H]-ketanserin were suspended in 10 ml Hydrofluor scintillation fluid (National Diagnostics, Manville, NJ) and equilibrated for a minimum of 6 hours prior to counting in a Beckman model LS-230 liquid scintillation counter.

Human platelet preparation

We drew whole blood from 4 healthy volunteers via a 20 gauge needle into vacuum glass tubes containing 100 μl of 15% K$_2$EDTA (Becton Dickinson Vacutainer Systems, Rutherford, NJ). Platelet rich-plasma (PRP) was obtained by centrifugation of blood at 180 x g for 15 minutes at room temperature (Geaney et al., 1984). Following centrifugation, we removed 1.4 ml of PRP to 1.5 ml conical polypropylene tubes containing 20 μl of M-1 embedding matrix (Lipshaw Mfg. Co., Detroit, MI). We centrifuged the tubes at 1200 x g for 8 minutes at room temperature. After discarding the supernatant, we slowly added 250 μl of M-1 embedding matrix on top of the pellet, together with a thin wooden stick. The tubes were frozen in isopentane on dry ice. We separated the pellets by lightly warming the tubes and pulling the stick.

In vitro autoradiography of 5-HT$_2$ receptors in human platelets

For the autoradiographic analysis of [^{125}I]-R-DOI binding sites, we cut the platelet pellets in 16 μm-thick consecutive sections in a cryostat at -20°C. The sections were thaw-mounted onto gelatin-coated glass slides and dried under vacuum at 4°C overnight.

We preincubated the sections at room temperature in 50 mM Tris-HCl buffer, pH 7.4, containing 4 mM CaCl$_2$ and 0.1% ascorbate, for 15 minutes. After preincubation, we labeled consecutive sections from a single platelet pel-

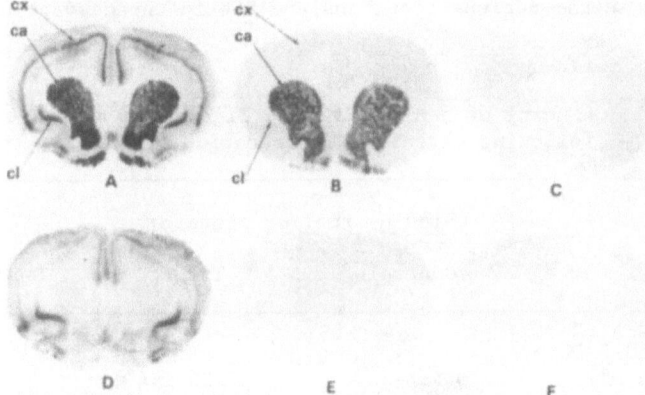

Fig. 1. (A-E) Cross-displacement of [^{125}I]-LSD and [^{125}I]-R-DOI binding in rat brain by R-DOB and LSD. [^{125}I]-LSD showed the highest binding density in the caudate nucleus, nucleus accumbens and olfactory tubercle (Fig. 1). The upper panel (A-C) shows (A) total binding of [^{125}I]-LSD; (B) displacement of [^{125}I]-LSD by R-DOB; and (C) displacement of [^{125}I]-LSD by LSD. The lower panel (D-F) shows (D) total binding of [^{125}I]-R-DOI; (E) displacement of [^{125}I]-R-DOI by R-DOB; and (F) displacement of [^{125}I]-R-DOI by LSD. The arrows in (A) and (B) indicate the cortex (cx), claustrum (cl) and caudate nucleus (ca).

let in vitro by incubation for 60 minutes in fresh buffer with addition to [125I]-R-DOI in concentrations ranging from 88 pM to 6.6 nM. We determined non-specific binding by incubation of adjacent sections under the same conditions with the addition of 1 μM ketanserin (McBride et al., 1987).

Following incubation, we rinsed the sections for 30 seconds three times in ice-cold buffer, followed by a dip in distilled water, and dried the sections as described above. The sections were exposed to [3H]-Ultrofilm, and the films developed as described above. Optical densities were obtained as above, and corrected by the protein concentrations of the same sections used in the receptor assay, determined by a new densitometric procedure (Miller et al., 1988).

RESULTS

Quantitative autoradiography

The highest density of [125I]-R-DOI binding was found in the claustrum, followed by the frontal cortex, nucleus accumbens and olfactory tubercle. Lower binding density was found in the caudate nucleus (Fig. 1).

[125I]-LSD showed the highest binding density in the caudate nucleus, nucleus accumbens and olfactory tubercle (Fig. 1).

Addition of 500 nM unlabeled LSD displaced more than 90% of the [125I]-LSD binding in all regions measured (Fig. 1C). This concentration of unlabeled LSD also displaced 97-99% of the [125I]-R-DOI binding in the cortex, claustrum, nucleus accumbens and olfactory tubercle (Fig. 1).

DOB displaced 90% of total [125I]-R-DOI binding in the claustrum and cortex, while displacing 91% of total binding in the nucleus accumbens and 47% in the caudate nucleus (Fig. 1). DOB displaced over 80% of the total [125I]-LSD binding in the claustrum and cortex, but only 48% in the olfactory tubercles, 44% in the nucleus accumbens and 26% in the caudate nucleus (Fig. 1).

Table 1. Binding of $5-HT_2$-selective ligands in rat cortex following chronic administration of (±)DOI

	(Binding fmol/mg protein)		
	Controls	Treated	% Change
[125I]-R-DOI (0.5 nM)	56 ± 8	25 ± 4*	-55
[125I]-R-DOI (5.0 nM)	282 ± 27	186 ± 31*	-34
[125I]-LSD (1.0 nM)	380 ± 59	282 ± 47*	-26
[3H]-Ketanserin (1.0 nM)	239 ± 22	152 ± 22*	-36

Data are the mean ± SEM of three independent experiments.
*$P < 0.05$ (Student's t-test) when compared to controls.

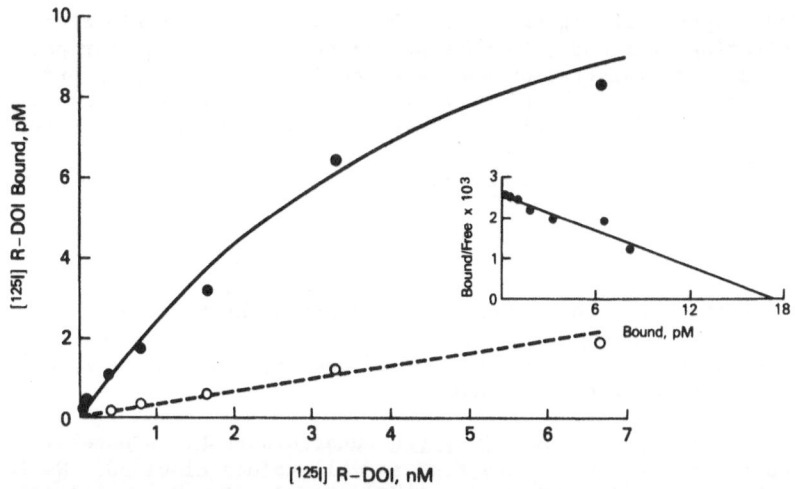

Fig. 2.   Saturation curve and Scatchard plot of $[^{125}I]$-R-DOI binding to human platelet pellet sections.   The Figure represents a typical experiment.   Each point is the average of duplicate determinations with consecutive tissue sections from a single human platelet pellet. Solid line: specific binding; broken line: nonspecific binding.

In vivo administration of (±)DOI

Chronic treatment with (±)DOI resulted in significant reductions in binding for all three radioligands in cortical membranes (Table 1).

Autoradiographic of DOI receptors in human platelets

The [125I]-R-DOI binding sites were homogeneously distributed throughout the platelet pellet sections (data not shown).   Non-specific binding, as determined in consecutive sections incubated in the presence of 1 μM ketanserin, was less than 20% of the total binding. $[^{125}I]$-R-DOI bound to a single class of sites in human platelets (Fig. 2), with a maximum binding capacity (B_{max}) of 100 ± 10 fmol/mg protein and a binding affinity (K_d) of 6.4 ± 0.7 nM.

DISCUSSION

The localization of $[^{125}I]$-R-DOI receptors reported here is in good agreement with that described earlier (Pazos et al., 1985) for rat brain 5-HT_2 receptors (McKenna et al., 1989b).   The finding of highest numbers of receptors for the hallucinogenic compound DOI in the claustrum is of interest.   The claustrum is involved in a multisensory link between the neocortex and limbic systems (Wilhite et al., 1986) and integrates the sensory input before its projection to the neocortex (Carey and Neal, 1986).

There are structural analogies between LSD and phenylalkylamine hallucinogens (Nichols et al., 1973).   Our autoradiographic studies demonstrate that a psychotomimetic phenylisopropylamine selectively displaces an anatomically discrete subset of the receptors occupied by the ergoline psychotomimetic LSD.   Our findings support the speculation that the psychotomimetic actions of ergoline and phenylalkylamine hallucinogens may be mediated at common receptors located in the claustrum and cortex (McKenna et al., 1987).

We analyzed binding of labeled DOI, LSD and ketanserin after chronic administration of a moderate dose of unlabeled DOI. A 4-hour period after the last drug administration was selected for our studies in order to minimize drug occupancy. The binding of all three radioligands was significantly reduced in chronically treated animals (McKenna et al., 1989a).

In vivo, 5-HT_2 agonist hallucinogens are able to downregulate 5-HT_2 receptors, and the administration of DOI, a hallucinogenic compound with 5-HT_2 agonist properties, downregulates both DOI and LSD binding sites. Both rats and humans develop rapid physiologic tolerance to hallucinogens following their repeated administration (Appel and Freedman, 1968). These observations, together with the present data, support the hypothesis that stimulation of 5-HT_2 postsynaptic receptors is crucial in the mechanism of action of different types of hallucinogenic drugs and may explain the appearance of rapid tolerance to these compounds.

We applied quantitative in vitro autoradiography to determine [^{125}I]-R-DOI binding in human platelets from a small volume of blood. We found a single class of binding sites for [^{125}I]R-DOI in platelets. Similar findings were reported in human platelets using [^{3}H]- or [^{125}I]-LSD and [^{3}H]-ketanserin (Geaney et al., 1984; Biegon et al., 1987; McBride et al., 1987). Because [^{125}I]-R-DOI, [^{3}H]-, [^{125}I]-LSD and [^{3}H]-ketanserin bind to 5-HT_2 receptors, these data can be interpreted as a demonstration of the presence of 5-HT_2 receptors in platelets (Geaney et al., 1984; Biegon, et al., 1987; Glennon et al., 1987; Johnson et al., 1987; McBride et al., 1987). However, LSD and phenylisopropylamine hallucinogens, including DOI, might have common binding sites different from 5-HT_2 receptors in brain (Pierce and Peroutka, 1989).

REFERENCES

Appel, J.B., and Freedman, D.X., 1968, Tolerance and cross-tolerance among psychotomimetic drugs, Psychopharmacologia, 13:267-274.

Biegon, A., Weizman, A., Karp, L., Ram, A., Tiano, S., and Wolff, M., 1987, Serotonin 5-HT_2 receptor binding on blood platelets - a peripheral marker for depression? Life Sci., 41:2485-2492.

Bradford, M.M., 1976, A rapid and sensitive method for quantitation of microgram quantities of protein utilizing the principle of protein dye binding, Anal. Biochem., 72:248-254.

Buckholtz, N.S., Freedman, D.X., and Middaugh, L.D., 1985, Daily LSD administration selectively decreases serotonin$_2$ receptor binding in rat brain, Eur. J. Pharmacol., 109:421-425.

Carey, R.G., and Neal, T.L., 1986, The rat claustrum: afferent and efferent connections with visual cortex, Brain Res., 329:185-193.

Fuller, R.W., 1986, Pharmacologic modification of serotonergic function: Drugs for the study and treatment of psychiatric and other disorders, J. Clin. Psych., 47:4-8.

Geaney, D.P., Schachter, M., Elliott, J.M., and Grahame-Smith, D.G., 1984, Characterization of [^{3}H]-lysergic acid diethylamide binding to a 5-hydroxytryptamine receptor on human platelet membranes, Eur. J. Pharmacol., 97:87-93.

Glennon, R.A., McKenney, J.D., Lyon, R.A., and Titeler, M., 1986, 5HT_1 and 5HT_2 binding characteristics of 1-(2,5-dimethoxy-4-bromophenyl)-2-aminopropane analogues, J. Med. Chem., 29:194-199.

Glennon, R.A., Seggel, M.R., Soine, W.H., Herrick-Davis, K., Lyon, R.A., and Titeler, M., 1988, [^{125}I]-1-(2,5-dimethoxy-4-iodophenyl)-2-aminopropane: an iodinated radioligand that specifically labels the agonist high-affinity state of 5HT_2 serotonin receptors, J. Med. Chem., 31:5-7.

Israel, A., Correa, F.M.A., Niwa, M., and Saavedra, J.M., 1984, Quantitative determination of angiotensin II binding sites in rat brain and pituitary gland by autoradiography, Brain Res., 322:341-345.

Jacobs, B.L., 1987, How hallucinogenic drugs work, Am. Sci., 75:386-392.

Johnson, M.P., Hoffman, A.J., Nichols, D.E., and Mathis, C.A., 1987, Binding to the serotonin 5HT$_2$ receptor by the enantiomers of [^{125}I]-DOI, Neuropharmacology, 26:1803-1806.

Mann, J.J., Stanley, M., McBride, A., and McEwen, B., 1986, Increased serotonin$_2$ and β-adrenergic receptor binding in the frontal cortices of suicide victims, Arch. Gen. Psych., 43:954-959.

McBride, P.A., Mann, J.J., Polley, M.J., Wiley, A.J., and Sweeney, J.A., 1987, Assessment of binding indices and physiological responsiveness of the 5-HT$_2$ receptor on human platelets, Life Sci., 40:1799-1809.

McKenna, D.J., Mathis, C.A., Shulgin, A.T., Sargent, T., 3rd, and Saavedra, J.M., 1987, Autoradiographic localization of binding sites for [^{125}I]-DOI, a new psychotomimetic radioligand, in the rat brain, Eur. J. Pharmacol., 137:289-290.

McKenna, D.J., Nazarali, A.J., Himeno, A., and Saavedra, J.M., 1989a, Chronic treatment with (±)DOI, a psychotomimetic 5-HT$_2$ agonist, down-regulates 5-HT$_2$ receptors in rat brain, Neuropsychopharmacology, 2:81-87.

McKenna, D.J., Nazarali, A.J., Hoffman, A.J., Nichols, D.E., Mathis, C.A., and Saavedra, J.M., 1989b, Common receptors for hallucinogens in rat brain: a comparative autoradiographic study using [^{125}I]-LSD and [^{125}I]-DOI, a new psychotomimetic radioligand, Brain Res., 476:45-56.

McKenna, D.J., and Saavedra, J.M., 1987, Autoradiography of LSD and 2,5-dimethoxyphenylisopropylamine psychotomimetics demonstrates regional, specific cross-displacement in the rat brain, Eur. J. Pharmacol., 142:313-315.

Miller, J.A., Curella, P., and Zahniser, N.R., 1988, A new densitometric procedure to measure protein levels in tissue slices used in quantitative autoradiography, Brain Res., 447:60-66.

Nichols, D.E., Barfnecht, C.F., Rusterholz, D.B., Benington, F., and Morin, R.D., 1973, Asymmetric systhesis of psychotomimetic phenylisoproylamines, J. Med. Chem., 16:480-483.

Pazos, A., Cortes, A., and Palacios, J.M., 1985, Quantitative autoradiographic mapping of serotonin receptors in the rat brain, II. Serotonin-2 receptors, Brain Res., 346:231-249.

Pierce, P.A., and Peroutka, S.J., 1988, Antagonism of 5-hydroxytryptamine$_2$ receptor-mediated phosphatidylinositol turnover by d-lysergic acid diethylamide, J. Pharmacol. Exp. Ther., 247:918-925.

Scott, J.A., and Crews, F.T., 1986, Down-regulation of serotonin$_2$ but not of beta-adrenergic receptors during chronic treatment with amitriptyline is independent of stimulation of serotonin$_2$ or beta-adrenergic receptors, Neuropharmacology, 25:1301-1306.

Shulgin, A.T., 1982, Chemistry of Psychotomimetics, in: "Handbook of Experimental Pharmacology", Vol. 55, Part III, Alcohol and Psychotomimetics, Psychotropic Effects of Central Acting Drugs, Hoffmeister, F., and Stille, G., eds., Springer-Verlag, Berlin, pp. 3-29.

Wilhite, B.L., Teyler, T.J., and Hendricks, C., 1986, Functional relations of the rodent claustral-entorhinal-hippocampal system, Brain Res., 365:54-60.

AMBIVALENCE ON THE MULTIPLICITY OF MAMMALIAN AROMATIC L-AMINO ACID

DECARBOXYLASE

M. Ebadi[1,2] and V. Simonneaux[1]

Departments of Pharmacology[1] and Neurology[2]
University of Nebraska College of Medicine
Omaha, Nebraska 68105
USA

INTRODUCTION

Hydroxytryptophan is decarboxylated to 5-hydroxytryptamine (serotonin) by aromatic L-amino acid decarboxylase which requires pyridoxal phosphate and is widely distributed throughout mammalian tissues, occurring most abundantly in the pineal gland, liver, kidney, adrenal medulla, and striatum. Earlier studies concluded that 3,4-dihydroxyphenylalanine decarboxylase (which catalyzes the decarboxylation of dopa, producing dopamine) and 5-hydroxytryptophan decarboxylase (which catalyzes the decarboxylation of 5-hydroxytryptophan, yielding serotonin) are the same enzyme, which the IUPAC Commission on Biomedical Nomenclature in 1972 named aromatic L-amino acid decarboxylase (EC 4.1.1.28). However, recent studies have questioned the validity of a single enzyme capable of decarboxylating both substrates. For example, since the pineal gland accumulates a large concentration of serotonin, melatonin and other indoleamines, it is assumed that the enzyme functions as a 5-hydroxytryptophan decarboxylase. On the other hand, since the striatum and the adrenal medulla accumulate mainly dopamine, norepinephrine and epinephrine, it is felt that the enzyme primarily decarboxylates dopa. Other factors dealing with the complexity of catalytic process are the results of reports revealing that both dopa decarboxylase and histidine decarboxylase exhibit complete immunochemical cross reactivity, suggesting the presence of similar antigenic recognition sites. Furthermore, a monoclonal antibody directed against the aromatic L-amino acid decarboxylases from various tissues revealed that the enzymes from the striatum, adrenal medulla, pineal gland, liver and kidney are indistinguishable with respect to immunological cross-reactivity and molecular size. In contrast to these findings, other evidence suggests that there exist numerous dissimilarities in the optimal pH and temperature conditions, kinetic parameters, affinity for pyridoxal phosphate, activation and inhibition by chemicals, and regional variation in the distribution of dopa and 5-hydroxytryptophan decarboxylases. Furthermore, in pyridoxine-deficient animals, hypothalamic serotonin content is significantly reduced without any changes in catecholamine levels. Moreover, since catecholamines and indoleamines are not synthesized in the liver and kidney, it is difficult to comprehend the reason for the presence of copious amounts of decarboxylases in these organs. In conclusion, although strong support for the concept of a single aromatic L-amino decarboxylase may be suggested, the presence of also non-uniform local regulatory mechanisms governing the differential synthesis

Kynurenine and Serotonin Pathways
Edited by R. Schwarcz et al., Plenum Press, New York, 1991

of catecholamines and indoleamines may be anticipated throughout mammalian tissues.

Discovery and importance of aromatic L-amino acid decarboxylase in mammalian tissues

The discovery of dopa decarboxylase in kidney by Holtz et al. (1938) prompted Blaschko (1939, 1945) to postulate an important role for this enzyme in the synthesis of catecholamines in mammalian tissues. Furthermore, it was postulated that this enzyme was able to decarboxylate all naturally occurring aromatic L-amino acids. However, the presence of a distinct histidine decarboxylase (EC 4.1.1.28) was identified initially in the stomach (Schayer, 1957) and confirmed and documented subsequently in other tissues (Weissbach, et al., 1961; Aures and Hakanson, 1971; Tran and Snyder, 1981). The importance of aromatic L-amino acid decarboxylase (EC 4.1.1.28) to catalyze the conversion of L-3,4-dihydroxyphenylalanine (L-dopa) to dopamine and of 5-hydroxytryptophan to serotonin is established. Furthermore, it is well known that this enzyme is also involved in the synthesis of trace amines catalyzing the conversion of tyrosine to tyramine, phenylalanine to phenylethylamine, and tryptophan to tryptamine (for review, see Boulton, 1978). Moreover, aromatic L-amino acid decarboxylase is involved in the synthesis of melatonin in the pineal gland and retina (see Ebadi, 1984 for review and references). Purification of the enzyme enhances its catalytic activity toward dopa and 5-hydroxytryptophan proportionately, showing that the different activities are concentrated pari passu with one another. Although no doubt remains about the significance of aromatic L-amino acid decarboxylase in neurobiology, an ongoing controversy has existed for decades about homogeneity of the decarboxylase(s), and whether aromatic L-amino acid decarboxylase is one enzyme decarboxylating both dopa or 5-hydroxytryptophan, or two enzymes decarboxylating dopa and 5-hydroxytryptophan respectively (for reviews, see Sims, 1974; Sourkes, 1977, 1987; Bowsher and Henry, 1986; Borri Voltattorni et al., 1987). In this report, pertinent information dealing with the ambivalent views on the multiplicity of aromatic L-amino acid decarboxylase will be highlighted.

AROMATIC L-AMINO ACID DECARBOXYLASE AS TWO ANALOGOUS BUT DISTINCT ENZYMES

Arguments in support of separate dopa decarboxylase and 5-hydroxytryptophan decarboxylase are A) the presence of aromatic L-amino acid decarboxylase in areas of the body with no apparent functions, B) differential kinetic properties of the two enzymes, C) dissimilar subcellular distribution of the two enzymes, D) non uniform distribution of the two enzymes in peripheral tissues and in the brain, E) non parallel changes in the activities of the enzymes in vitamin B6 deficiency, and F) gender differences in the concentration of serotonin but not dopamine.

Presence of L-amino acid decarboxylase in areas of the body with no apparent synthesis of biogenic amines

The activity of aromatic L-amino acid decarboxylase is nonuniform in the brain (Table 3) and varies dramatically in peripheral tissues (Table 4). The presence of an extremely high activity of the enzyme in the pineal gland corresponds presumably with the presence of high concentrations of serotonin and melatonin in this body. Similarly, the high activity of aromatic L-amino acid decarboxylase in the adrenal gland and corpus striatum corresponds with the high concentrations of catecholamines in these areas and other parts of the brain (Kuntzman et al., 1961; Lloyd and Hornykiewicz, 1972; Sims et al., 1973). On the other hand, next to the pineal gland, the highest activities of aromatic amino acid decarboxylase occur in the liver and kidney (Awapara et al., 1962; Christenson et al., 1970), areas of the body which do not synthesize and store indoleamines or catecholamines. Therefore, it is possible that L-dopa and 5-hydroxytryptophan may not be the endogenous substrates for the renal and hepatic decarboxylases. Recent immunohistochemical techniques using specific antidecarboxylase antiserum have shown that aromatic L-amino acid decarboxylase-immunoreactivity occurs in perikarya, axons, and fibers in the rat brain (Jaeger et al., 1983a,b, 1984; Jaeger, 1986). Furthermore, it was shown that some neurons may contain aromatic L-amino acid decarboxylase without monoamines (Jaeger et al., 1983b). In addition, aromatic L-amino acid decarboxylase-immunoreactive neurons in and around the cerebrospinal fluid containing neurons of the central canal do not contain dopamine or serotonin in the mouse and rat spinal cord (Nagatsu et al., 1988). Moreover, Kitahama et al. (1988), by investigating the distribution of aromatic L-amino acid decarboxylase-immunoreactive neurons in the cat hypothalamus, limbic areas and thalamus, and by using specific antiserum raised against porcine kidney decarboxylase, concluded that aromatic L-amino acid decarboxylase may play a more general role in physiological functions than had been considered.

Differential kinetic properties of dopa decarboxylase and 5-hydroxytryptophan decarboxylase in the brain

Sims et al. (1973) developed optimal assay conditions for dopa decarboxylase and 5-hydroxytryptophan decarboxylase in homogenates of rat brain. The results of this study showed that the two activities exhibited widely different optima for pH, temperature and substrate concentrations. The activity of 5-HTP decarboxylase was stimulated 2-fold by added pyridoxal-5-phosphate and was relatively resistant to antagonists of pyridoxal-phosphate. By contrast, the activity of dopa decarboxylase was stimulated 20-fold by added coenzyme and could be completely inhibited by carbonyl trapping agents (Table 1).

Dissimilar subcellular distribution of dopa decarboxylase and 5-hydroxytryptophan decarboxylase in brain

Sims et al. (1973) studied the relative subcellular distribution of dopa decarboxylase and 5-hydroxytryptophan decarboxylase in the rat brain. They showed that the activity of dopa decarboxylase was associated predominantly with the soluble fractions, whereas the activity of 5-hydroxytryptophan decarboxylase was distributed equally between soluble and particulate fractions (Table 2).

The subcellular distribution of the activity of dopa decarboxylase paralleled closely that of lactic acid dehydrogenase, whereas the subcellular

Table 1. Variation in kinetic parameters of dopa decarboxylase and 5-hydroxytryptophan decarboxylase in rat brain

| | Substrates | |
	Dopa	5-Hydroxytryptophan
pH	6.7	8.3
Temperature	38°C	30-42°C
Pyridoxal phosphate (1 mM)	15-fold stimulation	2-fold stimulation
K_m	600 μM	1.6 μM
Amino oxyacetic acid (1 mM)	Complete inhibition	No inhibition
L-Ascorbate (5 mM)	No inhibition	Inhibition

Data are taken from Sims et al. (1973).

Table 2. Subcellular distribution of brain decarboxylases

Fraction	Protein (% of H)	Dopa decarboxylase		5-HTP decarboxylase	
		RSA	% of H activity	RSA	% of H activity
Homogenate (H)	-	1.0	-	1.0	-
P₁ (crude nuclear)	32	0.40	12.8	0.80	23.0
P₂ (crude mitochondrial)	30	0.38	11.9	0.18	35.4
P₃ (microsomal)	14	0.15	2.4	0.52	7.4
S₃ (supernatant)	19	3.10	60.0	1.86	35.5
Recovery	95%		87.1%		101.3%
Sucrose gradient					
A (0.85 M; myelin)	4.5	0.10	0.4	0.20	1.0
B (1.2 M; synaptosomal)	8.3	0.34	3.9	1.54	15.4
C (Pellet; mitochondrial)	7.7	0.25	2.5	0.78	6.3
X (Intermediate zones)	3.2	0.54	1.6	0.93	2.9
Recovery (% of P₂)	79%		70.5%		72.0%

Enzymatic rates in the initial homogenate were: Dopa-D 5.4 μmol/h/g and 5-HTP-D 0.9 μmol/h/g. RSA: relative specific activity. Data are taken from Sims et al. (1973).

distribution of the activity of 5-hydroxytryptophan decarboxylase within the particulate fractions differed from that of succinic acid dehydrogenase (Sims et al., 1973).

Nonuniform distribution of aromatic L-amino acid decarboxylase in brain regions

Sims et al. (1973) who studied the regional distribution of dopa decarboxylase and 5-hydroxytryptophan decarboxylase in the rat brain, showed that the activity of dopa decarboxylase in the striatum was 11-fold higher than that in whole cerebellum (Table 3). Furthermore, the activity of 5-hydroxytryptophan decarboxylase relative to that of dopa decarboxylase was greater in hindbrain regions (medulla-pons and cerebellum).

Nonuniform distribution of aromatic L-amino acid decarboxylase in peripheral tissues

By using both fluorescence assay and HPLC - voltammetry technique, Rahman and colleagues (1981) studied the distribution of aromatic L-amino acid decarboxylase in eleven peripheral tissues in the rat using both L-dopa and 5-hydroxytryptophan as substrates. The results of this study revealed that pineal gland had the highest activity, followed by liver, kidney and adrenal glands. The heart and spleen exhibited low activities, whereas the salivary gland was devoid of any apparent aromatic amino acid decarboxylase (Table 4).

Using L-dopa and 5-hydroxytryptophan as substrates for decarboxylases in the rat pineal gland, adrenals and liver, Rahman et al. (1981) showed that the K_m values were similar in the adrenals and liver, with the K_m values for L-dopa being three-fold higher than that for 5-hydroxytryptophan. However,

Table 3. Regional distribution of Dopa-decarboxylase and 5-HTP-decarboxylase in rat brain

	Dopa decarboxylase	5-HTP decarboxylase
	(μmol/h/g wet weight)	
Corpus striatum	18.7 ± 1.1	2.2 ± 0.18
Mesencephalic tegmentum	9.0 ± 0.5	1.2 ± 0.07
Hypothalamus	8.6 ± 0.5	1.2 ± 0.05
Hippocampus	2.4 ± 0.2	0.34 ± 0.03
Frontal grey	3.1 ± 0.3	0.44 ± 0.03
Occipital grey	1.9 ± 0.1	0.36 ± 0.4
Midbrain colliculi	5.2 ± 0.3	0.98 ± 0.1
Midline medulla-pons	5.4 ± 0.4	1.05 ± 0.08
Lateral medulla-pons	3.8 ± 0.3	0.82 ± 0.05
Cerebellum	1.6 ± 0.1	0.71 ± 0.03

Data are taken from Sims et al. (1973).

the K_m values for both L-dopa and 5-hydroxytryptophan were significantly lower in the pineal gland.

Nonparallel changes in the concentration of serotonin and dopamine in pyridoxine deficiency

Since dihydroxyphenylalanine decarboxylase and 5-hydroxytryptophan de-carboxylase, respectively, have high and low affinities for pyridoxal phos-

Table 4. The distribution of aromatic L-amino acid decar-boxylase (AADC) in peripheral tissues

Tissues	AADC activity	
	(nmol/min/g wet weight)	
	L-Dopa as substrate	L-5-HTP as substrate
Pineal gland	1400 ± 139	277 ± 27
	1.39 ± 0.13[*]	0.27 ± 0.02[*]
Liver	444 ± 40	68.2 ± 6.9
Kidney	418 ± 39	105 ± 8.0
Adrenal glands	353 ± 27	43.5 ± 4.7
	18.3 ± 0.2[**]	2.1 ± 0.2[**]
Small intestine	56.8 ± 10.6	8.0 ± 0.8
Large intestine	44.8 ± 4.7	10.4 ± 1.2
Lung	31.9 ± 4.3	4.16 ± 0.48
Heart	11.9 ± 1.3	2.43 ± 0.11
Spleen	4.9 ± 0.1	1.3 ± 0.03
Blood serum	121 ± 11[***]	46.3 ± 5.7[***]
Salivary gland	ND	ND

[*]nmoles/min per pineal gland; [**]nmoles/min per pair of adrenal glands; [***]pmoles/min per ml serum; ND: Not detectable.
Data modified from Rahman et al. (1981).

Table 5. The effects of vitamin B6 deficiency on the synthesis of
dopamine, norepinephrine, serotonin and melatonin

| | Hypothalamus | | | Pineal gland |
	Dopamine (nmol/g)	Norepi- nephrine (nmol/g)	Serotonin (nmol/g)	Night time melatonin (pg/gland)
Pyridoxine- supplemented (control)	1.18 ± 0.07	2.17 ± 0.09	1.70 ± 0.20	1923 ± 452
Pyridoxine- deficient	1.25 ± 0.10	2.01 ± 0.10	1.00 ± 0.27[*]	1150 ± 139[*]

[*]Significantly different from controls (P < 0.05). Data are taken
from Siow and Dakshinamurti (1985) and from Dakshinamurti et al.
(1988).

phate, Siow and Dakshinamurti (1985) and Dakshinamurti et al. (1988) studied
the concentrations of dopamine, norepinephrine, serotonin and melatonin in
pyridoxine deficient and pyridoxine supplemented (control) rats. The results
showed that vitamin B6 deficiency reduced the concentrations of serotonin and
melatonin, but not of dopamine or norepinephrine (Table 5).

Sex differences in the concentration of serotonin in rat brain

Carlsson and Carlsson (1988) studied the concentrations of serotonin and
dopamine and their metabolites in the brain stem, hypothalamus, corpora striata, limbic forebrain and cortex of male (body weight 350-450 g) and female
(250-300 g) Sprague-Dawley rats. The results showed that the concentration
of serotonin was higher in the brains of female rats, whereas the concentration of dopamine was identical in the brains of both male and female rats.

AROMATIC L-AMINO ACID DECARBOXYLASE AS A UNITARY ENZYME

Much controversy continues to exist as to whether a single or two decarboxylases catalyze the decarboxylation of 3-4-dihydroxyphenylalanine and
5-hydroxytryptophan (for review, see Sourkes, 1977). In 1972, the concept of
a unitary enzyme had received the approval of the commission on biochemical
nomenclature, stating that the formerly known dopa decarboxylase be hence-
forth identified as aromatic L-amino acid decarboxylase. The argument in
favor of a unitary enzyme are A) singular catalytic action with enzymes from
diversified sources; B) lack of selectivity of decarboxylases to dissimilar
neurotoxins; and C) immunological cross reactivity of diversified decarboxylases.

Singular catalytic action with enzyme from diversified sources

The finding that the enzyme purified from pig kidney (Christenson et
al., 1970; Lancaster and Sourkes, 1972), guinea pig kidney (Srinivasan and
Awapara, 1978), rat kidney (Bowsher and Henry, 1983; Shirota and Fujisawa,
1988), human pheochromocytoma (Ichinose et al., 1985) and bovine brain
(Nishigaki et al., 1988) can catalyze the decarboxylation of dopa and
5-hydroxytryptophan support the concept of a unitary enzyme.

Lack of selectivity of decarboxylases to dissimilar neurotoxins

Dairman et al. (1975) administered to rats 5,6-dihydroxytryptamine or 6-hydroxydopamine intracisternally in order to effect a selective destruction of serotonin- or catecholamine-containing neurons. The L-dopa- and L-5-hydroxytryptophan-decarboxylating activities of the spinal cord and brain were then determined at several time intervals following this treatment. In both cases, the relative loss of L-dopa-decarboxylating activity was the same as the relative loss of L-5-hydroxytryptophan-decarboxylating activity. 5,6-dihydroxytryptamine treatment had little or no affect on catecholamine-containing neurons, and 6-hydroxydopamine did not affect serotonin-containing neurons. These data were interpreted to indicate that only one decarboxylase is involved in the biosynthesis of both serotonin and catecholamines in the rat CNS.

Immunological cross reactivity of diversified decarboxylases

The concept of a unitary enzyme is further supported by the observations that antisera against purified pig kidney enzyme (Christenson et al., 1972) and beef adrenal gland enzyme (Goldstein et al., 1973) inhibit the activities of dopa decarboxylase and 5-hydroxytryptophan decarboxylase to the same extent. Moreover, Shirota and Fujisawa (1988) purified to homogeneity from rat kidney aromatic L-amino acid decarboxylase. Furthermore, they showed that the amino acid composition of decarboxylase from rat kidney (Table 6) is similar to that of pig kidney (Christenson et al., 1970).

The purified preparation showed an activity of dopa decarboxylation of ~11,000 nmol/min/mg of protein at 37°C. Furthermore, the purified enzyme was capable of decarboxylating 5-hydroxytryptophan, tyrosine, tryptophan and phenylalanine.

Table 6. Amino acid composition of
aromatic L-amino acid decarboxylase from rat kidney

Amino acid	Residues/mole
Aspartic acid	54.8
Threonine	35.1
Serine	51.1
Glutamic acid	84.6
Proline	32.0
Glycine	55.0
Alanine	84.8
Valine	54.8
Methionine	19.4
Isoleucine	43.7
Leucine	87.0
Tyrosine	28.2
Phenylalanine	50.1
Lysine	34.4
Histidine	21.5
Arginine	57.1
Tryptophan	12.9
Half-cystine	21.2

Data are taken from Shirota and Fujisawa
(1988).

When the crude extracts from the striatum, adrenal medulla, pineal gland, liver and kidney were subjected to affinity chromatography on mono-clonal antibody-coupled Sepharose, each of the eluates showed one major pro-tein band corresponding in mobility to a molecular weight of 48,000 on SDS-polyacrylamide gel electrophoresis (Fig. 1).

Recapitulation and perspective

The mammalian aromatic L-amino acid decarboxylase, a dimer of identical subunits, is immunologically indistinguishable in several peripheral tissues and in the CNS (Nishigaki et al., 1988; Shirota and Fujisawa, 1988). Al-though this evidence provides strong support for the concept of a unitary en-zyme for aromatic L-amino acid decarboxylase, the possibility still exists that one may deal with heterogeneous groups of enzymes with identical molec-ular weights and immunological cross reactivity. Additional studies dealing with specific amino acid compositions and sequences of various peripheral and central decarboxylases are required to cast an unequivocable vote on either singularity or multiplicity of the enzymes(s). Furthermore, in order to learn the reasons for the differential kinetic properties of the enzyme, it

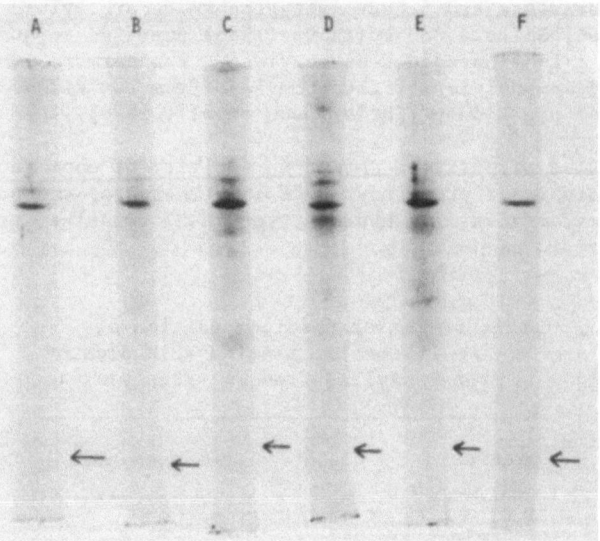

Fig. 1. SDS-polyacrylamide gel electrophoresis of aromatic L-amino acid decarboxylase purified from various rat tissues by using monoclonal antibody-coupled Sepharose. The crude extracts - 3.0 ml from the striatum (A), 2.5 ml from adrenal medulla (B), 0.5 ml from pineal gland (C), 0.4 ml from liver (D), and 0.5 ml from kidney (E), each of which contained ~100 units of dopa decarboxylase - were incubated for 17 h at 24°C with vigorous shaking with 0.1 ml of the monoclonal antibody-coupled Sepharose 4B, in the presence of 20 µg/ml each of the microbial protease inhibitors antipain, leupeptin, pepstatin, and chymostatin, respectively, and then packed into small columns. Each column was washed with 20 ml of 10 mM potassium phosphate buffer (pH 7.0), containing 0.5% Triton X-100 and 0.2 M NaCl and then with 1 ml of 10 mM potassium phosphate buffer (pH 7.0) and finally eluted with 0.3 ml of 0.1 M glycine buffer (pH 2.2). These eluates were dialyzed against 10 mM sodium phosphate buffer (pH 7.2), containing 1% SDS. An aliquot of 30-50 µl of each dialysate (A-E) and 0.4 µg of the purified enzyme (F) were subjected to electrophoresis on 7.5% polyacrylamide gels in the presence of SDS. The arrows indicate the bromophenol blue dye front. Data are taken from Shirota and Fujisawa (1988).

is necessary to characterize precisely the active site of the decarboxylase with respect to the number of substrates, number of binding site(s) for pyridoxal phosphate, and the precise mechanisms for the decarboxylation reactions. Furthermore, although it has been reported that a single gene codes for decarboxylases in neuronal and non-neuronal tissues (Albert et al., 1987), an extensive number of puzzling questions related to a few decarboxylases need to be clarified. A few unrelated examples are cited below:

1) Brain aromatic L-amino acid decarboxylase increases prenatally (Lamprecht and Coyle, 1972), whereas the peripheral decarboxylases increases postnatally (Awapara and Saines, 1975).

2) The aromatic L-amino acid decarboxylase from blowfly (Calliphorea vicina), which apparently has an identical molecular weight as kidney decarboxylase, shows very little activity toward 5-hydroxytryptophan (Fragoulis and Sekeris, 1975).

3) Due to the presence of inhibitors (Rahman et al., 1981), the salivary gland exhibits no decarboxylase activity.

4) Decarboxylation of dopa by erythrocytes occurs with both isomers of dopa nonenzymatically (Vogel, 1969; Dairman and Christenson, 1972). The reaction, which is catalyzed by oxyhemoglobin (Yambe and Lovenberg, 1972), does not result in the synthesis of dopamine (Mackowiak et al., 1972).

AKNOWLEDGEMENTS

The authors gratefully acknowledge the expert secretarial assistance of Mrs. L. Swigart and Mrs. D. Panowicz. This report was completed in part from a grant from USPHS ES03949.

REFERENCES

Albert, V.R., Allen, J.M., and Joh, T.H., 1987, A single gene codes for aromatic L-amino acid decarboxylase in both neuronal and non-neuronal tissues, J. Biol. Chem., 262:9404-9411.

Aures, D., and Hakanson, R., 1971, Histidine decarboxylase, in: "Methods in Enzymology", Vol. 17, Part B, Colowick, S.P., and Kaplan, N.O., eds., Academic Press, New York, pp. 667-677.

Awapara, J., and Saine, S., 1975, Fluctuations in dopa decarboxylase activity with age, J. Neurochem., 24:817-818.

Awapara, J., Sandman, R.P., and Hanly, C., 1962, Activation of dopa decarboxylase by pyridoxal phosphate, Arch. Biochem. Biophsys., 98:520-525.

Blaschko, H., 1939, The specific action of L-dopa decarboxylase, J. Physiol., 96:50P-51P.

Blaschko, H., 1945, The amino acid decarboxylase of mammalian tissues, Adv. Enzymol., 5:67-85.

Borri Voltattorni, C., Giartosio, A., and Turano, C., 1987, Aromatic L-amino acid decarboxylase from pig kidney, in: "Methods in Enzymology", Vol. 142, Kaufman, S., ed., Academic Press, New York, pp. 179-187.

Boulton, A.A., 1978, The tyramines: functionally significant biogenic amines or metabolic accidents? Life Sci., 23:659-672.

Bowsher, R.R., and Henry, D.P., 1983, Decarboxylation of p-tyrosine: a potential source of p-tyramine in mammalian tissues, J. Neurochem., 40:992-1002.

Bowsher, R.R., and Henry, D.P., 1986, Aromatic L-amino acid decarboxylase: biochemistry and functional significance, in: "Neuromethods. 5.
Neurotransmitter Enzymes", Boulton, A.A., Baker, G.B., and Yu, P.H.,
eds., Humana Press, Clifton, New Jersey, pp. 33-78.

Carlsson, M., and Carlsson, A., 1988, A regional study of sex differences in
rat brain serotonin, Prog. Neuropsychopharmacol. Biol. Psychiat., 12:53-
61.

Christenson, J.G., Dairman, W., and Udenfriend, S., 1970, Preparation and
properties of a homogeneous aromatic L-amino acid decarboxylase from hog
kidney, Arch. Biochem. Biophys., 141:356-367.

Christenson, J.G., Dairman, W., and Udenfriend, S., 1972, On the identity of
dopa decarboxylase and 5-hydroxytryptophan decarboxylase, Proc. Natl.
Acad. Sci. USA, 69:343-347.

Dairman, W., and Christenson, J.G., 1972, Dopa decarboxylating activity of
human erythrocytes, Fed. Proc., 31:590.

Dairman, W., Horst, W.D., Marchell, M.E., and Bautz, G., 1975, The proportionate loss of L-3,4-dihydroxyphenylalanine and L-5-hydroxytryptophan
decarboxylating activity in rat central nervous system following intra-
cisternal administration of 5,6-dihydroxytryptamine or 6-hydroxydopamine,
J. Neurochem., 24:619-623.

Dakshinamurti, K., Paulose, C.S., Viswanathan, M., and Siow, Y.L., 1988,
Neuroendocrinology of pyridoxine deficiency, Neurosci. Biobehav. Rev.,
12:189-193.

Ebadi, M., 1984, Regulation of the synthesis of melatonin and its significance to neuroendocrinology, in: "The Pineal Gland", Reiter, R.J., ed.,
Raven Press, New York, pp. 1-37.

Fragoulis, E.G., and Sekeris, C.E., 1975, Purification and characteristics of
DOPA-decarboxylase from the integument of Calliphora vicina larvae, Arch.
Biochem. Biophys., 168:15-25.

Goldstein, M., Anagnoste, B., Freedman, L.S., Roffman, M., Ebstein, R.P.,
Park, D.H., Fuxe, K., and Hökfelt, T., 1973, Characterization, localization and regulation of catecholamine synthesizing enzymes, in: "Frontiers in Catecholamine Research", Usdin, E., and Snyder, S., eds.,
Pergamon, New York, pp. 69-78.

Holtz, P., Heise, R., and Luedtke, K., 1938, Enzymic destruction of L-dopa by
the kidney, Arch. Exp. Path. Pharmakol., 191:87-118.

Ichinose, H., Kojima, K., Togari, A., Kato, Y., Parvez, S., Parvez, H., and
Nagatsu, T., 1985, Simple purification of aromatic L-amino acid decarboxylase from human pheochromocytoma using high-performance liquid chromatography, Anal. Biochem., 150:408-414.

Jaeger, C.B., 1986, Aromatic L-amino acid decarboxylation in the rat brain:
immunocytochemical localization during prenatal development, Neuroscience,
18:121-150.

Jaeger, C.B., Albert, V.R., Joh, T.H., and Reis, D.J., 1983a, Aromatic
L-amino acid decarboxylase in the rat brain: coexistence with vasopressin
in small neurons of the suprachiasmatic nucleus, Brain Res., 276:362-
366.

Jaeger, C.B., Ruggiero, D.A., Albert, V.R., Park, D.H., Joh, T.H., and Reis,
D.J., 1984, Aromatic L-amino acid decarboxylase in the rat brain: immunocytochemical localization in neurons of the rat brain stem, Neuroscience,
11:691-713.

Jaeger, C.B., Titelman, G., Joh, T.H., Albert, V.R., Park, D.H., and Reis,
D.J., 1983b, Some neurons of the rat central nervous system contain aromatic L-amino acid decarboxylase but not monoamines, Science, 219:1233-
1235.

Kitahama, K., Denoyer, M., Raynaud, B., Borri-Voltattorni, C., Weber, M., and
Jouvet, M., 1988, Immunohistochemistry of aromatic L-amino acid decarboxylase in the cat forebrain, J. Comp. Neurol., 270:337-353.

Kuntzman, R., Shore, P.A., Bogdanski, D., and Brodie, B.B., 1961, Micro-analytical procedures for fluorometric assay of brain DOPA-5HTP decarbox-ylase, norepinephrine and serotonin, and detailed mapping of decarboxylase activity in brain, J. Neurochem., 6:226-232.

Lamprecht, F., and Coyle, J.T., 1972, Dopa decarboxylase in the developing rat brain, Brain Res., 41:503-506.

Lancaster, G.A., and Sourkes, T.L., 1972, Purification and properties of hog kidney 3,4-dihydroxyphenylalanine decarboxylase, Can. J. Biochem., 50:791-797.

Lloyd, K.G., and Hornykiewicz, O., 1972, Occurrence and distribution of aro-matic L-amino acid (L-DOPA) decarboxylase in the human brain, J. Neuro-chem., 19:1549-1559.

Mackowiak, E.D., Hare, T.A., and Vogel, W.H., 1972, Measurements of aromatic L-amino acid decarboxylase - a technical comment, Biochem. Med., 6:562-567.

Nagatsu, I., Sakai, M., Yoshida, M., and Nagatsu, T., 1988, Aromatic L-amino acid decarboxylase-immunoreactive neurons in and around the cerebrospinal fluid-contacting neurons of the central canal do not contain dopamine or serotonin in the mouse and rat spinal cord, Brain Res., 475:91-120.

Nishigaki, I., Ichinose, H., Tamai, K., and Nagatsu, T., 1988, Purification of aromatic L-amino acid decarboxylase from bovine brain with monoclonal antibody, Biochem. J., 252:331-335.

Rahman, K., Nagatsu, T., and Kato, T., 1981, Aromatic L-amino acid decarbox-ylase activity in central and peripheral tissues and serum of rats with L-dopa and L-5-hydroxytryptophan as substrates, Biochem. Pharmacol., 30:645-649.

Schayer, R.W., 1957, Histidine decarboxylase of rat stomach and other mam-malian tissues, Am. J. Physiol., 189:533-536.

Shirota, K., and Fujisawa, H., 1988, Purification and characterization of aromatic L-amino acid decarboxylase from rat kidney and monoclonal anti-body to the enzyme, J. Neurochem., 51:426-434.

Sims, K.L., 1974, Biochemical characteristics of mammalian brain 5-hydroxy-tryptophan decarboxylase activity, Adv. Biochem. Psychopharmacol., 11:43-50.

Sims, K.L., Davis, G.A., and Bloom, F.E., 1973, Activities of 3,4-dihydroxy-L-phenylalanine and 5-hydroxy-L-tryptophan decarboxylases in rat brain: assay characteristics and distribution, J. Neurochem., 20:449-464.

Siow, Y.L., Dakshinamurti, K., 1985, Effect of pyridoxine deficiency on aromatic L-amino acid decarboxylase in adult rat brain, Exp. Brain Res., 59:575-581.

Sourkes, T.L., 1977, Structure and function of monoamine enzymes, in: "Enzymology of Aromatic Amino Acid Decarboxylase", Usdin, E., Weiner, N., and Youdim, M.B.H., eds., Marcel Dekker, New York, pp. 477-496.

Sourkes, T.L., 1987, Aromatic L-amino acid decarboxylase, in: "Methods in Enzymolology", Vol. 142, Kaufman, S., ed., Academic Press, New York, pp. 170-178.

Srinivasan, K., and Awapara, J., 1978, Substrate specificity and other pro-perties of dopa decarboxylase from guinea pig kidney, Biochim. Biophys. Acta, 526:597-604.

Tran, V.T., and Snyder, S.H., 1981, Histidine decarboxylase, J. Biol. Chem., 256:680-686.

Vogel, W.H., 1969, Non-enzymatic decarboxylation of dihydroxyphenylalanine, Naturwissenschaften, 56:462.

Weissbach, H., Lovenberg, W., and Udenfriend, S., 1961, Characteristics of mammalian histidine decarboxylating enzymes, Biochim. Biophys. Acta, 50:177-179.

Yamabe, H., and Lovenberg, W., 1972, Decarboxylation of 3-,4-dihydroxy-phenylalanine by oxyhemoglobin, Biochem. Biophys. Res. Comm., 47:733-739.

INTERRELATIONSHIPS OF DNA CYCLES AND MELATONIN SYNTHESIS

S.D. Wainwright

Biochemistry Department
Dalhousie University
Halifax, Nova Scotia B3H 4H7
Canada

Both the chick and the rat pineal glands convert tryptophan to melatonin by the same pathway, outlined by Klein et al. (1981). In both cases, the rate-limiting step is the acetylation of serotonin by the enzyme serotonin N-acetyltransferase (NAT), described by Voisin et al. (1984). Further, the pineals of both species exhibit circadian rhythms in the production of melatonin and in the level of NAT activity.

However, the mechanisms controlling these circadian rhythms differ markedly between the two species. The rhythms of the rat pineal gland are controlled by a complex adrenergic neural stimulation which, in turn, is regulated by a biological clock located in the suprachiasmatic nuclei of the brain (see Namboodiri, this volume).

In contrast, the pacemaker for these rhythms in the chick pineal is endogenous to the gland itself. The rhythms persist _in vivo_ after disruption of neural input to the gland by removal of the superior cervical ganglia (Ralph et al., 1975). They also persist _in vitro_ when isolated glands (Binkley et al., 1978a; Deguchi, 1979a; Kasal et al., 1979; Wainwright and Wainwright, 1979; Takahashi et al., 1980) or small portions of gland (Takahashi and Menaker, 1984) are incubated under appropriate conditions of culture. Moreover, levels of NAT activity change significantly in "anticipation" of changes of lighting in the schedule of illumination to which the birds are adjusted or entrained (Fig. 1). Rhythms in NAT activity (Deguchi, 1979b) and melatonin formation (Robertson and Takahashi, 1988; Zatz et al., 1988) have also been shown with cultures from dispersed pineal cells under conditions which limited growth of fibroblasts. These latter systems have also been used to demonstrate that the pineal gland itself possesses the property of entraining its rhythms to a new schedule of lighting.

However, the chick pineal pacemaker is probably not the entire clock controlling pineal rhythms. Removal of the superior cervical ganglia had no effect upon the rhythm of plasma melatonin content (a measure of pineal production) in birds kept under a normal lighting schedule. Yet the rhythm was not sustained in birds kept in constant darkness unless they were infused with norepinephrine for 12 h in each 24-h period (Cassone and Menaker, 1983). The norepinephrine depressed plasma melatonin rhythms and it has been suggested that the pineal gland, superior cervical ganglia and suprachiasmatic

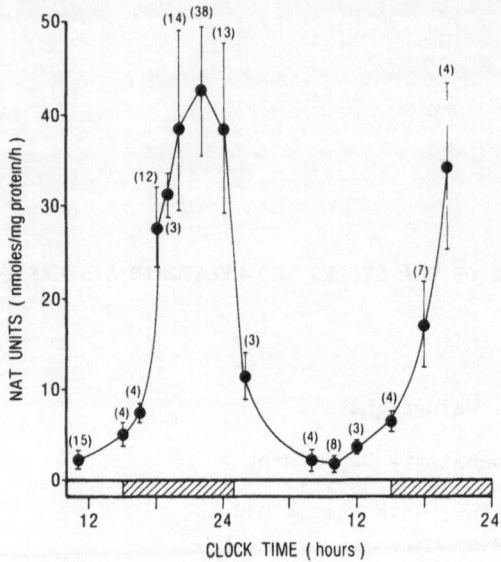

Fig. 1. Persistent NAT rhythm in culture. Glands were cultured in standard
medium and NAT assayed as described by Wainwright and Wainwright
(1979). Lighting conditions are indicated by the bar diagram, in
which the clear portion represents illumination at an intensity of
1000 lux.

nuclei constitute a self-sustaining neuro-endocrine loop (Cassone and Menaker, 1984).

Nevertheless, the pacemaker or "timekeeping" components of the clock appear to be restricted to the pineal gland itself.

The depression of plasma melatonin levels by norepinephrine reflects a
further difference between the pineal glands of the chick and rat. NAT activity and melatonin production in the rat pineal are stimulated by norepinephrine or agonists; and both α_1-and β-specific receptors are involved. In
contrast, NAT activity and melatonin production in the chick pineal are depressed in vivo (Binkley et al., 1975) and in vitro (Binkley et al., 1978b;
Deguchi, 1979c) by norepinephrine and agonists acting on α_2-specific receptors (Voisin and Collin, 1986; Pratt and Takahashi, 1987; Voisin et al.,
1987). Norepinephrine appears to prevent increase of NAT activity, rather
than affect the pineal pacemaker (Zatz and Mullen, 1990).

Our major interest is in the mechanism of the pacemaker(s) which controls the circadian rhythms of the chick pineal. We have probed the mechanism by studying the effects of supplements on the cycle in NAT activity. Results obtained with aphidicolin (Fig. 2) indicate that this agent probably
phase-shifts the cycle, i.e. resets the timing of the pineal pacemaker. When
the glands were cultured in continuous darkness (Fig. 2a), increase of activity was stimulated slightly and the time of peak activity was advanced at
least 2-h. When they were similarly cultured under normal diurnal lighting
(and therefore subject to dominant phase-shifting effects of changes in
lighting) the slight increase in maximal activity (Fig. 2b) was not accompanied by any change in time of peak activity.

We obtained similar results with methotrexate, cytosine arabinoside and
dideoxythymidine (Wainwright and Wainwright, 1984, 1986). Each of the effective agents shares the property of inhibiting DNA synthesis, but by three

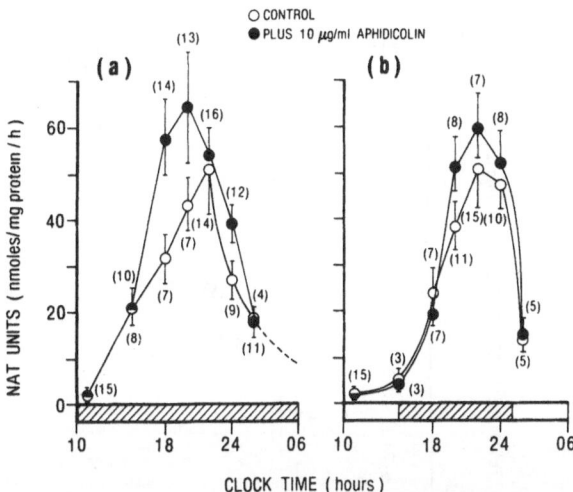

Fig. 2.   Glands were cultured with a standard medium ± aphidicolin and assayed for NAT activity as described by Wainwright and Wainwright
(1986). Lighting conditions are indicated by the bar diagrams.

different mechanisms. Due caution must always be given to the possibility of
unsuspected side-effects. Nevertheless, it seemed probable that the clock
controlling the cycle in NAT activity is in some way linked to the synthesis
of DNA.

Accordingly, we have commenced a study of the DNA metabolism of the intact chick pineal. Since it is highly inconvenient to use the procedure of
pulse-labelling under our experimental conditions, we began by studying the
cumulative incorporation of precursor.

We have demonstrated (Fig. 3) repetitive, or rhythmic, cycles of incorporation which are entrained to the light cycle under which the birds are
maintained (Wainwright and Wainwright, 1989a, and in preparation). Control
of this rhythm in thymidine incorporation appears to be independent of that
of the normal cell cycle. We found no evidence of selective labelling of a
portion of the chick genome. Rather, the incorporation corresponded to DNA
synthesis in a population of unsynchronized cells. Yet both the arrest and
renewal of thymidine incorporation were essentially synchronous (Wainwright
et al., 1989). Further, the rhythm in incorporation does not reflect rhythms
in either uptake of exogenous thymidine by the pineal cells (Wainwright and
Wainwright, 1989b), or in activity of the enzymes which convert nucelosides
to their phosphates (Wainwright and Wainwright, 1989c, and in preparation).

In pulse-labelling experiments (Table 1) we found a marked (but not
total) reduction of incorporation during the first dark period in culture
(Wainwright et al., 1989). The rate of pulse-labelling rose again at the
start of the next photoperiod but did not fall again during subsequent dark
periods (Wainwright and Wainwright, 1989c, and in preparation).

The rhythm in cumulative thymidine incorporation differs markedly from
those in NAT activity and melatonin formation in that it is unresponsive to
adrenergic stimulation (Wainwright and Wainwright, 1990).

However, the most important difference from the rhythms in NAT activity
and melatonin formation is that the bulk of the thymidine incorporated is in

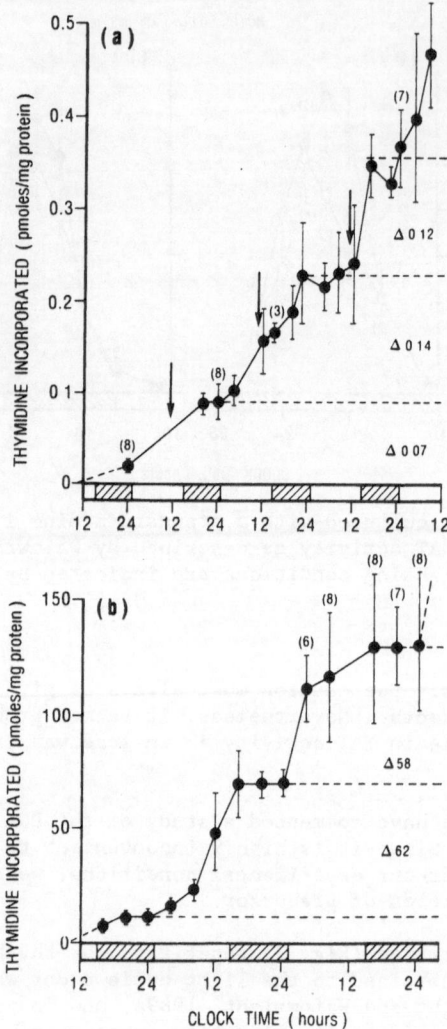

Fig. 3.   Repetitive cycles of incorporation.   (a) Cultures with standard
          medium plus 2.5 nM thymidine (total) renewed at times indicated by
          the arrows.   (b) Undisturbed cultures with standard medium plus
          2.5 μM thymidine and additional nucleosides.

stromal cells of the gland.   Only a very small number of pinealocytes are labelled after culture for up to 36 h with ^{3}H-thymidine (Wainwright et al.,
1989; Meiniel et al., unpublished).

These differences pose obvious questions about the relationship between
the mechanisms controlling these pineal rhythms.   At this stage, any attempt
to answer these questions must necessarily be highly speculative.   Moreover,
there are two further sets of observations which are highly relevant.

The first is simply that the nocturnal arrest of thymidine incorporation
is almost certainly not due to feedback inhibition by accumulated melatonin
(Table 2) or other indoles tested.   Serotonin at 1 μM tended to inhibit incorporation slightly (Table 2), but the effect was only significant in one of
three experiments.

Table 1. Thymidine incorporated in a 1-h pulse

Start of pulse		Thymidine (pmoles/mg/protein)	
Day	Hours	Experiment 1	Experiment 2
1	11:00	1.26 ± 0.32	1.29 ± 0.23
	20:00	0.44 ± 0.08	0.61 ± 0.10
2	8:00	3.42 ± 1.19	2.21 ± 0.29(3)
	20:00	3.25 ± 0.62	3.46 ± 0.92

Glánds were incubated with standard medium plus 1 μM of thymidine and other nucleosides (Wainwright and Wainwright, 1989a) to the time of start of pulse. They were were then transferred into the same medium plus 20 μCi/ml of ^{3}H-thymidine for 1 h. Lighting was diurnal (cf. legend to Fig. 2) and only infra-red illumination (with image converter) was used for transfers at 20:00 h. All values are means of 4 independent cultures except the one noted in parentheses.

The second point is more complex and originates in an apparent requirement for either some unidentified metabolite of thymidine, or a contaminant - in all samples of radioactive thymidine tested. The rate and extent of incorporation during each cycle depends upon the thymidine concentration of the medium (Fig. 3). Further, undisturbed pineals incubated under diurnal lighting with our current standard medium (which contains 1 μM of carrier thymidine) did not show renewal in a second cycle of incorporation. [Glands cultured in constant darkness show a more variable and complex response (e.g. Figs. 4 and 5). Incorporation does usually resume on the third day (sometimes after a substantial lag) and continues without interruption to a maximum level which may be considerably more than twice that attained in the previous normal cycle. We assume that this reflects enhaustion of endogenous reserves of non-radioactive thymidine and of other essential materials (Wainwright and Wainwright (1989c)]. However, corresponding cultures completed at least one additional cycle of incorporation when the concentration of thymidine was raised to 2.5 μM (Wainwright and Wainwright, 1989a). Yet, in both cases the amount incorporated represents less than 2% of the available thymidine (Wainwright et al., 1989; Wainwright and Wainwright, 1989a).

The only reported contaminant of radiothymidine is traces of 5-methyluridine, or ribothymidine. It seems improbable that this is the unidentified requirement as we only know of it as a minor base formed by methylation of transcribed tRNA precursors. Indeed, it had no effect upon thymidine incorporation under the standard cycle of illumination.

Yet, to our surprise, at 1 μM it appears to "uncouple" incorporation of thymidine in the dark from control by the pineal pacemaker (Fig. 4) without affecting the amount of precursor incorporated per day. It was inactive at 10 nM in the dark.

Since it seemed possible that this unanticipated action of methyluridine could be attributed to it serving as a donor of methyl groups, we examined the effect of adding 1 μM of choline to the medium. Again we observed (Fig. 5) uncoupling of incorporation from the pacemaker.

Table 2.   Effect of supplements on 1-h pulse incorporation

Supplement	Thymidine (pmoles/mg protein)	
	Control	Plus Supplement
Melatonin (1 μM)	1.01 ± 0.08 (3)	1.10 ± 0.08 (4)
	0.94 ± 0.18 (3)	1.19 ± 0.28 (4)
N-Acetylserotonin (1 μM)	1.08 ± 0.40 (4)	0.65 ± 0.10 (4)
	0.58 ± 0.09 (7)	0.55 ± 0.13 (8)
Serotonin (1 μM)	1.26 ± 0.32 (4)	0.83 ± 0.15 (4)
	1.08 ± 0.40 (4)	0.71 ± 0.13 (4)
	0.58 ± 0.09 (7)	0.48 ± 0.10 (7)*

Glands were cultured with standard medium plus 1 μM thymidine and
other nucleosides in the dark with or without 1 μM of the indicated
supplements. Each line represents a separate experiment and values
are means for the number of independent cultures indicated in parentheses. *Value significantly different from controls. (Anova,
$P < 0.05$).

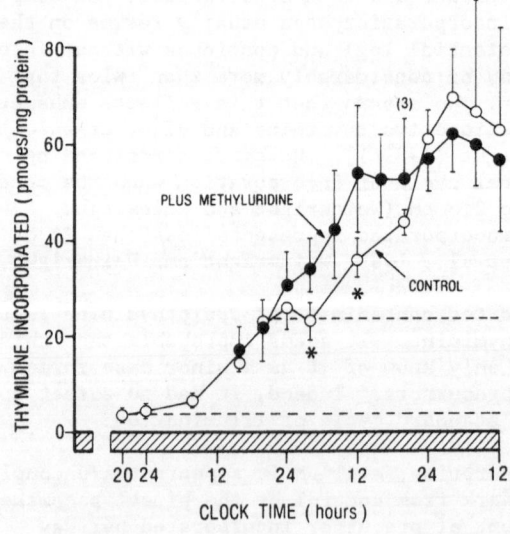

Fig. 4.   Uncoupling of incorporation from the clock
on day 3 of culture by 1 μM methyluridine.
Glands were cultured in the dark with or
without methyluridine. Values are the
mean ± SD of 4 separate cultures. *Significant difference. (Anova, $P < 0.05$).

I must add two things. One, our standard medium already contains approximately three times this amount of choline. Second, the rhythm in melatonin production by reaggregates of chick pineal cells reported by Zatz et al. (1988) was obtained with a medium containing roughly 10 times as much choline as ours. We suspect that the critical factor is the concentration of choline relative to other pineal constituents, rather than the absolute concentration in the medium. Nevertheless, this may also reflect a real difference in the mechanisms controlling incorporation of thymidine and level of NAT activity or melatonin formation.

Now we can return to the pacemakers controlling the rhythms in melatonin formation and thymidine incorporation, respectively. One controls the start of a dark-phase increase in melatonin formation by the pinealocytes, which is depressed by adrenergic stimulation and (at least under some conditions) is unaffected by relatively high concentrations of choline. The other controls the start of a light-phase incorporation of thymidine, which is unaffected by adrenergic stimulation, but uncoupled from its pacemaker by a moderate concentration of choline.

What is the relationship between these two pacemakers? There appear to be four possibilities:

1. The apparent complementarity of the two rhythms is entirely fortuitous. They are controlled by two distinct pacemakers which are totally independent of each other. (Coupling the pacemakers would amount to postulating one of the other possibilities);

2. Both rhythms are controlled by one (type of) pacemaker located exclusively in the pinealocytes;

3. Both rhythms are controlled by one (type of) pacemaker located exclusively in the fibroblasts of the stroma;

4. The complete (and fully efficient) pacemaker consists of a mutually-complementary aggregate containing both pinealocytes and stromal cells.

In most respects, the first of these possibilities is the easiest to en-

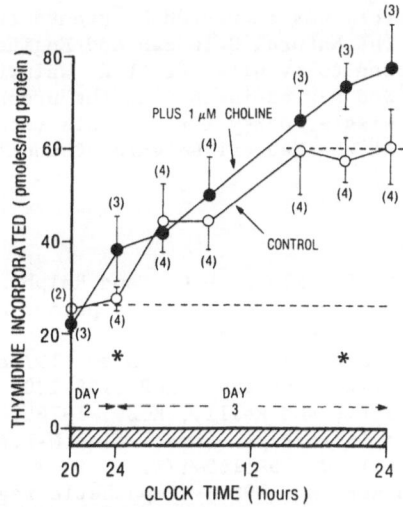

Fig. 5. Uncoupling of incorporation from the clock on day 3 of culture by 1 μM choline. Glands were cultured in the dark with or without choline. *Significant difference (Anova).

133

visage. However, it seems improbable that there is any selective advantage to retention of two distinct pacemakers in a tissue as small as the pineal. Certainly we are unaware of any precedent for favoring this possibility. Further, neither of the first two possibilities requires the conclusion that the cycle in NAT activity would be affected by various inhibitors of DNA synthesis as seen in our previous studies (Wainwright and Wainwright, 1984, 1986). They return us to the point of departure in our studies of thymidine incorporation.

On the other hand, neither of the last two possibilities leads to the expectation that dispersed pinealocytes resolved from most of the fibroblast population by adsorption to carrier beads would exhibit a circadian rhythm in melatonin formation (Robertson and Takahashi, 1988). However, the daily production of melatonin by these cultured pinealocytes (Robertson and Takahashi, 1988) appears to be less than 10% of that of a comparable number of pinealocytes in the intact cultured pineal (Takahashi et al., 1980).

We therefore currently accept the last of the four possibilities as our working hypothesis. There is a considerable body of evidence compatible with the hypothesis that at least a major portion of the "typical" biological clock or pacemaker is membranous. It seems possible that other components of the clock could be coupled to such a "core" portion through different membrane components and/or specific receptors.

It seems very probable that the clock controlling the rhythm in NAT activity will be located through fractionation of suspensions of dispersed pineal cells. However, it seems unlikely that the clock controlling the rhythm in DNA incorporation studied here can be located by the same type of approach without major modification. Dispersal of the pineal cells in a medium rich in serum will remove the constraints upon normal growth of all fibroblasts. The replication of DNA during this growth will probably totally mask any synthesis due to the cycle we have seen with intact pineal glands. We therefore look to other approaches for appropriate clues.

ACKNOWLEDGEMENTS

Work in my laboratory was supported by grants from the Medical Research Council of Canada and the Natural Sciences and Engineering Research Council of Canada. I am indebted to my wife, Dr. L.K. Wainwright, for her participation in these studies and her assistance in the preparation of this article. I am also indebted to Miss S. Dorey for valuable technical assistance. I am a Career Investigator of the Medical Research Council of Canada.

REFERENCES

Binkley, S., MacBride, S.E., Klein, D.C., and Ralph, C.L., 1975, Regulation of pineal rhythms in chickens: refractory period and nonvisual light perception. Endocrinology, 96:848-853.
Binkley, S.A., Riebman, J.B., and Reilly, K.B., 1978a, The pineal gland: a biological clock in vitro, Science, 202:1198-1201.
Binkley, S., Riebman, J.B., and Reilly, K.B., 1978b, Regulation of pineal rhythms in chickens: inhibition of dark-time N-acetyltransferase activity, Comp. Biochem. Physiol. C., 59:165-171.
Cassone, V.M., and Menaker, M., 1983, Sympathetic regulation of chicken pineal rhythms, Brain Res., 272:311-317.
Cassone, V.M., and Menaker, M., 1984, Is the avian circadian system a neuroendocrine loop? J. Exp. Zool., 232:539-549.
Deguchi, T., 1979a, Circadian rhythm of serotonin N-acetyltransferase activity in organ culture of chicken pineal gland, Science, 203:1245-1247.

Deguchi, T., 1979b, A circadian oscillator in cultured cells of chicken pineal gland, Nature, 282:94-96.

Deguchi, T., 1979c, Role of adenosine 3',5'-monophosphate in the regulation of circadian oscillation of serotonin N-acetyltransferase activity in cultured chicken pineal gland, J. Neurochem., 33:45-51.

Kasal, C.A., Menaker, M., and Perez-Polo, J.R., 1979, Circadian clock in culture: N-acetyltransferase activity of chick pineal glands oscillates in vitro, Science, 203:656-658.

Klein, D.C., Auerbach, D.A., Namboodiri, M.A.A., and Wheler, G.H.T., 1981, Indole metabolism in the mammalian pineal gland, in: "The Pineal Gland", Vol. 1, Reiter, R.J., ed., CRC Press, Boca Raton, pp. 199-227.

Pratt, B.L., and Takahashi, J.S., 1987, Alpha-2-adrenergic regulation of melatonin release in chick pineal cell cultures, J. Neurosci., 7:3665-3674.

Ralph, C.L., Binkley, S., MacBride, S.E., and Klein, D.C., 1975, Regulation of pineal rhythms in chickens: effects of binding, constant light, constant darkness, and superior cervical ganglionectomy, Endocrinology, 97:1373-1378.

Robertson, L.M., and Takahashi, J.S., 1988, Circadian clock in cell culture: I. Oscillation of melatonin release from dissociated chick pineal cells in flow-through microcarrier culture, J. Neurosci., 8:12-21.

Takahashi, J.S., Hamm, H., and Menaker, M., 1980, Circadian rhythms of melatonin release from individual superfused chick pineal glands in vitro, Proc. Natl. Acad. Sci. USA, 77:2319-2322.

Takahashi, J.S., and Menaker, M., 1984, Multiple redundant circadian oscillators within the isolated avian pineal gland, J. Comp. Physiol., 154A: 435-440.

Voisin, P., and Collin, J.-P., 1986, Regulation of chicken arylalkylamine N-acetyltransferase by postsynaptic α_2-adrenergic receptors, Life Sci., 39:2025-2032.

Voisin, P., Martin, C., and Collin, J.-P., 1987, α_2-Adrenergic regulation of arylalkylamine N-acetyltransferase in organ-cultured chick pineal gland: characterization with agonists and modulation of experimentally stimulated enzyme activity, J. Neurochem., 49:1421-1426.

Voisin, P., Namboodiri, M.A.A., and Klein, D.C., 1984, Arylamine N-acetyltransferase and arylalkylamine N-acetyltransferase in the mammalian pineal gland, J. Biol. Chem., 259:10913-10918.

Wainwright, S.D., and Wainwright, L.K., 1979, Chick pineal serotonin acetyltransferase: a diurnal cycle maintained in vitro and its regulation by light, Can. J. Biochem., 57:700-709.

Wainwright, S.D., and Wainwright, L.K., 1984, Phase-shifting of cycles in level of serotonin N-acetyltransferase activity and cyclic GMP content of cultured chick pineal glands by methotrexate, J. Neurochem., 43:364-370.

Wainwright, S.D., and Wainwright, L.K., 1986, Effects of some inhibitors of DNA synthesis and repair upon the cycle of serotonin N-acetyltransferase activity in cultured chick pineal glands, Biochem. Cell Biol., 64:344-355.

Wainwright, S.D., and Wainwright, L.K., 1989a, On the uptake and incorporation of thymidine by cultured chick pineal glands, J. Pineal Res., 6:169-178.

Wainwright, S.D., and Wainwright, L.K., 1989b, Rhythmic incorporation of thymidine by chick pineal glands in vitro, J. Pineal Res., 7:253-264.

Wainwright, S.D., and Wainwright, L.K., 1989c, A circadian rhythm in thymidine incorporation by the chick pineal in culture? Adv. Pineal Res., 3: 61-66.

Wainwright, S.D., and Wainwright, L.K., 1990, Thymidine incorporation by cultured chick pineal glands is not subject to adrenergic regulation, Biochem. Cell Biol., 68:145-147.

Wainwright, S.D., Wainwright, L.K., and Lucarotti, C.J., 1989, On thymidine incorporation by the cultured chick pineal gland, J. Pineal Res., 6:179-197.

Zatz, M., and Mullen, D.A., 1988, Norepinephrine, acting via adenylate cyclase, inhibits melatonin output but does not phase-shift the pacemaker in cultured chick pineal cells, Brain Res., 450:137-143.
Zatz, M., Mullen, D.A., and Moskal, J.R., 1988, Photoendocrine transduction in cultured chick pineal cells: effects of light, dark, and potassium on the melatonin rhythm, Brain Res., 438:199-215.

REGULATION OF MELATONIN SYNTHESIS IN THE OVINE PINEAL GLAND

M.A.A. Namboodiri, H. M. Valivullah, and J.R. Moffett

Department of Biology
Georgetown University
Washington, District of Columbia 20057
USA

INTRODUCTION

Based on studies in the rat, it is now generally believed that pineal serotonin N-acetyltransferase (NAT) activity is the major factor which regulates melatonin synthesis in the pineal gland and its concentration in the circulation on a circadian basis (Axelrod and Zatz, 1977; Klein et al. 1981). NAT activity in the rat increases 50-100-fold at night, causing increased synthesis of melatonin in the pineal gland and about a 10-fold increase in melatonin in the circulation (Reppert and Klein, 1980). The duration of increased melatonin in the circulation is proposed to act as the chemical signal that conveys information about the length of the night to the body. Norepinephrine released from the sympathetic nerve endings in the pineal gland causes the large increase in the NAT activity via a cAMP-dependent mechanism, with alpha and beta adrenergic receptors acting in concert (Klein, 1978).

Recent studies suggest that the above model, based on the rat, may not be applicable to other species. First, the large nocturnal increase in NAT activity characteristic of the rat is not observed in sheep and syrian hamster, two species in which the seasonal control of reproduction has been clearly demonstrated (Vanecek and Illnerova, 1982; Namboodiri et al., 1985b; Rudeen et al., 1985). Second, the nocturnal increase in sheep is extremely rapid, which may enable the system to reflect the length of the night more accurately than in the rat (Rollag and Niswender, 1976). These observations prompted us to examine other species to understand other potential mechanisms of regulation of pineal melatonin synthesis. We have selected sheep for our studies because the role of melatonin in the seasonal control of reproduction in sheep is clearly established (Bittman et al., 1983).

RESULTS

Rapid increase in melatonin synthesis in the ovine pineal gland at night

During the night, melatonin levels in the blood increase 5-10-fold with-

---

Abbreviations used: 5-HIAA: 5-hydroxyindoleacetic acid; 5-HT: 5-hydroxytryptamine; 5-HTOH: 5-hydroxytryptophol; 5-HTP: 5-hydroxytryptophan; 5-MIAA: 5-methoxyindoleacetic acid; 5-MTOH: 5-methoxytryptophol; NAS: N-acetylserotonin.

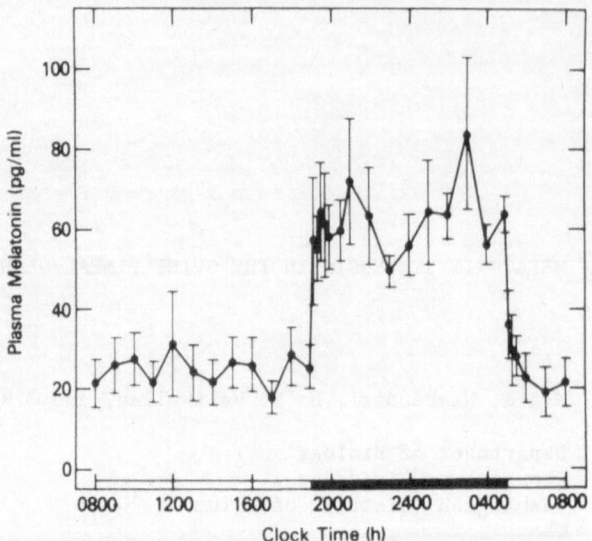

Fig. 1. Circadian rhythm in ovine plasma melatonin. The melatonin concentration in each plasma sample was obtained via an indwelling catheter and was determined in duplicates. Each point represents the
mean ± SEM of five sheep. The black bar represents the dark period
(from Sugden et al., 1985b).

in 30 min after light removal, and the increased level is maintained throughout the night (Fig. 1). We found the synthesis of pineal melatonin also to
be increased under this condition, as reflected in the increased activity of
NAT and increased levels of N-acetylserotonin (NAS) and melatonin in the pineal gland, measured 30 min after the onset of darkness (Table 1). No significant change in hydroxyindole-O-methyltransferase (HIOMT) activity was detected under this condition. Also, there were no detectable changes in serotonin or its oxidation products in the pineal gland (Table 2). These results
show that serotonin N-acetylation increases rapidly in the pineal gland at
night, leading to increased production of melatonin. This is consistent with
the proposal that increased production of melatonin in the pineal gland is
responsible for the rapid increase of circulating melatonin in sheep.

Table 1. Rapid increase in pineal melatonin synthesis at night

	NAT (nmoles/min/ mg protein)	NAS (ng/mg wet weight)	HIOMT (nmoles/min/ mg protein)	Melatonin (ng/mg wet weight)
Day (17:30h)	0.14 ± 0.07	0.02 ± 0.01	0.05 ± 0.01	0.12 ± 0.03
Night (18:30h)	0.43 ± 0.09[a]	0.36 ± 0.11[a]	0.07 ± 0.01	1.14 ± 0.30[a]

All values are the mean ± SEM of four to seven sheep (from Sugden et
al., 1985b). [a]Significantly different from day value ($P < 0.05$).

Table 2. Serotonin and oxidation metabolites before and after the onset
of darkness

	5-HT	5-HIAA	5-HTOH	5-MIAA	5-MTOH
			(ng/mg wet weight)		
Day (17:30h)	3.2 ± 0.9	19.6 ± 1.8	0.24 ± 0.01	1.6 ± 0.2	0.07 ± 0.02
Night (18:30h)	2.1 ± 0.5	18.3 ± 3.8	0.31 ± 0.09	2.5 ± 1.0	0.06 ± 0.01

Pineal indoles were measured using HPLC-EC. All values are the mean ± SEM
of four to seven sheep (from Sugden et al., 1985b).

Effect of a light pulse on the synthesis and concentrations of melatonin

A 30 min pulse of light given 4 hours after the onset of darkness decreased blood melatonin to daytime levels. Re-exposure to darkness following the light pulse returned blood melatonin to nighttime levels within 30-60
min (Table 3). Pineal melatonin showed an identical pattern. However, the
responses of NAT and NAS were less clear. NAT activity decreased significantly in response to the light pulse, but failed to increase on reexposure
to darkness. The changes in NAS, although consistent with the NAT response,
failed to reach adequate levels of significance under these conditions.
These results further support the view that changes in pineal melatonin are

Table 3. Effect of a 30 min light pulse on melatonin synthesis and
concentrations

	NAT	NAS	NIOMT	Melatonin	
	(nmoles/min/ mg protein)	(ng/mg wet weight)	(nmol/min/ mg protein)	Pineal (ng/mg wet weight)	Blood (pg/ml)
Day	1.1 ± 0.1	0.04 ± 0.01	0.14 ± 0.04	0.2 ± 0.0	15 ± 1.8
Night	3.1 ± 0.3[a]	0.21 ± 0.11	0.11 ± 0.02	1.5 ± 0.3[a]	120 ± 21.8[a]
Night + 30 min light	2.1 ± 0.2[b]	0.15 ± 0.07	0.14 ± 0.04	0.3 ± 0.1[b]	18 ± 2.5[b]
Night + 30 min light + 60 min dark	2.3 ± 0.4	0.26 ± 0.15	0.10 ± 0.04	1.6 ± 0.4[c]	98 ± 7.9[c]

The activities of NAT and HIOMT were measured in duplicate by incubating
(30 min, 37°C) 20 μl of the pineal homogenate in a total volume of 100 μl.
Pineal NAS was measured by HPLC-EC and melatonin by a RIA (from Namboodiri
et al., 1985b). [a]Significantly different from 'Day' (P < 0.05); [b]Signifi-
cantly different from 'Night' (P < 0.05); [c]Significantly different from
'Night + Light' (P < 0.05).

Table 4. Reciprocal relationship between serotonin oxidation and N-acetylation production in the pineal
gland during day and night

	Sheep		Rat	
	Day	Night	Day	Night
	(ng/mg wet weight)			
Tryptophan	4.0	4.3	6.0	7.0
5-HTP	0.09	0.05	ND	ND
5-HT	2.3	0.2	80	10
5-HIAA	12.9	2.6	6	2.5
5-HTOH	0.20	0.13	0.3	0.08
5-MIAA	7.6	0.3	0.06	0.005
5-MTOH	0.23	0.03	0.03	0.01
NAS	0.04	0.21	0.1	5.3
Melatonin	0.19	1.21	0.2	2.8

'Day' represents 4-5 h after lights on and 'Night' represents
4-5 h after lights off (from Namboodiri et al., 1985b). ND:
not detectable.

responsible for the changes in melatonin in the circulation. However, the
change in NAT activity alone is not sufficient to explain the altered rate of
melatonin production in the pineal gland.

**Reciprocal regulation of serotonin oxidation and N-acetylation in the pineal
gland during day and night**

During analysis of serotonin metabolites in the ovine pineal gland, it
was also observed that the oxidative metabolites of serotonin were generally
high during the day and low at night. The metabolites derived from N-acetylation, NAS and melatonin, showed the opposite pattern, being low during

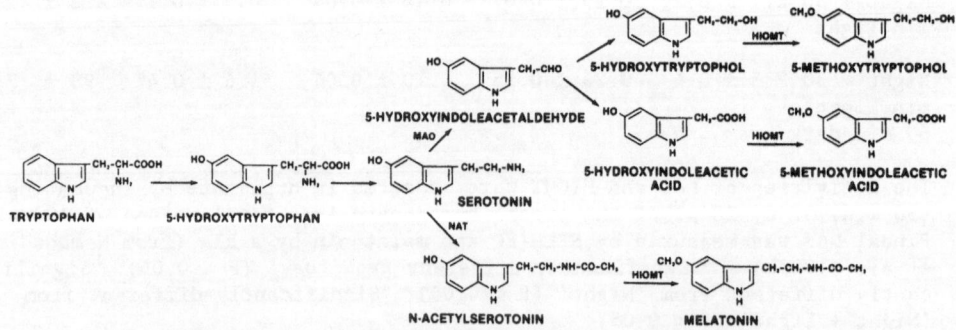

Fig. 2. Serotonin metabolism in the pineal gland.

Table 5. Effect of Prazosin (3 mg/animal) on melatonin synthesis in the ovine pineal gland

	NAT (nmol/min/ mg protein)	NAS (ng/mg wet weight)	HIOMT (nmol/min/ mg protein)	Melatonin (ng/mg wet weight)
Day	0.14 ± 0.07	0.02 ± 0.01	0.055 ± 0.006	0.12 ± 0.03
Night	0.43 ± 0.09[a]	0.36 ± 0.11[a]	0.073 ± 0.007	1.14 ± 0.30[a]
Night + Prazosin	0.33 ± 0.08	0.17 ± 0.04[a]	0.075 ± 0.011	0.48 ± 0.13[b]

Pineals from 'Day' (17:30h), 'Night' (18:30h) and 'Night + Prazosin' sheep were used for analysis. All values are the mean ± SEM of four to six sheep (from Sugden et al., 1985b). [a]Significantly different from the 'Day' group (P < 0.05); [b]Significantly different from the 'Night' group (P < 0.05).

the day and high at night. Such a reciprocal relationship between serotonin oxidation and N-acetylation products also exists in the rat pineal gland (Young and Anderson, 1982; Mefford et al., 1983; Table 4). No change in monoamine oxidase activity was detected between day and night, whereas NAT activity and serotonin N-acetylation increased at night. Therefore, it seems reasonable to propose that, in sheep as in the rat, increased serotonin N-acetylation at night is a major factor responsible for this reciprocal relationship. Reciprocal regulation of the two branches of the serotonin metabolic pathway by the activity of the first enzyme in one branch may constitute a novel regulatory mechanism in vertebrates (Fig. 2). Therefore, it will be important to determine if this mechanism is involved in serotonin metabolism elsewhere or in other metabolic pathways.

Table 6. Effect of cycloheximide (200 mg/sheep) on ovine pineal melatonin synthesis

	NAT (nmol/min/ mg protein)	NAS (ng/mg wet weight)	HIOMT (nmol/min/ mg protein)	Melatonin (ng/mg wet weight)
Day	0.14 ± 0.07	0.02 ± 0.01	0.055 ± 0.006	0.12 ± 0.03
Night	0.70 ± 0.09[a]	1.9 ± 0.53[a]	0.056 ± 0.012	1.61 ± 0.30[a]
Night + Cycloheximide	0.12 ± 0.04[b]	0.51 ± 0.10[b]	0.091 ± 0.013[b]	1.48 ± 0.32

Pineals from 'Day' (17:30h), 'Night' (18:30h) and 'Night + Cycloheximide' sheep were used for analysis. All values are the mean ± SEM of four to six sheep (from Namboodiri et al., 1985a). [a]Significantly different from the 'Day' group (P < 0.05); [b]Significantly different from the 'Night' group (P < 0.05).

141

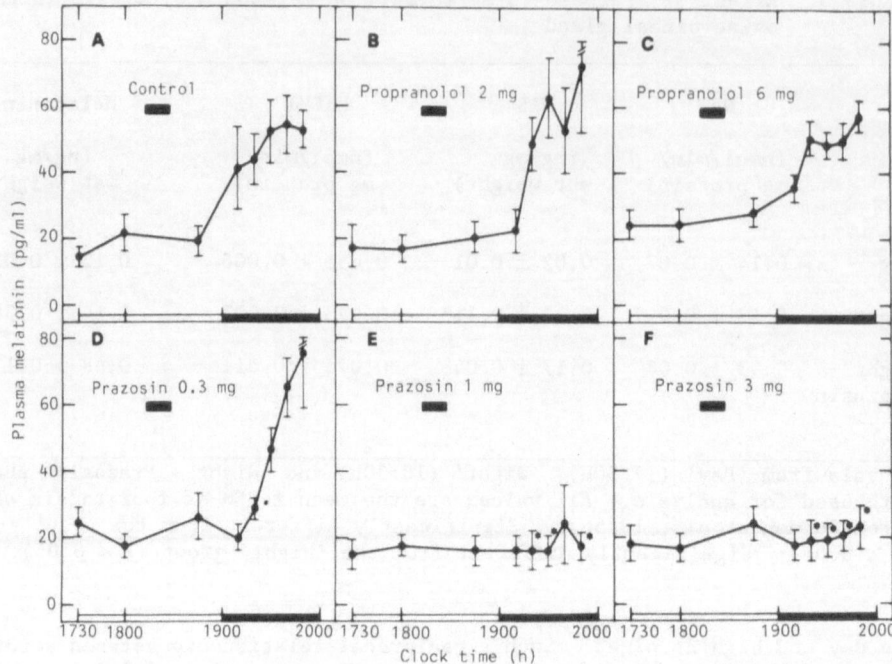

Fig. 3. Effect of prazosin and propranolol on the rise in melatonin after
dark onset. Each point represents the mean ± SEM melatonin concentration in plasma samples obtained via an indwelling catheter
during the final 90 min of light and the first 60 min of darkness.
The bar in each figure indicates the period during which the drugs
were injected (i.v.). A: N = 8; B, C and F: N = 5; D: N = 3;
E: N = 4. *Significantly different from control plasma melatonin
value at the same timepoint (P < 0.05) (from Sugden et al., 1985b).

It is not clear whether day/night changes in oxidative products of serotonin are reflected in their circulating levels, as in the case of melatonin.

Table 7. Effect of cycloheximide on pineal serotonin and metabolites

	5-HT	5-HIAA	5-HTOH	5-MIAA	5-MTOH
		(ng/mg wet weight)			
Day	3.2 ± 0.86	19.6 ± 1.8	0.29 ± 0.01	1.63 ± 0.21	0.07 ± 0.02
Night	2.1 ± 0.84	17.0 ± 1.7	0.42 ± 0.05	1.23 ± 0.68	0.03 ± 0.01
Night + Cycloheximide	3.2 ± 1.07	29.0 ± 5.7	0.65 ± 0.12	4.06 ± 1.30	0.11 ± 0.03[a]

Pineals from 'Day' (17:30h), 'Night' (18:30h) and 'Night + Cycloheximide'
sheep were used for analysis. All values are the mean ± SEM of four to six
sheep (from Namboodiri et al., 1985a). [a]Significantly different from 'Night'
group (P < 0.05).

Table 8. Effect of tryptophan administration on ovine
pineal indole concentrations

	Control	Treated
	(ng/mg tissue)	
Tryptophan	11.2 ± 3.8	64.7 ± 9.3*
5-HTP	0.09 ± 0.04	0.05 ± 0.04
5-HT	0.92 ± 0.25	0.79 ± 0.25
5-HIAA	8.7 ± 1.4	6.7 ± 1.2
5-HTOH	0.14 ± 0.02	0.17 ± 0.08
Indoleacetic acid	0.07 ± 0.03	1.66 ± 0.60*
Tryptophol	ND	0.04[a]
5-MIAA	1.77 ± 0.36	3.04 ± 0.78
5-MTOH	0.07 ± 0.001	0.05 ± 0.01
NAS	0.07 ± 0.02	0.08 ± 0.01
Melatonin	0.26 ± 0.05	0.28 ± 0.01

Sheep were dosed with vehicle or L-tryptophan (500 mg/kg,
i.p.) and killed 5 h later. Each value represents the mean
± SEM of each indole measured using HPLC-EC in three sheep
pineals (from Sugden et al., 1985a). [a]Not detectable in two
of three sheep; ND: not detectable; *P < 0.05 compared with
Controls.

Some of them, for example methoxytryptophol and tryptophol, are believed to
be biologically active molecules (Feldstein et al., 1970). Therefore, the
day/night changes of these metabolites in the pineal gland, and possibly in
the circulation, and their potential physiological significance are worthy of
further investigation.

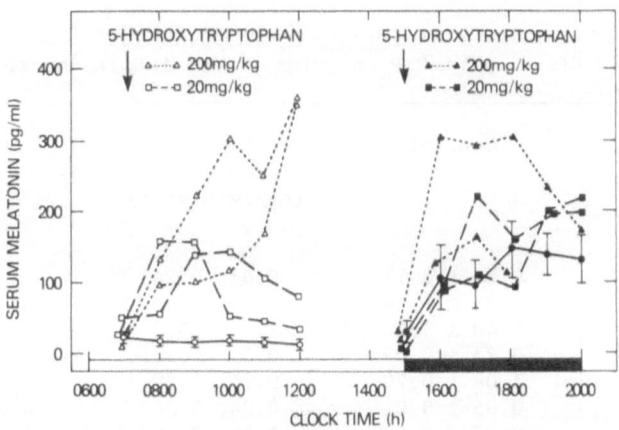

Fig. 4. Effect of 5-HTP administration on serum melatonin. Melatonin levels
were measured by RIA. The experiments were performed during November and December. Control values are the mean ± SEM of four animals. Experimental values are for individual animals (from Namboodiri et al., 1983).

Table 9. Effect of 5-HTP on ovine pineal indole
concentrations 2 h after treatment

	Control	Treated
	(ng/mg tissue)	
5-HTP	0.12 ± 0.01	11.5 ± 4.9
5-HT	4.1 ± 0.5	9.3 ± 2.0
5-HIAA	14.9 ± 0.6	37.2 ± 7.0
5-HTOH	0.15 ± 0.03	0.38 ± 0.12
5-MIAA	5.10 ± 1.00	6.49 ± 1.00
5-MTOH	0.12 ± 0.003	0.13 ± 0.02
NAS	0.07 ± 0.01	0.23 ± 0.04
Melatonin	0.27 ± 0.08	0.41 ± 0.08

Sheep were injected with 5-HTP (20 mg/kg, i.p.) or
saline at 8:00 a.m. and killed 2 h later. Pineal
indoles were measured by HPLC-EC. Values are the
mean ± SEM of three sheep (from Sugden et al.,
1985a).

Involvement of adrenergic receptors in ovine pineal melatonin synthesis

To further characterize the rapid increase in ovine pineal melatonin
synthesis, the involvement of adrenergic receptors was tested using specific
agonists and antagonists. Treatment with prazosin, an alpha-adrenergic antagonist, completely prevented the night time rise in pineal and blood melatonin (Fig. 3, Table 5). However, the night time increases in NAT activity
and NAS concentration in the pineal gland were not affected to any significant extent by the prazosin treatment (Table 5). Also, there were no significant changes in HIOMT activity, serotonin and its metabolites under this
condition. In contrast to the rat, even relatively high concentrations of
propranolol, a beta-adrenergic antagonist, failed to inhibit the nighttime
rise in melatonin in the circulation (Fig. 3).

Table 10. Effect of 5-HTP on ovine pineal indole concentrations

	Control	20 mg/kg	200 mg/kg
	(ng/mg tissue)		
5-HTP	0.20 ± 0.17	0.41, 0.24	4.9, 10.4
5-HT	6.73 ± 2.48	7.9, 6.4	61.1, 56.3
5-HIAA	18.40 ± 3.90	21.1, 30.1	180.6, 163.6
5-HTOH	0.23 ± 0.09	0.31, 0.33	2.61, 1.19
5-MIAA	1.08 ± 0.20	0.75, 2.39	5.62, 0.47
5-MTOH	0.05 ± 0.01	0.04, 0.06	0.08, 0.03
NAS	0.07 ± 0.02	0.23, 0.08	3.89, 8.50
Melatonin	0.25 ± 0.05	0.16, 0.20	1.22, 1.06

Sheep were injected i.p. with 5-HTP or saline and killed 5 h later.
Pineal glands were removed and indole concentrations determined by
HPLC-EC. Control values are the mean ± SEM of four sheep; 5-HTP
treated data are given as individual values (from Sugden et al.,
1985a).

We also attempted to stimulate pineal melatonin synthesis and blood melatonin by treatment with isoproterenol, a specific beta-adrenergic agonist, and with phenylephrine, a specific alpha-adrenergic agonist. On a few occasions, using both agonists separately, the rise in melatonin was rapid and about 10-fold, a pattern similar to that seen in vivo. However, animals often failed to respond, and the same animal failed to respond on different days. The reason for the erratic nature of this response remains unclear, so that no meaningful conclusions can be drawn from these experiments. Recently, Morgan et al. (1988) showed that melatonin production and release can be stimulated by treatment with norepinephrine in ovine pineal slices. In a superfusion system, norepinephrine stimulated melatonin release about 5-fold in 30 min, whereas in static cultures a similar increase in melatonin release was obtained in 3-4 hours. Also, dibutyryl cAMP as well as phorbol 12-myristate-13-acetate caused similar increases in melatonin release in static cultures. Further, conversion of radiolabelled tryptophan to melatonin was also stimulated by norepinephrine in static cultures. In unpublished studies, the same authors have obtained stimulatory effects on melatonin release by both isoproterenol and phenylephrine. In view of these in vivo and in vitro data, it seems highly likely that both alpha and beta-adrenergic receptors play significant roles in the regulation of melatonin production in the ovine pineal gland. The relative contribution of the two receptor systems and their respective mechanisms of action, however, need further investigation.

Is protein synthesis involved in the rapid nocturnal increase in ovine pineal melatonin synthesis?

Since the increase in melatonin synthesis at night in the ovine pineal gland was unusually rapid, it was of interest to test whether it involved protein synthesis. We found that treatment with cycloheximide, an inhibitor of protein synthesis, completely prevented the night time rise in NAT activity (Table 6). Consistent with this, NAS concentration also decreased significantly, although the inhibited level of NAS was still significantly higher than the day time level. Melatonin concentration in the pineal gland and in the blood did not decrease to a significant extent under this condition (Table 6). These results suggest that a mechanism independent of NAT may be involved in the nocturnal increase in melatonin synthesis.

Quite unexpectedly, a significant increase in HIOMT activity was detected on treatment with cycloheximide (Table 6). Consistent with this, significant increases in MIAA and MTOH were observed under this condition (Table 7). The nature of this increase in HIOMT activity remains unclear. However, a significant increase in HIOMT activity in the ovine pineal gland has not been observed previously. Further study of the cycloheximide effect will be valuable for understanding the regulatory mechanisms of HIOMT in the pineal gland and other tissues.

Serotonin availability in the regulation of melatonin synthesis

In search of the NAT-independent mechanism which may be involved in the regulation of ovine pineal melatonin synthesis, we tested the role of serotonin availability by administering the serotonin precursors tryptophan and 5-hydroxytryptophan. Tryptophan administration during the day increased pineal tryptophan levels about 6-fold, but had no effect on pineal serotonin, N-acetylserotonin or melatonin (Table 8). 5-Hydroxytryptophan administration, on the other hand, increased serotonin, N-acetylserotonin and melatonin in the pineal gland in a dose-dependent manner (Table 9, Table 10). Blood melatonin also increased under this condition (Fig. 4). NAT and HIOMT activities, however, did not change, indicating that the increase in melatonin is due to the increased serotonin acetylation caused by the increase in serotonin concentration. These results show that NAT is not saturated even during the day when pineal serotonin levels are higher than at night. Whether this mechanism is also involved in the rapid nocturnal increase in pineal melatonin synthesis can not be ascertained by the present results. However, the present observation that 5-HTP administration can increase melatonin levels in the circulation may have clinical implications because 5-HTP is used to treat neurological disorders such as myoclonus and depression which are suspected to involve serotonin deficiency in the central nervous system (Magnussen et al., 1978; Van Praag, 1981). The present results stress the need to consider the effect of 5-HTP on melatonin production in the interpretation of these clinical results.

SUMMARY AND CONCLUSIONS

The results presented here show clearly that the rapid nocturnal increase in circulating melatonin in sheep is associated with an equally rapid increase in melatonin production in the pineal gland. Pineal NAT activity and NAS concentration also increased under this condition, indicating that NAT activity is an important factor in this process. According to the current model for the regulation of pineal melatonin production in the rat, the large nocturnal increase in NAT activity is the major factor responsible for the daily rhythm in melatonin, as well as for the opposite rhythm in serotonin and its oxidation products in the pineal gland. Increased NAT activity in the pineal gland at night channels serotonin toward melatonin production at an enhanced rate, thereby causing pineal serotonin levels to drop. This, in turn, leads to decreased production of oxidative metabolites due to reduced substrate availability for MAO. Thus, the large increase in NAT activity acts as the key factor, directly or indirectly, for the generation of all rhythms in serotonin metabolites in the pineal gland, and possibly in the circulation. Such an exclusive role for NAT is doubtful in the sheep pineal gland for the following reasons: First, the nocturnal increase in NAT activity is relatively small (3-5-fold) compared to the rat (50-100-fold). Second, in at least two instances - a 30 min light pulse and prazosin treatment - there was a clear dissociation between melatonin production and NAT activity. This lack of correlation between NAT activity and melatonin production does not seem to be due to serotonin availability since serotonin levels did not change under the above conditions. Further, serotonin levels were found to decrease rather than increase when melatonin levels increased at night. Finally, our observation that melatonin production can be increased during the day by increasing serotonin levels in the pineal gland may reflect only the incompletely saturated nature of pineal NAT and may have little relevance for the physiological regulation of pineal melatonin production.

Even though NAT activity may not be the only factor responsible for the rhythmic production of serotonin metabolites in sheep pineal gland, activation of the serotonin → melatonin pathway seems to be the primary metabolic response involved in this regulation, since the rhythms in the serotonin metabolites are similar in both rat and sheep. In the rat, the activation of the serotonin → melatonin pathway is brought about exclusively by the increase in NAT activity. Assuming that the NAT activity measured in broken cell preparation reflects the in vivo situation - there are no reasons to think that this is not the case - it seems reasonable to propose that HIOMT activity also plays some role in this activation process in sheep. In fact, most of our data are consistent with a role for HIOMT. However, our inability to demonstrate a significant increase in HIOMT activity at night strongly argues against this possibility. Yet, it is tempting to speculate that HIOMT activity is regulated by allosteric or covalent modifications that are too labile to be detected in broken cell preparations. In fact, there are indications that HIOMT activity may be regulated by these mechanisms. First, bovine HIOMT purified to homogeneity was found to contain S-adenosyl-L-homocysteine, and HIOMT activity was increased by incubation with adenosylhomocysteinase, suggesting that HIOMT activity may be regulated in vivo by this end product (Deguchi and Barchas, 1971; Kuwano and Takahashi, 1980). Second, protein thiol/disulfide exchange mechanisms have been shown to regulate HIOMT activity in broken cell preparations (Sugden and Klein, 1987). Our observation that prazosin treatment decreases melatonin levels without affecting pineal NAS also suggests a potential control of melatonin production at the level of methylation by HIOMT. Finally, tyrosine hydroxylase, which is slowly and adrenergically regulated in pineal gland and other neural tissues via protein synthesis (like HIOMT in the rat pineal gland), is now known to be regulated in a rapid manner by phosphorylation. This type of regulation of tyrosine hydroxylase remained undetected or neglected during the early phase of research on tyrosine hydroxylase (Zigmond et al., 1989).

Thus, the most likely NAT-independent regulation of melatonin synthesis in sheep seems to occur at the level of HIOMT. Such reliance on two independent mechanisms instead of one for efficiency and accuracy may be an evolutionary adaptation, and may therefore be characteristic of all species exhibiting a small day/night rhythm in NAT activity. Thus, it may turn out that the large increase in NAT activity characteristic of the rat may be an exception rather than the rule in higher mammals.

ACKNOWLEDGEMENT

We thank Drs. D. Sugden, D.C. Klein, and I.N. Mefford who participated in the work described here. Preparation of this review was supported by a grant from the National Institutes of Health (DK 37024) to M.A.A.N.

REFERENCES

Axelrod, J., and Zatz, M., 1977, The beta-adrenergic receptors and the regulation of circadian rhythms in the pineal gland, in: "Biochemical Actions of Hormones", Vol. 4, Litwack, G., ed., Academic Press, New York, pp. 249-268.

Bitman, E.I., Dempsey, R.J., and Karsch, F.J., 1983, Pineal melatonin secretion drives the reproductive response to daylength in ewe, Endocrinology, 113:2276-2283.

Deguchi, T., and Barchas, J., 1971, Inhibition of transmethylation of biogenic amines by S-adenosylhomocysteine: enhancement of transmethylation by adenosylhomocysteinase, J. Biol. Chem., 246:3175-3181.

Feldstein, A., Chang, F.H., and Kucharski, J.M., 1970, Tryptophol, 5-hydroxytryptophol and 5-methoxytryptophol induced sleep in mice, Life Sci., 9:323-329.

Klein, D.C., Auerbach, D., Namboodiri, M.A.A., and Wheler, W.H.T., 1981, Indole metabolism in the mammalian pineal gland, in: "Pineal Gland: Anatomy and Biochemistry", Vol. 1, Reiter, R.J., ed., CRC Press, Boca Raton, pp. 199-227.

Klein, D.C., 1978, The pineal gland: a model of neuroendocrine regulation, in: "The Hypothalamus", Reichlin, S., Baldessarini, R.J., and Martin, J.D., eds., Raven Press, New York, pp. 303-327.

Kuwano, R., and Takahashi, Y., 1980, S-Adenosyhomocysteine is bound to pineal hydroxyindole-O-methyltransferase, Life Sci., 27:1321-1326.

Magnussen, I., Dupont, E., Engbaek, E., and Fine Olivarius, B. de, 1978, Post-hypoxic intention myoclonus treated with 5-hydroxytryptophan and an extracerebral decarboxylase inhibitor, Acta Neurol. Scand., 57:289-295.

Mefford, I.N., Chang, P., Klein, D.C., Namboodiri, M.A.A., Sugden, C., and Barchas J.D., 1983, Reciprocal day/night relationship between serotonin oxidation and N-acetylation products in the rat pineal gland, Endocrinology, 113:1582-1586.

Morgan, P.J., Williams, L.M., Lawson, W., and Riddoch, G., 1988, Stimulation of melatonin synthesis in ovine pineals in vitro, J. Neurochem., 50:75-81.

Namboodiri, M.A.A., Sugden, D., Klein, D.C., Grady, R., and Mefford, I.N., 1985a, Rapid nocturnal increase in ovine pineal N-acetyltransferase activity and melatonin synthesis: effects of cycloheximide, J. Neurochem., 45:832-835.

Namboodiri, M.A.A., Sugden, D., Klein, D.C., and Mefford, I.N., 1983, 5-Hydroxytryptophan elevates serum melatonin, Science, 221:659-661.

Namboodiri, M.A.A., Sugden, D., Klein, D.C., Tamarkin, L., and Mefford, I.N., 1985b, Serum melatonin and pineal indoleamine metabolism in a species with a small day/night N-acetyltransferase rhythm, Comp. Biochem. Physiol., 80B:731-736.

Reppert, S.M., and Klein, D.C., 1980, Mammalian pineal gland: basic and clinical aspects, in: "The Endocrine Functions of the Brain", Motta, M., ed., Raven Press, New York, pp. 327-337.

Rollag, M.D., and Niswender, G.D., 1976, Radioimmunoassay of serum concentrations of melatonin in sheep exposed to different lighting regimens, Endocrinology, 98:482-489.

Rudeen, P.K., Reiter, R.J., and Vaughan, M.K., 1975, Pineal serotonin N-acetyltransferase activity in four mammalian species, Neurosci. Lett., 1:225-229.

Sugden, D., and Klein, D.C., 1987, Inactivation of rat pineal hydroxyindole-O-methyltransferase by disulfide containing compounds, J. Biol. Chem., 262:6489-6493.

Sugden, D., Namboodiri, M.A.A., Klein, D.C., Grady Jr., R.K., and Mefford, I.N., 1985a, Ovine indoles: effect of 5-hydroxytryptophan administration, J. Neurochem., 44:769-772.

Sugden, D., Namboodiri, M.A.A., Klein, D.C., Pierce, J.E., Grady Jr., R., and Mefford, I.N., 1985b, Ovine pineal α-adrenoceptors: characterization and evidence for a functional role in the regulation of serum melatonin, Endocrinology, 116:1960-1967.

Van Praag, H.M., 1981, Management of depression with serotonin precursors, Biol. Psych., 16:291-310.

Vanecek, J., and Illnerova, H., 1982, Effect of light at night on the pineal rhythm and N-acetyltransferase activity in the syrian hamster Mesocricetus auratus, Experientia, 38:513-514.

Young, S.N., and Anderson, G.M., 1982, Factors affecting melatonin, 5-hydroxytryptophol, 5-hydroxyindoleacetic acid, 5-hydroxytryptamine and tryptophan in rat pineal gland, Neuroendocrinology, 35:464-468.

Zigmond, R.E., Schwarzschild, M.A., and Rittenhouse, A.R., 1989, Acute regulation of tyrosine hydroxylase by nerve activity and by neurotransmitters by phosphorylation, Ann. Rev. Neurosci., 12:415-461.

MELATONIN SYNTHESIS: MULTIPLICITY OF REGULATION

R. J. Reiter
Department of Cellular and Structural Biology
The University of Texas Health Science Center at San Antonio
San Antonio, Texas 78284
USA

INTRODUCTION

N-acetyl-5-methoxytryptamine, commonly known as melatonin, was initially thought to be synthesized exclusively in the pineal gland (Ebadi, 1984). However, subsequent examination of a variety of tissues indicates melatonin is very likely produced in the retina (Gern et al., 1978), the Harderian glands (Bubenik et al., 1976), the extra-orbital lacrimal glands (Mhatre et al., 1988), portions of the gastrointestinal tract (Vakkuri et al., 1985), and certain blood cells (Finocchiaro et al., 1988). The regulation of melatonin synthesis likely differs at these sites; the current brief survey will consider melatonin production exclusively in the mammalian pineal gland. Much of what is known concerning pineal melatonin production has been derived from studies in two species, the rat and the Syrian hamster (*Mesocricetus auratus*). Thus, the current review will summarize the findings primarily from these species and the assumption will be made that the regulation of the conversion of serotonin to melatonin is similar in the pineal gland of other mammals.

THE MELATONIN SYNTHETIC PATHWAY

A number of recent reviews consider the metabolic conversion of tryptophan to melatonin in the pinealocyte (Lewy, 1983; Ebadi, 1984; Reiter, 1988a), the endocrine unit of the pineal gland. As a consequence, the synthetic steps will be only briefly considered here. After its uptake from the systemic circulation, tryptophan is initially hydroxylated with the product of this conversion, 5-hydroxytryptophan, being decarboxylated to serotonin. Serotonin has a number of metabolic fates in the pinealocyte (Fig. 1). The one of current concern is its conversion to melatonin.

The synthesis of melatonin is particularly high during the dark phase of the light:dark (LD) cycle. Initially, serotonin undergoes N-acetylation in the presence of the enzyme serotonin N-acetyltransferase (NAT) with the resultant formation of N-acetylserotonin (Weissbach et al., 1961). N-acetylserotonin, in the presence of the enzyme hydroxyindole-O-methyltransferase (HIOMT), is then converted to melatonin (Axelrod and Weissbach, 1960). Once melatonin is produced, it rapidly escapes the cell into the systemic circulation; as a result, blood levels of the indole are generally taken as a reliable index of pineal melatonin production.

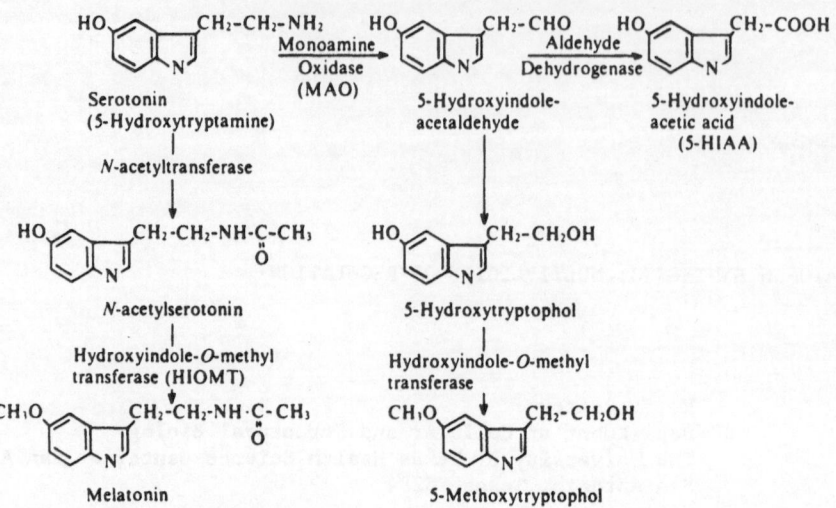

Fig. 1. Some of the metabolic fates of serotonin in the mammalian pineal gland. The product of one of these pathways, i.e. melatonin, is the chief secretory product of the gland.

The increased production of melatonin during the night, relative to the day, is determined by the release of norepinephrine (NE) (Zatz, 1981) from postganglionic sympathetic nerve endings which terminate in the gland and have their cell bodies in the superior cervical ganglia (Kappers, 1960). These ganglia are anatomically and functionally connected to the eyes by a series of neurons which include ganglion cells of the retina and parikarya in the suprachiasmatic nuclei (SCN) (Reiter, 1981). These latter nuclei are considered by some, but not all workers, to be the anatomical substrate for the master biological clock (Powell, 1988). Unquestionably, the SCN generate the nocturnal increase in pineal melatonin production (Moore and Klein, 1974).

The quantity of melatonin formed during darkness seems, in many cases, to be related to the activity of the acetylating enzyme (Ebadi, 1984). A nocturnal increase in NAT was initially described in the rat pineal gland (Klein and Weller, 1970) and, soon thereafter, this was observed in a variety of other rodents as well (Rudeen et al., 1985). The magnitude of the night-time rise in the activity of the acetylating enzyme varies from several to 100-fold depending on the species examined. Whereas the rhythm in pineal NAT is often considered to be universal feature of the mammalian pineal gland, it may be at least seasonally absent in some species (Reiter et al., 1987). When a nocturnal increase in NAT is absent, there also is no nighttime rise in pineal melatonin levels.

The final enzyme in the melatonin synthetic pathway, i.e., HIOMT, exhibits very little or no nighttime increase in activity (Ebadi, 1984; Reiter, 1988a). Presumably because of the high activity of this enzyme, however, any N-acetylserotonin formed is readily converted to melatonin which typically rises in the pineal gland at night (Reiter, 1988b). Because of its rapid release into the systemic circulation, melatonin concentrations in the blood follow a pattern like that in the pineal gland and rise at night (Reiter, 1986).

The pattern of nocturnal melatonin production and secretion by the pineal gland varies considerably among species (Reiter, 1988b). In some animals, high melatonin levels in the pineal gland are observed virtually throughout the dark period; conversely, at the other end of the spectrum, in some species high melatonin levels are restricted to a brief interval during the late dark phase. The pattern of the nocturnal melatonin rise seems to relate to the speed with which the intrapinealocyte synthetic machinery responds to NE released from the sympathetic nerve endings within the pineal gland (Gonzalez-Brito et al., 1988a).

POSTSYNAPTIC ADRENERGIC RECEPTORS

NE is a mixed β- and α-adrenergic receptor agonist, and is capable of interacting with α_1-, α_2-, β_1-, and β_2- receptors. Both α_1- and β_1-adrenoceptors are located on the pinealocyte membrane.

β_1-adrenergic receptors have been identified in the pineal gland of several species using a variety of ligands (Romero et al., 1975; Craft et al., 1985; Wilkinson and Wilkinson, 1985). The β-receptor density varies in a circadian manner in the pineal gland of the rat; however, reports differ in term of when peak receptor density occurs. Two basic circadian patterns of variation have been recorded. Initially, β-receptor density was reportedly greatest during the late light period with a marked decline in receptor number shortly after lights off (Romero et al., 1975; Wilkinson et al., 1987; Wirz-Justice, 1987); these observations were made when tritiated ligands for the β-receptor were employed and rats were under a LD cycle of 12:12. In this scheme, the minimal number of β-receptors on the rat pinealocyte occurred coincident with maximal melatonin production. Presumably, the drop in pineal β-receptor density is explained by the downregulation (densensitization, internalization?) of the receptors by endogenously released agonist, i.e. NE, at night.

Using iodinated β-receptor ligands, i.e. either iodocyanopindolol or pindolol, and studying binding over a 24 hour period in rats kept under LD 14:10, Reiter et al. (1985) and Gonzalez-Brito et al. (1988a) observed low β-receptor numbers during the day with an increase early in the dark phase and maximal receptor density at the time of maximal melatonin production, i.e. near the middle of the dark period. This rhythm in β-receptor density on the rat pinealocytes indicates an increased sensitivity of the rat pineal gland to NE stimulation at night. Indeed when subjected to test, the response of NAT and melatonin in the rat pineal gland to either physiologic or pharmacologic stimulation proved to be greatest during nighttime (Gonzalez-Brito et al., 1988b; Wu et al., 1988). Gonzalez-Brito and colleagues (1988b) feel that the circadian variation in β-receptor density in the rat pineal is determined primarily by NE released from the sympathetic nerve endings which terminate in the gland. This group has also shown that adrenalectomy, a procedure which removes a major source of circulating catecholamines, alters β-adrenoceptor density in the rat pineal gland in a predictable manner (Gonzalez-Brito et al., 1988d). This observation lends further support to the idea that the rhythms in the density of β-receptors on the rat pinealocyte are a function of released NE. Presumably, β-receptor density increases in the presence of low catecholaminergic activity while the stimulation of the receptors by locally released or circulating NE downregulates the receptors.

The difference in rat pineal β-adrenergic rhythms observed in the studies which used triated as opposed to iodinated ligands remains unexplained. The physiological significance of the rhythms in receptor density to pineal melatonin production depends on a constant uniform affinity of the receptors for the endogenous ligand, a condition that is not necessarily assured (Bevan et al., 1989).

The other species in which a time dependent variation in pinealocyte β-

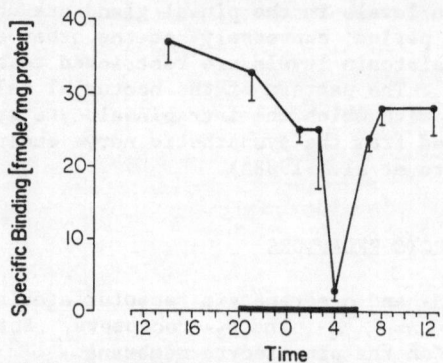

Fig. 2. Twenty-four hour rhythms in the binding of [^{125}I]PIN to crude membrane preparations of Syrian hamster pineal gland. The binding indicates the presence of β-adrenergic receptors. The period of darkness is indicated by the black bar on the time axis. From B. Pangerl et al. (1989).

adrenergic receptor density has been studied is the Syrian hamster (B. Pangerl et al., 1989). Using [^{125}I]-iodopindolol, these workers described a circadian variation in β-adrenoreceptors in the hamster pineal which also suggests that the melatonin rhythm in this species is determined by a change in receptor density. In the hamster, the specific binding of the ligand was high throughout the light phase and early dark phase. In the latter half of the night, however, specific binding decreased rapidly to a low point roughly 8 hours after darkness onset (Fig. 2). This drop presumably represented downregulation of the receptors by endogenously released NE since both light exposure at night (which prevents intrapineal NE release) or propranolol treatment (which prevents NE from acting on the receptors) overcame the late night reduction in specific binding of the ligand to the β-adrenoceptors (B. Pangerl et al., 1989). Since agonists normally downregulate β-adrenoceptors in cell membranes (Sibley and Lefkowitz, 1985), the 24 hour rhythms in receptor density in the pineal gland of the rat and Syrian hamster are most likely due to this phenomenon which is later followed by upregulation when the secretion of NE is discontinued.

The nighttime rise in pineal melatonin production in both the rat and Syrian hamster seems to be primarily mediated by β-adrenoceptors. On the other hand, α-receptor agonists amplify the rise in melatonin production that is induced by the administration of β-agonists (Klein et al., 1981; Santana et al., 1989). A search for α-receptors on pinealocyte membranes reveal they are present (Vanecek et al., 1985; A. Pangerl et al., 1989) and at least in the hamster, they exhibit a 24 hour rhythm (A. Pangerl et al., 1989). The ligand used to identify the α_1-receptors was [^{125}I]HEAT. In the hamster pineal, highest α-adrenoceptor density occurred during the day with lowest levels at about 24:00 h when lights went out at 20:00 h. It thus appears that the rhythms in β- and α-receptors in the hamster pineal are slightly out of phase since the nocturnal minima for these receptors occur at 04:00 h and 24:00 h, respectively. Whether these phase differences in receptor densities have physiological relevance remains unknown. Also, the importance of the 24 hour variation in α-adrenoceptor density on hamster pinealocyte production remains, for the most part, uninvestigated. It has been shown, however, that the administration of α-receptor antagonists to hamsters does little to interfere with the nocturnal rise in pineal melatonin production (Lipton et al., 1981), an effect not totally unexpected since only about 15% of the nocturnal rise in hamster pineal melatonin production seems to be a result of NE acting via the α_1-adrenoceptor (Nilsson and Reiter, 1989).

FACTORS INFLUENCING MELATONIN PRODUCTION

In general, the melatonin rhythm is relatively unperturbable. Indeed, its stability suggests that the rhythm plays an important role in the economy of the organism. Yet, there are some experimental perturbations which do alter the magnitude of the nocturnal melatonin rise.

Photoperiod

As already noted, the bulk of pineal melatonin formed during a given 24 hour period is synthesized during the night regardless of whether the animals exhibit a diurnal, nocturnal or crepuscular pattern of locomotor activity. Furthermore, it is known that the longer the daily dark period, (e.g., winter versus summer photoperiods), the longer the duration of elevated melatonin (Reiter, 1988b). Light exposure at unusual times, i.e., during the night, typically halts pineal melatonin production provided the light is of proper intensity and wavelength for the species being studied (Reiter, 1985). The sensitivity of the melatonin synthetic pathway to light varies markedly among species with the pineal gland of nocturnal animals usually being more sensitive than the pineal gland of diurnally active animals; the differential responses of the pineal gland to light intensity may be due to differences in the detector, i.e., the retinas, rather than in the pineal itself. The wavelengths of light that are most capable of suppressing pineal biosynthetic activity seem to be similar across mammalian species, with wavelengths in the 500-520 nm range (blue) being most suppressive (Reiter, 1985).

If mammals are acutely exposed to light at night, pineal NAT activity and melatonin levels drop precipitously (Rollag et al., 1980) with the half time in all cases being less than 10 min. Typically, within 20-30 min after light onset at night, basal daytime melatonin values are achieved. With the exception of the human, it seems that the melatonin supression response due to light exposure is an all-or-nothing phenomenon.

The drop in pineal melatonin production in light-exposed animals presumably is a consequence of the interruption of NE release from the intrapineal postganglionic sympathetic neurons. In the absence of this stimulation, the intracellular mechanisms that promote the conversion of serotonin to melatonin shut down, any residual melatonin escapes from the cells, and the concentration of the indole in the pinealocytes plummets. Although moonlight may be sufficiently bright to interrupt melatonin production in the pineal gland of some species, it is of the improper wavelength to do so (Brainard et al., 1984).

Stress/Exercise

The responses of the pineal gland to aversive stimuli and/or exercise seem to be very complex. It was originally surmised that stressful stimuli, which release catecholamines from a variety of sites, would not promote pineal melatonin production because the sympathetic nerve endings in the pineal gland took up excess circulating NE and thereby protected the pinealocytes from stimulation (Parfitt and Klein, 1976; Reiter, 1989)(Fig. 3). However, the uptake system may not be operative under certain conditions or it may be merely overwhelmed by large amounts of circulating catecholamines, since certain stressful encounters in rats are associated with very high levels of pineal NAT activity (Lynch et al., 1973). Indeed, there are many examples in the literature which show that experimental procedures that promote catecholamine release into the systemic circulation leads to an elevation of daytime melatonin synthesis in the pineal gland of rats (Reiter, 1989).

The application of stress/exercise during the nighttime, when pineal melatonin levels are already elevated, may have very different consequences

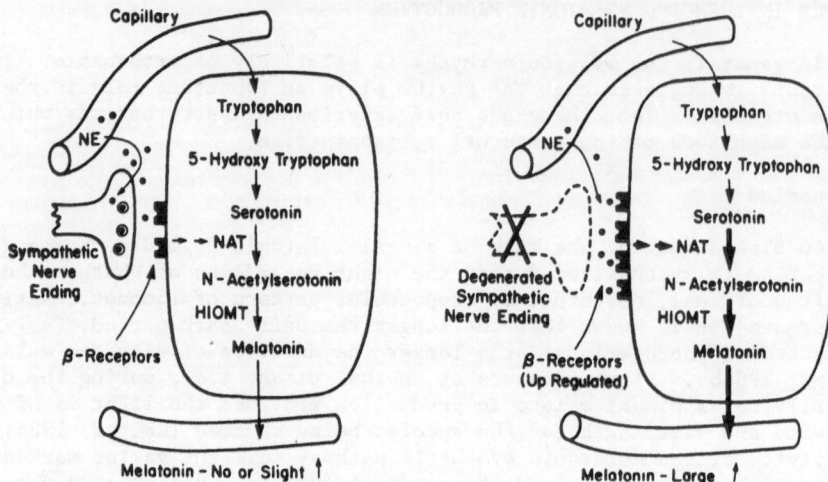

Fig. 3. Theoretical relationships between circulating norepinephrine (NE) and the β-receptors on the mammalian pinealocyte. Presumably, under usual conditions (left panel), excess circulating NE is taken up by the sympathetic nerve endings in the pineal; thus, melatonin production is not increased. This system, however, apparently is not operative in all cases since circulating NE sometimes promotes melatonin synthesis. The right panel illustrates the marked increase in melatonin production induced by circulating NE when the sympathetic nerve endings are absent. From Reiter (1989).

in terms of the levels of melatonin within the gland. The investigations of Troiani and colleagues (1988) show that forcing rats to swim during darkness is without effect on pineal NAT activity while it causes a precipitous drop in the melatonin content of the gland. Considering that the activity of the acetylating enzyme remains high, Troiani et al. (1988) assumed that melatonin production was not interrupted; rather pineal melatonin levels drop presumably because of the rapid discharge of the indole from the gland, a speculation supported by the observation that, as pineal melatonin levels plummet, circulating levels of the indole rise. These responses were not modified in adrenalectomized rats nor by any procedure which altered the neural information arriving at the pineal gland through the sympathetic nervous system. One speculation was that the forced exercise caused the release of atrial natriuretic factor from cardiac muscle which caused, by mechanisms that are not clear, a rapid efflux of pineal melatonin.

What is obvious from these studies is that either stress and/or exercise may, in fact, alter pineal melatonin production possibly by different mechanisms. It also is apparent that the responses of the pineal gland to these perturbations may depend on the functional status of the pineal gland at the time the stimulus is applied.

Hormones

Many endocrine glands are part of a closed loop system which allows feedback information to influence their secretory activity. Considering the marked influence of pineal hormonal products on reproductive physiology, it was suspected that gonadal steroids might influence the production of melatonin in the pineal gland. Investigations designed to examine this have generally revealed, however, that reproductively active hormones of pituitary origin or gonadally produced secretory products have a minor influence on the 24 hour melatonin rhythm. Indeed, gonadectomy in male rats causes no discernible change in the circadian rhythms of either pineal NAT activity or melatonin (Reiter et al., 1982). Most other endocrinectomies are equally inept in modifying the synthesis of melatonin within the pineal gland. One notable exception, however, is hypophysectomy, which significantly attenuates the nocturnal rises in the activity of the pineal serotonin acetylating enzyme as well as melatonin itself (Reiter et al., 1982). This reduction is generally believed to be a consequence of metabolic dysfunction produced by the loss of a variety of hormones rather than due to loss of a factor that specifically acts on the serotonin metabolism pathway.

Other factors

A variety of other factors may be instrumental in determining the quantity of melatonin produced during a 24 hour period (Reiter, 1989). There are, for example, certain strains of highly inbred mice which reportedly are devoid of a pineal melatonin rhythm because of a genetic absence of the two enzymes involved in its synthesis, i.e. NAT and HIOMT. If these melatonin deficient mice are crossbred with a strain that exhibits a normal circadian melatonin rhythm, the hybrids do contain measurable quantities of melatonin within their pineal glands.

Age plays a dominant role in the regulation of melatonin production. Certainly, in advanced age the amplitude of the nocturnal melatonin increase is much less than that in young animals, including humans (Reiter, 1986). The gradual reduction in melatonin production is speculated to be a consequence of either a reduction in the number of adrenergic receptors on the pinealocyte membrane or loss of sensitivity of the receptors associated with the aging process. In a few species, the pineal gland accumulates calcium deposits as aging progresses; whether these interfere with melatonin production is debatable.

FINAL COMMENT

The circadian melatonin rhythm is a very consistent feature in mammals suggesting that it has an important position in the economy of the organism. In general, the regular production of melatonin seems to ensure that an organism is daily and seasonally synchronized with its environment. In view of this, the stability of the melatonin cycle is essential and, thus, it must be carefully preserved and accurately regulated. Repeated measurements of the melatonin rhythm show that indeed it is very reproducible over time. A variety of factors influence the cyclic production of melatonin but, in general, the rhythm is rather unperturbable.

ACKNOWLEDGEMENTS

Work by the author was supported by grants from the National Science Foundation, the National Institutes of Health, and the Spanish-American Committee.

REFERENCES

Axelrod, J., and Weissbach, H., 1960, Enzymatic O-methylation of N-acetyl-serotonin to melatonin, **Science**, 131:1312.
Bevan, J.A., Bevan, R.D., and Shreeve, S.M., 1989, Variable receptor affinity hypothesis, **FASEB J.**, 3:1696-1704.

Brainard, G.C., Richardson, B.A., Hurlbut, E.C., Steinlechner, S., Matthews, S.A., and Reiter, R.J., 1984, The influence of various irradiances of artificial light, twilight and moonlight on the suppression of pineal melatonin content in the Syrian hamster, J. Pineal Res., 1:105-119.

Bubenik, G.A., Brown, G.M., and Grota, L.J., 1976, Immunohistochemical local-ization of melatonin in the rat Harderian gland, J. Histochem. Cytochem., 24:1173-1177.

Craft, C.M., Morgan, W.W., Jones, D.J., and Reiter, R.J., 1985, Hamster and rat pineal gland β-adrenoceptor characterization with iodocyanopindolol and the effect of decreased catecholamine synthesis on the receptor, J. Pineal Res., 2:51-66.

Ebadi, M., 1984, Regulation of the synthesis of melatonin and its signifi-cance to neuroendocrinology, in: "The Pineal Gland", Reiter, R.J., ed., Raven Press, New York, pp. 1-38.

Finocchiaro, L.M.E., Arzt, E.S., Fernandez-Castelo, S., Crescuolo, M., Finkielman, S. and Nahmod, V.E., 1988, Serotonin and melatonin synthesis in peripheral blood mononuclear cells: stimulation by interferon-γ as part of an immunomodulatory pathway, J. Interferon Res., 8:705-716.

Gern, W.A., Owens, D.W., and Ralph, C.J., 1978, The synthesis of melatonin in the trout retina, J. Exp. Zool., 206:263-270.

Gonzalez-Brito, A., Jones, D.J., Ademe, R.M., and Reiter, R.J., 1988a, Characterization and measurement of [^{125}I]-pindolol binding in individual rat pineal glands: existence of a 24 h rhythm in β-adrenergic receptor density, Brain Res., 438:108-114.

Gonzalez-Brito, A., Reiter, R.J., Menendez-Pelaez, A., Guerrero, J.M., Santana, C., and Jones, D.J., 1988b, Darkness-induced changes in nor-adrenergic input determine the 24 hour variation in beta-adrenergic recep-tor density in the rat pineal gland: in vivo physiological and pharmaco-logical evidence, Life Sci., 43:707-714.

Gonzalez-Brito, A., Reiter, R.J., Santana, C., Menendez-Pelaez, A., and Guerrero, J.M., 1988c, β-Adrenergic stimulation prior to darkness advances the nocturnal increase of Syrian hamster pineal melatonin synthesis, Brain Res., 475:393-396.

Gonzalez-Brito, A., Reiter, R.J., Tannenbaum, M.G., and Jones, D.J., 1988d, Adrenalectomy but not gonadectomy affects rat pineal β-adrenergic receptor density, Neurosci. Lett., 92:330-334.

Kappers, J.A., 1960, The development, topographical relations and innervation of the epiphysis cerebri in the albino rat, Z. Zellforsch., 52:163-215.

Klein, D.C., Sugden, D., and Weller, J.L., 1981, Postsynaptic α-adrenergic receptors potentiate the β-adrenergic stimulation of pineal serotonin N-acetyltransferase, Proc. Nat. Acad. Sci. USA, 80:599-603.

Klein, D.C., and Weller, J.L., 1970, Indole metabolism in the pineal gland: A circadian rhythm in N-acetyltransferase, Science, 169:1093-1095.

Lewry, A.J., 1983, Biochemistry and regulation of mamalian melatonin produc-tion, in: "The Pineal Gland", Relkin, R., ed., Elsevier, Amsterdam, pp. 77-128.

Lipton, J.S., Petterborg, L.J., and Reiter, R.J., 1981, Influence of propran-olol, phenoxybenzamine or phentolamine on the in vivo nocturnal rise of pineal melatonin levels in the Syrian hamster, Life Sci., 28:2377-2382.

Lynch, H.J., Eng, J.P., and Wurtman, R.J., 1973, Control of pineal indole biosynthesis by changes in sympathetic tone caused by factors other than environmental lighting, Proc. Nat. Acad. Sci. USA, 70:1704-1707.

Mhatre, M.C., van Jaarsveld, A.S., and Reiter, R.J., 1988, Melatonin in the lacrimal gland: first demonstration and experimental manipulation, Biochem. Biophys. Res. Commun., 153:1186-1192.

Moore, R.Y., and Klein, D.C., 1974, Visual pathways and the central neural control of a circadian rhythm in pineal serotonin N-acetyltransferase activity, Brain Res., 17-33.

Nilsson, K.J., and Reiter, R.J., 1989, In vivo stimulation of Syrian hamster pineal melatonin levels by isoproterenol plus phenylephrine is not accompanied by a commensurate large increase in N-acetyltransferase activity, **Neuroendocrin. Lett.**, in press.

Pangerl, A., Pangerl, B., Reiter, R.J., Vaughan, G.M., and Jones, D.J., 1989, Twenty-four hour variation of α_1-adrenergic receptors in the pineal gland of the male Syrian hamsters, **Brain Res.**, in press.

Pangerl, B., Pangerl, A., Reiter, R.J., and Jones, D.J., 1989, Circadian variation of β-adrenoceptor binding sites in the pineal gland of the Syrian hamster and prevention of the nocturnal reduction by light exposure or propranolol treatment, **Neuroendocrinology**, in press.

Parfitt, A.G., and Klein, D.C., 1976, Sympathetic nerve endings in the pineal gland protect against acute stress-induced increase in N-acetyltransferase activity, **Endocrinology**, 99:840-851.

Powell, E.W., 1988, The master clock illusion, **Chronobiologia**, 15:321-322.

Reiter, R.J., 1981, The mammalian pineal gland: Structure and function, **Amer. J. Anat.**, 162:287-313.

Reiter, R.J., 1985, Action spectra, dose response relationships, and temporal aspects of light's effects on the pineal gland, **Ann. N.Y. Acad. Sci.**, 453: 215-230.

Reiter, R.J., 1986, Normal patterns of melatonin levels in the pineal gland and body fluids of humans and experimental animals, J. Neural. Transm., (Suppl.), 21:35-54.

Reiter, R.J., 1988a, Neuroendocrinology of melatonin, in: "Melatonin - Clinical Perspectives", Miles, A., Philbrick, D.R.S., and Thompson, C., eds., Oxford University Press, Oxford, pp. 1-42.

Reiter, R.J., 1988b, Comparative aspects of pineal melatonin rhythms in mammals, in: "ISI Atlas of Science: Animal and Plant Sciences", Vol. 1, No. 2, Atkins, H., ed., ISI Press, Philadelphia, pp. 111-116.

Reiter, R.J., 1989, The pineal and its indole products: Basic aspects and clinical applications, in: "The Brain as an Endocrine Organ", Cohen, M.P., and Foa, P., eds., Springer, Vienna, pp. 96-149.

Reiter, R.J., Britt, J.H., and Armstrong, J.D., 1987, Absence of a nocturnal rise in either NE, NAT, HIOMT or melatonin in the pineal gland of the domestic pig kept under natural environmental conditions, **Neurosci. Lett.**, 81:171-176.

Reiter, R.J., Esquifino, A.I., Champney, T.H., Craft, C.M., and Vaughan, M.K., 1985, Pineal melatonin production in relation to sexual development in the male rat, in: "Paediatric Neuroendocrinology", Gupta, D., Borrelli, P., and Attanasio, A., eds., Croom and Helm, London, pp. 190-202.

Reiter, R.J., Trakulrungsi, W.K., Trakulrungsi, C., Vriend, J., Morgan, W.W., Vaughan, M.K., and Johnson, L.Y., 1982, Pineal melatonin production: endocrine and age effects, in: "Melatonin Rhythm Generating System", Klein, D.C., ed., Karger, Basel, pp. 143-154.

Rollag, M.D., Panke, E.S., Trakulrungsi, W., Trakulrungsi, C., and Reiter, R.J., 1980, Quantification of daily melatonin synthesis in the hamster pineal gland, **Endocrinology**, 106:232-236.

Romero, J.A., Zatz, M., Kebabian, J.W., and Axelrod, J., 1975, Circadian cycles in binding of ^{3}H-alprenolol to β-adrenergic receptor sites in rat pineal, **Nature**, 258:935-436.

Rudeen, D.K., Reiter, R.J., and Vaughan, M.K., 1975, Pineal serotonin N-acetyltransferase in four mammalian species, **Neurosci. Lett.**, 1:225-229.

Santana, C., Guerrero, J.M., Reiter, R.J., and Menendez-Pelaez, A., 1989, Role of postsynaptic α-adrenergic receptors in the β-adrenergic stimulation of melatonin production in the Syrian hamster pineal gland in organ culture, J. **Pineal Res.**, in press.

Sibley, D.R., Lefkowitz, R.J., 1985, Molecular mechanisms of receptor desensitization using the β-adrenergic receptor coupled to adenylate cyclase as a model, **Nature**, 317:124-129.

Troiani, M.E., Reiter, R.J., Tannenbaum, M.G., Puig-Domingo, M., Guerrero, J.M., and Menendez-Pelaez, A., 1988, Neither the pituitary gland nor the sympathetic nervous system is responsible for eliciting the large drop in elevated rat pineal melatonin levels due to swimming, J. Neural Transm., 74:149-160.

Vakkuri, O., Rintamäki, H., and Leppaluoto, J., 1985, Plasma and tissue concentrations of melatonin after midnight light exposure and pinealectomy in the pigeon, J. Endocrinol., 105:263-268.

Vanecek, J., Sugden, D., Weller, J., and Klein, D.C., 1985, Atypical synergistic α_1- and β-adrenergic regulation of adenosine 3-,5-monophosphate and guanosine 3',5'-monophosphate in rat pinealocytes, Endocrinology, 116:2167-2173.

Weissbach, H., Redfield, B.G., and Axelrod, J., 1961, The enzymatic acetylation of serotonin and other naturally occurring amines, Biochim. Biophys. Acta, 54:190-192.

Wilkinson, M., Joshi, M., Werstiuk, E.S., and Seggie, J., 1987, Lithium and rhythms of beta-adrenergic pineal gland and hypothalamus, Biol. Psych., 22:1191-1200.

Wilkinson, M., and Wilkinson, D.A., 1985, Beta-adrenergic ($[^3H]$CGP-12177) binding to brain slices and single intact pineal glands, Neurochem. Res., 10:829-839.

Wirz-Justice, A., 1987, Circadian rhythms in mammalian neurotransmitter receptors, Progr. Neurobiol., 29:219-259.

Wu, W., Chen, Y.C., and Reiter, R.J., 1988, Day-night differences in the response of the pineal gland to swimming stress, Proc. Soc. Exp. Biol. Med., 187:315-319.

Zatz, M., 1981, Pharmacology of the rat pineal gland, in: "The Pineal Gland, Vol. 1, Anatomy and Biochemistry", Reiter, R.J., ed., CRC Press, Boca Raton, pp. 225-242.

SESSION III

TRYPTOPHAN METABOLISM: KYNURENINES

THE ROLE OF TRYPTOPHAN AND KYNURENINE TRANSPORT IN THE CATABOLISM OF

TRYPTOPHAN THROUGH INDOLEAMINE 2,3-DIOXYGENASE

R.G. Knowles, N.A. Clarkson, C.I. Pogson, M.Salter, D.S. Duch[1], and M.P. Edelstein[1]

Biochemical Sciences
Wellcome Research Laboratories
Beckenham, Kent, BR3 3BS
UK

[1]Research Triangle Park, North Carolina 27709
USA

INTRODUCTION

    In 1967, indoleamine 2,3-dioxygenase (IDO) was first reported as being an enzyme capable of cleaving the indole ring of tryptophan (Scheme 1), which was distinct from the tryptophan 2,3-dioxygenase (TDO) found in the liver (Yamamoto and Hayaishi, 1967). Later studies revealed its presence in a range of tissues in mice and rats (Cook et al., 1980; Yoshida et al., 1980), and more recently in man (Yamazaki et al., 1985).

    IDO is markedly induced by interferonγ (IFNγ) and pathological conditions in which IFNγ production is stimulated (Yoshida et al., 1981a,b); this was the first hint that the physiological role of IDO might be connected with the biological effects of interferons. Between 1984 and 1986, Pfefferkorn and co-workers reported that the IFNγ growth suppression of Toxoplasma gondii growth in human fibroblasts is a consequence of the induction of IDO in these cells which then causes tryptophan depletion (Pfefferkorn, 1984; Pfefferkorn et al., 1986a,b). More recently, three separate research groups reported evidence that IDO induction and tryptophan depletion may also be responsible, at least in part, for the anti-tumor effects of IFNγ (de la Maza and Peterson, 1988; Ozaki et al., 1988; Takikawa et al., 1988). In vivo, the depletion of tryptophan following IDO induction could in principle occur at the intracellular, whole tumor or whole body level (Scheme 2). Brown et al. (1987) have shown a 63% decrease in the plasma tryptophan concentration in patients given a single dose of IFNγ, but it is not known whether more severe depletion of tryptophan occurs inside particular tissues or tumors or inside the cells in which IDO has been induced.

    We have carried out studies on tryptophan metabolism and transport and on the intracellular concentrations of tryptophan and kynurenine in cells in which IDO has been induced in order to elucidate the role of the plasma membrane transport of tryptophan and kynurenine in the anti-tumor effects of IFNγ.

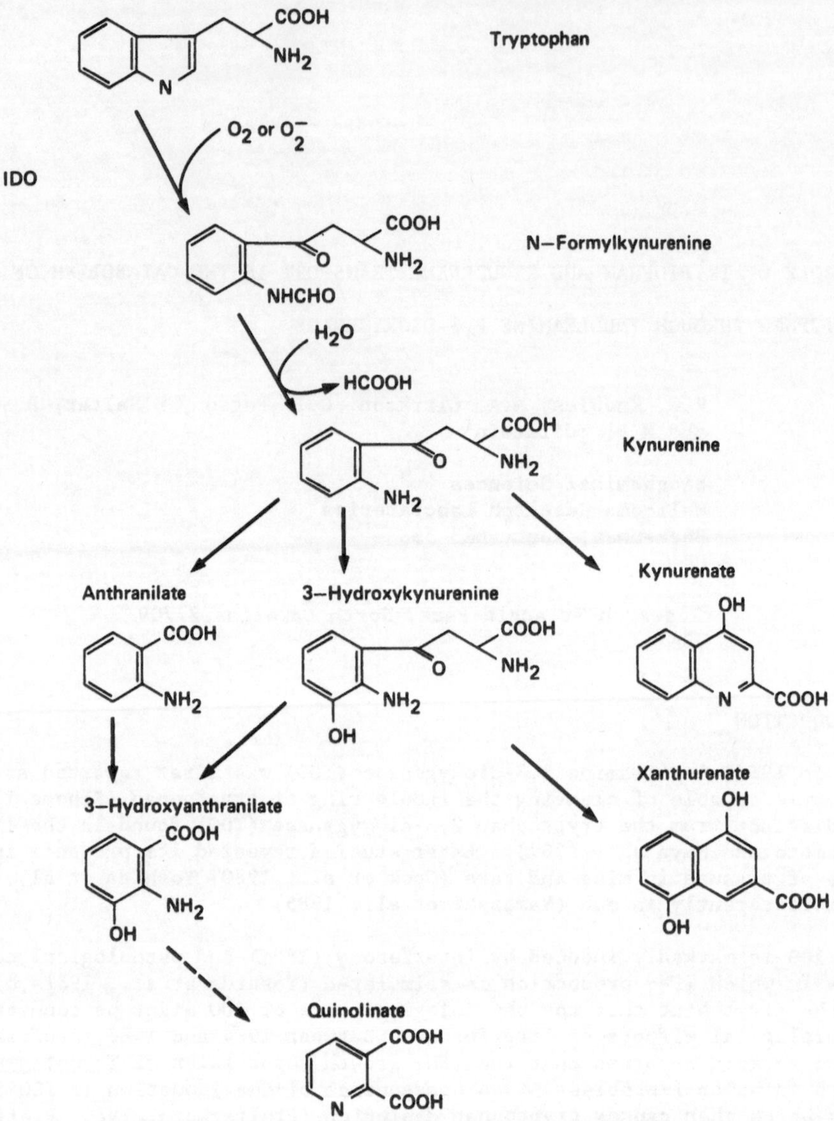

Scheme 1. The indoleamine 2,3-dioxygenase-initiated pathway of tryptophan metabolism.

RESULTS AND DISCUSSION

In studies of tryptophan catabolism through tryptophan dioxygenase in liver cells, we have observed a non-linear relationship between TDO activity and tryptophan catabolic flux: at high TDO activity we were able to show that the flux becomes limited by the rate of transport of tryptophan into the cells (Salter et al., 1986a,b). More recently we have found that when IDO is induced by exposure of MRC-5 cells to various concentrations of IFNγ, a non-linear relationship between IDO activity and tryptophan catabolism is also observed (Fig. 1). This suggests that induction of IDO by IFNγ can also result in tryptophan transport into cells becoming limiting (Knowles et al., 1989).

In order to test the hypothesis that a limitation of tryptophan cata-

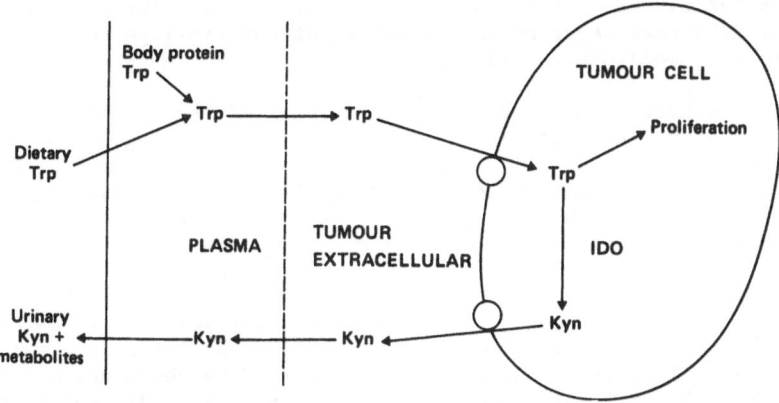

Scheme 2. Tryptophan metabolism to kynurenine by tumor cells in vivo.

bolism at the level of transport into cells could cause intracellular depletion of tryptophan, IDO was induced by various pretreatments of two human
tumor cell lines; the intracellular tryptophan and kynurenine concentrations
were then determined following a 30 min incubation in fresh medium containing
25 μM tryptophan (Fig. 2). The tryptophan concentration in WiDr cells was
greatly (> 95%) decreased by pretreatment with IFNγ, to a concentration not
differing significantly from zero. In THP-1 cells in contrast the intracellular tryptophan concentration was only moderately (< 50%) decreased by the
treatment giving the greatest induction of IDO (Fig. 2). Conversely, induction of IDO in WiDr cells resulted in a small but measurable concentration
of kynurenine in these cells whereas in THP-1 cells induction of IDO caused
the accumulation of very high intracellular concentrations of kynurenine
(Fig. 2, lower panels) despite the low extracellular concentration (< 5 μM).
We did not detect the further metabolites of kynurenine (anthranilate, 3-hy-
droxyanthranilate or 3-hydroxykynurenine) in either the WiDr or the THP-1
cells (Scheme 1). This is consistent with reports that kynurenine 3-hydroxy-
lase and kynureninase are present essentially only in liver and kidney, and

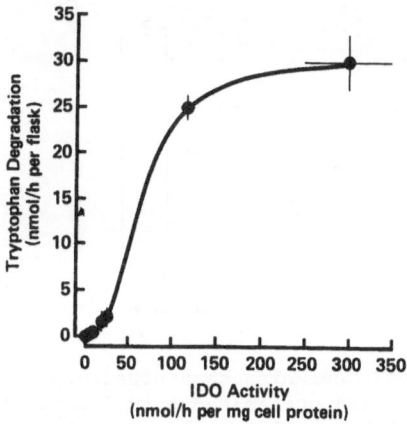

Fig. 1. Relationship between IDO activity and tryptophan degradation
in MRC-5 cells. The cells with pretreated for 2 days with 0-
4000 U/ml of IFNγ, and tryptophan degradation and IDO activity
were then determined on parallel flasks. The activity is
expressed as V_{max}, calculated from a K_m of 16 μM; the data are
shown as means ± SD.

not in other tissues containing IDO (Takikawa et al., 1986), although Werner et al. (1987) have reported the formation of anthranilate and 3-hydroxyanthranilate by some human cells in culture.

Because kynurenine concentrations above 1 mM cause inhibition of WiDr or THP-1 cell growth (results not shown), these results suggest that induction of IDO by IFNγ may inhibit cell proliferation by two distinct mechanisms: (a) intracellular tryptophan depletion and (b) intracellular kynurenine toxicity. Which of these two mechanisms predominates in any particular cell will depend on the relative activities of tryptophan entry, IDO and kynurenine efflux from the cell (Scheme 2).

We have studied tryptophan transport in WiDr adenocarcinoma cells. In these cells tryptophan transport is catalyzed by an L-type neutral amino acid transporter, since it is substantially inhibited by the model substrate 2-amino-bicyclo-[2.2.1]heptane-2-carboxylic acid (Fig. 3), and is Na^+-independent (data not shown). The transporter was found to concentrate tryptophan within the WiDr cells, and had a K_m for L-tryptophan of 35 μM (Fig. 4). Competition experiments showed that transport was highly stereospecific and that, of the neutral amino acids, only phenylalanine and tyrosine had a similar affinity to that of tryptophan. Leucine, isoleucine and valine all had K_i value for tryptophan transport more than 10 times greater than the K_m for tryptophan. L-kynurenine was a poor inhibitor of tryptophan transport, suggesting that this product of the IDO pathway may leave the WiDr tumor cells by a separate transporter.

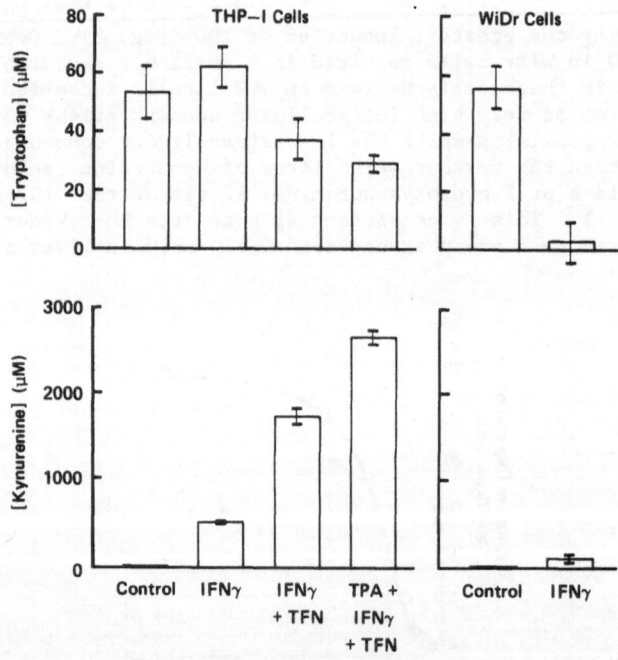

Fig. 2. Tryptophan and kynurenine concentrations in THP-1 and WiDr tumor cells. Cells were pretreated over 5 days: for 3 days in the absence or presence of 16 nM 12-0 tetradecanoylphorbol-13-acetate (TPA) and then for 2 days in the absence of presence of 1000 U/ml IFNγ or 10 ng/ml tumor necrosis factor (TNF). The cells were then harvested, suspended in fresh RPMI 640 medium and incubated for 30 min. Following centrifugation through silicone oil, intra- and extracellular tryptophan and kynurenine concentrations were determined by reverse-phase HPLC.

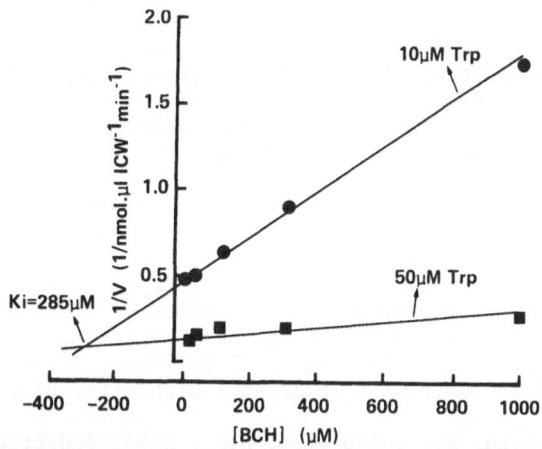

Fig. 3. Dixon plot of the inhibition of WiDr cell transport of L-tryptophan by BCH. WiDr cells were suspended in Earles balanced salts solution and preincubated for 10 min with 0-1000 μM 2-aminobicyclo-[2.2.1]-heptane-2-carboxylic acid (BCH) before addition of L-[¹⁴C]-tryptophan and incubation for 5s. Transport was terminated with 5 volumes of ice-cold papaverine (1 mM) and tryptophan (12.5 mM) in 0.9% NaCl, and the cells were centrifuged through silicone oil. Intra- and extracellular [¹⁴C]tryptophan radioactivity were then measured to determine the rate of tryptophan transport.

CONCLUSIONS

These results demonstrate that the transporters of both tryptophan and kynurenine appear to have important roles in the antiproliferative effects of IFNγ mediated by induction of IDO. We are looking for inhibitors of these transporters in order further to elucidate this antitumor mechanism.

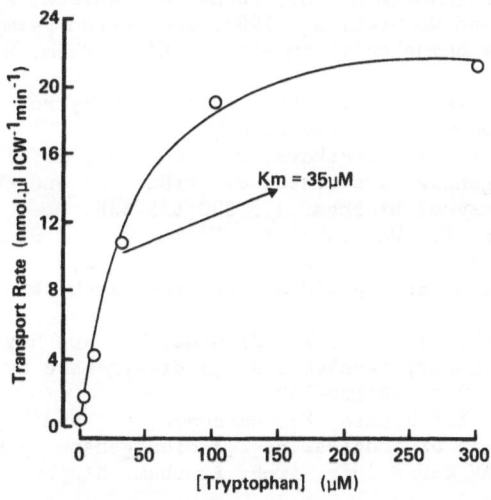

Fig. 4. Tryptophan concentration dependence of tryptophan transport into WiDr cells. Rates of tryptophan transport into WiDr cells were determined as described in Fig. 3, at 0-300 μM tryptophan.

REFERENCES

Brown, R.R., Borden, E.C., Sondel, P.M., and Lee, C.M., 1987, Effects of interferons and interleukin-2 on tryptophan metabolism in humans, in: "Progress in Tryptophan and Serotonin Research 1986", Bender, D.A., Joseph, M.H., Kochen, W., and Steinhart, H., eds., de Gruyter, Berlin, pp. 19-26.

Cook, J.S., Pogson, C.I., and Smith, S.A., 1980, Indoleamine 2,3-dioxygenase: a new, rapid, sensitive radiometric assay and its application to the study of the enzyme in rat tissues, Biochem. J., 189:461-466.

de la Maza, L.M., and Peterson, E.M., 1988, Dependence of the in vitro anti-proliferative activity of recombinant human γ-interferon on the concentration of tryptophan in culture medium, Cancer Res., 48:346-350.

Knowles, R.G., Salter, M., and Pogson, C.I., 1989, Tryptophan degradation and indoleamine 2,3-dioxygenase in MRC-5 fibroblasts, Biochem. Soc. Trans., 17:539-540.

Ozaki, U., Edelstein, M.P., and Duch, D.S., 1988, Induction of indoleamine 2,3-dioxygenase: a mechanism of the antitumor activity of interferonγ, Proc. Natl. Acad. Sci. USA, 85:1242-1246.

Pfefferkorn, E.R., 1984, Interferonγ blocks the growth of Toxoplasma gondii in human fibroblasts by inducing the host cells to degrade tryptophan, Proc. Natl. Acad. Sci. USA, 81:908-912.

Pfefferkorn, E.R., Eckel, M., and Rebhun, S., 1986a, Interferonγ suppresses the growth of Toxoplasma gondii in human fibroblasts through starvation for tryptophan, Mol. Biochem. Parasitol., 20:215-224.

Pfefferkorn, E.R., Rebhun, S., and Eckel, M., 1986b, Characterization of an indoleamine 2,3-dioxygenase induced by interferonγ in cultured fibroblasts, J. Interferon Res., 6:267:279.

Salter, M., Knowles, R.G., and Pogson, C.I., 1986a, Transport of the aromatic amino acids into isolated rat liver cells, Biochem. J., 233:499-506.

Salter, M., Knowles, R.G., and Pogson, C.I., 1986b, Quantification of the importance of individual steps in the control of aromatic amino acid metabolism, Biochem. J., 234:635-647.

Takikawa, O., Kuroiwa, T., Yamazaki, F., and Kido, R., 1988, Mechanism of interferonγ action, J. Biol. Chem., 263:2041-2048.

Takikawa, O., Yoshida, R., Kido, R., and Hayaishi, O., 1986, Tryptophan degradation in mice initiated by indoleamine 2,3-dioxygenase, J. Biol. Chem., 261:3648-3653.

Werner, E.R., Hirsch-Kauffmann, M., Fuchs, D., Hausen, A., Reibnegger, G., Schweiger, M., and Wachter, H., 1987, Interferon-gamma induced degradation of tryptophan by human cells in vitro, Biol. Chem. Hoppe-Seyler,368:1407-1412.

Yamamoto, S., and Hayaishi, O., 1967, Tryptophan pyrrolase of rabbit intestine, J. Biol. Chem., 242:5260-5266.

Yamazaki, F., Kuroiwa, T., Takikawa, O., and Kido, R., 1985, Human indolylamine 2,3-dioxygenase, its tissue distribution, and characterization of the placental enzyme, Biochem. J., 230:635-638.

Yoshida, R., Nukiwa, T., Watanabe, Y., Fujiwara, M., Hirata, F., and Hayaishi, O., 1980, Regulation of indoleamine 2,3-dioxygenase activity in the small intestine and epididymis of mice, Arch. Biochem. Biophys., 203: 343-351.

Yoshida, R., Imanishi, J., Oku, T., Kishida, T., and Hayaishi, O., 1981a, Induction of pulmonary indoleamine 2,3-dioxygenase by interferon, Proc. Natl. Acad. Sci. USA, 78:129-132.

Yoshida, R., Urade, Y., Nakata, K., Watanabe, Y., and Hayaishi, O., 1981b, Specific induction of indoleamine 2,3-dioxygenase by bacterial lipopolysaccharide in the mouse lung, Arch. Biochem. Biophys., 212:629-637.

KYNURENINASE AND KYNURENINE 3-HYDROXYLASE IN MAMMALIAN TISSUES

E. Okuno and R. Kido

Department of Biochemistry
Wakayama Medical College
Wakayama 640
Japan

INTRODUCTION

The major pathway of tryptophan catabolism in mammals is the kynurenine pathway. L-Tryptophan 2,3-dioxygenase, which is localized in liver and is induced by loading of tryptophan or treatment with glucocorticoids, initiates this pathway. Other organs, however, contain indoleamine 2,3-dioxygenase which is a different enzyme protein from tryptophan 2,3-dioxygenase but yields the same product, formylkynurenine, from tryptophan. Under normal conditions, formylkynurenine is mainly produced by liver tryptophan 2,3-dioxygenase. Kynurenine produced by formamidase is either hydroxylated to 3-hydroxykynurenine, hydrolytically cleaved to form anthranilic acid and alanine, or transaminated to produce kynurenic acid (Scheme 1).

Kynurenine 3-hydroxylase [L-kynurenine, NADPH:oxygen oxidoreductase (3-hydroxylating), EC 1.14.13.9] is a mixed function oxidase which catalyzes the 3-hydroxylation of L-kynurenine. This enzyme is localized in the outer mitochondrial membrane (Okamoto and Hayaishi, 1969). Because of increased ratios of kynurenine/hydroxykynurenine after the administration of cortisol or tryptophan, the hydroxylation of kynurenine is reported to be rate limitting _in vivo_ (Michael et al., 1964; Musajo and Benassi, 1964; Rose and

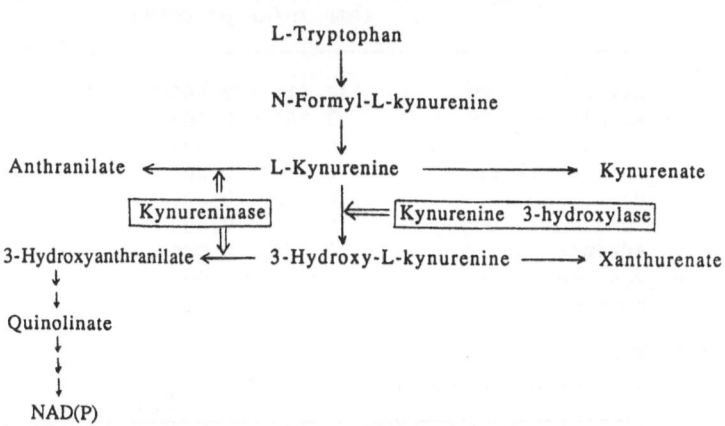

Scheme 1

McGinty, 1968; El-Zoghby et al., 1978; Bender and McCreanor, 1985). The enzyme kynureninase (L-kynurenine hydrolase, EC 3.7.1.3) (Takeuchi et al., 1980; Inada et al., 1984) catalyzes the hydrolysis of both L-kynurenine (yielding anthranilate) and L-3-hydroxykynurenine (yielding 3-hydroxyanthranilate) in mammals. The latter is an intermediate in the physiologically important pathway to nicotinamide nucleotides. Braunshtein et al. (1949) demonstrated a decrease in kynureninase activity in pyridoxine deficiency, showing that the enzyme requires pyridoxal 5'-phosphate as a coenzyme. In this paper, we describe the activities of kynureninase and kynurenine 3-hydroxylase in rat, human and suncus.

MATERIALS AND METHODS

Animals and human tissues

Male suncus [Suncus murinus (insectivora), 8 weeks old, 120-130 g body weight] were provided by our animal breeding room. They were maintained at about 20°C in a room with a 12 h light/12 h dark cycle. Food (MF laboratory chow, Oriental Yeast Co. Ltd., Tokyo, Japan) and water were available ad libitum.

Human tissues were obtained post mortem at autopsy from the Departments of Pathology and Legal Medicine at our Medical College.

Enzyme assay

Kynureninase activity was determined by measuring anthranilate fluorometrically in an HPLC system. The standard assay mixture contained 100 mM Tris-HCl buffer, pH 8.5, 20 μM pyridoxal 5'-phosphate, 2 mM L-kynurenine and enzyme solution in a final volume of 0.2 ml. Incubation was carried out at 37°C for 30 min and the reaction was stopped by adding 0.1 ml 1.0 M sodium acetate buffer, pH 4.0. In blank samples, kynurenine was added just before the addition of acetate buffer. For the standard, 500 pmol anthranilic acid was added to the assay mixture and treated identically. One ml ethyl acetate

Table 1. Organ distribution of kynureninase in rat

	Activity (nmol/h/mg protein)
Liver	12.811 ± 2.466
Kidney	1.369 ± 0.402
Spleen	0.519 ± 0.080
Intestinal muscle	0.035 ± 0.004
Heart	0.026 ± 0.009
Lung	0.020 ± 0.009
Adrenal	0.013 ± 0.006
Pancreas	0.008 ± 0.002
Muscle	0.004 ± 0.002
Brain	0.002 ± 0.001
Testis	ND
Intestinal mucosa	ND

Data are the mean ± SEM of five separate experiments. ND = Not detected.

Table 2. Organ distribution of kynureninase in human

	Activity (nmol/h/mg protein)	
	Case 1	Case 2
Liver	40.74	36.78
Spleen	5.04	3.06
Kidney	0.66	3.00
Lung	8.10	1.32
Adrenal	—	0.84
Ovary	—	0.78
Small intestine	1.26	—
Pancreas	0.90	0.12
Heart	0.54	0.12
Brain	0.30	—

Case 1: 51 old man who died of renal failure.
Case 2: 62 old woman who died of a heart attack.

was added and the newly formed anthranilic acid extracted after vigorous shaking (Vortex). The solution was the subjected to centrifugation (3,000 rpm, 5 min) and a 50 μl aliquot was injected into a Jasco Trirotor III HPLC system (Japan Spectro-scopic Co. Ltd., Tokyo, Japan) with a fluorometric detector. The detector was set at 314 nm for excitation and 420 nm for emission. A column (150 x 4.6 mm internal diameter) packed with TSK ODS gel (5 μm particle size) was used at a flow rate of 1.5 ml/min with 10 mM ammonium acetate solution containing 2% (v/v) methanol as the mobile phase. The column was kept at 40°C throughout.

Kynurenine 3-hydroxylase activity was in principle assayed by the method of Takikawa et al. (1986). The reaction mixture (100 μl) contained 100 mM Tris-HCl buffer, pH 8.5, 1% (w/v) Triton X-100, 5 mM FAD, 1 mM dithiothreitol, 1 mM kynurenine, 1 mM NADPH and tissue preparation. The reaction, at

Table 3. Organ distribution of kynureninase in suncus liver

	Kynureninase activity (nmol/h/mg protein)
Liver	1538 ± 126
Kidney	2.83 ± 0.46
Thymus	1.91 ± 0.17
Spleen	1.48 ± 0.14
Small intestine	0.67 ± 0.17
Heart	0.46 ± 0.14
Lung	0.32 ± 0.04
Whole brain	0.17 ± 0.03
Muscle	0.11 ± 0.02

Data are the mean ± SEM of four separate experiments.

37°C, was started by the addition of NADPH and was terminated after 30 min by addition of 20 μl 30% (w/v) trichloroacetic acid. For the blank, NADPH was added just before the addition of trichloroacetic acid. After centrifugation at 2,500 x g for 15 min, 3-hydroxykynurenine produced in the supernatant was analyzed by HPLC. The column (C_{18}; 5 x 150 mm) was eluted with 5% methanol and 95% water containing 5 mM ammonium acetate. 3-hydroxykynurenine was detected by UV absorbance at 340 nm.

Protein determination

Protein was measured by the method of Lowry et al. (1951) using bovine serum albumin as a standard.

RESULT AND DISCUSSION

Rat kynureninase

Male Sprague-Dawley rats (150-200 g) were killed by decapitation. The organs were quickly removed and homogenized (1:9, w/v) in 0.25 M sucrose using a Potter-Elvehjem homogenizer with a teflon pestle. The organ distribution of kynureninase was investigated. The highest activity was found in the liver, while the kidney and spleen contained activities of about 1/10 and 1/30 of that of liver, respectively. Some activity was also found in the pancreas, intestinal muscle, heart, lung, brain, adrenal and muscle (Table 1). Kynureninase in liver, kidney and spleen are likely to be the same protein as judged by physical and enzymic properties (Kawai et al., 1988). The molecular weight of rat kynureninase was estimated to be 95,000 and consisted of two identical subunits. The Michaelis constants were determined to be 240 μM for kynurenine and 13 μM for 3-hydroxykynurenine. The maximum velocity for 3-hydroxykynurenine was 11 times higher than that for kynurenine at pH 7.7 (Takeuchi et al., 1980).

The changes in kynureninase activity during development are illustrated in Fig. 1. During suckling, kynureninase activity in the liver was 5-6 nmol/h/mg protein. After weaning, the activity increased quickly and reached a peak (13 nmol/h/mg protein) on the 30th day. In the kidney, enzyme activity slowly increased after birth and reached a steady level (1.2 nmol/h/mg) by day 30. Enzyme activity in the spleen showed a similar ontogenetic pattern as the liver enzyme. Notably, the period between post-natal days 30 and 60 coincides with the period in which young rats have an increased requirement for NAD and with the most active feeding behavior. Liver, kidney and spleen also contain high NAD synthetase activity (Shibata et al., 1986) with a pattern of organ distribution similar to that of kynureninase activity, suggesting that kynureninase is of essence in NAD synthesis.

Human kynureninase

Kynureninase in human liver was described by Inada et al. (1984). The enzyme has a molecular weight of 130,000 and an isoelectric point of pH 5.9. The K_m values of the enzyme were 77 μM for 3-hydroxykynurenine and 1.0 mM for kynurenine. The enzyme was more active for 3-hydroxykynurenine than kynurenine, and its activity ratio was 15:1. These characteristics were similar to those of the rat enzyme. The organ distribution of kynureninase is shown in Table 2. The highest activity was found in the liver, and this activity activity is about 3 times higher than that of rat liver. High activity was also detected in spleen, kidney and lung. The low enzyme activity in the kidney of case 1 might be due to renal failure.

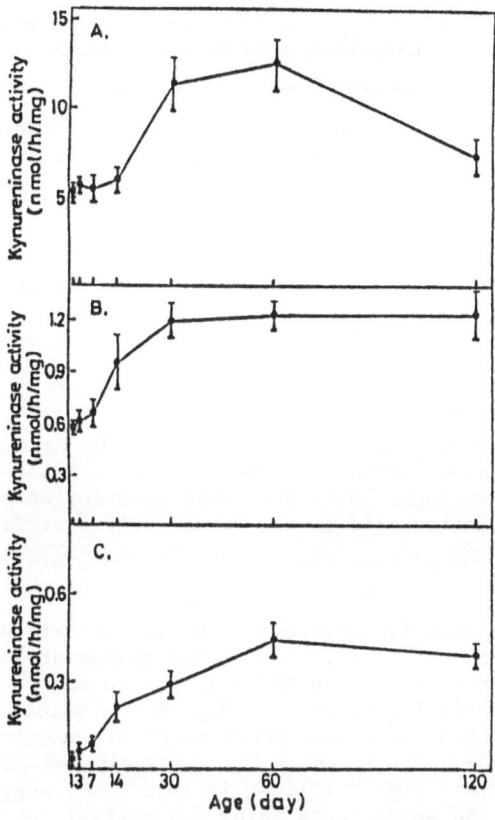

Fig. 1.   Developmental profile of kynureninase from rat liver (A), kidney (B) and spleen (C).   Data are the mean ± SEM of four separate experiments.

Suncus kynureninase

Suncus liver contains higher kynureninase activity than other mammalian livers, showing activity about 50 times that of pig liver (Tanizawa and Soda 1979), 100 times that of rat liver and 40 times that of human liver.   Other organs tested (kidney, thymus, spleen, intestine, heart, lung, brain and muscle) contained less than 1/300 of the liver activity (Table 3).   The molecular weight of the liver enzyme was estimated to be 110,000 and consisted of two identical subunits (Ishikawa et al., 1989). K_m values were 18 μM for 3-hydroxykynurenine and 250 μM for kynurenine.

The relationship between kynureninase activity and anthranilate levels in the serum is shown in Table 4.   Kinetic analysis of kynureninase (lower K_m and higher V_{max} values) shows that in the liver of this species, too, 3-hydroxykynurenine is a better substrate than kynurenine.   Moreover, Tanizawa and Soda have previously suggested that 3-hydroxykynurenine is probably the physiological substrate of kynureninase.   On the other hand, we found significant amounts of anthranilic acid in the serum which correlated with kynureninase (Table 4).

Rat kynurenine 3-hydroxylase

Rat kynurenine 3-hydroxylase has been purified by Nishimoto et al. (1975) and well characterized.   The molecular weight of the enzyme was est-

Table 4. Liver kynureninase activity and serum
anthranilic acid in rat, suncus and human

	Liver kynureninase (nmol/h/mg)	Serum anthranilic acid (μM)
Rat	12.8 ± 2.5[*]	0.25 ± 0.02
Suncus	1557 ± 416	1.96 ± 0.24
Human	38.9 ± 2.0	0.42 ± 0.03

[*]From Kawai et al. (1988). Data are the mean ± SEM
of 5 rats, 4 suncus and 2 humans. Five healthy male
volunteers (age: 35-41 yrs) were used for measurement
of anthranilic acid in the serum (different from liver
donors).

imated to be approximately 200,000 and the isoelectric point was determined
to be pH 5.4 (Nishimoto et al., 1975). The enzyme utilizes both NADPH and
NADH as an electron donor. When NADPH is the co-substrate, the K_m value for
kynurenine is slightly lower, and the V_{max} of the reaction slightly higher
than with NADH. The K_m value for NADPH is 25 μM, compared with 50 μM for
NADH (Shibata, 1978). The K_m value for kynurenine of partially purified
kynurenine hydroxylase from rat liver is 18 ± 6 μM, suggesting that under
normal conditions the enzyme acts below its maximal rate (Bender and
McCreanor, 1982).

The highest activities of kynurenine 3-hydroxylase were found in kidney
and liver, while spleen contained about 1/100 of the kidney activity. The
activities in brain, pancreas, heart, lung, intestine and adrenal gland were
less than 0.5 nmol/h/mg protein (Table 5). Developmental studies revealed
that the enzyme activity in liver reached a peak on the 14th day. On the
60th day, the tissue contained 50% of the highest activity. In kidney, the
enzyme activity increased after birth to the 120th day. One day old rats
showed very little enzyme activity in the spleen, but the activity gradually
increased after birth and reached a peak on the 60th day (Fig. 2).

Human kynurenine 3-hydroxylase

Human kidney and liver contained the highest kynurenine 3-hydroxylase
activity. Some activity was also detected In spleen and pancreas (Table 6).
This organ distribution is similar to that of the rat enzyme. The K_m values
for kynurenine were 125 μM with NADPH and 167 μM with NADH (Fig. 3). The
V_{max} of the reaction was slightly higher with NADPH. Compared to the rat
enzyme, activity was 1/5 and the K_m was about 7 times higher. In addition,
kynureninase activity in human liver was about 3 times higher than that of
rat liver. Taken together, these data suggest that in human liver kynurenine
is preferentially metabolized to anthranilic acid (Table 4).

Suncus kynurenine 3-hydroxylase

Kynurenine 3-hydroxylase activity was detected only in kidney and spleen,
while liver, pancreas, lung, small intestine, heart, brain and muscle contained no measurable activity (Table 7). Even in the particulate fraction of
liver, no kynurenine hydroxylase activity was detected. The activity in sun-

Table 5. Organ distribution of rat kynurenine 3-hydroxylase

	Activity (nmol/h/mg protein)
Kidney	155.3 ± 15.4
Liver	130.9 ± 20.9
Spleen	1.87 ± 0.77
Brain, pancreas, heart, lung, intestine, adrenal gland	< 0.5

Data are the mean ± SEM of five separate
experiments.

cus kidney was about 1/30 of that of rat kidney and about 1/5 of that of human kidney.

CONCLUSION

The major pathway of tryptophan is the kynurenine pathway. This pathway
provides a route for the total oxidation of tryptophan to acetyl Co-A, and is

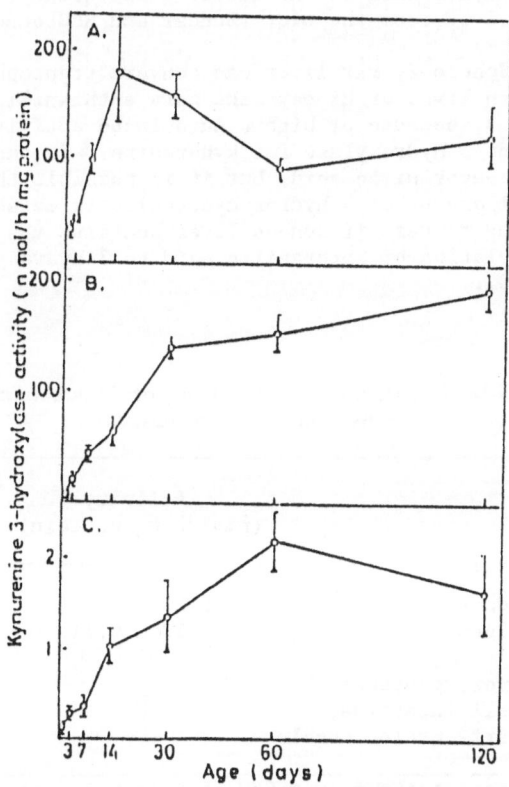

Fig. 2. Development profile of kynurenine 3-hydroxylase from rat liver (A),
kidney (B), and spleen (C). Data are the mean ± SEM of four
separate experiments.

Table 6. Organ distribution of human kynurenine 3-hydroxylase

| | Activity (nmol/h/mg protein) | | | |
	Case 1	Case 2	Case 3	Case 4
Liver	19.95	–	14.02	10.41
Kidney	20.84	31.33	–	–
Spleen	0.91	0.77	–	–
Pancreas	0.81	ND	–	–
Heart	ND	ND	–	–

Case 1: 27 old man who died of a traffic accident.
Case 2: 62 old woman who died of a heart attack.
Case 3: 51 old man who died of renal failure.
Case 4: 72 old man who died of stomach cancer.
ND: not detected

also the route for the de novo synthesis of nicotinamide nucleotide, NAD and NADPH. Animal studies suggest that de novo synthesis is a more important source of the coenzymes than is the direct utilization of nicotinamide and nicotinic acid derived from the diet (Bender and McCreanor, 1982).

As shown in Scheme 2, rat liver can convert tryptophan to nicotinamide nucleotides. Human liver or kidney make more anthranilic acid than the respective rat organs, because of high kynureninase activity and a high K_m value of kynurenine 3-hydroxylase for kynurenine. In suncus, most tryptophan is metabolized to anthranilic acid, but it is possible that in suncus kidney and spleen a small amount of 3-hydroxykynurenine is metabolized to NAD. It will be interesting to test if suncus liver contains the enzyme which catalyzes the hydroxylation of anthranilic acid to 3-hydroxyanthranilic acid (Ueda et al., 1978).

Table 7. Organ distribution of kynurenine 3-hydroxylase in suncus

	Activity (nmol/h/mg protein)
Kidney	5.58 ± 0.36
Spleen	2.30 ± 0.11
Liver, pancreas, lung, small intestine, heart, brain, muscle	ND

Data are the mean ± SEM of four separate experiments. ND: not detected.

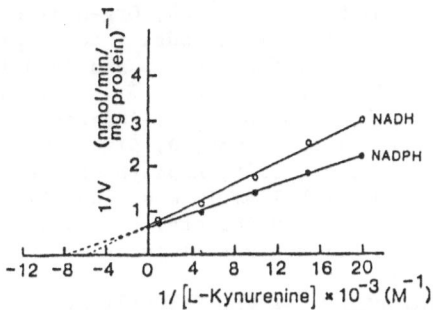

Fig. 3. Double reciprocal plots of the initial velocity of kynurenine 3-hydroxylase activity against L-kynurenine concentration with 4 mM β-NADPH (filled circles) or β-NADH (open circles) as co-substrate.

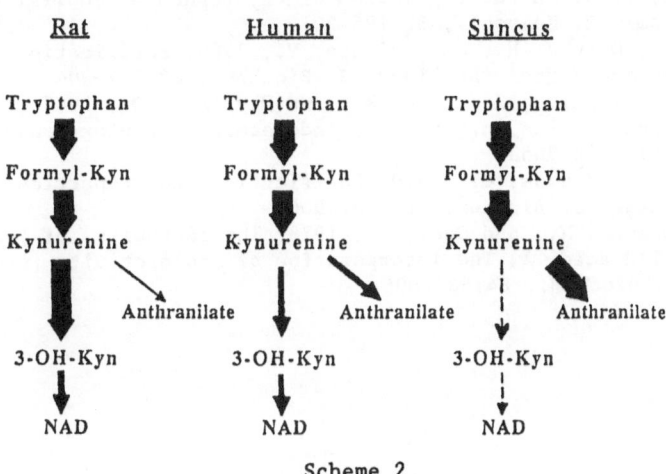

Scheme 2

REFERENCES

Bender, D.A., and McCreanor, G.M., 1982, The preferred route of kynurenine metabolism in the rat, Biochim. Biophys. Acta, 717:56-60.

Bender, D.A., and McCreanor, G.M., 1985, Kynurenine hydroxylase: a potential rate-limiting enzyme in tryptophan metabolism, Biochem. Soc. Trans., 13:441-443.

Braunshtein, A.E., Goryachenkova, E.V., and Paskhina, T.S., 1949, Enzymic formation of alanine from L-kynurenine and L-tryptophan and the role of vitamin B_6 in this process, Biokhimiya, 14:163-179.

El-Zoghby, S.M., El-Kholy, Z.A., Saad, A.A., Mostafa, M.H., Abdel-Tawab, G.A., El-Dardini, N., El-Kabariti, H., and Abdel-Rfea, A., 1978, The effect of environment on tryptophan metabolism via kynurenine in oral contraceptives user, Acta Vitaminol. Enzymol., 32:167-175.

Inada, J., Okuno, E., Kimura, M., and Kido, R., 1984, Intracellular localization and characterization of 3-hydroxykynureninase in human liver, Int. J. Biochem., 16:623-628.

Ishikawa, T., Okuno, E., Kawai, E., Kawai, J., and Kido, R., 1989, Organ distribution, purification and characterization of kynureninase in Suncus Murinus (Insectivore) and anthranilic acid level in the serum, Comp. Biochem. Physiol., 93B:107-111.

Kawai, J., Okuno, E., and Kido, R., 1988, Organ distribution of rat kynureninase and changes of its activity during development, **Enzyme**, 39:181-189.
Lowry, O.H., Rosebrough, N.J., Farr, A.L., and Randall, R.J., 1951, Protein
measurement with Folin phenol reagent, J. **Biol. Chem.**, 193:265-275.
Michael, A.F., Drummon, K.N., Doeden, D., Anderson, J.A., and Good, R.A.,
1964, Tryptophan metabolism in man, J. **Clin. Invest.**,43:1730-1746.
Musajo, L., and Benassi, C.A., 1964, Aspects of disorders of the kynurenine
pathway of tryptophan metabolism in man, **Adv. Clin. Chem.**, 7:63-135.
Nishimoto, Y., Takeuchi, F., and Shibata, Y., 1975, Isolation of L-kynurenine
3-hydroxylase from the mitochondrial outer membrane of rat liver, J.
Biochem., 78:573-581.
Okamoto, H., and Hayaishi, O., 1969, Solubilization and partial purification
of kynurenine hydroxylase from mitochondrial outer membrane and its
electron donors, **Arch. Biochem. Biophys.**, 131:603-608.
Rose, D.P., and McGinty, F., 1968, The influence of adrenocortical hormones
and vitamins upon tryptophan metabolism, **Clin. Sci.**, 35:1-9.
Shibata, K., Hayakawa, T., and Iwai, K., 1986, Tissue distribution of enzymes
concerned with the biosynthesis of NAD in rats, **Agric. Biol. Chem**, 50:
3037-3041.
Shibata, Y., 1978, On the regulation of tryptophan metabolism via kynurenine,
Acta Vitaminol. Enzymol., 32:195-207.
Takeuchi, F., Otsuka, H., and Shibata, Y., 1980, Purification and properties
of kynureninase from rat liver, J. **Biochem.**, 88:987-994.
Takikawa, O., Yoshida, R., Kido, R., and Hayaishi, O., 1986, Tryptophan
degradation in mice initiated by indoleamine 2,3-dioxygenase, J. **Biol.
Chem.**, 261:3648-3653.
Tanizawa, K., and Soda, K., 1979, Purification and properties of pig liver
kynureninase, J. **Biochem.**, 85:901-906.
Ueda, T., Otsuka, H., and Goda, K., 1978, The metabolism of [carboxyl-^{14}C]
anthranilic acid. I. The incorporation of radioactivity into NAD^+ and
$NADP^+$, J. **Biochem.**, 84:687-696.

RELATIONSHIPS BETWEEN PTERIDINE SYNTHESIS AND TRYPTOPHAN DEGRADATION

E.R. Werner, G. Werner-Felmayer, D. Fuchs, A. Hausen,
G. Reibnegger, and H. Wachter

Institute of Medical Chemistry and Biochemistry
University of Innsbruck
6020 Innsbruck
Austria

INTRODUCTION

In 1979, a fluorescent substance occurring in increased amounts in urinary specimens of patients suffering from viral infections or malignant diseases was characterized as D-erythro-neopterin (neopterin; Wachter et al. 1979). Extended studies of neopterin excretion in diseases reveaied that the enhanced excretion of neopterin is coupled to clinical conditions characterized by activated cell-mediated immunity (reviewed by Fuchs et al., 1988; Wachter et al., 1989). Studies investigating the cellular background of these observations in vitro confirmed that immunological activation of the host's peripheral blood mononuclear cells leads to release of neopterin into the culture medium. Surprisingly, large amounts of an unidentified fluorescent substance were always formed together with neopterin by the cells (Fuchs et al., 1982; Huber et al., 1983). The chemical identification of this substance as the tryptophan metabolite 3-hydroxyanthranilic acid (Werner et al., 1985a) initiated our work on the relationship between tryptophan degradation and pteridine synthesis, which is summarized in the present contribution.

INTERFERON-GAMMA INDUCES HUMAN MACROPHAGES TO DEGRADE TRYPTOPHAN

Detailed analysis of the observations in peripheral blood mononuclear cells indicated that the macrophages are the cells responsible for formation of neopterin (Huber et al., 1984) and 3-hydroxyanthranilic acid (Werner et al., 1985b). Interferon-gamma derived from activated T-lymphocytes could be demonstrated as a physiological inducer of the formation of both compounds by macrophages. Subsequently, we showed that the formation of 3-hydroxyanthranilic acid reflected a degradation of tryptophan by macrophages (Werner et al., 1987a). Metabolites formed from tryptophan were kynurenine, anthranilic acid and 3-hydroxyanthranilic acid. It was striking to see that the concentrations of neopterin in the supernatants of macrophages were strictly correlated with the extent of the degradation of tryptophan.

IMPACT OF INTERFERON-GAMMA ON THE TRYPTOPHAN METABOLISM OF HUMAN CELLS AND
CELL LINES OF VARIOUS TISSUE ORIGINS

    Cleavage of tryptophan induced by interferon-gamma is not specific for
macrophages, but had also been observed in a variety of other cultured human
cells (Pfefferkorn, 1984; Werner et al., 1987b; Ozaki et al., 1988; Takikawa
et al., 1988; Werner-Felmayer et al., 1989). In our experiments, all cells
reactive to interferon-gamma by enhanced MHC antigen expression were also
triggered to degrade tryptophan. We concluded therefore that induction of
tryptophan degradation by interferon-gamma is a general aspect of the action
of this lymphokine on human cells (Werner-Felmayer et al., 1989). In addition to cleavage of tryptophan, several of the cells also showed kynureninase

Pteridine and tryptophan metabolism of unstimulated human cells in vitro

	Mø	fibro	A 498	A 549	U 138	SkHep1	A 431	T24
neopterin supernatant								
pteridine synthesis								●
indoleamine-2,3-dioxygenase								●
kynureninase	●		●	●	●	●		
kynurenine-3-monooxygenase	●		●					

Pteridine and tryptophan metabolism of interferon gamma stimulated
human cells in vitro

	Mø	fibro	A 498	A 549	U 138	SkHep1	A 431	T24
neopterin supernatant	●	○			○		○	
pteridine synthesis	●	●	●	●	●	●	●	●
indoleamine-2,3-dioxygenase	●	●	●	●	●	●	●	●
kynureninase	●		●	●	●	●		
kynurenine-3-monooxygenase	●		●					

Fig. 1.   Schematic presentation of pteridine and tryptophan metabolism of
unstimulated (top) and interferon-gamma stimulated (bottom) human
cells in vitro. Mo/: peripheral blood derived macrophages; fibro:
normal dermal fibroblasts; A 498: kidney carcinoma; A 549: lung
carcinoma; U 138: glioblastoma; SkHep 1: liver carcinoma; A 431:
epidermoid carcinoma; T 24: bladder carcinoma. Open circles
indicate that a small amount of neopterin was detected in the
supernatant. Pteridine synthesis was observed by GTP-cyclohydrolase
activity, intracellular neopterin and biopterin concentrations (see
Fig. 2).

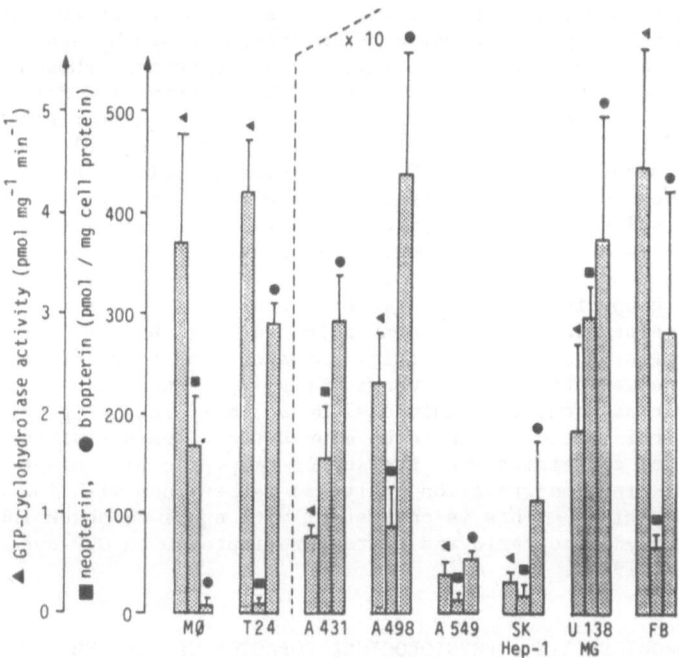

Fig. 2.   Pteridine synthesis in interferon-gamma stimulated human cells.
Mo/: peripheral blood derived macrophages; T 24: bladder carcinoma;
A 431: epidermoid carcinoma; A 498: kidney carcinoma; A 549: lung
carcinoma; SK Hep-1: liver carcinoma; U 138 MG: glioblastoma; FB:
normal dermal fibroblasts. Except for macrophages and T 24 cells,
values are magnified 10 fold (i.e. A 431 shows a GTP-cyclohydrolase
activity of 0.074 pmol/mg/min, neopterin of 15.4 pmol/mg and biopterin of 29.4 pmol/mg). Results are shown as mean ± SD of 4 observations. Cells were stimulated for 48 h with 250 U/ml human recombinant interferon-gamma. GTP-cylcohydrolase activity was assessed
according to Viveros et al. (1981). Pteridine concentrations were
determined by HPLC subsequent to oxidation with acidic iodine according to Werner et al. (1987c). Protein was determined by the
method of Bradford (1976).

activity and two of the cells kynurenine hydroxylase activity. However, only
the first step of the pathway, the indoleamine 2,3-dioxygenase reaction, was
induced by interferon-gamma. The kynurenine metabolizing activities were
constitutively present in the cells and remained unchanged by interferon-
gamma treatment (Werner-Felmayer et al., 1989; see also Fig. 1). The interferon treatment enhanced the V_{max} rather than the K_m of indoleamine 2,3-
dioxygenase in T 24 cells (Werner et al., 1988a).

PTERIDINE SYNTHESIS IN CELLS WITH INDOLEAMINE 2,3-DIOXYGENASE ACTIVITY

Since we had observed a close correlation of neopterin synthesis and
tryptophan cleaving activity in macrophages, we asked whether other cells
with indoleamine 2,3-dioxygenase activity also synthesized pteridines.
Pteridine synthesis is initiated by cleavage of guanosine 5'-triphosphate
(GTP) to 7,8-dihydroneopterin 3'triphosphate by the enzyme GTP-cyclohydrolase (reviewed by Nichol et al., 1985). This dihydroneopterintri-
phosphate is then converted to tetrahydrobiopterin, a cofactor of tryp-
tophan-5-, tyrosine-3- and phenylalanine-4-monooxygenase (Kaufman, 1986).

Fig. 1 schematically summarizes pteridine and tryptophan metabolism of eight human cell types responsive to interferon-gamma by enhanced MHC antigen expression. All cells with indoleamine 2,3-dioxygenase showed pteridine synthesis. In particular, T 24 cells showed both activities also when unstimulated (Fig. 1). Pteridine synthesis in interferon-gamma treated cells is demonstrated by GTP-cyclohydrolase activities, intracellular neopterin and biopterin concentrations (Fig. 2). All three quantities are 5 to 100 fold increased when compared to unstimulated controls. Except for T 24 cells, levels of unstimulated cells approximated the detection limits of the assays (not shown). About 35% of the neopterin occurred as the aromatic species, whereas more than 95% of the biopterin was in the tetrahydroform, the active hydroxylating species. Macrophages are peculiar in the synthesis of large amounts of neopterin (Fig. 1), confirming earlier observations (Huber et al., 1984). However, a small but significant amount of biopterin derivatives could also be detected in macrophages (Fig. 2). The ratio of neopterin to biopterin varies among the individual human cell lines (Fig. 2). Factors determining this ratio remain to be elucidated. The activities of GTP-cyclohydrolase correlated with the sum of neopterin plus biopterin (r = 0.94, P = 0.00019, linear correlation analysis) rather than with individual pteridine concentrations. This is consistent with equal feedback inhibiting potency of dihydroneopterin and tetrahydrobiopterin on GTP-cyclohydrolase (Shen et al., 1988).

IS TETRAHYDROBIOPTERIN A PHYSIOLOGICAL COFACTOR OF THE INDOLEAMINE 2,3-DIOXYGENASE REACTION?

The results presented in Figures 1 and 2 demonstrate that tetrahydrobiopterin is synthesized in cells with indoleamine 2,3-dioxygease activity. In experiments using indoleamine 2,3-dioxygenase purified from rabbit (Nishikimi, 1975) or mouse (Ozaki et al., 1986) tissues, tetrahydrobiopterin was claimed to act as cofactor of the indoleamine 2,3-dioxygenase reaction. Subsequently, highest enzyme activities were observed by using reduced flavin-mononucleotide ($FMNH_2$) as electron donor (Ozaki et al., 1987). We were unable, however, to confirm this stimulatory action of $FMNH_2$ in extracts of human T 24 cells (Werner et al., 1988a). On the other hand, addition of tetrahydrobiopterin to T 24 cell extracts supported the indoleamine 2,3-dioxygenase reaction in the presence of Methylene Blue (not shown) in a manner similar to observations on indoleamine 2,3-dioxygenase purified from rabbit intestine (Nishikimi, 1975).

If a cofactor such as tetrahydropterin were involved in the overall indoleamine 2,3-dioxygenase reaction, then it should be possible to replace the reducing agent ascorbic acid in crude cell extracts by NADH. NADH should then act by shuttling the cofactor from oxidized to reduced forms by means of a reductase. Further, the NADH dependent reaction, but not the ascorbic acid dependent reaction, should be impaired by an agent inhibiting this reductase. Results of such an experiment with T 24 cell extracts and Methotrexate, an inhibitor of dihydrofolate reductase, and in higher concentrations also of dihydropteridine reductase, are shown in Fig. 3. As expected, ascorbic acid can be replaced by NADH. Only the NADH dependent, but not the ascorbic acid dependent reaction is impaired by Methotrexate in doses known to inhibit dihydropteridine reductase (Fig. 3). These results demonstrate that a Methotrexate-sensitive reductase is involved in the stimulation of the indoleamine 2,3-dioxygenase reaction by NADH. We cannot exclude, however, that the used concentrations of Methotrexate used also have effects on reductases other than dihydropteridine reductase, e.g. on flavin-dependent reductases. A clear proof or disproof of tetrahydrobiopterin as the physiological cofactor of indoleamine 2,3-dioxygenase remains to be provided.

CONDITIONS STIMULATING TRYPTOPHAN DEGRADATION AND PTERIDINE SYNTHESIS IN
VITRO AND IN VIVO

We have shown above that we find induction of pteridine synthesis and
indoleamine 2,3-dioxygenase activity in human cells stimulated by interferon-
gamma. In peripheral blood mononuclear cells, in macrophages and in fibroblasts we investigated a number of immunomodulators alone or in combination
for their capacity to induce indoleamine 2,3-dioxygenase and pteridine synthesis (Werner-Felmayer et al., this volume). The results of this investigation showed that several agents are capable to cooperate with interferon-
gamma in the induction of indoleamine 2,3-dioxygenase. In all these experiments the magnitude of the pteridine synthesis correlated strictly with indoleamine 2,3-dioxygenase activity. As an example, dexamethasone augments
the induction of both, pteridine synthesis and indoleamine 2,3-dioxygenase by
interferon-gamma in fibroblasts. With dexamethasone, GTP-cyclohydrolase increased from 0.396 ± 0.020 to 0.507 ± 0.009 pmol mg^{-1} min^{-1} (mean $\pm$ SD, N = 3,
$P < 0.0001$, Student's t-test). Indoleamine 2,3-dioxygenase activities increased from 1.352 ± 0.110 to 2.037 ± 0.420 nmol mg^{-1} min^{-1} (mean $\pm$ SD, N = 3,
$P < 0.001$, Student's t-test). Similarly, changes in the indoleamine 2,3-
dioxygenase activity by variation of the extracellular tryptophan concentrations (Werner-Felmayer et al., 1989) are also accompanied by coherent changes
of pteridine synthesis (not shown). Thus, also minor regulations of indoleamine 2,3-dioxygenase activity are paralleled by changes in pteridine synthesis.

Increased excretion of tryptophan metabolites along the kynurenine pathway in certain diseases had already been observed more than thirty years ago

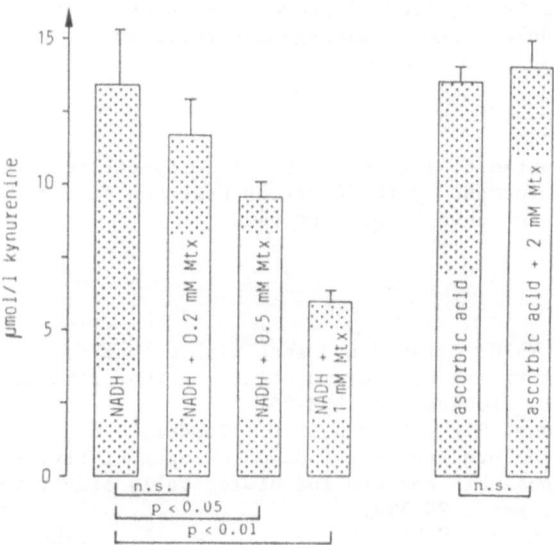

Fig. 3. Inhibition of the NADH dependent indoleamine 2,3-dioxygenase by
Methotrexate (Mtx) in crude extracts of T 24 cells. T 24 cells
were stimulated for 48 hours by 250 U/ml human recombinant interferon-gamma. Cell extracts were then freed from low molecular
weight substances by Sephadex G 25 gel filtration. The eluate was
incubated with 50 μM L-tryptophan, 10^4 U/ml catalase, 10 μM Methylene Blue, and with either 10 mM ascorbic acid or 1 mM NADH with or
without the indicated concentrations of Mtx for 10 min at 37°C. The
reaction was stopped by the addition of 20% perchloric acid and the
amount of kynurenine formed quantified by HPLC according to Werner
et al. (1988a). Results are shown as the mean $\pm$ SD of three experiments. Significance was calculated by Student's t-test.

(e.g. Musajo et al., 1955). These diseases are now known to be characterized by activated cell-mediated immunity. In clinical conditions with activated cell-mediated immunity, we also find elevated neopterin levels (Fuchs et al., 1988). Treatment of cancer patients with high doses of interferon-gamma led to induction of both pathways in vivo (Datta et al., 1987). A pronounced rise in neopterin concentrations was accompanied by a much smaller but distinct rise in biopterin concentrations in this study.

A topic of special interest of our work was the neopterin excretion in AIDS, which led us to conclude that in AIDS patients a pronounced stimulation of endogenous interferon-gamma synthesis occurred despite the observed immunodeficiency (reviewed by Fuchs et al., 1988). This prompted us to study tryptophan and kynurenine in a sample of AIDS patients. We found a significant alteration of serum tryptophan levels to accompany the observed synthesis of neopterin (Werner et al., 1988b).

CONCLUSION

Our data demonstrate that indoleamine 2,3-dioxygenase is the only enzyme activity in the degradation of tryptophan to 3-hydroxyanthranilic acid which is regulated by interferon-gamma. Together with indoleamine 2,3-dioxygenase, biosynthesis of tetrahydrobiopterin is induced in all cells investigated. Factors cooperating with interferon-gamma in regulating the activity of indoleamine 2,3-dioxygenase have parallel effects on pteridine synthesis. Although some observations seem to indicate the possibility of an involvement of tetrahydrobiopterin in the indoleamine 2,3-dioxygenase reaction, further evidence is required to prove or disprove a functional role of pteridines in the indoleamine 2,3-dioxygenase reaction.

ACKNOWLEDGEMENT

We are indebted to Ing. M. Broz for competent technical assistance. This work was supported by the Austrian Research Fund "Zur Förderung der wissenschaftlichen Forschung", project 6922.

REFERENCES

Bradford, M.M., 1976, A rapid and sensitive method for the quantitation of microgram quantities of protein utilizing the principle of protein-dye binding, Anal. Biochem., 72:248-254.

Datta, S.P., Brown, R.R., Borden, E.C., Sondel, P.M., and Trump, D.L., 1987, Interferon and interleukin 2 induced changes of tryptophan and neopterin metabolism: possible markers for biologically effective doses, Proc. Am. Assoc. Cancer Res., 28:338.

Fuchs, D., Hausen, A., Huber, C., Margreiter, R., Reibnegger, G., Spielberger, H., and Wachter, H., 1982, Pteridine secretion as a marker for the proliferation of alloantigen-induced lymphocytes, Hoppe-Seyler's Z. Physiol. Chem., 363:661-664.

Fuchs, D., Hausen, A., Reibnegger, G., Werner, E.R., Dierich, M.P., and Wachter, H., 1988, Neopterin as a marker for activated cell-mediated immunity: application in HIV-infection, Immunol. Today, 9:150-155.

Huber, C., Batchelor, J.R., Fuchs, D., Hausen, A., Lang, A., Niederwieser, D., Reibnegger, G., Swetly, P., Troppmair, Jr., and Wachter, H., 1984, Immune response associated production of neopterin: release from macrophages primarily under control of interferon-gamma, J. Exp. Med., 160: 310-316.

Huber, C., Fuchs, D., Hausen, A., Margreiter, R., Reibnegger, G., Spielberger, M., and Wachter, H., 1983, Pteridines as a new marker to detect
  human T-cells activated by allogeneic or modified self major histocompat-
  ibility complex (MHC) determinants, J. Immunol., 130:1047-1050.
Kaufman, S., 1986, The metabolic role of tetrahydrobiopterin, in: "Chemistry
  and Biology of Pteridines", Cooper, B.A., and Whitehead, V.M., eds., de
  Gruyter, Berlin, pp. 185-200.
Musajo, L., Benassi, C.A., and Parpajola, A., 1955, Isolation of kynurenine
  and 3-hydroxykynurenine from human pathological urine, Nature, 175:855-
  856.
Nichol, C.A., Smith, G.K., and Duch, D.S., 1985, Biosynthesis and metabolism
  of tetrahydrobiopterin and molybdopterin, Ann. Rev. Biochem., 54:729-764.
Nishikimi, M., 1975, A function of tetrahydropteridines as cofactors for
  indoleamine 2,3-dioxygenase, Biochem. Biophys. Res. Commun., 63:92-98.
Ozaki, Y., Edelstein, M.P., and Duch, D.S., 1988, Induction of indoleamine
  2,3-dioxygenase: a mechanism for the antitumor activity of interferon-
  gamma, Proc. Natl. Acad. Sci. USA, 85:1242-1246.
Ozaki, Y., Nichol, C.A., and Duch, D.S., 1987, Utilization of superoxide
  anion for the decyclization of L-tryptophan by murine epididymal indoleamine 2,3-dioxygenase, Arch. Biochem. Biophys., 257:207-216.
Ozaki, Y., Reinhard, J.F. Jr., and Nichol, C.A., 1986, Cofactor activity of
  dihydroflavin mononucleotide and tetrahydrobiopterin for murine epididymal
  indoleamine 2,3-dioxygenase, Biochem. Biophys. Res. Commun., 137:1106-
  1111.
Pfefferkorn, E.R., 1984, Interferon-gamma blocks the growth of toxoplasma
  gondii in human fibroblasts by inducing the host cells to degrade tryptophan, Proc. Natl. Acad. Sci. USA, 81:908-912.
Shen, R.S., Alam, A., and Zhang, Y., 1988, Inhibition of GTP-cyclohydrolase I
  by pterins, Biochim. Biophys. Acta, 965:9-15.
Takikawa, O., Kuroiwa, T., Yamazaki, F., and Kido, R., 1988, Mechanism of
  interferon-gamma action: characterization of indoleamine 2,3-dioxygenase
  in cultured human cells induced by interferon-gamma and evaluation of the
  enzyme mediated tryptophan degradation in its anticellular activity, J.
  Biol. Chem., 263:2041-2048.
Viveros, H.O., Lee, C.L., Abou-Donia, M.M., Nixon, J.C., and Nichol, C.A.,
  1981, Biopterin cofactor biosynthesis: independent regulation of GTP-
  cyclohydrolase in adrenal medulla and cortex, Science, 213:349-350.
Wachter, H., Fuchs, D., Hausen, A., Reibnegger, G., and Werner, E.R., 1989,
  Neopterin as a marker for the activation of cellular immunity: immuno-
  logic basis and clinical application, Adv. Clin. Chem., 27:81-141.
Wachter, H., Hausen, A., and Grassmayr, K., 1979, Increased urinary excretion
  of neopterin in patients with malignant tumors and with virus diseases,
  Hoppe Seyler's Z. Physiol. Chem., 360:1957-1960.
Werner, E.R., Bitterlich, G., Fuchs, D., Hausen, A., Reibnegger, G., Szabo
  G., Dierich, M.P., and Wachter, H., 1987a, Human macrophages degrade
  tryptophan upon induction by interferon-gamma, Life Sci., 41:273-280.
Werner, E. R., Fuchs, D., Hausen, A., Jaeger, H., Reibnegger, G., Werner-
  Felmayer, G., Dierich, M.P., and Wachter, H., 1988b, Tryptophan degrada-
  tion in patients infected by human immunodeficiency virus, Biol. Chem.
  Hoppe-Seyler, 369:337-340.
Werner, E.R., Fuchs, D., Hausen, A., Lutz, H., Reibnegger, G., and Wachter,
  H., 1985b, Interferon-gamma induced in vitro excretion of neopterin and
  3-hydroxyanthranilic acid by human macrophages, in: "Biochemical and
  Clinical Aspects of Pteridines", Wachter, H., Curtius, H.C., and
  Pfleiderer, W., eds., de Gruyter, Berlin, pp. 473-486.
Werner, E.R., Fuchs, D., Hausen, A., Reibnegger, G., and Wachter, T., 1987c,
  Simultaneous determination of neopterin and creatinine in serum with
  solid phase extraction and on-line elution liquid chromatography, Clin.
  Chem., 33:2028-2033.

Werner, E.R., Hirsch-Kaufmann, M., Fuchs, D., Hausen, A., Reibnegger, G.,
Schweiger, M., and Wachter, H., 1987b, Interferon-gamma induced
degradation of tryptophan by human cells in vitro, **Biol. Chem. Hoppe
Seyler**, 368:1407-1412.

Werner, E.R., Lutz, H., Fuchs, D., Hausen, A., Huber, C., Niederwieser, D.,
Pfleiderer, W., Reibnegger, G., Troppmair, J., and Wachter, H., 1985a,
Identification of 3-hydroxyanthranilic acid in mixed lymphocyte cultures,
Biol. Chem. Hoppe Seyler, 366:99-102.

Werner, E.R., Werner-Felmayer, G., Fuchs, D., Hausen, A., Reibnegger, G.,
and Wachter, H., 1988a, Influence of interferon-gamma and extracellular
tryptophan on indoleamine 2,3-dioxygenase activity in T 24 cells as
determined by a non-radiometric assay, **Biochem. J.**, 256:537-541.

Werner-Felmayer, G., Werner, E.R., Fuchs, D., Hausen, A., Reibnegger, G.,
and Wachter, H., 1989, Characteristics of interferon-induced tryptophan
metabolism in human cells in vitro, **Biochim. Biophys. Acta**, 1012:140-147.

Werner-Felmayer, G., Werner, E.R., Fuchs, D., Hausen, A., Reibnegger, G.,
and Wachter, H., Induction of indoleamine 2,3-dioxygenase in human cells
in vitro, this volume.

QUINOLINIC ACID AND KYNURENIC ACID IN THE MAMMALIAN BRAIN

R. Schwarcz and F. Du

Maryland Psychiatric Research Center
Baltimore, Maryland 21228
USA

INTRODUCTION

Over the last decade, the study of neuroexcitatory amino acids has become one of the most rapidly expanding areas of neuroscientific research.
Interest in this class of compounds was precipitated mainly by the realiza-
tion that metabolites such as glutamate and aspartate are major neurotransmitters in the central nervous system (Fonnum, 1984; Erecińska and Silver,
1990). However, it is now clear that excitatory amino acids not only play a
significant role in a wide array of brain processes such as synaptic plasticity and motor control but may also, as "excitotoxins", be causally involved
in the pathogenesis of various neuro-psychiatric diseases (Cavalheiro et al.,
1988). Thus, the pathological overstimulation of excitatory amino acid receptors situated in the neuronal membrane is now believed to result in neurodegeneration in cerebral hypoxia/ischemia, temporal lobe epilepsy and several
other diseases including chronic neurodegenerative disorders such as Huntington's disease (HD) (Schwarcz et al., 1984; Schwarcz and Meldrum, 1985; Rothman and Olney, 1986; Choi, 1988). Pharmacological probes have been used to
subdivide the receptors which mediate excitotoxic insults. These receptor
are linked to ion channels and are named after their model agonists N-methyl-
D-aspartate (NMDA), kainate and α-amino-3-hydroxy-5-methyl-4-isoxazolepro-
pionate (AMPA) (Watkins et al., 1990).

While glutamate and aspartate, which are present in the mammalian brain
in millimolar concentration, are likely to be responsible for many of the
physiological processes mediated by excitatory amino acid receptors (Cavalheiro et al., 1988), they are remarkably ineffective as neurotoxins in vivo.
For example, acute or chronic intracerebral application of concentrated solutions of either compound causes only very little damage in the rat striatum
and hippocampus, two brain regions known to be exquisitely sensitive to excitotoxic insults (Mangano and Schwarcz, 1983). In contrast, the discovery
of the potent convulsive (Lapin, 1978, 1981) and NMDA receptor-specific excitatory (Stone and Perkins, 1981) properties of the peripheral tryptophan
metabolite quinolinic acid (QUIN) suggested that compounds other than glutamate and aspartate may substitute for these two most prevalent amino acids
as ligands at neuronal receptor sites.

Soon after the establishment of its neuroexcitatory properties, QUIN was
shown to be substantially (approximately 100-fold) more potent than glutamate
as an excitotoxin after direct intracerebral injection into adult rats

(Schwarcz et al., 1983). Subsequently, the major characteristics of QUIN excitotoxicity in the rodent brain were rapidly elucidated both in vivo and, using organotypic or dissociated culture systems, in vitro (Whetsell and Schwarcz, 1983; Garthwaite and Garthwaite, 1987; Kim and Choi, 1987). Neuronal vulnerability to QUIN was found to increase with brain development (Foster et al., 1983) and to be profoundly heterogeneous between brain regions and among various neuronal populations of a given susceptible area of the brain (Schwarcz and Köhler, 1983). Taken together, the excitotoxic features of QUIN are unique when compared to all other known exogenous and endogenous excitotoxins and suggest that QUIN should be considered as a possible pathogen in human brain diseases (see below). A literature review of QUIN's neurotoxic effects in the mammalian brain has been published recently (Foster and Schwarcz, 1988).

In 1982, Perkins and Stone described for the first time the neuroinhibitory properties of kynurenic acid (KYNA). KYNA, too, interacts directly with central excitatory amino acid receptors, albeit less specifically than QUIN (Ganong et al., 1983). Thus, KYNA is a competitive antagonist at NMDA, kainate and AMPA sites and, in addition, preferentially blocks the glycine site which is closely associated with the NMDA receptor (Kessler et al., 1989). In pharmacological and physiological experiments, KYNA is now widely used as a non-selective blocker of excitatory amino acid receptor function. Moreover, in accordance with excitotoxic theory, i.e. that inhibition of receptor function should also provide neuroprotection against excitotoxic insults, KYNA was shown to prevent neurodegeneration in experimental animal models (Simon et al., 1986; Germano et al., 1987). Notably, KYNA is particularly powerful as an antagonist of QUIN neurotoxicity (Foster et al., 1984b). Like QUIN, KYNA qualifies conceptually as an endogenous modulator of the multiple physiological processes which are mediated by cerebral excitatory amino acid receptors. A dysfunctional KYNA system, alone or in concert with changes in brain QUIN, has therefore been speculatively linked to a host of neuropsychiatric diseases (see below).

METABOLISM OF KYNURENINES IN THE BRAIN

Work performed during the past decade has shown that both QUIN and KYNA are present in the mammalian brain in low concentration, several orders of magnitude lower than glutamate and aspartate (Wolfensberger et al., 1983; Moroni et al., 1984, 1988; Turski et al., 1988; Swartz et al., 1990). QUIN is rather evenly distributed in various brain regions and in different species, whereas the brain content of KYNA varies more substantially with brain areas and is dramatically different in rat and human. Thus, the human brain contains micromolar quantities of KYNA, whereas approximately 50 times less KYNA is found in the rat brain. Both QUIN and KYNA are present in cerebrospinal fluid (concentration in normal humans: 5-20 nM), i.e. in the extracellular compartment, and can therefore be expected to act upon excitatory amino acid receptors under physiological conditions (Schwarcz et al., 1988b; Swartz et al., 1990). Notably, no effective energy-driven uptake process appears to exist in the brain for either of the two metabolites, nor are efficient enzymes in place in cerebral tissue to catalyze their rapid degradation (Foster et al., 1984a; Turski and Schwarcz, 1988). This and their largely non-neuronal origin make it unlikely that QUIN and KYNA are major neurotransmitters in the central nervous system (cf. below).

Probably due to their polar nature, neither QUIN nor KYNA is capable of entering the brain from the periphery to a significant extent (Fukui et al., in press). Of the other kynurenine pathway metabolites tested, 3-hydroxyanthranilic acid, too, does not appear to enter the brain freely from the circulation, while anthranilic acid, probably a major bioprecursor of brain QUIN (Baran and Schwarcz, this volume), has easier access to the brain by passive diffusion (Fukui et al., in press). In contrast, kynurenine and 3-hydroxy-kynurenine are actively transported through the blood-brain barrier by the large neutral amino acid carrier which is also used, _inter alia_, by trypto-phan. Thus, peripherally produced kynurenine, when competing successfully with other endogenous amino acids for brain entry (Christensen, 1984), can be expected to influence substantially the synthesis and function of cerebral QUIN and KYNA (cf. Fig. 1). This is of particular relevance since the con-version rate of tryptophan to kynurenine in the brain is very low (Gál, 1974), making it unlikely that brain tryptophan constitutes a major parent compound of kynurenine pathway metabolites. Biochemical and immunohisto-chemical studies have therefore been performed to elucidate some of the issues of kynurenine neurobiology in the brain as they relate to QUIN and KYNA function.

Experiments using brain slices and ^{3}H-kynurenine have demonstrated the existence of two independent processes which are in place to facilitate the uptake of kynurenine into cells. One transporter, which is efficient and Na^+-independent, has been localized to glial cells (Speciale et al., 1989a; Speciale and Schwarcz, 1990) and is likely to be responsible for the rapid accumulation of kynurenine in astrocytes after it enters the brain. A sec-ond, much less effective, uptake process is Na^+-dependent and mediates the transport of kynurenine into neurons (Speciale and Schwarcz, 1990).

In spite of the fact that the essential features of QUIN and KYNA metab-olism in mammalian cells had been established for decades, very little infor-mation on their disposition in brain tissue was available at the time when their neuroactive properties became apparent (Gál and Sherman, 1980). In the periphery but not in the brain, QUIN and KYNA are known as integral elements of the enzymatic chain leading from tryptophan to NAD^+ (Fig. 1). The indi-vidual steps of the so-called kynurenine pathway, which accounts for the vast majority of tryptophan degradation outside the brain, are reasonably well understood, and several enzymes in the cascade have been purified from pe-ripheral tissues.

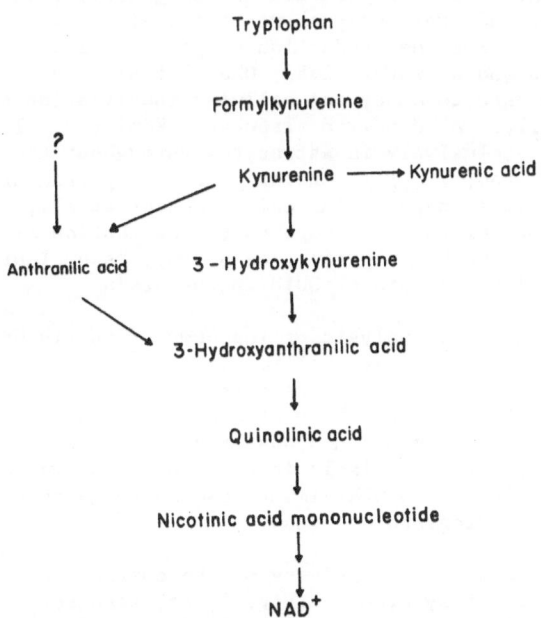

Fig. 1. Schematic representation of the kynurenine pathway in the mammalian brain.

Quinolinic acid

Studies of the enzymatic machinery of QUIN biosynthesis in the brain are generally in good agreement with the notion that astrocytic metabolism, fueled by accumulated kynurenine, plays a pivotal role in the function of brain kynurenines. While not all enzymes of the peripheral kynurenine pathway have been examined in sufficient detail to draw conclusions about their cellular localization in the brain, the entire cascade is clearly present in cerebral tissue (cf. below). Kynurenine-3-hydroxylase (Uemura and Hirai, this volume) and kynureninase (Kawai et al., 1988; Okuno et al., this volume) have very low activity in the rat brain, whereas other enzymes of kynurenine catabolism are remarkably active.

The two enzymes directly responsible for the biosynthesis and degradation of QUIN, respectively, 3-hydroxyanthranilic acid oxygenase (3HAO) and quinolinic acid phosphoribosyltransferase (QPRT), have been characterized quite extensively in the rat brain. Like peripheral organs, the brain has a much higher ability to synthesize QUIN than to break it down, as evidenced by the approximately 80 times higher V_{max} value of 3HAO. 3HAO and QPRT display similar distribution patterns between various brain areas, with the olfactory bulb containing the highest and hindbrain regions the lowest activities. During ontogeny, both enzymes show activity peaks during the second postnatal week, reaching 150-200% of adult levels (Foster et al., 1985b, 1986). Lesion studies, using surgical axotomy or excitotoxins in the striatum and the hippocampus, have revealed higher activities of both 3HAO and QPRT within days after the insult, i.e. during the period of pronounced reactive astrogliosis (Speciale et al., 1987; Schwarcz et al., 1989). These biochemical experiments provided the first evidence that at least a substantial portion of both enzymes reside in non-neuronal cellular elements, probably astrocytes. Taken together, the studies summarized here indicated a remarkable parallelism in several aspects of 3HAO and QPRT neurochemistry, suggesting a possible co-localization and co-regulation of the two enzymes in physiological and pathological situations.

3HAO and QPRT have been purified to homogeneity from rat liver and the identity of liver and brain enzymes was ascertained by biochemical, physico-chemical and, following the production of polyclonal antibodies, immunological means (Okuno and Schwarcz, 1985; Okuno et al., 1987). Purified antibodies were then used to study the cellular localization of both enzymes immunohistochemically. As detailed elsewhere (Köhler et al., 1988), 3HAO is contained almost exclusively in astrocytes throughout the rat brain (Figs. 2E, 3C,D). QPRT, too, is preferentially an astroglial enzyme, but discrete QPRT-positive neuronal populations exist (Köhler et al., 1987). The functional significance of this heterogeneous distribution of QPRT, which is even more pronounced in the human brain (see below), is unclear at present, but suggests a physiological role of QUIN in the brain.

Immunohistochemical analysis of lesioned brain tissue further confirmed a close association of QUIN metabolism with glial cells and is in full accordance with the biochemical assessments. For instance, QUIN-induced lesions in the rat hippocampus, which at a low toxic dose preferentially affect CA1 and some CA4/CA3 pyramidal neurons (Fig. 2B), cause a selective <u>increase</u> in 3HAO-immunoreactivity precisely in the area of nerve cell loss (Fig. 2F). A similar band of increased QPRT-immunoreactivity is seen in the QUIN-lesioned CA1 region (micrograph not shown).

Because of the very low activity of the enzymes preceeding 3HAO in the metabolic breakdown of kynurenine (Fig. 1; cf. also Baran and Schwarcz, this volume), most studies designed to investigate QUIN function in intact brain tissue have used 3-hydroxyanthranilic acid (3HANA) as QUIN's bioprecursor. QUIN synthesis from 3HANA is readily demonstrable in cerebral slices

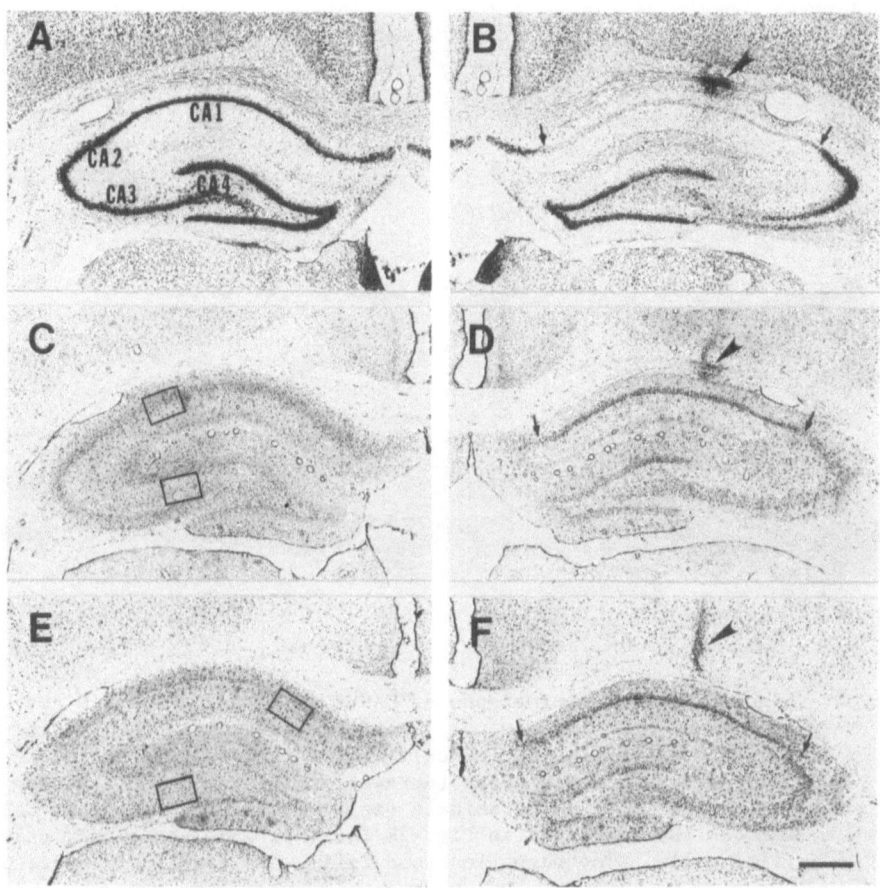

Fig. 2. Photomicrograph of coronal sections (20 μm) taken from the hippocampi of QUIN-lesioned rats. Adult male Sprague-Dawley rats were
anesthetized and a 30 gauge needle was inserted stereotaxically into
the right hippocampus (coordinates: 3.3 mm posterior to bregma, 2.1
mm from the midline and 2.9 mm below dura). 15 nmol QUIN, pH 7.4,
was administered over 10 min using a microinfusion pump. After 3
days, the animals were re-anesthetized and perfused transcardially
with saline followed by 500 ml of a fixative containing paraformaldehyde, lysine and periodate (McLean and Nakane, 1974). The brains
were post-fixed for another 3 hours and then washed in phosphate
buffer containing 15% sucrose. Serial cryostat sections were cut
coronally through the whole hippocampus and collected separately in
cold phosphate buffer. The first section of each series was stained
with thionin. The second and third sections were processed for KAT-
immunoreactivity (-i) and 3HAO-i, respectively (Köhler et al., 1988;
Okuno et al., 1990). A,C, and E: Photomicrographs from the contralateral (control) hippocampus, showing normal patterns of Nissl-
staining (A), KAT-i (C), and 3HAO-i (E). B,D, and F: Photomicrographs from the QUIN-lesioned hippocampus. B: Nissl-staining. Note
neurodegeneration in areas CA1 and CA4. Small arrows in B indicate
the borders of degeneration, which coincide with an obvious change
in the intensity of KAT-i (see D, small arrows) and 3-HAO-i (see F,
small arrows). D (KAT-i) and F (3HAO-i): Both immunoreactivities
are increased substantially in the lesioned CA1 field. Arrowheads
in B,D, and F indicate the needle track, passing through the cortex.
The magnifications in A-F are identical. Bar: 500 μm.

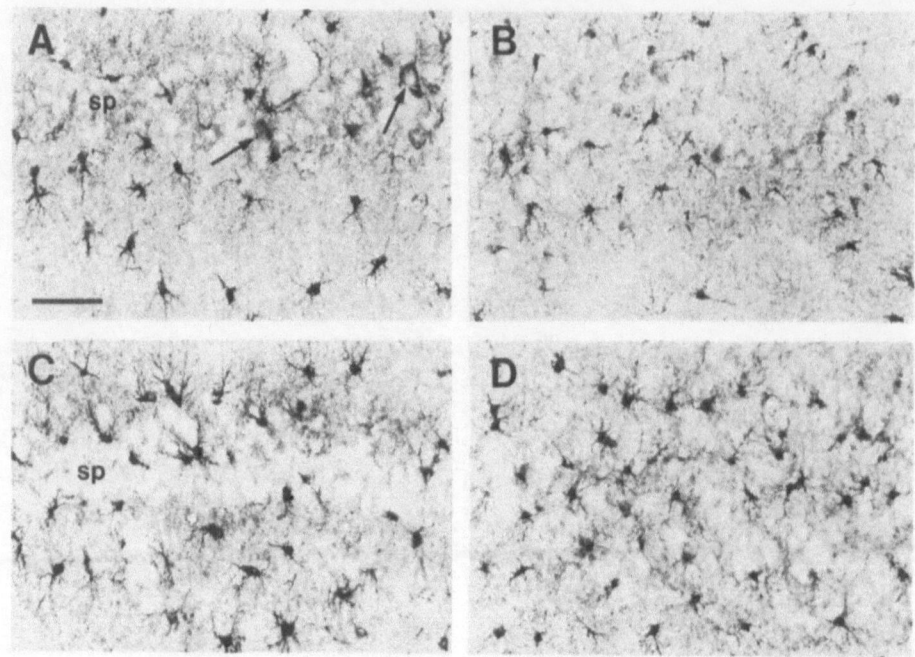

Fig. 3. High power photomicrographs of KAT-i (A,B) and 3HAO-i (C and D) from
the normal rat hippocampus. A and B: Portions of CA1 (A) and CA4
(B) as seen in Fig. 2C (boxes). KAT-i is mainly localized in glial
cells though a few neurons (arrows) exhibiting KAT-i can also be
seen in the stratum pyramidale (sp). C and D: Portions of CA1 (C)
and CA4 (D) indicated in Fig. 2E (boxes). 3HAO-i is only present in
glial cells. The magnifications in A-D are identical. Bar: 50 μm.

(Speciale and Schwarcz, this volume) and in the freely moving animal in vivo
(Speciale et al., 1989b). Notably, under identical experimental conditions
no significant production of QUIN can be shown when even high concentrations
of either tryptophan or kynurenine are supplied as bioprecursors (Speciale et
al., 1989b). Brain tissue, in contrast to peripheral tissue, is capable of
retaining de novo synthesized QUIN. This mechanism, which can be hypothesized to constitute a regulatory process controlling QUIN release into the
extracellular milieu, is compromised in lesioned, neuron-depleted tissue,
leading to a more rapid emergence of extracellular QUIN (Speciale and
Schwarcz, this volume). The altered disposition of QUIN following nerve cell
degeneration remains to be explored in greater detail and could play a role
in pathophysiological processes (see below).

Relatively little information is currently available on QUIN metabolism
in the human brain, but all evidence suggests that the features of QUIN function which have been elaborated in the rat are essentially also operant in
the human. As in rats, kynurenine-3-hydroxylase and kynureninase are present
in normal and pathological human brain tissue with low activities (unpublished data; cf. Vezzani et al., this volume). However, the existence of an efficient functional cascade from kynurenine to QUIN has not yet been firmly
established.

Like in the rat, 3HAO and QPRT have been purified to homogeneity from
the human liver, antibodies were produced, and the identity of liver and
brain enzymes has been ascertained using various methodologies (Okuno et al.,

1988; Okuno and Schwarcz, unpublished data). The V_{max} ratio between 3HAO and QPRT, similar to the rat, is approximately 100:1. However, anti-human 3HAO and QPRT antibodies very poorly recognize the respective rat enzymes (and vice versa). Moreover, the V_{max} values of both enzymes exceed those of the rat enzymes several fold. Both enzymes show regional variation in the human brain with up to 6-fold differences between the areas of highest and lowest activity (Foster et al., 1985a; Schwarcz et al., 1988a).

So far, immunohistochemical studies have been performed with anti-human QPRT antibodies, and detailed analyses of QPRT-immunoreactive cells have been conducted in the hippocampal formation (Du et al., 1990a) and the neostriatum (Du et al., in press). Like in the rat brain, the majority of immunoreactive cells are astrocytes. However, the human brain also harbors an astonishingly diverse array of QPRT-containing neurons. These neurons are present in virtually all brain areas examined to date. Their size and shape varies substantially between brain regions, but QPRT-positive neuronal populations appear to be remarkably homogeneous within a given area (Fig. 4). The organization of QPRT-positive cellular elements does not immediately suggest a functional commonality, but far too little is currently known about QUIN-related brain processes to generate reasonable inferences from these initial anatomical studies of QUIN's catabolic enzyme.

Kynurenic acid

KYNA biosynthesis in the brain has remained largely unexplored until very recently, though the presence of kynurenine aminotransferase (KAT) in the mammalian brain had been briefly reported by Minatogawa et al. (1974) (cf. also Kido, this volume). In the rat brain, a single enzyme appears to be responsible for the production of KYNA at physiological (micromolar) concentrations of kynurenine (Okuno, Schmidt et al., in press). However, at least one other enzyme, alanine aminotransferase, could conceivably play a significant anabolic role if kynurenine concentrations were to rise locally into the millimolar range (Okuno et al., 1990). In contrast to most known aminotransferases, KAT is remarkably substrate-specific. Thus, the enzyme prefers kynurenine not only to a wide spectrum of amino acids which are commonly excellent substrates for transamination, but also to 3-hydroxykynurenine, the parent compound of xanthurenic acid. KAT is differentially distributed between brain regions, with the olfactory bulb displaying highest and the cerebellum lowest activity of the areas studied to date (Okuno, Schmidt et al., in press).

The cellular localization of brain KAT was first explored biochemically. Thus, KAT activity was measured in the striatum at various timepoints after a QUIN-induced lesion. A biphasic response of KAT activity was noted: a modest but significant decrease (-20%) after two days was followed by a substantial increase (+87%) in KAT after 7 days. These data suggest the presence of at least two populations of KAT-containing cells in the rat striatum, one neuronal and excitotoxin-sensitive, and the other astroglial (Okuno, Schmidt et al., in press; cf. also Turski et al., 1989).

Immunohistochemical studies with anti-KAT antibodies (Okuno et al., 1990) confirmed the heterogeneous cellular localization of the enzyme. For instance, the vast majority of KAT-positive cells in the hippocampus are astrocytes (Figs. 2C and 3A,B). In the striatum, a small neuronal population containing KAT exists in addition to a large number of KAT-immunoreactive astrocytes. The detailed mapping of KAT-containing cells currently carried out in our laboratory demonstrates that discrete populations of KAT-containing neurons exist throughout the rat brain, though most cells harboring KAT are clearly astrocytic in nature. KAT-positive neurons are particularly observed in the brainstem and in deep cortical layers (Okuno et al., 1990).

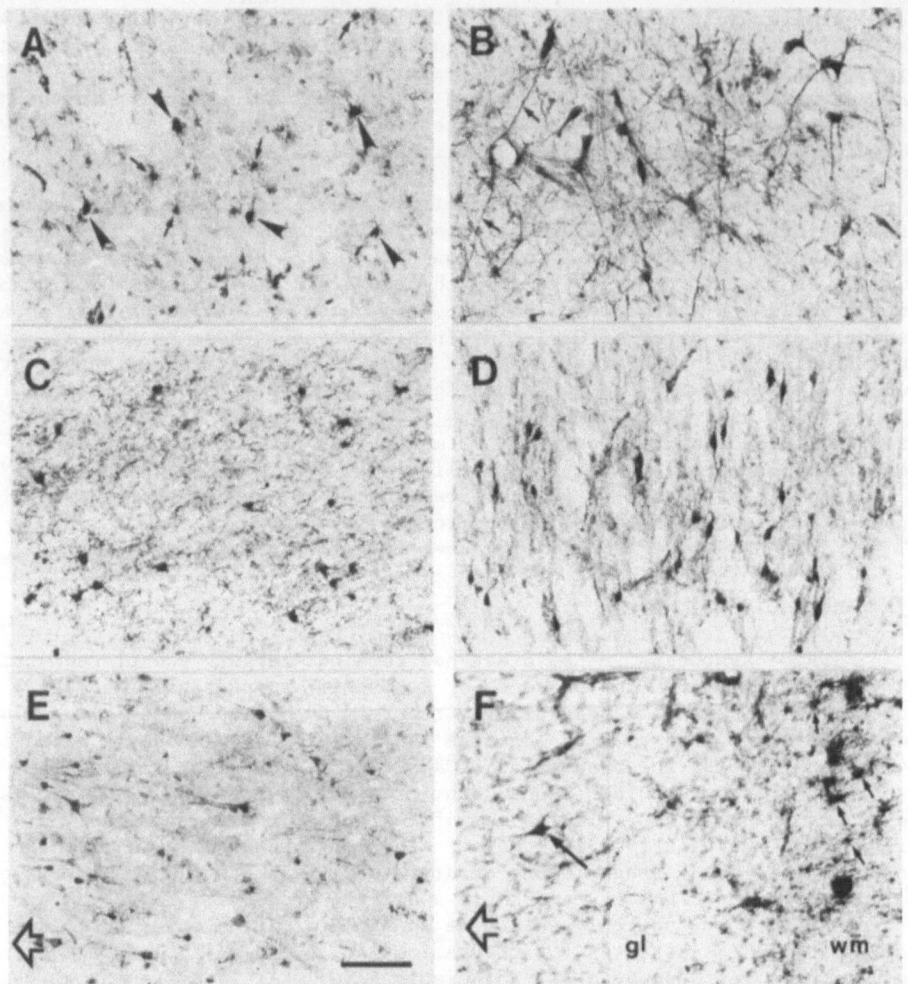

Fig. 4.   Photomicrographs of QPRT-immunostained sections taken from normal
human brain.   Tissue samples were obtained at autopsy and were fixed
in a fixative containing paraformaldehyde, lysine and periodate
(McLean and Nakane, 1974).   Following washes in phosphate buffer
containing 15% sucrose, the tissue was frozen and cut at 30 μm with
a cryostat.   Sections were then processed according to the avidin-
biotin-complex method for QPRT-i (cf. Du et al., 1990a).   A: Caudate
nucleus; QPRT-i is localized both in glial cells (some indicated by
small arrows) and in neurons (arrowheads).   QPRT-i neurons are mainly of medium size and closely resemble aspiny neurons seen in Golgi
preparations.   B: Medial segment of the globus pallidus, showing the
localization of QPRT-i in many large neurons.   Small arrows indicate
a possible axon.   C: Thalamus; QPRT-i is contained in a population
of small neurons.   D: Subiculum proper; most QPRT-i neurons are
fusiform or triangular in shape and are oriented perpendicular to
the molecular layer.   E: Olfactory cortex, showing QPRT-i neurons in
layers IV-VI.   F: Cerebellar cortex; QPRT-i is present in many glial
cells (small arrows) in white matter (wm).   Occasionally, a single
QPRT-i neuron (large arrow) can be observed in the granular layer
(gl).   Open arrows in E and F indicate the direction of the molecular layer.   The magnifications in A-F are identical.   Bar: 100 μm.

Therefore, a role of KYNA as a neurotransmitter in those neurons is certainly possible and remains to be investigated.

QUIN-induced degeneration and subsequent astrogliosis in the rat hippocampus result in an increased KAT-immunoreactive band at the site of nerve cell loss in the CA1 area, virtually identical in appearance to the reaction of 3HAO-positive cells (cf. Figs. 2D and F). Again, it will be necessary to evaluate the functional significance of the anatomically discrete change following a QUIN lesion and, in particular, to weigh the consequences of increased KAT- and 3HAO-activity against each other.

Experiments performed with tissue slices and by microdialysis in vivo have revealed salient features of KYNA function in the brain. Most importantly, the production of KYNA from kynurenine can be readily demonstrated in intact tissue, and KYNA can be recovered extracellularly after its liberation from KAT-containing cells (Turski et al., 1989; Speciale et al., 1990). As expected from immunohistochemical analyses and biochemical studies in brain homogenate, astrogliotic lesioned tissue has a far higher ability to produce KYNA under otherwise identical conditions (Wu and Schwarcz, 1990). Notably, the efflux of KYNA does not appear to be Ca^{2+}-dependent and occurs very rapidly after the compound is formed by KAT (Turski et al., 1989). However, KYNA release is regulated by mechanisms which are linked to neuronal activity, such that increased activity causes a decrease in extracellular KYNA concentration (Gramsbergen et al., this volume). This phenomenon has recently also be validated by measuring endogenous KYNA levels in the extracellular compartment and may therefore play a role in amplifying depolarizing effects mediated by ionotropic excitatory amino acid receptors (unpublished data). In this context, it is also noteworthy that in parallel with an increase in KYNA brain content with age (Moroni et al., this volume), both KAT activity and the production of KYNA from kynurenine increases up to three-fold between young adult (3 months) and old (2 years) brains (Schwarcz et al., 1988c and in preparation). At this point, the possible functional implications, e.g. a lack of memory function in the aged brain due to a lessened activity of hippocampal NMDA receptors, remain to be elucidated.

In spite of some evidence in peripheral organs that enzymatic mechanisms for the degradation of KYNA exist (Takahashi et al., 1956), no such catabolic events have yet been discovered in the brain (Turski and Schwarcz, 1988). In view of the lack of an efficient re-uptake process, it therefore seems that the action of extracellular KYNA in the brain is terminated by its transport out of the brain. A probenecid-sensitive such process has been identified (Moroni et al., this volume), but its precise role in the regulation of KYNA function remains to be investigated in greater detail.

In contrast to the rat brain, the human brain contains two enzymes capable of catalyzing the transamination of kynurenine to KYNA. The two proteins, which have been arbitrarily termed KAT I and KAT II, can be separated using a DEAE-Sepharose resin and differ with regard to co-factor preference, pH optimum and substrate specificity. Thus, KAT I is far more active with pyruvate than with α-ketoglutarate as the aminoacceptor, and shows the highest activity at pH 9.6 and is potently inhibited by glutamine. KAT II, on the other hand, is equipotent with either pyruvate or α-ketoglutarate as a co-factor, and has a pH optimum of 7.4 and cannot be inhibited even by very high (10 mM) concentrations of glutamine (Okuno et al., in press). Essential neurobiological questions regarding the two enzymes, for example their regional distribution and cellular localization in the brain, have as yet not been examined. Most importantly, it will be of interest to determine to what extent each of the two enzymes bears responsibility for KYNA production in the human brain.

QUINOLINIC ACID AND KYNURENIC ACID IN HUMAN BRAIN DISEASES

With the discovery of QUIN's and KYNA's excitotoxic and neuroprotective properties, respectively, it became apparent that both metabolites may play a role in the pathogenesis of neuro-psychiatric disorders (Schwarcz et al., 1983; Foster et al., 1984b). In particular, the excitatory amino acid receptor ligands QUIN and KYNA can be envisioned to participate as pathogens in a wide variety of central nervous system diseases which involve a malfunction of these receptors. The arguments for a central role of excitatory amino acid receptors in an ever growing list of brain diseases are in several cases quite compelling and have been amply reviewed (Table 1; Cavalheiro et al., 1988). Only recently, however, have attempts been made to examine directly the status of brain QUIN and KYNA in pathological human tissues and relevant experimental animal models. These studies were greatly facilitated by the development of specific and extremely sensitive methods for the determination of both QUIN (Heyes et al., 1988) and KYNA (Shibata, 1988; Swartz et al., 1990), as well as by the possibility to examine relevant enzymatic processes by novel radiochemical techniques (Foster et al., 1985a; Schwarcz et al., 1988a; Okuno et al., in press).

The hereditary neurodegenerative disease HD was the first to be speculatively linked to a pathological overabundance of QUIN. Intrastriatal injections of QUIN in the rat indeed mimick the neuropathological features of the disease in great detail (Schwarcz et al., 1983; Beal et al., 1989a). Thus, like in HD, the vast majority of striatal neurons degenerate after exposure to QUIN, whereas small populations of medium-sized aspiny (somatostatin- and neuropeptide Y-containing) and large cholinergic neurons survive (cf. also Davies and Roberts, 1988). To date, direct assessment of QUIN in HD victims has not revealed an unequivocal answer regarding its possible pathogenic role. QUIN levels are unchanged in HD tissue and cerebrospinal fluid (Reynolds et al., 1988; Schwarcz et al., 1988b). QUIN biosynthesis, on the other hand, is clearly impaired in HD, as indicated by large increases in 3HAO activity (Schwarcz et al., 1988a), possibly due to astrogliosis (see above). As judged from work in animals (Speciale and Schwarcz, this volume), this increase can be expected to translate into a higher ability of diseased tissue to produce extracellular QUIN if sufficient 3-hydroxyanthranilic acid is available. Implications for a role of QUIN in the slow progression of neurodegeneration in HD have been discussed (Schwarcz et al., 1987).

A potential involvement of QUIN in temporal lobe epilepsy was originally suggested by Lapin (1981). Indeed, as depicted in Fig. 2B, hippocampal CA1 and CA4/CA3 pyramidal cells are preferentially susceptible to QUIN; thus, the QUIN-lesioned hippocampus bears a very close resemblance to most hippocampi which are removed during the surgical treatment of drug-resistant temporal lobe epileptics. Moreover, QPRT activity is reportedly decreased in surgically excised tissue samples, possibly leading to local increases in tissue QUIN (Feldblum et al., 1988). The pattern of QPRT-immunoreactivity, too, is profoundly altered in the epileptic hippocampus: the hilar region, which in the normal brain is densely populated with QPRT-positive astrocytes (Du et al., 1990a), shows very little immunoreactivity in epileptic tissue. In contrast, QPRT-immunoreactivity, mainly associated with proliferated astrocytes, is substantially increased in the CA1 area of tissues that display neuronal degeneration (Du et al., 1990b). Clearly, the functional significance of this striking redistribution, and in particular its possible contribution to the initiation or propagation of seizure phenomena, remains to be explored.

Measurement of QUIN levels have provided evidence for changes in QUIN metabolism in a variety of other clinical situations which are associated with brain dysfunction or neurodegeneration. Thus, reports have appeared in support of increases in brain QUIN in hepatic encephalopathy (Moroni et al.,

1986), cerebral ischemia (Heyes and Nowak, 1990) and hypoglycemia (Heyes et al., 1990). Moreover, the recent finding by Heyes and colleagues (this volume) of substantial, dementia-related, increases in cerebrospinal fluid QUIN in AIDS victims indicates that viral disorders of the brain are linked to an overproduction of QUIN.

The status of KYNA in human brain diseases is merely beginning to be investigated. Conceptually, a hypofunctional KYNA system is equivalent to a hyperfunctional QUIN system and can be envisioned to result in excitotoxicity due to insufficient blockade of excitatory amino acid receptor(s). First studies in HD brain tissue have yielded contradictory results, i.e. KYNA synthesis appears to be decreased in the basal ganglia (Beal et al., 1989b) whereas the tissue concentration of KYNA is higher than normal in cortical samples (Connick et al., 1989). In HD as in other disease states, definitive answers are not likely to be deducible from the measurement of steady-state tissue levels alone, but must await thorough analyses including the assessment of more dynamic measures, e.g. KAT activity. More generally, a comprehensive study of the role of KYNA in pathological situation must take into consideration the possibility of concomitant changes in QUIN metabolism, which can conceivably attenuate or potentiate the effects of a dysfunctional KYNA system.

Table 1. Neurological and psychiatric diseases hypothetically
linked to a dysfunction in brain QUIN and/or KYNA

Neurodegenerative Disorders
    Huntington's disease
    Parkinson's disease
    Alzheimer's disease
    Olivopontocerebellar atrophy
    Amyotropic lateral sclerosis

Hypoxic/Ischemic Insults
    Stroke
    Cerebral palsy
    Transient global ischemia
    Traumatic brain injury

Hypoglycemia

Hepatic Encephalopathy

Seizure Disorders
    Myoclonus
    Temporal lobe epilepsy

Tremors and Spasticity

Psychiatric Disorders
    Schizophrenia
    Anxiety disorders

Viral Disorders
    AIDS dementia

ACKNOWLEDGEMENTS

We are indebted to Drs. A.C. Foster, M. Nakamura, E. Okuno, W. Schmidt, C. Speciale, and W.A. Turski for their important contributions to the work described in this article. We also wish to acknowledge the long-standing collaborations with the laboratories of Drs. E.D. French, C. Köhler, U. Ungerstedt, and W.O. Whetsell. This work was supported by USPHS grants NS 16102, NS 28236 and MH 44211, and a fellowship from the Huntington's Disease Society of America (to F.D.).

REFERENCES

Beal, M.F., Kowall, N.W., Swartz, K.J., Ferrante, R.J., and Martin, J.B., 1989a, Differential sparing of somatostatin-neuropeptide Y and cholinergic neurons following striatal excitotoxin lesions, Synapse 3:38-47.

Beal, M.F., Matson, W.R., Swartz, K.J., Gamache, P.S., and Bird, E.D., 1989b, Multicomponent analysis of tryptophan and tyrosine metabolism in Huntington's disease: evidence for reduced formation of kynurenic acid, Soc. Neurosci. Abstr., 15:328.8.

Cavalheiro, E.A., Lehmann, J., and Turski, L., eds., 1988, "Frontiers in Excitatory Amino Acid Research", A. Liss, New York.

Choi, D.W., 1988, Glutamate neurotoxicity and diseases of the nervous system, Neuron 1:623-634.

Christensen, H.N., 1984, Organic ion transport during seven decades: the amino acids, Biochim. Biophys. Acta, 779:255-269.

Connick, J.H., Carlà, V., Moroni, F., and Stone, T.W., 1989, Increase in kynurenic acid in Huntington's disease motor cortex, J. Neurochem., 52:985-987.

Davies, S.W., and Roberts, P.J., 1988, Sparing of cholinergic neurons following quinolinic acid lesions of the rat striatum, Neuroscience 26:387-393.

Du, F., Okuno, E., Whetsell, W.O. Jr., Köhler, C., and Schwarcz, R., 1990a, Distribution of quinolinic acid phosphoribosyltransferase in the human hippocampal formation and parahippocampal gyrus, J. Comp. Neurol., 295:71-82.

Du, F., Okuno, E., Whetsell, W.O. Jr., Köhler, C., and Schwarcz, R., Immunohistochemical localization of quinolinic acid phosphoribosyltransferase in the human neostriatum, Neuroscience, in press.

Du, F., Whetsell, W.O. Jr., Abou-Khalil, B., Blumenkopf, B., Okuno, E., and Schwarcz, R., 1990b, Immunohistochemical study of quinolinic acid catabolism in epileptic human hippocampus, Soc. Neurosci. Abstr., 16:138.11.

Erecińska, M., and Silver, I.A., 1990, Metabolism and role of glutamate in mammalian brain, Prog. Neurobiol., 35:245-296.

Feldblum, S., Rougier, A., Loiseau, H., Loiseau, P., Cohadon, F., Morselli, P.L., and Lloyd, K.G., 1988, Quinolinic phosphoribosyltransferase activity is decreased in epileptic human brain tissue, Epilepsia 29:523-529.

Fonnum, F., 1984, Glutamate: a neurotransmitter in mammalian brain, J. Neurochem., 42:1-11.

Foster, A.C., Collins, J.F., and Schwarcz, R., 1983, On the excitotoxic properties of quinolinic acid, 2,3-piperidine dicarboxylic acids and structurally related compounds, Neuropharmacology, 22:1331-1342.

Foster, A.C., Miller, L.P., Oldendorf, W.H., and Schwarcz, R., 1984a, Studies on the disposition of quinolinic acid after intracerebral or systemic administration in the rat, Exp. Neurol., 84:428-440.

Foster, A.C., and Schwarcz, R., 1988, Neurotoxic effects of quinolinic acid in the central nervous system, in: "Quinolinic Acid and other Kynurenines", Stone, T.W., ed., CRC Press, pp. 173-192.

Foster, A.C., Vezzani, A., French, E.D., and Schwarcz, R., 1984b, Kynurenic acid blocks neurotoxicity and seizures induced in rats by the related brain metabolite quinolinic acid, Neurosci. Lett., 48:273-278.

Foster, A.C., Whetsell, W.O. Jr., Bird, E.D., and Schwarcz, R., 1985a, Quinolinic acid phosphoribosyltransferase in human and rat brain: activity in
Huntington's disease and in quinolinate lesioned rat striatum, Brain Res.,
336:207-214.

Foster, A.C., White, R.J., and Schwarcz, R., 1986, Synthesis of quinolinic
acid by 3-hydroxyanthranilic acid oxygenase in rat brain tissue in vitro,
J. Neurochem., 47:23-30.

Foster, A.C., Zinkand, W.C., and Schwarcz, R., 1985b, Quinolinic acid phosphoribosyltransferase in rat brain, J. Neurochem., 44:446-454.

Fukui, S., Schwarcz, R., Rapoport, S.I., Takada, Y., and Smith, Q.R., Blood-
brain barrier transport of kynurenines: implications for brain synthesis
and metabolism, J. Neurochem., in press.

Gàl, E.M., 1974, Cerebral tryptophan-2,3-dioxygenase (pyrrolase) and its induction in rat brain, J. Neurochem., 30:607-613.

Gàl, E.M., and Sherman, A.D., 1980, L-kynurenine: its synthesis and possible
regulatory function in brain, Neurochem. Res., 5:223-239.

Ganong, A.H., Lanthorn, T.H., and Cotman, C.W., 1983, Kynurenic acid inhibits
synaptic and amino acid-induced responses in the rat hippocampus and
spinal cord, Brain Res., 273:170-174.

Garthwaite, G., and Garthwaite, J., 1987, Quinolinate mimics neurotoxic actions of N-methyl-D-aspartate in rat cerebellar slices, Neurosci. Lett.,
79:35-39.

Germano, I.M., Pitts, L.H., Meldrum, B.S., Bartkowski, H.M., and Simon, R.P.,
1987, Kynurenate inhibition of cell excitation decreases stroke size and
deficits, Ann. Neurol., 22:730-734.

Heyes, M.P., and Markey, S.P., 1988, Quantification of quinolinic acid in rat
brain, whole blood and plasma by gas chromatography and negative chemical
ionization mass spectrometry: effects of systemic L-tryptophan administration on brain and blood quinolinic acid concentrations, Anal. Biochem.,
174:349-359.

Heyes, M.P., and Nowak, T.S.J., 1990, Delayed increases in regional brain
quinolinic acid follow transient ischemia in the gerbil, J. Cereb. Blood
Flow Metab., 10:660-667.

Heyes, M.P., Pappagapiou, M., Leonard, C., Markey, S.P., and Auer, R.N.,
1990, Brain and plasma quinolinic acid concentrations in profound hypoglycemia, J. Neurochem., 54:1027-1033.

Kawai, J., Okuno, E., and Kido, R., 1988, Organ distribution of rat kynureninase and changes of its activity during development, Enzyme, 39:181-
189.

Kessler, M., Terramani, T., Lynch, G., and Baudry, M., 1989, A glycine site
associated with N-methyl-D-aspartic acid receptors: characterization and
identification of a new class of antagonists, J. Neurochem., 52:1319-1328.

Kim, J.P., and Choi, D.W., 1987, Quinolinate neurotoxicity in cortical cell
culture, Neuroscience, 23:423-432.

Köhler, C., Eriksson, L.G., Okuno, E., and Schwarcz, R., 1988, Localization
of quinolinic acid metabolizing enzymes in the rat brain. Immunohistochemical studies using antibodies to 3-hydroxyanthranilic acid oxygenase and
quinolinic acid phosphoribosyltransferase, Neuroscience, 27:49-76.

Köhler, C., Okuno, E., Flood, P.R., and Schwarcz, R., 1987, Quinolinic acid
phosphoribosyltransferase: preferential glial localization in the rat
brain visualized by immunocytochemistry, Proc. Natl. Acad. Sci. USA, 84:
3491-3495.

Lapin, I.P., 1978, Stimulant and convulsive effects of kynurenines injected
into brain ventricles in mice, J. Neural Transm., 42:37-43.

Lapin, I.P., 1981, Kynurenines and seizures, Epilepsia, 22:257-265.

Mangano, R.M., and Schwarcz, R., 1983, Chronic infusion of endogenous excitatory amino acids into rat striatum and hippocampus, Brain Res. Bull., 10:
47-51.

McLean, I., and Nakane, P., 1974, Periodate-lysine paraformaldehyde fixative:
a new fixative for immunoelectronmicroscopy, J. Histochem. Cytochem., 22:
1077-1083.

Minatogawa, Y., Noguchi, T., and Kido, R., 1974, Kynurenine pyruvate transaminase in rat brain, J. Neurochem., 23:271-272.

Moroni, F., Lombardi, G., Carlà, V., Lal, S., Etienne, P., and Nair, N.P.U.,
1986, Increase in the content of quinolinic acid in cerebrospinal fluid
and frontal cortex of patients with hepatic failure, J. Neurochem., 47:
1667-1671.

Moroni, I., Lombardi, G., Carlà, V., and Moneti, G., 1984, The excitotoxin
quinolinic acid is present and unevenly distributed in the rat brain,
Brain Res., 295:352-355.

Moroni, F., Russi, P., Lombardi, G., Beni, M., and Carlà, V., 1988, Presence
of kynurenic acid in the mammalian brain, J. Neurochem., 51:177-180.

Okuno, E., Du, F., Ishikawa, T., Tsujimoto, M., Nakamura, M., Schwarcz, R.,
and Kido, R., 1990, Purification and characterization of kynurenine-pyruvate aminotransferase from rat kidney and brain, Brain. Res., 534:37-44.

Okuno, E., Köhler, C., and Schwarcz, R., 1987, Rat 3-hydroxyanthranilic acid
oxygenase: purification from the liver and immunocytochemical localization
in the brain, J. Neurochem., 49:771-780.

Okuno, E., Nakamura, M., and Schwarcz, R., Two kynurenine aminotransferases
in rat brain, Brain Res., in press.

Okuno, E., Schmidt, W., Parks, D.A., Nakamura, M., and Schwarcz, R., Measurement of rat brain kynurenine aminotransferase at physiological kynurenine concentrations, J. Neurochem., in press.

Okuno, E., and Schwarcz, R., 1985, Purification of quinolinic acid phosphoribosyltransferase from rat liver and brain, Biochim. Biophys. Acta, 841:
112-119.

Okuno, E., White, R.J., and Schwarcz, R., 1988, Quinolinic acid phosphoribosyltransferase: purification and partial characterization from human
liver and brain, J. Biochem., 103:1054-1059.

Perkins, M.N., and Stone, T.W., 1982, An iontophoretic investigation of the
actions of convulsant kynurenines and their interaction with the endogenous excitant quinolinic acid, Brain Res., 247:184-187.

Reynolds, G.P., Pearson, S.J., Halket, J., and Sandler, M., 1988, Brain
quinolinic acid in Huntington's disease, J. Neurochem., 50:1959-1960.

Rothman, S.M., and Olney, J.W., 1986, Glutamate and the pathophysiology of
hypoxic-ischemic brain damage, Ann. Neurol., 19:105-111.

Schwarcz, R., Foster, A.C., French, E.D., Whetsell, W.O. Jr., and Köhler, C.,
1984, Excitotoxic models for neurodegenerative disorders, Life Sci., 35:
19-32.

Schwarcz, R., and Köhler, C., 1983, Differential vulnerability of central
neurons to quinolinic acid, Neurosci. Lett., 38:85-90.

Schwarcz, R., and Meldrum, B., 1985, Excitatory amino acid antagonists
provide a novel therapeutic approach to neurological disorders, Lancet,
2:140-143.

Schwarcz, R., Okuno, E., Speciale, C., Whetsell, Jr. W.O., and Köhler, C.,
1987, Neuronal degeneration in animals and man: the quinolinic acid
connection. in: "Neurotoxins and their Pharmacological Implications",
Jenner, P.G., ed., Raven, pp. 19-32.

Schwarcz, R., Okuno, E., and White, R.J., 1989, Basal ganglia lesions in the
rat: effects on quinolinic acid metabolism, Brain Res., 490:103-109.

Schwarcz, R., Okuno, E., White, R.J., Bird, E.D., and Whetsell, W.O. Jr.,
1988a, 3-Hydroxyanthranilic acid oxygenase activity is increased in the
brains of Huntington disease victims, Proc. Natl. Acad. Sci. USA, 85:
4079-4081.

Schwarcz, R., Tamminga, C., Kurlan, R., and Shoulson, I., 1988b, Cerebrospinal fluid levels of quinolinic acid in Huntington's disease and schizophrenia, Ann. Neurol., 24:580-582.

Schwarcz, R., Turski, W.a., and Gramsbergen, J.B.P., 1988c, Enhanced kynurenic acid synthesis in cortical and hippocampal slices from aged rats,
Soc. Neurosci. Abstr., 14:99.7.

Schwarcz, R., Whetsell, W.O. Jr., and Mangano, R.M., 1983, Quinolinic acid: an endogenous metabolite that causes axon-sparing lesions in rat brain, **Science**, 219:316-318.

Shibata, K., 1988, Fluorimetric micro-determination of kynurenic acid, an endogenous blocker of neurotoxicity, by high-performance liquid chromatography, J. **Chromatogr.**, 430:376-380.

Simon, R.P., Young, R.S.K., Stont, S., and Cheng, J., 1986, Inhibition of excitatory neurotransmission with kynurenate reduces brain edema in neonatal anoxia, **Neurosci. Lett.**, 71:361-364.

Speciale, C., Hares, K., Schwarcz, R., and Brookes, N., 1989a, High-affinity uptake of L-kynurenine by a Na^+-independent transporter of neutral amino acids in astrocytes, J. **Neurosci.**, 9:2066-2072.

Speciale, C., Okuno, E., and Schwarcz, R., 1987, Increased quinolinic acid metabolism following neuronal degeneration in the rat hippocampus, **Brain Res.**, 436:18-24.

Speciale, C., and Schwarcz, R., 1990, Uptake of kynurenine into rat brain slices, J. Neurochem., 54:156-163.

Speciale, C., Ungerstedt, U., and Schwarcz, R., 1989b, Production of extracellular quinolinic acid in the striatum studied by microdialysis in unanesthetized rats, **Neurosci. Lett.**, 104:345-350.

Speciale, C., Wu, H.-Q., Gramsbergen, J.B.P., Turski, W.A., Ungerstedt, U., and Schwarcz, R., 1990, Determination of extracellular kynurenic acid in the striatum of unanesthetized rats: effect of aminooxyacetic acid, **Neurosci. Lett.**, 116:198-203.

Swartz, K.J., During, M.J., Freese, A., and Beal, M.F., 1990, Cerebral synthesis and release of kynurenic acid: an endogenous antagonist of excitatory amino acid receptors, J. Neurosci., 10:2965-2973.

Takahashi, H., Kaihara, M., and Price, J.M., 1956, The conversion of kynurenic acid to quinaldic acid by humans and rats, J. **Biol. Chem.**, 223:705-708.

Turski, W.A., Gramsbergen, J.B.P., Traitler, H., and Schwarcz, R., 1989, Rat brain slices produce and liberate kynurenic acid upon exposure to L-kynurenine, J. Neurochem., 52:1629-1636.

Turski, W.A., Nakamura, M., Todd, W.P., Carpenter, B.K., Whetsell, W.O. Jr., and Schwarcz, R., 1988, Identification and quantification of kynurenic acid in human brain tissue, **Brain Res.**, 454:164-169.

Turski, W.A., and Schwarcz, R., 1988, On the disposition of intrahippocampally injected kynurenic acid in the rat, **Exp. Brain Res.**, 71:563-567.

Watkins, J.C., Krogsgaard-Larsen, P., and Honoré, 1990, Structure-activity relationships in the development of excitatory amino acid receptor agonists and competitive antagonists, **Trends Pharmacol. Sci.**, 11:25-33.

Whetsell, W.O. Jr., and Schwarcz, R., 1983, The organotypic tissue culture model of corticostriatal system used for examining amino acid neurotoxicity and its antagonism: studies of kainic acid, quinolinic acid and (-) 2-amino-7-phosphonoheptanoic acid, J. **Neural Transm.**, Suppl., 19:53-63.

Wolfensberger, M., Amsler, U., Cuénod, M., Foster. A.C., Whetsell, W.O. Jr., and Schwarcz, R., 1983, Identification of quinolinic acid in rat and human brain tissues, **Neurosci. Lett.**, 41:247-252.

Wu, H.-Q., and Schwarcz, R., 1990, Dual regulation of kynurenic acid production in rat hippocampus in vivo, Soc. Neurosci. Abstr., 16:200.11.

KYNURENATE FORMING ENZYMES IN LIVER, KIDNEY AND BRAIN

R. Kido

Department of Biochemistry
Wakayama Medical College
Wakayama 640
Japan

INTRODUCTION

Kynurenate was discovered by Liebig in 1853. The name he chose for this compound is still appropriate, since it occurs in higher concentration in the urine of dogs and closely related species than in the urine of other animals (Gordon et al., 1936; Jackson, 1939; Brown and Price, 1956). The discovery of tryptophan by Hopkins and Cole (1901) was followed shortly by the observation (Ellinger, 1904) that kynurenate was a metabolite of this amino acid. By the time the conversion of tryptophan-^{14}C to kynurenate-^{14}C was demonstrated (Heidelberger et al., 1949), it came as no surprise. On the other hand, the Kotake school found that kynurenine was an intermediary metabolite of tryptophan (Kotake and Iwao, 1931) and a metabolic precursor of kynurenate (Kotake, 1931).

In recent years, exogenously applied kynurenate has been shown to block synaptic neurotransmission in several parts of the nervous system (Ganong et al., 1983; Herrling, 1985) and to exert prominent antiexcitotoxic and anticonvulsive (Foster et al., 1984), neonatal anoxia/neuroprotective (Simon et al., 1986) and myorelaxant (Turski et al., 1985) effects in animal models. It is therefore appropriate to review kynurenate forming enzymes in the liver, kidney and brain as studied mainly in rats and humans.

For many years, it has been believed that the conversion of kynurenine to kynurenate is catalyzed exclusively by the enzyme kynurenine aminotransferase (EC 2.6.1.7; L-kynurenine:2-oxo-glutarate aminotransferase). In 1977, it was shown that this enzyme was identical with 2-aminoadipate aminotransferase (EC 2.6.1.39) in rat kidney (Tobes and Mason, 1977). More recently, it has been found that several different enzymes catalyze the transamination between kynurenine and various 2-oxoacids to form kynurenate in mammalian species (Minatogawa et al., 1974; Nakatani et al., 1974; Noguchi et al., 1975a; Harada et al., 1978).

RAT ENZYMES

In rat tissues, there are at least four types of enzymes which are capable of catalyzing the kynurenine-2-oxoacids transamination reaction to form kynurenate (Noguchi et al., 1973; Minatogawa et al., 1974; Nakatani et al.,

Table 1. Rat kynurenine aminotransferases

Enzyme No.	pI	pH Optimum	Aminoacceptor	Molecular weight	Identical with
1	5.2	9.0-0.3	Pyruvate	79,000	Gln T (EC 2.6.1.15) HPT-2 (EC 2.6.1.-)
2	6.6	6.0-6.5	2-Oxoglutarate	100,000	2-Aminoadipate T (EC 2.6.1.39)
3	8.0	8.0-8.5	Glyoxylate	80,000	AGT-1 (EC 2.6.1.44)
			2-Oxoglutarate		SPT (EC 2.6.1.51)
			Pyruvate		HPT-1 (EC 2.6.1.-)
4	9.4	8.0-8.5	2-Oxoglutarate	88,000	m-GOT (EC 2.6.1.1) m-TGT (EC 2.6.1.5)

1974; Noguchi et al., 1975a,b,c; Harada et al., 1978), designated Enymes 1, 2, 3 and 4. A summary of some of their properties is given in Table 1.

In the liver, all four enzymes exist. Enzyme 1 has a pI of 5.2 and a pH optimum between 9.0 and 9.3 (Noguchi and Kido, 1976). The enzyme is active with pyruvate as amino acceptor but not with 2-oxoglutarate, and utilizes various aromatic amino acids as amino donors. The apparent K_m values are about 1.4 mM and 1.3 mM for kynurenine and pyruvate, respectively (Noguchi and Kido, 1976). The molecular weight is estimated to be approximately 79,000 (Noguchi and Kido, 1976). This enzyme is identical with histidine-pyruvate aminotransferase isoenzyme 2 (EC 2.6.1.-) (Noguchi and Kido, 1976; Harada et al., 1978) and with glutamine-oxo acid aminotransferase (EC 2.6.1.15) (Noguchi et al., 1977). Enzyme 2 possesses a pI of 6.6 (Tobes and Mason, 1977) and a pH optimum between 6.0 and 6.5 (Noguchi et al., 1975a). It is specific for 2-oxoglutarate as amino acceptor (Tobes and Mason, 1977). The molecular weight is about 100,000 (Noguchi et al., 1975a). This is historically called kynurenine aminotransferase (EC 2.6.1.7) which is identical with 2-aminoadipate aminotransferase (EC 2.6.1.39) (Noguchi et al., 1975a; Tobes and Mason, 1975; Tobes and Mason, 1977). Enzyme 3 is inducible by glucagon (Noguchi et al., 1976a; Harada et al., 1978; Noguchi et al., 1978). It has a dimeric structure formed by identical subunits (Noguchi et al., 1976b). The molecular weight is approximately 80,000 (Noguchi et al., 1976b). The enzyme has a pI of 8.0 and a pH optimum between 8.0 and 8.5 (Harada et al., 1978). This enzyme is identical with histidine-pyruvate aminotransferase isoenzyme 1 (EC 2.6.1.-) (Noguchi and Kido, 1976; Harada et al., 1978), serine-pyruvate aminotransferase (EC 2.6.1.51) (Noguchi et al., 1976a,b) and alanine-glyoxylate aminotransferase (EC 2.1.6.44) (Noguchi et al., 1978). Enzyme 4 has a molecular weight of 88,000, a pH optimum between 8.0 and 8.5, and a pI of 9.4 (Noguchi et al., 1975a). This enzyme is identical with mitochondrial aspartate-2-oxoglutarate aminotransferase (EC 2.6.1.1) and mitochondrial tyrosine-2-oxoglutarate aminotransferase (EC 2.6.1.5) (Miller and Litwack, 1971; Noguchi et al., 1975a). Rat kidney contains Enzymes 1, 2

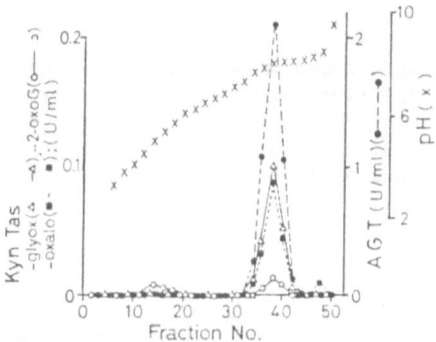

Fig. 1.   Isoelectric focusing profile of human liver extract.

and 4 (Noguchi et al., 1975c; Tobes and Mason, 1975; Tobes and Mason 1977). The brain contains Enzymes 1 and 4 (Minatogawa et al., 1974; Noguchi et al., 1975a).

HUMAN ENZYMES

In the liver there are two forms of kynurenate forming enzymes. The first is kynurenine-glyoxylate aminotranferase which is identical with alanine-glyoxylate aminotransferase and serine-pyruvate aminotransferase. The enzyme has a pI of 8.3, a pH optimum between 9.0 and 9.5, and a molecular weight of 90,000 with two identical subunits (Okuno et al., 1980). This enzyme corresponds to Enzyme 3. The second is kynurenine-2-oxoglutarate aminotransferase with a pI of 5.0 as shown in Fig. 1. This is identical with 2-amino-adipate aminotransferase which corresponds to Enzyme 2. In the kidney, no kynurenine aminotransferase activity is detectable (unpublished data).

Isoelectric focusing of human brain extract results in the appearance of two activity peaks of kynurenate forming enzymes (Fig. 2). One of them is that of glutamine phenylpyruvate aminotransferase with a pI of 5.2, coinciding with the activity of kynurenine pyruvate aminotransferase which has a pH optimum between 9.0 and 9.2 and may correspond to Enzyme 1. The second is the peak of kynurenine 2-oxoglutarate aminotransferase activity with a pI of 6.0 and a pH optimum between 8.0 and 8.5, coinciding with 2-aminoadipate aminotransferase activity which may correspond to Enzyme 2.

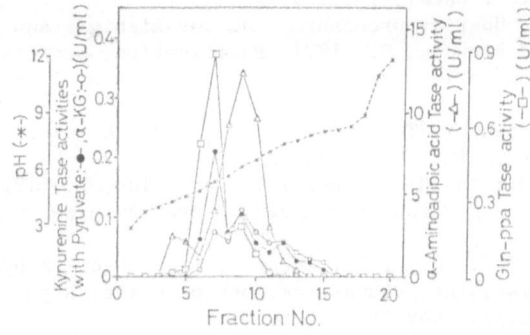

Fig. 2.   Isoelectric focusing profile of human brain extract.

CONCLUSION

Kynurenate forming enzymes were surveyed in the liver, kidney and brain of rats and humans. Rat liver contains all four types of enzymes, the kidney three types, i.e. Enzymes 1,2 and 4, and the brain two types, i.e. Enzymes 1 and 4. In humans, the liver contains two types, i.e. Enzymes 2 and 3. The kidney does not contain any of these enzymes. The brain harbors two types, possibly Enzymes 1 and 2.

REFERENCES

Brown, R.R., and Price, J.M., 1956, Quantitative studies on metabolites of tryptophan in the urine of the dog, cat, and man, J. Biol. Chem., 219:985-997.

Ellinger, A., 1904, Die Entstehung der Kynurensäure, Z. Physiol. Chem., 43:325-337.

Foster, A.C., Vezzani, A., French, E.D., and Schwarcz, R., 1984, Kynurenic acid blocks neurotoxicity and seizures induced in rats by the related brain metabolite quinolinic acid, Neurosci. Lett., 48:273-278.

Ganong, A.H., Lanthorn, T.H., and Cotman, C.W., 1983, Kynurenic acid inhibits synaptic and amino acid-induced reponses in the rat hippocampus and spinal cord, Brain Res., 27:170-174.

Gordon, W.G., Kaufman, R.E., and Jackson, R.W., 1936, The excretion of kynurenic acid by the mammalian organism. A method for the identification of small amounts of kynurenic acid, J. Biol. Chem., 113:125-134.

Harada, I., Noguchi, T., and Kido, R., 1978, Purification and characterization of aromatic-amino-acid-glyoxylate aminotransferase from monkey and rat liver, Hoppe-Seyler's Z. Physiol. Chem., 359:481-488.

Heidelberger, C., Gullberg, M.E., Morgan, A.F., and Lepkovsky, S., 1949, Tryptophan metabolism. I. The mechanism of the mammalian conversion of tryptophan into kynurenine, kynurenic acid, and nicotinic acid, J. Biol. Chem., 179:143-150.

Herrling, P.L. 1985, Pharmacology of the corticocaudate excitatory postsynaptic potential in the cat: evidence for its mediation by quisqualate or kainate receptors, Neuroscience, 14:417-426.

Hopkins, F.G., and Cole, S.W., 1901, On the proteid reaction of Adamkiewiez, with contributions to the chemistry of glyoxylic acid, Proc. Roy. Soc. London, 68:21-33.

Jackson, R.W., 1939, The excretion of kynurenic acid by members of various families of the order Carnivora, J. Biol. Chem., 131:469-478.

Kotake, Y., 1931, The mechanism of kynurenic acid formation in the organism, Z. Physiol. Chem., 195:158-166.

Kotake, Y., and Iwao, J., 1931, Studies on the intermediatary metabolism of tryptophan. I. Kynurenine, an intermediary metabolic product of tryptophan, Z. Physiol. Chem., 195:139-147.

Liebig, J., 1853, Über Kynurensäure, Justus Liebig's Ann. Chem., 86:125-126.

Miller, J.E., and Litwack, G., 1971, Purification, properties, and identity of liver mitochondrial tyrosine aminotransferase, J. Biol. Chem., 246:3234-3240.

Minatogawa, Y., Noguchi, T., and Kido, R., 1974, Kynurenine pyruvate transaminase in rat brain, J. Neurochem., 23:271-272.

Nakatani, M., Morimoto, M., Noguchi, T., and Kido, R., 1974, Subcellular distribution and properties of kynurenine transaminase in rat liver, Biochem. J., 143:303-310.

Noguchi, T., and Kido, R., 1976, Identity of kynurenine:pyruvate aminotransferase with histidine:pyruvate aminotransferase, Hoppe-Seyler's Z. Physiol. Chem., 357:649-656.

Noguchi, T., Minatogawa, Y., Okuno, E., and Kido, R., 1976a, Organ distribution of rat histidine-pyruvate aminotransferase isoenzymes, Biochem. J., 157:635-641.

Noguchi, T., Minatogawa, Y., Okuno, E., Nakatani, M., Morimoto, M., and Kido, R., 1975a, Purification and characterization of kynurenine-2-oxoglutarate aminotransferase from the liver, brain and small intestine in rats, **Biochem. J.**, 151:399-406.

Noguchi, T., Nakamura, J., and Kido, R., 1973, Kynurenine pyruvate transaminase and its inhibitor in rat intestine, **Life Sci.**, 13:1001-1010.

Noguchi, T., Nakatani, M., Minatogawa, Y., and Kido, R., 1975b, Cerebral aromatic aminotransferase, **J. Neurochem.**, 25:579-582.

Noguchi, T., Nakatani, M., Minatogawa, Y., Morimoto, M., and Kido, R., 1975c, Subcellular distribution and properties of kynurenine pyruvate transaminase in rat kidney, **Hoppe-Seyler's Z. Physiol. Chem.**, 356:1245 1250.

Noguchi, T., Okuno, E., and Kido, R., 1976b, Identity of isoenzyme 1 of histidine-pyruvate aminotransferase with serine-pyruvate aminotransferase, **Biochem. J.**, 159:607-613.

Noguchi, T., Okuno, E., and Kido, R., 1977, Identity of rat kidney histidine-pyruvate aminotransferase with glutamine-oxo-acid aminotransferase, **Biochem. J.**, 161:177-179.

Noguchi, T., Okuno, E., Takada, Y., Minatogawa, Y., Okai, K., and Kido, R., 1978, Characteristics of hepatic alanine-glyoxylate aminotransferase in different mammalian species, **Biochem. J.**, 169:113-122.

Okuno, E., Minatogawa, Y., Nakamura, M., Kamoda, N., Nakanishi, J., Makino, M., and Kido, R., 1980, Crystallization and characterization of human liver kynurenine-glyoxylate aminotransferase: Identity with alanine-glyoxylate aminotransferase and serine-pyruvate aminotransferase, **Biochem. J.**, 189:581-590.

Simon, R.P., Young, R.S.K., Stout, S., and Chen, J., 1986, Inhibition of excitatory neurotransmission with kynurenate reduces brain edema in neonatal anoxia, **Neurosci. Lett.**, 71:361-364.

Tobes, M.C., and Mason, M., 1975, L-Kynurenine aminotransferase and L-α-aminoadipate aminotransferase. I. Evidence for identity, **Biochem. Biophys. Res. Comm.**, 62:390-397.

Tobes, M.C., and Mason, M., 1977, α-Aminoadipate aminotransferase: purification, characterization, and further evidence for identity, **J. Biol. Chem.**, 252:4591-4599.

Turski, L., Schwarz, M., Turski, W.A., Klockgether, T., Sontag, K.-H., and Collins, J.F., 1985, Muscle relaxant action of excitatory amino acid antagonists, **Neurosci. Lett.**, 53:321-326.

REGULATION OF PYRIDINE NUCLEOTIDE COENZYME METABOLISM

K. Shibata[1], T. Hayakawa[2], H. Taguchi[3], and K. Iwai[4]

[1]Teikoku Women's University Moriguchi
Osaka 570

[2]Toyko University of Agriculture
Tokyo 156

[3]Mie University Tsu
Mie 514

[4]Kobe Women's University
Kobe 654
Japan

INTRODUCTION

Four NAD biosynthetic pathways are known (Henderson, 1983): (I) Nicotin-amide (Nam) → nicotinamide mononucleotide (NMN) →NAD; (II) nicotinic acid (NA) → nicotinic acid mononucleotide (NaMN) → nicotinic acid adenine dinuc-leotide (NaAD) → NAD; (III) Nam → NA → NaMN → NaAD → NAD; (IV) quinolinic acid (QA) → NaMN → NaAD → NAD.

In the catabolism of NAD, five pathways are known (Henderson, 1983): (I) NAD → Nam → N^1-methylnicotinamide (MNA) → N^1-methyl-2-pyridone-5-carboxamide (2-pyr); (II) NAD → Nam → MNA → N^1-methyl-4-pyridone-3-carboxamide (4-pyr); (III) NAD → Nam → Nam N-oxide; (IV) NAD → Nam → NA → nicotinuric acid (NuA); (V) NAD → Nam → NA → N^1-methylnicotinic acid (trigonelline, TRG).

Not all of these biosynthetic and catabolic pathways would be expected to function in mammalian tissues. Therefore, the functional metabolic path-ways of NAD in various rat tissues were investigated. To elucidate the func-tional pathways of NAD biosynthesis and catabolism, the tissue distribution of enzyme activities involved in NAD metabolism, and the tissue levels and the urinary excretion of Nam and NA and their metabolites were measured. The properties of certain enzymes were also examined to investigate the regula-tory mechanisms in NAD synthesis.

MATERIALS AND METHODS

Chemicals

[2,3,7,8-^{14}C]QA (0.23 Ci/mol) was purchased from the Daiichi Pure Chem-ical Co., Ltd. [7-^{14}C]NA (1.136 Ci/mol) and [7-^{14}C]Nam (55 Ci/mol) were ob-

tained from Amersham, England. NAD, NaAD, NMN, NuA, ATP and 5-phosphori-
bosyl-1-pyrophosphate (PRPP) were obtained from Sigma Chemical Co. Nam N-
oxide was purchased from Aldrich Chemical Co., Inc. MNA chloride and TRG
were obtained from Tokyo Kasei Kogyo Co., Ltd. 2-Pyr and 4-pyr were synthesized by the methods of Pullman and Colowick (1954) and Shibata et al.
(1988a), respectively. All other chemicals used were of the highest purity
available from commercial sources.

Animals and diets

Except where specified otherwise, male Sprague-Dawley rats, weighing
around 300 g, were used. A commercial stock diet (MF from the Oriental Yeast
Co., Ltd.) or a 20% casein diet (vitamin-free milk casein, 20%; α-cornstarch,
46%; sucrose, 23%; corn oil, 5%; mineral mixture, 5%; vitamin mixture, 1%)
and water were fed ad libitum.

Analyses

Nicotinamide phosphoribosyltransferase (NamPRTase), NMN adenylyltransferase, nicotinic acid phosphoribosyltransferase (NaPRTase), NAD synthetase,
nicotinamidase and quinolinic acid phosphoribosyltransferase (QPRTase) (Shibata et al., 1986), Nam methyltransferase, (Shibata, 1986) and 2-pyr and 4-
pyr forming MNA oxidase (Shibata et al. 1988b) were measured as described in
the literature. Nam, 2-pyr and 4-pyr (Shibata et al., 1988a), MNA (Shibata,
1987b), Nam N-oxide (Shibata, 1989a) and NA and NuA (Shibata, 1988b) were
measured by HPLC methods. NAD was measured by the method of Shibata and Murata (1986).

RESULTS AND DISCUSSION

I. Biosynthetic pathways of NAD

(1) Tissue distribution of enzymes in biosynthetic pathway I (Nam → NMN →
NAD).

Table 1 shows the tissue distribution of NamPRTase and NMN adenylyltransferase, which are involved in pathway I (Shibata et al., 1986). These
two enzyme activities were found in all tissues examined: liver, kidney,
small intestine, spleen, heart, brain, testis, thigh muscle, lung and pancreas. In humans, a similar finding would not be unreasonable, because Nam,
but not NA, is normally assimilated from food. The main foods supplying niacin are meat and fish, in which niacin is in the form of Nam. In rat, the
rate-limiting step in this pathway was considered to be the NamPRTase reaction because of the lower specific activity of this enzyme. NamPRTase activities from various tissues are inhibited by NAD as shown in Table 2. This
property would be important in regulating NAD biosynthesis.

(2) Tissue distribution of enzymes in biosynthetic pathway II (NA → NaMN →
NaAD → NAD).

Table 3 shows the tissue distribution of NaPRTase, NaMN adenylyltransferase and NAD synthetase, which are involved in pathway II (Shibata et al.,
1986). NaPRTase activity was observed only in liver, kidney, heart and pancreas. The activity of NaPRTase from hog liver is inhibited by the product
NAMN (Hayakawa et al., 1984a). It has been shown that NaMN adenylyltransferase and NMN adenylyltransferase are the same protein (Cantarow and Stollar, 1977). Therefore, NaMN adenylyltransferase exists in all the tissues
examined. NAD synthetase was detected in liver, kidney and spleen. All
three enzymes were found only in liver and kidney; thus, in the small intestine, spleen, heart, brain, testis, muscle, lung and pancreas, NA could not

Table 1. Biosynthetic pathway I in rats [Nam → NMN → NAD]

	NamPRTase	NMN adenylyltransferase	Pathway I
	(nmol/hr/g wet weight)		
Liver	102 ± 23	6352 ± 529	Yes
Kidney	21 ± 2	4179 ± 1106	Yes
Small intestine	1	1153 ± 544	Yes
Spleen	16	2378 ± 796	Yes
Heart	14	485 ± 65	Yes
Brain	9	1522 ± 35	Yes
Testis	10 ± 1	412 ± 124	Yes
Muscle	21	255 ± 147	Yes
Lung	10	1235 ± 351	Yes
Pancreas	2	1899 ± 388	Yes

Values for NamPRTase in liver, kidney and testis are means ± SD
of 4 rats; for other values, tissues from 4 rats were pooled for
an individual data point. Values for NMN adenylyltransferase
are means ± SD of 5 rats.

be used for synthesizing NAD. The blood NA level was below the limit of
detection (Shibata, 1988b).

(3) Tissue distribution of enzymes in biosynthetic pathway III (Nam → NA →
NaMN → NaAD → NAD).

Pathway III differs from pathway II only in the initial step, in which
the enzyme involved, nicotinamidase, was investigated (Shibata et al., 1986).
Its tissue distribution is shown in Table 4. Activity was detected only in
liver and small intestine. The activity in the small intestine may be attributable to intestinal microorganisms, since microorganisms with strong

Table 2. Effect of NAD on the activity of NamPRTase
from various rat tissues

	Relative activity	
	+ 0.2 mM NAD	+ 1.0 mM NAD
Liver	97	52
Kidney	79	41
Small intestine	58	22
Heart	62	27
Brain	75	16
Testis	70	2
Muscle	7	5
Lung	96	52
Pancreas	87	55
Stomach	75	50

Each value is expressed as % of control (value in the
absence of NAD) and is the mean of two separate experiments.

Table 3. Biosynthetic pathway II in rats [NA → NaMN → NaAD → NAD]

	NaPRTase	NaMN adenylyl-transferase	NAD synthetase	Pathway II
		(nmol/hr/g wet weight)		
Liver	102 ± 17	6352 ± 529	590 ± 181	Yes
Kidney	70 ± 4	4179 ± 1106	262 ± 44	Yes
Small intestine	N.D.	1153 ± 54	N.D.	No
Spleen	N.D.	2328 ± 796	170 ± 23	No
Heart	11 ± 1	485 ± 64	N.D.	No
Brain	N.D.	1522 ± 35	N.D.	No
Testis	N.D.	412 ± 124	N.D.	No
Muscle	N.D.	255 ± 147	N.D.	No
Lung	N.D.	1235 ± 351	N.D.	No
Pancreas	15 ± 2	1899 ± 388	N.D.	No

NaMN adenylyltransferase = NMN adenylyltransferase. Values are means ± SD of 3-5 rats. N.D. = not detected.

nicotinamidase activity have been reported in the gastrointestinal tract (Tanigawa et al., 1970). For the rat liver enzyme, it has been reported that its K_m value for Nam is around 0.1 M (Petrack et al., 1965) and that the concentration of free Nam in liver is around 0.1 mM (Dietrich et al., 1968). Therefore, nicotinamidase does not seem to have any physiological significance. Further, an analysis of the distribution of nicotinamidase in 150 different samples from living organisms revealed generally high activity in mushrooms, shellfish, microorganisms and plants, but none or very low activity in mammals (Taguchi et al., 1988). Accordingly, the reaction Nam → NA appears normally to be absent in mammals. This is borne out by the findings that urinary excretion of NA was not detected in rats (Shibata, 1988b) and humans (Shibata and Matsuo, 1989c) after physiological intake of niacin, although it was detected when a large amount of Nam (500 mg/kg body weight) was injected into rats (Shibata, 1989b). In mammals, it occurs only when a

Table 4. Biosynthetic pathway III in rats [NAM → NA → NaMN → NaAD → NAD]

	Nicotinamidase	Pathway III
	(nmol/hr/g wet weight)	
Liver	13 ± 5	Yes
Kidney	N.D.	No
Small intestine	18 ± 6	No
Spleen	N.D.	No
Heart	N.D.	No
Brain	N.D.	No
Testis	N.D.	No
Muscle	N.D.	No
Lung	N.D.	No
Pancreas	N.D.	No

See Tables 1 and 3 for other enzyme activities in pathway III. Values are means ± SD of 4 rats.

large amount of Nam is present in the liver as a remnant of evolution (see above). Some investigators (Ijichi et al., 1966; Tanigawa et al., 1970) have stated that gastrointestinal flora may play an important role in the overall metabolism of Nam and the pyridine nucleotides in mammals; but the finding that Nam, as well as NA, is utilized by germ-free rats (Lee et al., 1972) indicates that gastrointestinal microorganisms are not important in Nam metabolism. In the present study, nicotinamidase activity was detected in small intestine, but the NA formed could not be converted to NAD since low or no activities of NamPRTase and NAD synthetase were observed (Table 3). In the liver, the synthesis of NAD by pathway III would be limited to times of high intake of Nam, through the observed activities of NaPRTase, NaMN adenylyltransferase and NAD synthetase.

(4) Tissue distribution of enzymes in biosynthetic pathway IV (QA → NaMN → NaAD → NAD).

Pathway IV differs from pathway II only in the initial step. The tissue distribution of the enzyme involved, QPRTase, was therefore investigated. As shown in Table 5, enzyme activity was detected only in liver and kidney (Shibata et al., 1986). QA is synthesized mainly from the essential amino acid tryptophan (Trp). When free Trp was added to a Trp-limited and niacin-free diet, the levels of NAD and niacin in liver but not in kidney increased depending on the level of added Trp (Shibata and Murata, 1982). These results may be attributable to the fact that aminocarboxymuconate-semialdehyde decarboxylase activity is considerably higher in the kidney than in the liver (Ikeda et al., 1965), with the result that QA is hardly synthesized from Trp in the kidney. Accordingly, pathway IV would function only in the liver.

(5) NAD and total Nam levels in various tissues.

Table 6 shows NAD (Shibata, 1987a) and total Nam (Shibata et al., 1987) levels in various tissues.

II. NaMN synthesis in liver

The above results indicate that NA and QA are not precursors of NAD in non-hepatic tissues. In non-hepatic tissues, the only precursor of NAD is Nam. Therefore, when the intake of Nam is limited, the function of the liver, where NA (pathway II) and QA (pathway IV) are converted to Nam, is very important. It has been reported that pathway I does not operate in the liver under physiological conditions because the initial step, the NamPRTase reaction, is subject to end-product inhibition by pyridine nucleotides. Under conditions, therefore, only pathways II and IV functions. It is of interest to investigate the ratio of the amount of NaMN formed from NA (NaPRTase, pathway II) and QA (QPRTase, pathway IV). In the reconstituted assay system for physiological conditions (QA, 872 μM; NA, 50 μM; PRPP, 25 μM; MgCl$_2$, 3 mM; ATP, various concentrations; purified NaPRTase (Hayakawa et al., 1984a), 8.4 units; purified QPRTase (Taguchi and Iwai, 1975), 55 units) (Hayakawa et al., 1984b), all of the NaMN was formed by QPRTase in the absence of ATP. As the ATP concentration was raised, NaMN formation from NA (by NaPRTase) increased. At 0.5 mM ATP, about 70% of NaMN was formed from QA and 30% from NA. This effect of ATP is obviously attributable to the fact that ATP greatly lowers the K_m value of NaPRTase for PRPP (from 220 μM to 13 μM) (Hayakawa et al., 1984b). In this connection, the K_m value of PRPP for hog liver QPRTase was about 25 μM at pH 7.4 (Shibata and Iwai, 1980). These results strongly suggest that the ratio of formation of NaMN by NaPRTase and QPRTase in liver is controlled by the availability of PRPP for the two enzymes. Since its level in liver is around 1 μmol/g wet weight, ATP may contribute significantly to the salvage biosynthetic pathway of NAD. In practice, the supply of NA would normally be limited, because niacin in meat and fish, its main source, is in the form of Nam (not NA), and the nicotinamidase reaction

Table 5. Biosynthetic pathway IV in rats [QA → NaMN
→ NaAD → NAD]

	QPRTase	Pathway IV
	(nmol/hr/g wet weight)	
Liver	352 ± 80	Yes
Kidney	193 ± 38	Yes
Small intestine	N.D.	No
Spleen	N.D.	No
Heart	N.D.	No
Brain	N.D.	No
Testis	N.D.	No
Muscle	N.D.	No
Lung	N.D.	No
Pancreas	N.D.	No

See Table 3 for other enzyme activities in pathway IV.
Values are means ± SD of 5 rats.

does not seem to operate during physiological niacin intake. Accordingly,
with a normal diet, NaMN is formed mainly from QA, which is synthesized from
the essential amino acid Trp. Thus, pathway IV functions in the liver. This
pathway plays an important role in supplying Nam to non-hepatic tissues.
But, it is also true that liver can efficiently synthesize NAD from NA when
NA is supplied. This NAD can be further converted in the liver to Nam, which
is a universal precursor of NAD in animal tissues.

III. Catabolic pathways of NAD

(1) NAD-degrading activities in tissues (common initial step in the five
catabolic pathways; NAD → Nam).

The tissue distribution of NAD-degrading activity was estimated by measuring the residual NAD content in tissues, stored by wrap-film at 37°C.

Table 6. NAD and total Nam levels in various
rat tissues

	NAD	Total Nam
	(nmol/g wet weight)	
Liver	753 ± 27	1259 ± 94
Kidney	616 ± 67	1061 ± 78
Small intestine	219 ± 54	453 ± 34
Spleen	144 ± 29	504 ± 29
Heart	599 ± 58	1047 ± 36
Brain	271 ± 58	457 ± 29
Testis	154 ± 9	241 ± 13
Muscle	574 ± 45	677 ± 31
Lung	106 ± 20	391 ± 50
Pancreas	233 ± 38	352 ± 16
Blood	95 ± 7	136 ± 10

Values are means ± SD of 5 rats.

Table 7 shows the results (Shibata, 1987a). After 1 hr, the NAD content decreased to around 70% in liver, kidney, brain, small intestine, testis, spleen, lung and heart. In heart and muscle, it decreased more. After 5 hr, the NAD content in each tissue decreased to around 20%. Only blood NAD content did not decrease, indicating that NAD-degrading activity is not present in blood.

(2) Tissue distribution of enzymes in catabolic pathway I (NAD → Nam → MNA → 2-pyr).

Table 8 shows the tissue distribution of Nam methyltransferase (Shibata, 1986) and 2-pyr forming MNA oxidase (Shibata, 1989c) in rats. Nam methyltransferase activity was detected only in liver and kidney. The specific activity was 4-fold higher in liver than in kidney and the tissue weight of liver, too, was 5-fold higher. Thus, the main site of methylation of Nam is the liver. For the rat liver enzyme, the activity was inhibited by 80% in the presence of 50 μM MNA (Shibata et al., unpublished data). The liver MNA content is around 20 nmol/g wet weight when rats are fed a normal diet, however its level increases to around 100 nmol/g wet weight when rats are fed a low-protein diet (Shibata, 1988a). 2-Pyr forming MNA oxidase was also detected mainly in the liver. The daily urinary excretions of MNA and 2-pyr were around 1 μmol and 0.8 μmol, respectively, when a normal diet was fed to rats (Shibata and Matsuo, 1989a). These findings indicate pathway I (NAD → Nam → MNA → 2-pyr) functions in the liver.

(3) Tissue distribution of enzymes in catabolic pathway II (NAD → Nam → MNA → 4-pyr).

Pathway II differs from pathway I only in the final step, for which Table 9 shows the tissue distribution of 4-pyr forming MNA oxidase (Shibata, 1989c). Strong 4-pyr-forming MNA oxidase activity was detected in liver. The ratio of 4-pyr-forming activity to 2-pyr-forming activity was about 7. The daily urinary excretion of 4-pyr was around 8 μmol when a normal diet was fed (Shibata and Matsuo, 1989a). Thus, catabolic pathway II (NAD → Nam → MNA → 4-pyr) would function mainly in the liver. The 4-pyr-forming MNA oxidase activity greatly decreases when the amino acids intake is not accurate (Table 10)(Shibata and Matsuo, 1989a,b). In particular, liver MNA level and the urinary excretion of MNA increase when rats are fed a low-protein diet, since the reaction MNA → 4-pyr is reduced.

Table 7. NAD-degrading activity in rat tissues (common initial
         step in the five catabolic pathways; NAD → Nam)

	0 hr	1 hr	5 hr
	%	%	%
Liver	100	67 ± 9	11 ± 3
Kidney	100	67 ± 10	11 ± 4
Small intestine	100	62 ± 6	29 ± 22
Spleen	100	72 ± 16	33 ± 16
Heart	100	37 ± 2	16 ± 9
Brain	100	62 ± 6	7 ± 9
Testis	100	68 ± 6	24 ± 4
Muscle	100	48 ± 8	14 ± 7
Blood	100	100	100

Values are means ± SD of 3 rats.

Table 8. Catabolic pathway I in rats [NAD → Nam → MNA → 2-pyr]

	Nam methyl-transferase	2-Pyr-forming MNA oxidase	Pathway I
	(nmol/hr/g wet weight)		
Liver	282 ± 47	647 ± 89	Yes
Kidney	77 ± 10	71 ± 27	Yes
Small intestine	N.D.	N.D.	No
Spleen	N.D.	N.D.	No
Heart	N.D.	N.D.	No
Brain	N.D.	N.D.	No
Testis	N.D.	N.D.	No
Muscle	N.D.	N.D.	No
Lung	N.D.	N.D.	No
Pancreas	N.D.	N.D.	No
Blood	N.D.	N.D.	No

Values are means ± SD of 5 rats.

(4) Urinary excretion of Nam N-oxide (end product of catabolic pathway III; NAD → Nam → Nam N-oxide).

The daily urinary excretion of Nam N-oxide is about 0.2 μmol when rats are fed a normal diet (Shibata, 1989a). Therefore, catabolic pathway III (NAD → Nam → Nam N oxide) functions in rats, but this pathway is not a main catabolic pathway of NAD.

(5) Urinary excretion of NA and NuA (metabolites of catabolic pathway IV; NAD → Nam → NA → NuA).

Urinary excretion of NA and NuA is not detected when rats were fed a normal diet (Shibata, 1988b). However, when a large amount of Nam (500 mg/kg of body weight) is injected into rats, the urinary excretion of NA and NuA was observed (Shibata, 1989b). Therefore, pathway IV (NAD → Nam → NA → NuA)

Table 9. Catabolic pathway II in rats [NAD → Nam → MNA → 4-pyr]

	4-Pyr-forming MNA oxidase	Pathway II
	(nmol/hr/g wet weight)	
Liver	3285 ± 1151	Yes
Kidney	239 ± 22	Yes
Small intestine	N.D.	No
Spleen	N.D.	No
Heart	N.D.	No
Brain	N.D.	No
Testis	N.D.	No
Muscle	N.D.	No
Lung	N.D.	No
Pancreas	N.D.	No

See Table 8 for Nam methyltransferase ativity. Values are means ± SD of 5 rats.

Table 10. Effect of dietary protein levels on 4-pyr-forming MNA oxidase in rats

Diet	4-Pyr-forming MNA oxidase activity
	(nmol/hr/g wet weight)
20% Casein	2877 ± 359
10% Casein	177 ± 68
- -	
40% Soy protein isolate	2986 ± 679
20% Soy protein isolate	378 ± 24
- -	
10% Soy protein isolate + 0.25% threonine + 0.31% methionine	2379 ± 799
10% Soy protein isolate	65 ± 33

Male rats of the Wistar strain (6 weeks old) were used.
Values are means ± SD of 5 rats.

functions only when excess Nam is supplied.

(6) Tissue distribution of enzymes in catabolic pathway V (NAD → Nam → NA → TRG).

NA methyltransferase activity can not be detected in the liver or kidney of various mammals, but is present in mushrooms, shellfish and plants (Taguchi et al., 1987). This distribution is similar to that of nicotinamidase (Taguchi et al., 1988). Thus, in mushrooms, shellfish and plants, catabolic pathway V (NAD → Nam → NA → TRG) functions actively; in mammals, it does not function.

(7) MNA, 2-pyr and 4-pyr levels in various tissues.

Table 11 shows the MNA, 2-pyr and 4-pyr levels in various rat tissues (Shibata, 1988a). These metabolites of Nam were only detected in liver, although the urinary excretion of NMA, 2-pyr and 4-pyr was high.

CONCLUSION

All of the tissues examined can synthesize NAD from Nam (pathway I; Nam → NMN → NAD). The biosynthesis of NAD from NA (pathway II; NA → NaMN → NaAD → NAD) and QA (synthesized from Trp: pathway IV; QA → NaMN → NaAD → NAD) seems to be limited to the liver. In pathway III (Nam → NA → NaMN → NaAD → NAD), the nicotinamidase reaction appears not to function during physiological niacin intake. Therefore, the role of the liver as a supplier of Nam, derived from NA and QA (Trp), is very important from the viewpoint of niacin nutrition. In hepatic tissue, NAD is mainly synthesized from Trp via QA. The biosynthesis of NAD from NA is also conceivable, but this pathway would not operate in vivo since the nicotinamidase reaction does not occur during

Table 11. MNA, 2-pyr and 4-pyr levels in various rat tissues

	MNA	2-pyr	4-pyr
	(nmol/g wet weight)		
Liver	19 ± 7	Trace	Trace
Kidney	N.D.	N.D.	N.D.
Small intestine	N.D.	N.D.	N.D.
Spleen	N.D.	N.D.	N.D.
Heart	N.D.	N.D.	N.D.
Brain	N.D.	N.D.	N.D.
Testis	N.D.	N.D.	N.D.
Muscle	N.D.	N.D.	N.D.
Lung	N.D.	N.D.	N.D.
Pancreas	N.D.	N.D.	N.D.
Blood	N.D.	N.D.	N.D.

Values are means ± SD of 5 rats.

physiological niacin intake. Only when NA is supplied, does this pathway function effectively to convert NA to Nam via NAD. The ratio of the amounts of NaMN formed by NaPRTase and QPRTase in liver would be controlled by the availability of PRPP for the two enzymes. Furthermore, QPRTase activity is inhibited by PRPP. Therefore, the regulation of NAD synthesis seems to be related to purine and pyrimidine nucleotide biosynthesis.

The catabolic pathway of Nam functions mainly in the liver. The metabolic fate of Nam in the liver is therefore determined by two pathways: biosynthetic pathway I (Nam → NMN → NAD) and catabolic pathway II (NAD → Nam → MNA → 4-pyr). When the animal's nutritional status is good (i.e. the tissue NAD level is normal), NamPRTase activity is subject to end-product inhibition by NAD, and excess Nam is mainly metabolized by catabolic pathway II. The liver MNA level is very low under good nutritional conditions, because the 4-pyr-forming MNA oxidase is very active. In contrast, when nutritional status is poor (i.e. tissue NAD level is low), Nam methyltransferase activity is subject to product inhibition by MNA, which accumulates in the liver as a results of extremely low 4-pyr-forming MNA oxidase activity. Nam is then re-utilized for synthesis of NAD. Accordingly, the Nam-NAD-level in the whole body seems to be regulated by the liver MNA level. The regulation of NAD metabolism in hepatic and non-hepatic tissues is summarized in Fig. 1.

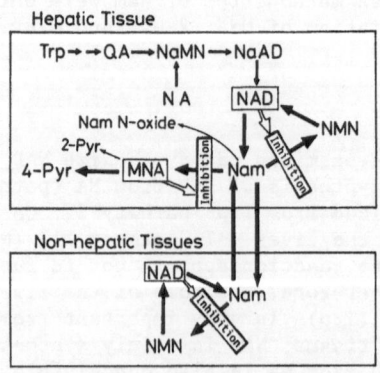

Fig. 1. Regulation of NAD metabolism in rats.

REFERENCES

Cantarow, W., and Stollar, B.D., 1977, Nicotinamide mononucleotide adenylyl-transferase, a nonhistone chromatin protein, **Arch. Biochem. Biophys.**, 180:26-34.

Dietrich, L.S., Muniz, O., and Powanda, M., 1968, NAD synthesis in animal tissues, **J. Vitaminol.**, 14:123-129.

Hayakawa, T., Shibata, K., and Iwai, K., 1984a, Purification and some properties of nicotinate phosphoribosyltransferase from hog liver, **Agric. Biol. Chem.**, 48:445-453.

Hayakawa, T., Shibata, K., and Iwai, K., 1984b, Nicotinate phosphoribosyltransferase from hog liver: regulatory effect of ATP at physiological concentrations of 5-phosphoribosyl-1-pyrophosphate, **Agric. Biol. Chem.**, 48:455-460.

Henderson, L.M., 1983, Niacin, **Ann. Rev. Nutr.**, 3:289-307.

Ijichi, H., Ichiyama, A., and Hayaishi, O., 1966, Studies on the biosynthesis of nicotinamide adenine dinucleotide. III. Comparative in vivo studies on nicotinic acid, nicotinamide and quinolinic acid as precursors of nicotinamide adenine dinucleotide, **J. Biol. Chem.**, 241:3701-3707.

Ikeda, M., Tsuji, H., Nakamura, S., Ichiyama, A., Nishizuka, Y., and Hayaishi, O., 1965, Studies on the biosynthesis of nicotinamide adenine dinucleotide. II. A role of picolinic carboxylase in the biosynthesis of nicotinamide adenine dinucleotide from tryptophan in mammals, **J. Biol. Chem.**, 240:1395-1401.

Lee, Y.C., McKenzie, R.M., Gholson, R.K., and Raica, N., 1972, A comparative study of the metabolism of nicotinamide and nicotinic acid in normal and germ-free rats, **Biochim. Biophys. Acta**, 264:59-64.

Petrack, B., Greengard, P., Craston, A., and Sheppy, F., 1965, Nicotinamide deamidase from mammalian liver, **J. Biol. Chem.**, 240:1725-1730.

Pullman, M.E., and Colowick, S.P., 1954, Preparation of 2- and 6-pyridones of N^1-methylnicotinamide, **J. Biol. Chem.**, 206:121-127.

Shibata, K., 1986, Nutritional factors affecting the activity of liver nicotinamide methyltransferase and urinary excretion of N^1-methylnicotinamide in rats, **Agric. Biol. Chem.**, 50:1489-1493.

Shibata, K., 1987a, The metabolism of niacin in each organ and the biological method for assessing the nutritional status in the rat, **Vitamins**, 61:39-56.

Shibata, K., 1987b, Ultramicro-determination of N^1-methylnicotinamide in urine by high-performance liquid chromatography, **Vitamins**, 61:599-604.

Shibata, K., 1988a, Micro-determination of nicotinamide and its metabolites by high-performance liquid chromatography, **Vitamins**, 62:225-233.

Shibata, K., 1988b, Simultaneous measurement of nicotinic acid and its major metabolites, nicotinuric acid in blood and urine by a reversed-phase high-performance liquid chromatography, **Agric. Biol. Chem.**, 52:2973-2976.

Shibata, K., 1989a, High-performance liquid chromatographic measurement of nicotinamide N-oxide in urine after extraction with chloroform, **Agric. Biol. Chem.**, 53:1329-1331.

Shibata, K., 1989b, Fate of excess nicotinamide and nicotinic acid differs in rats, **J. Nutr.**, 119:892-895.

Shibata, K., 1989c, Tissue distribution of N^1-methyl-2-pyridone-5-carboxamide- and N^1-methyl-4-pyridone-3-carboxamide-forming N^1-methylnicotinamide oxidase in rats, **Agric. Biol. Chem.**, 53:3355-3356.

Shibata, K., Hayakawa, T., and Iwai, K., 1986, Tissue distribution of the enzymes concerned with the biosynthesis of NAD in rats, **Agric. Biol. Chem.**, 50:3037-3041.

Shibata, K., and Iwai, K., 1980, Effect of 5-phosphoribosyl-1-pyrophosphate from hog kidney and hog liver, **Agric. Biol. Chem.**, 44:2785-2791.

Shibata, K., Kawada, T., and Iwai, K., 1988a, Simultaneous microdetermination of nicotinamide and its major metabolites, N^1-methyl-2-pyridone-5-carboxamide and N^1-methyl-4-pyridone-3-carboxamide, by high-performance liquid chromatography, **J. Chromat.**, 424:23-28.

Shibata, K., and Matsuo, H., 1989a, Effect of supplementing low protein diets with the limiting amino acids on the excretion of N^1-methylnicotinamide and its pyridone in rats, J. Nutr., 119:896-901.

Shibata, K., and Matsuo, H., 1989b, Effect of dietary soy protein isolate level on the ratio of N^1-methyl-2-pyridone-5-carboxamide plus N^1-methyl-4-pyridone-3-carboxamide to N^1-methylnicotinamide excretion in rats, Agric. Biol. Chem., 53:1003-1007.

Shibata, K., and Matsuo, H., 1989c, Correlation between niacin equivalent intake and urinary excretion of its metabolites, N^1-methylnicotinamide, N^1-methyl-2-pyridone-5-carboxamide, and N^1-methyl-4-pyridone-3-carbox-amide, in humans consuming a self-selected food, Am. J. Clin. Nutr., 50: 114-119.

Shibata, K., Matsuo, H., and Iwai, K., 1987, Non-uniform decrease of nicotin-amide in various tissues of rats fed on a niacin-free and tryptophan-limited diet, Agric. Biol. Chem., 51:3429-3430.

Shibata, K., and Murata, K., 1982, Niacin requirement depending on tryptophan level of diet in rat, Vitamins 56:469-477.

Shibata, K., and Murata, K., 1986, Blood NAD as an index of niacin nutrition, Nutr. Int., 2:177-181.

Shibata, K., Taguchi, H., and Iwai, K., 1988b, Effects of dietary protein levels on the enzyme activities involved in tryptophan-niacin metabolism in rats, Agric. Biol. Chem., 52:3165-3167.

Shibata, K., Taguchi, H., Nishitani, H., Okumura, K., Shimabayashi, Y., Matsushita, N., and Yamazaki, H., 1989, End product inhibition of the ac-tivity of nicotinamide phosphoribosyltransferase from various tissues of rats by NAD, Agric. Biol. Chem., 53:2283-2284.

Taguchi, H., Inamori, K., Muto, H., Okumura, K., and Shimabayashi, Y., 1988, Distribution and fundamental properties of nicotinamidase, Vitamins, 62: 399-406.

Taguchi, H., and Iwai, K., 1975, The isolation and physico-chemical proper-ties of crystalline quinolinate phosphoribosyltransferase from hog liver, Agric. Biol. Chem., 39:1493-1500.

Taguchi, H., Yamada, H., Ishihara, N., Okumura, K., and Shimabayashi, Y., 1987, Distribution and fundamental properties of nicotinate methyltrans-ferase, Vitamins, 61:355-360.

Tanigawa. Y., Shimoyama, M., Murashima, R., Ito, T., Yamaguchi, Y., and Ueda, I., 1970, The role of microorganisms as a function of nicotinamide de-amidation, Biochim. Biophys. Acta, 201:394-397.

SESSION IV

BIOLOGICAL EFFECTS: SEROTONIN AND METATONIN

SEROTONIN: ITS ROLE AND RECEPTORS IN ENTERIC NEUROTRANSMISSION

M.D. Gershon

Department of Anatomy and Cell Biology
Columbia University, College of Physicians and Surgeons
New York, New York 10032
USA

INTRODUCTION: ENTERIC STORES OF 5-HYDROXYTRYPTAMINE

The gastrointestinal tract represents the largest depot of 5-hydroxy-tryptamine (5-HT) in the body (Erspamer, 1966). Within the gut, 5-HT is found both in enteroendocrine (EC) cells in the epithelium of the mucosa (Erspamer, 1966; Nilsson et al., 1985) and in neurons of the myenteric plexus (Costa et al., 1982; Gershon, 1982; Furness and Costa, 1987). The 5-HT-containing EC cells develop from embryonic endoderm, while the neurons develop from the neural crest (Le Douarin, 1982). Neither the roles played in gastrointestinal physiology by EC cell 5-HT nor those of serotonergic neurons have been ascertained; however, 5-HT is probably involved in the regulation of gastrointestinal motility. A variety of stimuli have been reported to release 5-HT from EC cells, including mucosal pressure (Bülbring and Lin, 1958; Bülbring and Crema, 1959) and activation of splanchnic or vagus nerves (Ahlman et al., 1976; Ahlman and Dahlström, 1983; Gronstad et al., 1985). Pressure applied to the mucosa of the gut also initiates the peristaltic reflex, a response that can also be elicited by mucosal but not serosal application of 5-HT (Bülbring and Crema, 1958; Bülbring and Lin, 1958). It has thus been proposed that EC cells are pressure receptors and that they release 5-HT, not into the lumen of the bowel, but to the lamina propria, where intrinsic primary afferent nerve fibers that are sensitive to 5-HT are located. If this hypothesis is correct, then enteric neuronal 5-HT receptors would have to be present in the lamina propria. Enteric serotonergic neurons are interneurons, innervating serotonergic and other ganglion cells in both enteric plexuses (Dreyfus et al., 1977; Furness and Costa, 1982, 1987; Gershon and Sherman, 1987). Enteric neuronal 5-HT receptors, therefore, should also be found with enteric ganglia.

ENTERIC ACTIONS OF 5-HT

A wide variety of actions of 5-HT have been described in gastrointestinal preparations. In fact, so many different gastrointestinal responses to 5-HT have been reported that it has been difficult to determine which of these are of physiological significance. For example, Cl-secretion by cells of the crypt epithelium is stimulated by 5-HT (Donowitz et al., 1980; Hardcastle et al., 1981; Cooke and Carey, 1985;). This response is at least partially the result of stimulation of submucosal neurons by 5-HT (Cassuto et

al., 1982; Cooke and Carey, 1985; Cooke, 1987). Smooth muscle is activated
by 5-HT directly (Gaddum and Picarelli, 1957; Vane, 1957; Gershon, 1967) and
indirectly, via neurons. Actions of 5-HT on enteric neurons include stimulation of extrinsic (Paintal, 1964; Wallis et al., 1982; Lew and Longhurst,
1986; Ireland and Tyers, 1987) and intrinsic sensory neurons (Bülbring and
Crema, 1958, 1959; Bülbring and Lin, 1959), activation of those cholinergic
neurons that excite smooth muscle (Gaddum and Picarelli, 1957; Bülbring and
Lin, 1958; Brownlee and Johnson, 1963; Harry, 1963; Gershon, 1967; Drakontides and Gershon, 1968; Vizi and Vizi, 1978; Costa and Furness, 1979; Kamikawa and Shimo, 1983), and excitation of intrinsic non-adrenergic, non-cho-
linergic neurons, which relax smooth muscle (Gaddum and Picarelli, 1957;
Bülbring and Gershon, 1967; Gershon, 1967; Furness and Costa, 1973, 1979).
Presynaptic inhibition of the release of acetylcholine (ACh) from terminals
that form nicotinic synapses on neurons has also been described (North et
al., 1980; Sanger, 1985). Physiological responses in which an involvement of
5-HT has been implicated include relaxation of the stomach (Bülbring and
Gershon, 1967) or lower esophageal sphincter in response to vagal stimulation
(Rattan and Goyal, 1978), post-tetanic excitation of myenteric neurons
(Dingeldine and Goldstein, 1976), ascending excitation (Furness and Costa,
1973; Costa and Furness, 1976), descending inhibition of vagal excitation of
the colon (Julé, 1980), cholera toxin-induced mucosal secretion (Cassuto et
al., 1982), and regulation of the migrating myoelectric complex (Ormsbee et
al., 1984; Davidson and Pilot, 1986).

One reason that it has been difficult to define precisely the function
of gastrointestinal 5-HT is that specific antagonists of actions at particular 5-HT receptors have not been available until recently. Receptors for
5-HT in the gut were classified for the first time by Gaddum and Picarelli
(1957) who defined "M" and "D" subtypes. Responses at M receptors were those
blocked by morphine, while responses at D receptors were those blocked by dibenzyline. Unfortunately, both morphine and dibenzyline were subsequently
discovered to antagonize 5-HT responses physiologically, not through a direct
action on 5-HT receptors; nevertheless, because morphine inhibits the release
of ACh and ACh is the major excitatory neurotransmitter, morphine inhibits
intestinal contractions in response to activation of enteric neurons by 5-HT
(Paton, 1957; Schaumann, 1957; Schulz and Cartwright, 1974). Dibenzyline
non-specifically blocks smooth muscle responsiveness to 5-HT and other agents
(Boyd et al., 1963; Day and Vane, 1963; Iversen, 1963; Nishi and North 1973).
Despite its lack of direct relevance to 5-HT receptors, the classification of
Gaddum and Picarelli was useful, because it distinguished neural effects of
5-HT from those due to actions on smooth muscle. It should be pointed out,
however, that interference with the release of the mediator of the final common motor neuron in the excitatory neural pathway to enteric smooth muscle
can mask effects of 5-HT on neurons situated proximal to the final output
neuron in the myenteric plexus. Since the myenteric plexus contains micro-
circuits that are capable of mediating reflexes, even in the absence of input
from the CNS (Wood, 1987; Gershon et al., 1989), a great deal of enteric neuronal activity would be expected to occur proximal to the final common excitatory motor neuron. A clear picture of the neural actions of 5-HT thus cannot be discerned by investigating only the excitatory output of the system.
Studies of the contraction of the guinea pig ileum, however, do just this.

In contrast to studies of muscle contractivity, electrophysiological experiments with intracellular microelectrodes reveal the actions of 5-HT on
individual neurons. Electrophysiology has now been used to define enteric
neuronal 5-HT receptors (Mawe et al., 1986; Branchek et al., 1988a; Galligan
et al., 1988; Surprenant and Crist, 1988). This approach has been pursued in
association with investigations of the binding of ^{3}H-5-HT to its receptors.
^{3}H-5-HT binding can be assayed either by rapid filtration of enteric membranes or by radioautography (Branchek et al., 1984a,b, 1988a; Gershon et
al., 1985). The type of response evoked by 5-HT has been found to depend on

the type of enteric neuron from which a recording is made. There are four classes of myenteric neuron (Holman et al., 1972; Nishi and North, 1973; Hirst et al., 1974; Wood, 1987). Type I/S cells have a high input resistance, spike repeatedly when injected with depolarizing current, manifest anodal break excitation, and generate no action potentials in the presence of tetro-dotoxin. Type II/AH cells have a lower input resistance than type I/S cells, do not spike repeatedly when injected with depolarizing current, do not manifest anodal break excitation (except when synaptically excited), and display tetrodotoxin-resistant somal spikes. Most characteristically, they exhibit prolonged hyperpolarizing afterpotentials (the AH). Excitatory post-synaptic potentials (EPSPs) but not action potentials, are seen in type III/ NS neurons. A fourth type of cell is excitable when initially impaled, but eventually behaves like a type II/AH neuron. Types I/S and II/AH have been reported to respond to 5-HT. One response in II/AH cells is a long lasting depolarization (the "slow response") that is associated with a decrease in membrane conductance and diminution of hyperpolarizing afterpotentials (Wood and Meyer, 1979; Johnson et al., 1980a,b; Takaki et al., 1985a; Surprenant and Crist, 1988). These effects of 5-HT have been attributed to inhibition of the Ca^{2+}-activated K^+ conductance that gives rise to the AH. During a slow response to 5-HT, II/AH neurons become hyperexcitable. A second effect of 5-HT is a short-lived depolarization (the "fast response") of myenteric or submucosal neurons. In contrast to the slow response, the fast response is associated with a rise membrane conductance (Takaki et al., 1985a; Mawe et al., 1986; Surprenant and Crist, 1988). Finally, a hyperpolarization has also been seen in response to 5-HT. The hyperpolarization is associated with a decrease in input resistance.

SUBTYPES OF 5-HT RECEPTOR ON ENTERIC NEURONS

Two dipeptides, N-hexanoyl- and N-acetyl-5-hydroxytryptophyl-5-hydroxy-tryptophan amide (5-HTP-DP) have been reported to antagonize slow responses to 5-HT (Takaki et al., 1985a; Mawe et al., 1986). These dipeptides do not inhibit those responses to ACh (which are muscarinic) or to substance P, which are similar to the slow response to 5-HT. Neither does 5-HTP-DP affect fast EPSPs or nicotinic responses to ACh (Takaki et al., 1985a; Mawe et al., 1986). In constrast to 5-HTP-DP, ICS 205-930, a $5-HT_3$ receptor antagonist, does not inhibit slow responses to 5-HT. ICS 205-930, but not 5-HTP-DP, however, blocks fast responses to 5-HT (Mawe et al., 1986; Surprenant and Crist, 1988). The slow response to 5-HT is mimicked by 5- and 6-hydroxyindalpine and responses to the hydroxylated indalpines are antagonized by 5-HTP-DP, but not by ICS 205-930. The fast response to 5-HT is mimicked by 2-methyl-5-HT and this action of 2-methyl-5-HT is blocked by ICS 205-930. These results show that the slow and fast responses to 5-HT are mediated by different subtypes of 5-HT receptor. The receptor that mediates slow responses to 5-HT has been called the $5-HT_{1P}$ site, while that which mediates the fast response, is a $5-HT_3$ receptor (Mawe et al., 1986; Branchek et al., 1988a; Surprenant and Crist, 1988; Gershon et al., 1989). Very recently, BRL 24924, a substituted benzamide related in structure to metoclopramide, has been found to inhibit slow responses to 5-HT; therefore, BRL 24924 is a physiological antagonist of $5-HT_{1P}$ receptor-mediated responses (Branchek et al., 1988b; Mawe et al., 1989). Unlike 5-HTP-DP, BRL 24924 is also an antagonist at $5-HT_3$ receptors; thus, BRL 24924 inhibits fast responses to 5-HT when applied at a 10-fold higher concentration than that which blocks slow responses (Mawe et al., 1989). Another substituted benzamide, zacopride (and particularly one of its steroisomers, AHR 4965), which is also an antagonist at $5-HT_3$ receptors, mimics the slow response of 5-HT (Wade et al., 1989). Since this action of zacopride is antagonized by desensitization of 5-HT receptors and by BRL 24924, the drug appears to have an agonist effect at $5-HT_{1P}$ receptors. The hyperpolarizing response to 5-HT is not mimicked by hydroxylated indalpines or 2-methyl-5-HT (Mawe et al., 1986); nor is it antagonized by 5-HTP-DP (personal observation) or by ICS 205-930 (Galligan et al., 1988). The $5\text{-}HT_{1A}$ agonist, 8-hydroxy-2(di-n-propyl-amino)tetralin (8-OH-DPAT) and the non-specific $5\text{-}HT_1$ agonist 5-carboxamidotryptamine, however, do mimic the hyper-polarizing response to 5-HT, which is thus likely to be mediated by a $5\text{-}HT_{1A}$ receptor (Galligan et al., 1988). 8-OH-DPAT, like 5-HT itself, has an additional effect and presynaptically inhibits ACh release (Surprenant and Crist, 1988). The hyperpolarizaing and presynaptic responses to 5-HT are antagonized by spiperone, but spiperone is not a specific 5-HT antagonist; therefore, effects of spiperone do not adequately define subtypes of 5-HT receptor. In summary, electrophysiological studies imply that $5\text{-}HT_{1P}$, $5\text{-}HT_3$, and $5\text{-}HT_{1A}$ receptors exist in the ENS and each is responsible for mediating a different response of enteric neurons to 5-HT.

Enteric neuronal $5\text{-}HT_{1P}$ receptors can be characterized by radioligand binding techniques using either $^3H\text{-}5\text{-}HT$ or $^3H\text{-}5$-hydroxyindalpine (5-OHIP) (Branchek et al., 1984b; Gershon et al., 1985; Branchek and Gershon, 1987; Branchek et al., 1988a) as well as by electrophysiology. The binding of $^3H\text{-}5\text{-}HT$ to myenteric membranes is saturable and dissociable ($K_D \approx 3$ nM). The Hill coefficient is 0.96; therefore, there is probably one high affinity enteric $^3H\text{-}5\text{-}HT$ binding site without positive or negative cooperativity. Similarly, a single high affinity binding site ($K_D = 7.6$ nM) has been detected with the $5\text{-}HT_{1P}$ agonist, $^3H\text{-}5\text{-}OHIP$, as the radioligand (Branchek et al., 1988a). The binding of $^3H\text{-}5\text{-}HT$ or $^3H\text{-}5\text{-}OHIP$ to enteric membranes is not antagonized by substances known to bind to receptors for other neurotransmitters, to $5\text{-}HT_{1A,B \text{ or } C}$, $5\text{-}HT_2$ or $5\text{-}HT_3$ receptors, or to sites responsible for transmembrane transport of $^3H\text{-}5\text{-}HT$. In contrast, the binding of $^3H\text{-}5\text{-}HT$ is antagonized by compounds that act as antagonists or agonists at $5\text{-}HT_{1P}$ receptors, such as 5-HTP-DP and 5- or 6-hydroxyindalpine (Brancheck et al., 1984b, 1988b; Takaki et al., 1985a; Mawe et al., 1986). The high affinity $^3H\text{-}5\text{-}HT$ binding sites in isolated enteric neuronal membranes, therefore, are likely to be $5\text{-}HT_{1P}$ receptors.

Radioautography has been utilized to visualize specific $^3H\text{-}5\text{-}HT$ and $^3H\text{-}5\text{-}OHIP$ binding sites in the gut (Branchek et al., 1984b, 1988a; Gershon et al., 1985; Mawe et al., 1986; Branchek and Gershon, 1987). The binding of these radioligands in radioautographs has characteristics that are identical to those found by rapid filtration (Branchek et al., 1984b, 1988a; Gershon et al., 1985; Mawe et al., 1986; Branchek and Gershon, 1987); therefore, radioautographs probably demonstrate the $5\text{-}HT_{1P}$ receptor. Initial radioautographic studies were done with a dry mount technique in order to prevent movement of bound $^3H\text{-}5\text{-}HT$ during processing. This type of method does not provide the resolution needed to locate $^3H\text{-}5\text{-}HT$ binding sites on a cellular level. More recently, radioautography has been done following the coupling of bound $^3H\text{-}5\text{-}HT$ to tissue protein with glutaraldehyde. This step of postfixation does not extract the label or cause it to diffuse; however, it prevents translocation of $^3H\text{-}5\text{-}HT$ during the radioautographic procedure. Glutaraldehyde fixation also makes it possible to dip $^3H\text{-}5\text{-}HT$-labeled sections in liquified photographic emulsion, instead of apposing them to film as done with dry mounts, thereby enhancing resolution. Radioautography reveals $^3H\text{-}5\text{-}HT$ and $^3H\text{-}5\text{-}OHIP$ binding sites ($5\text{-}HT_{1P}$ receptors) in the myenteric and submucosal plexuses and also in the lamina propria, underlying the mucosal epithelium (Branchek and Gershon, 1987; Branchek et al., 1988a). The binding sites in the lamina propria are located exactly where a subepithelial plexus of nerve fibers is also found; moreover, these lamina propria $^3H\text{-}5\text{-}HT$ binding sites are virtually absent from segments of aganglionic bowel from ls/ls mutant mice (Brancheck et al., 1984a). The almost complete absence of $5\text{-}HT_{1P}$ receptors in the lamina propria of aganglionic bowel of the ls/ls mouse, implies strongly that these receptors are normally located on the mucosal processes derived from intrinsic nerve fibers, which are missing in the mutant tissue. On the other hand, the $^3H\text{-}5\text{-}HT$ binding sites in the lamina propria do not disappear following destruction of extrinsic sensory nerve fibers by neonatal administration of capsaicin or after chemical sympathectomy with 6-hydroxy-dopamine. They also remain following acute administration of 5,7-dihydroxy-tryptamine to destroy enteric serotonergic axons. The lamina propria ^{3}H-5-HT binding sites, therefore, appear not to be auto-receptors and they are not located in significant numbers on extrinsic sensory or sympathetic axons. The localization of 5-HT$_{1P}$ receptors on intrinsic nerve fibers supports the hypothesis that 5-HT may participate in the initiation of the peristaltic reflex (Bülbring and Crema, 1958, 1959; Bülbring and Lin, 1959). If 5-HT were to be released by pressure from enteroendocrine cells, it could reach and activate 5-HT$_{1P}$ receptors on intrinsic sensory nerve fibers in the lamina propria and thereby trigger the peristaltic reflex. This hypothesis could be tested in the future by determining the effect of the newly characterized 5-HT$_{1P}$ antagonists on the peristaltic reflex.

5-HT$_{1P}$ receptors are the only one of the three subtypes of 5-HT receptor that have been found on enteric neurons that has been shown to mediate a physiological action of 5-HT. Slow EPSPs are evoked in myenteric type II/AH neurons by repetitive stimulation of an interganglionic fiber tract (Wood and Mayer, 1979; Takaki et al., 1985a). Slow responses of the same neurons to 5-HT mimic these slow EPSPs. This mimicry does not itself establish that 5-HT is a mediator of slow EPSPs, because other putative neurotransmitters, including substance P and ACh (through muscarinic receptors), also evoke re-sponses that resemble a slow EPSP; however, 100% of neurons that display a slow EPSP in response to fiber tract stimulation are covered by multiple serotonergic synapses (Erde et al., 1985). Furthermore, slow EPSPs evoked by fiber tract stimulation, as well as slow responses to 5-HT, are blocked by the 5-HT$_{1P}$ receptor antagonists, 5-HTP-DP (Takaki et al., 1985a) and BRL 29424 (Branchek et al., 1988b; Mawe et al., 1989), which do not interfere with responses of myenteric type II/AH neurons to substance P or ACh. In ad-dition, endogenous 5-HT is released in active form from enteric serotonergic neurons by tryptamine (Takaki et al., 1985b). When the myenteric plexus is exposed to tryptamine the released 5-HT rapidly desensitizes 5-HT$_{1P}$ recep-tors, and responses of II/AH neurons to both fiber tract stimulation and 5-HT are lost. If exposure to tryptamine is prolonged, 5-HT is depleted from the tissue. When 5-HT stores are exhausted, slow EPSPs cannot be evoked by fiber tract stimulation, but 5-HT$_{1P}$ receptors recover from desensitization and slow responses to 5-HT are regained. 5-HT thus is one of the transmitters responsible for the mediation of slow EPSPs in the myenteric plexus; this ef-fect is mediated by 5-HT$_{1P}$ receptors. It is important to note that the 5-HT-mediated slow EPSP is the slow EPSP that is initiated by fiber tract stimula-tion. Since serotonergic axons are prominent in fiber tracts, but are spare and scattered in ganglia, it is likely that slow EPSPs, evoked by focal intraganglionic stimulation instead of by stimulating fiber tracts, are medi-ated not by 5-HT, but by another neurotransmitter, such as substance P.

The ability of substituted benzamides to interact with 5-HT$_{1P}$ receptors has not been adequately explained. Neither zacopride nor BRL 24924 displace the binding of ^{3}H-5-HT from enteric neuronal membranes (Branchek et al., 1988b and personal observation), yet zacopride is physiologically an agonist (Wade et al., 1989) and BRL 24924 an antagonist (Mawe et al., 1989) of 5-HT$_{1P}$-mediated responses. These observations suggest that zacopide and BRL 24924 act, not at the ligand recognition site of the 5-HT$_{1P}$ receptor, but on the receptor-effector coupling mechanism. ^{3}H-Zacopride binding to enteric neuronal membranes is displaced by BRL 29424, but not by 5-HT, suggesting that these compounds may bind to a common benzamide site that is distinct from the 5-HT$_{1P}$ receptor. ICS 205-930 produces a relatively slight displace-ment of ^{3}H-zacopride binding, which is expected since zacopride, like ICS 205-930, is a 5-HT$_3$ antagonist; nevertheless, the fact that displacement of ^{3}H-zacopride by ICS 205-930 is much less than that of BRL 24924 suggests that the density of 5-HT$_3$ receptors in the bowel is probably not great and that the binding of ^{3}H-zacopride in the gut to a benzamide site is more substantial than its binding to 5-HT$_3$ receptors. No specific binding of the 5-HT$_{1A}$ receptor agonist, ^{3}H-8-OH-DPAT, can be detected by rapid filtration assays. Although ^{3}H-8-OH-DPAT binding, which can be displaced by a 1000-fold excess of 5-HT can be detected by radioautography when very long exposures are used, the pattern of the ^{3}H-8-OH-DPAT labeling is diffuse and not, like that of ^{3}H-5-HT or^{3}H-5-OHIP, segregated in ganglia and the lamina propria.

The physiological significance of the 5-HT$_{1P}$ receptor has been investigated using substituted benzamides as tools. BRL 24924 and 5-HTP-DP are both gastrokinetic and increase the rate of emptying of a ^{51}Cr-labelled liquid meal from the murine stomach (Branchek et al., 1988b; Mawe et al., 1989). In contrast, the stereoisomer of zacopride, AHR 4965, decreases the rate of gastric emptying in the same assay (Wade et al., 1989). 5-HT$_3$ receptor antagonists, ICS 205-930 and AHR 4964 (a stereoisomer of zacopride that does not act as an agonist at 5-HT$_{1P}$ receptors) do not affect the rate of gastric emptying (Branchek et al., 1988b). These observations suggest, therefore, that with respect to the emptying of a liquid meal from the stomach of the mouse, 5-HT$_{1P}$ antagonists enhance emptying, 5-HT$_{1P}$ agonists inhibit emptying, and 5-HT$_3$ antagonists are without effect. Since the actions of two stereoisomers of zacopride (AHR 4965 and 4964) on gastric emptying are different despite the fact that both are 5-HT$_3$ receptor antagonists, the actions of the benzamides on gastric emptying is probably not due to their action at 5-HT$_3$ receptors. This conclusion is supported by the different actions of gastric emptying of ICS 205-930 and BRL 24924. ICS 205-930 is a far more potent 5-HT$_3$ antagonist than is BRL 24924, yet only BRL 24924 is gastrokinetic in this liquid meal assay. On the basis of these observations it is suggested that intrinsic inhibitory neurons of the murine stomach are tonically driven by serotonergic axons acting on 5-HT$_{1P}$ receptors. Blocking this serotonergic excitatory drive to neurons in a relaxant pathway by 5-HTP-DP or BRL 24924 may explain the gastrokinetic effects of these compounds. Enhancement of this drive by AHR 4965 could explain the delay in gastric emptying caused by this drug.

ABSTRACT

Enteric neural 5-HT receptors were analyzed and related to possible physiological actions of 5-HT. Receptors were identified electrophysiologically with intracellular microelectrodes and by studies of the binding of radioligands. Radioligand binding was assessed by rapid filtration and by radioautography. Three subtypes of 5-HT receptor, 5-HT$_{1P}$, 5-HT$_3$, and 5-HT$_{1A}$, were identified. 5-HT$_{1P}$ receptors were found to mediate slow depolarizations of myenteric neurons that were associated with a decrease in membrane conductance. These responses were inhibited by 5-HTP-DP and by BRL 24924 and mimicked by 5- and 6-hydroxyindalpine. 5-HT$_{1P}$ receptors were labeled with high affinity by ^{3}H-5-HT and were located on both submucosal and myenteric neurons and on processes of intrinsic neurons in the lamina propria. Serotonergic EPSPs were found to be mediated by 5-HT$_{1P}$ receptors; it is postulated that 5-HT$_{1P}$ receptors may be involved in initiation of the peristaltic reflex and in the regulation of gastic empyting. 5-HT$_3$ receptors have been shown to be responsible for fast depolarizations of myenteric and submucosal neurons associated with a rise in membrane conductance. These responses are antagonized by ICS 205-930 and mimicked by 2-methyl-5-HT. 5-HT$_{1A}$ receptors have been reported by others to mediate hyperpolarizing responses of myenteric neurons associated with a rise in membrane conductance. Hyperpolarizing responses are also elicited by the 5-HT$_{1A}$ agonist, 8-OH-DPAT. No physiological role has yet been identified for 5-HT$_3$ or 5-HT$_{1A}$ receptors in the ENS.

REFERENCES

Ahlman, H., and Dahlström, A., 1983, Vagal mechanisms controlling serotonin release from the gastrointestinal tract and pyloric motorfunction, J. Auton. Nerv. Syst., 9:119-140.

Ahlman, H., Lundberg, J., Dahlström, A., and Kewenter, J., 1976, A possible vagal adrenergic release of serotonin from enterochromaffin cells in the cat, Acta Physiol. Scand., 98:366-375.

Bornstein, J., North, R.A., Costa, M., and Furness, J.B., 1984, Excitatory synaptic potentials due to activation of neurons with short projection in the myenteric plexus, Neuroscience, 11:723-731.

Boyd, H., Burnstock, G., Campbell, G., Jowett, A., O'Shea, J., and Wood, M., 1963, The cholinergic blocking action of adrenergic blocking agents in pharmacological analysis of autonomic innervation, Br. J. Pharmacol. Chemother., 20:418-435.

Branchek, T.A., and Gershon, M.D., 1987, Development of neural receptors for serotonin in the murine bowel, J. Comp. Neurol., 258:597-610.

Branchek, T.A., Kates, M., and Gershon, M.D., 1984b, Enteric receptors for 5-hydroxytryptamine, Brain Res., 324:107-118.

Branchek, T., Mawe, G., and Gershon, M.D., 1988a, Characterization and localization of a peripheral neural 5-hydroxytryptamine receptor subtype with a selective agonist, ^{3}H-5-hydroxyindalpine, J. Neurosci., 8:2582-2595.

Branchek, T., Mawe, G., and Gershon, M.D., 1988b, Actions of BRL 24924 on enteric neurons: Role of 5-HT$_{1P}$ receptors, Proc. Symp. Cardiovascular Pharmacology of Serotonin, Amsterdam, The Netherlands.

Branchek, T., Rothman, T., and Gershon, M.D., 1984a, Serotonin receptors on the processes of intrinsic enteric neurons: Reduction in the aganglionic bowel of the ls/ls mouse, Soc. Neurosci. Abstr., 10:1097.

Brownlee, G., and Johnson, E.S., 1963, The site of the 5-hydroxytryptamine receptor on the peristaltic reflex, Br. J. Pharmacol., 21:306-322.

Bülbring, E., and Crema, A., 1958, Observations concerning the action of 5-hydroxytryptamine on the peristaltic reflex, Br. J. Pharmacol., 13:444-457.

Bülbring, E., and Crema, A., 1959, The release of 5-hydroxytryptamine in relation to pressure exerted on the intestinal mucosa, J. Physiol. (London), 146:381-407.

Bülbring, E., and Gershon, M.D., 1967, 5-Hydroxytryptamine participation in the vagal inhibitory innervation of the stomach, J. Physiol. (London), 192:823-846.

Bülbring, E., and Lin, R.C.Y., 1958, The effect of intraluminal application of 5-hydroxytryptamine and 5-hydroxytryptophan on peristalsis, the local production of 5-hydroxytryptamine and its release in relation to intraluminal pressure and propulsive activity, J. Physiol. (London), 140:381-407.

Cassuto, J., Jodal, M., Tuttle, R., and Lundgren, O., 1982, 5-Hydroxytryptamine and cholera secretion, Scand. J. Gastroenterol., 17:695-703.

Cooke, H.J., 1987, Neural and humoral regulation of small intestinal electrolyte transport, in: "Physiology of the Gastrointestinal Tract", Vol. 2, Johnson, L.R., ed., Raven Press, New York, pp. 1307-1350.

Cooke, H.J., and Carey, H.V., 1985, Pharmacological analysis of 5-hydroxytryptamine actions on guinea pig ileal mucosa, Eur. J. Pharmacol., 111:329-337.

Costa, M., and Furness, J.B., 1976, The peristaltic reflex: an analysis of the nerve pathways and their pharmacology, Naunyn-Schmiedeberg's Arch. Pharmacol., 294:47-60.

Costa, M., and Furness, J.B., 1979, The sites of action of 5-HT in nerve muscle preparations from guinea-pig small intestine and colon, Br. J. Pharmacol., 65:237-248.

Costa, M., Furness, J.B., Cuello, A.C., Verhofstad, A.A.J., Steinbusch, H.W.J., and Elde, R.P., 1982, Neurons with 5-hydroxytryptamine-like immunoreactivity in the enteric nervous system: their visualization and reactions to drug treatment, **Neuroscience**, 7:351-363.

Davidson, H.I., and Pilot, M.A., 1986, Does endogenous neuronal 5-hydroxytryptamine influence canine intestinal motility, **J. Physiol. (London)**, 376:49P.

Day, M., and Vane, J.R., 1963, An analysis of the direct and indirect actions of drugs on the isolated guinea-pig ileum, **Br. J. Pharmacol. Chemother.**, 20:150-170.

Dingeldine, R., and Goldstein, A., 1976, Effect of synaptic transmission blockade on morphine action in the guinea pig myenteric plexus, **J. Pharmacol. Exp. Ther.**, 196:97-106.

Donowitz, M., Tai, Y.-H., and Asarkof, N., 1980, Effect of serotonin on active electrolyte transport in rabbit ileum, gall bladder, and colon, **Amer. J. Physiol.**, 239:G463-G472.

Drakontides, A.B., and Gershon, M.D., 1968, 5-HT receptors in the mouse duodenum, **Br. J. Pharmacol.**, 33:480-492.

Dreyfus, C.F., Sherman, D., and Gershon, M.D., 1977, Uptake of serotonin by intrinsic neurons of the myenteric plexus grown in organotypic tissue culture, **Brain Res.**, 128:109-123.

Erde, S., Sherman, D., and Gershon, M.D., 1985, Morphology of the serotonergic innervation of physiologically identified cells of the guinea pig myenteric plexus, **J. Neurosci.**, 5:617-633.

Erspamer, V., 1966, Occurrence of indolealkylamines in nature, in: "Handbook of Experimental Pharmacology, Vol. 19, 5-Hydroxytryptamine and Related Indolealkylamines", Erspamer, V., ed., Springer, New York, pp. 132-181.

Furness, J.B., and Costa, M., 1973, The nervous release and the action of substances which affect intestinal muscle through neither adrenoreceptors nor cholineroreceptors, **Phil. Trans. Roy. Soc. Series**, B265:123-133.

Furness, J.B., and Costa, M., 1982, Neurons with 5-hydroxytryptamine-like immunoreactivity in the enteric nervous system: their projections in the guinea pig small intestine, **Neuroscience**, 7:341-350.

Furness, J.B., and Costa, M., 1987, "The Enteric Nervous System", Churchill, Livingston, New York, pp. 65-69.

Gaddum, J.H., and Picarelli, Z.P., 1957, Two kinds of tryptamine receptor, **Br. J. Pharmacol. Chemother.**, 12:323-328.

Galligan, J.J., Sukrprenant, A., Tonini, M., and North, R.A., 1988, Differential localization of $5-HT_1$ receptors on myenteric and submucosal neurons, **Am. J. Physiol.**, (Gastrointest. Liver Physiol. 18), 255:G603-G611.

Gershon, M.D., 1967, Effects of tetrodotoxin on innervated smooth muscle preparations, **Br. J. Pharmacol.**, 29:259-279.

Gershon, M.D., 1982, Enteric serotonergic neurons, in: "Biology of Serotonergic Neurotransmission", Osborne, N., ed., Wiley, New York, pp. 363-399.

Gershon, M.D., Mawe, G., and Branchek, T., 1989, 5-Hydroxytryptamine and enteric neurons, in: "The Peripheral Actions of 5-HT", Fozard, J.R., ed., Oxford Press, UK, pp. 247-264.

Gershon, M.D., and Sherman, D.L. 1987, Noradrenergic innervation of serotonergic neurons in the myenteric plexus, **J. Comp. Neurol.**, 259:193-210.

Gershon, M.D., Takaki, M., Tamir, H., and Branchek, T., 1985, The enteric neural receptor for 5-hydroxytryptamine, **Experientia**, 41:863-868.

Gronstad, K.O., DeMagistris, L., Dahlström, A., Nilsson, O., Price, B., Zinner, M.J., Jaffe, B.M., and Ahlman, H., 1985, The effects of vagal nerve stimulation on edoluminal release fo serotonin and substance P into the feline small intestine, **Scand. J. Gastroenterol.**, 20:163-169.

Hardcastle, J., Hardcastle, P.E.T., and Redfern, J.S., 1981, Action of 5-hydroxytryptamine on intestinal transport in the rat, **J. Physiol. (London)**, 320:41-55.

Harry, J., 1963, The action of drugs on the circular muscle strip from the guinea pig isolated ileum, **Br. J. Pharmacol. Chemother.**, 20:399-417.

Hirst, G.D.S., Holman, M.E., and Spence, I., 1974, Two types of neurons in the myenteric plexus of duodenum in the guinea pig, J. Physiol. (London), 236:303-326.

Holman, M.E., Hirst, G.D.S., and Spence, I., 1972, Preliminary studies of the neurons of Auerbach's plexus using intracellular microelectrodes, Aust. J. Exp. Biol. Med., 50:795-801.

Ireland, S.J., and Tyers, M.B., 1987, Pharmacological characterization of 5-hydroxytryptamine-induced depolarization of the rat isolated vagus nerve, Br. J. Pharmacol., 90:229-238.

Iversen, L.L., 1963, Uptake of noradrenalin by the isolated perfused rat heart, Br. J. Pharmacol. Chemother., 21:523-537.

Johnson, S.M., Katayama, Y., and North, R.A., 1980a, Multiple actions of 5-hydroxytryptamine on myenteric neurons of the guinea-pig ileum, J. Physiol. (London), 304:459-479.

Johnson, S.M., Katayama, Y., and North, R.A., 1980b, Slow synaptic potentials in neurons of the myenteric plexus, J. Physiol. (Lond.), 301:505-516.

Julé, Y., 1980, Nerve-mediated descending inhibition in the proximal colon of the rabbit, J. Physiol. (London), 159:361-368.

Kamikawa, Y., and Shimo, Y., 1983, Indirect action of 5-hydroxytryptamine on the isolated muscularis mucosa of the guinea pig oesophagus, Br. J. Pharmacol., 78:103-110.

Le Douarin, N.M., 1982, "The Neural Crest", Cambridge, Cambridge University Press.

Lew, W.Y.W., and Longhurst, J.C., 1986, Substance P, 5-hydroxytryptamine and bradykinin stimulate abdominal visceral afferents, Am. J. Physiol., 250: R465-R473.

Mawe, G.M., Branchek, T., and Gershon, M.D., 1986, Peripheral neural serotonin receptors: Identifications and characterization with specific agonists and antagonists, Proc. Nat. Acad. Sci. USA, 83:9799-9803.

Mawe, G.M., Branchek, T., and Gershon, M.D., 1988, Blockade of 5-HT-mediated enteric slow EPSPs by BRL 24924: Gastrokinetic effects, Am. J. Physiol., (Gastrointest. Liver Physiol.), in press.

Nilsson, O., Ericson, L.E., Dahlström, A., Steinbusch, H.W.M., and Ahlman, H., 1985, Subcellular localization of serotonin immunoreactivity in rat enterochromaffin cells, Histochemistry, 82:351-361.

Nishi, S., and North, R.A., 1973, Intracellular recording from the myenteric plexus of the guinea pig ileum, J. Physiol. (London), 231:471-491.

North, R.A., Henderson, C., Katayama, Y., and Johnson, S.M., 1980, Electrophysiological evidence of presynaptic inhibition of acetylcholine release by 5-hydroxytryptamine in the enteric nervous system, Neuroscience, 5: 581-586.

Ormsbee, H.S., Silver, D.A., and Hardy, F.E., 1984, Effects of 5-hydroxytryptamine on the migrating myoelectric complex in the canine intestine, J. Pharmacol. Exp. Ther., 231:436-440.

Paintal, A.S., 1964, Effects of drugs on vertebrae mechanoreceptors, Pharmacol. Rev., 16:341-380.

Paton, W.D.M., 1957, The action of morphine and related substances on contraction and on acetylcholine output on coaxially stimulated guinea-pig ileum, Br. J. Pharmacol. Chemother., 12:119-127.

Rattan, S., and Goyal, R.K., 1978, Evidence of 5-HT participation in vagal inhibitory pathway to opossum LES, Am. J. Physiol., 234:E273-E276.

Sanger, G.J., 1985, Three different ways in which 5-hydroxytryptamine can affect choline activity in guinea-pig isolated ileum, J. Pharm. Pharmacol., 37:584-586.

Schaumann, W., 1957, Inhibition by morphine of the release of acetylcholine from the intestine of the guinea pig, Br. J. Pharmacol., 12:115-118.

Schulz, R., and Cartwright, C., 1974, Effect of morphine on serotonin release from the myenteric plexus of the guinea pig, J. Pharmacol. Exp. Ther., 190:420-430.

Surprenant, A., and Crist, J., 1988, Electrophysiological characterization of functionally distinct 5-HT receptors on guinea-pig submucous plexus, **Neuroscience**, 24:283-295.

Takaki, M., Branchek, T., Tamir, H., and Gershon, M.D., 1985a, Specific antagonism of enteric neural serotonin receptors by dipeptides of 5-hydroxytryptophan: evidence that serotonin is a mediator of slow synaptic excitation in the myenteric plexus, J. Neurosci., 5:1769-1780.

Takaki, M., Mawe, G.M., Barasch, J., and Gershon, M.D., 1985b, Physiological responses of guinea-pig myenteric neurons secondary to the release of endogenous serotonin by tryptamine, **Neuroscience**, 16:223-240.

Vane, J.R., 1957, A sensitive method for the assay of 5-hydroxytryptamine, **Br. J. Pharmacol. Chemother.**, 12:344-349.

Vizi, V.A., and Vizi, E.S., 1978, Direct evidence for acetylcholine releasing effect of serotonin in the Auerbach's plexus, J. **Neural Transm.**, 42:127-138.

Wade, P.R., Branchek, T.A., Mawe, G.M., and Gershon, M.D., 1990, Use of stereoisomers of Zacopride to distinguish between 5-HT receptor subtypes: an intracellular study of myenteric neurons and gastric emptying, Proc. **N.Y. Acad. Sci.**, in press.

Wallis, D.I., Stansfeld, C.E., and Nash, N.L., 1982, Depolarizing responses recorded from nodose ganglion cells in the rabbit evoked by 5-hydroxytryptamine and other substances, Neuropharmacology, 21:31-40.

Wood, J.D., 1987, Physiology of enteric neurons, in: "**Physiology of the Gastrointestinal Tract**", Johnson, L.R., ed., Vol. 1, 2nd Edition, Raven Press, New York, pp. 1-41.

Wood, J.D., and Mayer, C.J., 1979, Serotonergic activation of tonic-type enteric neurons in guinea pig small bowel, J. Neurophysiol., 422:582-593.

MECHANISMS OF SEROTONERGIC AFFECT CONTROL

G.A. Kennett[1]

Dept. of Neurochemistry
Institute of Neurology
London

[1]Beecham Pharmaceuticals Research Division
Harlow, Essex
UK

INTRODUCTION

Early pharmacological studies suggested that 5-hydroxytryptamine (5-HT) might be involved in the regulation of mood and in particular of anxiety. Thus, lesions of serotonergic pathways using the specific 5-HT neurotoxins 5,6- or 5,7-dihydroxytryptamine (5,7-DHT) were reported to result in anxiolytic actions in various conflict paradigms (Pelham et al., 1977; Tye et al., 1977), although not all studies have confirmed these findings (Commissaris et al., 1981). The dorsal raphe may be of particular importance in this response since the injection of 5,7-DHT into this nucleus causes anxiolytic actions in either the conflict (Thiebot et al., 1984) or social interaction tests (File et al., 1978). The 5-HT synthesis inhibitor parachlorophenylalanine (PCPA) was also reported to have anxiolytic actions in both conflict and social interaction anxiety models (Robichaud and Sledge, 1969; Geller and Blum, 1970; File and Hyde, 1978), an effect reported to be reversed in the conflict test by the 5-HT precursor 5-hydroxytryptophan (5-HTP) (Stein et al., 1973). However, significant anxiolytic effects of PCPA have not been universally reported (Blakely and Parker, 1973; Cook and Sepinwall, 1975; Pelham et al., 1975; Kilts et al., 1982). Studies using receptor antagonists to reduce 5-HT function have produced even less clear results, probably due to the complexity of 5-HT receptor pharmacology and the lack of truly selective drugs. Nevertheless, some receptor antagonists have been reported to have anxiolytic actions. Thus the non-specific 5-HT antagonists methysergide, cyproheptadine and cinanserin (Hoyer, 1988) were reported by some to have anxiolytic actions in conflict tests (Graeff and Schoenfield, 1970; Stein et al., 1973; Geller et al., 1974; Cook and Sepinwall, 1975; Becker, 1986) but not by others (Commissaris and Rech, 1982; Kilts et al., 1982; Becker, 1986; Deacon and Gardner, 1986). Another non-specific 5-HT antagonist metergoline (Hoyer, 1988), has also been reported ineffective in conflict (Deacon and Gardner, 1986) and the social interaction test (File, 1981).

Treatments increasing 5-HT function were also suggestive. Electrical stimulation of the median raphe causes behavioral inhibition resembling fear (Graeff and Schoenfield, 1970), while the re-uptake inhibitors cause non-specific behavioral inhibition in both conflict (Kilts et al., 1982) and social interaction (File and Tucker, 1986) tests. 5-HTP exerts a pro-conflict effect (Kilts et al., 1982), but this may depend upon a peripheral action. The 5-HT releaser fenfluramine had no effect in the conflict test (Kilts et al., 1982), though it has been reported to exert anxiogenic-like effects in the elevated x maze (Chopin and Briley, 1987).

Finally, benzodiazepines reduce 5-HT turnover (Corrodi et al., 1971; Saner and Pletscher, 1979; Haefely et al., 1981; Pratt et al., 1985), reduce the rate of 5-HT neuronal firing (Pratt et al., 1979; Trulson et al., 1982; Laurent et al., 1983) and reduce 5-HT release from nerve terminals (Soubrie et al., 1983). Indeed, benzodiazepines can elicit anxiolytic responses in the conflict test following intra-raphe administration (Thiebot et al., 1982).

Recent advances in radioligand binding studies have led to the identification of at least six 5-HT receptor subtypes in the rat brain: $5-HT_{1A}$, $5-HT_{1B}$, $5-HT_{1C}$, $5-HT_{1D}$, $5-HT_2$ and $5-HT_3$. The identification of these subtypes may lead to a better understanding of the role of 5-HT in anxiety and may explain the conflicting and unclear nature of earlier studies.

$5-HT_{1A}$ Receptors and anxiety

One observation to emerge from the newly identified pharmacology was the clinically effective non-benzodiazepine anxiolytic buspirone (Goa and Ward, 1986) had high affinity for $5-HT_{1A}$ receptors (Peroutka, 1985) and acted as a $5-HT_{1A}$ agonist, reducing 5-HT turnover (Hjorth and Carlsson, 1982) and the firing rate of 5-HT neurons (Trulson and Trulson, 1986) and inducing hypothermia (Goodwin et al., 1985), hyperphagia (Dourish et al., 1986a) and ACTH release (Gilbert et al., 1988). Like benzodiazepines, buspirone and other $5-HT_{1A}$ agonists including the specific $5-HT_{1A}$ agonist 8-hydroxy-2-(di-n-propylamino) tetralin (8-OH-DPAT) (Gozlan et al., 1983; Middlemiss and Fozard, 1983) exert anxiolytic effects in the majority of studies using Vogel or Geller-Seifter conflict tests (Table 1). However, unlike benzodiazepines, their efficacy is not universally reported in the social interaction test while many studies suggest an anxiogenic profile in the elevated X-maze test (Table 1). This has led to suggestions that either these tests do not measure anxiety but some other property sensitive to benzodiazepines, or that the $5-HT_{1A}$ agonists are not truly effective anxiolytics.

Since early pharmacological studies (outlined above) had predicted that treatments enhancing 5-HT release might be anxiogenic, it was suggested that $5-HT_{1A}$ agonists might exert their anxiolytic effects by stimulating the $5-HT_{1A}$ cell body autoreceptor (Dourish et al., 1986a; Traber and Glaser, 1987) and hence reducing 5-HT neuronal firing (Sprouse and Aghajanian, 1987) and 5-HT release in terminal regions (Sharp et al., 1988). Indeed, 5-HT infusion into the dorsal raphe had been reported to produce anxiolytic effects in the conflict test which are prevented by prior lesioning of the raphe 5-HT cell bodies (Thiebot et al., 1982). Recently, this explanation for the actions of the $5-HT_{1A}$ agonists has been supported by the ability of intra-raphe infusion of 8-OH-DPAT, buspirone and ipsapirone to induce anxiolysis in both conflict and social interaction tests (Higgins et al., 1988). Since buspirone was ineffective in the social interaction test when administered peripherally, this may suggest that the drug's anxiolytic effects in the raphe can be exposed by non-specific actions elsewhere, perhaps as a result of its neuroleptic properties (Stanton et al., 1981) or through stimulation of postsynaptic $5-HT_{1A}$ receptors.

The above actions of $5-HT_{1A}$ agonists are observed shortly after acute administration of the drugs. However, studies on an animal model of depression suggest that the drugs can also exert long term effects. In this model, rats exposed to 2 h restraint stress show deficits in locomotion and increased defaecation in a 5 min open field test on the following day. These stress-induced deficits are reversed by chronic antidepressant treatment (Kennett et al., 1987a). Single injection of large doses of 8-OH-DPAT, buspirone, ipsapirone (Kennett et al., 1987a) and gepirone (Kennett et al., submitted) 2 h after the end of restraint attenuate these stress-induced locomotor deficits one day later. Since the direct responses to these drugs are relatively transient (Dourish et al., 1986b), it seemed likely that the effect might be mediated by the induction of a persistant neurochemical change. As all the drugs share a high affinity for $5-HT_{1A}$ sites, their actions on pre- and post- synaptic $5-HT_{1A}$ receptors were studied. Single injections of 8-OH-DPAT were found to selectively reduce the density of $[^3H]$-8-OH-DPAT binding sites in the raphe but not in caudate or frontal cortex when determined one day later (Beer et al., in press), suggesting that $5-HT_{1A}$ pre- but not postsynaptic sites were down-regulated, K_d values being unaltered (Table 2). Evidence from two models of presynaptic $5-HT_{1A}$ function suggests that this down-regulation is of significance. Firstly, the ability of 8-OH-DPAT to reduce 5-HT metabolism 30 min after administration (Hjorth et al., 1982; Hutson et al., 1986; Kennett et al., 1987b) is attenuated one hour after prior treatment with either 8-OH-DPAT (Kennett et al., 1987b) or gepirone (Kennett et al., submitted). Secondly, the ability of 8-OH-DPAT to induce hyperphagia (Dourish et al., 1986b; Hutson et al., 1988) is also attenuated following administration of 8-OH-DPAT, buspirone, ipsapirone (Kennett et al., 1987b) and gepirone (Kennett et al., submitted) one day earlier. However, the 8-OH-DPAT-induced 5-HT syndrome, a model of $5-HT_{1A}$ postsynaptic function (Tricklebank et al., 1984), was unaffected (Kennett et al., 1987b, submitted). The ability of $5-HT_{1A}$ agonists to down-regulate raphe autoreceptors has also been shown electrophysiologically following gepirone treatment (Blier and De Montigny, 1987).

It might be supposed that a reduction of raphe autoreceptor sensitivity would lead to a loss of negative feedback control and hence enhanced 5-HT release upon stimulation. The observation that electrical stimulation of the dorsal raphe leads to enhanced accumulation of 5-HIAA in some rat brain regions (particularly the frontal cortex) of rats treated one day earlier with 8-OH-DPAT (Beer et al., in press) or gepirone (Kennett et al., submitted) is consistent with this supposition, although Blier and De Montigny (1987) did not report elevated hippocampal postsynaptic activation following stimulation of raphe axons after chronic pre-treatment with gepirone. However, mediation of the antidepressant-like effects of 8-OH-DPAT by presynaptic mechanisms may be inferred by its prevention in rats pretreated with the 5-HT synthesis inhibitor PCPA (Beer et al., in press).

Mediation of the antidepressant-like actions of $5-HT_{1A}$ agonists by $5-HT_{1A}$ autoreceptor desensitization is consistent with reports that a number of conventional antidepressants have similar effects after chronic administration (De Montigny and Blier, 1984; Blier and De Montigny, 1985) and with evidence that $5-HT_{1A}$ agonists may be clinically effective antidepressants (Goldberg and Finnerty, 1979; Schweizer et al., 1986; Amsterdam et al., 1987; Cott et al., 1988). It is also consistent with evidence of 5-HT hypofunction commonly observed in depressives (Van Praag et al., 1989).

The above evidence argues in favor of two mechanisms by which 5-HT agonists can alter mood, one an immediate effect dependent on autoreceptor stimulation and reduced 5-HT neuronal firing and the second dependent on autoreceptor desensitization and perhaps paradoxically increased 5-HT function. However, clinical data on the anxiolytic actions of $5-HT_{1A}$ agonists does not favor a mechanism dependent solely on immediate drug effects. Thus, many clinical trials note a delayed onset of anxiolytic potency of about 2 weeks compared to either diazepam or placebo (Table 2). Preliminary evidence sug-

Table 1. Effects of 5-HT agonists in animal models of anxiety

Animal model	8-OH-DPAT	Buspirone	Ipsapirone	Gepirone
Conflict test	Engel et al., 1984(+) Petersen, 1985(+) Green and Hodges, 1986(+) Higgins et al., 1988(+) Deacon and Gardner, 1986(o)	Geller and Hartman, 1982(+) Riblet et al., 1982(+) Sanger et al., 1985(+) Taylor et al., 1985(+) Eison et al., 1986(+) Merlo Pich and Samanin, 1986(+) Higgins et al., 1988(+) Gardner, 1986(o) Budhram et al., 1986(o) Goldberg et al., 1983(o)	Traber et al., 1985(+) Amrick and Bennett, 1982(+) Higgins et al., 1988(+) Deacon and Gardner, 1986(o)	Eison et al., 1986(+)
Social interactions	Higgins et al., 1988(+)	Higgins et al., 1988(+) File, 1985(o) Gardner and Guy, 1985(o)	Higgins et al., 1988(+)	
Elevated X-maze	Johnson et al., 1986(o) File et al., 1987(o) Critchley and Handley, 1987(-)	Chopin and Briley, 1987(o) File et al., 1987(o) Pellow et al., 1987(-)	Critchley and Handley, 1988(+) Johnson et al., 1986(o) Chopin and Briley, 1987(o) File et al., 1987(o) Pellow et al., 1987(-)	

+ = Anxiolytic; o = No effect; - = Anxiogenic

Table 2. Clinical reports on buspirone and gepirone in anxiety reporting slow onset of action

Double-blind placebo controlled	Patient population	Average daily dose (mg)	Findings	Reference
no (double blind without placebo)	100 patients with generalized anxiety disorder	buspirone 16.5 mg diazepam 15 mg	Buspirone tended to have slower onset of action over two weeks but slightly more effective over four-six weeks	Feighner et al., 1982
yes	131 patients with anxiety rating of > 15 on HAM A scale	buspirone 10-30 mg diazepam 10-30 mg	Buspirone no better than placebo over first two weeks	Wheatley, 1982
yes	36 patients with generalized anxiety disorder, panic disorder or agrophobia with panic attacks	buspirone 25.5 mg diazepam 14 mg	Buspirone no different from placebo after one week Diazepam effective	Tyrer and Owen, 1984
yes	51 patients with 1° anxiety symptoms	buspirone 10 mg diazepam 10 mg	Buspirone less effective than diazepam over two weeks but as effective over four-six weeks	Tyrer et al., 1985
yes	66 outpatients with generalized anxiety disorder	buspirone 16.5 mg diazepam 13 mg	Diazepam had earlier onset over two weeks but equieffective after four weeks	Jacobsen et al., 1985
yes	60 outpatients with generalized anxiety	buspirone 5-40 mg diazepam 5-4 mg	Buspirone less effective over two weeks	Pecknold et al., 1985
yes	41 patients with HAM A scores > 18	gepirone 15-60 mg	No difference between gepirone and placebo over first two weeks	Cott et al., 1988

Table 3. Effect of 5-HT$_{1A}$ agonist pretreatment on social interaction
24 h later

	(All points are means ± SEM)	
Pretreatment	Total interaction time (sec)	Total locomotion (squares crossed)
Saline	72.5 ± 16.5	63.3 ± 16.3
8-OH-DPAT (1 mg/kg, s.c.)	171.1 ± 21.9** (+ 136%)	88.0 ± 15.3
Saline	172.5 ± 9.8	117.7 ± 5.6
Gepirone (8 mg/kg, s.c.)	246.6 ± 8.9** (+ 43%)	124.6 ± 7.8

Significantly different from saline treated group. **$P < 0.01$ by two-tailed Student's t-test.

gests that this is not a unique property of buspirone but is also observed
following gepirone treatment (Cott et al., 1988). It is therefore conceivable that the mechanism underlying the antidepressant-like actions of 5-HT$_{1A}$
agonists also underlies their clinical efficacy. One argument against this
notion is the discrepancy in dose and time course of these effects. However,
in a recent study low doses of gepirone were only found to induce antidepressant-like actions in the rat model and autoreceptor subsensitivity following
subchronic administration (Kennett et al., submitted), suggesting that the
even lower doses might achieve the same effect over 2 weeks in the clinic.
The hypothesis is also directly supported by the observed anxiolytic actions
of a single administration of 8-OH-DPAT and gepirone in the social interaction test one day later (Table 3).

5-HT$_{1C}$ Receptors and anxiety

A second mechanism by which 5-HT can interact with anxiety may be revealed by the actions of 1,3-chlorophenylpiperazine (mCPP) and 1-3-(tri-
fluoro-methyl)-phenylpiperazine (TFMPP). Those drugs have previously been
considered to be 5-HT$_{1B}$ agonists (Lucki and Frazer, 1982; Martin and Sanders-
Bush, 1982; Sills et al., 1984; Asarch et al., 1985). However, analysis of
their hypolocomotor (Kennett and Curzon, 1988a) and hypophagic (Kennett and
Curzon, 1988b, 1990) effects leads to the conclusion that they are mediated
by 5-HT$_{1C}$ receptor stimulation. This is consistent with evidence that both
drugs stim-ulate 5-HT$_{1C}$-mediated phosphoinositide hydrolysis (Conn and
Sanders-Bush, 1987) albeit weakly and that they have 10-fold greater
affinities for 5-HT$_{1C}$ than for 5-HT$_{1B}$ receptors (Hoyer, 1988).

Since anxiety has been reported in humans given mCPP (Mueller et al.,
1985; Charney et al., 1987), the effect of these drugs on a rat model of anxiety, the social interaction test (File and Hyde, 1978), was examined. Both
mCPP and TFMPP dose-dependently reduced total interaction time in the test
but only the highest dose of mCPP used (1 mg/kg, i.p.) reduced locomotion
(Kennett et al., 1989). This profile of action had been interpreted as indicating anxiogenesis (File and Hyde, 1978). The ability of chronic pretreatment with the benzodiazepine anxiolytic chlordiazepoxide to prevent the effect of mCPP in the social interaction test also argues for an anxiogenic
mechanism. Pharmacological evaluation of the action of mCPP suggests mediation by 5-HT$_{1C}$ receptor stimulation (Kennett et al., 1989) since it was
blocked by three non-specific antagonists with high affinity for 5-HT$_{1C}$ sites

(Hoyer, 1988), metergoline, mianserin and cyproheptadine (Kennett et al., 1989), but not (Kennett et al., 1989) by the 5-HT$_2$ antagonists ketanserin (Leysen et al., 1981) or ritanserin (Leysen et al., 1985). Despite the latter drug's reportedly high affinity for the 5-HT$_{1C}$ receptor (Hoyer, 1988), we have found its ID$_{50}$ in blocking mCPP-induced hypophagia to be considerably higher than the dose used in this study (Kennett and Curzon, 1990). The effect of mCPP was also not blocked (Kennett et al., 1989) by the 5-HT$_{1A}$, 5-HT$_{1B}$ and β-receptor antagonists cyanopindolol and (-) propranolol (Hoyer, 1988) nor (Kennett et al., 1989) by the specific 5-HT antagonist ICS 205 930 (Richardson et al., 1985). Indeed, mCPP is reportedly an antagonist of 5-HT$_3$ receptors on the vagal nerve (Kilpatrick et al., 1988).

The above results indicate that mCPP induces anxiety in rats as in humans. In rats, at least, it apparently does so by activating 5-HT$_{1C}$ receptors. Whether these effects are secondary to non-specific stress or a more direct response to mCPP remains to be elicited as does its site of action. However, it is of interest that several 5-HT$_{1C}$ antagonists with high affinity for 5-HT$_{1C}$ receptors, metergoline, cyproheptadine, methysergide and cinanserin (Hoyer, 1988) have all been reported to have some anxiolytic properties (see Chopin and Briley, 1987, for review). It is also of interest that mCPP-induced anxiety is greater in patients with obsessive compulsive disorder (Zohar and Insel, 1987) or panic disorder (Charney et al., 1987; Kahn et al., 1988).

In conclusion, 5-HT may have opposing effects on affect. Anxiogenic mechanisms may include the activation of 5-HT$_{1C}$ (Kennett et al., 1989), 5-HT$_2$ (Colpaert et al., 1985) or 5-HT$_3$ (Jones et al., 1988) receptors. While down-regulation of 5-HT$_{1A}$ receptors and conceivably enhanced 5-HT release may underlie the clinically effective anxiolysis produced by chronically administered 5-HT$_{1A}$ agonists, their acute actions in animal models are probably caused by reducing 5-HT release onto postsynaptic 5-HT receptors associated with anxiogenic properties. The ability of 5-HT re-uptake inhibitors to down-regulate 5-HT$_{1A}$ autoreceptors (De Montigny and Blier, 1984; Blier and De Montigny, 1985) may also underlie their reported anxiolytic efficacy (Kahn et al., 1988).

REFERENCES

Amrick, C.L., and Bennett, D.A., 1986, A comparison of the anticonflict activity of the serotonin agonists and antagonists in rat, Soc. Neurosci. Abstr., 12:907.
Amsterdam, J.D., Berwich, N., Potter, L., and Rickels, K., 1987, Open trial of gepirone in the treatment of major depressive disorder, Curr. Ther. Res., 41:185-193.
Asarch, K.E., Ransom, R.W., and Shih, J.S., 1985, 5-HT$_{1A}$ and 5-HT$_{1B}$ selectivity of two phenylpiperazine derivatives: evidence for 5-HT heterogeneity, Life Sci., 36:1265-1273.
Becker, H.C., 1986, Comparison of the effects of the benzodiazepine midazolam and three serotonin antagonists on a consummatory conflict paradigm, Pharmacol. Bioch. Behav., 24:1057-1064.
Beer, M., Kennett, G.A., Stahl, S.M., and Curzon, G., 1989, A single dose of 8-OH-DPAT reduces raphe binding of [^{3}H]-8-OH-DPAT and increases the effect of raphe stimulation on 5-HT metabolism, Eur. J. Pharmacol., in press.
Blakely, T.A., and Parker, L.F., 1973, Effects of parachlorophenylalanine on experimentally induced conflict behavior, Pharmacol. Bioch. Behav., 1: 609-613.
Blier, P., and De Montigny, C., 1985, Serotonergic but not noradrenergic neurons in rat central nervous system adapt to long term treatment with monoamine oxidase inhibitors, Neuroscience, 16:949-955.

Blier, P., and De Montigny, C., 1987, Modification of 5-HT neuron properties by sustained administration of the 5-HT$_{1A}$ agonist gepirone: electrophysiological studies in the rat brain, **Synapse**, 1:470-478.

Budhram, P., Deacon, R., and Gardner, C.R., 1986, Some putative non-sedating anxiolytics in a conditioned licking conflict, **Brit. J. Pharmacol.**, 88:331P.

Charney, D.S., Woods, S.W., Goodman, W.K., and Heninger, G.R., 1987, Serotonin function in anxiety II Effects of the serotonin agonist MCPP in panic disorder patients and healthy subjects, **Psychopharmacology**, 92:14-21.

Chopin, P., and Briley, M., 1987, Animal models of anxiety: the effect of compounds that modify 5-HT neurotransmission, **Trends Pharmacol. Sci.**, 8:383-386.

Colpaert, F.C., Meert, T.F., Niemegeers, C.J.E., Janssen, P.A.J., 1985, Behavioural and 5-HT antagonist effects of ritanserin: a pure and selective antagonist of LSD discrimination in rat, **Psychopharmacology**, 86:45-54.

Commissaris, R.L., Lyness, W.H., and Rech, R.H., 1981, The effects of lysergic acid diethylamide (LSD), 2,5-dimethoxy-4-methyl-amphetamine (DOM), pentobarbital and methaqualone on punished responding in control and 5,7-dihydroxytryptamine-treated rats, **Pharmacol. Biochem. Behav.**, 14:617-623.

Commissaris, R.L., and Rech, R.H., 1982, Interactions of metergoline with diazepam, quipazine and hallucinogenic drugs on a conflict behavior in the rat, **Psychopharmacology**, 76:282-285.

Conn, P.J., and Sanders-Bush, E., 1987, Relative efficacies of piperazines at the phosphoinositide hydrolysis-linked serotonergic (5-HT$_2$ and 5-HT$_{1C}$) receptors, **J. Pharmacol. Exp. Ther.**, 242:552-557.

Cook, L., and Sepinwall, J., 1975, Behavioral analysis of the effects and mechanisms of action of benzodiazepines, in: "Mechanisms of Action of Benzodiazepines", Costa, E., and Greengard, P., eds., Raven Press, New York, pp. 1-28.

Corrodi, H., Fuxe, K., Lidbrink, P., and Olson, L., 1971, Minor tranquilizers, stress and central catecholamine neurons, **Brain Res.**, 29:1.

Cott, J.M., Kurtz, N.M., Robinson, D.S., Lancaster, S.P., and Copp, J.E., 1988, 5-HT$_{1A}$ ligand with both antidepressant and anxiolytic properties, **Psychopharmacol. Bull.**, 24:164-167.

Critchley, M.A.E., and Handley, S.L., 1987, Effects in the X-maze anxiety model of agents acting at 5-HT$_1$ and 5-HT$_2$ receptors, **Psychopharmacology**, 93:502-506.

Critchley, M.A.E., and Handley, S.L., 1989, Dorsal raphe lesions abolish effects of 8-OH-DPAT and ipsapirone in X-maze, **Brit. J. Pharmacol.**, 96:309P.

Deacon, R., and Gardner, C.F., 1986, Benzodiazepine and 5-HT ligands in a rat conflict test, **Brit. J. Pharmacol.**, 88:330P.

De Montigny, C., and Blier, P., 1984, Effects of antidepressant treatments of 5-HT neurotransmission: electrophysiological and clinical studies, in: "Frontiers in Biochemical and Pharmacological Research in Depression", Usdin E., et al., eds., Raven Press, New York, pp. 223-239.

Dourish, C.T., Hutson, P.H., and Curzon, G., 1986a, Putative anxiolytics 8-OH-DPAT, buspirone and TURQ 7821 are agonists at 5-HT$_{1A}$ autoreceptors in the raphe nuclei, **Trends Pharmacol. Sci.**, 7:212-214.

Dourish, C.T., Hutson, P.H., Kennett, G.A., and Curzon, G., 1986b, 8-OH-DPAT-induced hyperphagia: its neural basis and possible therapeutic relevance, **Appetite**, Suppl., 7:127-140.

Eison, A.S., Eison, M.S., Stanley, M., and Riblet, L.A., 1986, Serotonergic mechanisms in the behavioral effects of buspirone and gepirone, **Pharmacol. Biochem. Behav.**, 24:701-707.

Engel, J.A., Hjorth, S., Sensson, K., Carlsson, A. and Liljequist, 1984, Anticonflict effect of the putative serotonin receptor agonist 8-OH-DPAT, **Eur. J. Pharmacol.**, 105:365-368.

Feighner, J.P., Meredith, C.H., and Hendrickson, S.A., 1982, A double blind comparison of buspirone and diazepam in outpatients with generalized anxiety disorder, J. Clin. Psychiat., 43:103-107.

File, S.E., 1981, Behavioral effects of serotonin depletion, in: "Metabolic Disorders of the Nervous System", Clifford Rose, E., ed., Pitmans, London, pp. 429-445.

File, S.E., 1985, Animal models for predicting clinical efficacy of anxiolytic drugs: social behavior, Neuropsychobiology, 13:55-62.

File, S.E., and Hyde, J.R.G., 1978, Can social interaction be used to measure anxiety?, Brit. J. Pharmacol., 62:19-24.

File, S.E., Hyde, J.R.G., and Macleod, N.K., 1979, 5,7-dihydroxytryptamine lesions of dorsal and median raphe nuclei and performance in the social interaction test of anxiety and in a home aggression test, J. Affect. Disord., 1:115-122.

File, S.E., Johnston, A.L., and Pellow, S., 1987, Effects of compounds acting at CNS 5-hydroxytryptamine systems on anxiety in the rat, Brit. J. Pharmacol., 90:265P.

File, S.E., and Tucker, J.C., 1986, Behavioral consequences of antidepressant treatment in rodents, Neurosci. Biobehav. Rev., 10:123-134.

Gardner, C.R., 1986, Recent developments in 5-HT-related pharmacology of animal models of anxiety, Pharmacol. Biochem. Behav., 24:1479-1485.

Gardner, C.R., and Guy, A.P., 1985, Pharmacological characterization of a modified social interaction model of anxiety in the rat, Neuropsychobiology, 13:194-201.

Geller, I., and Blum, K., 1970, The effects of 5-HTP on parachlorophenylalanine (pCPA) attenuation of conflice behavior, Eur. J. Pharmacol., 9:319-324.

Geller, I., and Hartmann, R.J., 1982, Effects of buspirone on operant behaviour of laboratory rats and cynomologous monkeys, J. Clin. Psychiat., 43:25-32.

Geller, I., Hartmann, R.J., and Croy, D.J., 1974, Attenuation of conflict behavior with cinanserin, a serotonin antagonist: reversal of the effect with 5-hydroxytryptophan and α-methytryptamine, Res. Commun. Chem. Pathol. Pharmacol., 7:165-175.

Gilbert, F., Brazell, C., Tricklebank, M.D., and Stahl, S.M., 1988, Activation of the $5\text{-}HT_{1A}$ receptor subtype increases rat plasma ACTH concentration, Eur. J. Pharmacol., 147:431-439.

Goa, K.L., and Ward, A., 1986, Buspirone: a preliminary review of its pharmacological properties and therapeutic efficacy as an anxiolytic, Drugs, 32:114-129.

Goldberg, H.L., and Finnerty, R.J., 1979, The comparative efficacy of buspirone and diazepam in the treatment of anxiety, Am. J. Psychiat., 136:1184-1187.

Goldberg, M.E., Salama, A.I., Patel, J.B., and Malick, J.B., 1983, Novel nonbenzodiazepine anxiolytics, Neuropharmacology, 22:1499-1508.

Goodwin, G.M., De Souza, R.J., and Green, A.R., 1985, The pharmacology of the hyptothermic response in mice to 8-hydroxy-2-(di-N-propylamino) tetralin (8-OH-DPAT): a model of presynaptic 5-HT function, Neuropharmacology, 24:1187-1194.

Gozlan, H., El Mestikawy, S., Pichat, L., Glowinski, J., and Hamon, M., 1983, Identification of presynaptic serotonin autoreceptors using a new ligand: H-PAT, Nature, 305:140-142.

Graeff, F.G., and Schoenfeld, R.I., 1970, Tryptamine mechanisms in punished and non-punished behavior, J. Pharmacol. Exp. Ther., 173:277-283.

Green, S., and Hodges, H., 1986, The lateral amygdala and benzodiazepine effects on conflict behavior in rats, Psychopharmacology, 89:5S.

Haefely, W., Pieri, L., Plc, P., and Schaffner, R., 1981, General pharmacology and neuropharmacology of benzodiazepine derivatives, in: "Handbook of Experimental Pharmacology", Vol. 55, Hoffheister, F., and Stille, G., eds., Springer, Berlin, pp. 13-262.

Higgins, G.A., Bradbury, A.J., Jones, B.J., and Oakley, N.R., 1988, Behavioural and biochemical consequences following activation of 5-HT$_1$-like and GABA receptors in the dorsal raphe nucleus of the rat, **Neuropharmacology**, 27:993-1001.

Hjorth, S., Carlsson, A., Lindberg, P., Sanchez, D., Wilkstrom, H., Arvidsson, L.E., Hacksell, U., and Nilsson, J.L.G., 1982, 8-Hydroxy-2-(di-n-propylamino)-tetralin, 8-OH-DPAT, a potent and selective simplified ergot congener with central 5-HT-receptor stimulating activity, J. **Neural Transm.**, 55:169-188.

Hjorth, S., and Carlsson, A., 1982, Buspirone: effects on central monoaminergic transmission - possible relevance to animal experimental and clinical findings, **Eur. J. Pharmacol.**, 83:299-303.

Hoyer, D., 1988, Functional correlates of 5-HT recognition sites, J. **Receptor Res.**, 8:59-81.

Hutson, P.H., Dourish, C.T., and Curzon, G., 1986, Neurochemical and behavioural evidence for mediation of the hyperphagic action of 8-OH-DPAT by 5-HT cell body autoreceptors, **Eur. J. Pharmacol.**, 129:347-352.

Hutson, P.H., Dourish, C.T., and Curzon, G., 1988, Evidence that the hyperphagic response to 8-OH-DPAT is mediated by 5-HT$_{1A}$ receptors, **Eur. J. Pharmacol.**, 150:361-366.

Jacobson, A.F., Dominguez, R.a., Goldstein, B.A., and Steinbook, R.M., 1985, Comparison of buspirone and diazepam in generalized anxiety disorder, **Pharmacotherapy**, 5:290-296.

Johnson, A.L., Pellow, S., and File, S.E., 1986, The effects of selective antagonists for 5-hydroxytryptamine receptor subtypes in a test of anxiety in the rat, **Soc. Neurosci. Abstr.**, 12:907.

Jones, B.J., Costall, B., Domeney, A.M., Kelly, M.E., Naylor, R.J., Oakley, N.R., and Tyers, M.B., 1988, The potential anxiolytic activity of GR 38032F, a 5-HT$_2$-receptor antagonist, **Brit. J. Pharmacol.**, 93:985-993.

Kahn, R.S., Van Praag, H.M., Wetzler, S., Asnis, G.M., and Barr, G., 1988, Serotonin and anxiety revisited, **Biol. Psychiat.**, 23:189-208.

Kennett, G.A., and Curzon, G., 1988a, Evidence that mCPP may have behavioral effects mediated by central 5-HT$_{1C}$ receptors, **Brit. J. Pharmacol.**, 94:137-147.

Kennett, G.A., and Curzon, G., 1988b, Evidence that hypophagia induced by mCPP and TFMPP requires 5-HT$_{1C}$ and 5-HT$_{1B}$ receptors; hypophagia induced by RU 24969 only requires 5-HT$_{1B}$ receptors, **Psychopharmacology**, 96:93-100.

Kennett, G.A., Dourish, C.T., and Curzon, G., 1987a, Antidepressant-like action of 5-HT$_{1A}$ agonists and conventional antidepressants in an animal model of depression, **Eur. J. Pharmacol.**, 134:265-274.

Kennett, G.A., Marcou, M., Dourish, C.T., and Curzon, G., 1987b, Single administration of 5-HT$_{1A}$ agonists decreases 5-HT$_{1A}$ presynaptic but not postsynaptic receptor mediated responses: relationship to antidepressant-like action, **Eur. J. Pharmacol.**, 138:53-60.

Kennett, G.A., Whitton, P., Shah, K., and Curzon, G., 1989, Anxiogenic-like effects of mCPP and TFMPP in animal models are opposed by 5-HT$_{1C}$ receptor antagonists, **Eur. J. Pharmacol.**, 164:445-454.

Kennett, G.A., Whitton, P., and Curzon, G., 1990, ID$_{50}$ values of antagonists vs. mCPP-induced hypophagia and 5-HT$_2$ mediated head shakes indicate 5-HT$_{1C}$ sites mediate the hypophagia, **Br. J. Pharmacol.**, in press.

Kennett, G.A., Whitton, P., and Curzon, G., Effects of gepirone on a depression model and on indices of pre and postsynaptic 5-HT function, **Eur. J. Pharmacol.**, submitted.

Kilpatrick, G.J., Jones, B.J., and Tyers, M.B., 1988, Identification and distribution of 5-HT$_3$ receptors in rat brain using radioligand binding, **Nature**, 330:746-748.

Kilts, C.D., Commissaris, R.L., Cordon, J.J.; and Rech, R.H., 1982, Lack of central 5-hydroxytryptamine influence on the anticonflict activity of diazepam, **Psychopharmacology**, 78:156-164.

Laurent, J.P., Margold, M., Hunkel, V., and Haefely, W., 1983, Reduction by two benzodiazepines and pentobarbitone of the multi unit activity in substantia nigra, hippocampus nucleus locus coeruleus and dorsal raphe nucleus of 'encephale isole' rats, Neuropharmacology, 22:501-511.

Leysen, J.E., Awouters, F., Kennis, L., Laduron, P.M., Vanderberk, J., and Janssen, P.A.J., 1981, Receptor binding profile of R41 468, a novel antagonist of 5-HT receptors, Life Sci., 28:1015-1022.

Leysen, J.E., Gommeren, W., Van Gompel, P., Wynants, J., Janssen, P.A.J., and Laduron, P.M., 1985, Receptor binding properties in vitro and in vivo of ritanserin, a very potent and long acting serotonin-S2 antagonist, Mol. Pharmacol., 27:600-611.

Lucki, I., and Frazer, A., 1982, Behavioral effects of indole and piperazine type serotonin receptor agonists, Soc. Neurosci. Abstr., 8:101.

Martin, L.L., and Sanders-Bush, E., 1982, Comparison of the pharmacological characteristics of 5-HT$_1$ and 5-HT$_2$ binding sites with those of serotonin autoreceptors which modulate serotonin release, Naunyn-Schmiedeberg's Arch. Pharmacol., 321:165-170.

Merlo Pich, E., and Samanin, R., 1986, Disinhibitory effects of buspirone and low doses of sulpiride and haloperidol in two experimental anxiety models in rats: possible role of dopamine, Psychopharmacology, 89:125-130.

Middlemiss, D.N., and Fozard, J.R., 1983, 8-Hydroxy-2-(di-n-propylamino)-tetralin discriminates between subtypes of the 5-HT$_1$ recognition site, Eur. J. Pharmacol., 90:151-153.

Mueller, E.A., Murphy, D.L., and Sunderland, T., 1985, Neuroendocrine effects of m-chlorophenylpiperazine, a serotonin agonist, in humans, J. Clin. Endocrinol. Metab., 61:1179-1184.

Pecknold, J.C., Familamiri, P., Chang, H., Wilson, R., Acarcia, J., and McClure, D.J., 1985, Buspirone: anxiolytic? Prog. Neuropsychopharmacol. Biol. Psychiat., 9:639-642.

Pelham, R.W., Osterberg, A.C., Thibault, L., and Tankella, T., 1975, Interactions between plasma cortisone and anxiolytic drugs on conflict behaviour in rats, Presented at 4th Int. Congr. Soc. Psychoneuroendorcrinol., Aspen, Colorado.

Pellow, S., Johnston, A.L., and File, S.E., 1987, Selective agonists and antagonists for hydroxytryptamine receptor subtypes and interactions with yohimbine and FG 7142 using elevated plus-maze test in the rat, J. Pharm. Pharmacol., 39:917-928.

Peroutka, S.J., 1985, Selective interaction of novel anxiolytics with 5-hydroxytryptamine$_{1A}$ receptors, Biol. Psych., 20:971-979.

Petersen, E.N., and Scheel-Krüger, J., 1985, 5-HT receptor-mediated anticonflict effects in a benzodiazepine-sensitive part of the amygdala, Proc. 5th European Winter Conference on Brain Research, Varslesclaux.

Pratt, J., Jenner, P., Reynolds, E.H., and Marsden, C.D., 1979, Clonazepam induces decreased serotonergic activity in the mouse brain, Neuropharmacology, 18:791-799.

Pratt, J., Jennifer, P., and Marsden, C.D., 1985, Comparison of the effects of benzodiazepines and other anticonvulsant drugs on synthesis and utilization of 5-HT in mouse brain, Neuropharmacology, 24:59-68.

Riblet, L.A., Taylor, D.P., Eison, M.S., and Stanton, H.C., 1982, Pharmacology and neurochemistry of buspirone, J. Clin. Psychiat., 43:11-16.

Richardson, B.P., Engel, S., Donatsch, P., and Stadler, P.A., 1985, Identification of serotonin m-receptor subtypes and their specific blockade by a new class of drugs, Nature, 316:126-131.

Robichaud, R.C., and Sledge, K.L., 1969, The effects of p-chlorophenylalanine on experimentally induced conflict in the rat, Life Sci., 8:965-969.

Saner, A., and Pletscher, A., 1979, Effects of diazepam on cerebral 5-hydroxytryptamine synthesis, Eur. J. Pharmacol., 55:315-318.

Sanger, D.J., Joly, D., and Zivkovic, B., 1985, Behavioral effects of non-benzodizepine anxiolytic drugs: a comparison of CGS 9896 and zopiclone with chlordiazepoxide, J. Pharmacol. Exp. Ther., 232:831-837.

Schweizer, E.E., Amsterdam, J., Rickels, K., Kaplan, M., and Droba, M.,
1986, Open trial of buspirone in the treatment of major depressive disorder, Psychopharmacol. Bull., 22:183-185.

Sharp, T., Bramwell, S., Maskell, L., and Grahame-Smith, D.G., 1988, 5-HT$_1$
agonists reduce 5-HT release in rat hippocampus in vivo as determined by
brain microdialysis, Brit. J. Pharmacol., 93:94P.

Sills, M.a., Wolfe, B.B., and Frazer, A., 1984, Determination of selective
and non-selective compounds for the 5-HT$_{1A}$ and 5-HT$_{1B}$ receptor subtypes
in rat frontal cortex, J. Pharmacol. Exp. Ther., 231:480-487.

Soubrie, P., Blas, C., Ferron, A., and Glowinski, J., 1983, Chlordiazepoxide
reduces in vivo serotonin release in the basal ganglia of encephale isole
but not anaesthetized cats: evidence for a dorsal raphe site of action, J.
Pharmacol. Exp. Ther., 226:526-532.

Sprouse, J.S., and Aghajanian, G.K., 1987, Electrophysiological responses
of dorsal raphe neurons in 5-HT$_{1A}$ and 5-HT$_{1B}$ agonists, Synapse, 1:3-9.

Stanton, H.C., Taylor, D.P., and Riblet, L. A., 1981, Buspirone - an anxio-
selective drug with dopaminergic action, in: "The Neurobiology of the
Nucleus Accumbens", Chronister, R.B., and DeFrance, J.F., eds., Haer
Institut, Brunswick/Main, pp. 316-321.

Stein, L., Wise, C.D., and Berger, B.D.., 1973, Antianxiety action of
benzodiazepines: decrease in activity of serotonergic neurones in the
punishment system, in: "Benzodiazepines", Garattini, S., Mussini, E., and
Randell, L.O., eds., Raven Press, New York, pp. 299-326.

Taylor, D.P., Eison, M.S., Riblet, L.A., and Vandermaelen, C.P., 1985,
Pharmacological and clinical effects of buspirone, Pharmacol. Biochem.
Behav., 23:687-694.

Thiebot, M.H., Hamon, M., and Soubrie, P., 1982, Attenuation of induced-
anxiety in rats by chlordiazepoxide: role of raphe dorsalis benzodiazepine binding sites and serotoninergic neurons, Neuroscience, 7:2287-
2294.

Thiebot, M.H., Soubrie, P., Hamon, M., and Simon, P., 1984, Evidence against
the involvement of serotonergic neurons in the antipunishment activity of
diazepam in the rat, Psychopharmacology, 82:355-359.

Traber, J., and Glaser, T., 1987, 5-HT$_{1A}$ receptor-related anxiolytics,
Trends Pharmacol. Sci., 8:432-437.

Traber, J., Glaser, T., Spencer, D.G., Schuurman, T., Zilles, K., and
Schleicher, A., 1985, Behavioral pharmacology and autoradiographic dis-
tribution of a novel anxiolytic: TVQ 7821, 4th World Congress on
Biological Psychiatry, Philadelphia.

Tricklebank, M.D., Forler, C., and Fozard, J.R., 1984, The involvement of
subtypes of the 5-HT receptor and of catacholeaminergic systems in the
behavioral response to 8-hydroxy-2-(di-n-propylamino) tetralin in the
rat, Eur. J. Pharmacol., 106:271-282.

Trulson, M.E., Preussler, D.W., Howell, G.A., and Fredrickson, C.J., 1982,
Raphe unit activity in freely moving cats: effects of benzodiazepines,
Neuropharmacology, 21:1050-1080.

Trulson, M.E., and Trulson, T.J., 1986, Buspirone decreases the activity of
serotonin-containing neurons in the dorsal raphe in freely-moving cats,
Neuropharmacology, 25:1263-1266.

Tye, N.C., Everitt, B.J., and Iversen, S.D., 1977, 5-hydroxytryptamine and
punishment, Nature, 268:741-743.

Tyrer, P., Murphy, S., and Owen, R.T., 1985, The risk of pharmacological
dependence with buspirone, Brit. J. Clin. Pract., 39 Suppl. 38:91-93.

Tyrer, P., and Owen, R., 1984, Anxiety in primary care: is short term drug
treatment approximate? J. Psychiat. Res., 18:73-78.

Van Praag, H.M., Kahn, K.S., Asnis, G.M., Wetzler, S., Brown, S.L., Bleich,
A., and Korn, M.L., 1989, Beyond nosology in biological psychiatry: 5-HT
disturbances in mood aggression and anxiety disorders, in: "New Concepts
in Depression", Vol. 2, Macmillan Press, London, pp. 53-81.

Wheatley, D., 1982, Buspirone: multicentre efficacy study, J. Clin.
Psychiat., 43:92-94.

Zohar, J., and Insel, T.R., 1987, Obsessive-compulsive disorder: psycho-
  biological approaches to diagnosis, treatment and pathophysiology, Biol.
  Psych., 22:667-687.

    ... ...

NEUROENDOCRINE RESPONSES TO L-TRYPTOPHAN AS AN INDEX OF BRAIN SEROTONIN

FUNCTION: EFFECT OF WEIGHT LOSS

I. Anderson and P. Cowen

MRC Clinical Pharmacology Unit and
University Department of Psychiatry
Research Unit
Littlemore Hospital
Oxford OX4 4XN
UK

INTRODUCTION

Over the last two decades the relationship between plasma tryptophan (TRP) and brain serotonin (5-hydroxytryptamine, 5-HT) synthesis and function has been extensively investigated and many of the factors influencing the transport of TRP into the brain have been elucidated. This is of particular relevance in psychiatry because of longstanding interest in the involvement of 5-HT in processes such as the control of appetite and macronutrient selection, mood and sleep. If physiological alterations in plasma TRP can result in altered brain 5-HT function, this is of potential etiological and therapeutic importance in psychiatric conditions such as depression and eating disorders.

Two questions recur when the effect of changes of plasma TRP on brain 5-HT function are considered:

1. Do the measurable alterations in brain 5-HT and its principal metabolite 5-HIAA after manipulations of plasma TRP reflect changes in 5-HT neurotransmission (i.e. 5-HT synaptic release)?

2. Are the changes in plasma TRP seen under physiological conditions large enough to influence brain 5-HT transmission?

This paper considers the use of TRP infusion in humans as a probe of brain 5-HT function and then examines the effect of dieting on plasma TRP and brain 5-HT function. The implications of these studies for the questions above are discussed.

BACKGROUND

TRP is the amino acid precursor of 5-HT and is actively transported across the blood-brain barrier (Oldendorf and Szabo, 1976). The rate limiting step in 5-HT synthesis is the hydroxylation of brain TRP to 5-hydroxytryptophan by tryptophan hydroxylase (Fernstrom and Wurtman, 1971; Young and Gauthier, 1981; Fernstrom, 1988) which occurs specifically in 5-HT neurons

(Moir and Eccleston, 1968). The amount of 5-HT synthesized is therefore dependent on brain TRP concentration, which in turn is dependent on the rate
of TRP transport across the blood-brain barrier. The factors that influence
this transport are the plasma TRP concentration (Fernstrom and Wurtman,
1971), the amount of free TRP compared to albumin bound TRP (Knott et al.,
1973) and the ratio of TRP to other amino acids competing for the same membrane carrier (Fernstrom and Wurtman, 1972). Which factor is predominant appears to depend on the experimental condition studied and it seems likely
that all three contribute to the physiological regulation of TRP entry into
the brain.

It is well established that administration of TRP to rats increases
brain 5-HT and 5-HIAA (Fernstrom and Wurtman, 1971; Grahame-Smith, 1971) but
whether this affects 5-HT synaptic release remains controversial, particularly under physiological conditions. It has been shown that TRP decreases 5-HT
cell body firing (Trulson and Jacobs, 1976) which might be expected to offset
any increase in the amount of 5-HT released with each impulse. In apparent
support of this, TRP alone does not produce the 5-HT behavioral syndrome (although it does after inhibition of monoamine oxidase) (Grahame-Smith, 1971).
However the hypotensive effect of TRP in spontaneously hypertensive rats does
appear to be a central effect and is inhibited by parachlorophenyl-alanine
and metergoline (a TRP hydroxylase inhibitor and 5-HT receptor antagonist respectively) and enhanced by the 5-HT uptake inhibitor, fluoxetine (Fernstrom,
1988). This indicates that the hypotensive effect of TRP is likely to be mediated through increasing the release of brain 5-HT. In addition, high doses
of TRP have been reported to increase prolactin release in rats (Mueller et
al., 1976) which is also consistent with enhanced 5-HT neurotransmission.

NEUROENDOCRINE EFFECTS OF L-TRYPTOPHAN IN HUMANS

In humans, administration of intravenous TRP in doses of 5-10 g over 15-
30 minutes stimulates the secretion of prolactin and growth hormone (Charney
et al., 1982; Cowen et al., 1985). It has been demonstrated that TRP administration increases CSF and cortical TRP and 5-HIAA in patients undergoing
neurological investigation and psychosurgery (Eccleston et al., 1970; Gillman
et al., 1981) and accordingly the assumption has been that the hormone responses are caused by increased 5-HT release in the hypothalamus which then
acts on postsynaptic 5-HT receptors to stimulate prolactin and growth hormone
releasing factors. This has led many groups, including our own, to use TRP
infusion as a way of dynamically assessing brain 5-HT function. However, the
question has remained as to whether these hormone responses are mediated by
increased 5-HT release. The specificity of TRP as a 5-HT challenge has been
questioned because large amounts of TRP will not only increase the synthesis
of 5-HT but also tryptamine and other potentially active compounds (Young et
al., 1980; Young and Gauthier, 1981). van Praag et al. (1986) have also suggested that a large excess of plasma TRP will compete with tyrosine for
transport into the brain and will therefore decrease the brain levels of tyrosine and hence the synthesis of dopamine and noradrenaline. These changes
may change catecholaminergic neurotransmission with possible effects on hormone secretion (e.g. lowered dopaminergic function leading to increased prolactin secretion).

In order to investigate whether or not TRP-induced hormone increases
were mediated through 5-HT mechanisms we carried out a series of drug interaction studies in normal male volunteers. The results are summarized in
Table 1.

We were able to strikingly enhance the prolactin and growth hormone responses by pretreatment with a single oral dose of the relatively selective
5-HT reuptake inhibitor, clomipramine (Anderson and Cowen, 1986). In a

Table 1. Prolactin and growth hormone responses to TRP infusion: effects of pretreatment

Drug	Action	Hormone response	
		PRL	GH
Clomipramine	5-HT uptake inhibitor	↑	↑
Metergoline	$5\text{-}HT_1 + 5\text{-}HT_2$ antagonist	↓	→
Ritanserin	$5\text{-}HT_2$ antagonist	↑	→
BRL 43694	$5\text{-}HT_3$ antagonist	→	→

↑ - response enhanced; ↓ - response blocked; → - response unchanged.

series of antagonist studies we were unable to antagonize the TRP induced increase in growth hormone but the prolactin response was abolished by the non-selective 5-HT antagonist, metergoline (McCance et al., 1987), but not by the selective $5\text{-}HT_2$ antagonist, ritanserin (Charig et al., 1986), or the selective $5\text{-}HT_3$ antagonist, BRL 43694 (Anderson et al., 1988). Our conclusion is that the prolactin response to TRP is due to increased 5-HT release and further that it is likely to be mediated by activation of postsynaptic $5\text{-}HT_1$ receptors. Although clomipramine enhanced the growth hormone response, our inability to antagonize it using 5-HT antagonists leaves open the possibility that 5-HT is not involved in this response.

An interesting effect was seen with ritanserin. Not only was the prolactin response not inhibited, it was significantly increased. We interpret this as indicating an interaction between $5\text{-}HT_1$ and $5\text{-}HT_2$ receptors for which there is also evidence from animal studies (Lakoski and Aghajanian, 1985).

DIETING STUDIES

We have employed a neuroendocrine challenge strategy to investigate 5-HT function in depressed patients. Weight loss is a common component of the depressive syndrome and we were concerned that this might in itself alter the hormonal responses to TRP. In order to investigate this we have embarked on a series of normal volunteer studies in which subjects are asked to undertake a calorie restricting diet (1000 kcal for women, 1200 kcal for men). The diet was devised with the help of the Department of Nutrition and Dietetics, Oxford University, Oxford, England, and consisted of approximately 31% protein, 44% carbohydrate and 25% fat. The diet was relatively high in animal protein so that the absolute protein intake was equivalent to the subjects usual (non-dieting) intake. Diaries were kept by the subjects principally to enhance compliance and a detailed analysis of these were not carried out. Subjects were tested before dieting commenced and then towards the end of the third week of their diet. Women were tested at the same stage of their menstrual cycle before and after dieting by delaying the onset of dieting by one week after the pre-diet tests. Weight loss has been comparable in the three studies we present (Table 2).

In our initial study, 6 women and 6 men were tested with TRP (100mg/kg). All the women had enhanced prolactin responses after dieting while there was

Table 2. Weight loss after dieting

	Males			Females		
Study	Age	BMI[*]	Weight loss/kg(%)	Age	BMI[*]	Weight loss/kg(%)
1[a] (6F,6M)	32 ± 4	24.5 ± 2.4	5.0 ± 0.6 (6.2 ± 0.6)	30 ± 7	23.6 ± 1.8	2.8 ± 0.8 (4.5 ± 1.0)
2[b] (11F)	-	-	-	27 ± 6	23.2 ± 1.3	3.1 ± 0.7 (5.1 ± 1.2)
3 (9F,5M)	34 ± 7	25.8 ± 1.3	4.8 ± 0.8 (6.0 ± 1.3)	31 ± 5	24.6 ± 1.5	3.4 ± 1.0 (5.0 ± 1.4)

Values are the mean ± SD. [*]Body Mass Index [weight in kg divided by
(height in meters)]^2]. Normal weight = 20-25; Moderate obesity = 26-30.
a: Goodwin et al., 1987; b: Anderson et al., 1989.

no change in the men's response even though they had lost more weight (Table
3) (Goodwin et al., 1987). This was not due to an alteration in the disposition of plasma TRP which had the same plasma profile in men and women before and after dieting. Baseline prolactin was not altered.

However, it is well know that dieting and weight loss can alter pituitary hormone secretion (Fichter et al., 1984), not necessarily through a
5-HT mechanism, so it was necessary to control for possible effects of weight
loss on prolactin secretion itself, and on the inhibitory dopaminergic control of prolactin secretion. In a second dieting study in women we measured
the prolactin response to metoclopramide, 5 μg/kg, a dopamine antagonist, and
to thyrotropin releasing hormone (TRH), 0.1 μg/kg, which releases prolactin
by direct stimulation of pituitary lactotrophs (Anderson et al., 1989). We
chose doses which gave submaximal prolactin responses comparable to those
seen with TRP. Eleven women received metoclopramide and 8 received TRH.
Weight loss was comparable to the first study but there was no alterations in
the prolactin responses to either challenge (Figs. 1 and 2; Table 3).

Taken together, these two studies suggest that dieting may alter brain
5-HT function in women without a similar effect in men. In order to investi-

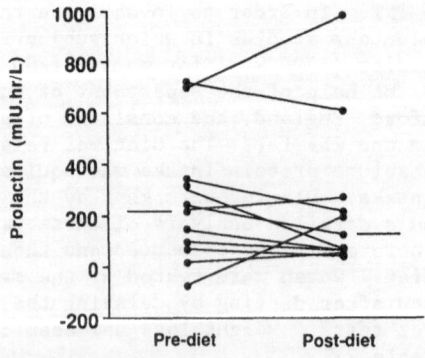

Fig. 1. Individual and median prolactin responses (AUC) to metoclopramide
infusion (5 μg/kg) in eleven female volunteers before and after a
three-week 1000 kcal diet.

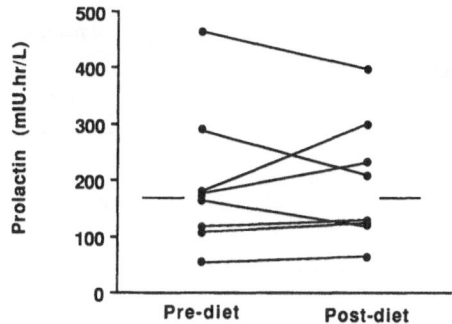

Fig. 2.   Individual and median prolactin responses (AUC) to TRH infusion
          (0.1 μg/kg) in eight female volunteers before and after a three-
          week 1000 kcal diet.

gate the possible mechanism of this effect, we measured fasting plasma amino
acid levels in blood samples taken from the dieters before the neuroendocrine
tests.   Fasting plasma TRP levels fell significantly in both diet studies.
This occurred equally in men and women (Fig. 3).   There were no changes in
plasma concentrations of branch chain amino acids or tyrosine and consequently the ratio of TRP to competing amino acids also fell significantly after
dieting.   Plasma free TRP was not measured but plasma levels of lactate, butyrate and free fatty acids were unchanged by dieting (Dr. G. Goodwin, personal communication).   We are untaking a further dieting study in men and
women.   Nine women and 5 men have completed the study.   We have confirmed
that the prolactin response in TRP is enhanced after dieting in women but not
in men (Fig. 4).   Fasting plasma total TRP has been analyzed in the women and
in agreement with the previous studies there is a decrease of about 20% after
dieting (47.3 ± 5.8 before dieting falling to 38.4 ± 5.6 μmol/1, mean ± SD,
paired T = 3.56, df = 8, P = 0.0074).   There is a non-significant negative
correlation between the increase in the prolactin response and the change in
plasma TRP (Fig. 5).

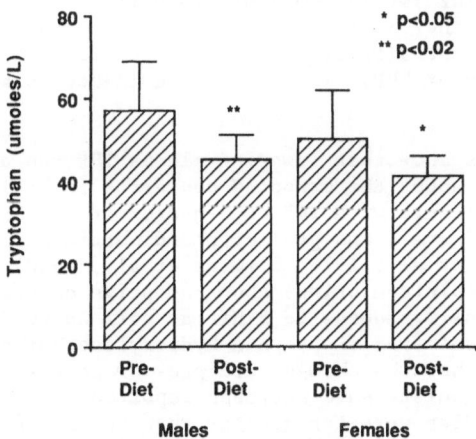

Fig. 3.   Fasting plasma tryptophan (mean ± SD) in eight male and eight
          female volunteers before and after dieting. Plasma tryptophan fell
          significantly in both sexes. (Paired t-test, two tailed).

249

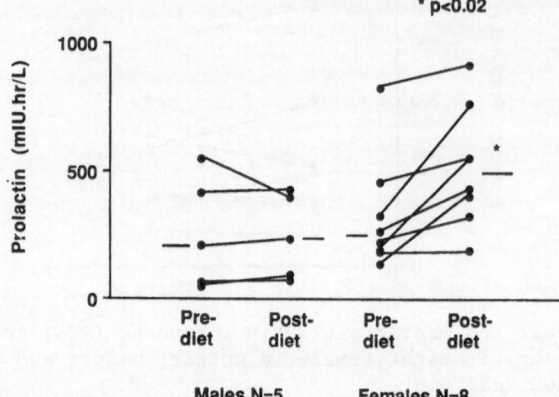

Fig. 4. Individual and median prolactin responses (AUC) to TRP infusion in
         five male and eight female volunteers before and after dieting.
         There was a significant enhancement of the response after dieting in
         females. (Wilcoxon signed-rank test, two-tailed).

DISCUSSION

     Moderate dieting selectively lowered plasma TRP in our male and female
volunteers. However, only the female volunteers showed enhanced prolactin
secretion to TRP infusion, a response which we have shown is likely to be mediated through 5-HT$_1$ receptors. This increase in prolactin response cannot
be explained by an alteration in dopamine function or prolactin secretion by
the pituitary because the prolactin responses to metoclopramide and TRH are
not altered. This suggests that the change in the prolactin response to TRP
must be due either to an alteration in TRP transport or metabolism or in the
functioning of 5-HT pathways. The altered response cannot be explained pure-
ly by kinetic factors as the disposition of TRP after TRP infusion was not
altered by dieting and also the prolactin response was not increased in men
in spite of lowered plasma total TRP (Dr. G. Goodwin, personal communication). The finding of lowered plasma TRP after dieting indicates a possible
mechanism by which dieting might influence brain 5-HT function. Our hypothesis is that lowered plasma TRP (and its ratio to competing amino acids) leads
to decreased brain TRP levels and 5-HT synthesis. This in turn leads to compensatory changes either in the 5-HT neurone or at the level of the postsynaptic receptor (supersensitivity) which are revealed when the system is challenged with infusion of TRP. Is there any evidence to support this hypothesis?

     Studies in rats show that lowering plasma TRP sub-chronically using a
low TRP diet brings about the expected decreases in brain TRP, 5-HIAA and 5-
HT. Subsequent challenge with TRP or 5-hydroxytryptophan (5-HTP) results in
enhanced brain 5-HT levels and prolactin and corticosterone levels compared
to controls (Gil-Ad et al., 1976; Clemens et al., 1980). These results demonstrate that TRP depletion brings about adaptive changes in the 5-HT system.
The enhanced hormonal responses to 5-HTP as well as to TRP indicate that
these are 'downstream' from tryptophan hydroxylase but at present it is not
possible to say whether the changes are pre- or postsynaptic as studies with
postsynaptic 5-HT agonists have not been reported. The changes in plasma TRP
seen in the rat studies were far greater than the 15-20% seen in our human
studies. This raises the issue mentioned at the beginning of the paper as
to whether changes of plasma TRP within the physiological range can alter
5-HT function. In other words can the modest effect on dieting on plasma TRP
levels account for the enhanced prolactin responses to TRP infusion?

Table 3. Effect of dieting on prolactin secretion after challenge
tests

Study	Test		N	Prolactin	AUC	(mIUH/1)
				Pre-diet		Post-diet
1[a]	TRP (100mg/kg)	Males	6	114 (0-863)	100	(18-700)
		Females	6	244 (5-454)	859	(153-1826)[*]
2[b]	Metoclopramide (5μg/kg)	Females	11	216 (-73-719)	124	(69-983)
	TRH (0.1μg/kg)	Females	8	169 (54-405)	169	(62-399)

Results are Median (range). Prolactin response is area under the
response curve minus baseline value (AUC). a: Goodwin et al., 1987;
b: Anderson et al., 1989; [*]P < 0.025 (Wilcoxon signed-mark test, two-
tailed).

This question is directly addressed in a recent study by Delgado et al.
(1989). Volunteers were given a low TRP diet (either 200 or 700 mg/day)
which maintained calorie intake. Subjects had a 7 g TRP infusion before and
after 10 days of the diet. Total plasma TRP levels fell significantly by
about 15% on both diets. Comparable to our findings with calorie restricting
subjects, there was a significant overall enhancement of the prolactin response after the 200 mg TRP diet, with a non-significant increase after the
700 mg TRP diet. Analysis of males and females separately revealed that this
was more pronounced in females - in fact only the females demonstrated a
significant increase in prolactin response to TRP. One difference from our
studies, however, was that females showed a greater decrease in plasma TRP
than men after both TRP restricting diets.

We can conclude therefore that the evidence points towards the decrease
in plasma TRP in our female dieters causing the increased prolactin response

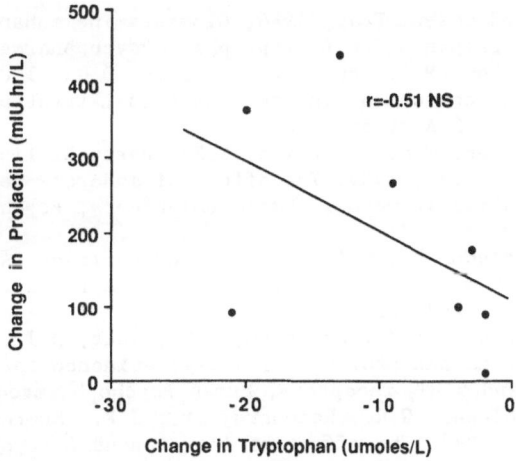

Fig. 5. Individual changes in prolactin response to TRP infusion plotted
against change in fasting plasma TRP values for eight female
dieters. (Change = post-dieting value - pre-diet value). There is
a non-significant negative correlation.

to TRP infusion through an alteration in the functioning of brain 5-HT systems. The reason for the sex difference is not clear, but one possible explanation is the increased turnover in 5-HT seen in females in a number of species including humans (Young et al., 1980; Carlsson et al., 1985). A modest decrease in precursor availability may not compromise 5-HT function in men due to the lower demand and turnover, but be sufficient to compromise 5-HT function in females.

CONCLUSIONS

In this paper we have shown that TRP infusion in humans stimulates hormone secretion through stimulation of 5-HT receptors. This indicates that alterations in TRP supply to the brain can alter 5-HT synaptic release, although it has to be admitted that this occurs under pharmacological, rather than physiological, conditions during the TRP test.

However, we have also shown that a modest alteration in plasma TRP brought about by calorie restriction may be sufficient to bring about a detectable alteration in the functioning of brain 5-HT neurones in women. Dieting of the kind utilized in our studies has a high prevalence among young women and so it appears that a common physiological situation can alter plasma TRP and affect brain 5-HT function. This has considerable, although at this stage speculative, implications for psychiatric disorders, in particular for eating disorders, where a reduction in brain 5-HT function could contribute to disordered appetitive behavior (the diet-binge-purge cycle) and to the associated mood changes.

ACKNOWLEDGEMENTS

We thank J. Robinson, R. Hockney, M. Scholes, and J. Moss for expert nursing care and L. Eeley for help with devising the diet. The studies were supported by the Medical Research Council and Oxford Regional Health Authority. IMA is an MRC Training Fellow.

REFERENCES

Anderson, I.M., and Cowen, P.J., 1986, Clomipramine enhances prolactin and growth hormone responses to L-tryptophan, **Psychopharmacology**, 89:131-133.

Anderson, I.M., Cowen, P.J., and Grahame-Smith, D.G., 1988, The effect of BRL 43694 on the neuroendocrine responses to L-tryptophan infusion, J. **Psychopharmacol.**, 2:Abstract.

Anderson, I.M., Crook, W.S., Gartside, S.E., Parry-Billings, M., Newsholme, E.A., and Cowen, P.J., 1989, The effect of moderate weight loss on prolactin secretion in normal female volunteers, **Psychiat. Res.**, 29:161-167.

Carlsson, M., Svensson, K., Eriksson, E., and Carlsson, A., 1985, Rat brain serotonin: biochemical and functional evidence for a sex difference, J. **Neural Trans.**, 63:297-313.

Charig, E.M., Anderson, I.M., Robinson, J.M., Nutt, D.J., and Cowen, P.J., 1986, L-tryptophan and prolactin release: evidence for interaction between $5\text{-}HT_1$ and $5\text{-}HT_2$ receptors, **Human Psychopharmacol.**, 1:93-97.

Charney, D.S., Heninger, G.R., Reinhard, Jr., J.F., Sternberg, D.E., and Hafstead, K.M., 1982, The effect of intravenous L-tryptophan on prolactin and growth hormone and mood in healthy subjects, **Psychopharmacology**, 77:217-222.

Clemens, J.A., Bennett, D.R., and Fuller, R.W., 1980, The effect of a tryptophan-free diet on prolactin and corticosterone release by serotonergic stimuli, **Horm. Metab. Res.**, 12:35-38.

Cowen, P.J., Gadhui, H., Gosden B., and Kolakowska T., 1985, Responses of prolactin and growth hormone to L-tryptophan infusion: effects in normal subjects and schizophrenic subjects receiving neuroleptics, **Psychopharmacology,** 86:164-169.

Delgado, P.L., Charney, D.S., Price, L.H., Landis, H., and Heninger, G.R., 1989, Neuroendocrine and behavioral effects of diet tryptophan restriction in healthy subjects, **Life Sci.,** in press.

Eccleston, D., Ashcroft, G.W., Crawford, T.B.B., Stanton, J.B., Wood, D., and McTurk, P.H., 1970, Effect of tryptophan administration on 5-HIAA in cerebrospinal fluid in man, J. Neurol. Neurosurg. Psychiatr., 33:269-272.

Fernstrom, J.D., and Wurtman, R.J., 1971, Brain serotonin content: physiological dependence on plasma tryptophan levels, **Science,** 173:149-152.

Fernstrom, J.D., and Wurtman, R.J., 1972, Brain serotonin content: physiological regulation by plasma neutral amino acids, **Science,** 178:414-416.

Fernstrom, J.D., 1988, Tryptophan availability and serotonin synthesis in brain, in: "**Amino Acid Availability and Brain Function in Health and Disease**", Huether, G., ed., Springer, Berlin, pp. 137-146.

Fichter, M.M., Pirke, K.-M., and Holsboer, F., 1984, Weight loss causes neuroendocrine disturbances: experimental study in healthy starving subjects, **Psychiat. Res.,** 17:61-72.

Gil-Ad, I., Zambotti, F., Carruba, M.O., Vicentini, I., and Muller, E.E., 1976, Stimulatory role for brain serotonergic system on prolactin secretion in the male rat, Proc. Soc. **Exper. Biol. Med.,** 151:512-518.

Gillman, P.K., Bartlett, J.R., Bridges, P.K., Hunt, A., Patel, A.J., Kantamaneni, A.J., and Curzon, G., 1981, Indolic substances in plasma, CSF and frontal cortex of human subjects infused with saline or L-tryptophan, J. Neurochem., 37:410-417.

Goodwin, G.M., Fairburn, C.G., and Cowen, P.J., 1987, Dieting changes serotonergic function in women, not men: implications for the aetiology of anorexia nervosa, Psychol. Med., 17:839-842.

Grahame-Smith, D.G., 1971, Studies in vivo on the relationship between brain tryptophan, brain 5-HT synthesis and hyperactivity in rats treated with a monoamine oxidase inhibitor and L-tryptophan, J. **Neurochem.,** 18:1053-1066.

Knott, P.J., and Curzon, G., 1973, Free tryptophan in plasma and brain tryptophan metabolism, **Nature,** 239:452-453.

Lakoski, J.M., and Aghajanian, G.K., 1985, Effects of ketanserin on neuronal responses to serotonin in the prefrontal cortex, lateral geniculate and dorsal raphe nucleus, **Neuropharmacology,** 24:265-273.

McCance, S.L., Cowen, P.J., Waller, H., and Grahame-Smith, D.G., 1987, The effect of metergoline on endocrine responses to L-tryptophan, J. **Psychopharmacol.,** 1:90-94.

Moir, A.T.B., and Eccleston, D., 1968, The effects of precursor loading on the cerebral metabolism of 5-hydroxyindoles, J. Neurochem., 15:1093-1108.

Mueller, G.P., Twohy, C.P., Chen, H.T., Advis, J.P., and Meites, J., 1976, Effects of L-tryptophan and restraint stress on hypothalamic and brain serotonin turnover, and pituitary TSH and prolactin release in rats, Life Sci., 18:715-724.

Oldendorf, W.H., and Szabo, J., 1976, Amino acid assignment to one of three blood-brain barrier amino acid carriers, Am. J. Physiol., 230:94-98.

Trulson, M.E., and Jacobs, B.L., 1976, Dose response relationship between systemically administered L-tryptophan or L-5-hydroxytryptophan and raphe unit activity in the rat, **Neuropharmacology,** 15:339-344.

van Praag, H.M., Asnis, G., and Zukin, S., 1986, Peripheral hormones: a window on the central MA? Psychopharmacol. Bull., 22:565-570.

Young, S.N., and Gauthier, S., 1981, Effect of tryptophan administration on tryptophan, hydroxyindoleacetic acid and indoleacetic acid in human lumbar and cisternal cerebrospinal fluid, J. Neurol. Neurosurg. Psychiat., 44:323-328.

Young, S.N., Gauthier, S., Anderson, G.M., and Purdy, W.C., 1980, Tryptophan, 5-hydroxyindoleacetic acid and indoleacetic acid in human cerebrospinal fluid: interrelationships and the influence of age, sex, epilepsy and anticonvulsant drugs, J. Neurol. Neurosurg. Psychiat., 43:438-445.

MELATONIN RECEPTORS IN THE CENTRAL NERVOUS SYSTEM

M.L. Dubocovich

Department of Pharmacology
Northwestern University Medical School
Chicago, Illinois 60611
USA

INTRODUCTION

The hormone melatonin is synthesized primarily in the pineal gland of vertebrates from its precursor serotonin (5-hydroxytryptamine). Serotonin is N-acetylated by a relatively specific N-acetyltransferase to yield N-acetyl-serotonin, which in turn is methylated by hydroxyindole-O-methyltransferase to melatonin (5-methoxy-N-acetyltryptamine) (Axelrod, 1974).

The synthesis and secretion of melatonin from the pineal gland are inhibited by environmental light and thus exhibit a circadian rhythm in which the highest levels are seen during the dark period (Klein, 1979). Melatonin regulates a number of processes in vertebrates, including the modulation of neural and endocrine processes that are cued by the daily change in photoperiod (Cardinali, 1981; Goldman, 1983), circadian rhythms (Takahashi and Zatz, 1985) and retinal physiology (Dubocovich, 1988c).

In photoperiodic mammals such as sheep and hamsters, seasonal changes in day length (photoperiod) regulate reproductive function, body weight, pelage color, metabolism and behavior (Goldman and Darrow, 1983; Holthorf et al., 1985; Vitale et al., 1985; Bartness and Goldman, 1988). The neuroendocrine effects of melatonin leading to reproductive changes are believed to occur in the central nervous system (CNS) (Glass and Lynch, 1981, 1982; Hasting et al., 1985) through activation of melatonin receptor sites, primarily within the hypothalamus (Duncan et al., 1989; Vanecek and Jadislow, 1989).

Evidence suggests that melatonin may be effective in regulating circadian rhythmicity in vertebrates (Underwood and Goldman, 1987). However, in rodents pinealectomy does not affect the period of free running circadian rhythms or the rate of reintrainment following phase shifts of the light cycle (Cheung and McCormack, 1982). Melatonin is effective in synchronizing disturbed circadian rhythms in mammals including man (Arendt et al., 1986; Cassone et al., 1986; Arendt, 1988). These effects of melatonin appear to be due to an action of this hormone within the hypothalamus since lesions of the rat suprachiasmatic nucleus prevent the entraining effects of melatonin (Cassone et al., 1986).

Melatonin, once thought to be an unique pineal product, has been recently found in the retina (Cardinali, 1981). In the retina, melatonin is synthesized from its precursor serotonin and secreted in a diurnal rhythm with peak levels during the dark period (Iuvone and Besharse, 1983).

Evidence suggests that melatonin secreted from the photoreceptors or other indoleamine-containing neurons may function as a modulator of photoreceptor outer segment disc shedding and phagocytosis (Besharse and Dunnis, 1983), melanosome aggregation in pigmented ephithelium (Pang and Yew, 1979), cone photoreceptor motor movement (Pierce and Besharse, 1985), and dopaminergic activity (Dubocovich, 1983; Dubocovich et al., 1985; Dubocovich, 1988c).

Here, I will review the presence of melatonin receptors modulating dopamine release from retina, and the pharmacological characteristics, localization and function of melatonin receptors in the central nervous system.

MELATONIN RECEPTORS IN RETINA

Melatonin and the related 5-methoxy-N-acetyltryptamine are potent inhibitors of the calcium-dependent release of dopamine from retina. In the

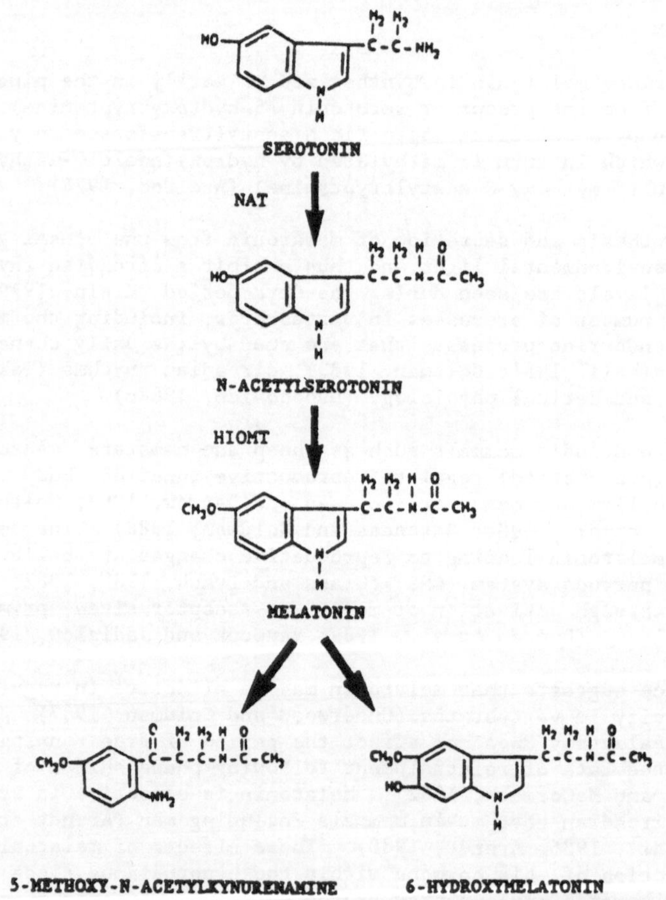

Fig. 1. Synthetic and metabolic pathways of melatonin. NAT: N-acetyl-transferase; HIOMT: hydroxyindole-O-methyltransferase; Serotonin: 5-hydroxytryptamine (5HT); N-acetylserotonin: N-acetyl-5-hydroxy-tryptamine (N-A-5HT); Melatonin: N-acetyl-5-methoxytryptamine (ML).

256

chicken and rabbit retina, melatonin selectively inhibits the calcium-dependent release of dopamine elicited by electrical stimulation or high potassium through activation of a site possessing the pharmacological and functional characteristics of a specific melatonin receptor (Fig. 2A) (Dubocovich, 1983, 1985, 1988b; Dubocovich et al., 1985; Dubocovich and Takahashi, 1987). While melatonin does not modify the spontaneous outflow of radioactivity, picomolar concentrations of the hormone inhibit the calcium-dependent release of dopamine (Fig. 2A) in a concentration-dependent manner (Dubocovich, 1983, 1985). Melatonin is about 1,000 times more potent than its precursor N-acetylserotonin, while the neurotransmitter serotonin does not affect either the spontaneous or field stimulation-evoked release of ^{3}H-dopamine from rabbit retina (Fig. 2A). Moreover, serotonin antagonists such as spiperone, methysergide or methiothepin did not change the response to melatonin (Dubocovich, 1983, 1985, 1988b), suggesting that the site activated by melatonin in the retina is pharmacologically distinct from a serotonin receptor. The inhibitory effect of melatonin is also unaffected by the inhibitors of the neuronal uptake of dopamine, specific antagonists of α-adrenergic, dopamine, serotonin and opiate receptors, ruling out a possible effect of this hormone on other presynaptic receptors (Dubocovich, 1983).

The potent inhibitory effect of melatonin on dopamine release was mimicked by other 5-methoxyindoles possessing an acetamidoethyl group on carbon-3 (i.e., ethyl-N-acetyl) (Fig. 1). Structure-activity relationship studies on the melatonin receptor demonstrated that the most potent inhibitors (agonists) of dopamine release from rabbit and chicken retina are 5-methoxy-N-acetyltryptamines such as 2-iodomelatonin, 6-chloromelatonin and 6,7-dichloro-2-methylmelatonin (Dubocovich, 1985, 1988b; Dubocovich and Takahashi, 1987). Melatonin analogs with substitutions on position 6 (i.e., 6-hydroxymelatonin, 6-methoxymelatonin) are about 40 times less potent than melatonin in activating the presynaptic melatonin receptors of rabbit retina (Dubocovich, 1985).

Melatonin and related agonists inhibit the calcium-dependent release of dopamine through activation of a melatonin receptor since these effects are blocked by the competitive melatonin receptor antagonist luzindole (N-0774; 2-benzyl-N-acetyltryptamine) (Dubocovich, 1988a,b). Tested alone, luzindole did not modify the spontaneous or stimulation-evoked release of tritium in concentrations as high as 10 μM from rabbit retinal pieces labeled in vitro with ^{3}H-dopamine (Dubocovich, 1988b; Fig. 3A). Fig. 3A shows that luzindole (1 μM) completely antagonizes the inhibition of the calcium-dependent release of dopamine elicited by melatonin (10 pM - 10 nM). At concentrations of 0.1, 1 and 10 μM, luzindole shifted the concentration effect curve for melatonin (IC_{50} = 40 pM) to the right. The affinity of luzindole for the retinal melatonin receptor is 20 nM (Dubocovich, 1988b). In the rabbit retina, luzindole does not modify the inhibition of dopamine release elicited through activation of either D_2 dopamine autoreceptors by apomorphine or α2 adrenoceptors by clonidine (Dubocovich, 1983, 1988b) (Fig. 3b). Further evidence for luzindole's selectivity is provided by binding studies, in which luzindole did not affect the binding of specific radioligands to monoamine receptor subtypes, muscarinic, adenosine-1 or benzodiazepine receptors (Dubocovich, 1988a). These results with luzindole further support our hypothesis that melatonin inhibits the calcium-dependent release of dopamine from retina through activation of melatonin receptor sites.

Metabolites of melatonin formed either in the peripheral or central nervous system are potent inhibitors of the calcium-dependent release of dopamine from retina. The metabolites of melatonin, 6-hydroxymelatonin (IC_{50}: 1.6 nM), which is formed in the liver, and N-acetyl-5-methoxykynurenamine (IC_{50}: 10 nM) which is formed in the central nervous system (Kopin et al., 1961; Hirata et al., 1974) are potent activators of melatonin receptors sites in retina (Dubocovich, 1985; Dubocovich and Takahashi, 1987). Other metab-

olites of melatonin formed by demethylation (i.e. N-acetylserotonin) and de-
acetylation (i.e. 5-methoxytryptamine) (Leone and Silman, 1984) also interact with melatonin receptors. These metabolites activate melatonin receptors
in retina since the inhibitory effect is blocked by luzindole.

PHARMACOLOGY PROFILE OF 2-[^{125}I]-IODOMELATONIN BINDING SITES IN THE CENTRAL
NERVOUS SYSTEM

High affinity melatonin receptor sites are currently being characterized
and localized in the mammalian central nervous system using the new radioligand 2-[^{125}I]-iodomelatonin (Vakkuri et al., 1984). This radioligand selectively labels melatonin sites, since drugs which interact with either
serotonin, dopamine and α- and β-adrenergic receptors are less potent than
melatonin analogs in competing for this site (Dubocovich and Takahashi, 1987;
Duncan et al., 1988, 1989). The specific binding of 2-[^{125}I]-iodomelatonin
in brain and retina membranes fulfills all the criteria for binding to a receptor site, i.e. the binding is stable, reversible, saturable, and of high
affinity (Dubocovich and Takahashi, 1987; Duncan et al., 1988, 1989).

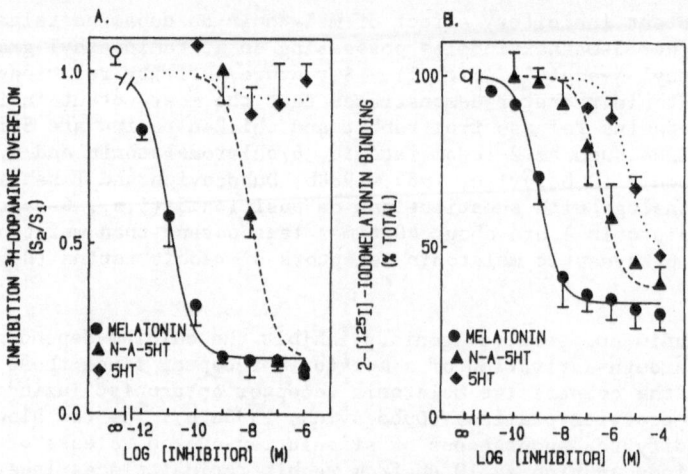

Fig. 2A. Effect of melatonin, N-acetylserotonin and serotonin on the
calcium-dependent release of ^{3}H-dopamine from rabbit retina.
Retinas from albino rabbits were dissected, labeled with ^{3}H-
dopamine and superfused as previously described (Dubocovich,
1985). Ordinate: ^{3}H-overflow as the percentage of total
tissue radioactivity released by field stimulation above the
spontaneous levels of release. Results are expressed as the
ratio obtained between the second (S$_2$) and the first (S$_1$)
stimulation periods within the same experiment. The calcium-
dependent release of ^{3}H-dopamine was elicited by 1-min period
electrical stimulation of 3 Hz (20 mA, 2 msec). Abscissa:
molar concentrations of melatonin (IC$_{50}$ = 9 pM), N-acetylserotonin (N-A-5HT; IC$_{50}$ = 8.6 nM) and serotonin (5HT). Data are
the mean ± SEM of 3 to 8 experiments per group.

B. Competition curves for the inhibition of 2-[^{125}I]-iodomelatonin
binding by melatonin, N-acetylserotonin and serotonin in rabbit
retinal membranes. Washed rabbit retinal membranes were incubated with 2-[^{125}I]-iodomelatonin (50-100 pM) and various concentrations of melatonin (K$_i$ = 1.03 nM), N-acetylserotonin
(N-A-5HT; K$_i$ = 2.6 nM) and serotonin (5HT; K$_i$ = 3.7 μM). Values
are mean of at least 3 independent determinations. (From
Dubocovich, 1988b).

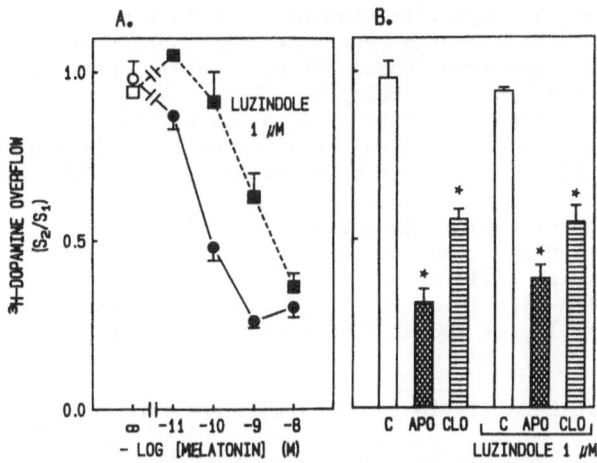

Fig. 3. Luzindole antagonizes the melatonin-induced inhibition of dopamine release from rabbit retina. Retinas from albino rabbits were dissected, labeled with ^{3}H-dopamine and superfused as previously described (Dubocovich, 1985). Ordinates: ^{3}H-dopamine overflow as a percentage of total tissue radioactivity released by field stimulation (3 Hz, 2 min, 20 mA, 2 msec) above the spontaneous levels of release. Results are expressed as the ratio obtained between the second (S_2) and the first (S_1) stimulation periods within the same experiment. Panel A: The abscissa represents the molar concentration of melatonin (logarithmic scale) added 20 min before the second period of stimulation (S_2) (closed symbols). Control: (O); 1 μM luzindole (□ , ■) was added 40 min before the first period of stimulation (S_1) and remained present throughout. Panel B: Control: (□); apomorphine (▨ , APO 0.1 μM) and clonidine (▤ , CLO 1 μM) were added 20 min before the second stimulation (S2). Where indicated, luzindole (1 μM) was added 40 min before the second stimulation (S_1), and remained present throughout. Data are the mean ± SEM. (From Dubocovich, 1988b).

2-[^{125}I]-Iodomelatonin shows low picomolar affinity for the melatonin binding site of retina, and nanomolar affinity for the site of the hamster brain (Table 1). As shown in Table 1, the affinity of [^{125}I]-iodomelatonin for the melatonin site of chicken retina and brain increase with the temperature of incubation of the membranes (Dubocovich et al., 1987). Melatonin and related indoles inhibited 2-[^{125}I]-iodomelatonin binding in retinal membranes with the same order of potency as that found for inhibition of dopamine release from chicken and rabbit retina (Dubocovich and Takahashi, 1987) (Fig. 2). In retina, N-acetylserotonin and 6-methoxymelatonin are less potent than melatonin in inhibiting dopamine release and competing for 2-[^{125}I]-iodomelatonin binding (Dubocovich and Takahashi, 1987; Dubocovich, 1988a) (Fig. 2). By contrast, in Syrian and Djungarian hamster brain membranes both N-acetylserotonin and 6-methyoxymelatonin are equipotent with melatonin to compete for 2-[^{125}I]-iodomelatonin binding (Duncan et al., 1988, 1989).

Based on binding and functional studies, two melatonin sites, the ML-1 and ML-2, have been postulated (Dubocovich, 1988a). ML-1 receptor sites which show picomolar affinity for 2-[^{125}I]-iodomelatonin have been extensively characterized in rabbit and chick retina where they function to inhibit dopamine release (Dubocovich, 1985, 1988a,b,c; Dubocovich and Takahashi, 1987). The ML-1 and ML-2 sites possess distinct pharmacological characteristics, e.g., N-acetylserotonin and 5-methoxymelatonin show low affinity for the ML-1 site, while they are equipotent with melatonin on the ML-2 site

(Dubocovich, 1988a). Specific binding sites in the rat median eminence
(Vanecek et al., 1987), ovine pars tuberalis (Morgan et al., 1989) and Djungarian hamster hypothalamus (Duncan et al., 1989) appear to possess the same
picomolar affinity and pharmacological characteristics seen for the ML-1
receptor in the retina (Dubocovich, 1985, 1988a,b). Activation of ML-1 sites
inhibits dopamine release and may be involved in mediating the photoperiodic
changes in reproduction induced by melatonin (Dubocovich, 1988a).

Our recent work using Djungarian and Syrian hamster brain membrane homogenates revealed a second type of specific melatonin site, ML-2, which
shows low nanomolar affinity for 2-[^{125}I]-iodomelatonin and uneven regional
distribution with high levels in the hypothalamus (Duncan et al., 1988,
1989). The function of the ML-2 site is yet unknown.

AUTORADIOGRAPHIC LOCALIZATION OF MELATONIN RECEPTOR SITES IN DISCRETE BRAIN
AREAS

Autoradiographic studies have revealed discrete locations of 2-[^{125}I]-
iodomelatonin binding sites in brains of a variety of species. In particular, two hypothalamic areas, the median eminence and suprachiasmatic nuclei
of hamster, rat and human brain (Vanecek et al., 1987; Reppert et al., 1988;
Weaver et al., 1988; Duncan et al., 1989) were shown to contain melatonin receptor binding which may represent the physiological sites where melatonin
acts to regulate and synchronize circannual rhythms of reproduction in photoperiodic animals as well as other mammalian circadian and seasonal rhythms.
The median eminence is known to exhibit photoperiodic monoamine changes that
are pineal-dependent (Steger et al., 1984; Glass et al., 1988). The density
of 2-[^{125}I]-iodomelatonin binding sites decreases in the median eminence and

Table 1. Affinity and number of 2-[^{125}I]-iodomelatonin binding sites
in membranes from various central nervous system tissues

Tissue	N	K_d (pM)	B_{max} (fmol/ mg protein)
A.			
Rabbit retina	3	353 ± 70	8.5 ± 3.9
Chicken retina	5	434 ± 56	74 ± 13.6
Chicken brain	3	344 ± 24	57.6 ± 10.1
Syrian hamster brain	4	3,300 ± 500	110.2 ± 13.4
Djungarian hamster brain	4	1,500 ± 300	293 ± 38
B.			
Chicken retina	5	58 ± 0.9	59 ± 9.3
Chicken brain	3	176 ± 43	19 ± 2.2
Djungarian hamster hypothalamus	3	43 ± 5.1	2.9 ± 0.5

The affinity and number of 2-[^{125}I]-iodomelatonin binding sites to
membranes of various central nervous system tissues were determined
by Scatchard analysis. Membranes from all tissue were incubated with
various concentrations of 2-[^{125}I]-iodomelatonin (0.02-12 nM) at 0°C
(A) or 25°C (B). Data were obtained from Dubocovich and Takahashi,
1987; Dubocovich et al., 1987; Dubocovich, 1988b; Duncan et al., 1988,
1989; Chung and Dubocovich (unpublished).

anterior pituitary of Syrian hamsters maintained in short photoperiod (Vanecek and Jadislow, 1989).

The suprachiasmatic nuclei, which serve as an endogenous oscillator for the circadian time keeping, have also been proposed as target sites for the effects of melatonin on entraining circadian rhythms or regulating seasonal reproductive changes (Glass and Lynch, 1982; Cassone et al., 1986). The density of 2-[^{125}I]-iodomelatonin binding sites exhibits diurnal rhythm in the rat suprachiasmatic nucleus (Laitinen et al., 1989). Peripheral melatonin administration inhibits metabolic activity in the rat suprachiasmatic nucleus (Cassone et al., 1987). Melatonin administration near the suprachiasmatic nucleus inhibited reproductive function in white-footed mice (Glass and Lynch, 1982; Glass and Knott, 1987). However, whether the suprachiasmatic nuclei are necessary for the reproductive effects of melatonin is unclear, since lesions of the suprachiasmatic nucleus prevents the reproductive effects of melatonin in some, but not all, studies (Bittman et al., 1979; Rusak, 1980).

In addition to the median eminence and suprachiasmatic nucleus, Djungarian brain showed high levels of specific 2-[^{125}I]-iodomelatonin binding in discrete regions of the thalamus [paraventricular nucleus and reuniens nucleus], and the pars tuberalis of the anterior pituitary (Duncan et al., 1989). A similar distribution of binding sites has been reported in Syrian hamster after pinealectomy (Williams et al., 1989). Recent reports have shown 2-[^{125}I]-iodomelatonin binding sites in the pars tuberalis of Syrian hamster, rat and sheep (Morgan et al., 1989; Williams et al., 1989). Photoperiod-dependent changes in morphology and TSH-like immunoreactivity have been reported in the pars tuberalis of Djungarian hamster (Wittdowski et al., 1984, 1988). The biological importance of melatonin sites in the paraventricular nucleus and reuniens nucleus of the thalamus is unclear. These nuclei may act as relay stations, receiving input from the olfactory tubercle, thalamus, and striatum, and sending projections to the hypothalamus and the limbic system (Newman and Winans, 1980a,b).

ROLE OF ENDOGENOUS MELATONIN IN MODULATING CENTRAL NERVOUS SYSTEM FUNCTION

Evidence suggests that melatonin modulates monoaminergic systems in the brain and that these actions may underlie a number of physiological effects including photoperiodic-induced changes in reproductive function (Goldman and Darrow, 1983; Stetson and Watson-Whitmyre, 1984). Hypothalamic areas, in which dopamine, norepinephrine and/or serotonin are known to regulate gonadotropin release, appear to be of particular relevance for mediating reproductive effects. In hamsters, short photoperiod-induced testicular regression is correlated with reduced levels and turnover of norepinephrine and dopamine in hypothalamic areas including the median eminence (Steger et al., 1982, 1984). This effect is pineal-dependent. In the C3H/HeN mouse, 6-chloromelatonin affected hypothalamic norepinephrine levels after inhibition of synthesis and this effect was blocked by in vivo administration of luzindole (Fang and Dubocovich, 1990). In rat hypothalamic areas, melatonin inhibits dopamine release in vitro and this effect shows a diurnal change in sensitivity (Zisapel et al., 1983, 1985). The best characterized melatonin-monoamine interaction is the inhibition of dopamine release from both rabbit and chicken retina where melatonin acts via well-defined presynaptic receptors of the ML-1 type (Dubocovich, 1983, 1985, 1988a; Dubocovich and Takahashi, 1987).

Exposure of animals to light, which inhibits the synthesis of melatonin, or removal of the pineal gland, are methods classically used to antagonize the physiological effects of endogenous melatonin (Darrow and Goldman, 1983; Tamarkin et al., 1985). Therefore, a melatonin receptor antagonist, by

blocking its receptor in target tissues, would be expected to be more effective than pinealectomy in antagonizing the effects of endogenous melatonin
since this hormone is also secreted by extrapineal tissues (Cardinali, 1981).

Pharmacological and behavioral experiments support the concept that the
melatonin receptor antagonist luzindole blocks the effect of activation of
melatonin receptors in vivo. In this regard, Fang and Dubocovich (1990) recently found that luzindole antagonizes the 6-chloromelatonin-induced rever-
sion of the decrease in hypothalamic norepinephrine levels elicited by α-
methyl-p-tyrosine in the C3H/HeN mouse. In addition, luzindole, administered
at midnight when the levels of melatonin in the C3H/HeN mouse are high, further decreases the levels of hypothalamic norepinephrine following α-methyl-
p-tyrosine administration (Dubocovich and Fang, 1990). These results suggest
that in brain luzindole blocks the activation of melatonin receptor sites by
endogenous melatonin.

Luzindole shows antidepressant-like activity in the behavioral despair
test conducted on the C3H/HeN mouse, possible by blocking the effect on endogenous melatonin (Mogilnicka and Dubocovich, 1987; Dubocovich et al.,
1990). In the behavioral despair test, typical and atypical antidepressants
are known to decrease the duration of immobility during swimming. Luzindole
decreases the time of immobility during swimming in the C3H/HeN mouse, possible by blocking the effects of endogenous melatonin, because the action of
the melatonin receptor antagonist was more pronounced at midnight, when the
levels of melatonin in the pineal gland are high (Dubocovich et al., 1990).

Recently, several groups of investigators have used bright light to
treat chronobiologic sleep and mood disorders in humans (Lewy et al., 1987).
Patients with seasonal affective disorders, a syndrome characterized by re-
current depressions that occur annually at the same time of the year, have
been successfully treated by lengthening their winter days with bright artificial light (Lewy et al., 1987). If the therapeutic effects of light are
related to the suppression of melatonin secretion, the melatonin receptor
antagonist luzindole may be useful in treating chronobiologic disorders involving changes in the pattern of melatonin secretion. Moreover, melatonin
receptor antagonists may be useful to treat neuroendocrine disturbances associated with alterations in melatonin secretion and rhythmicity.

ACKNOWLEDGEMENTS

I would like to thank Ms. V. James-Houff for excellent secretarial assistance. Recent work was supported by USPHS grants MH 42922 and DK 38607.

REFERENCES

Arendt, J., 1988, Melatonin, Clin. Endocrinol., 29:205-229.
Arendt, J., Aldhous, M., and Markus, V., 1986, Alleviation of jet lag by
    melatonin: preliminary results of controlled double blind trials, Brit.
    Med. J., 292:1170.
Axelrod, J., 1974, The pineal gland: a neurochemical transducer, Science,
    184:1341-1348.
Bartness, T.J., and Goldman, B.D., 1988, Peak duration of serum melatonin and
    short day responses in adult Siberian hamsters, Am. J. Physiol., 255:R812-
    R-822.
Besharse, J.C., and Dunis, D.A., 1983, Methoxyindoles and photoreceptor
    metabolism: activation of rod shedding, Science, 219:1341-1343.
Bittman, E.L., Goldman, B.D., and Zucker, I., 1979, Testicular responses to
    melatonin are altered by lesions of the suprachiasmatic nuclei in golden
    hamster, Biol. Reprod., 21:647-656.

Cardinali, D.P., 1981, Melatonin: a mammalian pineal hormone, **Endocrine Rev.**, 2:327-346.

Cassone, V.M., Chesworth, M.J., and Armstrong, S.M., 1986, Dose-dependent entrainment of rat circadian rhythms by daily injection of melatonin, J. **Biol. Rhythms**, 1:219-229.

Cassone, V.M., Roberts, M.H., and Moore, R.V., 1987, Melatonin inhibits metabolic activity in the rat suprachiasmatic nuclei, **Neurosci. Lett.**, 81:29-34.

Cheung, P.W., and McCormack, C.E., 1982, Failure of pinealectomy or melatonin to alter circadian activity rhythm of the cat, **Am. J. Physiol.**, 242:R261-R264.

Dubocovich, M.L., 1983, Melatonin is a potent modulator of dopamine release in the retina, **Nature**, 306:782-784.

Dubocovich, M.L., 1985, Characterization of a retinal melatonin receptor, J. **Pharmacol. Exp. Ther.**, 234:395-401.

Dubocovich, M.L., 1988a, Luzindole (N-0774): a novel melatonin receptor antagonist, J. **Pharmacol. Exp. Ther.**, 246:902-910.

Dubocovich, M.L., 1988b, Pharmacology and function of melatonin receptors, **FASEB J.**, 2:2765-2773.

Dubocovich, M.L., 1988c, Role of melatonin in retina, in: "**Progress in Retinal Research**", Vol. 8, Osborne, N.N., and Cheder, G.J., eds., Pergamon Press, Oxford, pp. 129-151.

Dubocovich, M.L., Lucas, R.C., and Takahashi, J.S., 1985, Light-dependent regulation of dopamine receptors in mammalian retina, **Brain Res.**, 335: 321-325.

Dubocovich, M.L., Mogilnicka, E., and Areso, P.M., 1990, Antidepressant-like activity of the melatonin receptor antagonist luzindole (N-0774) in the mouse behavioral despair test, **Eur. J. Pharmacol.**, 182:313-325.

Dubocovich, M.L., Schabauer, A., Murphy, R., and Takahashi, J.S., 1987, Melatonin receptor binding sites in chicken retina: effect of temperature, **Pharmacologist**, 29:166.

Dubocovich, M.L., and Takahashi, J.S., 1987, Use of 2-[^{125}I]-iodomelatonin to characterize melatonin binding sites in chicken retina, **Proc. Natl. Acad. Sci. USA**, 84:3916-3920.

Duncan, M.J., Takahashi, J.S., and Dubocovich, M.L., 1988, 2-[^{125}I]Iodomelatonin binding sites in hamster brain membranes: pharmacological characteristics and regional distribution, **Endocrinology**, 122:1825-1833.

Duncan, M.J., Takahashi, J.S., and Dubocovich, M.L., 1989, Characteristics and autoradiographic localization of 2-[^{125}I]iodomelatonin-binding sites in Djungarian hamster brain, **Endocrinology**, 125:1011-1018.

Fang, J.M., and Dubocovich, M.L., 1990, Activation of melatonin receptors retarded the depletion of norepinephrine from hypothalamus of C3H/HeN mouse, J. **Neurochem.**, 55:76-82.

Glass, J.D., Ferreira, S., and Deaver, D.R., 1988, Photoperiodic adjustments in hypothalamus amines, gonadotropin-releasing hormone, and betaendorphin in the white-footed mouse, **Endocrinology**, 123:1119-1127.

Glass, J.D., and Knotts, L.K., 1987, A brain site for the antigonadal action of melatonin in the white-footed mouse (Peromyscus leucopus): involvement of the immunoreactive GnRH neuronal system, **Neuroendocrinology**, 46:48-55.

Glass, J.D., and Lynch, G.R., 1981, Melatonin: identification of site of antigonadal action in mouse brain, **Science**, 214:821-823.

Glass, J.D., and Lynch, G.R., 1982, Evidence for a brain site of action in the white-footed mouse Peromyscus leucopus, **Neuroendocrinology**, 34:1-6.

Goldman, B.D., 1983, The physiology of melatonin in mammals, in: "**Pineal Research Reviews**", Reiter, R.J., ed., Alan R. Liss, New York, pp. 145-182.

Goldman, B.D., and Darrow, J.M., 1983, The pineal gland and mammalian photoperiodism, **Neuroendocrinology**, 37:386-396.

Hasting, M.H., Roberts, A.C., and Herbert, J., 1985, Neurotoxic lesions of the anterior hypothalamus disrupt the photoperiodic but not the circadian system of the syrian hamster, **Neuroendocrinology**, 40:316-324.

Hirata, F., Hayaishi, O., Tokuyama, T., and Senboy, S., 1974, In vitro and in vivo formation of two new metabolites of melatonin, J. Biol. Chem., 249: 1311-1313.

Holtorf, A.P., Heldmaier, G., Thiele, G., and Steinlechner, S., 1985, Diurnal changes in sensitivity to melatonin in intact and pinealectomized Djungarian hamsters: effects on thermogenesis, cold tolerance, and gonads, J. Pineal Res., 2:393-403.

Iuvone, P.M., and Besharse, J.C., 1983, Regulation of indoleamine N-acetyl-transferase activity in the retina: effects of light and dark, protein synthesis inhibitors, and cyclic nucleotide analogs, Brain Res., 273:111-119.

Klein, D.C., 1979, Circadian rhythms in the pineal gland, in: "Endocrine Rhythms", Krieger, D.T., ed., Raven Press, New York, pp. 203-223.

Kopin, I.J., Pare, C.M.B., Axelrod, J., and Weissbach, H., 1961, The fate of melatonin in animals, J. Biol. Chem., 236:3072-3075.

Laitinen, J.T., Castren, E., Vakkuri, O., and Saavedra, J.M., 1989, Diurnal rhythms of melatonin binding in the rat suprachiasmatic nucleus, Endocrinology, 124:1585-1587.

Leone, R.M., and Silman, R.E., 1984, Melatonin can be differentially metabolized in the rat to produce N-acetylserotonin in addition to 6-hydroxy-melatonin, Endocrinology, 114:1825-1832.

Lewy, A.J., Sack, R.O., and Miller, L.S., 1987, Antidepressant and circadian phase-shifting effects of light, Science, 235:352-354.

Mogilnicka, E., and Dubocovich, M.L., 1987, Effect of melatonin receptor antagonist luzindole (N-0774) in the mouse behavioral despair test, Soc. Neurosci. Abstr., 13:1039.

Morgan, P.J., Williams, L.M., Davidson, G., Lawson, W., and Howell, E., 1989, Melatonin receptors on ovine pars tuberalis: characterization and auto-radiographical localization, J. Neuroendocrinol., 1:1-4.

Newman, R., and Winans, 1980a, Experimental study of the ventral striatum of the golden hamster. I. Neuronal connections of the accumbens, J. Comp. Neurol., 191:167.

Newman, R., and Winans, 1980b, An experimental study of the ventral striatum of the golden hamster. II. Neuronal connections of the olfactory tubercle, J. Comp. Neurol. 191:193.

Pang, S.F., and Yew, D.T., 1979, Pigment aggregation by melatonin in the retinal pigment epithelium and choroid of guinea pig, Cavia Porcellus, Experientia, 35:231-233.

Pierce, M.E., and Besharse, J.C., 1985, Circadian regulation of retinomotor movements. I. Interaction of melatonin and dopamine in the control of cone length, J. Gen. Physiol., 86:671-689.

Reppert, S.M., Weaver, D.R., Rivkees, S.A., and Stopa, E.G., 1988, Putative melatonin receptors in a human biological clock, Science, 242:78-81.

Rusak, B., 1980, Suprachiasmatic lesions prevent an antigonadal effect of melatonin, Biol. Reprod., 22:148-154.

Steger, R.W., Bartke, A., and Goldman, B.D., 1982, Alterations in neuro-endocrine function during photoperiod induced testicular atrophy and recrudescence in the golden hamster, Biol. Reprod., 26:437-444.

Steger, R.W., Bartke, A., Matt, K.S., Soares, M.J., and Talamantes, F., 1984, Neuroendocrine changes in male hamsters following photo-stimulation, J. Exp. Zool., 229:467-474.

Stetson, M.H., and Watson-Whitmyre, M., 1984, Physiology of the pineal and its hormone melatonin in annual reproduction in rodents, in: "The Pineal Gland", Reiter, R.J., ed., Raven Press, New York, pp. 109-153.

Takahashi, J.S., and Zatz, M., 1982, Regulation of circadian rhythmicity, Science, 217:1104-1111.

Tamarkin, L., Baird, C.J., and Almeida, O.F.X., 1985, Melatonin: A coordinating signal for mammalian reproduction? Science, 227:714-720.

Underwood, H., and Goldman, B.D., 1987, Vertebrate circadian and photo-periodic systems. Role of the pineal gland and melatonin, J. Biol. Rhythms, 2:279-315.

Vakkuri, O., Leppaluoto, J., and Vuolteenaho, O., 1984, Development and validation of a melatonin radioimmunoassay using radioiodinated melatonin as tracer, **Acta Endocrinol.**, 106:152-157.

Vanecek, J., 1988, Melatonin binding sites, J. **Neurochem.**, 51:1436-1440.

Vanecek, J., and Jansky, L., 1989, Short days induce changes in specific melatonin binding in hamster median eminence and anterior pituitary, **Brain Res.**, 477:387-390.

Vanecek, J., Pavlik, A., and Illnerova, H., 1987, Hypothalamic melatonin receptor sites revealed by autoradiography, **Brain Res.**, 435:359-362.

Vitale, P., Darrow, J.M., Duncan, M.J., Shustak, C., and Goldman, B.D., 1985, Effects of photoperiod and pinealectomy on body weight and daily torpor in Djungarian hamsters: gonad-dependent and gonad-independent effects, J. **Endocrinol.**, 106:367-375.

Weaver, D.R., Namboodiri, M.A.A., and Reppert, S.M., 1988, Iodinated melatonin mimics melatonin action and reveals discrete binding sites in fetal brain, **FEBS Lett.**, 228:123-127.

Williams, L.M., Hasting, M.H., and Morgan, P.J., 1989, 2-[^{125}-I]-Iodomelatonin binding sites in the rat brain: the effect of long-term pinealectomy, J. **Endocrinol.**, 121:192.

Wittkowski, W., Hewing, M., Hoffmann, K., Bergmann, M., and Fechner, J., 1984, Influence of photoperiod on the ultrastructure of the hypophysial pars tuberalis of the Djungarian hamster, Phodopus Sungorus, **Cell Tissue Res.**, 238:213-216.

Zisapel, N., Egozi, Y., and Laudon, M., 1983, Inhibition by melatonin of dopamine release from rat hypothalamus in vitro: variations with sex and the estrous cycle, **Neuroendocrinology**, 37:41-71.

Zisapel, N., Egozi, Y., and Laudon, M., 1985, Circadian variations in the inhibition of dopamine release from adult and newborn rat hypothalamus by melatonin, **Neuroendocrinology**, 40:102-108.

MELATONIN INTERACTION WITH THE BENZODIAZEPINE-GABA RECEPTOR COMPLEX IN

THE CNS

L. Niles

Department of Biomedical Sciences
Division of Neuroscience
McMaster University
Hamilton, Ontario L8N 3Z5
Canada

INTRODUCTION

The major inhibitory neurotransmitter, γ-aminobutyric acid (GABA) acts on a postsynaptic receptor complex which includes GABA and benzodiazepine (BZ) recognition sites and a chloride ionophore (Turner and Whittle, 1983). Pharmacological, biochemical and neurophysiological studies suggest that BZ agonists, such as diazepam, exert their central effects by enhancing low-affinity GABA binding (DeFeudis, 1983; Olsen and Venter, 1986).

The pharmacological effects of the pineal hormone, melatonin, are similar to those produced by psychoactive agents such as the BZ's and barbiturates. Melatonin has been reported to induce sedation and to act as an anticonvulsant in humans and experimental animals (Romijn, 1978). Pharmacological doses of this hormone reduced spiking activity and seizure frequency in epileptic patients (Anton-Tay, 1974) and produced sedative and hypnotic effects in normal subjects (Anton-Tay, 1971) and rodents (Holmes and Sugden, 1982; Sugden, 1983).

Similarities in the effects of melatonin with those of the BZ's, coupled with its ability to inhibit [^{3}H]diazepam binding (Marangos et al., 1981), suggested that its psychopharmacological effects are mediated by BZ receptor sites. In attempting to determine whether melatonin also modulates central GABAergic function, as has been reported for BZ's, we earlier examined its effects on GABA binding in vitro and in vivo (Coloma and Niles, 1988; Niles et al., 1987). Recently, we have studied melatonin's effects on the binding of the labelled cage convulsant, t-[^{35}S]-butylbicyclophosphoro-thionate ([^{35}S]TBPS), which binds to sites on the GABA-gated chloride ionophore. The results of these studies and their implications in clarifying the mechanism(s) underlying the pharmacological effects of melatonin are presented in this review.

EFFECTS OF MELATONIN ON DIAZEPAM BINDING

Inhibition of [^{3}H]diazepam binding

Inhibition experiments indicated that melatonin inhibits [^{3}H]diazepam

Kynurenine and Serotonin Pathways
Edited by R. Schwarcz et al., Plenum Press, New York, 1991

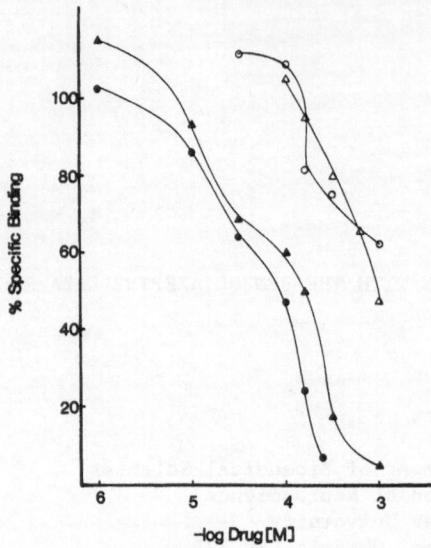

Fig. 1. Inhibition of [^{3}H]diazepam (1.25 nM) binding by melatonin
(closed symbols) and its precursor, N-acetylserotonin (NAS,
open symbols) in bovine (●, ○) and human (▲, △) cortical
membranes. Means of triplicate determinations are presented.

binding with IC_{50}'s of 67 μM and 45 μM in human and bovine cortex respectively. The melatonin precursor, N-acetylserotonin was much less potent with
corresponding IC_{50}'s of 795 μM and 950 μM (Fig. 1). In rat brain membranes,
melatonin was less effective as a displacer, with an IC_{50} of ~500 μM, while
its brain metabolite, N-acetyl-5-methoxykynurenamine (AMK), had an IC_{50} of
100 μM (Fig. 2). These affinities of melatonin and AMK for BZ receptor sites
in the rat brain are similar to those previously reported by Marangos et al.
(1981). The higher affinity found for melatonin in human and bovine brain,
as compared with rat brain, may be related to species differences. However,
we have observed that melatonin's inhibitory potency is increased after
membranes are frozen and extensively washed. Therefore, it is possible that
melatonin's greater affinity for diazepam binding sites in human and calf
brain was due to the long-term freezing of these tissues, while rat brain
tissues were usually frozen for less than one month.

Saturation binding studies conducted in human cerebral cortex in the absence and presence of melatonin, indicated that this hormone competitively
inhibits [^{3}H]diazepam binding as reflected by a significant decrease in binding affinity with no change in binding site density (Table 1).

Protection of diazepam receptors

Studies in bovine cortical membranes indicated that preincubation with
melatonin protected BZ receptors against heat-induced inactivation. In the
presence of 1 mM melatonin, about 90% of [^{3}H]diazepam binding was retained,
while in its absence, about 55% of binding activity was lost following heat
treatment (Table 2). This protective effect of melatonin is presumably due
to occupation of BZ receptor sites and is consistent with its ability to inhibit BZ binding as discussed above.

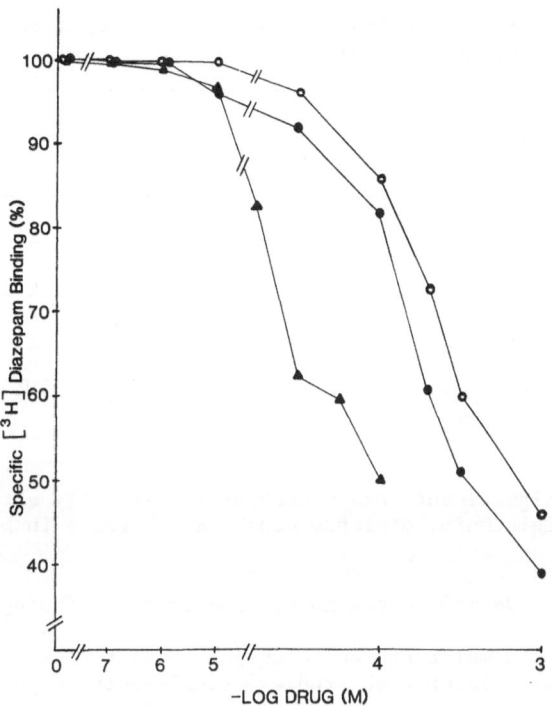

Fig. 2. Inhibition of [³H]diazepam binding by melatonin in fresh (○)
and frozen (●) membranes and by N-acetyl-5-methoxykynurenamine
in frozen (▲) membranes. Means of triplicate determinations
are presented.

MODULATION OF GABA RECEPTORS

In vivo effects of melatonin

     Chronic injection of melatonin, in increasing doses ranging from 1-5
mg/kg for three weeks, produced significant changes in GABA binding. Both
high- and low-affinity binding was increased in forebrain membranes from
melatonin-treated animals. However, the dominant effects were on low-affinity sites which showed an enhancement in binding of about 84% as compared
with an increase of 28% in high-affinity binding (Fig. 3).

Effects of melatonin on [³H]muscimol binding

     Preincubation of rat cortical membranes with melatonin (100 µM) caused a
significant increase of about 96% in the density of GABA receptor sites labelled by the tritiated GABA$_A$ agonist, [³H]muscimol (Table 3). Melatonin
also caused a concomitant decrease of more than 3-fold in binding affinity.
These effects were completely abolished by incubation with Triton X-100,
which significantly enhanced binding in both control and melatonin treated
membranes (Table 3).

Effects of melatonin and diazepam on GABA binding

     An examination of GABA$_A$ receptor binding in rat cortical synaptic mem-
branes revealed that preincubation with melatonin caused a significant increase in the density of low-affinity GABA receptor sites while decreasing
binding affinity (Table 4). Preincubation of membranes with diazepam also

Table 1. Effects of melatonin on [^{3}H]diazepam binding in human cerebral cortex

Treatment	K_d (nM)	B_{max} (fmol/mg protein)
Control	9.6 ± 0.7	776 ± 170
Melatonin (50 nM)	7.9 ± 0.5	693 ± 125
Melatonin (250 μM)	12.6 ± 0.8*	514 ± 90

Means of 4 experiments conducted in duplicate are presented. *P < 0.05 vs. control.

resulted in a significant enhancement of low-affinity site density. In contrast with melatonin, diazepam caused an increase in binding affinity (Table 5).

Effects of chloride ion on the modulation of GABA binding

Although both melatonin and diazepam consistently enhanced low-affinity binding of GABA in saturation studies, single-point experiments produced inconsistent results. Addition of NaCl to single-point assays enhanced the ef-

Table 2. Melatonin protection of ^{3}H-diazepam binding sites in bovine cortex against heat inactivation

Treatment	Specific binding	
	fmol/mg Protein	% of unheated control
Control	145	---
Control and heat	65	45
Melatonin (1.0 mM)	131	90
Melatonin (5.0 mM)	135	93

Cortical membranes were preincubated at 0°C for 30 min with or without melatonin dissolved in 0.4% methanol (final concentration) and 50 mM Tris-HCl buffer (pH 7.4). Controls were also preincubated with methanol in buffer to control for possible effects on binding. Following preincubation, all groups (except the unheated control), were incubated at 60°C for 15 minutes. Membranes were then washed 15 times to remove melatonin and incubated with 1.25 nM ^{3}H-diazepam at 0°C for 30 minutes. Nonspecific binding was measured in the presence of 5 μM non-radioactive diazepam as previously reported (Niles et al., 1987). Means of triplicate determinations are presented.

Table 3. Effects of melatonin on [³H]muscimol binding in rat brain synaptic membranes with or without Triton X-100 treatment

Tissue treatment	K_d (nM)		B_{max} (fmol/mg protein)	
	Control	Melatonin	Control	Melatonin
None	6.9 ± 0.9	25 ± 1.8*	1372 ± 20	2689 ± 14*
Triton X-100	3.5 ± 0.2	3.9 ± 0.3	5316 ± 214	5306 ± 189

Frozen cortical membranes were thawed and treated with 0.05% Triton-X 100 as previously reported (Coloma and Niles, 1988). Membranes were incubated with [³H]muscimol (0.3 to 80 nM) at 0° for 30 in following preincubation with or without melatonin (100 μM) at 0-4° for 60 min. Means ± SEM of four experiments conducted in duplicate are presented. *P < 0.05 vs. control. Data from Coloma and Niles (1988).

fects of both melatonin and diazepam on GABA binding. Maximal enhancement of GABA binding by diazepam and melatonin were found at NaCl concentrations of 50 mM and 150 mM respectively (Fig. 4).

Dose response studies in the presence of 150 mM NaCl indicated that melatonin caused a maximal enhancement of about 60% of GABA binding with an EC_{50} of ~10 μM. Diazepam caused a maximal increase of about 70% with an EC_{50} of ~1 μM (data not shown).

Effects of melatonin on TBPS binding

Melatonin inhibited [³⁵S] TBPS binding in rat brain with an IC_{50} of about 2000 μM. Addition of GABA (1 μM) to incubates caused a six-fold increase in

Table 4. Effects of melatonin (100 μM) on [³H]GABA binding in rat brain

Treatment	K_{d1} (nM)	B_{max1} (fmol/mg protein)	K_{d2} (nM)	B_{max2} (fmol/mg protein)
Control	38 ± 4	637 ± 123	200 ± 19	3661 ± 348
Melatonin	28 ± 3	864 ± 132	394 ± 45*	8097 ± 817*

Frozen forebrain membranes were thawed, incubated at 37° for 30 min, and washed three times in the assay buffer before use in binding experiments. Membranes were preincubated with or without melatonin at 0-4°C for 60 min. Saturation binding of [³H]GABA was determined using a fixed concentration of radioligand and a range of nonradioactive GABA concentrations (2.5 to 1000 nM). Means ± SEM of four experiments conducted in triplicate are presented. *P < 0.01 vs. control. Data from Coloma and Niles (1988).

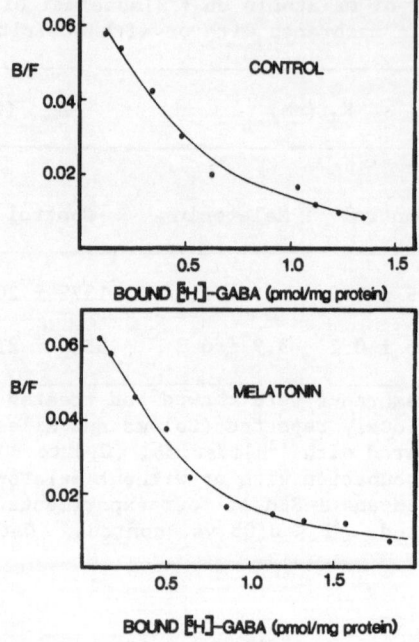

Fig. 3. Scatchard plots of saturation binding of [³H]GABA in rat
forebrain membranes following chronic injection of vehicle or
melatonin. Dissociation constants (K_d, nM) and receptor
densities (B_{max}, pmol/mg protein) for high- and low-affinity
GABA binding were: Control: K_{d1} = 27, B_{max1} = 517; K_{d2} = 612,
B_{max2} = 1960; Melatonin: K_{d1} = 38, B_{max1} = 662; K_{d2} = 1700,
B_{max2} = 3600. Means of triplicate determinations are presented.

melatonin's affinity (IC_{50} = ~300 μM) as shown in Fig. 5.

A comparison of the effects of melatonin, related tryptamines and catecholamines on TBPS binding in the presence of GABA is presented in Table 6.
Although compounds bearing an ethyl N-acetyl substituent tended to reduce
binding, only melatonin significantly inhibited binding by about 36%, indicating the structural specificity of this action.

Table 5. Effects of diazepam (100 μM) on [³H]GABA binding in rat
brain

Treatment	K_{d1} (nM)	B_{max1} (fmol/mg protein)	K_{d2} (nM)	B_{max2} (fmol/mg protein)
Control	38 ± 6	810 ± 96	191 ± 23	1883 ± 183
Diazepam	17 ± 2[*]	731 ± 196	100 ± 7[**]	3645 ± 363[**]

Assays were carried out as described in Table 4. Means ± SEM of six
experiments conducted in triplicate are presented. [*]P < 0.01 vs.
control, [**]P < 0.001 vs. control. Data from Coloma and Niles (1988).

Saturation binding studies in fresh unwashed forebrain membranes indicated that melatonin's effect is due to a significant decrease in the density of TBPS binding sites which is associated with an increase in binding affinity (Table 7).

Target sites of melatonin action

In attempting to confirm that melatonin's pharmacological effects are due to a direct action on BZ receptors, its effects were examined in the presence of a central BZ receptor antagonist, Ro15-1788 (flumazenil) (Haefely, 1989). Preliminary findings from single-point experiments indicate that flumazenil blocks the effects of both melatonin and diazepam on GABA binding in the rat cerebellum (data not shown), supporting a direct pharmacological action of melatonin on central-type BZ receptors. However, saturation binding studies in the presence and absence of flumazenil are required to confirm that the effects of melatonin are mediated by central-type BZ receptors.

There is evidence that melatonin binds with about 6-fold higher affinity to non-central type BZ receptors as compared with central-type receptors (Marangos et al., 1982). Moreover, these non-central type BZ receptors, which are present in the brain, have recently been found to mediate the modulatory effects of various drugs on [^{35}S]TBPS binding (Gee, 1987). Thus, it is possible that melatonin's pharmacologic effects involve central and other BZ receptors in the CNS.

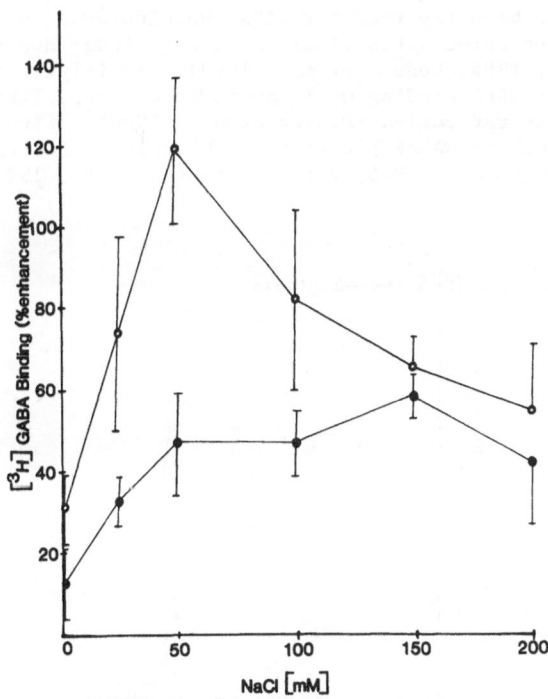

Fig. 4.   Effects of melatonin (500 μM; ●) and diazepam (100 μM; ○) on [^{3}H]GABA binding in rat cortex as a function of chloride ion concentration. Means ± SEM of four experiments conducted in triplicate are presented.

Table 6. Effects of melatonin and other compounds on [^{35}S]TBPS binding in rat cortex

	Specific binding	
Drug	CPM	%
None	3487 ± 490	100
Melatonin	2221 ± 395	64
N-Acetylserotonin	3164 ± 231	91
5-Methoxytryptamine	3402 ± 600	98
N-Acetyl-5-methoxykynurenamine	3288 ± 315	94
N-Acetyltryptamine	3021 ± 253	87
5-Hydroxytryptamine	3536 ± 513	101
Dopamine	3659 ± 538	105
Norepinephrine	3670 ± 531	105

Synaptosomal membranes were incubated with ~2 nM of [^{35}S]TBPS and 500 μM of each drug in the presence of 1 μM GABA. Means ± SEM of three experiments are presented.

CONCLUSIONS

The evidence presented indicates that micromolar concentrations of melatonin alter the characteristics of GABA$_A$ receptors and the associated chloride ionophore binding sites for TBPS. A similar allosteric modulation of TBPS sites has been reported for GABA, barbiturates and various drugs which act on BZ receptor sites (Squires et al., 1983; Honore and Drejer, 1985; Gee et al., 1986; Concas et al., 1988). Anxiolytic agents, such as diazepam, inhibit TBPS binding while anxiogenic drugs, like the β-carbolines, enhance binding in rat cortex (Concas et al., 1988). Since TBPS is a convulsant which inhibits GABAergic acity by blocking GABA-regulated chloride channels (Tehrani et al., 1986; Van Renterghem et al., 1987), the ability of

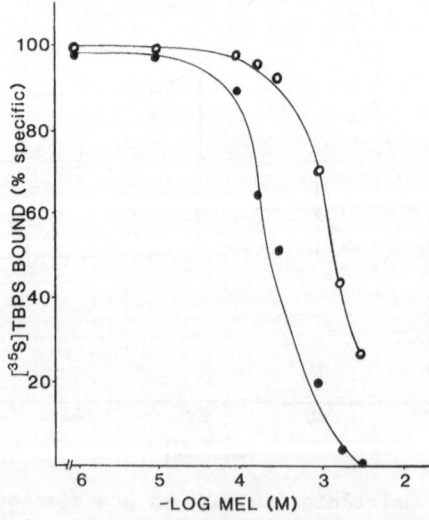

Fig. 5. Effects of melatonin on [^{35}S]TBPS binding in rat brain in the presence (●) or absence (○) of GABA.

Table 7. Effects of melatonin (100 μM) on [^{35}S]TBPS
binding in rat forebrain

Treatment	K_d (nM)	B_{max} (fmol/mg protein)
Control	296 ± 10	6.1 ± 0.3
Melatonin	222 ± 26[*]	4.0 ± 0.3[**]

Fresh homogenates were centrifuged at 18,000 x g for 20
min at 4°C. The pellets were resuspended in 10-20 volumes of 50 mM Tris-citrate buffer (pH 7.4 at 25°C) for
binding assays. Following incubation at 25°C for 90
minutes, bound radioactivity was separated by rapid
vacuum-filtration. Means ± SEM of three experiments
conducted in triplicate are presented. [*]P < 0.05;
[**]P < 0.01 vs. control.

anxiolytic drugs to inhibit its binding is consistent with the GABA-positive
actions of these drugs. Similarly, melatonin's inhibitory effects on TBPS
binding is in keeping with its enhancing effects on GABA binding and supports
our view that the psychopharmacological effects of this hormone involve
facilitation of central GABAergic activity (Niles et al., 1987; Coloma and
Niles, 1988).

Further studies are necessary to determine whether the observed effects
of melatonin are functionally relevant in terms of modulating GABA-gated
chloride ionophore activity in the brain.

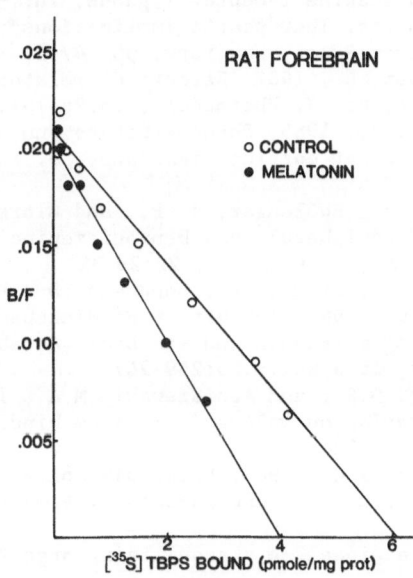

Fig. 6. Scatchard plots of [^{35}S]TBPS binding in fresh unwashed
forebrain membranes in the presence (●) or absence (○)
of 100 μM melatonin. Assays were conducted with 2-380 nM
of isotopically diluted radioligand as described in Table 7.
Means of triplicate measurements from one representative
experiment are presented.

ACKNOWLEDGEMENTS

This work was supported by the Ontario Mental Health Foundation and MRC, Canada. The excellent secretarial assistance of Ms. N. Miksza is acknowledged.

REFERENCES

Anton-Tay, F., 1974, Melatonin: effects on brain function, **Adv. Biochem. Psychopharmacol.**, 11:315-324.

Anton-Tay, F., Diaz, J.L. and Fernandez-Guardiola, A., 1971, On the effect of melatonin upon human brain: its possible therapeutic implications, **Life Sci.**, 10:841-850.

Coloma, F.M., and Niles, L.P., 1988, Melatonin enhancement of [^{3}H]-γ-aminobutyric acid and [^{3}H]muscimol binding in rat brain, **Biochem. Pharmacol.**, 37:1271-1274.

Concas, A., Serra, M., Atsoggiu, T., and Biggio, G., 1988, Foot-shock stress and anxiogenic ß-carbolines because of t-[^{35}S]butylbicyclo-phosphoro-thionate binding in the rat cerebral cortex, an effect opposite to anxiolytics and γ-aminobutyric acid mimetics, **J. Neurochem.**, 51:1868-1876.

DeFeudis, F.V., 1983, Psychoactive agents and GABA-receptors, **Pharmacol. Res. Comm.**, 15:29-39.

Gee, K.W., 1987, Phenyquinolines PK8165 and PK9084 allosterically modulate [^{35}S]t-butybicyclophosphorothionate binding to a chloride ionophore in rat brain via a novel Ro5 4864 binding site, **J. Pharmacol. Exp. Ther.**, 240: 747-753.

Gee, K.W., Lawrence, L.J., and Yamamura, H.I., 1986, Modulation of the chloride ionophore by benzodiazepine receptor ligands: influence GABA and ligand efficacy, **Mol. Pharmacol.**, 30:218-225.

Haefely, W.E., 1989, Pharmacology of the allosteric modulation of GABA$_A$ receptors by benzodiazepine receptor ligands, in: "Allosteric Modulation of Amino Acid Receptors: Therapeutic Implications", Barnard, E.A., and Costa, E., eds., Raven Press, New York, pp. 47-69.

Holmes, S.W., and Sugden, D., 1982, Effects of melatonin on sleep and neuro-chemistry in the rat, **Br. J. Pharmacol.**, 76:95-101.

Honoré, T., and Drejer, J., 1985, Phenobarbitone enhances [^{35}S]TBPS binding to extensively washed rat cortical membranes, **J. Pharm. Pharmacol.**, 37: 928-929.

Marangos, P.J., Patel, J., Boulenger, J.-P., and Clark-Rosenberg, R., 1982, Characterization of peripheral-type benzodiazepine binding sites in brain using [^{3}H]Ro5-4864, **Mol. Pharmacol.**, 22:26-32.

Marangos, P.J., Patel, J., Hirata, F., Sondhein, D., Paul, S.M., Skolnick, P., and Goodwin, F.K., 1981, Inhibition of diazepam binding by tryptophan derivatives including melatonin and its brain metabolite N-acetyl-5-methoxykynurenamine, **Life Sci.**, 29:259-267.

Niles, L.P., Pickering, D.S., and Arciszewski, M.A., 1987, Effects of chronic melatonin administration on GABA and diazepam binding in rat brain, **J. Neural Transm.**, 70:117-124.

Olsen, R.W., and Venter, J.C., 1986, "Benzodiazepine-GABA Receptors and Chloride Channels: Structural and Functional Properties", Alan R. Liss, New York.

Romijn, H.J., 1978, The pineal, a tranquilizing organ? **Life Sci.**, 23:2257-2274.

Squires, R.F., Casida, J.E., Richardson, M., and Saederup, E., 1983, ^{35}S-t-Butylbicyclophosphorothionate binds with high affinity to brain-specific sites coupled to γ-aminobutyric-A and ion recognition sites, **Mol. Pharmacol.**, 23:326-336.

Sugden, D., 1983, Psychopharmacological effects of melatonin in mouse and rat, **J. Pharmacol. Exp. Therap.**, 227:587-591.

Tehrani, M.H.J., Vaidyanathaswamy, R., Verkade, J.G., and Barnes, E.M. Jr., 1986, Interaction of t-butylbicyclophosphorothionate with γ-aminobutyric acid-gated chloride channels in cultured cerebral neurons, J. Neurochem., 46:1542-1548.

Turner, A.J., and Whittle, S.R., 1983, Biochemical dissection of the γ-aminobutyrate synapse, Biochem. J., 209:29-41.

Van Renterghem, C., Bilbe, G., Moss, S., Smart, T.G., Constanti, A., Brown, D.A., and Barnard, E.A., 1987, GABA receptors induced in xenopus oocytes by chick brain mRNA: evaluation of TBPS as a use-dependent channel blocker, Mol. Brain Res., 2:21-31.

SESSION V

BIOLOGICAL EFFECTS: TRYPTOPHAN AND KYNURENINES

EFFECTS OF TRYPTOPHAN 2,3-DIOXYGENASE INHIBITORS IN THE RAT

M. Salter[1], R.M. Beams[2], M.A.E. Critchley[3], H.F. Hodson[2],
R. Iyer[2], R.G. Knowles[1], D.J. Madge[2], and C.I. Pogson[1]

[1]Biochemical Sciences

[2]Medicinal Chemistry

[3]Pharmacology
The Wellcome Research Laboratories
Beckenham, Kent BR3 3BS
UK

INTRODUCTION

Tryptophan is an essential amino acid and is therefore, under normal conditions, only supplied net from the diet (see Fig. 1). Although tryptophan is metabolized through several pathways in the body, it is thought that the catabolism of tryptophan in the liver through the kynurenine pathway is of greatest quanitative significance (Young et al., 1978). However, under certain conditions, enzymes which control other pathways elsewhere in the body, such as indoleamine 2,3-dioxygenase, are induced to such an extent that their respective pathways become quantitatively significant (Brown et al., 1987; Knowles et al., 1989). Under normal conditions, the concentration of tryptophan in the blood will be regulated by the activity of the kynurenine pathway of the liver (Knowles et al., 1989) because there is little regulation of dietary input of tryptophan, apart from substrate supply. Until recently, it has been thought that the activity of the kynurenine pathway of the liver is controlled exclusively by its first step, tryptophan 2,3-dioxygenase (TDO); however, studies with isolated liver cells have shown that significant control resides in the transport of the amino acid across the plasma membrane (Salter et al., 1985; Salter et al., 1986a). Upon induction of TDO, control moves from TDO to transport until transport becomes the major controlling step in the pathway (Salter et al., 1986a). Tryptophan is transported across the liver plasma membrane by two transport systems, systems L and T (Salter et al., 1986b). Because these systems are not subject to regulation, apart from competition effects of other amino acids (Salter et al., 1986b), changes in plasma tryptophan will usually be caused by changes in TDO activity.

Tryptophan is transported across the blood-brain barrier and into the serotonergic neurones where it is hydroxylated and then decarboxylated to 5-hydroxytryptamine (5-HT; for review see Pogson et al., 1989). The hydroxylation of tryptophan by tryptophan hydroxylase is thought to be the major limiting step for the synthesis of 5-HT with the enzyme catalying hydroxylation at or below its K_m for tryptophan in vivo. The rate of transport of tryptophan across the blood-brain barrier and then into the serotonergic neu-

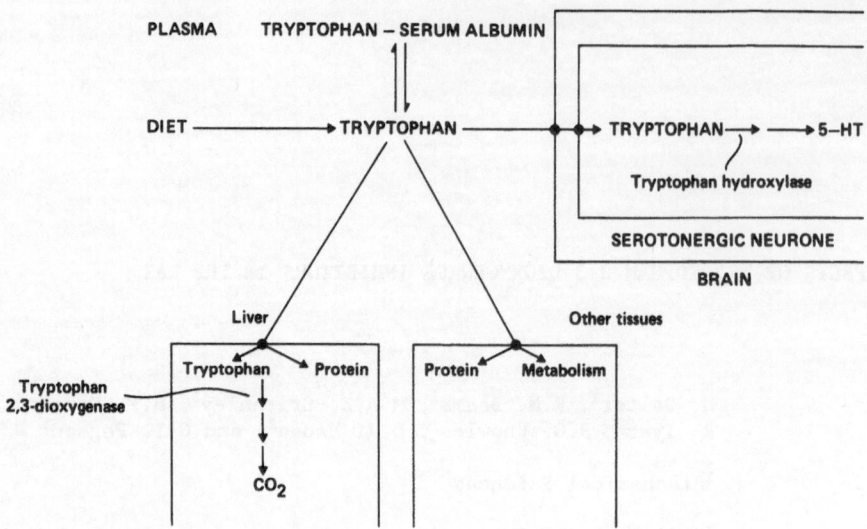

Fig. 1. Tryptophan metabolism in the body.

rons is very fast compared to the rate of 5-HT synthesis, so that it is unlikely that these steps will significantly control the synthesis of 5-HT in the brain (reviewed in Pogson et al., 1989). It has been suggested many times that the synthesis of 5-HT may be regulated by amino acids that compete for the same transporter as tryptophan; however, the relatively fast rate of transport makes it unlikely that direct kinetic competitive effects of these amino acids would significantly affect flux. Transport of tryptophan into synaptosomes, however, has been shown to be energy-linked (Knowles and Pogson, 1984); depletion of the energy available to maintain the gradient of $tryptophan_{in}$: $tryptophan_{out}$ by energy-linked accumulation of other amino acids may well affect this tryptophan gradient and thus the rate of 5-HT synthesis (Pogson et al., 1989).

Many data support the hypothesis that increases in plasma tryptophan will increase brain tryptophan and therefore the synthesis of 5-HT in the brain (Young, 1986; Pogson et al., 1989). The synthesis of 5-HT in the brain appears to be almost solely dependent upon the concentration of free tryptophan in the blood and therefore displacement of tryptophan from serum albumin (thereby increasing free tryptophan) increases the synthesis of 5-HT in the brain (Spano et al., 1974; Salter, Knowles, and Pogson, unpublished observations). The rate of tryptophan catabolism in the liver, however, appears to be determined not only the the concentration of the free pool of tryptophan but by the concentration of the bound pool as well (Salter, Knowles, and Pogson, unpublished observations). This is a consequence of the buffering of the free pool by the bound pool as the free pool is catabolized, and thus depleted, as the blood passes through the liver.

It has been thought for some time that the conditions of depression and, more recently, anxiety are associated with altered serotonergic functioning in the brain (Yaryura-Tobias and Bhagavan, 1977; Young, 1986; Hjorth et al., 1987). Many antidepressants are believed to elevate functional levels of 5-HT and recently 5-HT$_{1A}$ agonists have been marketed as an alternative to benzodiazapines in the treatment of anxiety (Taylor, 1988). Tryptophan has been used for many years as an antidepressant with equivocal results (for review see Young, 1986); however, it does appear to have antidepressant potential in mild to moderate depression (Thomson et al., 1982). There are few data available on the use of tryptophan to treat anxiety apart from a small study with patients suffering from Obsessive-Compulsive Disorders, in which tryptophan showed significant therapeutic efficacy (Yaryura-Tobias and Bhagavan, 1977). The fact that tryptophan may be therapeutically active in both of these conditions is not too surprising, considering the large overlap in the symptomatology of depression and anxiety.

Tryptophan is metabolized rapidly in many species including man (Moller, 1981; Young, 1986 and references therein) and the equivocal action of tryptophan in the treatment of depression has been suggested to be due to this rapid catabolism and therefore limited increase in 5-HT (Young, 1986). Because of this rapid catabolism of tryptophan through the liver kynurenine pathway, the doses of tryptophan used are already large and cannot reasonably be increased. It has been suggested that inhibition of this catabolism would lead to consistent elevation of blood and brain tryptophan and brain 5-HT and thus lead to greater efficacy of action. The TDO inhibitors nicotinamide and allopurinol were shown to have little or no effect upon the clearance of a tryptophan load (Green et al., 1980); however, these negative effects may well have been due to the low potency of these compounds in vivo, rather than the possibility that TDO does not significantly regulate tryptophan clearance.

We have developed potent inhibitors of TDO to help us elucidate this enzyme's role in tryptophan homeostasis and hence the synthesis of 5-HT in the brain.

RESULTS AND DISCUSSION

Fig. 2. shows the effect of racemic TDO inhibitor 64C87 on isolated TDO activity and on flux through TDO in isolated liver cells (in the presence and absence of phenylalanine). It can be seen that the IC_{50} of the inhibitor (with the isolated enzyme) is approximately 2 μM, and, because the concentration of tryptophan (35 μM) is well below the K_m of TDO for tryptophan (approximately 250 μM), this will approximate to the K_i. Single isomers of other TDO inhibitors have K_i's for TDO as low as 400 nM (results not shown). The lower apparent potency of 64C87 on flux through TDO in the liver cells can be explained by the fact that the control of tryptophan catabolism, under these conditions, is shared between transport and TDO. In the presence of phenylalanine (an inhibitor of transport), control is situated mainly in the transport of tryptophan across the liver plasma membrane and therefore inhibition of TDO becomes less effective under these conditions. Thus, it is clear that 64C87 can cross the liver plasma membrane and is therefore able to inhibit flux through the kynurenine pathway (as well as the activity of TDO in cell homogenates).

A 100 mg/kg dose of tryptophan (i.p.) was rapidly catabolized (Fig. 3), as has previously been observed in a large number of studies in the rat. Administration of a racemic mixture of 64C87 (60 mg/kg, 10 minutes after the tryptophan) produced a significant decrease in the rate of elimination of the tryptophan load (Fig. 3). At the peak concentration of tryptophan it is clear that there is a steady-state from 30 minutes to 120 minutes. From the known rate of tryptophan input (all of the tryptophan from the exogenous administration should already have entered the bloodstream at 30 minutes) and the kinetics of TDO for tryptophan, it can be shown that TDO is greatly inhibited over this steady-state period.

Fig. 4 shows the effect of 64C87 on endogenous levels of plasma tryptophan. A 60 mg/kg dose of 64C87 (given intraperitoneally or orally) produced significant increases in plasma tryptophan over 6 hours. If it is assumed that, at steady-state, tryptophan removal is equal to its output (essentially

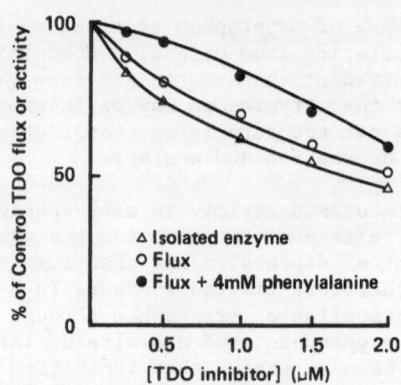

Fig. 2. The effect of TDO inhibitor 64C87 on flux through or activity of
TDO.

through TDO), then the rate of input can be simply determined by calculating
the output through TDO. If tryptophan is distributed in all the body water,
then the rate of TDO (and thus input) at 10 μM free tryptophan concentration will be approximately 8 μM/hour. With total inhibition of TDO one could
therefore expect an increase in plasma tryptophan of 8 μM/hour, which is similar to that seen in Fig. 4. The most important site of tryptophan elevation
for brain 5-HT synthesis is obviously the brain; Fig. 5 shows a repeat of
the experiment in Fig. 4, this time measuring brain tryptophan instead of
plasma tryptophan. 64C87 increased levels of brain tryptophan significantly,
in parallel with the increases seen in the plasma, over a period of 6 hours.

Table 1 shows the effect of 64C87 (60 mg/kg, intraperitoneally) or tryptophan (400 mg/kg, intraperitoneally) on the brain content of tryptophan and
5-hydroxyindoleacetic acid (5-HIAA) 2 hours after administration. 64C87 produced a 2-fold increase in brain tryptophan and 5-HIAA, of which the latter
is indicative of elevated 5-HT synthesis. The very large dose of tryptophan
(400 mg/kg) produced, as expected, a large increase in brain tryptophan but
only a modest further increase in 5-HIAA; consistent with the kinetics of
tryptophan hydroxylase for tryptophan and the major limiting role of trypto-
tophan hydroxylase for 5-HT synthesis.

Evidence supporting the hypothesis that 64C87 is increasing plasma and

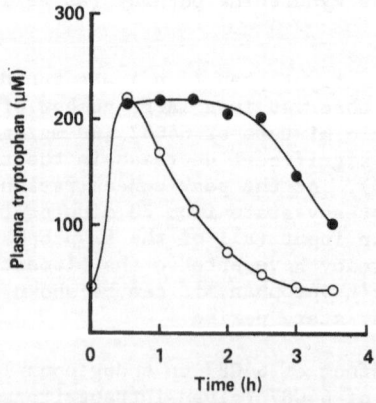

Fig. 3. The effect of 64C87 on the metabolism of a 100 mg/kg dose of
tryptophan; (O) control, (●) plus 64C87.

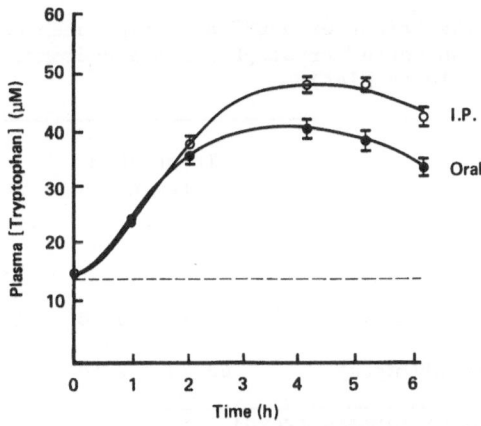

Fig. 4.   The effect of 64C87 on plasma tryptophan.

brain tryptophan by inhibition of TDO is presented in Fig. 6. Rats were injected (intraperitoneally) with L-[ring-2-^{14}C]tryptophan, which is metabolized by TDO and indoleamine 2,3-dioxygenase to unlabelled kynurenine and [^{14}C]-formate. Under normal conditions, most of the [^{14}C]-formate is produced by TDO rather than indoleamine 2,3-dioxygenase; production of [^{14}C]-formate is therefore an indicator of TDO activity in vivo. Fig. 6 shows a time-dependent increase in plasma [^{14}C]-formate after administration of the radiolabelled tryptophan. Subsequent administration of 64C87 (60 mg/kg, intraperitoneally) prevented this increase in [^{14}C]-formate over the 2 hour period of observation, a finding consistent with a large degree of inhibition of TDO in vivo.

The results show that we have a potent TDO inhibitor that is active in vivo as well as in vitro and that significant inhibition of TDO activity in vivo leads to increases in plasma and brain tryptophan and an increase in the synthesis of brain 5-HT. Relatively large doses of 64C87 are needed for two reasons: (1) the compound used was present as a racemic mixture of which only one isomer appeared to be active; (2) the inhibitor binds strongly to serum albumin and therefore large amounts of the compound are not in the free pool in the plasma. From this work, it would appear that previous attempts to regulate plasma tryptophan levels, by inhibition of TDO, have failed not because of the importance (or lack of it) of TDO in regulating plasma trypto-

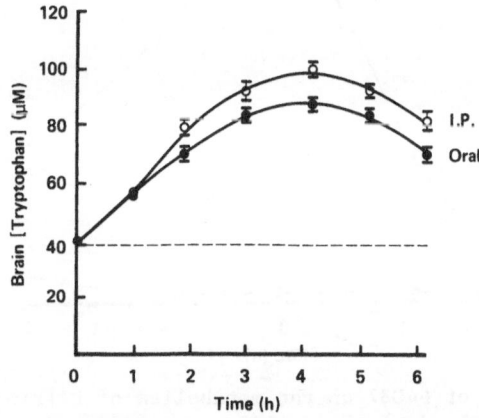

Fig. 5.   The effect of 64C87 on brain tryptophan.

Table 1. The effect of 64C87 and tryptophan on the brain content of tryptophan and 5-hydroxyindoleacetic acid (5-HIAA)

	Tryptophan (μg/g)	5-HIAA (μg/g)
Control	2.10 ± 0.15	0.85 ± 0.10
64C87	4.20 ± 0.40	1.62 ± 0.20
Tryptophan (400 mg/kg)	63.50 ± 2.50	2.51 ± 0.35

Means ± SEM (3 animals/group).

phan but because of the low potency or pharmacokinetics of the inhibitors used. We have shown that inhibition of TDO leads to elevation of plasma tryptophan without the requirement of exogenous tryptophan. With continued administration of such a TDO inhibitor, it may be possible to maintain chronic increases in plasma tryptophan and therefore chronic elevation of serotonergic function. TDO activity has recently been found in rat skin (Naito et al., 1989) but it is not clear at present whether this will have any significant quantitative importance for tryptophan catabolism.

We have investigated the role of 5-HT in anxiety by comparing the effect of 5-hydroxytryptophan (5-HTP) with the classical anxiolytic, diazepam (1 mg/kg, subcutaneously) in the Vogel conflict model of anxiety (Fig. 7). 5-HTP produced a significant decrease in anxiety (increasing licks) at 100 mg/kg, whereas diazepam decreased anxiety but not significantly. This effect of

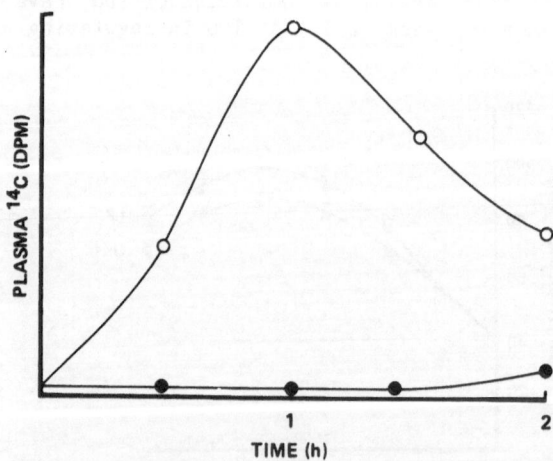

Fig. 6. The effect of 64C87 on the metabolism of L-[ring-2-^{14}C]tryptophan; (O) control, (●) plus 64C87.

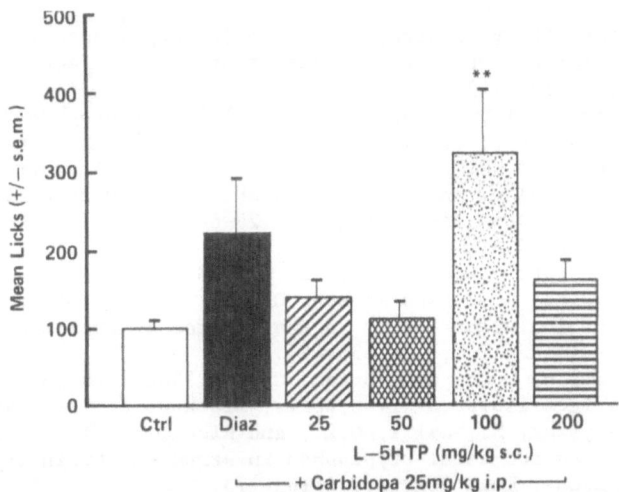

Fig. 7. The effect of diazepam and 5-hydroxytryptophan in the Vogel conflict model of anxiety. **P < 0.005 (Mann-Whitney test).

5-HT is consistent with previous work (Hjorth et al., 1987) and shows for the first time a comparison of the effects of elevation of serotonergic function and the effects of a benzodiazepine on anxiety. The dose response with 5-HTP was slightly different from previous reports (Hjorth et al., 1987) and is probably explained by the less effective dose of peripheral decarboxylase inhibitor (carbidopa) used in this study. We have also obtained preliminary evidence that tryptophan is also an anxiolytic in the Vogel conflict model of anxiety (results not shown). It is likely that elevation of serotonergic function is a valid anxiolytic mechanism of comparable efficacy to that of the classical benzodiazepines when given acutely in the Vogel conflict model of anxiety.

ACKNOWLEDGEMENTS

We would like to thank M. Lowe for typing the manuscript.

REFERENCES

Brown, R.R., Borden, E.C., Sondel, P.M., and Lee, C., 1987, Effects of interferons and interleukin-2 as tryptophan metabolism in humans, in: "Progress in Tryptophan and Serotonin Research 1986", Bender, D.A., Joseph, M.H., Kochen W., and Steinhart, H., eds., de Gruyter, Berlin, pp. 19-26.
Green, A.R., Aronson, J.K., Curzon, G., and Woods, H.F., 1980, Metabolism of an oral tryptophan load, Br. J. Pharmacology, 10:611-618.
Hjorth, S., Soderpalm, B., and Engel, J.A., 1987, Biphasic effect of 5-HTP in the Vogel conflict model, Psychopharmacology, 92:96-99.
Knowles, R.G., and Pogson, C.I., 1984, Characteristics of tryptophan accumulation by isolated rat forebrain synaptosomes, J. Neurochem., 42:663-669.
Knowles, R.G., Pogson, C.I., and Salter, M., 1989, Application of control analysis to the study of amino acid metabolism, in: "Control of Metabolic Processes", Cornish-Bowden, A., Ricard, J., Westerhoff, H.V., and Goldbeter, A., eds., Plenum, New York, pp. 377-384.

Moller, S.E., 1981, Pharmacokinetics of tryptophan, renal handling of kynurenine and the effect of nicotinamide on its appearance in plasma and urine following L-tryptophan loading of healthy subjects, **Eur. J. Clin. Pharmacol.**, 21:137-142.

Naito, J., Ishiguro, I., and Nagamura, Y., 1989, Tryptophan 2,3-dioxygenase activity in rat skin, **Arch. Biochem. Biophys.**, 270:236-241.

Pogson, C.I., Knowles, R.G., and Salter, M., 1989, The control of aromatic amino acid catabolism and its relationship to neurotransmitter amine synthesis, **Crit. Rev. Neurobiol.**, pp. 29-64.

Salter, M., Bender, D.A., and Pogson, C.I., 1985, Leucine and tryptophan metabolism in rats, **Biochem. J.**, 225:277-280.

Salter, M., Knowles, R.G., and Pogson, C.I., 1986a, Quantification of the importance of individual steps in the control of aromatic amino acid metabolism, **Biochem. J.**, 234:635-647.

Salter, M., Knowles, R.G., and Pogson, C.I., 1986b, Transport of the aromatic amino acids into isolated liver cells, **Biochem. J.**, 233:499-506.

Spano, P.F., Szyszka, K., Galli, C.L., and Ricci, A., 1974, Effect of clofibrate on free and total tryptophan in serum and brain tryptophan metabolism, **Pharmacol. Res. Commun.**, 6:163-173.

Taylor, D.P., 1988, Buspirone, a new approach to the treatment of anxiety, **FASEB J.**, 2:2445-2452.

Thomson, J., Rankin, H., Ashcroft, G.W., Yates, C.M., McQueen, J.K., and Cummings, S.W., 1982, The treatment of depression in general practice: a comparison of L-tryptophan, amitriptyline and combination of L-tryptophan and amitriptyline with placebo, **Psychol. Med.**, 12:741-751.

Yaryura-Tobias, J.A., and Bhagavan, H.N., 1977, L-tryptophan in obsessive-compulsive disorders, **Am. J. Psychiat.**, 134:1298-1299.

Young, S.N., 1986, The clinical psychopahrmacology of tryptophan, in: "Nutrition and the Brain", Wurtman, R.J., and Wurtman, J.J., eds., Raven Press, New York, pp. 49-88.

Young, S.N., St. Arnaud-McKenzie, D., and Sourkes, T.L., 1978, Importance of tryptophan pyrrolase and aromatic amino acid decarboxylase in the catabolism of tryptophan, **Biochem. Pharmacol.**, 27:763-767.

BIDIRECTIONAL RELATIONSHIPS BETWEEN TRYPTOPHAN AND SOCIAL BEHAVIOR IN VERVET

MONKEYS

M.J. Raleigh[1,2,3] and M.T. McGuire[1,2]

[1]Department of Psychiatry and Biobehavioral Sciences
UCLA School of Medicine
Los Angeles, California 90024

[2]Nonhuman Primate Laboratory
Sepulved Veterans Administration Medical Center
Sepulveda, California 91304

[3]Neurobiochemistry Laboratory
Brentwood Veterans Administration Medical Center
Los Angeles, California 90049
USA

INTRODUCTION

Exogenous tryptophan induces a wide variety of behavioral effects in socially-living vervet monkeys. Tryptophan administration produces dose-dependent increases in grooming, proximity to group members, and other affiliative behaviors (Raleigh et al., 1980). In stable social groups, tryptophan administration also reduces aggressive, submissive, and retaliatory aspects of agonistic behavior (McGuire and Raleigh, 1985). Pharmacological and physiological studies suggest that tryptophan's effects are mediated by central serotonin (McGuire et al., 1982; Raleigh et al., 1985). The diversity of the behavioral effects, together with the diffuse distribution of central serotonergic projections and the heterogeneity of serotonergic receptors make it unlikely that each of these distinct behavioral effects is caused by the action of serotonin alone on some final common motor pathway. Rather, enhanced central serotonergic neurotransmission appears to promote both the mood and cognitive states that in turn facilitate the expression of quiescent, calm bahaviors (Raleigh et al., 1988).

In addition to documenting the behavioral effects of tryptophan, we have observed that social and environmental factors modulate the effects of tryptophan. In stable social groups, social rank influences the behavioral effects of tryptophan. For example, relative to low-ranking males, dominant males are more responsive to tryptophan. They exhibit behavioral responses to lower doses of tryptophan than do subordinate animals and the magnitude of their behavioral alterations is larger than are those of subordinate animals. Besides social rank, an animal's response to tryptophan is influenced by group stability, the individual's arousal level, and the behavioral options available at the time of treatment (Kraemer, 1985; Raleigh et al., 1985; Chamberlain et al., 1987; Young, 1987).

Kynurenine and Serotonin Pathways
Edited by R. Schwarcz et al., Plenum Press, New York, 1991

The preceding observations point to a bidirectional relationship between tryptophan and behavior. While tryptophan can profoundly affect a monkey's behavior, behavioral history and environmental setting influence the form and magnitude of these behaviors as well as their physiological concomitants. The present chapter reviews two aspects of this bidirectional approach. One concerns the selectivity and sensitivity of social behavioral measures which provide a rich data base for assessing the effects of tryptophan and its metabolites. The other addresses the extent to which environmental alterations can shape both the magnitude and the direction of tryptophan's behavioral effects. Prior to presenting these data, we will briefly describe selected features of vervet monkeys and their behavior that make them well-suited for these types of investigations.

VERVET MONKEYS AND THEIR BEHAVIOR

At least three aspects of vervet monkeys make them exceedingly useful for studies of the interrelationships between monoaminergic function and social behavior. Behaviors such as social development, migration, intergroup behavior, and dominance subordinance relationships have been well-documented in both captive and free-ranging settings. Further, nearly all of the species-typical behavioral patterns observed in the wild also occur in captive groups living in large outdoor enclosures (Cheney et al., 1986; Hauser and Fairbanks, 1988). For instance, in both captive and free-ranging settings, matrilineal relationships exert pronounced influences on affiliative and agonistic behavior. Matrilineally related animals are more likely to groom each other, support each other in fights, and be in close proximity (Fairbanks, 1988). Similarly, in both free-ranging and captive settings animals are strikingly aggressive to adults who are not members of their groups (Keddy, 1986). In free-ranging settings, males, but not females, migrate out of their natal groups and in captivity males are much more tolerant of analogous changes in group membership than are females (Cheney and Seyfarth, 1983). Signs of severe stress such as cage stereotypies or incessant pacing are not observed in large outdoor enclosures. Consequently, captive vervet monkey groups provide a reproducible setting for examining the interplay between etiologically relevant behaviors and monoaminergic function.

Vervets are among the most successful primates. Morphologically generalized, they are exceedingly adaptive and are distributed throughout sub-Saharan Africa. Except for humans, they are the most numerous of primate species. Living in a wide variety of niches, they have been appropriately called "opportunistic omnivores" (Struhsaker, 1967). Both single- and multimale groups have been repeatedly observed (McGuire, 1974). Their adaptability has allowed them to flourish on at least three Eastern Caribbean islands since their importation in the 17th century. Because of their abundance, vervets represent a species in which both correlative and experimental investigations can be conducted. For instance, vervets have been used to evaluate the extent to which peripheral physiological measures (e.g. platelet serotonin receptors) mirror central indices (e.g. cortical S_2 receptors) of serotonergic function (Brammer et al., 1987). In endangered or less flexible species such concerns could not be addressed directly.

The third feature is vervets' relative reliance on social skills rather than physical prowess. Adult females weigh about 75% as much as males and can successfully lead aggressive coalitions against males. Indeed, in free-ranging settings, females may strongly influence the entry of a migrating male into a new group. Thus for males the development of social skills needed for the formation and maintenance of enduring, heterosexual affiliative relationships are critical to survival. These skills are also important among captive males where, for example, the initial phase in achieving high rank involves forming affiliative alliances with females (Raleigh and

Table 1. Morphological and behavioral concomitants of male
         social status

	Dominant/Subordinate
Weight (19)	94 ± 15
Canine length (9)	108 ± 19
Testes volume (9)	92 ± 18
Crown-rump length (19)	109 ± 10
Aggressivity (13)	71 ± 8*
Appropriate targets (13)	217 ± 23*
Affiliative behavior (13)	169 ± 13*

Data are the ratio (times 100) of dominant to subordinate
males for selected morphological and behavioral measures.
Numbers in parentheses indicate the number of groups used in
each comparison. From each such group, the dominant male was
compared to an age matched subordinate male. *Significant
difference between the dominant and subordinate groups
(P < 0.05). Data are the mean ± SEM.

McGuire, 1989).

Table 1 shows that rank among males is unrelated to such physical features as weight, canine length, testicular volume, and body length. None of
these morphometric measurements correlate with dominance and they do not differ between dominant and subordinate males by more than 9%. In contrast to
the absence of physical differences, Table 1 also shows that behavioral style
differed between high and low ranking animals. Aggressivity, the rate at
which individuals initiate fights, was not linked to high rank. Indeed subordinate males are more aggressive than dominant males. Further, dominant
males are less likely to attack inappropriate targets such as females or immature animals. Affiliative behavior differs with rank. Dominant males are
approximately twice as likely as subordinate animals to engage in affiliative
behavior with females.

SENSITIVITY AND SELECTIVITY OF SOCIAL BEHAVIOR

A number of studies have shown that tryptophan and related compounds can
specifically effect selected social behaviors. Tryptophan treatment does not
universally inhibit all behaviors: some are enhanced, others reduced, and
still others unaltered. In addition, social behavior is sensitive to low
doses of tryptophan. Because of this, investigations of the social behavioral effects of tryptophan can use doses producing relatively few side effects.
The observed behavioral consequences are not secondary to stereotypies or
other toxicological disruptions: rather, species-typical, normal behaviors
are influenced and these alterations appear to represent the physiological
actions of tryptophan and its metabolites on the neural bases of social
behavior.

Both the sensitivity and selectivity of social behavior were illustrated
in an investigation that compared the effects of tryptophan and 5-HTP (Raleigh, 1987). Both compounds are precursors to serotonin, and their administration increases central and peripheral serotonin concentrations. Among socially living vervet monkeys, both treatments augment grooming and other affiliative behaviors. However, tryptophan and 5-HTP result in opposite effects on aggression and viligance as well as differing effects on locomotion
and eating (Raleigh et al., 1980).

We sought to determine whether these differences were due to 5-HTP's effects on catecholaminergic neurons. Many distinct lines of evidence indicate that exogenous 5-HTP can enter catecholaminergic neurons and be converted there to serotonin by aromatic amino acid decarboxylase (Ebadi and Simonneaux, this volume). The resulting serotonin may displace catecholaminergic transmission. In contrast, while tryptophan enters both catecholaminergic and serotonergic neurons it can only be converted to 5-HTP and thence to serotonin in serotonergic neurons. The distribution of serotonin synthesized from exogenous tryptophan closely resembles that of endogenous serotonin while the distribution of serotonin following exogenous 5-HTP is more widespread (Moir and Eccleston, 1968).

To test whether some of 5-HTP's induced behaviors resulted from 5-HTP's impact on catecholamines, we evaluated the effects of several doses of 5-HTP and tryptophan alone and in combination with fluoxetine and desipramine (DMI). Fluoxetine is a fairly selective 5-HT reuptake inhibitor while DMI is a relatively specific catecholaminergic reuptake inhibitor. Thus, if the contrasts between tryptophan and 5-HTP arise from 5-HTP's effects on catecholamines, concurrent 5-HTP and DMI should enhance these differences while concurrent fluoxetine and 5-HTP should diminish them.

As Table 2 shows, tryptophan produced dose-dependent decreases in aggression, vigilance, and locomotion. At the highest tested dose, all three behaviors occurred at less than 60% of the baseline rates. The moderate and high doses of tryptophan increased eating to 128% and 156% of baseline, respectively. Table 2 also shows that 5-HTP had strikingly different behavioral effects than tryptophan: it increased aggression and vigilance and did not alter locomotion or eating.

Table 3 shows that concurrent fluoxetine reduced the differences between

Table 2. Effects of tryptophan and 5-HTP on social behavior

			Tryptophan	5-HTP
Aggression	-	Low	102	101
	-	Moderate	81	157
	-	High	56	193
Vigilance	-	Low	98	102
	-	Moderate	66	158
	-	High	42	258
Locomotion	-	Low	105	99
	-	Moderate	61	95
	-	High	39	106
Eating	-	Low	96	111
	-	Moderate	128	96
	-	High	156	105

Six animals received tryptophan at 10, 20, or 40 mg/kg (low, moderate, or high) doses and six received 20, 40, and 80 mg/kg 5-HTP. For each behavior, the low, moderate, and high rows represent the rate of that behavior relative to vehicle (times 100). For example, the moderate dose of tryptophan reduced aggression to 81 percent of vehicle. All behaviors were significantly altered by tryptophan treatment. 5-HTP significantly altered aggression and vigilance.

Table 3. Effects of concurrent fluoxetine and DMI on 5-HTP induced behavioral alterations

FL	DMI	Aggression	Vigilance	Locomotion	Rating
High	None	-46	-49	-14	+51
Low	None	-42	-36	-10	+ 9
None	None	Increase	Increase	None	None
None	Low	+32	+34	- 3	+ 8
None	High	+61	+127	+56	-30

The FL and DMI columns indicate the dose of fluoxetine or DMI given concurrently with 40 mg/kg of 5-HTP. High and low doses of fluoxetine and DMI were 1.0 and 0.5 mg/kg and 3 and 1.5 mg/kg, respectively. For each behavior, numbers represent the percentage difference between the rate under concurrent treatment regimes and 5-HTP alone. Thus, for aggression, concurrent high fluoxetine reduced the effects of 5-HTP by 46%. Concurrent low dose of fluoxetine diminished aggression by 42%. The direction of the behavioral effects of 5-HTP alone is indicated qualitatively in the third row.

5-HTP and tryptophan. For instance, concurrent high and low doses of fluoxetine reduced the vigilance induced by 40 mg/kg of 5-HTP by 49% and 36%, respectively. Conversely, DMI enhanced the differences between 5-HTP and tryptophan. For example, when given with 40 mg/kg of 5-HTP, concurrent low and high DMI augmented vigilance by 34% and 127%, respectively. Concurrent fluoxetine and DMI had similar dose-dependent effects on the high dose of 5-HTP.

The differences between 5-HTP and tryptophan are centrally mediated and not due to peripheral effects of tryptophan or 5-HTP. Nor do pharmacokinetic differences explain the 5-HTP/tryptophan differences. 5-HTP and DMI produce identical increases in aggression and vigilance, and concurrent DMI augmented the effects of 5-HTP. In contrast, fluoxetine antagonized the 5-HTP effects on these behaviors. Thus, DMI increased and fluoxetine reduced the differences between 5-HTP and tryptophan. For behaviors in which catecholaminergic and serotonergic agents interact antagonistically, 5-HTP produced different effects than tryptophan and the greater the catecholamine contribution the larger the differences between 5-HTP and tryptophan. It seems reasonable to conclude that the different behavioral effects of 5-HTP and tryptophan result from 5-HTP's effects on central catecholaminergic systems (McGuire and Raleigh, 1987).

A second set of investigations that also underscores the usefulness of social behavior in psychopharmacology is summarized qualitatively in Table 4. This study evaluated the effects of several doses of tryptophan and tyrosine on the behavior of subordinate males. As in other studies, tryptophan resulted in dose-dependent increases in affiliative behaviors including grooming, approaching and being in proximity. Tryptophan reduced avoiding, being solitary, and aspects of aggressing. Individual behaviors such as eating, locomoting and resting were also altered by tryptophan. Table 4 shows that tyrosine treatment produced a different pattern of behavioral alterations. As described elsewhere (Raleigh et al., 1988), these two treatments had opposite effects on grooming, resting, locomoting, being vigilant, aggressing and differing effects on eating avoiding, and submitting. The data in Table 4 show that species-typical social behaviors are useful and sensitive endpoints for pharmacological studies. Direct, quantitative observation of social behavior may be a particularly sensitive procedure for documenting the specific effects of tryptophan in humans as well.

Table 4. Qualitative effects of tryptophan and tyrosine
on social behavior

Behavior	Tryptophan	Tyrosine
Aproach	Increase	Increase
Groom	Increase	Decrease
Rest	Increase	Decrease
Eat	Increase	No effect
Locomote	Decrease	Increase
Avoid	Decrease	No effect
Be vigilant	Decrease	Increase
Be solitary	Decrease	Decrease
Huddle	No effect	No effect
Aggress	Decrease	Increase

This Table summarizes the effects of several doses of tryptophan (10,20 and 40 mg/kg) and tyrosine (15, 30 and 60 mg/
kg) on social behavior. An increase or a decrease indicates
that the treatment produced significant alterations in the
rate or duration of that behavior.

ENVIRONMENTAL AND BEHAVIORAL CONSTRAINTS

The physiological and behavioral consequences of tryptophan are shaped
by environmental and behavioral conditions. Tryptophan does not induce in-
variant effects on social behavior. Rather, its effects are tempered by pre-
existing behavioral and environmental conditions. Among humans, this envir-
onmentally-induced variability is clearly shown in studies of the effects of
tryptophan on cortisol, growth hormone, and prolactin. For example, Lehnert
and Beyer (this volume) reported that in adult men the effects of tryptophan
on cortisol and prolactin are affected by time and test setting. Many human
studies have been conducted on depressed, hypertensive or other clinical populations. In these studies, both the study and referent groups often have
unusual diets, sleep patterns or activity levels (Cowen, 1988); consequently,
the extent to which these findings generalize to normative settings is uncertain.

Two sets of investigations of monkeys show that the social influences on
the behavioral and physiological effects of tryptophan might be very widespread. One social factor that strongly shapes the effects of tryptophan and
other compounds is dominance status. Relative to subordinate males, dominant
individuals are hyper-responsive to tryptophan. This status linked difference is state dependent in that alterations in status are accompanied by corresponding changes in tryptophan responsivity. Formerly subordinate animals
who have become dominant respond to tryptophan in a characteristically dominant animal fashion.

At a phenomenological level, status-linked differences are also apparent
for other compounds that primarily affect other neurotransmitters systems.
For instance, among squirrel monkeys the effects of alcohol change with social status and season (Winslow and Miczek, 1988).

Descriptively, these studies point to the influences of social factors
on the effects of tryptophan and other compounds. Identification of status-
linked differences are critical if the range of normative responses is to be
understood. Obviously the more fully the underlying mechanisms can be delineated, the more informative such status-linked differences become. Status-
linked differences in the effects of tryptophan may arise from pharmacokin-

etic differences. They could reflect differences in the activity of tryptophan hydroxylase or of tryptophan pyrrolase. These (or other) potential mechanisms may result in varying amounts of tryptophan being converted to brain
serotonin. Status-linked differences in response to tryptophan could also
arise at the level of post-synaptic receptor. Either the number or the affinity of such receptors for serotonin could differ. Clearly these possible
mechanistic explanations are far from exhaustive. Determination of the mechanisms that support these status-linked differences in animals may point to
new directions for the investigations of the bases for individual differences
in the effects of tryptophan in humans.

A second set of environmental/behavioral factors impacting on the effects of tryptophan are the conditions under which animals are housed. In
socially living animals, tryptophan reliably represses some types of elicited
aggression. A paradigm that affects an animal's sensitivity to tryptophan
involves briefly moving an animal from his social group and placing him in a
neutral enclosure with an unfamiliar conspecific. Under these conditions,
monkeys who are living in normative social groups will direct a substantial
amount of species-typical aggression to the other, unfamiliar animal. As
shown in Table 5, in such socially living animals, the rate of such elicited
aggressive behavior can be dramatically diminished by tryptophan administration.

In a preliminary cross-over study we have evaluated the impact of social
isolation on the effects of tryptophan on this type of elicited aggression.
Under isolation conditions, animals lived in standard primate cages, appeared
healthy and could see, hear, but not touch, other vervets. This housing did
not induce stereotypies or other behavioral signs of severe stress. The animals were tested in the behavioral paradigms described above. In the vehicle
condition isolated animals aggressed at the same rates as when they were members of social groups. Table 5 also shows that for isolated animals tryptophan doubled the rate of aggression. For isolated animals, tryptophan produced diametrically opposed effects to what was seen in these animals when
they lived in social groups. Thus, tryptophan enhanced aggression to approx-

Table 5. Housing conditions constrain the effects of
          tryptophan on stranger-elicited aggression

	Vehicle	Tryptophan
Social	100	65 ± 12
Isolated	109 ± 14	216 ± 23

Subjects were studied as members of social groups as
when living in isolation. In each setting, they were
examined following vehicle and tryptophan (20 mg/kg)
administration. There was substantial intersubject
variability in each of the four conditions. For example, when socially living animals received vehicle, the
rate of aggression directed to the stranger ranged from
4 to 61 events/hour. For clarity of presentation, each
subjects score in the social living vehicle condition
was assigned a relative value of 100 and other cells
represent the mean ± SEM of aggression relative to this
value. Among socially living animals, tryptophan treatment reduced aggression directed to an unfamiliar conspecific. In contrast, when housed in individual cages
the same treatment increased aggressive behavior.

imately twice baseline values. These data indicate that social setting strongly affects the impact of tryptophan on behavior. The differences are in the direction and not merely the magnitude of the behavioral effects.

In sum, behavioral and housing conditions profoundly affect the behavioral consequences of tryptophan. Among monkeys living in species representative social groups, social status modifies the effects of tryptophan. Similarly, housing primates in isolation dramatically alters the effects of tryptophan on aggression. In view of these and other observations, sweeping statements about tryptophan having inhibitory or facilitory effects on agressive or other behavior are inexact. The impact of tryptophan cannot be adequately described without delineating an animal's social and housing conditions.

CONCLUSIONS

Vervet monkeys are well suited for examining behavior-physiology relationships. Their behavior is well characterized in free-ranging and captive settings, they are abundant, they live in complex social groups, and they develop enduring social relationships. These features make them ideally suited for evaluating the functional consequences of alterations in tryptophan. They are also appropriate for studies that examine the role of central mechanisms and because of their phylogenetic proximity to humans behavior-physiology relationshipss found in vervet monkeys are likely to generalize to humans.

This chapter has underscored the utility of social behavioral measures for evaluating the impact of an individual's history and the current environmental setting on the behavioral effects of tryptophan and other compounds. Many previous investigations of the behavioral effects of tryptophan and its metabolites have utilized individually housed animals and relied on learning and conditioning tasks. While these investigations have generated vast amounts of data, their relationship to the social challenges individuals encounter is problematic. For humans as well as monkeys, the formation and maintenance of friendships, the adroit handling of aggressive interactions, and the selection of an appropriate mate are critically important challenges. Among vervets, the social behaviors involved in these and other patterned relationships can be quantified and are sensitive to low doses of tryptophan. Social behavioral measures can be used to differentiate between closely related compounds such as tryptophan and 5-HTP and may prove useful in evaluating the effects of other metabolites of tryptophan such as quinolinic or kynurenic acid.

Finally, this chapter has shown that monkeys deprived of social interactions may show abherent responses to tryptophan. Many potentially psychoactive compounds are screened by evaluating their effects of conditioned performances in individually caged animals. However, such data may not be predictive of the behavioral or cognitive effects of the compounds administered to socially living animals. Environmental conditions may limit the degree to which one can generalize about tryptophan's behavioral effects, and a central problem to be addressed concerns the mechanisms by which environmental factors alter tryptophan-elicited behaviors.

ACKNOWLEDGEMENTS

This research is supported in part by grants from the United States Public Health Service, the Research Service of the Veterans Administration, and the Giles W. and Elise G. Mead Foundation. Useful comments were provided by A. Yuwiler and expert technical assistance was provided by G. Morton, D. Bergin, N. Kimble, and D. Diekman.

REFERENCES

Brammer, G.L., McGuire, M.T., and Raleigh, M.J., 1987, Similarity of 5HT$_2$ receptor sites in dominant and subordinate vervet monkeys, Pharmac. Biochem. Behav., 27:701-705.

Chamberlain, B., Ervin, F.R., Pihl, R.O., and Young, S.N., 1987, The effect of raising or lowering tryptophan levels on aggression in vervet monkeys, Pharmacol. Biochem. Behav., 28:503-510.

Cheney, D.L., and Seyfarth, R.M., 1983, Nonrandom dispersal in free-ranging vervet monkeys: social and genetic consequences, Amer. Natur., 122:392-412.

Cheney, D.L., Seyforth, R., and Smuts, B., 1986, Social relationships and social cognition in nonhuman primates, Science, 234:1361-1366.

Cowen, P.J., 1988, Neuroendocrine responses to tryptophan as an index of brain serotonin function, in: "Amino Acid Availability and Brain Function in Health and Disease", Huether, G., ed., Springer, Berlin, pp. 285-290.

Ebadi, M., and Simonneaux, Ambivalence on the multiplicity of mammalian aromatic L-amino acid decarboxylase, this volume.

Fairbanks, L.A., 1988, Vervet monkey grandmothers: effect on mother-infant relationships, Behavior, 104:176-188.

Hauser, M.H., and Fairbanks, L.A., 1988, Mother-offspring conflict in vervet monkeys: variation in response to ecological conditions, Anim. Behav., 36:802-813.

Keddy, A.C., 1986, Female mate choice in vervet monkeys (Cercopithecus aethiops sabaeus), Am. J. Primatol., 10:125-134.

Kraemer, G.W., 1985, The primate social environment, brain neurochemical changes and psychopathology, Trends Neurosci., 8:339-340.

Lehnert, H., and Beyer, J., Cardiovascular and endocrine properties of L-tryptophan in combination with various diets, this volume.

McGuire, M.T., 1974, The St. Kitts green monkey, Contrib. Primatol., 1:1-199.

McGuire, M.T., and Raleigh, M.J., 1985, Serotonin-behavior interactions in vervet monkeys, Psychopharm. Bull., 21:458-463.

McGuire, M.T., and Raleigh, M.J., 1987, Serotonin, social behavior, and aggression on vervet monkeys, in: "Psychopharmacology of Aggression", Mos J., and Brain, P.F., eds., Martinus Nijhoff, Dardrecht, The Netherlands, pp. 207-222.

McGuire, M.T., Raleigh, M.J., and Brammer, G.L., 1982, Sociopharmacology, Ann. Rev. Pharm. Toxicol., 27:643-661.

Moir, A.T.B., and Eccleston, D., 1968, The effects of precursor loading in the cerebral metabolism of 5-hydroxyindoles, J. Neurochem., 15:1093-1108.

Raleigh, M.J., 1987, Differential behavioral effects of tryptophan and 5-hydroxytryptophan in vervet monkeys, influence of catecholaminergic systems, Psychopharmacology, 93:44-50.

Raleigh, M.J., Brammer, G.L., McGuire, M.T., and Yuwiler, A., 1985, Dominant social status facilitates the behavioral effects of serotonergic agonists, Brain Res., 348:274-282.

Raleigh, M.J., Brammer, G.L., Yuwiler, A., Flannery, J.W., McGuire, M.T., and Geller, E., 1980, Serotonergic influences on the social behavior of vervet monkeys (Cercopithecus aethiops sabaeus), Exp. Neurol., 68:322-334.

Raleigh, M.J., and McGuire, M.T., Female influence of male dominance acquisition in captive vervet monkeys (Cercopithecus aethiops sabaeus), Animal Behav., 38:59-67.

Raleigh, M.J., McGuire, M.T., and Brammer, G.L., 1988, Behavioral and cognitive effects of altered tryptophan and tyrosine supply, in: "Amino Acid Availability and Brain Function in Health and Disease", Huether, G., ed., Springer, Berlin, pp. 299-308.

Struhsaker, T.T., 1967, Behavior of vervet monkeys, Univ. Calif. Publ. Zool., 82:1-64.

Winslow, J.T., and Miczek, K.A., 1988, Androgen dependency of alcohol effects on aggressive behavior: a seasonal rhythm in high-ranking squirrel monkeys, Psychopharmacology, 95:92-98.

Young, S.N., 1987, The effect of altering tryptophan levels on human mood and behavior, in: "Progress in Tryptophan and Serotonin Research 1986", Bender, D.A., Joseph, M.H., Kochen, W., and Steinhart, H., eds., de Gruyter, Berlin, pp. 225-228.

THE REGULATION OF BRAIN KYNURENIC ACID CONTENT: FOCUS ON INDOLE-3-PYRUVIC

ACID

F. Moroni, P. Russi, V. Carlà, G. De Luca[1], and V. Politi[1]

Department of Pharmacology
University of Florence
Firenze

[1]Polifarma S.p.A.
Roma
Italy

INTRODUCTION

The actual interest in the role that kynurenic acid (KYNA), one of the first identified metabolic products of tryptophan, (Ellinger et al., 1904 in Heidelberger et al., 1949; see Fig. 1) may have in physiology or pathology stems from at least three groups of observations. First, KYNA antagonizes in a non-competitive manner excitatory amino acid (EAA) receptors (Perkins and Stone, 1982; Moroni et al., 1986); second, it prevents the excitotoxic actions of the related tryptophan (TRP) metabolite quinolinic acid (QUIN) (Foster et al., 1984) and reduces neuronal damage after anoxic and ischemic brain insults (Germano et al., 1987); and third, it is present in mammalian biological fluids and in the central nervous system (Moroni et al., 1988b).

The mechanism whereby kynurenate antagonizes the EAA receptors has been extensively studied during the last few years. It is now accepted that concentrations of KYNA in the order of 10 μM displace glycine (GLY) from its strychnine-insensitive binding sites and competitively antagonize GLY actions on the NMDA receptor ion channel complex (see Johnson and Ascher, 1987; Moroni et al., 1989). Higher concentrations of KYNA also affect kainate and quisqualate receptors (Kemp et al., 1988). The existence of this endogenous antagonist of the EAA receptors in the mammalian brain, associated with the suggestion that an abnormal function of EAA mediated neurotransmission may play a pathogenetic role in different neurological disorders such as epilepsy, neurodegenerative diseases, stroke, metabolic encephalopathy etc., prompted us to investigate how KYNA synthesis and degradation are regulated both in physiology and pathology.

THE PRESENCE AND DISTRIBUTION OF KYNURENIC ACID IN THE BRAIN AND OTHER ORGANS IN THE RAT

The availability of an original method based on ion exchange chromatography and HPLC (Carlà et al., 1988; Moroni et al., 1988b) allowed us to identify KYNA in the brain and other organs of several animal species including humans. Table 1 shows KYNA distribution in the rat CNS compared to that

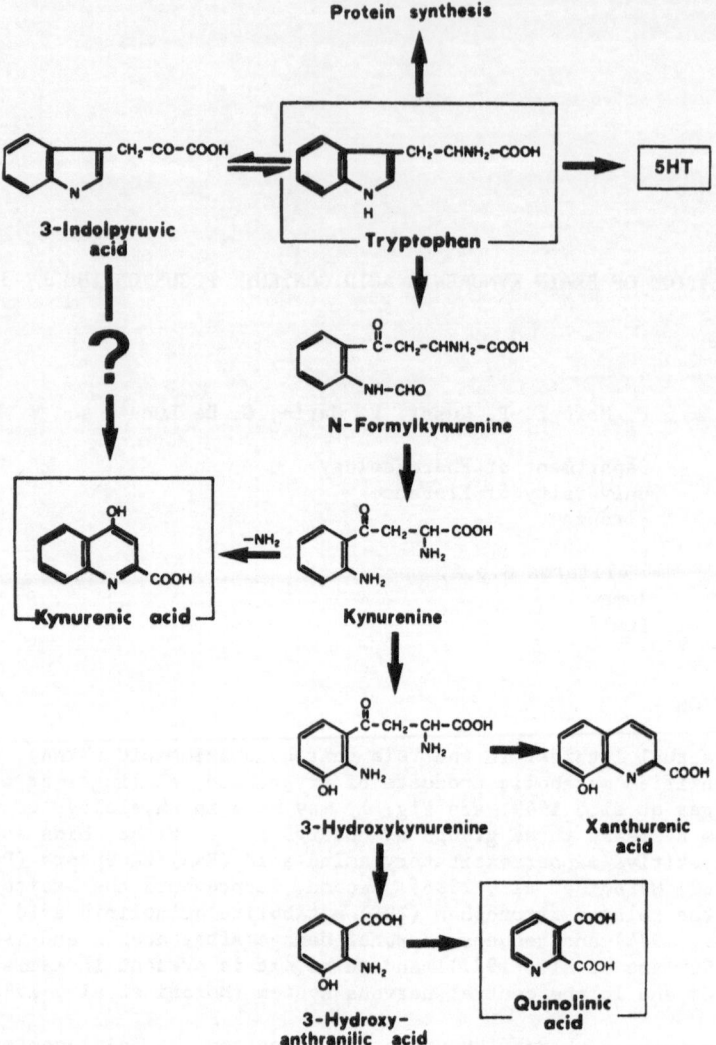

Fig. 1. The proposed pathways of tryptophan metabolism.

of QUIN and 5HT. On the average, both QUIN and 5HT, two of the other electrophysiologically active TRP metabolites, are approximately 100 times more concentrated than KYNA, whose brain content is similar to that of tryptamine, another extremely active TRP metabolite (Saavedra and Axelrod, 1972; Phillips et al., 1974).

The regional distribution of brain KYNA shows clear differences from that of QUIN or 5HT and indicates that KYNA is more concentrated in the brainstem than in other brain areas. Thus, in spite of the common precursor (TRP), different compartments are probably available for the storage of these molecules. The average concentration of KYNA in the brain of 2 month-old rats was 20 ± 2.5 picomol/g tissue, which was significantly lower than the concentration in the liver (90 ± 9.3), kidney (250 ± 23), heart (45 ± 5.1) and blood (30 ± 2.3). However, it should be noted that brain KYNA concentrations significantly increased with the age of the rats. Fig. 2 reports how these changes occur in the different organs.

Table 1. The regional distribution of kynurenic and quinolinic acid
and 5HT in the brain of 2-month old rats

	KYNA	QUIN	5-HT
Cortex	16 ± 3.1	1500 ± 112	1200 ± 110
Hippocampus	19 ± 2.2	900 ± 97	1300 ± 120
Striatum	25 ± 2.3	800 ± 63	1600 ± 210
Diencephalon	10 ± 1.4	850 ± 81	2500 ± 320
Cerebellum	13 ± 1.3	780 ± 73	500 ± 55
Brain stem	27 ± 1.8	810 ± 75	2400 ± 200

Values are picomol/g tissue and are the mean ± SEM of at least seven
animals.

In focussing on the brain (Fig. 3), it appears that a rapid increase of
KYNA content occurs during the first 80 days of life: 2-month old animals had
a brain KYNA content approximately 20 times higher than those 7 days old.
After sexual maturity, however, KYNA content in the brain increased much more
slowly, reaching, at 18 months of age, values 3 times higher than those found
in 3 month-old rats. It is not easy to clarify whether or not these large
changes of brain KYNA concentration may have functional implications (Moroni
et al., 1988a). Considering the importance of EAA-mediated neurotransmission
in different phases of development and the aging process, it is not unreasonable to propose that the almost complete absence of KYNA in the brains of
newborn animals could be useful to allow glutamate or other endogenous agonists of glutamate receptors to play their roles both as trophic factors able
to regulate neuronal growth and development (Pearce et al., 1987) and/or as
excitotoxic agents able to cause a reduction in the number of neurons during
maturation of the central nervous system. The extremely low levels of this
endogenous antagonist of the EAA receptors in the very first period of postnatal life could also partially explain the particular susceptibility to convulsions and to glutamic acid neurotoxicity of the brains of newborn animals.

On the other hand, the accumulation of KYNA in the brains of aged rats
could help explain some age-related pathophysiological changes. Memory loss
and impairment of acquisition of several tasks are common observations in
aged animals. EAA antagonists affect learning and memory in a similar manner

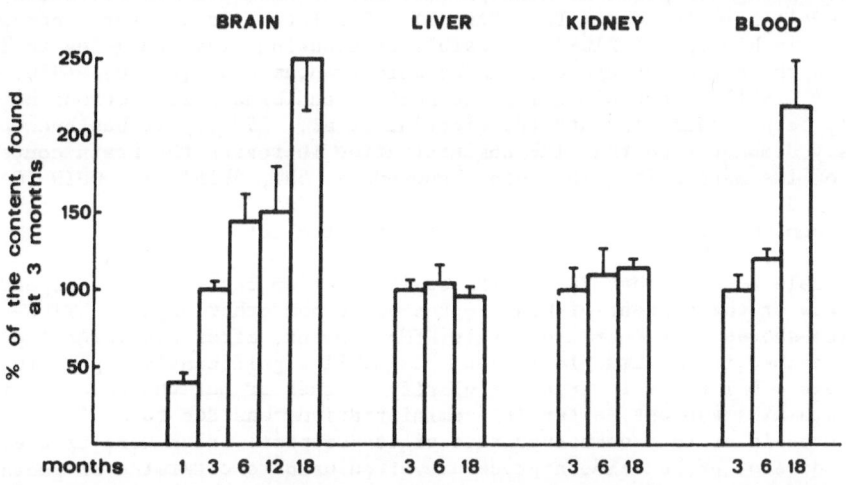

Fig. 2. Changes of KYNA content in rat organs at different ages.

301

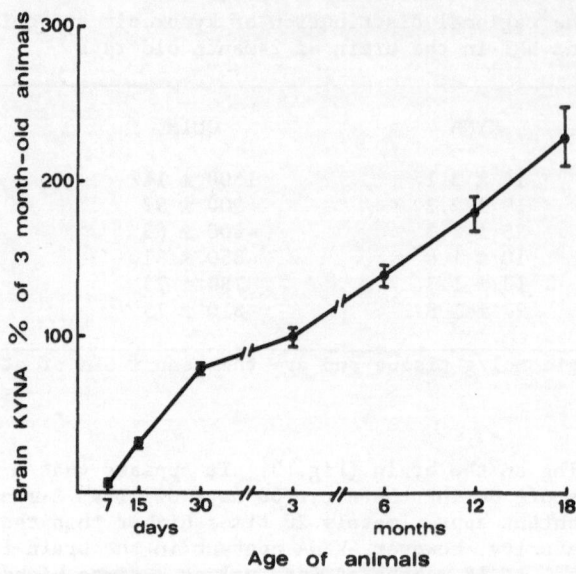

Fig. 3.   Developmental and aging changes of brain KYNA content.

(Morris et al., 1986).   Other interesting spontaneous changes in brain KYNA
concentrations occur at various times of the day.   Fig. 4 shows that rat
brain KYNA concentrations are two times higher during daytime than at night,
suggesting a possible role for KYNA in biological rhythms and in sleep induction or maintenance.

THE MODULATION OF BRAIN KYNA CONTENT: EFFECTS OF INDOLE-PYRUVIC ACID

In view of the potential pharmacological interest in antagonizing EAA
receptors in different pathological situations, we thought it interesting to
investigate whether and how the administration of possible precursors of KYNA
or the induction or activation of the enzymes possibly leading to its synthesis could result in an increase in brain content and utilization of this compound.   Among the possible KYNA precursors, we focussed our attention on TRP
and indole-pyruvic acid (IPA) (Fig. 1).   The latter is a natural compound
present in biological fluids, possibly originating from and going to TRP
through the action of aromatic amino acid transaminase (Kuroda, 1950; Millard
and Gál, 1971).   Several of its biochemical and behavioral actions have recently been studied in rats (Bacciottini et al., 1987).   It has been previously demonstrated that TRP administration increases the brain content of
most of its metabolites including kynurenine, 5HT, 5HIAA, and QUIN (Gál et
al., 1978; Moroni et al., 1984) and that IPA administration increases the
brain content of 5HT, 5HIAA and TRP (Bacciottini et al., 1987).

Table 2 shows that TRP or IPA administration caused a dose-dependent
increase of the content of KYNA in the brain and other organs.   Time-course
studies showed that increased brain KYNA content, after 250 mg/kg i.p. of
IPA, reached its maximum in 1 h and was still significantly higher than in
controls 4 h later.   In order to clarify whether or not the observed increase
of brain KYNA content, after IPA administration, was due to an increased rate
of synthesis or to a decreased rate of its disposal, rats were treated with a
large dose of probenecid, a procedure often used to determine an index of the
rate of synthesis of acidic neurochemicals (Neff et al., 1967).   Fig. 5 shows
that IPA-induced accumulation of brain KYNA was significantly higher in pro-

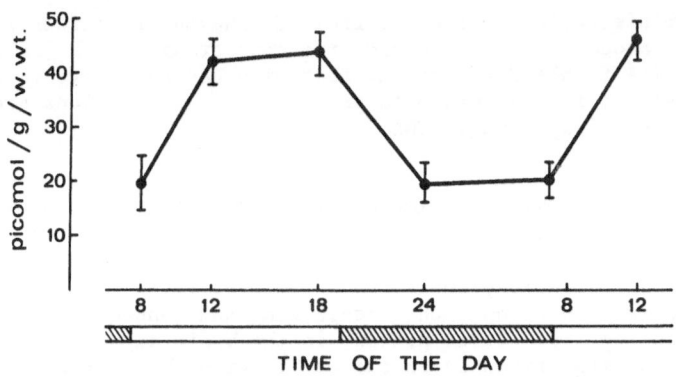

Fig. 4. Circadian rhythm of brain KYNA content.

benecid-treated animals than in controls, thus suggesting that IPA indeed increased the rate of brain KYNA synthesis. The figure also shows that brain
KYNA disposal probably occurs through a probenecid-sensitive mechanism.

The increased formation of brain KYNA after the administration of IPA
could occur because IPA is metabolized to TRP and subsequently to KYNA. By
measuring, in the same brains, the content of both TRP and KYNA and by calculating the ratio TRP/KYNA after the administration of the two precursors, it
was possible to show that IPA administration resulted in an increased brain
content of both TRP and KYNA, but that the ratio TRP/KYNA was significantly
lower after IPA than after TRP (Table 3).

Another approach used to clarify whether IPA had to be transaminated to
TRP in order to induce KYNA formation was the use of ^{3}H-uniformly labelled
IPA (1.1 mCi/mmol), administered i.c.v. at the dose of 20 μCi, and the measurement of the specific radioactivity of TRP, KYNA and other metabolites.
Table 4 shows that the specific activity found in KYNA following IPA injection was approximately 10 times higher than that found in TRP, thus suggesting that the metabolic pathway indicated in Fig. 1 with a question mark is
operant _in vivo_.

Table 2. Effects of tryptophan or IPA on the KYNA content of the
brain and other organs

	Brain	Liver	Kidney	Heart	Blood
Saline	16 ± 3	89 ± 10	282 ± 20	87 ± 8	20 ± 3
IPA 100	22 ± 2*	480 ± 35**	–	70 ± 5	–
IPA 250	30 ± 1**	827 ± 41**	2593 ± 180**	181 ± 2**	130 ± 9**
IPA 500	96 ± 10*	1980 ± 160**	–	–	–
TRP 100	23 ± 3*	455 ± 40**	1250 ± 110**	95 ± 15*	–
TRP 250	33 ± 4**	–	–	–	–

Values are expressed in picomol/g net weight and are the mean ± SEM of
at least seven animals. IPA and TRP were administered intraperitoneally at the indicated doses (mg/kg). *P < 0.05, **P < 0.01 (ANOVA
and Dunnett's test).

IPA administration caused a series of behavioral actions which were compatible with a dampening of EAA-mediated neurotransmission (Table 5). However, since IPA itself does not interact with EAA receptors (our unpublished
observations), it is reasonable to propose that such actions are due to its
metabolites, possibly mainly KYNA.

MODULATION OF KYNA CONTENT IN THE MAMMALIAN BRAIN: EFFECTS OF AGENTS ABLE TO
INDUCE INDOLEAMINE 2,3-DIOXYGENASE.

Indoleamine 2,3-dioxygenase (IDO, E.C.1.13.11.17) is a heme containing
enzyme which is able to open the indole ring and consequently to produce kynurenines (see Fig. 1). It is present in the brain and other organs and its
activity may be significantly increased "in vivo" by bacterial toxins or viral infections (Yoshida et al., 1979). It has been previously proposed that
the mechanisms whereby infections increase IDO activity involve the neosyn-
thesis of interferon which also strongly stimulates kynurenine production in
vitro (Werner et al., 1988). Fig. 6 shows that animals injected i.v. with
bacteria (Escherichia coli, 1,000,000 isolated from the clinic) or with bacterial lipopolysaccharides (400 μg/kg, i.v.) have brain KYNA concentrations
which are significantly higher than those of age-matched controls. No significant changes were found in the liver, kidney or lung. Thus, the possibility exists that in situations characterized by an elevated production of
interferon (such as infective states), brain KYNA levels are increased and
that some of the neurological or psychiatric symptomatology (sleepiness,
weakness, obnubilation of consciousness, depression, etc.) associated with
these pathological states could be ascribed to KYNA-induced changes in EAA
neurotransmission.

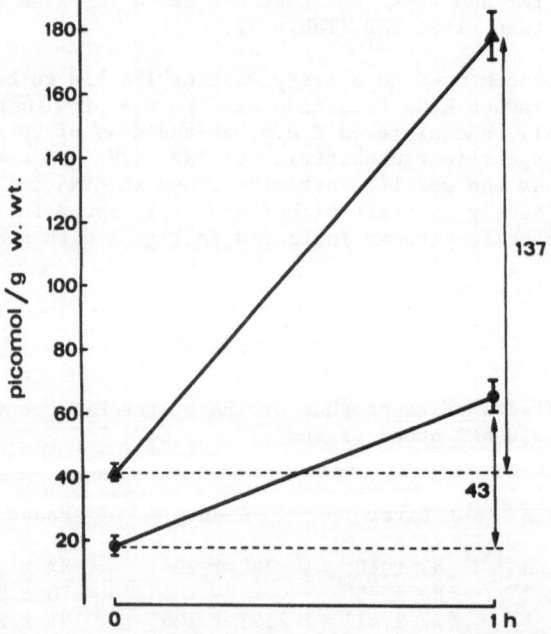

Fig. 5.   Effects of probenecid (200 mg/kg, i.p.) on IPA-induced
increase of brain KYNA content. IPA (250 mg/kg, i.p.) was
administered 60 min before probenecid. At least 6 animals
per group were used; vertical bars are SEM. The numbers on
the right portion of the Figure represent the probenecid-
induced KYNA accumulation and are expressed as picomol/g wet
weight/h.

Table 3. Relationship between the brain content of tryptophan and
         KYNA

	Brain TRP	Brain KYNA	Ratio
Saline	21.6 ± 2.4	28.4 ± 2.9	760
TRP (i.p., 250 mg/kg)	321 ± 28**	56.4 ± 4.2**	5700
IPA (i.p., 250 mg/kg)	160 ± 10**	72.2 ± 6.1**	2200

Values are the mean ± SEM of at least 6 animals and are expressed as
nmol/g wet weight for TRP and picomol/g wet weight for KYNA. The
brain KYNA content in these experiments was higher than that reported
in Table 1 because the animals were slightly older (see Fig. 3).

CONCLUSIONS

    The data here presented indicate that it is possible to significantly in-
crease the brain concentration of KYNA, a modulator of EAA neurotransmission.
Preliminary observations suggest that the induced changes of brain KYNA concentrations are associated with behavioral and biochemical modifications compatible with the expected KYNA induced dampening of the EAA receptor function.   However, it should be noted that brain KYNA content is relatively low
(in the nanomolar range), while "in vitro" experiments indicate that micromolar concentrations of KYNA are necessary in order to antagonize EAA receptors.   On the other hand, it should also be kept in mind that KYNA is an allosteric modulator of the NMDA receptor ion channel complex and that its action is strictly dependent upon the local concentration of glycine (Moroni et
al., 1989).

Table 4. Incorporation of labelled [^{3}H]-indole-
         pyruvic acid injected i.c.v. into
         various tryptophan metabolites

	cpm/g tissue
Brain alkaline extracts	200,000
Brain acid extracts	23,000

	cpm/μmol
KYNA	2,600
TRP	380
5-HT	4,000
5-HIAA	2,320

The specific activity of KYNA was obtained from
alkaline extract, after purification and HPLC
monitoring of both the endogenous content and
the radioactivity. Similarly, the specific activity of TRP, 5HT and 5HIAA were obtained after
HPLC of the acid extract and measurement of the
radioactivity in each peak.

Table 5. Behavioral and neurochemical effects of IPA
(20-500 mg/kg) administration to rats or mice

1)   Decreases locomotor activity

2)   Potentiates alcohol actions

3)   Potentiates barbiturate actions

4)   Possesses analgesic activity

5)   Decreases blood pressure in spontaneously
     hypertensive rats (repeated administration)

6)   Antagonizes NMDA induced convulsions

7)   Antagonizes audiogenic convulsions
     in DBA/2 mice

8)   Decreases norepinephrine turnover rate

9)   Increases 5HT turnover rate

Since it is difficult to evaluate the synaptic concentration of the different agonists and modulators of EAA receptors, including KYNA, it is certainly possible to hypothesize functional meanings to the observed changes of
the tissue content of this TRP metabolite.

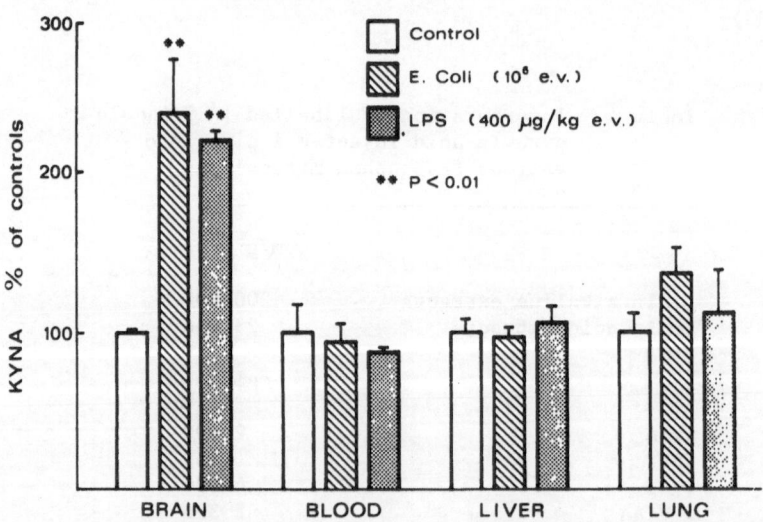

Fig. 6. Increase of brain KYNA content 24 h after the administration of
bacteria or of lipopolysaccharides. Note that KYNA content in the
lung, liver and blood is not changed by these treatments.

REFERENCES

Bacciottini, L., Pellegrini-Giampietro, D., Bongianni, F., De Luca, G., Beni, M., Politi, V., and Moroni, F., 1987, Biochemical and behavioural studies on indole-pyruvic acid: a keto analogue of tryptophan, Pharmacol. Res. Comm., 19:803-807.

Carlà, V., Lombardi, G., Beni, M., Russi, P., Moneti, G., and Moroni, F., 1988, Identification and measurement of kynurenic acid in the rat brain and other organs, Anal. Biochem., 169:89-94.

Foster, A.C., Vezzani, A.M., French, E.D., and Schwarcz, R., 1984, Kynurenic acid blocks neurotoxicity and seizures in rats by the related brain metabolite quinolinic acid, Neurosci. Lett., 48:273-278.

Gál, E.M., Young, R.B., and Sherman, A.D., 1978, Tryptophan loading: consequent effects on the synthesis of kynurenine and 5-hydroxyndoles in rat brain, J. Neurochem., 31:237-244.

Germano, I.M., Pitts, L.H., Meldrum, B.S., Bartkowski, H.M., and Simon, R.P., 1987, Kynurenate inhibition of cell excitation decreases stroke size and deficits, Ann. Neurol., 22:730-734.

Heidelberger, C., Guldberg, M.E., Morgan, A.F., and Lepkovsky, S., 1949, Tryptophan metabolism I: concerning the mechanism of the mammalian conversion of tryptophan into kynurenine, kynurenic acid and nicotinic acid, J. Biol. Chem., 179:143-150.

Johnson, J.W., and Ascher, P., 1987, Glycine potentiates the NMDA response in cultured mouse brain neurones, Nature, 325:529-531.

Kemp, J.A., Foster, A.C., Leeson, P.D., Priestley, T., Tridgett, R., Iversen, L.L., and Woodruff, G.N., 1988, 7-Chlorokynurenic acid is a selective antagonist at the glycine modulatory site of the N-methyl-D-aspartate receptor complex, Proc. Natl. Acad. Sci. USA, 85:6547-6550.

Kuroda, Y., 1950, A contribution to the metabolism of tryptophan, J. Biochem., 37:91-97.

Millard, A., and Gál, E.M., 1971, The contribution of 5-hydroxyindole-pyruvic acid to cerebral 5-hydroxyindole metabolism, Int. J. Neurosci., 1:211-218.

Moroni, F., Lombardi, G., Carlà, V., and Moneti, G., 1984, The excitotoxin quinolinic acid is present and unevenly distributed in the rat brain, Brain Res., 295:352-355.

Moroni, F., Luzzi, S., Franchi-Micheli, S., and Zilletti, L., 1986, The presence of NMDA-type receptors for glutamic acid in the guinea pig myenteric plexus, Neurosci. Lett., 68:57-62.

Moroni, F., Pellegrini-Giampietro, D.E., Alesiani, M., Cherici, G., Mori, F., and Galli, A., 1989, Glycine and kynurenate modulate the glutamate receptors present in the myenteric plexus and in cortical membranes, Eur. J. Pharmacol., 163:123-126.

Moroni, F., Russi, P., Carlà, V., and Lombardi, G., 1988a, Kynurenic acid is present in the rat brain and its content increases during development and aging processes, Neurosci. Lett., 94:145-150.

Moroni, F., Russi, P., Lombardi, G., Beni, M., Carlà, V., 1988b, Presence of kynurenic acid in the mammalian brain, J. Neurochem., 51:177-180.

Morris, R.G.M., Andersen, E., Lynch, G.S., and Baudry, M., 1986, Selective impairment of learning and blockade of long term potentiation by an NMDA antagonist, Nature, 319:774-776.

Neff, N.H., Tozer, T.N., and Brodie, B.B., 1967, Application of steady-state kinetics to studies of the transfer of 5-HIAA from brain to plasma, J. Pharmacol. Exp. Ther., 158:214-218.

Pearce, I.A., Cambray-Dekin, M.A., and Burgoyne, R.D., 1987, Glutamate acting on NMDA receptors stimulates neurite outgrowth from cerebellar granule cells, FEBS Lett., 233:143-147.

Perkins, M.N., and Stone, T.W., 1982, An iontophoretic investigation of the actions of convulsant kynurenines and the interaction with the endogenous excitant quinolinic acid, Brain Res., 247:184-187.

Phillips, S.R., Durden, D.A., and Boulton, A.A., 1974, Identification and distribution of tryptamine in the rat, Can. J. Biochem., 52:477-481.
Saavedra, J.M., and Axelrod, J., 1972, A specific and sensitive enzymatic essay for tryptamine in tissues, J. Pharmacol. Exp. Ther., 182:363-369.
Werner, E.R., Werner-Felmayer, G., Fuchs, D., Hausen, A., Reibnegger, G., and Wachter, H., 1988, Influence of interferon-gamma and extracellular tryptophan on indoleamine 2,3-dioxygenase activity in T24 cells as determined by a non-radiometric assay, Biochem. J., 256:537-541.
Yoshida, R., Urade, Y., Tokuda, M., and Hayaishi, O., 1979, Induction of indoleamine 2,3-dioxygenase in mouse lung during infection, Proc. Natl. Acad. Sci. USA, 76:4084-4086.

PHYSIOLOGICAL ROLE OF 3-HYDROXYKYNURENINE AND XANTHURENIC ACID UPON

CRUSTACEAN MOLTING

Y. Naya, M. Ohnishi, M. Ikeda, W. Miki, and K. Nakanishi

Suntory Institute for Bioorganic Research
Shimamoto-cho
Mishima-gun
Osaka 618
Japan

ABSTRACT

Studies with crabs (*Charybdis japonica*) and crayfish (*Procambarus clarkii*) revealed that the tryptophan metabolites, 3-hydroxy-L-kynurenine (3-OH-K) and xanthurenic acid (XA), common secretory products of the X-organsinus gland complex of eyestalks from several decapods, regulated the molting of crustaceans in species-nonspecific fashion. Injection of 3-OH-K to the eyestalk-ablated crayfish delayed the onset of the first molt and lengthened the interval between the first and second molts. These lines of evidence were in accord with previous accounts of the so-called "molt inhibiting hormone" (MIH) effect. Removal of eyestalks caused a change in the conversion capacity of exogenous 3-OH-K to XA in the hemolymph. The peak in transformation capacity was followed by a peak in the titer of 20-hydroxyecdysone or molting hormone. Moreover, the seasonal profiles of the XA and ecdysone titers in *Charybdis japonica* exhibited a staggered relationship in the tissues tested. The ratio of XA to 3-OH-K, which is expected to indicate the apparent 3-OH-Kase activity, fluctuated seasonally and locally. When the Y-organ with the adhering tissues (Y-organ complex or YOC) was incubated during the period of high XA titer, the YOC produced 100 times more ecdyson than before incubation. It is suggested that ecdysteroidogenesis *in situ* was suppressed during this period by XA, but incubation of the YOC lead to a dramatic acceleration in ecdysone synthesis by overriding this inhibitory effect. XA profoundly repressed ecdysteroidogenesis in the YOC culture. Thus, XA is the ecdysone biosynthesis inhibitor (EBI) and 3-OH-K the precursor in crustaceans. An interfering effect of XA to a biocatalyst cytochrome P-450 system was postulated for the inhibition mechanism of ecdysteroidogenesis.

INTRODUCTION

Hormones may function in the homeostatic regulation of physiological processes and/or may serve as signals for the initiation or regulation of developmental events. In crustaceans, the postembryonic developmental molting is believed to be controlled (Kleinholz, 1985; Skinner, 1985) by at least two major types of hormones; namely, the ecdysteroids or molting hormones (MH), and the so-called "molt-inhibiting hormone" (MIH). More recent studies indicate that a juvenile hormone (JH) may also be involved in

Table 1. The progress of "MIH" study

ES removal promotes molting and premature ecdysis: Zeleny, 1905.

ES reimplantation reverses premature ecdysis: "MIH"?: Brown and Cunningham, 1939.

"MIH" produced in X-organ inhibits release of ecdysone from Y-organ: Passano, 1960.

"MIH" is characterized as proteinaceous substance: Rao, 1965.

Cultured Y-organs secrete ecdysone (pro-MH); sole source of 20-hydroxy-ecdysone (MH): Chang and O'Connor, 1977.

"MIH" may be indole alklamine (*Pandalus jorani*): Soyez and Kleinholz, 1977.

"MIH" is a 61 amino acid peptide (*Carcinus maeas*): Webster and Keller, 1986.

Two similar "MIH" peptides (MW ~8700, *Homarus americanus*): Chang et al., 1987.

Xanthurenic acid and 3-hydroxy-L-kynurenine (crabs and crayfish) inhibit ecdysteroidogenesis: Naya et al., 1988.

---

developmental events (Borst et al., 1987; Laufer et al., 1987).

The Y-organ is the primary, if not the sole, physiological source of ec-dysone, which is converted to 20-hydroxyecdysone (a major molting hormone) in specific tissues peripheral to the Y-organ. A progressive increase in the titer of 20-hydroxyecdysone initiates the events generically referred to as molting. The onset of crustacean molting is controlled by "MIH", which is released from the X-organ-sinus gland complexes in the eyestalks (ES); eye-stalk ablation causes precocious ecdysis by accelerating the secretion of the molting hormone by the Y-organs. Reimplantation of ES returns ecdysis to the normal molt cycle. The injection of ES-extract delays the molting of eye-stalk-ablated animals. With the advances in Y-organ culture and the develop-ment of ecdysteroid RIA in the mid-1970s, it has been demonstrated that "MIH" extracted from ES decreases the basal rate of ecdysone secretion. The rates of ecdysone synthesis or secretion by the Y-organ are temporally varied; the critical period is controlled not only by "MIH" but rather by multiple fac-tors (Kleinholz, 1985; Skinner, 1985; Borst et al., 1987; Laufer et al., 1987). Each hormone must precisely interplay amongst hormones involved in molting events. The significant approaches of "MIH" study are summarized in the Table 1.

Although little is known in the literature, analogy to Insecta (Smith, 1985) suggests that the ecdysteroidogenesis in Crustacea involves a putative cerebral peptide or Y-organotropic hormone and hemolymph proteins (a sterol carrier protein) (Watson and Spaziani, 1985; Spaziani, 1988). In addition, the ecdysteroidogenesis is probably mediated by the actions of protein kinase C (Mattson and Spaziani, 1987) and of a biocatalyst cytochrome P-450. The cytochrome P-450s are known to be involved in the biosynthetic pathway of ecdysone in Insecta (Kappler et al., 1988).

We recently found (Naya et al., 1988) that two tryptophan metabolites, 3-hydroxy-L-kynurenine (3-OH-K) and xanthurenic acid (XA), function as ecdy-sone biosynthesis inhibitors (EBIs). Metabolites of tryptophan are known to exhibit physiologically important activities, but the physiological role of XA was found for the first time. We present evidence which indicates that 3-OH-K and XA inhibit ecdysteroidogenesis as homeostatic regulators of crabs (*Charybdis japonica*) and crayfish (*Procambarus clarkii*) in a species-non-specific fashion.

MATERIALS AND BIOASSAY

Several species of crabs (Naya et al., 1988), *Callinectes sapidus*, *Portunus trituberculatus*, *Charybdis japonica*, *Geothelphusa dehaai*, and crayfish (Naya et al., 1989), *Procambarus clarkii*, collected off different coasts of the USA and Japan were used in this study.

The Y-organ culture (Chang and O'Connor, 1977) established by O'Connor was modified (Naya et al., 1988) for the conventional bioassay system as follows. The Y-organs with their adherent tissues, the so-called Y-organ complex (YOC) were excised from the crabs. The YOC was homogenized in buffer (0.17 M KH_2PO_4/$NaHCO_3$, pH 7.1) and centrifuged (7,700 x g, 20 min). The supernatant containing a multiple enzymatic system and substrates was divided into several aliquots, which were incubated at 37°C for 20 hr with gentle stirring in the presence or absence (control) of ES-extract (see below). Upon termination of the incubation, each aliquot was lyophilized and extracted with methanol to determine the amount of ecdysone produced. The ecdysone was quantified by HPLC with UV detection at 247 nm (ERC-ODS-1161, ERMA, Tokyo) (MeOH:CH_3CN:H_2O, 28:10:62) in the presence of an internal standard, 14,15-dehydro-20-hydroxyecdysone; the detection limit was 3 ng. Ecdysone produced by the YOC, and the pertrimethylsilyl derivative of ecdysone were shown to be identical to the authentic compounds not only by HPLC but also by combined gas chromatography and mass spectrometry (GC-MS).

It is known that the level of ecdysone secretion varies widely between species and also in the same species between the individual stages of the molting cycle (Kleinholz, 1985; Skinner, 1985). Consequently, our procedure using YOC homogenates rather than individual Y-organs was effective for leveling the variations among animals, and hence for a reduction in the number of control experiments. The protocol of the bioassay system is illustrated in Fig. 1.

ISOLATION OF "MIH"

To isolate inhibitor(s) of ecdysone biosynthesis (Naya et al., 1988), eyestalks were excised from animals immobilized on ice. Ablated-eyestalks

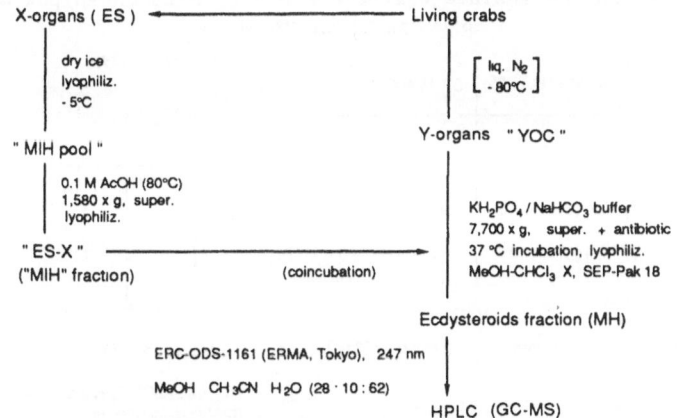

Fig. 1. Bioassay system for "MIH"/MH activity.

were frozen with dry-ice and lyophilized for storage as the "MIH" pools. The stored ES were extracted with 0.1 M acetic acid (100°C, 10 min) and centrifuged (1580 x g, 15 min). The supernatant was lyophilized to give the crude ES-extract (ES-X), which was assayed for inhibitory activity on ecdysone production. The inhibitory activity manifested itself even when the ES-X and YOC were taken from different species of crabs or crayfish. The inhibitor(s) thus appears to be species-nonspecific between ES and YOC donors.

The bioassay resulted in the isolation of 700 µg of the active principle (Fig. 2), which was identical to 3-OH-K in all respects (UV, CD, EI-MS, NMR, electrophoresis and amino acid analysis). The authentic L-enantiomer was as active as the isolated natural compound, but the synthetic DL-compound was less potent. These experiments suggested that XA, which we believe to be responsible for the inhibition of ecdysteroidogenesis, is a metabolite of the L-enanthiomer of 3-OH-K. For the isolation of 3-OH-K and XA, the stored ES were extracted with 0.17 M ammonium acetate buffer (pH 5). The extract was centrifuged (1580 x g, 15 min) and the supernatant was placed on an Asahipak GS 320 column (Asahi Chem. Ind.). 3-OH-K was eluted with 0.17 M ammonium acetate and XA with 10% acetonitrile/0.17 M ammonium acetate.

XA inhibited ecdysteroidogenesis in vitro more potently than 3-OH-K. The ED_{50} of ES-X (Callinectes sapidus) was about 1 mg. In this amount of ES-X, which was equivalent to 1/25 of ES, 136 ng of 3-OH-K and 78 ng of XA were detected. Even when assuming a 100% conversion of 3-OH-K into XA, this combined amount did not account for the full potency of ES-X; the detailed dose response study is still under investigation. HPLC analysis revealed the presence of 3-OH-K (27 ng) and XA (128 ng) in the X-organ itself, which was taken from a random selecting animal.

ENZYMATIC TRANSFORMATION OF 3-OH-K to XA

The body fluid (Naya et al., 1988) extracted from Charybdis japonica with 0.17 M KH_2PO_4/$NaHCO_3$ buffer (pH 7.1) was treated with a large excess of casein (to inhibit proteolysis) and further with saturated aqueous ammonium sulfate. After centrifugation, the residue was dialyzed for 24 hr against water. The non-dialyzed fraction was passed through Sephadex G-200 (Pharmacia) with the buffer and used as a crude enzyme preparation. This crude enzyme preparation was incubated with 3-OH-K in the buffer (0.17 M KH_2PO_4/$NaHCO_3$, pH 7.1) at 37°C for 3 hr. Upon termination of the incubation, the mixture was lyophilized and dissolved in 0.1 M ammonium hydroxide, then dialyzed overnight against 0.05 M ammonium hydroxide (Cellulose dialyzer tubing, VT-801, Nacalai Tesque). The dialyzable solution was lyophilized, and the quantity of XA in the residue was determined by HPLC (Asahipak GS-320, 247 nm

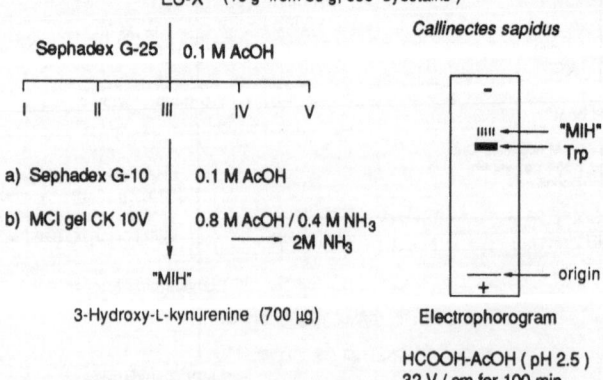

Fig. 2. Isolation of "MIH" from eyestalks of Callinectes sapidus.

detection) under the conditions described above. Transformation of L-3-OH-K into XA was thus accomplished in 60% yield. On the other hand, the transformation of D- and DL- compounds resulted in 1% and 9% yields, respectively; the transformation was L-enantiospecific but inhibited by the D-enantiomer. The crude enzyme preparations obtained from ES and YOC were also capable of transforming 3-OH-K into XA.

IN VIVO EFFECTS OF 3-OH-K and XA UPON CRAYFISH MOLTING

Eyestalk ablation and forced autotomy of chelae induced precocious ecdysis in *Procambarus clarkii* (Naya et al., 1989), which received daily injections of crustacean saline (100 μl/animal, control); eyed animals that received the same treatment did not molt for 22 days. Each of 10 animals in another group received daily injections of 3-OH-K solution in saline (30 ng/100 μl) at 24 hr intervals after the operation; this delayed by 2 days the onset of the first molt and lengthened by 3 days the interval between the first and second molts (Fig. 3). On the other hand, daily injection of XA solution in saline (10 and 30 ng/100 μl) induced no significant effects regarding the molt inhibiting properties before the animals died.

PHYSIOLOGICAL CHANGE INDUCED BY EYESTALK ABLATION

An equal volume of hemolymph was drawn from the abdomen of each of the 15 crayfish (bilaterally destalked and forced to autotomize at day 0) at 24 hr intervals beginning on day 1. The hemolymph collected on each day was combined and diluted with an equal volume of buffer solution (0.17 M KH_2PO_4/ $NaHCO_3$, pH 7.1) containing a large excess of casein. The mixture was stirred at 0°C for 1 hr, and then centrifuged at 1580 x g for 20 min. Aliquots of the supernatant were incubated at 37°C for 1 hr with gentle stirring in the presence or absence (control) of exogenous 3-OH-K. Upon termination of the incubation, each aliquot was worked up according to the procedure outlined above for XA analysis on HPLC.

The capacity of hemolymph to convert 3-OH-K to XA peaked after a period of several days when the animals were de-eyestalked at day 0. The conversion capacity varied according to animal size. The enzyme that converts 3-OH-K into XA was adopted as the apparent 3-OH-Kase. Although it is not known whether exogenous 3-OH-K is metabolized along the same lines as the endogenous substrate, the peak in the apparent 3-OH-Kase activity was followed by a peak in the titer of 20-hydroxyecdysone. A decrease in the enzymatic activity or in the XA titer presumably signals the timing for a sharp increase in the titer of ecdysone, and subsequently of 20-hydroxyecdysone, the active molting hormone, in hemolymph. The transient increase in the apparent 3-OH-

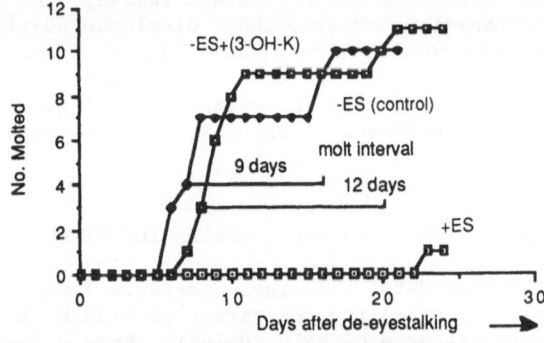

Fig. 3. Effect of 3-OH-K injection on crayfish *Procambarus clarkii*.
-ES: de-eyestalk animal, +ES: eyed animal.

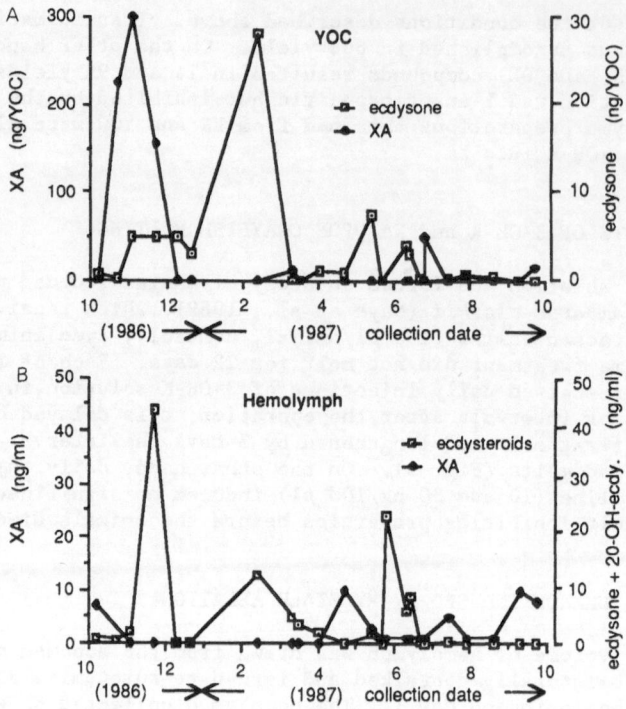

Fig. 4 A,B.   Staggered correlation between XA and ecdysteroids.

Kase activity induced by eyestalk ablation was not accompanied by a change in the total amount of the endogenous aminotransferase substrates. After removal of proteinaceous compounds in the hemolymph, the total concentration of aminotransferase substrates was determined by ninhydrin colorimetry and amino acid analysis. The apparent 3-OH-Kase in the crude enzyme preparation was somewhat inhibited by a general aminotransferase substrate, L-glutamic acid, though inhibition by other amino acids was not investigated.

Interestingly, 5-hydroxytryptophan (10^{-7} M) can provide excitatory input to "MIH"-containing (producing) neurosecretory cells (Mattson and Spaziani, 1985). This result is possibly caused by a feed-back regulation among the tryptophan metabolites. An injection of 20-hydroxyecdysone given to crayfish *Procambarus simulans* (Lowe et al., 1968) 24 hr after eyestalk removal did not induce molting as effectively as an injection given 48 hr or 72 hr after operation. Twenty-four hours after eyestalk removal, the Y-organs of crabs *Cancer antennarius* (Spaziani et al., 1982) displayed 30-times higher uptake of cholesterol-^{14}C than the Y-organs of unablated premolts. These reports, together with our results, indicate that de-eyestalking not only induces acute and long-term abnormalities in several physiological parameters but also directly affects the biochemical processes associated with late premolt stages.

SEASONAL VARIATION OF 3-OH-K, XA AND ECDYSTEROIDS TITERS

For each collection date, six crabs *Charybdis japonica* (Naya et al., 1989) were randomly selected, and the titers of 3-OH-K, XA, ecdysone and 20-hydroxyecdysone were measured by HPLC analysis above. The quantities of each compound in the ES, YOC and hemolymph were recorded for 12 months. Staggered correlations between the XA and ecdysteroids titers were found in the YOC

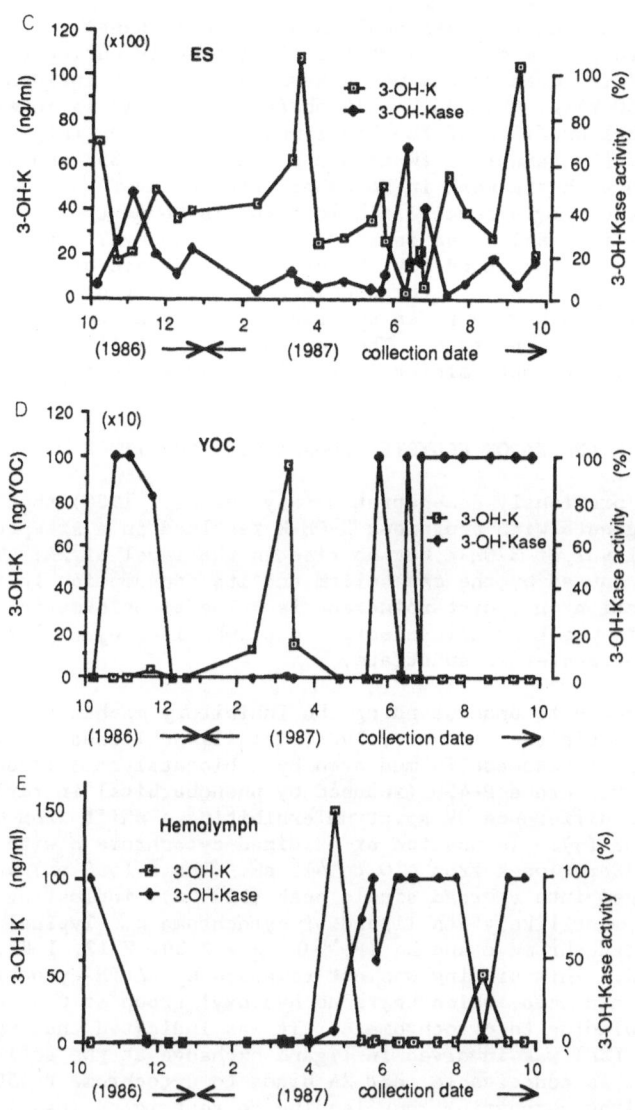

Fig. 4 C-E. The apparent 3-OH-Kase activity and the unchanged 3-OH-K titers.

(Fig. 4A) and hemolymph (Fig. 4B); a staggered correlation was also found in eyestalk-ablated crabs (see above). When the YOC was incubated in November/ December, the period of high XA titer (Fig. 4A), the ecdysone titer increased 100-fold as the inhibitory effect of XA was overridden. In February, the period following the high titer of XA, the ecdysone level rose to ca. 30 ng/ YOC. Incubation of the YOC during this period reduced the ecdysone level of 3 ng/YOC. 3-Hydroxy-L-kynurenine, which was in high concentration prior to incubation (ca. 1000 ng/YOC, Fig. 4D; 150 ng/ml of hemolymph, Fig. 4E), pre- sumably was the agent responsible for inhibition of ecdysteroidogenesis. Thus, incubation accelerated the metabolism of ecdysone, and XA generated from 3-OH-K during incubation suppressed ecdysteroidogenesis. The above evi- dence suggested the presence of a putative 3-OH-K receptor at the Y-organ.

The apparent 3-OH-Kase activity, estimated as the ratio [XA]/[3-OH-K] +

[XA], was an indicator of seasonal physiological change in the animals, in vivo aminotransferase capacity varied significantly during the year. The apparent 3-OH-Kase capacity was not always temporally coincident between the ES (Fig. 4C), YOC (Fig. 4D) and hemolymph (Fig. 4E). It is interesting to note that the annual profiles of the hemolymph lipoproteins (high-density type) in the crab *Cancer antennarius* (Watson and Spaziani, 1985) show variation in concentration with the peak in November being an average four-fold higher than the lows of early summer. Cholesterol is bound to lipoproteins of this class (Spaziani, 1988) to be carried and taken up by Y-organ cells. This report convincingly accounted for the dramatic increase of ecdysone titer upon YOC incubation (see above); the accelerated activity of the multi-enzyme complex in the YOC in November was sychronous with the increase in the concentration of carried substrate. The concentration of lipoprotein-bound cholesterol may be a rate-determining factor in ecdysteroidogenesis.

ROLE OF 3-OH-K AND XA ON ECDYSTEROIDOGENESIS INHIBITION

We have previously demonstrated (Naya et al., 1988) that incubation of the YOC homogenate with exogenous 3-OH-K resulted in a steep decrease (ca. 90%) in the level of 3-OH-K but no rise in the level of XA. The loss of XA was probably caused by the catabolism and its consumption in the course of inhibitory action on ecdysteroidogenesis. The transformation capacity of 3-OH-K into XA was L-enentioselective but inhibited by L-glutamic acid, a general aminotransferase substrate.

Our approach to understanding the inhibitory mechanism incorporated the working hypothesis (Naya et al., 1988; details published elsewhere) that the ecdysteroidogenic cascade is mediated by a biocatalyst. Incubation of the biocatalyst cytochrome P-450 (induced by phenobarbital in rat) with XA resulted in the difference UV spectrum exhibiting a shift from 417 to 445 nm of the Soret peak (γ). Incubation of oxidized cytochrome c with XA produced a shift of the Soret peak from 410 to 415 nm. The a-(550 nm) and b-bands (520 nm) were merged into a broad single peak (540 nm), indicating that hydroxyl group is the most likely 6th ligand of cytochrome c. Typical substrate binding between cytochrome c and XA (Fe^{+3}-O^-, g = 2.30, 2.12, 1.88, 77°K) was observed by ESR. This binding was not reversed by NADPH-cytochrome c reductase. Kynurenic acid, which bears no hydroxyl group at C-8, did not show evidence of binding to cytochrome c. It was indicated that the hydroxyl group at C-8 (XA) was involved in ligand exchange at the active site of cytochrome c. It is conceivable that XA binds to cytochrome P-450, probably at the site of iron porphyrin, thus leading to restricted inactivation of the

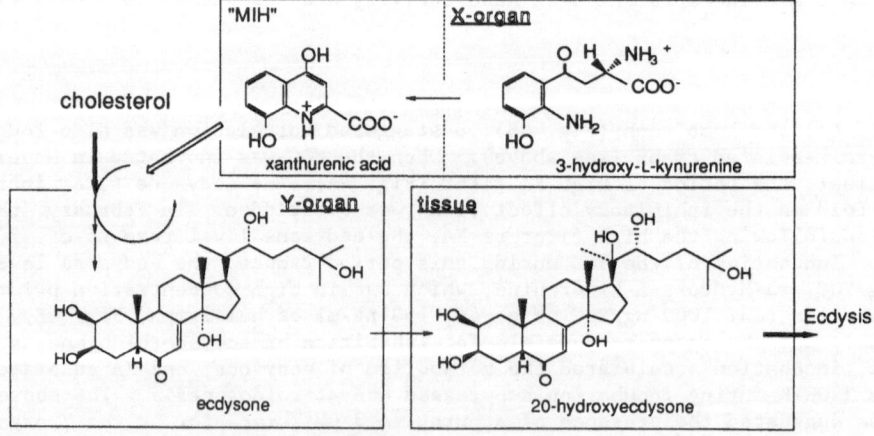

Fig. 5. Regulation in crustacean molting.

316

biocatalyst. As a result of this inactivation, ecdysteroidogenesis is antagonized.

Whereas XA seemed to inhibit ecdysone biosynthesis, incubation of XA with the body fluid, which was capable of hydroxylation of C-20 of ecdysone, revealed no interference in the hydroxylation. Experiments with insect tissues (Dr. T. Fujita, personal communication) and ecdysone 20-monooxygenase from insect sources (Dr. S. L. Smith, personal communication) also demonstrated that neither 3-OH-K nor XA inhibited the 20-hydroxylation of ecdysone, suggesting diversity among the oxygenase involved in ecdysteroidogenesis. No evidence of interaction between 3-OH-K and the biomolecules was found, either.

When eyestalk-ablated lobsters received two injections of 3-OH-K or XA (4 μg/animal) no significant effect was observed (Dr. E. Chang, personal communication). In this study, the "injection effect" was defined the log of the pre-injection ecdysteroid concentration divided by the log of the post-injection value. The ecdysteroid concentration before and after injection was determined by radioimmunoassay. However, we observed that daily 3-OH-K injection into eyestalk-ablated crayfish delayed by 2-3 days a significant increase of the 20-hydroxyecdysone titer and subsequent shedding of the exoskeleton (see above). The average weight of gastrolith pairs (from 10 animals) on the day of ecdysis was increased by 3-OH-K-injection (470 ± 130 mg) as compared to saline-injection (338 ± 114 mg); the lengthening of the proecdysial period caused this increase. The injection effect, which was connected with the quick metabolic turnover of 3-OH-K and 3-OH-Kase capacity, varied as a function of injection timing and animal size.

Our results, taken together, indicate that circulating 3-OH-K in the hemolymph was actively transported into the target tissue YOC, where it was enzymatically converted into XA to suppress ecdysteroidogenesis mediated by a biocatalyst (illustrated in the Fig. 5).

ACKNOWLEDGEMENTS

We thank Professor Y. Umebachi (Kanazawa University) for a gift of 3-hydroxy-L-kynurenine, Dr. S. Senoh (SUNBOR) for a gift of the synthetic DL-compound, Professor Y. Sugiura (Kyoto University) for the measurement of ESR, Professor I. Morishima (Kyoto University) for the measurement of paramagnetic NMR, Dr. T. Isawhita (SUNBOR) for NMR studies, Professor Y. Funae (Osaka City University) for a gift of phenobarbital-induced cytochrome P-450, and Professor H. Sonobe (Konan University) for helpful suggestions and contributions toward the crayfish experiments. We also thank Iwami Fishing Association for the collection of crabs in 1986-1988.

REFERENCES

Borst, D.W., Laufer, H., Landau, M., Chang, E.S., Hertz, W.A., Baker, F.C., and Schooley, D.A., 1987, Methyl farnesoate and its role in crustacean reproduction and development, Insect Biochem., 37:1123.
Brown, F., and Cunningham, O., 1939, Influence of the sinus gland of crustaceans on normal viability and ecdysis, Biol. Bull., 77:104.
Chang, E.S., Bruce, M.J., and Newcomb, R.W., 1987, Purification and amino acid composition of a peptide with molt-inhibiting activity from the lobster, Homarus americanus, Gen. Comp. Endocrinol., 65:56.
Chang, E.S., and O'Connor, J.D., 1977, Secretion of α-ecdysone by crab Y-organs in vitro, Proc. Natl. Acad. Sci. USA, 74:615.

Kappler, C., Kabbouh, M., Hetru, C., Durst, F., and Hoffmann, J.A., 1988, Characterization of three hydroxylases involved in the final steps of biosynthesis of the steroid hormone ecdysone in Locusta migratoria (Insecta, Orthoptera), J. Steroid Biochem., 31:891.

Kleinholz, L.H., 1985, Biochemistry of crustacean hormones, in: "The Biology of Crustacea", Vol. 9, Bliss, D.E., and Mantal, L.H., eds., Academic Press, Florida, pp. 463-522.

Laufer, H., Landau, M., Homola, E., and Borst, D.W., 1987, Methyl farnesoate: its site of synthesis and regulation of secretion in a juvenile crustacean, Insect Biochem., 17:1129.

Lowe, M.E., Horn, D.H.S., and Galbraith, M.N., 1968, The role of crustecdysone in the moulting crayfish, Experientia, 24:518.

Mattson, M.P., and Spaziani, E., 1985, 5-Hydroxytryptamine mediates release of molt-inhibiting hormone activity from isolated crab eyestalk ganglia, Biol. Bull., 169:246.

Mattson, M.P., and Spaziani, E., 1987, Demonstration of protein kinase C activity in crustacean Y-organs, and partial definition of its role in regulation of ecdysteroidogenesis, Mole. Cell. Endocrinol., 49:159.

Naya, Y., Kishida, K., Sugiyama, M., Murata, M., Miki, W., Ohnishi, M., and Nakanishi, K., 1988, Endogenous inhibitor of ecdysone synthesis in crabs, Experientia, 44:50.

Naya, Y., Miki, W., Ohnishi, M., Ikeda, M., and Nakanishi, K., 1989, Endogenous xanthurenic acid as a regulator of the crustacean molt cycle, Pure Appl. Chem., 61:465.

Passano, L.M., 1960, Molting and its control, in: "The Physiology of Crustacea", Vol. 1, Waterman, T.H., ed., Academic Press, New York, pp. 473-536.

Rao, K.R., 1965, Isolation and partial characterization of the molt-inhibiting hormone of the crustacean eyestalk, Experientia, 21:593.

Skinner, D.M., 1985, Molting and regeneration, in: "The Biology of Crustacea", Vol. 9, Bliss, D.E., and Mantel, L.H., eds., Academic Press, Florida, pp. 43-146.

Smith, S.L., 1985, Regulation of ecdysteroid titer: synthesis, in: "Comprehensive Insect Physiology, Biochemistry and Pharmacology", Kerkut, G.A., and Gilbert, L.I., eds., Pergamon Press, Oxford, pp. 295-334.

Soyez, D., and Kleinholz, L.H., 1977, Molt-inhibiting factor from the crustacean eyestalk, Gen. Comp. Endocrinol., 31:233.

Spaziani, E., 1988, Serum high-density lipoprotein in the crab, Cancer antennarius, Stimpson: II. annual cycles, J. Exp. Zool., 246:315.

Spaziani, E., Ostedgaard, L.S., Vensel, W.H., and Hegmann, J.P., 1982, Effects of eyestalk removal in crabs: relation to normal premolt, J. Exp. Zool., 221:323.

Watson, R.D., and Spaziani, E., 1985, Effects of eystalk removal on cholesterol uptake and ecdysone secretion by crab (Cancer antennarius) Y-organs in vitro, Gen. Comp. Endocrinol., 57:360.

Webster, S.G., and Keller, R., 1986, Purification, characterization and amino acid composition of the putative molt-inhibiting hormone (MIH) of Carcinus maenas (Crustacea, Decapoda), J. Comp. Physiol. B., 156:617.

Zeleny, C., 1905, Compensatory regulation, J. Exp. Zool., 2:1.

KYNURENINES IN THE REGULATION OF BEHAVIOR IN INSECTS

E. Savvateeva

Pavlov Institute of Physiology
Leningrad 199164
USSR

INTRODUCTION

The site and the mechanism of kynurenines effects have drawn much attention after the demonstration of their possible role as endogenous modulators of the NMDA receptor, which are involved in the mediation of basic functions and disorders in the central nervous system, such as learning, memory, ischemia, epilepsy and some degenerative diseases in a variety of species (Stone and Connick, 1985). Since genetics lend a powerful tool to the neurosciences, the mutational dissection of the kynurenine pathway of tryptophan metabolism (KPTM) seems to be fruitful in the elucidation of the mechanisms and behavioral manifestations of the kynurenines. It is noteworthy that biochemical genetics date back to 1941, when the eye-color mutants of Drosophila were used by Beadle and Tatum to demonstrate the exact sequence of reactions via the KPTM.

Since then, it has been proven that insects present a very convenient model for the study of the effects of kynurenines, because a great number of species has spontaneous mutations in the homologous genes coding for the enzymes of the successive stages of the KPTM. When ommochromes, the final products of KPTM, are the screening pigments of an insect eye, all the mutational blocks are manifested in the phenotype. Moreover, the mutations lead to a complete loss of all metabolites of the KPTM subsequent to a mutational block, and to an excess amount of a metabolite preceding the block. However, tryptophan metabolism in insects is blocked at the stage of 3-hydroxyanthranilic acid so that further metabolites, such as picolinic acid, quinolinic acid and nicotinamide are not synthesized (Linzen, 1974). Since the mutations are homologous, one can study the manifestations of the same genes while varying the species for the convenience of an experimental design. Therefore, the honey bee and Drosophila were chosen as the main objects in our studies, the latter permitting the construction of genotypes with two adjacent mutational blocks, which prohibit a substance injected or fed to metabolize further via the KPTM.

This report presents a review of the studies, carried out in two genetic laboratories at the Pavlov Institute of Physiology.

MUTATIONS OF:

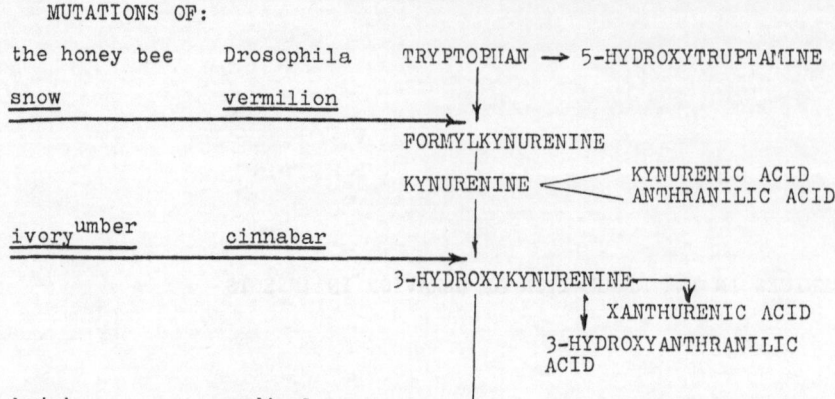

Fig. 1. Mutations of the kynurenine pathway of tryptophan metabolism in the honey bee and Drosophila.

MATERIALS AND METHODS

The effects of the following mutations were studied (Fig. 1): 1. Mutations in the homologous structural genes for tryptophan pyrrolase, vermilion in Drosophila (v) and snow (s) in the honey bee, which block the conversion of tryptophan to formylkynurenine, thus leading to an excess of unbound tryptophan and an absence of the subsequent metabolites of the pathway: 2. Mutations in the homologous structural genes for kynurenine hydroxylase, cinnabar (cn) in Drosophila and ivory^{umber} (i^u) in the honey bee, which lead to an excess of kynurenine (Linzen, 1974; Dustmann, 1975), the major difference between the two is the 10-15-fold excess of kynurenine in the i^u mutants and the only 1.5-fold excess in cn, since kynurenine is largely metabolized to kynurenic acid, which can not be detected in heads of the wild type flies (Ferré, 1983); 3. Mutations of the honey bee brick (bk) and chartreuse^{red} (ch^r). The first blocks the formation of ommochromes from 3-hydroxykynurenine and thus causes an excess of the latter and of xanthurenic acid. The second mutation causes the deposition of the redundant 3-hydroxykynurenine in granulated form in the pigment cells of the eyes. Mutations cardinal (cd) with a similar mode of action exists in Drosophila (Linzen, 1974).

The eyes in all mutants of Drosophila have the same bright red color, resulting from the absence of brown pigments and the presence of red pigments, drosopterins. In the honey bee, in contrast to Drosophila, the eye pigments are only ommochromes, which are responsible for the black eye color. Therefore the snow mutants have white eyes, while those of other mutants have a reddish tint and are pigmented to a different extent dependent on the ratio of ommochromes and 3-hydroxykynurenine.

The mutations of Drosophila melanogaster were brought to one genetic background of the wild type strain Canton-S. The flies were maintained on standard yeast-raisin mdeia at 25°C and a 12-hr cycle of light and darkness. Mutants of the honey bee were introduced on the genetic background of Apis mellifera Liguistica and were generously provided by Professors Laidlow and Bishop (Bee Research Center, USA), and by Professor Woyke, Poland. Experimental hives with an observation glass wall (inside temperature: 30°C) were kept in closed chambers at 22°C, artificially illuminated with diffused light to permit gathering of the mutants, which were unable to return into the hive

320

because of the defective orientation. All behavioral tests were carried out in the wild type bees and in homozygotes and heterozygotes for a mutation, which, being the offsprings of one heterozygous queen, inseminated with sperm of hemizygous drones, are semisibs. This permits the evaluation of an effect of a mutation in a one-bee family and the contribution of a sensory defect (since the eyes of heterozygotes contain ommochromes, they are phenotypically normal).

The behavioral and neurophysiologic performances studied in the honey bee were rhythm of dance (signal behavior), learning of free-moving and immobilized insects, both with food reward and olfactory stimuli (odors of clove and lavender), thresholds of neuro-muscular excitability and the rate of ether narcotization (Lopatina et al., 1977, 1985); in Drosophila, we examined olfactory learning with electroshock reinforcement (Quinn et al., 1974), spontaneous neuronal activity in the cervical connective (Smirnov and Ponomarenko, 1981) and stereotypic locomotor behavior in the immobilized insects.

The doses of injected kynurenines were adjusted to their contents in mutants, so that an injection could produce a phenocopy of a mutation (DL-kynurenine sulfate (Ferak): 10 mg/insect; DL-tryptophan (Reachim): 5 mg/insect; kynurenic acid (Sigma): 1-4 x 10^{-4} M; xanthurenic acid (Loba Chemie), 2-3 x 10^{-4} M; anthranilic acid (Serva): 10^{-4} M; 3-hydroxykynurenine (Sigma): 10^{-4} M; injected volume: 1-2 ml (honey bee) and 0.4 ml (Drosophila)).

EXPERIMENTS IN THE HONEY BEE

In early stages of our study, it was shown that mutation snow produces marked changes in a number of behavioral and neurophysiologic characteristics. This mutational block of the initial step of the kynurenine pathway led to a rather poor performance of signal behavior (dance), in rhythm and a reduction in the number of dancing bees (Kuzmina et al., 1975; Lopatina et al., 1976). This effect of the mutation could not be ascribed to the defects in vision or orientation, since the heterozygous bees with normal eyes showed a similar depression of the signal behavior, though less pronounced in magnitude. The scores of the wild type bees, heterozygotes and homozygotes, for the mutation showed a clear dependence on the dosage of the mutant gene. In the same dosage-dependent manner, the mutation reduced the excitability of the neuro-muscular apparatus both at the imaginal and the larval stage. Influence on excitability was mostly expressed in the range of short-duration (Kuzmina et al., 1975). Thus, the mutation affects function, independent of vision, and does so at an ontogenetic stage, at which sensory defects are not yet manifested. The snow mutation of the honey bee, as the vermilion in Drosophila, also significantly reduced the time of etherization. This effect, too, could not be ascribed to a sensory defect. The mutation also affected the bioelectric activity of the 2nd thoracic ganglion and reduced the spike frequency both in homozygotes and heterozygotes. Thus, the snow mutation, which inactivates tryptophan pyrrolase and therefore cuts off the KPTM, exerts an wide pleiotropic effect on behavior, inhibiting functional activity of the nervous system at different levels. Its influence on behavior, independent of the sensory defect and proportional to the gene dosage, indicates that one normal allele of the gene, coding for tryptophan pyrrolase, is sufficient for the improvement of the sensory defect, but not sufficient for the normal activity of the nervous system. A disturbance of tryptophan metabolism similar to that produced by the mutation is known in the case of the human disease tryptophanuria, which is characterized by an excess of urinary tryptophan and by mental retardation. If the manifestation of the mutation could in some respect resemble the disease, it might be reasonable to pose the question which biochemical changes, caused by the mutation, could produce such pronounced inhibitory effect on behavior and neurophysiology. We therefore developed a program for further analysis, involving 1) mutations that block subsequent stages of the KPTM and 2) pharmacological probes to correct the disturbances produced by the mutations.

Since the effect of inactivating tryptophan pyrrolase on the nervous system may be a result either of an excess of tryptophan, or of a deficit of kynurenines, tryptophan and kynurenine were injected into bees of different genotypes. Administration of DL-tryptophan reduced the rhythm of dance to the level of the snow heterozygotes in the wild type bees (Kuzmina et al., 1977). The fact that the homozygous level was not attained is indicative of the possible role of an absence of kynurenines together with the excess of tryptophan. Therefore, DL-kynurenine sulfate was injected into the snow hetero- and homozygotes. The injection raised the rhythm of dance in heterozygotes to the wild type level, while in homozygotes this effect could be achieved with twice as much kynurenine.

The effects of exogenously applied tryptophan on all the characteristics studied appeared to be similar to that of the snow mutation (Kuzmina et al., 1977). On the other hand, injections of kynurenine eliminated the effects of snow. The effects of exogenous kynurenine mimic those of an excess of endogenous kynurenine in the ivory^u mutant to a great extent. The mutation leads to a drastic rise in neuro-muscular excitability, both in homo- and heterozygotes, while the mutation brick (accumulation of 3-hydroxykynurenine) decreases the excitability. Mutants ch^r, too, with an excess of 3-hydroxykynurenine deposited in the eyes, shows decreased excitability, too. Thus, endogenously accumulated kynurenine that produces the same effects as the injected one, while 3-hydroxykynurenine is not involved in the excitatory effect of the i^u mutation.

Our further efforts were concentrated on the study of the role of kynurenines in the learning process. The rate of acquisition and transformation of conditioned reflexes on olfactory stimuli with food reward was studied in free-moving and in immobilized honey bees. In free-moving bees, the snow mutation decreased the rate of alteration of a conditioned stimulus (CS) in a dose-dependent manner. Injection of kynurenine normalized the mutant performance, with homozygotes needing twice the dose of heterozygotes. This indicates that the substance could interfere with the process of inhibition. However, excess amounts of kynurenine (either of endogenous origin, accumulated in the mutants i^u, or exogenous, injected to the wild type bees) have no effect on that type of conditioning. At the same time, conditioning in the fixed honey bees, after being subjected to immobilization stress, is much better in the i^u mutants than in the wild type. This effect of the mutation was even more pronounced following monotonous prolonged presentations (50-55) of the same CS and could be reproduced by the injections of kynurenine (Table 1). This type of conditioning is impaired in the mutants snow, but could not be improved by the injections of kynurenine.

The remarkable feature of this series of experiments with the immobilized i^u mutants is the profound impairment of the process of inhibition, manifested in the significant slowing of the rate of conditioning following the alteration of the CS (from positive to negative). A similar, but less strong, effect could be produced by the injection of kynurenine to the wild type bees. The mutants devoid of kynurenines exceed the wild type in the rate of transforming the positive reflex into the negative. In this respect, the performance of the immobilized mutants is like that of the free moving bees.

These data suggest that kynurenine can increase the behavioral response to external stimulation (either in new situation, which promotes an orienting-exploratory reaction, or under the stress of immobilization) and influence the process of inhibition. Moreover, the excess of kynurenine in the mutants leads to a functional maturation of the nervous system, which is

Table 1. Effects of KPTM mutations and of the injection of DL-tryptophan
(5 mg/insect) and DL-kynurenine (10 mg/insect) in the honey bee

Genotype	ivory[u]		wild type	snow	
Substance injected		KYN	TRY		KYN
Thresholds of excitability					
Larva	-14	-20	n.d.	+70	N
Imago	-60	-33	+42	<u>+104</u>	+12
Spontaneous bioelectric activity					
Total	N	+119	-30	<u>-62</u>	N
30-70 mV	N	+119	-40	-82	N
70-100 mV	-85	-35	N	N	-52
100 mV	+340	N	N	N	N
Conditioning					
Free-moving					
Number of CR(+)	N	N	N	N	N
Number of CR(-)	N	N	+17	-38	N
Immobilized					
Number of CR(-)	+50	+43	N	-33	-34
number of CR(+)	+150	+66	-53	-42	-40
number of UC	+15	+13	N	-39	-38
Foraging and signal behavior					
Rhythm of visiting of a food source	+16	n.d.	n.d.	-18	n.d.
Reinforcements before the 1st dance	-13	n.d.	n.d.	-133	n.d.
% of dancing bees	N	n.d.	-56	-58	n.d.
Rhythm of dance	+10	+16	-28	<u>-36</u>	N

Data are expressed as a percentage of wild type levels. Dose-dependent
figures are underlined. N: wild type level; n.d.: not determined.

achieved at earlier ontogenetic stages than in wild type bees. The study of
conditioning in the ontogenesis of the immobilized mutants has shown that the
mutation of i^u permits to reach the definitive level of conditioning on the

third day of imago life, while wild type bees usually reach it on the 7th day, and the snow mutants only on the 10th day (Lopatina et al., 1985).

The effects of 3-hydroxykynurenine are very similar to those of kynurenine, since the brick mutation affects the behavioral performances studied in the same manner, as i^u. The major difference is that the effects of bk are manifested only early during ontogenesis. From the 15th day of imago life, the mutants show strong depression of the functional activity of the nervous system, which is expressed in a drastic decrease in conditioning and in the rhythm of dance. One of the possible reasons lies in the cytotoxic and carcinogenic effects of 3-hydroxykynurenine (Linzen, 1974). Notably, the homologous mutation of Drosophila, cardinal, significantly shortens the life span of the flies (Kamyshev, 1980).

Injections of kynurenic acid to wild type bees revealed its inhibitory effects on conditioning in the immobilized insects. At the same time, kynurenic acid facilitated the process of transforming positive into negative reflexes. It also decreased the neuro-muscular excitability and increased the time of ether narcotization, the most effective concentration being 3×10^{-4} M (Table 2). These data demonstrate that kynurenic acid produces an inhibitory effects in insects as well.

Injection of xanthurenic acid caused similar effects, albeit of slightly lesser magnitude.

Though all the experimental paradigms for behavioral testing were chosen to avoid the vision defects of the mutants, they can be used to elucidate the involvement of kynurenines involvement in sensory perception. It has been shown that snow mutants hae an altered electroretinogram and an increased sensitivity to light (Chesnokova et al., 1981). In addition to the described defects of the snow mutants, and especially their ontogenetic delay in reach-

Table 2. Conditioned and unconditioned reflexes in the honey bee following the injections of saline (S), kynurenic (K) and xanthurenic (X) acids

Concentration (M)	1×10^{-4}		2×10^{-4}		3×10^{-4}		4×10^{-4}	
	S	K	S	K	S	K	S	K
Conditioned reactions	36.9	32.6	71.7	64.2	60.2	45.9*	63.1	59.7
Unconditioned reactions	87.6	87.0	83.9	80.8	86.7	78.9**	88.9	86.7
			S	X	S	X		
Conditioned reactions			58.9	62.2	63.7	56.5*		
Unconditioned reactions			85.4	88.9*	89.2	82.5		

Data are expressed as percent of reactions following injections. *$P < 0.05$; **$P < 0.01$ as compared to saline injected animals.

ing the definitive level of conditioning, it has been noticed that they form the queen's suite, which is twice as large as that of the wild type, later during ontogenesis. These facts are indicative of a pleiotropic effect of the mutation on the olfactory system of the mutants, and especially on the sensitivity to pheromones (Lopatina et al., in press). The homozygous and heterozygous snow have a decreased sensitivity of their antennal receptors to the queen's pheromone on the 3rd-7th day of imago life. This is manifested in a decreased amplitude of their electroantennogram following presentation of the pheromone extract in an air-stream. At the same time, sensitivity to the pheromone of the sting apparatus is quite normal. These data present an opportunity to use KPTM mutants for the elucidation of the involvement of kynurenines in olfaction. This is of interest, since in mammals the activities of kynurenine pathway enzymes are surprisingly high in the olfactory bulb (Stone and Connick, 1985).

The demonstration of a participation of kynurenines in behavioral regulation in the honey bee proved the utility of the mutants. However, pharmaco-genetic analysis in the honey bee is difficult because it is not possible to construct the appropriate genotypes by means of artificial insemination. In contrast, Drosophila, with homologous mutations of the KPTM, is readily amenable for genetic manipulations. This allows the production of genotypes with two adjacent mutational blocks which inhibit the degradation of exogenously administered compounds.

EXPERIMENTS IN DROSOPHILA

Studies of the neuronal activity of Drosophila mutants recorded in the cervical connective between the brain and the thoracic ganglion have shown that immobilized flies, fixed in wax on their backs, have different average frequencies and patterns of spontaneous neuronal activity (Smirnov and Ponomarenko, 1981) (Fig. 2). Mutation vermilion, homologous to snow, produced a 1.5-fold reduction in frequency, while cinnabar, homologous to i^u, led to a proportional increase, and cardinal had no notable effect. However, the wild type Drosophila had bursts of activity, lasting for 2 sec with the 8-10 sec intervals; mutation vermilion diminished burst duration to 1 sec without altering the interval between them, and mutation cinnabar prolonged both bursts and intervals.

A similar pattern of activity could be recorded for the spontaneous movements of legs of the fixed normal flies. However, prolonged observation revealed fading with time of this sterotypic behavior. This could be ap-

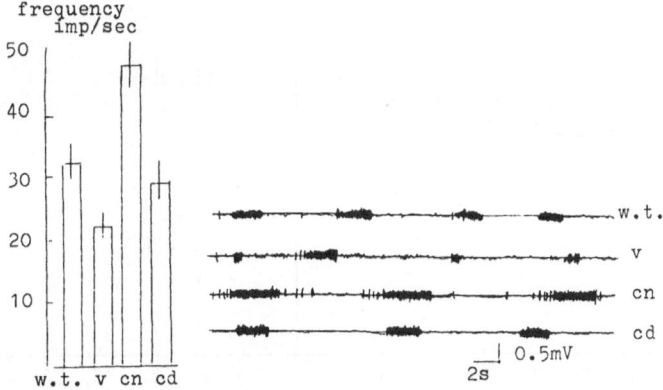

Fig. 2. Average frequency of impulse activity and patterns of spontaneous neuronal activity recorded in the cervical connective in Drosophila mutants (w.t.: wild type).

proximated by a curve, which was typical for the phases of stress reactions and showed clear-cut stages of activation and of adaptation. This finding has prompted a more detailed study, in which an index of this stereotypic behavior (ratio of time of activity to a sum of time of activity and time of rest) was measured in the mutants (Smirnov and Kamyshev, in press). Vermilion mutants, devoid of any kynurenines, could not switch down to the stage of adaptation, while cardinal mutants did not have the stage of activation. At the same time the performance of the cinnabar mutants resembled that of the wild type but had a more pronounced activation stage. This indicates that certain kynurenines might play a trigger role in switching the behavior during immobilization stress. Thus, kynurenine or kynurenic acid were injected into the double mutants v.cn, (in which metabolism was inhibited) and 3-hydroxykynurenine was injected into the double mutants cn.cd. While kynenine had a slight promoting effect on the switch from activation to adaptation, kynurenic acid was a strong promoter of the switch, and 3-hydroxykynurenine inhibited the primary reaction to stress, which is manifested as an activation (Fig.3).

As demonstrated earlier (Savvateeva, 1978), the cinnabar mutants, as well as the homologous i^u, are better learners than the wild type, while the

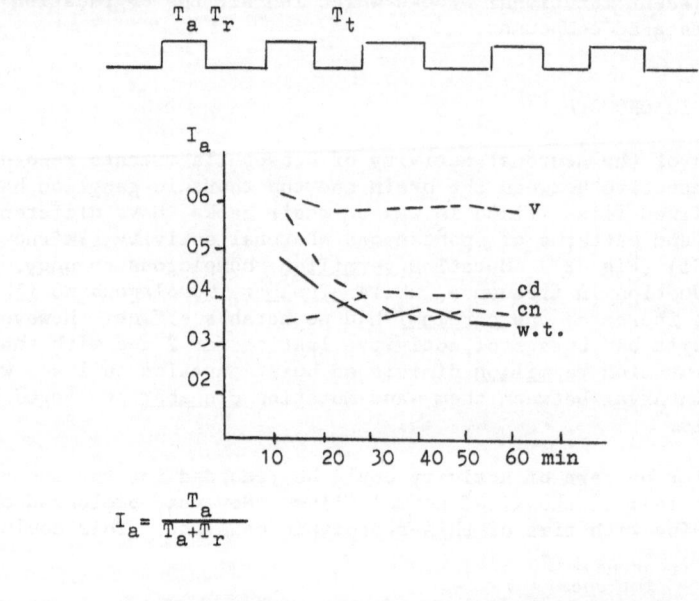

$$I_a = \frac{T_a}{T_a + T_r}$$

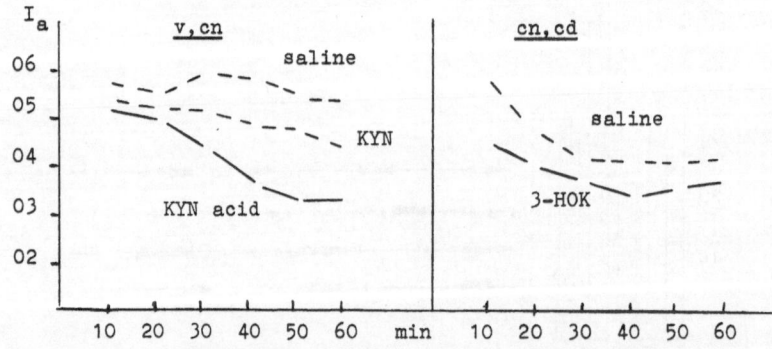

Fig. 3. Indices of activity of stereotypic locomotor behavior in Drosophila mutants and following the injections of kynurenines. KYN: kynurenine, KYN acid: kynurenic acid, 3-HOK: 3-hydroxykynurenine.

other kynurenine pathway mutations shown quite normal learning. Injection of kynurenine into the double mutants v.cn also increased learning ability (Savvateeva and Kamyshev, unpublished data). The only other bright learners in Drosophila are the agnostic mutants which show increased activity of calmodulin (CaM) (Savvateeva et al., 1985). Since CaM can initiate long term potentiation, mediated by NMDA receptors, in the mammalian hippocampus (Reymann et al., 1988), it is tempting to ask if kynurenines could produce such a variety of behavioral effects in other species by a modulation of CaM functions. In turn, CaM may trigger intracellular second messenger systems. This would be in line with the widespread notion that kynurenines are endogenous modulators of the NMDA receptors. Freezing and thawing procedures and detergent treatment, which diminish binding at Cl^-/Ca^{2+}-dependent glutamate-binding sites, reveal quinolinic and kynurenic acid binding in hippocampal membranes (French et al., 1984). Many Ca^{2+}-dependent binding proteins carry Ca^{2+}-binding domains homologous to that of CaM, thus allowing the binding of CaM modulators or antagonists. Among the latter are tricyclic antidepressants and some dicyclic or multi-cyclic agents. Lapin's structure-activity relationships for kynurenines (1983) should therefore be amended to include the role of the dicyclic structure, since the side chain of kynurenine allows it to attain nearly bicyclic structure.

Taken together, we therefore wish to advance a hypothesis of a "Calmodulin button", which assumes that the kynurenines with a dicyclic structure (kynurenine, kynurenic acid, 3-hydroxykynurenine, xanthurenic acid) might be the endogenous modulators of CaM. Upon binding with CaM, these substances could fix its conformation in a configuration which is advantageous for the activation of adenylate cyclase or phosphodiesterase (PDE), and which would be responsible for the inhibitory or stimulatory effects of kynurenines. They could bind to NMDA receptors as well, provided the latter bear a Ca^{2+}-binding sites homologous to that of CaM. Direct assessment of the hypothesis may be achieved by measuring the effects of kynurenines in an in vitro bioassay of CaM activation of PDE. We have therefore recently begun to study PDE activity in Drosophila kynurenine mutants and examined direct effects of kynurenines on the CaM activation of PDE. Preliminary data demonstrate that kynrenic acid (10^{-4}-10^{-5} M) has a stronger inhibitory effect than kynurenine, producing 35% inhibition of PDE. This may be sufficient to produce its modulatory effects.

CONCLUSION

Study of the neuroactive effects of kynurenines in the honey bee and Drosophila demonstrates their involvement in the regulation of a balance between the excitatory and the inhibitory processes in insect CNS, and in the formation of the definitive level of behavioral activity and of thresholds of sensory perception. The data suggest that kynurenine and 3-hydroxykynurenine are endogenous stimulators and kynurenic and xanthurenic acids inhibitors of the nervous system. Anthranilic acid appears to be indifferent to the behavioral and neurologic characters studied.

REFERENCES

Beadle, G.W., and Tatum, E.L., 1941, Genetic control of developmental reactions, Amer. Natur., 75:107-116.
Chesnokova, E.G., Polyanovsky, A.D., and Grybakin, F.G., 1981, The influence of mutations snow and laranja on the morphophysiologic characteristics of the compound eye of the honey bee, Dokl. Acad. Sci. USSR., 256:1503-1506.
Dustmann, J.H., 1975, Pigment studies on several eye-color mutants of the honey bee, Insect Biochem., 5:429-445.

Ferré. J., 1983, Accumulation of kynurenic acid in the cinnabar mutant of Drosophila melanogaster as revealed by thin-layer chromatography, **Insect Biochem.**, 13:289-294.

French, E.D., Foster, A.C., Vezzani, A., and Schwarcz, R., 1984, Quinolinate and kynurenate, two endogenous tryptophan metabolites with potential links to epileptic disorders, **Clin. Neuropharmacol.**, Suppl., 7:456-457.

Kamyshev, N.G., 1980, Life span and its relation to locomotor activity in Drosophila mutants of the metabolic pathway tryptophan-ommochromes, **Dokl. Acad. Sci. USSR**, 253:1476-1480.

Kuzmina, L.A., Lopatina, N.G., Nikitina, I.A., Ponomarenko, V.V., and Saifutdinova, Z.N., 1975, The effect of the mutant genes snow and laranja, controlling the tryptophan metabolism, on signal behavior of the honey bee, **Dokl. Acad. Sci. USSR**, 222:463-465.

Kuzmina, L.A., Lopatina, N.G., and Ponomarenko, V.V., 1977, On the biochemical mode of influence of the snow mutation on the neurologic characters of the honey bee, **Dokl. Acad. Sci. USSR**, 237:955-957.

Lapin, I.P., 1983, Structure-activity relationships of kynurenine, diazepam and some putative endogenous ligands of the benzodiazepine receptors, **Neurosci. Biobehav. Rev.**, 7:107-118.

Linzen, B., 1974, The tryptophan-ommochrome pathway in insects, **Insect Physiol.**, 10:117-246.

Lopatina, N.G., Marshin, V.G., Nikitina, I.A., Ponomarenko, V.V., Smirnova, G.P., and Savvateeva, E.V., 1976, The effects of several mutations on behavioral and neurophysiologic characters of insects, **Jurn. Vysh. Nervn. Deyat.**, 26:785-791.

Lopatina, N.G., Marshin, V.G., Ponomarenko, V.V., Smirnova, G.P., and Sogrin, B.V., 1977, The study of a neurophysiologic character - a rate of ether narcotization - in relation to insect behavior, **Genetika**, 13:1767-1772.

Lopatina, N.G., Ponomarenko, V.V., and Chesnokova, E.G., 1985, Tryptophan and its metabolites in the nervous system functions and behavior of the honey bee, **Jurn. Evol. Bioch. Physiol.**, 21: 25-32.

Quinn, W.G., Harris, W.A., and Benzer, S., 1974, Conditioned behavior in Drosophila melanogaster, **Proc. Natl. Acad. Sci. USA**, 71:708-712.

Reymann, K., Trey, U., Jork, R., and Matthies, H., 1988, Polymyxin B, an inhibitor of protein kinase C, prevents the maintenance of synaptic long-term potentiation in hippocampal CA1 neurons, **Brain Res.**, 440:305-314.

Savvateeva, E.V., 1977, Comparative study of learning abilityl in Drosophila strains selected on neurophysiologic characters and in mutants, **Dokl. Acad. Sci. USSR**, 235:1430-1432.

Savvateeva, E.V., Peresleny, I.V., Ivanushina, V., and Korochkin, L.I., 1985, Expression of adenylate cyclase and phosphodiesterase in development of temperature-sensitive mutants with impaired metabolism of cAMP in Drosophila melanogaster, **Develop. Genet.**, 5:159-172.

Smirnov, V.B., and Ponomarenko, V.V., 1981, The effect of mutatations blocking the kynurenine pathway of tryptophan metabolism on the neuronal activity of Drosophila melanogaster, **Dokl. Acad. Sci. USSR**, 258:489-491.

Stone, T.W., and Connick, J.H., 1985, Quinolinic acid and other kynurenines in the central nervous system, **Neuroscience**, 15:597-617.

EFFECTS OF QUINOLINIC AND KYNURENIC ACIDS ON CENTRAL NEURONS

T.W. Stone and J.H. Connick

Department of Pharmacology
University of Glasgow
Scotland

QUINOLINIC ACID

Interest in the cellular basis of the actions of kynurenines in the central nervous system can be dated to 1981 when a range of cyclic analogs of the excitatory amino acids glutamate and aspartate were tested on the excitability of neurons in the cerebral cortex of anesthetised rats (Stone and Perkins, 1981; Stone, 1984). The earliest observations were of an excitatory action of quinolinic acid on cortical neurons, an action which clearly provided some explanation for the convulsant effects of quinolinic acid reported previously by Lapin and his colleagues (see Lapin, 1989). The excitation is summarized in histogram form in Fig. 1 which also reveals one of the major properties of quinolinic acid excitation also discovered in 1981. This is the sensitivity of the response to antagonism by compounds known to be active at the population of dicarboxylic amino acid receptors which show a preference for activation by N-methyl-D-aspartate (NMDA). Thus, in Fig. 1, the cell recorded in this case in the hippocampus in vivo was excited by applications of quinolinic acid, NMDA and quisqualic acid all applied by microiontophoresis from glass micropipettes on to a single neuron in the CA1 region of the hippocampal formation. The iontophoresis of 2-amino-5-phosphonopentanoic acid (2AP5) is then able to prevent excitation in response to both quinolinic acid and NMDA, the response to quisqualic acid being unchanged.

More recent experiments have confirmed that this profile of selective NMDA activation is also observed in neurons in the striatum, cerebellum and brain stem of anesthetized rats.

A second early observation which aroused interest in quinolinic acid as an important centrally active neurotransmitter or neuromodulator candidate was the finding that cells in different parts of the central nervous system responded differently to quinolinic acid and either NMDA or glutamate. Early comparison of sensitivities of neurons was made by using doses which would allow an equilibrium or plateau of cell-firing rate to be achieved. The relative potencies of compounds could then be estimated by comparing the relative size of plateau firing rates for a given dose of agonist or conversely the different doses required to achieve the same plateau of firing provided that this was submaximal. In the case of microiontophoretic applications of agonists, this comparison is dependent on the assumption that the rate of release of compounds from micropipettes reflected in the transport number is approximately equal (Stone, 1985). For the compounds of interest in this case the transport numbers have not yet been identified, and the estimates of relative potency may eventually need to be corrected by an appropriate factor. Nevertheless the relative sensitivities of neurons to agonists can be compared provided that the same microelectrode system is used since the transport number for a given compound is constant for a given microelectrode barrel.

In the cerebral cortex, the doses of glutamate or NMDA and quinolinate were adjusted so that the responses were approximately equal in magnitude. The electrode being used was then removed from the cortex and reinserted into the exposed dorsal horn of the spinal cord of the same animal. Despite the constancy of the anesthetic concentration, the drug concentrations and injection characteristics from the one micropipette, it was clear that in the spinal cord NMDA was still able to activate cells whereas quinolinic acid was much less potent. Indeed on most cells in the spinal cord quinolinic acid was unable to produce an excitation even using 200 nA of ejecting current applied for two minutes, i.e. over forty times the dose used to produce excitation in the cerebral cortex (Perkins and Stone, 1983).

The regional variability in the relative sensitivities of neurons to quinolinic acid and NMDA as well as to quinolinic acid and glutamate strongly suggests that there may be sub-types of receptor only one of which is sensitive to quinolinic acid, for example in the cerebral cortex, whereas the spinal receptor is sensitive only to NMDA. In our earlier studies, the conclusion was reached that the former receptor, sensitive to quinolinic acid and NMDA was present in the cerebral cortex, hippocampus and the striatum whilst the latter receptor, sensitive to NMDA only, was present in the spinal cord and cerebellum (Perkins and Stone, 1983). This suggestion led us to propose a model of the NMDA receptor in which two sub-types are present as illustrated in Fig. 2. The receptor sub-type which is not sensitive to quinolinic acid is shown as responding to another NMDA-like agonist, ibotenic acid. The evidence for this assertion is rather weak but is based primarily on the failure of kynurenic acid to prevent ibotenic acid-induced neurodegeneration in the striatum (Foster et al., 1984) and the finding that kynurenic acid and 2AP5 exhibit different pA_2 values against ibotenate compared with NMDA and quinolinic acid (Burton et al., 1988).

OTHER KYNURENINES - RECENT FINDINGS

Besides quinolinic acid, several other members of the kynurenine pathway from tryptophan are also able to produce seizures in conscious animals. These include kynurenine and kynurenic acid (Lapin, 1989). We have attempted to determine the cellular mechanism of action of these compounds which would explain their convulsant properties. Kynurenine itself has no effect on central neurons when applied by microiontophoresis in vivo (Perkins and Stone, 1982) or when perfused on brain slices in vitro at concentrations up to 5 mM. Kynurenine does however possess a glycine-like side chain and glycine is well-known to be an inhibitory neurotransmitter particularly in the hindbrain and spinal cord regions of the central nervous system. Experiments were therefore performed in the hindbrain in which glycine was applied by microiontophoresis to produce depressions of the spontaneous firing rate of unidentified neurons. Kynurenine then applied for up to 5 minutes in duration with high ejecting currents (100-200 nA) failed not only to affect basal firing rates but also failed to modify the glycine inhibitory responses. Similarly, L-kynurenine did not modify the inhibition of cells of the hindbrain or cerebral cortex produced by the local iontophoresis of gamma aminobutyric acid (GABA) (Stone, unpublished observations). Pinelli et al. (1985), however, have reported an ability of kynurenine to displace GABA-related ligands from their binding sites in central nervous system neuronal membranes, and we have therefore reinvestigated this potential interaction using hippocampal

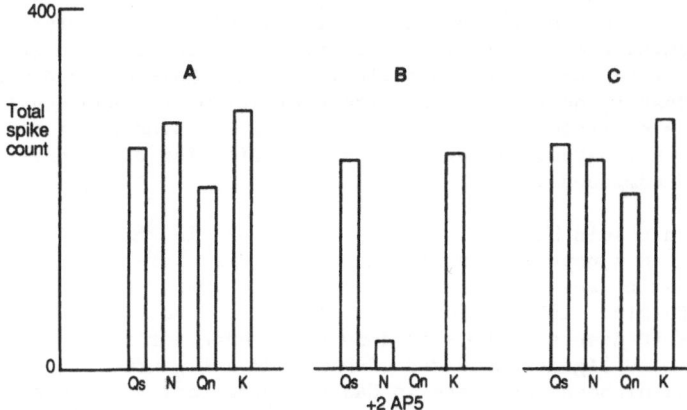

Fig. 1.   Histogram illustrating the excitatory activity of 4 agonists on a
         cortical neuron recorded in an anesthetized rat. Quisqualic acid
         (Qs), N-methyl-D-aspartate (N), quinolinic acid (Qn) and kainic acid
         (K) were applied by microiontophoresis and the total number of
         spikes elicited over background was determined. During panel B, the
         NMDA antagonist 2-amino-5-phosphonopentanoic acid (2AP5) was admini-
         stered simultaneously, causing a block of responses not only to
         NMDA, but also to quinolinate.

slices in which both GABA and kynurenine could be applied at known concentrations to the slice. Fig. 3 illustrates the depression of evoked potential
size recorded from populations of CAl pyramidal neurons by perfusion with
GABA at a concentration of 2 mM. Additional superfusion with kynurenine at
concentrations up to 10 mM had no effect either on the baseline level of the
evoked potentials or on the depression of those potentials by GABA.

KYNURENIC ACID - RECENT FINDINGS

     The other kynurenine of major interest is kynurenic acid and this com-
pound, like kynurenine, has been found not to modify the spontaneous firing
rates of neurons in the anesthetized rat brain in several regions (Stone and
Burton, 1988). Neither does it affect the inhibitory responses of hindbrain
neurons to glycine or of hippocampal neurons in vivo or in vitro to GABA
(unpublished). Early experiments in the cerebral cortex, however, indicated
that kynurenic acid was able to antagonize responses to excitatory compounds
including quinolinic acid, NMDA kainate and quisqualic acid (Perkins and
Stone, 1982). In the cortex, no clear differentiation could be achieved be-

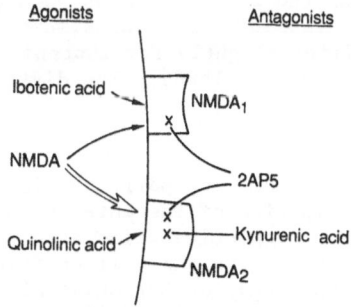

Fig 2.   Schematic of the proposed subdivision of NMDA receptors into two
        subtypes, an NMDA1 site responding to ibotenate and weakly to NMDA,
        and an NMDA2 site responding to quinolinate.

tween these various agonists, and kynurenic acid was therefore cast as a non-selective amino acid antagonist. It did not affect the sensitivity to ace-tylcholine or carbachol on neocortical neurons, however, and subsequent work has shown that it does not affect the excitation of hippocampal pyramidal cells by the excitatory peptide neuropeptide Y. Kynurenate does therefore demonstrate a substantial degree of selectivity for dicarboxylic amino acids and related compounds.

Kynurenic acid has now become a major tool in the analysis of central nervous system synaptic transmission mechanisms where the involvement of an excitatory amino acid or related substance is suspected (Ganong et al., 1983; Stone and Perkins, 1984). In some regions, it is a particularly useful compound since it does seem to show some preference for antagonizing responses to NMDA-related ligands compared with non-NMDA ligands. This is particularly clear in the hippocampus where in CA3 pyramidal cells kynurenic acid appears to be a selective antagonist of NMDA at concentrations which have no effect on sensitivity to kainic acid or quisqualic acid (Ganong and Cotman, 1986; Stone, 1988). In addition, it has been found that when recording the activi-ty of cells in the hippocampal formation in vivo kynurenic acid is able to suppress selectively the sensitivity to quinolinic acid at a time when re-sponses to NMDA are relatively unchanged (Perkins and Stone, 1985). This ob-servation on 8 of 20 neurons would be entirely consistent with the view ex-pressed above that there exist sub-types of NMDA receptor only one of which is sensitive to quinolinic acid (Fig. 2). Kynurenic acid would then be re-garded as a preferential antagonist at that population of NMDA sites which is sensitive to quinolinic acid.

More recently, we have been involved in examining the pharmacology of kynurenic acid in more detail. In order to obtain firm evidence for or against the hypothetical receptor model illustrated in Fig. 2, we performed an extensive quantitative analysis of responses of populations of pyramidal neurons in the mouse cerebral cortex. Slices of cortex were prepared and placed in two-compartment baths, the division between the two compartments consisting of a high-resistance grease seal as described by Harrison and Sim-monds (1985) and Burton et al. (1987). This system is basically analogous to the sucrose gap system used for many years in the analysis of electrophysio-logical changes in isolated nerve or smooth muscle. It allows recordings of the standing DC potential to be made between two regions of neurons, in this case the cell bodies of pyramidal neurons, located to one side of the grease gap, and the axons of those cells passing in the corpus callosum white matter which are located on the other side of this barrier. The slices are perfused continuously with drug solution. The addition of depolarizing agents to that solution then produces a DC shift lasting several minutes in duration, which can easily be measured and quantified in order to produce dose-response curves for agonists. The results from this work are summarized in Table 1. Kynurenic acid as well as the selective NMDA antagonist 2AP5 were able to block all three agonists tested (NMDA, quinolinic acid and ibotenic acid) but with pA_2 values which differ slightly for ibotenic acid compared with the other two agents (Burton et al., 1988). The difference is not great but it is statistically significant and is again consistent with the model proposed in Fig. 2.

A further extension of the work performed in this study allowed us to determine the molecular kinetics of the interaction between agonist and an-tagonist molecules and the respective amino acid receptors. By construc-ting Schmild plots and determining the Hill coefficients for the dose-re-sponse data obtained in the study of Burton et al. (1988) it was possible to propose a bimolecular interaction, that is 2 molecules of the agonists being required to activate the receptors. Surprisingly, substitution of the rele-vant Hill slopes into the Schild equations then allowed an assessment of the molecular kinetics of antagonist action; this yielded the result that 2

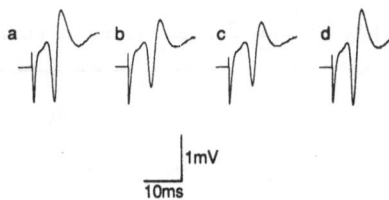

Fig. 3. Single records of orthodromic population spike potentials record in
area CA1 of the hippocampal slice. Record a is a control; b
illustrates the effect of adding 2 mM GABA into the perfusion
medium. There is an approximately 20% reduction in spike sizes.
Record c illustrates the effect of the same concentration of GABA
added onto a perfusion solution containing 2 mM L-kynurenine.
Kynurenine itself did not change the population potential and there
is no significant enhancement or blockade of the GABA response.

molecules of the antagonists 2AP5 and kynurenic acid were also required to
block receptors (Williams et al., 1988). This at first sight seems surprising since it might be expected that the combination of a single molecule of
antagonist with the receptor would be sufficient to prevent the binding of an
agonist molecule. However, it is conceivable that a single molecule of antagonist binds rather loosely to the receptor in such a way that it can easily be displaced by an incoming agonist molecule, and it is only when two
molecules of antagonist have acted on the receptor and locked it into an
agonist resistant conformation that true antagonism is obtained.

    While kynurenic acid has now been used in a wide range of preparations
to block the responses to exogenously applied amino acids or synaptic transmission in which amino acids are felt to be involved (see Stone and Burton,
1988 for review), it is only recently that great excitement has been directed
to the detailed mechanism of this antagonistic action. In 1987, Johnson and
Ascher showed that perfusion of cultured neurons with a control solution led
to a progressive decline in the sensitivity of those cells to locally applied
NMDA. This phenomenon they attributed to the washout of a factor or factors
required for NMDA receptor activation, and following analyses of the perfusate solution for a range of compounds including amino acids revealed that

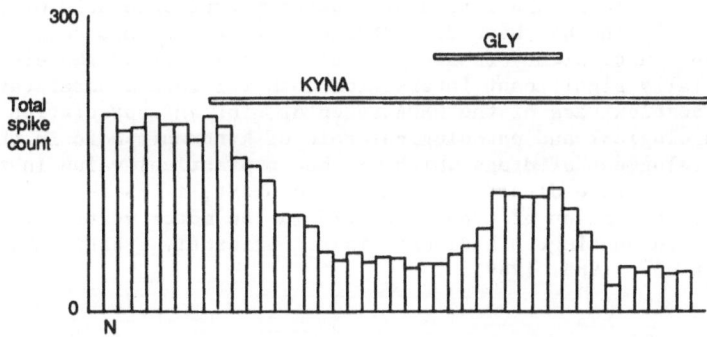

Fig. 4. Each bar of this histogram represents the total spike count elicited
by microiontophoretic applications of NMDA every 40 seconds to a
neuron in the CA1 region of the hippocampal slice. The addition of
kynurenic acid (0.2 mM) into the perfusion medium reduces the size
of those responses, but the additional perfusion with glycine (1 mM)
causes a partial reversal of the blockade.

Table 1. pA$_2$ Values for amino acid antagonists

	pA$_2$ values	
Agonist	Kynurenate	2AP5
NMDA	3.65	4.92
Quinolinate	3.65	4.98
Ibotenate	3.89	5.05

this compound might be glycine. Addition of glycine to the solution washing over the cultured cells from a micropipette indeed rapidly restored and maintained the sensitivity of those neurons to NMDA. It has become clear that the activation of the NMDA receptor or its associated ion channels requires micromolar concentrations of glycine acting at a strychnine-resistant site, i.e. a site pharmacologically distinct from that responsible for neuronal inhibition in the hindbrain and spinal cord. It now appears that kynurenic acid is able to act partly by displacing glycine from that strychnine-resistant binding site so that the action of kynurenic acid is due partly to a competitive interaction with incoming NMDA at the receptor site and partly a non-competitive action exerted by glycine displacement deep within the receptor channel complex. This interaction has now been studied in several areas of brain, and Fig. 4 illustrates its occurrence in slices of rat hippocampus. Here a neuron in the CA1 pyramidal cell layer is being recorded and NMDA then applied by microiontophoresis to that cell. Perfusion with kynurenic acid prevents that excitation but additional perfusion with glycine restores the size of the hippocampal excitation towards control levels. This mutual antagonism and interaction between kynurenic acid and glycine would be expected if kynurenic acid acted partly at the glycine binding site.

As mentioned earlier in this chapter, kynurenine has a glycine-like side-chain which led us to examine its ability to block glycine-induced inhibitions in the rat hindbrain. Clearly, it is also worth considering the possibility that kynurenine might interact with the strychnine-resistant glycine site associated with the NMDA receptor. However, experiments recently completed show that kynurenine does not prevent the glycine potentiation of NMDA responses in hippocampal or neocortical slices at concentrations up to 10 mM. Nevertheless, the interaction between kynurenic acid or 7-chlorokynurenate and glycine has been demonstrated not only on neuronal excitability but also on neurotoxicity (Shalaby et al., 1989). It is therefore a robust and potentially significant interaction both for future understanding of the molecular complexities of the NMDA receptor, for an appreciation of the possible physiological and pathological role of kynurenic acid in the brain, and for the development of drugs which may be of practical value in the treatment of neurodegenerative disorders. In this context, it is interesting to note that the concentration of kynurenic acid in the motor cortex of patients who have died with Huntington's disease is significantly greater than in controls (Connick et al., 1988, 1989).

ACKNOWLEDGEMENTS

We are grateful to Action Research for the Crippled Child, the S.E.R.C. and Wellcome Trust for support.

REFERENCES

Burton, N.R., Smith, D.A.S., and Stone, T.W., 1987, The mouse neocortical slice preparation and the pharmacology of excitatory neocortical amino acids, **Comp. Biochem. Physiol.**, 88:47-55.

Burton, N.R., Smith, D.A.S., and Stone, T.W., 1988, A quantitative pharmacological analysis of excitatory amino acid receptors in the mouse neocortex in vitro, **Br. J. Pharmacol.**, 93:693-702.

Connick, J.H., Carlà, V., Moroni, F., and Stone, T.W., 1989, Increase in kynurenic acid in Huntington's disease motor cortex, **J. Neurochem.**, 52:985-987.

Connick, J.H., Stone, T.W., Carlà, V., and Moroni, F., 1988, Increased kynurenic acid in Huntington's disease, **Lancet**, 2:1373.

Foster, A.C., Vezzani, A., French, E.D., and Schwarcz, R., 1984, Kynurenic acid blocks neurotoxicity and seizures induced in rats by the related brain metabolite quinolinic acid, **Neurosci. Lett.**, 48:273-278.

Ganong, A.H., and Cotman, C.W., 1986, Kynurenic acid and quinolinic acid act at NMDA receptors in the rat hippocampus, **J. Pharmacol. Exp. Therap.**, 236:293-299.

Ganong, A.H., Lanthorn, T.H., and Cotman, C.W., 1983, Kynurenic acid inhibits synaptic and acidic amino acid induced responses in the rat hippocampus and spinal cord, **Brain Res.**, 273:170-174.

Harrison, N.L., and Simmonds, M.A., 1985, Quantitative studies on some antagonists of NMDA in slices of rat cerebral cortex, **Br. J. Pharmacol.**, 84:381-392.

Johnson, J.W., and Ascher, P., 1987, Glycine potentiates the NMDA responses in cultured mouse brain neurons, **Nature**, 325:529-531.

Lapin, I.P., 1989, Behavioral and convulsant effects of kynurenines, in: "Quinolinic Acid and the Kynurenines", Stone, T.W., ed., CRC Press, Boca Raton, pp. 193-212.

Perkins, M.N., and Stone, T.W., 1981, An iontophoretic investigation of the actions of convulsant kynurenines and their interaction with the endogenous excitant quinolinic acid, **Brain Res.**, 247:184-187.

Perkins, M.N., and Stone, T.W., 1983, Pharmacology and regional variations of quinolinic acid evoked excitations in the rat CNS, **J. Pharmacol. Exp. Therap.**, 226:551-556.

Perkins, M.N., and Stone, T.W., 1985, Actions of kynurenic acid and quinolinic acid in the rat hippocampus in vivo, **Exp. Neurol.**, 88:570-579.

Pinelli, A., Govoni, S., Ossi, C., Battaini, F., Caimi, B.R., and Trivulzio, S., 1985, Kynurenine may directly interact with GABA receptors in brain, **Pharmacology**, 30:255-258.

Shalaby, I., Chenard, B., and Prochniak, M., 1989, Glycine reverses 7-chlorokynurenate block of glutamate neurotoxicity in cell culture, **Eur. J. Pharmacol.**, 160:309.

Stone, T.W., 1984, Excitant activity of methyl derivatives of quinolinic acid on rat cortical neurons, **Br. J. Pharmacol.**, 81:175-181.

Stone, T.W., 1985, "Microiontophoresis and Pressure Ejection", Wiley, Chichester.

Stone, T.W., 1988, Comparison of kynurenic acid and 2APV suppression of epileptiform activity in rat hippocampal slices, **Neurosci. Lett.**, 84:234-238.

Stone, T.W., and Burton, N.R., 1988, NMDA receptors and ligands in the vertebrate CNS, **Progr. Neurobiol.**, 30:333-368.

Stone, T.W., and Perkins, M.N., 1981, Quinolinic acid: a potent endogenous excitant at amino acid receptors in rat CNS, **Eur. J. Pharmacol.**, 72:411-412.

Stone, T.W., and Perkins, M.N., 1984, Actions of amino acids and kynurenic acid in the primate hippocampus, **Neurosci. Lett.**, 52:335-340.

Williams, T.L.M., Smith, D.A.S., Burton, N.R., and Stone, T.W., 1988, Amino acid pharmacology in neocortical slices: evidence for bimolecular actions from an extension of the Hill and Gaddum-Schild equations, **Br. J. Pharmacol.**, 95:805-810.

BEHAVIORAL ABNORMALITIES IN RATS AFTER SINGLE TREATMENT WITH QUINOLINIC ACID

DURING EARLY ONTOGENESIS

I.P. Lapin, V.L. Kozlovsky, and O.G. Kenunen

Laboratory of Psychopharmacology
Bekhterev Psychoneurological Research Institute
Leningrad 193019
USSR

INTRODUCTION

Neonatal brain injuries which occur under hypoxia, ischemia, hypogly-cemia and trauma (Rothman and Olney, 1986; Simon et al., 1986; Ikonomidou et al., 1988; McIntosh et al., 1988) have recently been suggested to be mediated through activation of excitatory amino acid systems. In particular, neuro-destructive changes in the developing brain are thought to be due to stimu-lation of N-methyl-D-aspartate (NMDA) receptors (Simon et al., 1986; McDonald et al., 1988).

Quinolinic acid (QUIN), a neuroactive metabolite of tryptophan, is an endogenous agonist of NMDA receptors (Perkins and Stone, 1983). Although there is no evidence so far that the concentration of QUIN in the brain is increased under and/or after those circumstances common for a developing or-ganism, i.e. hypoxia, ischemia, trauma, etc., such elevations are certainly possible. For that reason, we have suggested to answer the question whether an increase in the concentration of QUIN in early ontogenesis is of function-al importance, and what kind of behavioral after-effects that increase may have.

Using a single systemic administration of QUIN we have tried to mimic a situation when the level of QUIN in the organism, both in the periphery and in the CNS, is elevated only once and for a short time.

METHODS

Eighty five albino rats were used for the study. QUIN (Sigma) was in-jected i.p. in a dose of 50 mg/kg (pH = 7). Controls from the same litter received i.p. saline on the same days as the respective experimental groups. Injections were made on postnatal days 2-3, 5-6 and 11-12 (Groups 1, 2 and 3), which were chosen according to defined periods of rat brain development (Dmitrieva, 1969; Olenev, 1978). Thus, the period between days 1 and 5 is characterized by intensive development of the medulla oblongata and the mid-brain. From day 6 to 10, massive differentiation of neurons of the telence-phalon takes place. During days 11-20, maturation of the cortex is particu-larly progressing although the overall rate of brain development is already markedly reduced.

Kynurenine and Serotonin Pathways
Edited by R. Schwarcz et al., Plenum Press, New York, 1991

Behavioral studies were started three weeks after the birth.

Motor activity, locomotion and rearing, were measured once a week during 45-60 days in an open field (diameter: 1 m). A single rat was placed at the center. Each session lasted 2 min.

Conditioned reflex of passive avoidance (CRPA) was studied in a chamber divided into light and dark compartments. The latter had an electric grid floor. A rat was placed in the light compartment, and the latency of entering into the dark compartment was recorded with a stop-watch. 10 sec after entering the dark compartment, the rat was punished (15 Hz, 35 V) every 1 sec with 1 sec intervals until it left the compartment. Duration of the session was 100 sec. After 24 hours, rats were tested again, and preservation of the conditioned reflex was quantitated by determining the number of rats entering the dark compartment and the time spent in the light compartment.

"Darkness preference" reflex was studied in the same chamber. A compartment was regarded as preferred if a rat spent more than 50 per cent of the session in it.

Conditioned reflex of active avoidance (CRAA) was studied in a chamber divided into two halves connected via a window (diameter: 50 cm). A ring was used as a conditioned signal. 5-7 sec after the ring, electric current was given to the grid floor of this part of the chamber (intensity of the current was chosen individually according to the threshold of pain sensitivity measured by appearance of vocalization response). 5 sec after transition of the rat into the opposite half of the chamber, the combination "ring-current" was repeated. During the first session (10 combinations), only latency of the escape reaction was measured. During the next six sessions, the time of the first avoidance reaction and the number of conditioned reactions during a session were recorded (total number of combinations: 70). After 24 hours, CRAA was measured again. During the second session, the rat received 60 combinations. The total number of conditioned responses during the second day was compared with that during the first day. Control and treated rats were always tested on the same day.

Pain threshold was measured by means of electrical stimulation from the grid floor according to the vocalization response.

"Pentylenetetrazole titration" was made by means of subbcutaneous injections (10 mg/kg) every 15 min until generalized clonic-tonic seizures appeared. The threshold convulsant dose of pentylenetetrazol was calculated.

Catalepsy was measured in haloperidol (3 mg/kg, i.p.)-treated rats. A 3-score scale was used: 1 - one foreleg is placed on a platform 3 cm high, 2 - reared rat with one foreleg placed on a 12 cm high platform, 3 - rat standing on forelegs and one hindleg placed on a 12 cm high platform. The time of fixation of a posture was 15 sec. The degree of catalepsy in a group was determined as the ratio of the number of scores observed to the maximal possible number (6) of scores.

Stereotypy was induced by D,L-amphetamine (7 mg/kg, i.p.). Latency and duration of stereotypies were recorded.

Histological examination of the brains were performed in the Brain Institute in Moscow 3 months after treatment.

RESULTS

Motor activity during 6 weeks of observation did not differ significantly between controls and QUIN-treated rats. However, there was a marked tendency to more intensive locomotion. Thus, in groups 1, 2 and 3, it was + 13%, + 24% and + 14% vs. the respective controls. In the same groups, rearing increased by 21%, 34% and 6%, respectively.

Darkness preference appeared to be normal in all groups when it was first tested on day 25 of postnatal ontogenesis. Only in rats of group 2 (treated with QUIN on day 5-6) a longer stay in a dark compartment (84.9 ± 1.5 sec) and a shorter latency to entering into it (15.1 ± 1.5 sc) was registered vs. respective controls (73.8 ± 3.8 and 24.9 ± 3.7 sec; $P < 0.05$).

CRPA was studied on day 40 of ontogenesis. In 12 of 14 rats of Group 2 (treated with QUIN on day 5-6), CRPA was not developed (Table 1).

CRAA was studied in 3-3.5 months old rats. In group 2, the number of correct conditioned responses on the second day of the experiment was lower than in controls (Table 2).

Pain thresholds were higher in QUIN-treated rats of group 2: 21,8 ± 1,0 vs. 18.7 ± 0,9 V ($P < 0.05$). In other groups, no differences were observed.

Generalized clonic-tonic seizures were only observed in 4 of 10 rats of Group 3 3-4 min after QUIN injection.

"Pentylenetetrazole titration" thresholds did not differ in any group. Mean threshold convulsant doses were 56.0 ± 2.1 (controls) and 56.9 ± 3.8 mg/kg (QUIN-treated rats).

Haloperidol catalepsy and amphetamine stereotypy were virtually identical in all groups.

Histological examination did not reveal any marked degeneration in the hippocampus in any of the three groups studied (preliminary data).

Thus, behavioral abnormalities were found only in Group 2, i.e. in rats treated with QUIN on postnatal day 5-6.

Table 1. Effect of treatment with quinolinic acid (QUIN) on the conditioned reflex of passive avoidance (CRPA) in 40 day old rats

| | Day of treatment with QUIN (50 mg/kg) | | |
	2-3	5-6	10-11
Control rats			
Total	17	17	10
With CRPA in 24 h	11	10	4
Time spent in dark (sec)	76.2	54.7	30.2
QUIN-treated rats			
Total	16	14	10
With CRPA in 24 h	9	2**	4
Time spent in dark (sec)	67.0	68.8	60.0*

*$P < 0.05$, **$P < 0.01$ vs. controls.

Table 2. Effect of treatment with quinolinic acid (QUIN) on behavior
in a shuttle box (conditioned reflex of active avoidance:
CRAA) in 3-3.5 months old rats

| | Day of treatment with QUIN (50 mg/kg) | | |
	2-3	5-6	10-11
Control rats			
Latency of avoidance (sec)	8.2 ± 1.2	12.6 ± 2.2	9.6 ± 2.3
Number of conditioned responses			
1st experiment	2.5 ± 1.0	3.9 ± 1.1	0.6 ± 0.4
2nd experiment	6.0 ± 1.7	11.9 ± 1.1	8.4 ± 3.2
Treated rats			
Latency of avoidance (sec)	12.9 ± 3.2	14.1 ± 2.3	7.4 ± 1.9
Number of conditioned responses			
1st experiment	2.0 ± 0.6	3.6 ± 0.8	2.8 ± 0.8[*]
2nd experiment	4.6 ± 1.5	6.4 ± 1.7[**]	7.8 ± 2.5

The significantly lower number of conditioned reflexes in controls of
group 3 (0.6 ± 0.4) may be an artifact since those rats did not differ
from other controls on the second day of testing. Data are the mean ±
SEM. [*]$P < 0.05$; [**]$P < 0.01$.

DISCUSSION

Central administration of QUIN in adult animals induces generalized
clonic-tonic seizures and local neuro-destruction. These effects are related
to the stimulation of excitatory amino acid receptors (Perkins and Stone,
1983). QUIN does not penetrate into the brain after systemic administration
and therefore does not induce seizures or degeneration of neurons. In rats
younger than 12-15 days of postnatal ontogenesis, i.p. injected QUIN presumably penetrates into the brain due to an immature blood-brain barrier. In
our experiments, i.p. administered QUIN induced seizures only in 11-12 day
old rats. In 2-3 and 5-6 days old rats, seizures were not seen. This observation agrees with data on the appearance and development of QUIN-induced
seizures in rats beginning on postnatal days 7-8 (Lapin, 1978).

Lack of convulsant effect of QUIN in groups 1 and 2 may be related to
different reasons. It seems probable that in 2-3 days old rats the sensitivity of brain tissue to QUIN is low, resulting not only in an absence of
seizure activity but also of all other behavioral abnormalities. However, in
5-6 day old rats numerous distant behavioral effects of QUIN were found.
Lack of the convulsant effect may be related to the functional immaturity of
the CNS and inability to develop seizures. This age group has marked neuronal sensitivity to the neurodestructive action of excitatory amino acids
(Simon et al., 1986; McDonald et al., 1988). In 11-12 day old rats, in spite
of a convulsant response to QUIN, distant behavioral effects of QUIN were not
observed. This dissociation suggests a dissimilarity between this age group
and group 2 and adults in convulsant and behavioral mechanisms involved in
the action of QUIN.

The appearance of QUIN-induced seizures only in 11-12 days old rats is in accordance with the assumption that a brain structure other than the hippocampus (presumably the nucleus caudatus) is the trigger structure for those seizures (Lapin, 1988). Newborn rats do not differ from adults in the number of pyramidal cells in hippocampus (Olenev, 1978). In contrast, the caudate differentiates only in 11-12 day old rats (Olenev, 1978). However, without further data on the functional maturation of those two brain structures any assumption based on morphological findings only is merely tentative.

Because no neurodegenerative changes in hippocampus were found in any group tested (preliminary data), one may suppose that behavioral after-effects of QUIN are mainly of a functional nature. Thus, it appears that the abnormalities of postnatal development induced by QUIN are determined by both the sensitivity of the brain tissue to QUIN and the period of CNS maturation when an increased level of QUIN happens.

It is of particular interest that direct injection of QUIN into the striatum of 7 day old rats, in contrast to adult rats, does not result in neurodegeneration (Foster et al., 1983). These data disagree with the idea of a leading role of NMDA receptors in the development of neonatal destructions in the CNS (Simon et al., 1986; McDonald et al., 1988) and also, with data on the NMDA-mimetic action of QUIN (Rothman and Olney, 1986).

Data obtained in the present study suggest that even a single short-lasting increase in the level of QUIN at some stages of ontogenesis can cause distant behavioral abnormalities in the developing organism. This may be related to the minimal brain dysfunction syndrom, the hyperactivity syndrom, and the pathology of memory and learning. Antagonists of QUIN seem to be promising in the prevention and treatment of those disorders.

REFERENCES

Dmitrieva, N.I., 1969, Development of brain and spinal cord in postnatal ontogenesis in laboratory rats, in: "Development of animal brain", Olenev, S.N., ed., Nauka, Leningrad, pp. 132-144.

Foster, A.C., Collins, J.F., and Schwarcz, R., 1983, On the excitotoxic properties of quinolinic acid, 2,3-piperidine dicarboxylic acids and structurally related compounds, Neuropharmacology, 22:1331-1342.

Ikonomidou, C., Friedrich, G., Salles, S., Labruyere, J., Price, M.T., and Olney, J.W., 1988, Glutamate-like damage in infant brain produced by hypobaric-ischemic conditions, in: "Frontiers in Excitatory Amino Acid Research", Cavalheiro, E.A., Lehmann, J., and Turski, L., eds., Alan R. Liss, New York, pp. 657-660.

Lapin, I.P., 1978, Convulsions and tremor in immature rats after intraperitoneal injection of kynurenine and metabolites, Pharmacol. Res. Comm., 10:81-84.

Lapin, I.P., 1988, Kynurenines and behavior, in: "Frontiers in Excitatory Amino Acid Research", Cavalheiro, E.A., Lehmann, J., and Turski, L., eds., Alan R. Liss, New York, pp. 605-611.

McDonald, J.W., Silverstein, F.S., and Johnston, M.V., 1988, Neurotoxicity of N-methyl-D-aspartate is markedly enhanced in developing rat central nervous system, Brain Res., 459:200-203.

McIntosh, T.K., Soares, H., Hayes, R.L., and Simon, R.P., 1988, The N-methyl-D-aspartate receptor antagonist MK-801 prevents edema and improves outcome after experimental traumatic brain injury in rats, Neurochem. Internat., Suppl. 1, 12:40.

Olenev, S.N., 1978, "Developing Brain", Nauka, Leningrad.

Perkins, M.N., and Stone, T.W., 1983, Quinolinic acid: regional variations in neuronal sensitivity, **Brain Res.**, 259:172-176.

Rothman, S.M., and Olney, J.W., 1986, Glutamate and the pathophysiology of hypoxic-ischemic brain damage, **Ann. Neurol.**, 19:705-711.

Simon, R.P., Young, P.S.K., Stout, S., and Cheng, J., 1986, Inhibition of excitatory neurotransmission with kynurenate reduces brain edema in neonatal anoxia, **Neurosci. Lett.**, 71:361-364.

SESSION VI

NUTRITION

TRYPTOPHAN NUTRITION AND METABOLISM: AN OVERVIEW

J.C. Peters

The Procter & Gamble Company
Miami Valley Laboratories
Cincinnati, Ohio 45239
USA

Tryptophan (TRP) was the first amino acid to be recognized as being essential for normal growth of young animals when Wilcock and Hopkins (1906) and later Osborne and Mendel (1914) observed its ability to stimulate weight gain in mice and rats when added to low TRP rations. Subsequent studies in a variety of species confirmed that TRP was essential for normal growth and, furthermore, was required for maintenance of nitrogen equilibrium in mature animals. Some years after those early animal studies, Rose and collaborators (1957) demonstrated that TRP was an essential amino acid for human nutrition.

Tryptophan's role in maintaining normal physiologic function goes beyond its role as a substrate for tissue protein synthesis. Tryptophan has been suggested to play a unique role in regulating protein synthesis in the liver, and has been shown to affect protein synthesis in other tissues in a fashion that appears unrelated to its function as a precursor amino acid. Tryptophan gives rise to a wide array of metabolites involved in a variety of aspects of normal nutrition and metabolism. For example, picolinic acid, a product of TRP's oxidative metabolism, is involved in normal intestinal absorption of zinc. Another TRP metabolite, quinolinic acid, is involved in the regulation of gluconeogenesis. Tryptophan can also contribute to the body's pool of the nicotinamide nucleotides through its metabolic conversion to niacin. Finally, TRP is the precursor of several neuroactive compounds including serotonin (5-hydroxytryptamine, 5-HT), which functions as a neurochemical substrate for a variety of normal behavioral and neuroendocrine functions. In light of the neurotransmitter precursor function of TRP it is not surprising that many of the effects of treatments or conditions which severely alter TRP nutrition and metabolism are expressed as behavioral effects reflecting altered central nervous system function.

The purpose of this brief overview is to highlight some of the major functions of TRP in the body, to outline key pathways of TRP utilization and to discuss some of the mechanisms involved in the integration of whole-body TRP metabolism and the responses of those systems to variations in diet.

TRYPTOPHAN OCCURRENCE AND REQUIREMENT

Tryptophan is the least abundant amino acid in most proteins (Block and Weiss, 1956) accounting for roughly 1% to 1.5% of the total amino acids in typical plant and animal proteins, respectively. Despite its scarcity, it is rarely the most limiting amino acid for maintenance or growth when the dietary amino acid source is from naturally occurring proteins. For example the corn protein zein is nearly devoid of TRP, yet lysine is limiting for growth in this protein since the content of lysine in zein in the lowest of any of the essential amino acids in relation to its requirement.

It is important to keep in mind however, that while TRP may not be limiting for growth in some poor quality proteins, its supply may be insufficient for normal functioning of other pathways which depend on an adequate supply of this amino acid.

As might be expected due to TRP's scarcity in dietary proteins, the requirement of mammals for TRP is correspondingly the lowest among the indispensable amino acids. In adult man, the minimal daily requirement for TRP has been estimated to be 250 mg/day in males and 160 mg/day in females (Harper, 1977). In human infants, the requirement for growth is roughly 12-40 mg/kg. The recommended dietary allowance for protein in adult man ranges from 44 g/day for women to 56 g/day for men, an amount that would supply between 500 and 700 mg/day of TRP if the protein was of high quality.

PATHWAYS OF TRYPTOPHAN UTILIZATION

Protein synthesis

Tryptophan is one of 20-22 amino acids required for the synthesis of tissue proteins. In an average adult male at nitrogen equilibrium, approximately 225-250 grams of protein are synthesized each day (approx. 3 g/kg/day, Young et al., 1983). If TRP is assumed to represent 1.5% of the total amino acids in tissue protein, then approximately 3.5 grams of TRP would be utilized daily for protein synthesis. Thus, despite the fact that in an adult at nitrogen balance, no net accretion of body protein takes place, an appreciable flux of amino acids flows through this pathway each day. In the case of TRP, this amounts to more than 15 times the minimum intake requirement and more than 3 times the average daily intake of TRP in well nourished individuals, making this pathway quantitatively the most significant in the utilization of TRP.

It has long been recognized that protein synthesis in the whole animal is sensitive to nutritional factors, including the supply of energy and the amount and pattern of amino acids provided in relation to amino acid requirements. Over the past 30 years, a variety of studies have yielded information suggesting a unique role for TRP in the regulation of protein synthesis in a number of tissues including liver (Sidransky et al., 1984), muscle (Lin et al., 1988) and brain (Blazek and Shaw, 1978). Much of the work focusing on the role of TRP has come from studies of protein synthesis in the liver and suggests that TRP may act at several different points in the overall process.

In early studies of the effect of amino acids on hepatic protein synthesis, Munro and associates (1975) observed that tube-feeding fasted rats a complete mixture of amino acids caused a shift in the ribosomal pattern of liver from lighter to heavier aggregates. This response did not occur when animals were force-fed a complete amino acid mixture devoid of TRP. Furthermore, the response to the amino acid mixture including TRP was not influenced by treatment of the animals with Actinomycin D, suggesting that the effect of TRP was most likely at a post-transcriptional step in protein synthesis. In a systematic study by Pronczuk et al. (1968), the response of liver ribosomes was studied when fasted animals were fed 10 different amino acid mixtures, each one lacking in a single indispensable amino acid. These workers observed that impairment of polyribosome aggregation occurred only when TRP was omitted from the amino acid mixture. In view of the fact that the free TRP content of serum and tissues is the lowest of any indispensable amino acid (Munro, 1970), these findings raised the possibility that TRP might be the limiting amino acid for protein synthesis under conditions of fasting.

However, these studies did not establish whether or not TRP was unique in its effect on protein synthesis. In subsequent investigations, Pronczuk et al. (1970) found that when animals were fed threonine or isoleucine imbalanced diets, which depleted tissue pools of these amino acids, hepatic polyribosomes were disaggregated and were stimulated by addition of the limiting amino acid to the meal. Ip and Harper (1974) extended these findings in studies of rats fed a threonine-imbalanced diet. Feeding animals a threonine imbalanced diet for 7 days resulted in hepatic ribosomes that were largely disaggregated. Oral administration of threonine caused ribosomes to reaggregate and stimulated the incorporation of ^{14}C-leucine into tissue proteins. Administration of TRP to threonine-depleted animals did not improve protein synthesis, suggesting that liver protein synthesis was sensitive to the supply of the amino acid most limiting in the tissue.

Collectively, these studies established that TRP's ability to affect hepatic polyribosomal aggregation was not a unique effect of this amino acid on the protein synthetic machinery. However, the observation that TRP is normally the least abundant amino acid in the liver free amino acid pool when animals are fed nutritionally adequate diets, and the finding that the TRP-tRNA content of liver falls more rapidly during food deprivation than do the t-RNAs of other indispensable amino acids (Rogers, 1976), suggests that TRP may be an important effector of hepatic protein synthesis under many physiological circumstances.

In studies in which rats or mice were given solutions containing single amino acids, Sidransky and coworkers (1971) found that administering TRP alone stimulated ribosome aggregation and protein synthesis in liver while giving isoleucine, methionine or threonine alone did not. A somewhat lesser response was observed with certain TRP metabolites including 5-HT, 5-hydroxy-tryptophan (5-HTP), indole and 3-hydroxyanthranilic acid. This effect was still intact in adrenalectomized animals and thus could not be attributed to an effect of adrenal corticosteroid secretion. Protein synthesis was stimulated both in fed and fasted animals when TRP was given, and thus was apparently not due simply to increasing tissue TRP content.

Sidransky and associates have investigated the mechanism of this response to TRP and have found that TRP administration affects a number of aspects of hepatic RNA metabolism, including DNA-dependent RNA polymerase activity, polyribosomal RNA and nuclear RNA synthesis, cytoplasmic poly(A) and poly(A)-mRNA concentrations, nucleocytoplasmic translocation of poly(A)-mRNA and levels of nucleoside triphosphatase activity in the nuclear envelope (Sidransky et al., 1984). These workers have hypothesized that TRP can stimulate hepatic protein synthesis by at least two mechanisms: 1) increasing the synthesis of mRNA, and 2) increasing nucleocytoplasmic translocation of mRNA, which would increase the supply of message to locations in the cell where translation occurs. Recent evidence from this group indicates that TRP's effect involves its specific binding to a nuclear membrane glycoprotein (Sidransky et al., 1984).

The serotonin pathway

The conversion of TRP to 5-HT occurs in several tissues throughout the body including the enterochromaffin cells of the gut, blood platelets and the central nervous system. In the central nervous system, 5-HT functions as a neurotransmitter and is believed to be involved in a variety of normal brain functions. For example, serotoninergic neurons are thought to participate in regulating pain perception, aggressive behavior, sleep, and appetite (Sved, 1983). Furthermore, the serotoninergic system plays an important role in certain neuroendocrine systems (Fernstrom, 1981).

Based on measurements in man of urinary excretion of the major endproduct of 5-HT metabolism, 5-hydroxyindoleacetic acid (5-HIAA), it can be estimated that roughly 3.6 mg of 5-HT are turned over each day (Udenfriend et al., 1955). This represents the conversion of an equivalent molar amount of TRP to 5-HT, which corresponds to the utilization of less than 1% of dietary TRP intake. The proportion of total urinary 5-HIAA arising from 5-HT turnover in the central nervous system compared to that stemming from other sources such as the gut is probably variable but has been estimated to be 10% in the rat and as much as 30% in man (Bosnan, 1978).

Serotonin synthesis in brain occurs via a two-step reaction beginning with hydroxylation of L-TRP by the enzyme TRP hydroxylase, to form 5-hydroxytryptophan (5-HTP). This reaction is followed by the decarboxylation of 5-HTP to 5-HT carried out by aromatic L-amino acid decarboxylase. The principal product of 5-HT degradation is 5-HIAA, which is formed by the sequential action of monoamine oxidase and aldehyde dehydrogenase.

The reaction catalyzed by TRP hydroxylase is rate limiting in brain and regulates the flux of TRP through the 5-HT pathway. Studies in animals have shown that the Km of TRP hydroxylase for its substrate TRP is about 50 μM, which is close to the normal concentration of TRP in brain (Kaufman, 1974). Thus, 5-HT synthesis under normal conditions is controlled by the availability of TRP to serotoninergic neurons. The supply of TRP to 5-HT releasing cells is in turn dependent on many factors, some of which are influenced by diet composition and the previous nutritional status of the animal. Because 5-HT synthesis is sensitive to precursor supply, the possibility exists that a number of behavioral functions dependent on serotoninergic neuronal activity may be sensitive to variations in TRP availability to the brain. Thus, despite the quantitative insignificance of this pathway in terms of whole-body TRP disposal, derangements in this pathway have the potential to influence a wide variety of normal biological functions dependent on 5-HT mediated neurotransmission.

At least three factors are important in determining the supply of TRP to brain, and hence 5-HT synthesis. These include: 1) the plasma TRP concentration, 2) the plasma concentrations of other large neutral amino acids (LNAA), which compete with TRP for uptake into brain, and 3) the extent of binding of TRP to serum albumin, which can influence the pool of unbound TRP that interacts with the amino acid carrier mechanism situated at the blood-brain barrier. Each of these factors, in turn, can be influenced by the nutritional and hormonal status of the animal, and by interorgan relationships in the metabolism of amino acids.

For example, the concentration of TRP in plasma is a function not only of dietary TRP intake, but of the extent of removal of TRP from blood by body tissues. Since there is little net utilization of TRP by non-hepatic tissues owing to the relatively limited capacity of these tissues to oxidize TRP (Miller, 1962), the liver is the most important organ influencing plasma TRP concentration. The extent of removal of TRP by the liver following a meal is influenced by several factors at least one of which is the extent to which TRP stimulates the activity of TRP oxygenase, the principal enzyme regulating TRP oxidation. The direct relationship between dietary TRP intake and TRP catabolism will tend to limit its entry into the general circulation following a meal.

The plasma concentrations of the principal LNAA (LNAA = leucine, isoleucine, valine, tyrosine and phenylalanine) which compete with TRP for uptake into brain, are also influenced by the extent of their uptake and net utilization by tissues. Tyrosine and phenylalanine are similar to TRP in that the liver is the primary site for their metabolism (Miller, 1962). The branched chain amino acids (BCAA) on the other hand escape significant liver metabolism and are taken up and metabolized predominately by skeletal muscle (Harper et al., 1984). Because of this, following a protein-containing meal the BCAA rise more in peripheral blood than the other LNAA. The fact that the BCAA's rise in peripheral blood in proportion to their content in the diet, while the increase in other indispensable amino acids is blunted by liver metabolism, means that these amino acids dominate the effect of the LNAA's as a group on brain TRP uptake.

Finally, the binding of TRP to serum albumin, a phenomenon first described by McMenamy and Oncley (1958), can under some circumstances influence the carrier mediated transport of TRP across blood-brain barrier. In normal fed animals, the proportion of total plasma TRP bound to albumin is about 85-90%. This equilibrium can be shifted under conditions which raise plasma non-esterified fatty acid (NEFA) concentrations, such as during fasting or stress (McMenamy, 1965). This is because NEFA's compete with TRP for binding sites on the albumin molecule; therefore, when the concentration of NEFA's rise, TRP is displaced from the albumin molecule raising the concentration of free TRP.

A number of investigations have been carried out to determine the relative importance of albumin binding and of the fraction of plasma TRP which exists in the unbound or free form, on the uptake of TRP by brain under various conditions. Studies designed to mimic physiological situations in which the equilibrium between free and bound TRP was perturbed by altering plasma NEFA concentrations have shown that brain TRP concentration is not affected by changes in plasma free TRP concentration (Fernstrom et al., 1975), but is more closely predicted by the plasma ratio of TRP/LNAA. Other studies in which various drugs were used to displace TRP from albumin and increase the plasma free pool, have led to the opposite conclusion, namely that the size of the plasma free pool of TRP is the most important determinant of TRP supply to the brain (Bloxam et al., 1980). Although there is no universal agreement on this point, it is probable that under normal circumstances the binding of TRP to albumin has some influence (even if small) on the carrier-mediated transport of TRP into brain.

Nutritional effects on brain 5-HT synthesis

It has been known for many years that diets deficient in TRP lead to depletion of brain 5-HT and hence to disturbances in 5-HT-mediated brain function (Gál and Drewes, 1962). However, it wasn't appreciated until the work of Fernstrom and Wurtman that variations in plasma and brain TRP and brain 5-HT synthesis could occur under normal physiologic circumstances.

In the early 1970's, Fernstrom and Wurtman (1971a,b, 1972) began a systematic investigation of the relationship between TRP supply and 5-HT synthesis under a variety of circumstances. In their initial studies, they found that injecting rats with a small dose of TRP (12.5 mg/kg), only 5% of the daily intake of an adult rat, led to significant increases in both brain TRP and 5-HT content. They later observed that giving fasted rats a single injection of insulin rapidly elevated serum and brain TRP concentrations and produced a corresponding rise in brain 5-HT content. Giving fasted rats a single meal of a protein-free, high carbohydrate diet, produced the same effect as did insulin treatment, indicating that the response occurred under normal physiological conditions. Subsequent studies revealed that diet-induced increases in brain 5-HT content actually reflected an increased rate of 5-HT synthesis.

The ability of various treatments to alter brain TRP and brain 5-HT was not simply a function of their effects on serum TRP however, but turned out to be dependent on the concentration of TRP in plasma relative to the concentrations of other large neutral amino acids that compete with TRP for transport into brain via a common carrier (Pardridge, 1977). For example, it was found that in overnight fasted rats fed a meal containing 18-24% protein, plasma TRP concentrations were considerably higher than in animals fed a protein-free meal, yet brain TRP and 5-HT concentrations were unchanged compared to fasting (Fernstrom and Wurtman, 1972). Furthermore, feeding fasted rats a 40% protein meal actually decreased brain TRP and 5-HT content, despite the high TRP content of the meal (Fernstrom and Faller, 1978). Other studies showed that feeding animals diets in which the LNAA competitors were omitted resulted in a large increase in brain TRP and 5-HT compared to the response of animals fed a complete amino acid mixture having the same level of TRP (Fernstrom and Wurtman, 1972). Collectively, these observations demonstrated that changes in brain TRP and 5-HT concentrations were directly related to changes in the plasma ratio of TRP to LNAA (TRP/LNAA), and that alterations in this ratio induced by diet were predictive of changes in brain 5-HT synthesis.

Based on the pioneering work of Fernstrom and Wurtman, other workers began to investigate the possibility that diet-induced changes in brain TRP content and 5-HT synthesis might be involved in the control of normal animal feeding behavior. In long-term studies in weanling rats that were allowed to self-select between high and low protein diets, Ashley and Anderson (1975) observed a strong inverse correlation between chronic cumulative protein intake and the plasma TRP/LNAA ratio. These authors proposed that the plasma TRP/LNAA ratio, and hence 5-HT synthesis were involved in regulating protein intake and selection in rats. The results of these and other studies by Anderson and associates (Anderson, 1979) led to the hypothesis that changes in brain TRP and 5-HT brought about by single meals constituted a behavioral feedback loop by which animals regulated the selection of protein and carbohydrate. According to their hypothesis, ingestion of a protein-free or low-protein diet would increase the plasma ratio of TRP/LNAA and brain 5-HT which would cause a shift in diet selection toward a diet having a higher protein content, a move that presumably would restore the premeal level of brain 5-HT. Their hypothesis also predicted that high protein diets should reduce the plasma TRP/NAA ratio and 5-HT, and should shift diet selection toward a lower protein ration.

Peters and Harper (1985, 1987a) carried out a series of investigations of the effects of dietary protein content on the plasma TRP/LNAA ratio, brain TRP and brain 5-HT concentrations. In rats that were previously adapted to 20% casein diets, consumption of single meals of diets containing from 0 to 55% of casein led to increases in plasma TRP concentrations that were proportional to dietary protein level (i.e., TRP intake). The plasma TRP/LNAA ratio, however, remained unaffected by differences in dietary protein level across a wide range from 10% to 55% of casein. Likewise, brain TRP, 5-HT and 5-HIAA concentrations were unchanged over the range of increasing dietary protein levels. Feeding rats the protein-free diet on the other hand led to an elevation in the plasma TRP/LNAA ratio and brain 5-HT, despite causing a reduction in the absolute concentration of TRP in blood. This finding was consistent with the earlier work of Fernstrom and Wurtman (1971b). The results of these acute studies indicated that neither the plasma TRP/LNAA ratio nor brain 5-HT are likely to play an important role in directing protein intake or selection in rats when the dietary choices offered have protein contents within the range compatible with optimum growth. However, the data raise the possibility that elevations in brain TRP and 5-HT caused by consumption of a protein-free diet may act as a signal to alter food intake or selection.

The response of TRP and 5-HT to a protein-free, high-carbohydrate meal has been suggested to be primarily an effect of the carbohydrate on insulin secretion (Fernstrom and Wurtman, 1971b). According to this idea, the carbohydrate load induces the secretion of insulin which raises the plasma level of TRP/LNAA by stimulating uptake of the competing LNAA into muscle while having a relatively lesser effect on tissue uptake of TRP. If this effect were due entirely to the secretion of insulin per se, then one would predict that the TRP/LNAA ratio would also be elevated by a diet containing 5% or 10% protein since such a diet is still high in carbohydrate and would most likely lead to an equivalent stimulation of insulin secretion. In fact, diets containing substantial amounts of both protein and carbohydrate stimulate insulin release to a greater extent than do diets having carbohydrate alone (Spiller et al., 1987). Thus, as has been pointed out previously (Harper and Peters, 1989), the response of rats to meals lacking in protein would seem to be a response to protein deficiency rather than a response to the high level of carbohydrate. The potential of such a response to act as a signal to modify food intake or selection would then seem to make sense as it could function as a mechanism that would allow animals to avoid diets that are incompatible with survival.

It is noteworthy that the response of plasma TRP, the TRP/LNAA ratio and brain 5-HT are different when rats are fed different levels of protein for many days compared to the acute responses described above. When young rats were allowed to adapt to diets containing from 5% to 75% of casein for 11 days, plasma TRP did not rise in proportion to dietary protein level, but remained within a fairly narrow range when the diet contained between 15% and 75% of casein (Peters and Harper, 1985). Animals fed the 5% and 10% protein diets had lower plasma TRP concentrations reflecting the inadequacy of the diet to support rapid growth. The plasma TRP/LNAA ratio however, did show a relationship with dietary protein content and was significantly inversely correlated with dietary casein level. Furthermore, brain TRP, 5-HT and 5-HIAA showed a similar relationship, and although the absolute changes in their concentrations were small in comparison to the effects brought about by single meals, they declined in proportion to increasing dietary protein level.

An explanation for the apparently opposite effects of dietary protein content in the short and long term on plasma TRP, the TRP/LNAA ratio, brain TRP and 5-HT has been discussed in detail elsewhere (Harper and Peters, 1989). The different responses of animals fed increments of protein in the long-term compared to single meals of the same diets, results from adaptive responses in the activities of several amino acid degrading enzymes which increase over time in rats fed high protein diets. The activities of the enzymes responsible for degrading the BCAA do not increase with increasing dietary protein content to the same extent as do enzyme systems which dispose of other dietary indispensable amino acids (Harper et al., 1970, 1984). Thus, after several days of high protein diets, the plasma concentrations of most amino acids, including TRP, return to near normal owing to an increased capacity for their oxidation, but concentrations of the BCAA remain elevated. Because the BCAA quantitatively represent the largest fraction of plasma LNAA, the plasma TRP/LNAA ratio, brain TRP and 5-HT would be expected to decline as the protein content of the diet was increased.

The oxidative pathway via kynurenine

The oxidation of TRP via the kynurenine pathway is quantitatively the most significant route of TRP disposal in the body, accounting for 95% or more of daily TRP metabolism. Tryptophan catabolism via this route leads to a number of end products including the nicotinamide nucleotides and acetyl-CoA which can be further oxidized to CO_2 to yield energy. Furthermore, several intermediates in this pathway have been shown to play other important roles in nutrition and metabolism. Picolinic acid, the product of the non-enzymatic cyclization of aminomuconic semialdehyde, is involved in normal intestinal uptake of zinc (Evans and Johnson, 1980). Another pathway intermediate, quinolinic acid, can form a chelate with iron, which has been shown by Lardy and coworkers (Veneziale et al., 1967) to be involved in the physiological regulation of gluconeogenesis.

Whole-body TRP oxidation occurs almost exclusively in the liver (Miller, 1962) owing to the localization of TRP oxygenase in that tissue. There is however, some capacity in other visceral and peripheral tissues to degrade TRP. Hayaishi and coworkers (1980) have identified an indole oxygenase that is distributed widely throughout the body which can also oxidize TRP. This enzyme differs from the liver enzyme in that it uses superoxide anion instead of molecular oxygen as the oxidizing agent, and its substrate specificity is much broader than that of TRP oxygenase. Indole oxygenase will oxidize D-TRP as well as L-TRP and also demonstrates activity toward 5-hydroxytryptophan, tryptamine, and 5-HT.

Before TRP can be utilized by liver tissue, it must first gain entry into hepatic cells, a process influenced by extra-cellular factors. For example, the uptake of TRP by the liver, while not sensitive to competitive inhibition by other LNAA, appears to be influenced by the binding of TRP to serum albumin. Smith and Pogson (1980) have shown that TRP oxidation by isolated hepatocytes is significantly reduced when albumin is included in the incubation medium.

L-tryptophan-2,3-oxygenase (tryptophan oxygenase) is the first and rate controlling enzyme in TRP degradation and its activity regulates the overall flux of TRP through the oxidative pathway. Tryptophan oxygenase is somewhat unique among enzymes catalyzing degradation of essential amino acids in that both the amount and activity of the enzyme are controlled by its substrate, TRP. Knox and Mehler (1951) were the first to observe that the activity of rat liver TRP oxygenase was stimulated by TRP. Later, Greengard and Feigelson (1961) discovered that heme acts as a cofactor for the enzyme and is required for conversion of the apoenzyme to the active holoenzyme. In a series of investigations that followed, it was shown that TRP affects the activity of TRP oxygenase both by increasing heme saturation of the enzyme (thus increasing the level of active holoenzyme) as well as by stabilizing the enzyme against degradation (Schimke et al., 1965).

Tryptophan oxygenase is also responsive to various hormones, the most studied of which are the corticosteroids. Administration of hydrocortisone to rats induces liver TRP oxygenase activity (Schimke et al., 1964), an effect that has been shown to result from synthesis of new apoenzyme protein. Furthermore, induction of enzyme activity by glucocorticoids was shown to be additive with the effect of TRP (Schimke et al., 1964). Tryptophan oxygenase is also subject to feedback inhibition by many intermediates and products of the oxidative pathway, including NADH and NADPH (Badawy, 1977).

The product of TRP oxygenase, formylkynurenine, under conditions of normal flux of TRP through the oxidative pathway, is converted to kynurenine which is metabolized further to acroleyl aminofumarate. An important branch point occurs at this step in the pathway which commits the carbon skeleton either to complete oxidation to CO_2 or to synthesis of niacin and NAD. The rate limiting step in the branch of the pathway leading to complete oxidation is catalyzed by picolinic carboxylase, while the alternate pathway leading to NAD involves the non-enzymatic cyclization of acroleyl aminofumarate to yield quinolinic acid. The partitioning of acroleyl aminofumarate between the two possible metabolic routes appears to depend on the substrate supply in relation to the K_m and capacity of the carboxylase. Thus, significant conversion of TRP to niacin only occurs when the capacity of picolinic carboxylase becomes limiting (Bender, 1982). One of the most well known species differences in the TRP to niacin conversion is in the cat, in which picolinic carboxylase has both a high activity and high capacity to metabolize acroleyl aminofumarate. In this species, virtually no conversion of TRP to niacin takes place (Ikeda et al., 1965).

The efficiency of conversion of TRP to niacin has been a subject of several investigations. Horwitt et al. (1956) originally reported a conversion efficiency of 60:1 in young men, meaning that ingestion of 60 mg of TRP was required for each 1 mg of niacin formed. Subsequent work has shown values for the efficiency of conversion to be as high as 122:1, depending on the amount of TRP provided by the diet (Nakagawa et al., 1969). In a recent study of young adult men, Patterson et al. (1980) showed that when TRP intake was increased from 245 to 845 mg/day, urinary N^1-methyl-nicotinamide excretion increased from 5.4 to 17.1 μmole/24 hours while N^1-methyl-2-pyridone-5-carboxamide output increased from 11.7 to 38.6 μmole/24 hours. Based on the urinary excretion of these niacin metabolites, the authors calculated an efficiency of conversion of TRP to niacin which averaged 72:1, although there was a trend toward increased conversion efficiency at the higher TRP intakes. In general, it appears that conversion of TRP to niacin depends on the surplus of dietary TRP provided in excess of the body's needs for maintenance of nitrogen equilibrium and for the synthesis of other important molecules such as 5-HT (Bender, 1982).

INTEGRATION AND REGULATION OF WHOLE-BODY TRYPTOPHAN UTILIZATION

The utilization of TRP by the whole-body, like the utilization of other amino acids, is subject to relatively few points of regulation (Harper, 1974). Unlike certain other essential dietary nutrients, such as calcium or iron whose whole body utilization is regulated primarily at the level of absorption, there is little regulation of amino acid metabolism exerted at the level of amino acid absorption from the intestine. The digestibility of most naturally occurring proteins is high and amino acid uptake by the gut is nearly quantitative. Excretion of amino acids, including TRP, by the kidney is also not a significant route of amino acid disposal owing to the efficient reabsorption of amino acids in the renal tubule. Amino acid excretion by the kidney can occur, but only when blood amino acid levels are extremely high such as might be encountered in some inborn errors of metabolism or other pathological states. In the young animal, utilization of TRP and other amino acids for protein synthesis is substantial, but in mature animals, there is no net increase in body protein or free amino acid contents. Also, unlike the situation for carbohydrate and fat, there is no storage pool in the body for amino acids that are ingested in excess of immediate needs. Therefore, even in the young animal, protein synthesis, while sensitive to total amino acid supply, is not a major site of regulation of amino acid disposal. The low Km's of the amino acid synthetases insure that amino acid needs for protein synthesis are met even under conditions of low amino acid intake. Thus, in both growing and mature animals, once the body's demand for amino acids to supply protein synthesis or for the synthesis of other biologically important molecules such as 5-HT have been met, the surplus amino acids are oxidized to provide energy or are converted to fat.

It is evident then, that the primary mechanism by which utilization of amino acids is regulated is at the level of amino acid catabolism. The flux of TRP and other essential amino acids through their respective degradative pathways is determined largely by the characteristics of the various enzymes involved in their metabolism (Krebs, 1982), and by factors that determine the accessibility of the various amino acids to the enzymatic machinery. The availability of substrates for metabolism can be influenced by interorgan relationships in the transport of substrates and by factors that modify uptake of substrates into the tissues in which the appropriate enzyme systems are located. In the case of TRP, the liver is the major organ responsible for its degradation, which means that hepatic TRP disposal must be responsive to fluctuations in dietary TRP supply. The localization of TRP degrading enzymes in liver also means that hepatic TRP disposal is an important regulator of TRP supply to the rest of the body.

The rate controlling enzyme in TRP catabolism, TRP oxygenase, is not saturated with its substrate under normal conditions (Krebs, 1972). Because the K_m of the enzyme is considerably above the normal tissue TRP concentration, TRP degradation will increase with increasing availability of substrate. The supply of TRP in turn is determined by intake of TRP from the diet, and the extent to which TRP is bound to serum albumin. Liver extraction and degradation of TRP will determine the rise in peripheral blood TRP concentration after a meal which determines the availability of TRP to extrahepatic tissues. As pointed out previously, the extent to which peripheral blood TRP concentration increases after a protein-containing meal depends not only on the amount of TRP ingested, but also on the adaptive metabolic state of the animal (i.e., the capacity to degrade surplus amino acids).

The K_m of the LNAA transport system in muscle is considerably higher than the average plasma concentrations of the neutral amino acids. Thus, uptake into this tissue is relatively insensitive to substrate competition effects and should be determined primarily by the plasma TRP concentration relative to the Km of the muscle transporter for TRP. As discussed previously, the situation for brain is different because the K_m of the neutral amino acid transport carrier for TRP and the other LNAA is low, and approximates the normal plasma concentration of TRP (Pardridge, 1977). Therefore, uptake of TRP into brain, which regulates 5-HT synthesis, is influenced by the concentrations in blood of the other LNAA.

Because there are significant differences in the location and extent of metabolism of the LNAA as a function of dietary amino acid supply, the conversion of TRP to serotonin is influenced by the interaction between different organ systems (esp. muscle and liver) in the metabolism of TRP and the competing LNAA. Furthermore, the binding of TRP to serum albumin offers an additional mechanism by which TRP availability to tissues can be regulated.

The possible survival advantage that albumin binding of TRP may confer upon the animal is not readily apparent. At low TRP intakes, binding of TRP to albumin may protect TRP from hepatic catabolism, insuring an adequate supply to peripheral tissues. In the periphery, the low K_m for TRP uptake into brain compared to the K_m for transport into muscle would help to insure adequate TRP for normal 5-HT synthesis and hence normal brain function. The albumin-bound pool of TRP may therefore serve as a buffer which protects the brain from TRP and 5-HT depletion when dietary intake of TRP is low.

Under normal conditions, when animals are maintained on relatively low protein diets, the activities of most amino acid degrading enzymes are low, although activities are sufficient to dispose of amino acids in excess of requirements. However, when the protein content of the diet is very high, or when individual indispensable amino acids are given individually in large amounts above the requirement, the capacity of the animal to catabolize the surplus amino acids may be exceeded, and those amino acids present in excess accumulate in blood and tissues (Harper et al., 1970). When animals are offered only a single diet of fixed composition, accumulation of amino acids, including TRP, in body fluids and tissues is associated with depressed food intake. Alternatively, if animals are allowed to self-select their diet, elevated tissue amino acid levels lead to altered diet preference, such that animals choose a diet that will restore amino acid concentrations toward normal. These responses describe an additional mechanism by which animals can regulate whole-body amino acid metabolism; regulation through alterations in food intake and diet selection (Harper, 1974).

If animals are fed high protein diets for several days or longer, activities of several amino acid catabolizing enzymes increase, including those for degrading TRP. Increased catabolic capacity favors a reduction in the blood concentrations of amino acids toward normal, and food intake is restored to near control levels (Anderson et al., 1968). The ability of amino acid catabolic enzymes to adjust to amino acid intake over time allows the animal to maintain plasma and tissue amino acid concentrations within an acceptable range despite wide fluctuations in dietary protein content. However, under some circumstances the capacity of these systems may still be exceeded and plasma and tissue amino acid concentrations would remain elevated. In this situation, food intake would remain depressed and normal function and possibly survival would become threatened.

It is clear that modification of food intake is a mechanism involved in the regulation of metabolism of all of the indispensable amino acids, and is not unique for TRP (Harper et al., 1970). The mechanism by which changes in feeding behavior are brought about by alterations in amino acid intake and metabolism is however, not known. The question of specificity of the relationship between TRP metabolism and utilization and food intake is complicated by the fact that TRP (unlike most other dietary amino acids) can be converted to 5-HT, a neurotransmitter involved in the central nervous system control of feeding behavior (Blundell, 1977). Serious doubts have been raised as to whether diet-induced changes in brain TRP and 5-HT concentrations play any significant role in the control of normal food intake and selection (Peters and Harper, 1987b). Further work in this area is needed to identify those conditions under which feeding behavior might be specifically controlled by diet-induced alterations in brain TRP supply and 5-HT synthesis, and whether such a mechanism might play a more general role in regulating whole-body amino acid metabolism.

REFERENCES

Anderson, G.H., 1979, Control of protein and energy intake: role of plasma amino acids and brain neurotransmitters, Can J. Physiol. Pharmacol., 57:1043-1057.

Anderson, H.L., Beneverga, N.J., and Harper, A.E., 1968, Associations among food and protein intake, serine dehydratase, and plasma amino acids, Am. J. Physiol., 214:1008-1013.

Ashley, D.V.M., and Anderson, G.H., 1975, Correlation between the plasma tryptophan to neutral amino acid ratio and protein intake in the self-selecting weanling rat, J. Nutr., 105:1412-1421.

Badawy, A.A.-B., 1977, The functions and regulation of tryptophan pyrrolase, Life Sci., 21:755-768.

Bender, D.A., 1982, Biochemistry of tryptophan in health and disease, Molec. Aspects Med., 6:101-197.

Blazek, R., and Shaw, D.M., 1978, Tryptophan availability and brain protein synthesis, Proc. Brit. Assoc. Psychopharmacol., 17:1065-1068.

Block, R.J., and Weiss, K.W., 1956, "Amino Acid Handbook", The Ryerson Press, Toronto.

Bloxam, D.L., Tricklebank, M.D., Patel, A.J., and Curzon, G., 1980, Effects of albumin, amino acids, and clofibrate on the uptake of tryptophan by the rat brain, J. Neurochem., 34:43-49.

Blundell, J.E., 1977, Is there a role for serotonin (5-hydroxytryptamine) in feeding?, Int. J. Obes., 1:15-42.

Bosnan, T., 1978, Serotonin metabolism, in: "Serotonin in Health and Disease", Vol. 1, Essman, W.B., ed., SP Medical and Scientific Books, New York, pp. 181-300.

Evans, G.W., and Johnson, E.C., 1980, Zinc absorption in rats fed a low protein diet and a low protein diet supplemented with tryptophan or picolinic acid, J. Nutr., 110:1076-1080.

Fernstrom, J.D., 1981, Dietary precursors and brain neurotransmitter formation, Ann. Rev. Med., 32:413-425.

Fernstrom, J.D., and Faller, D.V., 1978, Neutral amino acids in the brain: Changes in response to food ingestion, J. Neurochem., 30:1531-1538.

Fernstrom, J.D., Hirsch, M.J., Madras, B.K., and Sudarsky, L., 1975, Effects of skim milk, whole milk and light cream on serum tryptophan binding and brain tryptophan concentrations, J. Nutr., 105:1359-1362.

Fernstrom, J.D., and Wurtman, R.J., 1971a, Brain serotonin content: Physiological dependence on plasma tryptophan levels, Science, 173:149-152.

Fernstrom, J.D., and Wurtman, R.J., 1971b, Brain serotonin content: Increase following ingestion of a carbohydrate diet, Science, 174:1023-1025.

Fernstrom, J.D., and Wurtman, R.J., 1972, Brain serotonin content: Physiological regulation by plasma neutral amino acids, Science, 178:414-416.

Gál, E.M., and Drewes, P.A., 1962, Studies on the metabolism of 5-hydroxy-tryptamine (serotonin). II. Effect of tryptophan deficiency in rats, Proc. Soc. Exptl. Biol. Med., 110:368-371.

Greengard, O., and Feigelson, P., 1961, The activation and induction of rat liver tryptophan pyrrolase in vivo by its substrate, J. Biol. Chem., 236:158-161.

Hayaishi, O., 1980, Newer aspects of tryptophan metabolism, in: "Biochemical and Medical Aspects of Tryptophan Metabolism", Hayaishi, O., Ishimura, Y., and Kido, R., eds., Elsevier/North-Holland, Amsterdam, pp. 15-30.

Harper, A.E., 1974, Control mechanisms in amino acid metabolism, in: "The Control of Metabolism", Sink, J.D., ed., The Pennsylvania State University Press, University Park, pp. 49-71.

Harper, A.E., 1977, Human amino acid and nitrogen requirements as the basis for evaluation of nutritional quality of protein, in: "Food Proteins", Whitaker, J.R., and Tannenbaum, S.R., eds., Avi Publishing Co., Inc., Westport, pp. 363-386.

Harper, A.E., Benevenga, N.J., and Wohlhueter, R.M., 1970, Effects of ingestion of disproportionate amounts of amino acids, Physiol. Rev., 50: 428-558.

Harper, A.E., Miller, R.H., and Block, K.P., 1984, Branched chain amino acid metabolism, Ann. Rev. Nutr., 4:409-454.

Harper, A.E., and Peters, J.C., 1989, Protein intake, brain amino acids and serotonin and protein self-selection, J. Nutr., 119:677-689.

Horwitt, M.K., Harvey, C.C., Rothwell, W.S., Cutler, J.L., and Haffron, D., 1956, Tryptophan-niacin relationship in man, J. Nutr., 60:1-43.

Ikeda, M., Tsuji, H., Nakamura, S., Ichiyama, A., Nishizuka, Y., and Hayaishi, O., 1965, Studies on the biosynthesis of nicotinamide adenine dinucleotide: (ii) a role of picolinic carboxylase in the biosynthesis of nicotinamide adenine dinucleotide from tryptophan in mammals, J. Biol. Chem., 240:1395-1401.

Ip, C.Y., and Harper, A.E., 1974, Liver polysome profiles and protein synthesis in rats fed a threonine-imbalanced diet, J. Nutr., 104:252-263.

Kaufman, S., 1974, Properties of pterin-dependent aromatic amino acid hydroxylases, in: "Aromatic Amino Acids in the Brain", Wolstenholme, G.E.W., and Fitzsimons, D.W., eds., Elsevier/North-Holland, Amsterdam, pp. 85-108.

Knox, W.E., and Mehler, A.H., 1951, The adaptive increase of the tryptophan peroxidase-oxidase system of liver, Science, 113:237-240.

Krebs, H.A., 1972, Some aspects of the regulation of fuel supply in omnivorous animals, Adv. Enz. Regul., 10:397-420.

Lin, F.D., Smith, T.K., and Bayley, H.S., 1988, A role for tryptophan in regulation of protein synthesis in porcine muscle, J. Nutr., 118:445-449.

McMenamy, R.H., 1965, The binding of indole analogues to human serum albumin: effects of fatty acids, J. Biol. Chem., 240:4235-4243.

McMenamy, R.H., and Oncley, J.L., 1958, The specific binding of L-tryptophan to serum albumin, J. Biol. Chem., 233:1436-1447.

Miller, L.L., 1962, The role of liver and the non-hepatic tissues in the
regulation of free amino acid levels in the blood, in: "Amino Acid Pools",
Holden, J.T., ed., Elsevier, New York, pp. 708-721.

Munro, H.N., 1970, Free amino acid pools and their role in regulation, in:
"Mammalian Protein Metabolism", Vol. 4, Munro, H.N., ed., Academic Press,
New York, pp. 299-386.

Munro, H.N., Hubert, C., and Baliga, B.S., 1975, Regulation of protein syn-
thesis in relation to amino acid supply - a review, in: "Alcohol and
Abnormal Protein Biosynthesis", Rothschild, M.A., Oratz, M., and
Schreiber, S.S., eds., Pergamon Press, New York, pp. 33-66.

Nakagawa, I., Takahashi, T., Suzuki, T., and Masana, Y., 1969, Effect in man
of the addition of tryptophan or niacin to the diet on the excretion of
their metabolites, J. Nutr., 99:325-330.

Osborne, T.B., and Mendel, L.B., 1914, Amino acids in nutrition and growth,
J. Biol. Chem., 17:325-349.

Pardridge, W.M., 1977, Kinetics of competitive inhibition of neutral amino
acid transport across the blood-brain barrier, J. Neurochem., 28:103-108.

Patterson, J.I., Brown, R.R., Linkswiler, H., and Harper, A.E., 1980, Excretion of tryptophan-niacin metabolites by young men: effects of tryptophan,
leucine, and vitamin B_6 intakes, Am. J. Clin. Nutr., 33:2157-2167.

Peters, J.C., and Harper, A.E., 1985, Adaption of rats to diets containing
different levels of protein: effects on food intake, plasma and brain
amino acid concentrations and brain neurotransmitter metabolism, J. Nutr.,
115:382-398.

Peters, J.C., and Harper, A.E., 1987a, Acute effects of dietary protein on
food intake, tissue amino acids, and brain serotonin, Am. J. Physiol.,
252:R902-R914.

Peters, J.C., and Harper, A.E., 1987b, A skeptical view of the role of central serotonin in the selection and intake of protein, Appetite, 8:206-
210.

Pronczuk, A.W., Baliga, B.S., Triant, J.W., and Munro, H.N., 1968, Comparison of the effect of amino acid supply on hepatic polysome profiles in
vivo and in vitro, Biochim. Biophys. Acta, 157:204-206.

Pronczuk, A.W., Rogers, Q.R., and Munro, H.N., 1970, Liver polysome patterns
of rats fed amino acid imbalanced diets, J. Nutr., 100:1249-1258.

Rogers, Q.R., 1976, The nutritional and metabolic effects of amino acid
imbalances, in: "Protein Metabolism and Nutrition", Cole, D.J.A., ed.,
Butterworths, London.

Rose, W.C., 1957, The amino acid requirements of adult man, Nutr. Abstr.
Rev., 27:631-647.

Schimke, R.T., Sweeney, E.W., and Berlin, C.M., 1964, An analysis of the
kinetics of rat liver tryptophan pyrrolase induction: the significance of
both enzyme synthesis and degradation, Biochem. Biophys. Res. Comm.,
15:214-219.

Schimke, R.T., Sweeney, E.W., and Berlin, C.M., 1965, The roles of synthesis
and degradation in the control of rat liver tryptophan pyrrolase, J. Biol.
Chem., 240:322-331.

Sidransky, H., Murty, C.N., and Verney, E., 1984, Nutritional control of protein synthesis: studies relating to tryptophan-induced stimulation of
nucleocytoplasmic translocation of mRNA in rat liver, Am. J. Pathol.,
117:298-309.

Sidransky, H., Verney, E., and Sarma, D.S.R., 1971, Effect of tryptophan on
polyribosomes and protein synthesis in liver, Am. J. Clin. Nutr., 24:779-
785.

Smith, S.A., and Pogson, C.I., 1980, The metabolism of L-tryptophan by isolated rat liver cells: effect of albumin binding and amino acid competition on oxidation of tryptophan by tryptophan 2,3-dioxygenase, Biochem.
J., 186:977-986.

Spiller, G.A., Jensen, C.D., Patterson, T.S., Chuck, C.S., Whittam, J.H.,
and Scala, J., 1987, Effect of protein dose on serum glucose and insulin
response to sugars, Am. J. Clin. Nutr., 46:474-480.

Sved, A.F., 1983, Precursor control of the function of monoaminergic neurons, in: "Nutrition and the Brain", Vol. 6, Wurtman, R.J., and Wurtman, J.J., eds., Raven Press, New York, pp. 223-275.

Udenfriend, S., Titus, E., and Weissbach, H., 1955, The identification of 5-hydroxy-3-indoleacetic acid in normal urine and a method for its assay, J. Biol. Chem., 216:499-505.

Veneziale, C.M., Walter, P., Kneer, N., and Lardy, H.A., 1967, Influence of L-tryptophan and its metabolites on gluconeogenesis in the isolated perfused liver, Biochemistry, 6:2129-2138.

Willcock, E.G., and Hopkins, F.G., 1906, The importance of individual amino acids in metabolism; observations on the effect of adding tryptophan to a diet in which zein is the sole nitrogenous constituent, J. Physiol. (London), 35:88-102.

Young, V.R., Munro, H.N., Matthews, D.E., and Bier, D.M., 1983, Relationship in energy metabolism to protein metabolism, in: "New Aspects of Clinical Nutrition", Kleinberger, G., and Deutsch, E., eds., Karger, Basel, pp. 44-73.

COMPARISON OF TRYPTOPHAN METABOLISM IN VIVO AND IN ISOLATED HEPATOCYTES FROM

VITAMIN B6 DEFICIENT MICE

D.A. Bender, E.N.M. Njagi, and P.S. Danielian

Department of Biochemistry
University College London
London WC1E 6BT
UK

Tryptophan dioxygenase meets the traditional criteria for the rate-limiting enzyme of the kynurenine pathway of tryptophan metabolism. From the concentrations of substrates in the liver and published values of K_m and V_{max} (Bender et al., 1975; Bender and McCreanor, 1985; Takikawa et al., 1986), the steady-state rates of activity can be calculated. As shown in Table 1, such calculations show that tryptophan dioxygenase has the lowest activity of the pathway under basal conditions. Furthermore, it is regulated by a variety of mechanisms, including: induction of mRNA synthesis by glucocorticoid hormones; increased translation of mRNA by glucagon; stabilization of the enzyme protein against catabolism by both tryptophan and the heme cofactor; and feedback inhibition and repression by the reduced nicotinamide nucleotide coenzymes NADH and NADPH.

Although tryptophan dioxygenase has a lower activity than either of the following two enzymes, kynurenine hydroxylase and kynureninase, both kynurenine and hydroxykynurenine accumulate in tissues and blood and are excreted in the urine. This suggests that these two subsequent steps in the pathway may, at least under some conditions, provide secondary rate-limiting steps in tryptophan metabolism. Knox (1953) noted that the induction of tryptophan dioxygenase by glucocorticoids, or stabilization by loading doses of tryptophan, increases its activity to such an extent that kynureninase now has the lowest activity in the pathway. Bender and McCreanor (1985) suggested that kynurenine hydroxylase has a lower activity than kynureninase, and may also be rate-limiting under such conditions. Table 2 shows that the administration of a loading dose of tryptophan results in a very considerable increase in the activity of tryptophan dioxygenase, so that kynureninase or kynurenine hydroxylase may indeed become rate-limiting.

As shown in Table 3, treatment of ovariectomized rats with estrone sulphate results in a sufficient reduction in the activity of kynurenine hydroxylase for the V_{max} of this step to become the lowest in the pathway (Bender and Totoe, 1984). Similarly, estrogen conjugates are competitive inhibitors of kynureninase (Bender and Wynick, 1981). As a result of these two effects, treatment with estrogens results in considerably increased urinary excretion of kynurenine and hydroxykynurenine, as well as kynurenic acid and xanthurenic acid, which are normally minor side-metabolites (Rose and Braidman, 1971).

As shown in Table 4, the administration of a test dose of tryptophan to

Table 1. Activities of enzymes of the tryptophan oxidative pathway
          rat liver

	V_{max} (nmol/min)	[Substrate] (nmol/g)	Steady state (nmol/min)
Tryptophan dioxygenase	10.0	6.7	0.32
Kynurenine hydroxylase	22.9	3.6	0.96
Kynureninase	33.1	4.9	1.30

Enzyme activities (expressed per g wet weight) were assayed in vitro
under V_{max} conditions. Steady state concentrations of substrates in
rat liver and calculated steady state activity are based on published
K_m values. (Modified from: Bender et al., 1975; Bender and McCreanor,
1985; Takikawa et al., 1986).

human beings results in a considerable increase in the excretion of both kynurenine and hydroxykynurenine. Urinary kynurenine increases more markedly
than hydroxykynurenine, so that the ratio of kynurenine:hydroxykynurenine
rises from approximately 0.3 under basal conditions to 1.3-1.6 after a tryptophan load (Michael et al., 1964). Treatment with cortisol results in a
further increase in the excretion of kynurenine and hydroxykynurenine, and
again kynurenine increases more than hydroxykynurenine (Rose and McGinty,
1968). This suggests that the activity of kynurenine hydroxylase is more
limiting than that of kynureninase under these conditions.

Further evidence that kynurenine hydroxylase may be rate-limiting in
vivo comes from data reported by El-Zoghby et al. (1978). Table 4 shows the
basal excretion of kynurenine and hydroxykynurenine in pre-pubertal girls,
mature and post-menopausal women. The ratio of urinary kynurenine:hydroxy-
kynurenine rises sharply in sexually mature women, then falls post-menopaus-
ally. This suggests that endogenous estrogens impair the activity of kynurenine hydroxylase in the same way as do administered estrogens. It is noteworthy that in areas where tryptophan and niacin intake is marginally inadequate, women show a two-fold greater incidence of pellagra than do men; in
rats administration of estrogens results in impairment of nicotinamide nucleotide synthesis from tryptophan, and lower tissue concentrations of NAD(P)
(Bender and Totoe, 1984).

Kynureninase is a pyridoxal phosphate (vitamin B_6) dependent enzyme, and

Table 2. Activity of tryptophan dioxygenase in rat liver after
          administration of 500 mg tryptophan/kg in intact and
          adrenalectomized rats

	Control	Tryptophan load
Intact animals	14.6 ± 2.25	134 ± 24.1
Adrenalectomized	6.8 ± 0.92	67 ± 10.9

Data are expressed in nmol kynurenine formed/h/g liver and are
the mean ± SEM of 4 animals per group (Bender et al., 1983).

Table 3. Activities of enzymes of the tryptophan oxidative pathway in liver homogenates from ovariectomized rats under basal conditions and after the administration of 3 mg/kg diet estrone sulphate for 7 days

	Control	Estrone
Tryptophan dioxygenase	8.8 ± 4.2	7.2 ± 3.9
Kynurenine hydroxylase	22.9 ± 6.45	6.7 ± 0.78[*]
Kynureninase	33.1 ± 0.65	25.9 ± 1.15[*]
3-Hydroxyanthranilate oxygenase	113 ± 3.2	102 ± 3.3[*]

Data are expressed as nmol product formed/min/g liver and show the mean V_{max} ± SEM for 4-8 animals. [*]$P < 0.001$ (t-test). (Modified from Bender et al., 1983 and Bender and Totoe, 1984).

vitamin B_6 deficiency results in severely impaired activity. Deficient people and experimental animals excrete higher than normal amounts of kynurenic and xanthurenic acids, kynurenine and hydroxykynurenine, both under basal conditions and, much more markedly, after a loading dose of tryptophan. The excretion of kynurenic and xanthurenic acids after a dose of tryptophan became a standard method for assessing vitamin B_6 nutritional status (Lepkovsky and Nielson, 1942; Coursin, 1964; Allegri et al., 1978). However, the

Table 4. Urinary excretion of kynurenine (Kyn) and 3-hydroxykynurenine (3HK) by human beings under various conditions

		Kyn	3HK	Ratio Kyn/3HK	Reference
Children	basal	8	36	0.29	(a)
	+ 100 mg/kg Trp	132	98	1.35	
Adult men	basal	22	78	0.28	
	+ 100 mg/kg Trp	643	472	1.26	
Adult women	basal	17	69	0.24	
	+ 100 mg/kg Trp	1351	835	1.62	
Adult men	5 g Trp load	108	103	1.03	(b)
	+ cortisol	469	237	2.04	
Girls 8-12	basal	6.6	12.7	0.52	(c)
Women 23-40	basal	5.1	1.1	4.66	
Women 50-65	basal	4.6	3.2	1.40	

Data represent μmol Kyn or 3HK excreted over a 24 hour period. Modified from data reported by (a) Michael et al., 1964, (b) Rose and McGinty, 1968, and (c) El-Zoghby et al., 1978.

validity of the tryptophan load test as an index of vitamin B_6 nutritional status has been discredited by a number of studies. Women taking estrogens as oral contraceptives show abnormalities of tryptophan metabolism (Rose and Braidman, 1971), although other indices of vitamin B_6 nutritional status are unaffected (Bender, 1987). Estrogen conjugates inhibit kynureninase and estrogens impair kynurenine hydroxylase (Bender and Wynick, 1981; Bender and Totoe, 1984). It is most probable that the abnormalities of tryptophan metabolism are due to these effects. Coon and Nagler (1969) demonstrated abnormal tryptophan metabolism in a large number of patients with widely differing clinical conditions, whose only common feature was that they were sick, under stress, and hence secreting more cortisol than normal. This would increase tryptophan dioxygenase activity. Again, the results suggest that kynureninase and kynurenine hydroxylase can provide rate-limiting steps in tryptophan metabolism.

A more sophisticated approach to the identification of which enzymes are important in metabolic regulation comes from the determination of Control Coefficients by flux analysis. As shown in Table 5, studies of the metabolism of [^{14}C]tryptophan in isolated hepatocytes show that the Control Coefficients of kynureninase (< 0.004) and kynurenine hydroxylase (< 0.037) are negligible. Under basal conditions the Control Coefficient of tryptophan dioxygenase is most important, while after the induction of the enzyme by cortisol it is the uptake of tryptophan into the cells which is most important. It is only in vitamin B_6 deficiency, when the activity of kynureninase is severely impaired, that this enzyme has a significant Control Coefficient (Salter and Pogson, 1985; Salter et al., 1986).

There is thus an obvious discrepancy between in vivo studies and the results of flux analysis in isolated hepatocytes. In vivo, the induction of tryptophan dioxygenase by glucocorticoid hormones; the inhibition of kynureninase by estrogen conjugates (Bender and Wynick, 1981); the impairment of kynurenine hydroxylation by estrogens (Bender and Totoe, 1984; Bender and McCreanor, 1985) and the inhibition of kynureninase by leucine and of kynurenine hydroxylase by 2-oxo-isocaproate (Magboul and Bender, 1983), all result in increased urinary excretion of kynurenine and its metabolites. Estrogen administration and feeding a high leucine diet also result in decreased production of $^{14}CO_2$ from [ring-2-^{14}C]tryptophan in vivo (Bender, 1983a,b). These results all suggest that in intact animals kynureninase and/or kynurenine hydroxylase are rate-limiting, despite their negligible Control Coefficients as determined in isolated hepatocytes (Salter and Pogson, 1985; Salter et al., 1986).

Table 5. Control Coefficients of tryptophan metabolism in rat hepatocytes isolated from control animals and after induction of tryptophan dioxygenase by cortisol

	Basal	+ Cortisol
Tryptophan transport	0.25	0.75
Tryptophan dioxygenase	0.75	0.25
Kynureninase	< 0.004	
Kynurenine hydroxylase	< 0.037	

Data are taken from Salter et al. (1986).

In order to investigate the discrepancy between studies of metabolic flux in isolated cells and studies in vivo, the metabolism of tryptophan has been assessed in vitamin B_6 deficient mice by measuring both the formation of $^{14}CO_2$ from [^{14}C]tryptophan and the excretion of tryptophan metabolites, and in hepatocytes from the same animals, by the metabolism of [^{14}C]tryptophan and the synthesis of tryptophan metabolites.

METHODS

Male mice were maintained for 4 weeks from weaning on a vitamin B_6 free diet or the same diet supplemented with 5 mg pyridoxine/kg (Symes et al., 1984). Urine was collected for 24 h for measurement of tryptophan metabolites: kynurenine and methyl pyridone carboxamide colorimetrically; kynurenic and xanthurenic acids, hydroxykynurenine and N^1-methyl nicotinamide fluorimetrically (Bender and Olufunwa, 1988).

The ability to metabolize [^{14}C]tryptophan was assessed by the formation of $^{14}CO_2$ after intraperitoneal injection of a tracer dose of [ring-2-^{14}C]-, [methylene-^{14}C]- or [U-^{14}C]tryptophan, undiluted by non-radioactive amino acid. Exhaled air was passed through 2-methoxy-methylamine to trap $^{14}CO_2$ for scintillation counting; the trapping agent was changed at 10 min intervals for 2 h.

Hepatocytes were prepared by perfusion with collagenase. Before the perfusion commenced, a sample of blood was withdrawn from the inferior vena cava for measurement of plasma tryptophan, and the red cell aspartate aminotransferase activation coefficient as an index of vitamin B_6 nutritional status. Hepatocytes were incubated in Krebs-Ringer phosphate-bicarbonate buffer containing glucose, lactate, pyruvate and glutamine (Salter and Pogson, 1985) and either [^{14}C]tryptophan or non-radioactive tryptophan over the range 0.06-8 mM (Bender and Olufunwa, 1988). In incubations using non-radioactive tryptophan, the formation of NAD(P) was measured in the cell pellet; N^1-methyl nicotinamide and total niacin (nicotinamide plus nicotinic acid) in the incubation medium; and kynurenine, hydroxykynurenine and xanthurenic acid in the cells plus incubation medium, as described previously (Bender and Olufunwa, 1988). Incubations were performed using [ring-2-^{14}C]-, [methylene-^{14}C]- and [U-^{14}C]tryptophan; total metabolism was determined by the formation of $^{14}CO_2$ and non-aromatic tryptophan metabolites.

RESULTS

The vitamin B_6 deficient animals grew more slowly than the controls, although since, by chance, the controls were initially lighter than those fed the vitamin B_6 free diet, the final weights of the two groups were not significantly different (control 30.80 ± 0.73 g; deficient 28.60 ± 1.00 g). There was a significant deficiency of vitamin B_6 in the experimental group, with an erythrocyte aspartate aminotransferase activation coefficient of 1.44 ± 0.10 compared with the control value of 1.09 ± 0.07. The plasma concentration of tryptophan was slightly, but not significantly, higher in the deficient animals (76.40 ± 8.54 $\mu mol/l$) than in the controls (72.70 ± 5.27 $\mu mol/l$).

As shown in Table 6, in the mice maintained for 4 weeks on the vitamin B_6 free diet there was the expected increase in the basal urinary excretion of kynurenine, hydroxykynurenine and xanthurenic acid (without a loading dose of tryptophan), and reduction in N^1-methyl nicotinamide and methyl pyridone carboxamide.

As shown in Fig. 1, the production of $^{14}CO_2$ from [methylene-^{14}C]- and [U-

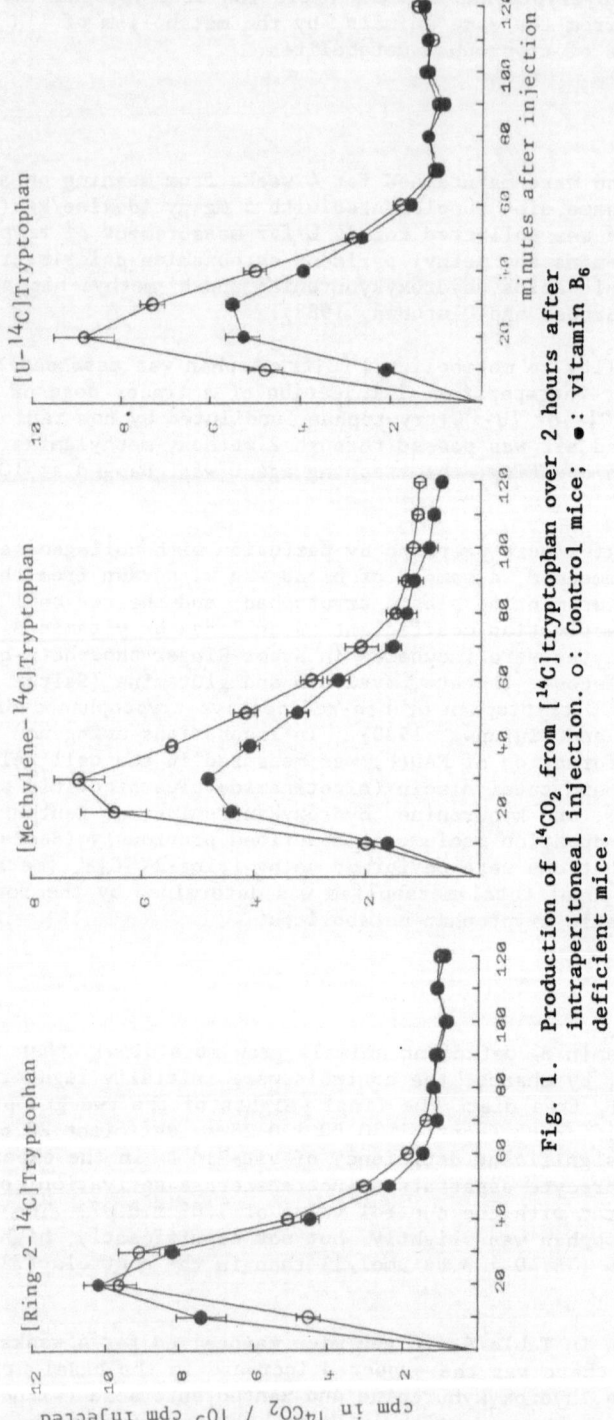

Fig. 1. Production of $^{14}CO_2$ from [^{14}C]tryptophan over 2 hours after intraperitoneal injection. o: Control mice; ●: vitamin B_6 deficient mice.

Table 6. Basal urinary excretion of tryptophan metabolites by control and vitamin B_6 deficient mice

	Control	B_6 deficient
Kynurenine	0.81 ± 0.058	0.96 ± 0.081
Hydroxykynurenine	0.072 ± 0.005	0.158 ± 0.009[*]
Kynurenic acid	135 ± 10.4	128 ± 11.0
Xanthurenic acid	50 ± 3.9	93 ± 17.4[*]
N-methyl nicotinamide	0.24 ± 0.029	0.16 ± 0.026[*]
Methyl pyridone carboxamide	3.13 ± 0.393	0.58 ± 0.148[*]

Data (μmol excreted over a 24 hour period) are the mean ± SEM of 5 mice per group. [*]$P < 0.001$ (t-test).

^{14}C]tryptophan in vivo was also reduced in deficient animals. The production of $^{14}CO_2$ from [ring-2-^{14}C]tryptophan was not significantly affected by vitamin B_6 status.

As shown in Fig. 2, the incubation of hepatocytes with [^{14}C]tryptophan showed slightly increased metabolism of all three positional isomers in cells from deficient animals, as assessed by the sum of $^{14}CO_2$ and non-aromatic tryptophan metabolites. This was significant by two-way analysis of variance only for [ring-2-^{14}C]tryptophan ($P < 0.083$).

As shown in Fig. 3, hepatocytes from vitamin B_6 deficient mice, showed increased formation not only of kynurenine, hydroxykynurenine and xanthurenic acid, but also of NAD(P), niacin and N^1-methyl nicotinamide.

DISCUSSION

The vitamin B_6 deficient mice showed the expected elevation of urinary kynurenine, hydroxykynurenine and xanthurenic acid, even in the absence of a loading dose of tryptophan. This was accompanied by reduced excretion of N-methyl nicotinamide and methyl pyridone carboxamide. These results, and the reduced production of $^{14}CO_2$ from [methylene-^{14}C]- and [U-^{14}C]tryptophan, are compatible with impairment of the activity of kynureninase, and impaired synthesis of NAD(P).

The results of isolated hepatocytes are at variance with those in the intact animals. There was an increase in the metabolism of all three positional isomers of [^{14}C]tryptophan in cells isolated from the vitamin B_6 deficient mice, with no evidence of impairment of kynureninase. While there was increased formation of kynurenine, hydroxykynurenine and xanthurenic acid in cells from vitamin B_6 deficient animals, they also formed more NAD(P), niacin and N^1-methyl nicotinamide than did the cells from control animals.

The results in isolated cells suggest that there was increased entry of tryptophan into the oxidative pathway in cells from deficient mice. This could be the result either of an increase in the activity of the tryptophan transport mechanism in the cells from vitamin B_6 deficient animals, or an increase in tryptophan dioxygenase. Vitamin B_6 deficiency increases the sen-

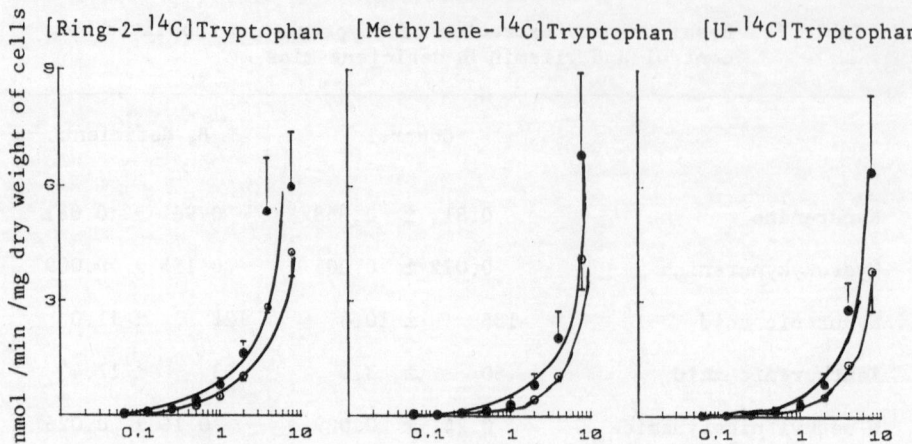

[Ring-2-^{14}C]Tryptophan   [Methylene-^{14}C]Tryptophan   [U-^{14}C]Tryptophan

Fig. 2.   The metabolism of [^{14}C]tryptophan ($^{14}CO_2$ + non-aromatic metabolites)
in isolated hepatocytes. Abscissa: tryptophan (mmol/l) in incuba-
tion.   Points show nmol metabolized/min/mg dry weight of cells,
mean and SEM for 5 animals in each group. O : Control mice; ● :
vitamin B_6 deficient.

sitivity of target tissues to steroid hormone action (Symes et al., 1984;
Bowden et al., 1986), and it is possible that the sensitivity to endogenous
glucocorticoid hormones is also increased in vitamin B_6 deficiency, resulting
in increased induction of tryptophan dioxygenase.

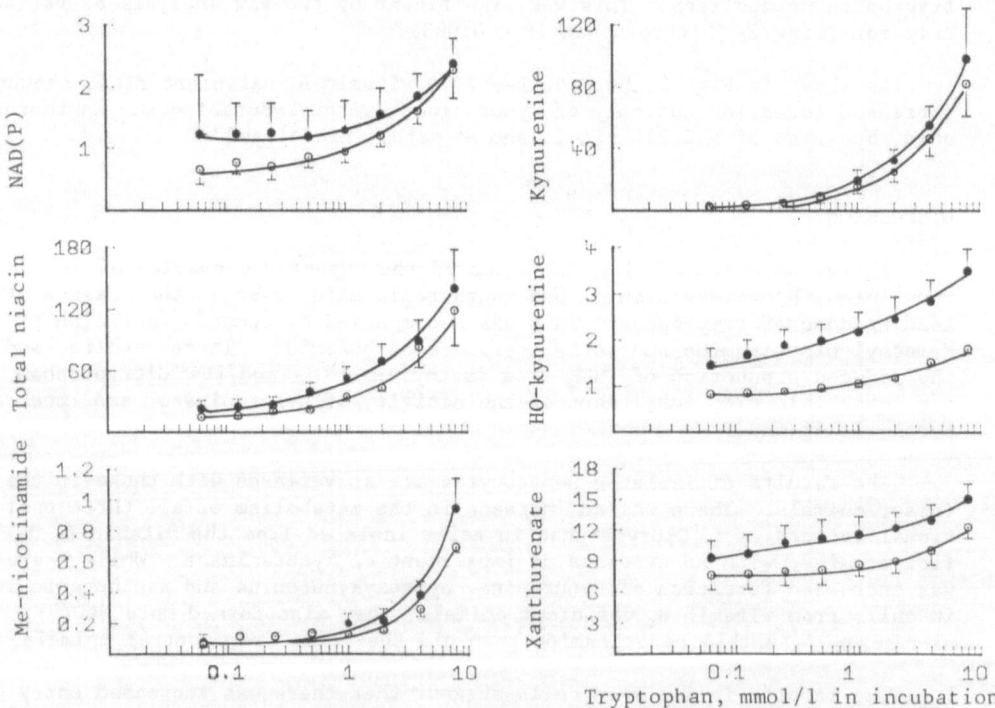

Fig. 3.   The metabolism of tryptophan in isolated hepatocytes.   Points show
nmol product formed/mg dry weight of cells, mean and SEM for 5
animals in each group. O: Control;   ● : vitamin B_6 deficient.

The discrepancy between the experiments in vivo and the studies with isolated hepatocytes suggests that some artefact may be induced during the preparation of hepatocytes by perfusion with collagenase. Although tryptophan transport is an important regulatory factor in isolated cells (Salter and Pogson, 1985; Salter et al., 1986), it is not known whether transport is similarly limiting in the intact liver. Alternatively, it may be that extra-hepatic metabolism of tryptophan, initiated by indoleamine dioxygenase, is more important than has been believed hitherto. In this case, the isolated hepatocyte may not be a useful model for studies of this kind.

ACKNOWLEDGEMENTS

E.N.M.N. was supported by a British Council Technical Development Training Award. P.S.D. was supported by a vacation studentship from the Wellcome Trust.

REFERENCES

Allegri, G., Costa, C., and de Antoni, A., 1978, A further contribution to the choice of dose for tryptophan load test, **Acta Vitaminol. Enzymol.** 32:163-166.

Bender, D.A., 1983a, Effects of a dietary excess of leucine on the metabolism of tryptophan in the rat: a mechanism for the pellagragenic action of leucine, **Br. J. Nutr.**, 50:25-32.

Bender, D.A., 1983b, Effects of oestradiol and vitamin B_6 on tryptophan metabolism in the rat: implications for the interpretation of the tryptophan load test for vitamin B_6 nutritional status, **Br. J. Nutr.**, 50:33-42.

Bender, D.A., 1987, Estrogens and vitamin B_6, actions and interactions, **Wld. Rev. Nutr. Diet.**. 51:140-188.

Bender, D.A., Armstrong, A.J., Monkhouse, C.R., and Richardson, J.P., 1975, Changes in pancreatic tryptophan in the rat in response to fasting: the effect of beta-cytotoxic agents and variation in the oestrous cycle, **Pflügers Archiv.**, 356:245-251.

Bender, D.A., Laing, A.E., Vale, J.A., Papadaki, M., and Pugh, 1983, Effects of estrogen administration on tryptophan metabolism in rats and in menopausal women receiving hormone replacement therapy, **Biochem. Pharmacol.**, 32:843-848.

Bender, D.A., and McCreanor, G.M., 1985, Kynurenine hydroxylase: a potential rate-limiting enzyme in tryptophan metabolism, **Biochem. Soc. Trans.**, 13:441-443.

Bender, D.A., and Olufunwa, R., 1988, Utilization of tryptophan, nicotinamide and nicotinic acid as precursors for nicotinamide nucleotide synthesis in isolated rat liver cells, **Br. J. Nutr.**, 59:279-287.

Bender, D.A., and Totoe, L., 1984, Inhibition of tryptophan metabolism by estrogens in the rat - a factor in the aetiology of pellagra, **Br. J. Nutr.**, 51:199-224.

Bender, D.A., and Wynick, D., 1981, Inhibition of kynureninase by oestrone sulphate: an alternative explanation for abnormal results of tryptophan load tests in women receiving estrogenic steroids, **Br. J. Nutr.**, 45:269-275.

Bowden, J.-F., Bender, D.A., Coulson, W.F. and Symes, E.K., 1986, Increased uterine uptake and nuclear retention of [^{3}H]oestradiol through the oestrous cycle and enhanced end-organ sensitivity to estrogen stimulation in vitamin B_6 deficient rats, **J. Ster. Biochem.**, 25:359-365.

Coon, W.W. and Nagler, E., 1969, The tryptophan load test for pyridoxine deficiency in hospitalized patients, **Ann. N.Y. Acad. Sci.**, 166:30-43.

Coursin, D.B., 1964, Recommendations for standardization of the tryptophan load test, **Amer. J. Clin. Nutr.**, 14:56-61.

El-Zoghby, S.M., El-Kholy, Z.A., El-Sewedy, S.M., and Adbel-Tawab, G.A., 1978, The spontanenous urinary excretion of tryptophan metabolites "via kynurenine" in women with regards to the prepuberty, sexual maturity and menopause, **Acta Vitaminol. Enzymol.**, 32:155-158.

Knox, W.E., 1953, The relation of liver kynureninase to tryptophan metabolism in pyridoxine deficiency, **Biochem. J.**, 53:379-385.

Lepkovsky, S., and Nielson, E., 1942, A green pigment-producing compound in urine of pyridoxine deficient rats, **J. Biol. Chem.**, 144:135-138.

Magboul, B.I., and Bender, D.A., 1983, The effects of a dietary excess of leucine on the synthesis of nicotinamide nucleotides in the rat, **Br. J. Nutr.**, 49:321-324.

Michael, A.F., Drummond, K.N., Doeden, D., Anderson, J.A., and Good, R.A., 1964, Tryptophan metabolism in man, **J. Clin. Invest.**, 43:1730-1746.

Rose, D.P., and McGinty, F., 1968, The influence of adrenocortical hormones and vitamins upon tryptophan metabolism in man, **Clin. Sci.**, 35:1-9.

Rose, D.P., and Braidman, I.P., 1971, Excretion of tryptophan metabolites as affected by pregnancy, contraceptive steroids and steroid hormones, **Amer. J. Clin. Nutr.**, 24:673-683.

Salter, M., and Pogson, C.I., 1985, The role of tryptophan-2,3-dioxygenase in the hormonal control of tryptophan metabolism in isolated rat liver cells, **Biochem. J.**, 229:499-504.

Salter, M., Knowles, R.G., and Pogson, C.I., 1986, Quantification of the importance of individual steps in the control of aromatic amino acid metabolism, **Biochem. J.**, 234:635-647.

Symes, E.K., Bender, D.A., Bowden, J.-F., and Coulson, W.F., 1984, Increased target tissue uptake of, and sensitivity to, testosterone in the vitamin B_6 deficient rat, **J. Ster. Biochem.**, 20:1089-1093.

Takikawa, O., Yoshida, R.R., Kido, R., and Hayaishi, O., 1986, Tryptophan degradation in mice initiated by indoleamine 2,3-dioxygenase, **J. Biol. Chem.**, 261:3648-3653.

EFFECTS OF THE DIET AND OTHER METABOLIC PHENOMENA ON BRAIN TRYPTOPHAN UPTAKE

AND SEROTONIN SYNTHESIS

J.D. Fernstrom

Department of Psychiatry
University of Pittsburgh School of Medicine
Western Psychiatric Institute and Clinic
Pittsburgh, Pennsylvania 15213
USA

INTRODUCTION

In this short article, information will be briefly reviewed that ties
serotonin (5HT) synthesis in brain to the availability of its amino acid precursor, tryptophan (TRP). In addition, the physiologic factors influencing
brain TRP uptake will also be discussed, which include food ingestion, starvation, stress and exercise. Since available data linking these physiologic
phenomenon to brain TRP and 5HT synthesis are in many cases incomplete, suggestions will be made concerning future research directions. The hope is
that further investigation will resolve some of the perplexing issues in this
complex area.

TRYPTOPHAN AVAILABILITY AND SEROTONIN SYNTHESIS IN BRAIN

Less than five years after serotonin's (5HT) discovery in the central
nervous system (Twarog and Page, 1953), investigators were already studying
whether changes in the dietary intake of its amino acid precursor, TRP, could
influence brain 5HT levels (Zbinden et al., 1958). There was no clearly
stated biochemical rationale at that time as to why one would expect dietary
TRP manipulations to influence 5HT levels. This oversight in logic was not
remedied until Lovenberg et al. (1968) provided the first set of kinetic constants for tryptophan hydroxylase, which suggested that the enzyme might not
be fully saturated at normal brain TRP concentrations. These results provided a biochemical context for interpreting the earlier dietary results, as
well as the results of pharmacologic studies showing rapid increases in brain
5HT levels following TRP injection (Hess and Doepfner, 1961; Ashcroft et al.,
1965). The observed increases and decreases in brain 5HT presumably paralleled changes in 5HT synthesis rate, which was a direct reflection of the
degree of substrate saturation of tryptophan hydroxylase, the enzyme catalyzing the rate-limiting step in the synthetic pathway. This notion, that
TRP hydroxylation rate and 5HT synthesis in brain respond directly to changes
in local TRP levels has now stood the test of time for over 25 years. Other
potential control mechanisms have been proposed from time-to-time [e.g., end-
product inhibition (Macon et al., 1971), only to be discarded later (Millard
et al., 1972)].

Since 5HT production increased rapidly following single injections of TRP, even at very small doses (Fernstrom and Wurtman, 1971a), it seemed to some that TRP uptake into brain from the blood must readily influence the brain pool of TRP used in 5HT synthesis. The implication of this recognition was that any physiologic factor that could influence blood TRP levels should be directly able to alter brain TRP levels and 5HT production. Because of an interest in amino acid metabolism, we studied the effects of one such physiologic factor, insulin (Fernstrom and Wurtman, 1971b). Insulin modifies the levels of most amino acids in blood, including TRP, and thus might be predicted to alter brain TRP levels as well. Such was indeed the case: when fasted rats received an injection of insulin (or consumed a carbohydrate meal to induce insulin secretion), blood and brain TRP levels rose rapidly, as did 5HT synthesis. This result clearly suggested that physiologically-induced (carbohydrate ingestion) changes in blood and brain TRP levels were large enough to alter 5HT production. Later studies focused more generally on the ingestion of foods, and explored in particular whether protein ingestion would alter blood TRP levels, and subsequently influence brain TRP levels and 5HT production. A clear effect was obtained, but the result was unexpected. The ingestion of proteins produced rapid and marked increments in blood TRP level, but unlike the carbohydrate meal, produced no increase in brain TRP level or 5HT production. The explanation for these effects ultimately underscored an important physiologic concept; viz., that amino acids are transported across the blood-brain barrier by saturable and competitive carrier mechanisms. In the case of TRP, the carrier is shared among several large neutral amino acids [LNAA] (Oldendorf, 1971). These LNAA rise substantially in blood when protein is consumed, offsetting any competitive advantage for brain uptake accorded to TRP by its increment in blood following protein ingestion. Hence, TRP uptake does not increase following a protein meal. The LNAA (but not TRP) fall when carbohydrates are consumed, a well-known action of insulin (Fernstrom and Wurtman, 1972). Hence, in this latter case, competitive transport greatly favors increased TRP uptake.

Such findings suggested that the competitive transport of TRP into brain could be generalized to predict in any situation how changes in blood LNAA levels would influence brain TRP levels and 5HT synthesis. Notwithstanding, other investigators developed an entirely different model for the regulation of brain TRP uptake (and 5HT formation). This model was based on the fact that [a] TRP circulates in blood 90% associated with serum albumin (McMenamy et al., 1957), and [b] TRP binding is competitive with serum non-esterified fatty acids [NEFA] (McMenamy and Oncley, 1958). By this formulation, raising or lowering serum NEFA levels would alter TRP binding to albumin, and thus the blood level of free (i.e., that not bound to albumin) TRP. The free TRP pool was presumed to be the fraction of TRP in blood accessible to brain transport sites, and was thus said to dictate the rate at which TRP would be taken across the blood-brain barrier. For example, the injection of drugs that unbind TRP from albumin was shown to raise brain TRP level; the interpretation was that the increase in the blood free TRP pool increased brain TRP uptake (Gessa and Tagliamonte, 1974). However, though some experiments involving drugs or "physiologic" treatments seem to support these notions, they do not hold in studies in which diet was used to modify serum NEFA levels. For example, serum NEFA levels are altered rapidly and substantially by changing the amount of fat consumed in a meal, but such changes produce no alterations in brain TRP levels (Madras et al., 1974; Fernstrom et al., 1975). Apparently, the variations in serum NEFA that occur in dietary contexts are too small to cause remarkable changes in brain TRP uptake (Yuwiler et al., 1977). Hence, dietary fat content does not appear to have any important impact on brain TRP uptake (or on 5HT synthesis).

TRYPTOPHAN, OTHER LARGE NEUTRAL AMINO ACIDS AND APPETITE CONTROL

The observation that the ingestion of carbohydrates or proteins could rapidly alter blood LNAA levels and cause predictable changes in brain TRP levels and 5HT formation has led to the postulation that the brain might use the [serum LNAA → brain TRP → 5HT synthesis] cascade to monitor protein and carbohydrate intake in the diet. In one formulation, chronic protein selection in the diet was linked to this model (Ashley and Anderson, 1975; Li and Anderson, 1987). The animal presumably modulated protein intake over days and weeks indirectly by monitoring the changes the foods produced in the serum LNAA pattern, and thus brain TRP uptake and 5HT synthesis. For this model to work, however, it is necessary that a chronic difference in protein intake produce a clear difference in serum LNAA (including TRP) levels, brain TRP uptake and 5HT production. Otherwise, the brain is not receiving the needed biochemical information by this route with which to decide what level of dietary protein is being consumed. In several recent studies, however, the serum LNAA pattern, brain TRP level and 5HT production were all found to experience no variations over wide ranges of chronic dietary protein intake, at several times during the 24-hr period (Fernstrom et al., 1985; Peters and Harper, 1985; Gustafson et al., 1986). Such results indicate that control mechanisms governing chronic protein selection in the diet cannot involve monitoring serum LNAA and brain TRP uptake, since no such changes occur. Other diet-induced perturbations must signal the brain regarding chronic level of protein intake.

Another formulation is that the meal-to-meal selection of dietary carbohydrates is monitored in brain via meal-induced changes in serum LNAA, brain TRP and 5HT. In this model, the carbohydrate-induced increment in 5HT synthesis, and the absence of such changes when even small amounts of protein are consumed, provide a signal to brain via the 5HT neuron that one or the other macronutrient has just been consumed. The brain is said to use this information to govern the appetite for carbohydrates (and protein) at the next meal (Wurtman, 1982). Much recent interest has been generated by this notion, because the idea is so attractive. The complete model is reputed to function as follows: when a rat consumes a carbohydrate meal, brain TRP and 5HT rise, leading to an increase in 5HT release. The effect of increased 5HT release is to suppress carbohydrate intake. Hence, at the next meal, less carbohydrate and more protein is ingested, which tends to lower brain TRP levels and 5HT production. As a consequence, 5HT release declines, and the signal to suppress carbohydrate intake is moderated. At the next meal, therefore, carbohydrate intake rises. As this negative feedback loop cycles, it is postulated to maintain carbohydrate intake around a fixed level or setpoint (see Fernstrom, 1988).

The model builds on two main data bases. The first is that showing how single meals of protein vs. carbohydrate influence brain TRP uptake and 5HT formation. The second is a set of pharmacologic results reputing to show that agents that stimulate 5HT receptors selectively suppress appetite for carbohydrates (Wurtman and Wurtman, 1979, 1989). The latter result, however, is currently a subject of debate (e.g., Kanarek, 1987; Fernstrom, 1988). The drug most commonly studied is fenfluramine, which has many actions on the 5HT neuron, all directed toward increasing 5HT receptor stimulation. Fenfluramine stimulation reputedly reduces selectively the elective intake of carbohydrates by rats (Wurtman and Wurtman, 1979). But the drug has now been tested in a number of laboratories, and has not consistently been found to suppress carbohydrate intake (Orthen-Gambill and Kanarek, 1982; McArthur and Blundell, 1983). In fact, from study to study, it appears that the drug shows no particular macronutrient specificity. Without a specific effect of such drugs to reduce carbohydrate intake, the negative feedback loop cannot work (Fernstrom, 1987). Data continue to appear, however, suggesting macronutrient selectivity in the effects of such 5HT drugs (Shor-Posner et al.,

1986; Kim and Wurtman, 1988), and the resolution of this point is an issue of considerable current interest. If there is no selective suppression of carbohydrates by 5HT drugs, then the proposed model certainly cannot work (Fernstrom, 1988). Even if such pharmacologic data were ultimately obtained by many labs, however, they alone may not save the model, because it is not at all clear that carbohydrate intake is a regulated commodity: [a] there is no clear requirement for this macronutrient (National Academy of Sciences, 1978), [b] rats show great variability in their ingestion of carbohydrate (Leathwood and Ashley, 1983; Yokogoshi et al., 1986), and [c] rats will not defend a particular level of carbohydrate intake, when they must work to obtain the macronutrient (Ashley, 1985).

PLASMA TRYPTOPHAN BINDING AND METABOLIC STRESS

Fasting/starvation and stress have been reported to increase brain TRP levels and 5HT production. These effects were tied to increments in serum free TRP levels. A common metabolic feature of stress and starvation is elevated serum NEFA levels, which are said to produce the rise in serum free TRP [by displacing TRP from albumin] (Knott and Curzon, 1972; Tagliamonte et al., 1973; Kennett and Joseph, 1981; Kennett et al., 1986). Recently, exercise has been a new focus of study, since it also causes a significant rise in serum NEFA levels. Treadmill exercise in rats has been found to raise serum NEFA levels, serum free TRP levels, and brain TRP and 5-hydroxyindoles (Chaouloff et al., 1985, 1986). While these data seem to provide good support for serum TRP binding to albumin as a factor governing TRP uptake into brain, at least under these conditions (no data have appeared disproving this possibility), even in these situations the importance of competitive LNAA transport at the blood brain-barrier has been affirmed experimentally: in each of these paradigms, the increment in brain TRP levels is blocked either by an injection of valine (an LNAA competitor of TRP) or the ingestion of a high-protein meal [which contains all LNAA] (Kennett and Joseph, 1981; Chaouloff et al., 1985; Kennett et al., 1986). Together, these data suggest that changes in TRP binding to albumin could be a factor influencing the alterations in brain TRP levels observed in conditions of substantial metabolic demand, but further experimentation is needed to work out and validate the details.

GENERAL DISCUSSION

There is a clear need for continued work on the relationship of diet and metabolism to brain TRP uptake and 5HT synthesis. Many interesting observations have been made over the years, but too few have been followed up sufficiently to enable firm conclusions to be drawn regarding their physiologic importance. First, additional studies are needed to resolve the conflicting views on the role of meal-induced changes in brain TRP and 5HT in the control of macronutrient selection. At present, it is at best unclear if 5HT drugs selectively and consistently reduce carbohydrate intake, and this issue should be resolved. And, if the notion of carbohydrate intake regulation is to survive, convincing and consistent evidence is needed demonstrating that the rat actually regulates its intake of this macronutrient. The resolution of these issues will determine the fate of the present hypothesis regarding the control of carbohydrate intake.

Much additional work is also needed to understand fully the conditions under which diet and food intake do and do not alter brain TRP and 5HT. Recent data suggest that chronic variations in protein intake do not influence brain TRP (or 5HT), despite large differences in the input of amino acids from the diet. The absence of such effects may reflect normal metabolic adaptation to handle widely differing amino acid loads (i.e., the induction of

372

amino acid metabolizing enzymes; see Peters and Harper, 1985). Metabolic adaptation therefore becomes a potential modifier of dietary effects; too little is known about this interaction to be able to predict the ultimate impact on brain TRP uptake in all chronic dietary and metabolic contexts. In the single meal context, no information is available regarding changes in brain TRP and 5HT that occur following the ingestion of a single meal in a non-fasted rat, though current models of macronutrient intake regulation implicitly assume continuing responsivity of serum LNAA levels and brain TRP uptake at every meal and snack. Is this likely to be true? An answer is presently unavailable. But since circulating insulin levels before lunch and dinner and before snacks are substantially higher than at breakfast (Fernstrom et al., 1978), because of recent food intake, it is unclear at best if the insulin contribution to meal-related changes in the serum LNAA pattern (and thus brain TRP uptake) would be the same at all meals. If not, then the data on single meal effects on brain TRP and 5HT, which involve premeal fasting, cannot be extrapolated to all meal and snack contexts. An answer is needed.

Further work on stress, starvation and exercise is also needed, to understand more fully the extent to which serum TRP binding to albumin might account for the changes observed in brain TRP and 5-hydroxyindoles. Changes in serum NEFA levels and consequently TRP binding to albumin do not seem to be important factors influencing brain TRP levels in the dietary context (Fernstrom et al., 1975; Yuwiler et al., 1977), but perhaps they are in these more robust metabolic contexts. If so, then these models would be quite useful in examining more fully the interaction between TRP binding to albumin and LNAA competition in determining net TRP uptake into brain. What would the effects of single meals be, for example, in the exercised vs the sedentary animal? Would a carbohydrate meal fail to raise brain TRP and 5HT in the exercised rat, since it is already elevated? Or would ingestion of a protein meal by the exercised rat produce a fall in brain TRP and 5HT, where no such reduction would be seen in the sedentary rat? Such possibilities (and others), of course, suggest that it may be appropriate to view the relationship of brain TRP supply to 5HT synthesis in a broader metabolic context, rather than in narrow dietary confines. A variety of metabolic factors, including diet, can presumably modify the serum LNAA pattern, and perhaps the association of TRP to albumin, and produce changes in brain TRP uptake and 5HT synthesis. How these factors actually interact in the normal (and abnormal) life settings of the animal to impact on brain TRP uptake and on 5HT production is currently unknown, and will be an object of future investigation.

Finally, one might ask in general why 5HT synthesis should be so vulnerable to changes in TRP supply. By way of comparison, acetylcholine synthesis is also somewhat responsive to diet-related changes in the supply of its substrate, choline (Cohen and Wurtman, 1976; Hirsch and Wurtman, 1978). But ACh synthesis is relatively insensitive to precursor supply, when compared to 5HT synthesis: very large choline doses are needed to raise ACh production, whereas small doses of TRP stimulate 5HT production. For ACh, the sluggishness of the response is probably related to the fact that choline is so effectively conserved by the cholinergic neuron, once ACh is released and hydrolyzed: an active, high-affinity reuptake mechanism exists at the cholinergic nerve terminal, which aggressively reabsorbs choline and reutilizes it for ACh synthesis (Jope, 1979). No such mechanism of conservation can exist for 5HT, of course, since once a molecule of TRP has been converted to 5HT, it cannot reappear as a substrate molecule. However, a neglected issue has been whether or not active uptake of TRP by the neuron itself can influence the TRP pool used in 5HT synthesis. A transport mechanism for TRP is said to exist at the neuronal membrane (Grahame-Smith and Parfitt, 1970; Denizeau and Sourkes, 1977), and some data in synaptosomal preparations suggest its activity varies directly with neuronal depolarization (Bruinvels and Moleman,

1980). If this is also the case in vivo, such plasticity might be particularly important in situations in which TRP demand is great (e.g., during
sustained depolarization of 5HT neurons), while exogenous supply is low
[e.g., if a dietary TRP deficiency exists, or excessive amounts of other LNAA
are being consumed (Ramanamurthy and Srikantia, 1970; Fernstrom and Lytle,
1976)]. An important future direction is therefore determining if active
transport of TRP into the neuron can influence or moderate physiologically
the vagaries in TRP supply to the brain, particularly in situations where 5HT
synthesis is increased (such as in stress and starvation).

ACKNOWLEDGEMENTS

Some of the studies described in this review were supported by the
National Institutes of Health (HD 24730).

REFERENCES

Ashcroft, G.W., Eccleston, D., and Crawford, T.B.B., 1965, 5-hydroxyindole
    metabolism in rat brain. A study of intermediate metabolism using the
    technique of tryptophan loading; I. Methods, J. Neurochem., 12:483-492.
Ashley, D.V.M., 1985, Factors affecting the selection of protein and carbo-
    hydrate from a dietary choice, Nutr. Res., 5:555-571.
Ashley, D.V.M, and Anderson, G.H., 1975, Correlation between the plasma
    tryptophan to neutral amino acid ratio and protein intake in the self-
    selecting weanling rat, J. Nutr., 105:1412-1421.
Bruinvels, J., and Moleman, P., 1980, Enhancement of tryptophan uptake by
    divalent cations in the absence of sodium ions, J. Neurochem., 34:1065-
    1070.
Chaouloff, F., Elghozi, J.L., Guezennec, Y., Laude, D., 1985, Effects of
    conditioned running on plasma, liver and brain tryptophan and on brain
    5-hydroxytryptamine metabolism of the rat, Br. J. Pharmacol., 86:33-41.
Chaouloff, F., Kennett, G.A., Serrurrier, B., Merino, D., and Curzon, G.,
    1986, Amino acid analysis demonstrates that increased plasma free trypto-
    phan causes the increase of brain tryptophan during exercise in the rat,
    J. Neurochem., 46:1647-1650.
Cohen, E.L., and Wurtman, R.J., 1976, Brain acetylcholine: control by dietary
    choline, Science, 191: 561-562.
Denizeau, F., and Sourkes, T.L., 1977, Regional transport of tryptophan in
    rat brain, J. Neurochem., 28:951-959.
Fernstrom, J.D., 1987, Food-induced changes in brain serotonin synthesis: is
    there a relationship to appetite for specific macronutrients? Appetite,
    8:163-182.
Fernstrom, J.D., 1988, Issues and Opinions: Tryptophan, serotonin and carbo-
    hydrate appetite: will the real carbohydrate craver please stand up! J.
    Nutr., 118:1417-1419.
Fernstrom, J.D., Arnold, M.A., Wurtman, R.J., Hammarstrom-Wiklund, B., Munro,
    H.N., and Davidson, C.S., 1978, Diurnal variations in plasma insulin con-
    centrations in normal and cirrhotic subjects. Effects of dietary protein,
    J. Neural Transm. Suppl., 14:133-142.
Fernstrom, J.D., Fernstrom, M.H., Grubb, P.E., and Volk, E.A., 1985, Absence
    of chronic effects of dietary pontrol of brain tryptophan level and sero-
    tonin synthesis, Adv. Biochem. Psychopharmacol., 11:119-131.
Grahame-Smith, D.G., and Parfitt, A.G., 1970, Tryptophan transport across the
    synaptosomal membrane, J. Neurochem., 17:1339-1353.
Gustafson, J.M., Dodds, S.J., Burgus, R.C., and Mercer, L.P., 1986, Predic-
    tion of brain and serum free amino acid profiles in rats fed graded levels
    of protein, J. Nutr., 116:1667-1681.
Hess, S.M., and Doepfner, W., 1961, Behavioral effects and brain amine con-
    tent in rats, Arch. Int. Pharmacodyn., 134:89-99.

Hirsch, M.J., and Wurtman, R.J., 1978, Lecithin increases acetylcholine concentrations in rat brain and adrenal gland, **Science**, 202:223-225.

Jope, R.S., 1979, High affinity choline transport and acetyl CoA production in brain and their roles in the regulation of acetylcholine synthesis, **Brain Res.**, 180:313-344.

Kanarek, R.B., 1987, Neuropharmacological approaches to studying diet selection, in: "Amino Acids in Health and Disease: New Perspectives", Kaufman, S., ed., Alan R. Liss, New York, pp. 383-401.

Kennett, G.A., Curzon, G., Hunt, A., and Patel, A.J., 1986, Immobilization decreases amino acid concentrations in plasma but maintains or increases them in brain, J. **Neurochem.**, 46:208-212.

Kennett, G.A., and Joseph, M.H., 1981, The functional importance of increased brain tryptophan in the serotonergic response to restraint stress, **Neuropharmacology**, 20:39-43.

Kim, S.H., and Wurtman, R.J., 1988, Selective effects of CGS 10686B, dl-fenfluramine or fluoxetine on nutrient selection, **Physiol. Behav.**, 42:319-322.

Knott, P.J., and Curzon, G., 1972, Free tryptophan in plasma and brain tryptophan metabolism, **Nature**, 239:452-453.

Leathwood, P.D., and Ashley, D.V.M., 1983, Strategies of protein selection by weanling and adult rats, **Appetite**, 4:97-112.

Li, E.T.S., and Anderson, G.H., 1987, Amino acids in food intake and selection, in: "Amino Acids in Health and Disease: New Perspectives", Kaufman, S., ed., Alan R. Liss, New York, pp. 345-367.

Lovenberg, W., Jequier, E., and Sjoerdsma, A., 1968, Tryptophan hydroxylation in mammalian systems, in: "Advances in Pharmacology", Vol. 6A, Garattini S., and Shore, P.A., eds., Academic Press, New York, pp. 21-36.

McArthur, A.R., and Blundell, J.E., 1983, Protein and carbohydrate self-selection: modification of the effects of fenfluramine and amphetamine by age and feeding regimen, **Appetite**, 4:113-124.

McMenamy, R.H., Lund, C.C., and Oncley, J.L., 1957, Unbound amino acid concentrations in human blood plasmas, J. **Clin. Invest.**, 36:1672-1679.

McMenamy, R.H., and Oncley, J.L., 1958, Specific binding of tryptophan to serum albumin, J. **Biol. Chem.**, 233:1436-1447.

Macon, J.B., Sokoloff, L., and Glowinski, J., 1971, Feedback control of rat brain 5-hydroxytryptamine synthesis, J. **Neurochem.**, 18:323-331.

Madras, B.K., Cohen, E.L., Messing, R., Munro, H.N., and Wurtman, R.J., 1974, Relevance of free tryptophan in serum to tissue tryptophan concentrations, **Metabolism**, 23:1107-1116.

Millard, S.A., Costa, E., and Gál, E.M., 1972, On the control of brain serotonin turnover rate by end product inhibition, **Brain Res.**, 40:545-551.

National Academy of Sciences, 1978, "Nutrient requirements of domestic animals, Number 10: Nutrient requirements of laboratory animals, third revised edition". National Academy of Sciences, Washington.

Oldendorf, W.H., 1971, Brain uptake of radiolabeled amino acids, amines, and hexoses after arterial injection, **Am. J. Physiol.**, 221:1629-1639.

Orthen-Gambill, N., and Kanarek, R.B., 1982, Differential effects of amphetamine and fenfluramine on dietary self-selection in rats, **Pharmacol. Biochem. Behav.**, 16:303-309.

Peters, J.C., and Harper, A.E., 1985, Adaptation of rats to diets containing different levels of protein: effects on food intake, plasma and brain amino acid concentrations and brain neurotransmitter metabolism, J. **Nutr.**, 115:382-398.

Ramanamurthy, P.S.V., and Srikantia, S.G., 1970, Effects of leucine on brain serotonin, J. **Neurochem.**, 17:27-32.

Shor-Posner, G., Grinker, J.A., Marinescu, C., Brown, O., and Leibowitz, S.F., 1986, Hypothalamic serotonin in the control of meal patterns and macronutrient selection, **Brain Res. Bull.**, 17:663-671.

Tagliamonte, A., Biggio, G., Vargiu, L., and Gessa, G.L., 1973, Free tryptophan in serum controls brain tryptophan levels and serotonin synthesis, **Life Sci.**, 12:277-287.

Twarog, B.M., and Page, I.H., 1953, Serotonin content in some mammalian tissues and urine and a method for its determination, Am. J. Physiol., 175:157-161.

Wurtman, R.J., 1982, Nutrients that modify brain function, Sci. Amer., 246: 50-59.

Wurtman, J.J., and Wurtman, R.J., 1979, Drugs that enhance central serotoninergic transmission diminish elective carbohydrate consumption by rats, Life Sci., 24:895-904.

Wurtman, R.J., and Wurtman, J.J., 1989, Carbohydrates and depression, Sci. Amer., 260:68-75.

Yokogoshi, H., Theall, C.L., and Wurtman, R.J., 1986, Selection of dietary protein and carbohydrates by rats: changes with maturation, Physiol. Behav., 36:979-982.

Yuwiler, A., Oldendorf, W.H., Geller, E., and Braun, L., 1977, Effect of albumin binding and amino acid competition on tryptophan uptake into brain, J. Neurochem., 28:1015-1023.

Zbinden, G., Pletscher, A., and Studer, A., 1958, Alimentäre Beeinflussung der enterochromaffinen Zellen und des 5-Hydroxytryptamin Gehaltes von Gehirn and Darm, Z. Ges. Exptl. Med., 129:615-620.

EFFECTS OF TRYPTOPHAN AND OF 5-HYDROXYTRYPTAMINE RECEPTOR SUBTYPE AGONISTS ON FEEDING

G. Curzon

Institute of Neurology
Queen Square
London WC1N 3BG
UK

INTRODUCTION

Most research on the relationships between tryptophan (TRP), 5-hydroxy-tryptamine (5-HT) and feeding comes from laboratories centered on either neurochemistry or pharmacology. Neurochemists have usually studied how feeding affects brain TRP concentration and the synthesis and availability of 5-HT. Pharmacologists have focussed on the use of 5-HTergic drugs to control appetite. Consideration of possible relationships between the neurochemical and pharmacological findings leads to other important questions, e.g. do effects of feeding on TRP and on 5-HT function have roles in the normal control of appetite and in its disorders such as anorexia nervosa. Research in this area has been stimulated in recent years by evidence for numerous 5-HT receptor subtypes and by the availability of drugs with selectivity towards them. The main subject of this chapter will be how some of these drugs affect feeding in the rat. While it is easy to show that many 5-HTergic drugs affect food intake in laboratory animals it is less easy to be sure that appetite changes are involved. However, the fact that the drug most commonly used clinically to decrease appetite (fenfluramine, Ponderax) acts by releasing 5-HT (Rowland and Carlton, 1986), encourages the belief that other 5-HTergic drugs can also directly influence appetite in the rat and do not decrease food intake merely, for example as a result of general malaise. Before describing research on this topic however, the effects of feeding on brain TRP and of TRP administration on feeding must be considered.

EFFECTS OF FEEDING ON BRAIN TRP AND 5-HT

Rat brain tryptophan has long been known to increase after a large high carbohydrate - low protein meal, due to a fall in plasma concentrations of amino acids which compete with the transport of TRP to the brain (Fernstrom and Wurtman, 1972; Fernstrom, this volume). It is fascinating that a similar increase occurs on overnight fasting due to the liberation of TRP from plasma albumin so that it becomes more available to the brain (Knott and Curzon, 1972; Gillman et al., 1981; Sarna et al., 1984). Evidence (discussed by Curzon, 1988) suggests that human brain TRP concentration is not markedly affected by normal acute changes of dietary intake but this evidence derives from data obtained on relatively few subjects studied over short periods. As values obviously range more widely within larger groups or over longer periods of study, it is possible that subgroups exist with brain TRP concentrations that are highly sensitive to dietary influences.

Although such TRP changes can clearly influence 5-HT synthesis, how much they might affect the availability of the transmitter to its receptors remains uncertain. There is still no convincing evidence from (for example) in vivo dialysis studies suggesting that giving TRP, even at pharmacological dosage, increases 5-HT at receptor sites. Nevertheless, as described below, there is much evidence that TRP and 5-HTergic drugs affect feeding. Furthermore, Schwartz et al. (1989) have used in vivo dialysis to show that feeding increases extracellular 5-HT in the lateral hypothalamus of the rat and suggest that this may play a part in controlling meal duration. As TRP was not measured in these experiments, it is not clear whether or not TRP changes were implicated in the increase of 5-HT.

The above findings, however, suggest that feeding may lead to central TRP and 5-HT changes which normally mediate its termination and that inappropriate relationships between feeding or food deprivation and changes of brain indole chemistry and 5-HT function could be involved in disorders of appetite such as anorexia nervosa.

EFFECT OF L-TRP ON FEEDING

Indications that 5-HTergic drugs affected feeding (Blundell, 1977) and that brain 5-HT synthesis could be increased by L-TRP (Fernstrom and Wurtman, 1971) encouraged investigation of its effects of feeding. Early work on rats by Weinberger et al. (1978) failed to reveal changes but this may have been because the amino acid was given after food deprivation as White et al. (1988) found that intake was decreased under other conditions, i.e. on the first day of exposure to a diet to which L-TRP was added. Similar results were obtained on human subjects by Hroboticky et al. (1985). The association of anorexia in cancer patients with high plasma free TRP and CSF TRP concentrations (Cangiano et al., 1986) may therefore be of causal significance.

Evidence also suggests that L-TRP influences food choice. Thus, either acute (Li and Anderson, 1984) or chronic (White et al., 1988) administration to rats within meals decreased subsequent carbohydrate intake and increased protein intake. Similarly, 1 g L-TRP given to humans in a high protein meal decreased carbohydrate intake in a later, free choice meal (Blundell and Hill, 1987). Conversely, if plasma TRP was decreased by giving a TRP deficient amino acid mixture to humans then protein intake in a later, free choice meal was decreased (Young et al., 1988). These findings are certainly consistent with an important role for 5-HT in feeding though it must be remembered that the influence of other pathways of TRP metabolism on feeding remains largely unexplored. However, emerging evidence, summarized below, on the effects of 5-HT receptor subtype agonists on feeding strengthens interpretation in terms of 5-HT.

EFFECTS OF 5-HT AGONISTS ON FEEDING

5-HT$_{1A}$ agonists

The putative 5-HT$_{1A}$ agonists 8-hydroxy-2-(di-n-propylamino)tetralin (8-OH-DPAT) (Dourish et al., 1985; Bendotti and Samanin, 1987), buspirone, ipsapirone (Dourish et al., 1986b), gepirone (Gilbert and Dourish, 1987) and LY165163 (Hutson et al., 1987) increase food intake in free feeding rats. Some typical data are shown in Table 1. As the anorexic effects of non-selective, indirect 5-HT agonists (e.g. fenfluramine and 5-hydroxytryptophan) and the hyperphagic effects of the 5-HT synthesis inhibitor p-chlorophenyl-

Table 1.   Effects of 5-HT$_{1A}$ agonists on food
            intake in freely feeding rats

Treatment (s.c.)	Food intake (g ± SEM)
	Two hours
Saline	0.8 ± 0.3
8-OH-DPAT, 0.125 mg/kg	2.2 ± 0.4
8-OH-DPAT, 0.250 mg/kg	2.7 ± 0.4
Saline	0.4 ± 0.1
LY165163, 0.5 mg/kg	1.5 ± 0.5
LY165163, 1 mg/kg	2.3 ± 0.3
	Four hours
Saline	1.6 ± 0.4
Gepirone, 4 mg/kg	3.4 ± 0.4
Gepirone, 8 mg/kg	4.4 ± 0.3
Saline	1.5 ± 0.3
Buspirone, 4 mg/kg	3.2 ± 0.5
Ipsapirone, 5 mg/kg	2.2 ± 0.5

Data from references given above.   Effects of
drugs were all $P < 0.05$.

alanine (pCPA) suggest that 5-HT inhibits food intake (Blundell, 1977; Rowland and Carlton, 1986) it was proposed (Dourish et al., 1986b) that 8-OH-
DPAT causes hyperphagia by activating presynaptic 5-HT receptors so that 5-HT
release from terminals is decreased. This mechanism is supported by the absence of 8-OH-DPAT- (Dourish et al., 1986a) and LY165163- (Hutson et al.,
1987) induced hyperphagia after depletion of 5-HT by pCPA.

Evidence is against presynaptic 5-HT$_{1A}$ receptors on 5-HT terminals (Middlemiss, 1984) but indicates a high density of these sites on 5-HT cell bodies in the raphe nuclei (Verge et al., 1985; Weissman-Nanopoulos et al.,
1985). Injecting 8-OH-DPAT into the raphe increased feeding (Hutson et al.,
1986).

8-OH-DPAT has high affinity for 5-HT$_{1A}$ sites but also has lower affin-
ity for 5-HT$_{1B}$, 5-HT$_{1C}$, 5-HT$_{1D}$, 5-HT$_2$ (Hoyer, 1988) and α_2-adrenergic (Middlemiss, 1987) sites. However, the effects of various antagonists points
strongly to the hyperphagic effect being mediated at 5-HT$_{1A}$ sites (Table 2).

Is the effect of 8-OH-DPAT on feeding behaviorally specific or part of a
general activation of motor and consummatory behavior due to a widespread decrease of 5-HT at terminals? The latter possibility is not unreasonable as
raphe lesions can enhance dopamine-dependent behavior such as locomotion and
gnawing (Costall and Naylor, 1974). Therefore 8-OH-DPAT might conceivably
increase feeding secondarily to a release of these behavioral components.
Indeed it has been suggested that it elicits eating merely as a consequence
of gnawing (Chaouloff et al., 1988; Montgomery et al., 1988). However, in
the study of Dourish et al. (1985), animals given a choice between food pellets and wood blocks chose to eat the pellets. Subsequently, Dourish et al.
(1988) reported that both 8-OH-DPAT and gepirone dose-dependently increased

Table 2. Effects of antagonists on the hyperphagic response to
8-OH-DPAT (1 mg/kg, s.c.) in normally fed rats

Treatment	Blockade	Interpretation
Metergoline, 5 mg/kg	yes	5-HT receptor
Ketanserin, 2.5 mg/kg	no	Not 5-HT$_2$
ICS-205-930, 1 mg/kg	no	Not 5-HT$_3$
(-) Pindolol, 4 mg/kg	yes	5-HT$_{1A}$, 5-HT$_{1B}$, or β
(+) Pindolol, 4 mg/kg	no	Not β
Spiperone, 0.05 mg/kg	yes	5-HT$_{1A}$ or DA
Haloperidol, 0.1 mg/kg	no	Not DA or α_1
Idazoxan, 3 mg/kg	no	Not α_2

From Hutson et al. (1988).

consumption of a liquid diet.

As TRP, which might increase 5-HT at postsynaptic receptors, decreases carbohydrate intake (Li and Anderson, 1986; Blundell and Hill, 1987; White et al., 1988) it is of interest that 8-OH-DPAT, which decreases 5-HT at these sites (Hutson et al., 1989) had the opposite effect, increasing carbohydrate intake by rats given a carbohydrate/protein choice (Sarna, Whitton, Curzon and Leathwood, unpublished). These findings are consistent with the hypothesis that the choice depends on 5-HT function (Wurtman and Wurtman, 1979). It seems unlikely however, that this mechanism operates normally (Peters and Harper, 1987; Curzon, 1988), even though it may occur as a result of drug treatment. Also, whether the resultant changes of 5-HT receptors influence discrimination between carbohydrate and protein as such or some unrelated sensory difference between the characteristics of the diets has been questioned (Booth, 1987; Fernstrom, 1987). Nevertheless, pharmacological evidence in favor of discrimination is substantial.

5-HT$_{1B}$ and 5-HT$_{1C}$ agonists

The above data implies that decreased 5-HT at postsynaptic sites is responsible for the hyperphagic effects of the 5-HT$_{1A}$ agonists. The questions arise: which 5-HT receptor subtypes are involved; where do they occur? It has been shown that stimulation of postsynaptic 5-HT$_{1B}$ receptors by RU 24969 causes hypophagia (Kennett et al., 1987; Hutson et al., 1988a). Also 1-(3-chlorophenyl)piperazine (mCPP) and 1-[3-(trifluoromethyl)-phenyl]piperazine (TFMPP) have substantial affinity for the 5-HT$_{1B}$ (Sills et al., 1984; Asarch et al., 1985) 5-HT$_{1C}$ and 5-HT$_2$ sites and caused hypophagia in normally fed (Kennett et al., 1987) and food-deprived rats (Samanin et al., 1979). Results are given in Table 3.

RU 24969 also causes hyperactivity (Green et al., 1984; Tricklebank et al., 1986) whilst mCPP and TFMPP cause hypoactivity which appears to be mediated by central 5-HT$_{1C}$ receptors (Kennett and Curzon, 1988a). It therefore seemed possible that mCPP- and TFMPP-induced hypophagia also involved stimulation of 5-HT$_{1C}$ receptors. Indeed, preliminary evidence (Kennett and Curzon, 1988a) suggested that they mediated the hypophagia seen on placing mCPP and TFMPP-treated rats for 20 min in a novel cage containing food. Furthermore, putative 5-HT$_{1C}$ antagonists caused hyperphagia in similar experiments. Responses to RU 24969, mCPP and TFMPP were therefore investigated under conditions more appropriate to the study of hypophagia, i.e., using animals previously deprived of food and measuring intake over longer periods.

Table 3. Effects of $5\text{-HT}_{1B}/5\text{-HT}_{1C}$ agonists on food intake in freely feeding rats

Treatment (i.p.)	Food intake (g/4 h $\pm$ SEM)
Saline	1.14 $\pm$ 0.18
RU 24969, 1 mg/kg	0.55 $\pm$ 0.12
RU 24969, 2 mg/kg	0.30 $\pm$ 0.10
RU 24969, 5 mg/kg	0.07 $\pm$ 0.16
Saline	1.21 $\pm$ 0.16
mCPP, 1 mg/kg	0.31 $\pm$ 0.17
mCPP, 5 mg/kg	0.15 $\pm$ 0.03
TFMPP, 1 mg/kg	0.52 $\pm$ 0.07
TFMPP, 5 mg/kg	0.15 $\pm$ 0.03

Data from Kennett et al. (1987). Effects of drugs were all $P < 0.05$.

The effects of various receptor antagonists shown in Tables 4 and 5 indicated that mCPP (and probably also TFMPP) caused hypophagia by direct stimulation of 5-HT_{1C} receptors although functional 5-HT_{1B} receptors were also needed, but that RU 24969 acted solely by stimulation of 5-HT_{1B} receptors. Since completing this work, Hoyer (1988) reported that mCPP and TFMPP bind somewhat more strongly to 5-HT_{1C} receptors than to 5-HT_{1B} sites.

Affinities of the antagonists for non-5-HT receptors are unable to explain their effects on hypophagias due to RU 24969 or mCPP. Thus, although mianserin (but not metergoline or mesulergine) has high affinity for histamine H_1 receptors (Leysen et al., 1981; Closse, 1983), reduced histaminergic

Table 4. Effects of antagonists (s.c.) on the hypophagic response to RU 24969 (5 mg/kg, i.p.)

Treatment	Blockade	Interpretation
Metergoline, 5 mg/kg	yes	5-HT receptor
Ketanserin, 0.2 mg/kg	no	Not 5-HT_2, α_1
ICS 205930, 1 mg/kg	no	Not 5-HT_3
Mianserin, 5 mg/kg	no	Not 5-HT_{1C}, 5-HT_2, H_1
Mesulergine, 0.2 mg/kg	no	Not 5-HT_{1C}, 5-HT_2, DA
1-Naphthylpiperazine, 2 mg/kg	no	Not 5-HT_{1C}, 5-HT_2
($\pm$) Cyanopindolol, 8 mg/kg	yes	5-HT_{1A}, 5-HT_{1B}, β
(−) Pindolol, 2 mg/kg	yes	5-HT_{1A}, 5-HT_{1B}, β
(+) Pindolol, 2 mg/kg	no	Not β
Spiperone, 0.10 mg/kg	no	Not 5-HT_{1A}, 5-HT_2, DA

Data mostly on food-deprived rats (Kennett and Curzon, 1988a); some on freely feeding rats (Kennett et al., 1987).

Table 5. Effects of antagonists (s.c.) on the hypophagic response to mCPP (5 mg/kg, i.p.)

Treatment	Blockade	Interpretation
Metergoline, 5 mg/kg	yes	5-HT receptor
Ketanserin, 0.2 mg/kg	no	Not 5-HT$_2$, α_1
Ritanserin, 0.6 mg/kg	no	Not 5-HT$_2$,
ICS 205930, 1 mg/kg	no	Not 5-HT$_3$
Mianserin, 5 mg/kg	yes	5-HT$_{1C}$, 5-HT$_2$, H$_1$
Mesulergine, 0.2 mg/kg	yes	5-HT$_{1C}$, 5-HT$_2$, DA
1-Naphthylpiperazine, 2 mg/kg	yes	5-HT$_{1C}$, 5-HT$_2$,
($\pm$) Cyanopindolol, 8 mg/kg	yes	5-HT$_{1A}$, 5-HT$_{1B}$, β
Propanolol, 16 mg/kg	yes	5-HT$_{1A}$, 5-HT$_{1B}$, β
Idazoxan, 1 mg/kg	no	Not α_2

Data on food-deprived rats (Kennett and Curzon, 1988b).

activity reduces food intake (Menon et al., 1971). Mediation by α_2 receptors for which mCPP and mianserin have some affinity (Clineschmidt et al., 1979; Smith and Suckow, 1985) is also unlikely, since the specific α_2 antagonist idazoxan did not oppose mCPP-induced hypophagia. The α_2 antagonist phentolamine (Samanin et al., 1979) is also ineffective. Lastly, mCPP has little affinity for dopamine receptors (Invernizzi et al., 1981).

A further comment on Tables 4 and 5 is necessary. Thus, cyproheptadine (10 mg/kg; not shown), unlike other 5-HT$_{1C}$ antagonists, did not block the hypophagic effect of either drug, although it antagonized mCPP-induced hypolocomotion (Kennett and Curzon, 1988a). Also, unlike other 5-HT$_{1A}$ and 5-HT$_{1B}$ antagonists, (-)pindolol (2 mg/kg; not shown) did not block mCPP-induced hypophagia even though it prevented hypophagia due to RU 24969.

Despite the above anomalies, evidence from antagonist experiments that mCPP and TFMPP cause hypophagia by action at 5-HT$_{1C}$ receptors appears strong. However, these sites and 5-HT$_2$ sites have much in common. Thus, most antagonists with high affinity for 5-HT$_{1C}$ sites also bind strongly to 5-HT$_2$ sites (Hoyer, 1988) while activation of both sites induces phosphoinositide hydrolysis (Conn and Sanders-Bush, 1987). Furthermore, it was recently proposed that activation of 5-HT$_2$ sites led to hypophagia, as the reduction of feeding on giving the 5-HT releaser fenfluramine was blocked by high doses of ketanserin (Hewson et al., 1988). The latter drug and two other putative 5-HT$_2$ antagonists, LY 53857 and 1-naphthylpiperazine (1-NP), also blocked the hypophagic effect of 1-(2,5-dimethoxy-4-iodophenyl)-2-aminopropane (DOI) (Schechter and Simansky, 1988). These findings, together with the above similarities between 5-HT$_2$ and 5-HT$_{1C}$ sites, might be considered to cast doubt on our evidence that activation of 5-HT$_{1C}$ sites causes hypophagia. However, since 1-NP is also a potent 5-HT$_{1C}$ receptor antagonist (Conn and Sanders-Bush, 1987; Hoyer, 1988) and both DOI and LY53857 have high affinity for the 5-HT$_{1C}$ site (Hoyer, 1988), the evidence for the 5-HT$_2$ involvement relies heavily upon the selectivity of ketanserin.

Therefore, the in vivo antagonist potencies of ketanserin, ritanserin, mianserin and 1-NP against 5-HT$_2$-mediated head twitch behavior (Bedard and

Pycock, 1977; Yap and Taylor, 1983) and mCPP-induced hypophagia were compared (Kennett and Curzon, unpublished). Table 6 shows some results. Log transformed ratios of the ID_{50} values obtained correlated significantly with the corresponding ratios of the affinities (r = 0.984, df2, P < 0.02). This strengthens our evidence that mCPP-induced hypophagia is mediated by $5\text{-}HT_{1C}$ receptors. The method of relating the in vivo effects of drugs to their in vitro binding affinities eliminates uncertainties due to differences of metabolism of different drugs as such differences would comparably affect ability to inhibit both hypophagic and head twitch behavior.

Table 6 shows that the ratios of the ID_{50} values are greater than the ratios of affinities. This presumably largely reflects differences between the in vivo and in vitro circumstances. Other very recent data is also consistent with $5\text{-}HT_1$ rather than $5\text{-}HT_2$ sites mediating hypophagia. Thus, a dose of ritanserin causing 50% occupation of $5\text{-}HT_2$ sites had no effect on D-fenfluramine-induced hypophagia while 4 times the dose caused about 50% inhibition of the hypophagia (Samanin et al., 1989). This suggests mediation by $5\text{-}HT_{1C}$ sites as affinity constants (Hoyer, 1988) indicate that the above increase of ritanserin dosage would half saturate $5\text{-}HT_{1C}$ receptors but have little effects on either $5\text{-}HT_{1A}$ or $5\text{-}HT_{1B}$ sites.

Are RU 24969 or mCPP and TFMPP-induced hypophagias secondary to other behavioral effects? It seems unlikely that hypophagia due to RU 24969 is secondary to its hypolocomotor effect, as haloperidol (Green et al., 1984; Tricklebank et al., 1986; Kennett et al., 1987) and (+) pindolol (Tricklebank et al., 1986) blocks this but not the hypophagia (Kennett et al., 1987). Conversely, metergoline blocks the hypophagia (Kennett et al., 1987; Kennett and Curzon, 1988b) but potentiates the hyperlocomotion (Green et al., 1984; Tricklebank et al., 1986). Also, infusing RU 24969 into the hypothalamus caused hypophagia only (Hutson et al., 1988a).

Similar arguments suggest a dissociation between mCPP (and TFMPP)-induced hypophagia and hypoactivity, since both (±) cyanopindolol and (-) propanolol did not alter the activity, yet prevented the hypophagia (Kennett and Curzon, 1988a,b). Also, TFMPP caused hypophagia but not hypoactivity on hypothalamic infusion (Hutson et al., 1988a). Lastly, while the $5\text{-}HT_{1C}$ antagonists mianserin, 1-NP, cyproheptadine and mesulergine opposed both mCPP-induce hypophagia and hypoactivity, they only increased food intake but not activity (Kennett and Curzon, 1988a,b) when given alone.

As cyproheptadine, 1-NP and mianserin increased food uptake by normally fed but not by food-deprived rats, the effect may either be rate dependent or due to effects on appetite rather than satiety. This may explain contradic-

Table 6. ID_{50} values for inhibition of mCPP-induced hypophagia and carbidopa + 5-HTP-induced head twitches

Drug	ID_{50} (mg/kg) vs. hypo	vs. TW	ID_{50} hypo ID_{50} (TW)	$5\text{-}HT_2$ affinity[1] $5\text{-}HT_{1C}$ affinity
Ketanserin	12.9	0.036	358	71.4
Ritanserin	4.6	0.19	24.4	4.11
Mianserin	2.6	0.12	17.5	1.2
1-NP	1.03	1.75	0.59	0.10

hypo = hypophagia, TW = head twitch, 1-NP = 1-naphthylpiperazine.
[1]Ratio calculated from Hoyer (1988).

tory reports on the effect of cyproheptadine on rat food intake (Baxter et al., 1970; Ghosh and Parvathy, 1973).

One possibility to be considered is that 5-HT_{1B} and 5-HT_{1C} agonists cause hypophagia due to drug-induced malaise. This was investigated (Kennett and Curzon, 1988c) using the antiemetic drug trimethobenzamide (TMB). Although rats cannot vomit, this is a rational procedure as TMB prevents both the putative malaise-induced hypophagic response to cholecystokinin (Moore and Deutsch, 1985) and lithium chloride-induced taste aversion (Coil et al., 1978). However, while it prevented the hypophagic response of rats to acetyl salicylate (a known emetic in man and dogs) it did not affect hypophagia due to RU 24969, mCPP and TFMPP. These drugs therefore probably do not cause hypophagia by a malaise-dependent mechanism.

In view of the finding by Shor-Posner et al. (1986) that infusing 5-HT or norfenfluramine into the paraventricular nucleus (PVN) of the hypothalamus causes hypophagia, it seemed likely that RU 24969, mCPP and TFMPP had a similar effect at this site. As briefly mentioned above, this was shown (Hutson et al., 1988). Thus, 0.5, 1.0 and 2.0 μg of both RU 24969 and TFMPP caused dose dependent hypophagia on PVN infusion. Similar infusion of 8-OH-DPAT was without effect.

SEX DIFFERENCE IN 5-HT DEPENDENT HYPOPHAGIA

Investigations on sex differences in the hypophagic effects of stress have recently led us to similar studies on hypophagic 5-HT agonists. In the former work a single 2 h restraint stress caused hypophagia over the next 24 h. This effect was not merely a result of the gastric lesions which can result from restraint as the histamine-H_2 antagonist ranitidine decreased their number but not the hypophagia (Donohoe et al., 1987). On repeating the restraint daily for 5 days, food intake returned to normal in male but not in female rats (Kennett et al., 1986; Haleem et al., 1987). It is also reported that chronic fenfluramine infusion has a greater hypophagic effect in female rats than in males (Rowland, 1986).

Subsequently, it was shown that the hypophagic effects of RU 24969 and mCPP were greater in female than in male rats if they had been deprived of food for 24 h but not in freely feeding animals (Haleem, 1988). These results suggest the existence of a 5-HT dependent hypophagic mechanism which is more active in food deprived female rats than in males. This could conceivably have a role in the greater incidence of anorexia nervosa in women.

SUMMARY

Feeding or food withdrawal can affect the supply of tryptophan to the brain and hence (in some circumstances) 5-HT synthesis therein. Also fenfluramine which releases 5-HT to postsynaptic receptors suppresses appetite and there are reports that tryptophan can have a similar effect. Furthermore, feeding is reported to release hypothalamic 5-HT. Therefore 5-HT could have a role in the normal termination of feeding and perhaps also in disorders of appetite. The recognition of various 5-HT receptor subtypes has stimulated research in this area. We have now investigated the involvement of the subtypes in the pharmacological control of feeding.

Thus, 5-HT_{1A} agonists (8-OH-DPAT, buspirone, gepirone etc.) stimulate intake in freely feeding rats, probably by activating autoreceptors on the cell bodies of 5-HT neurons so that 5-HT release at terminals is decreased. The hyperphagia is not explicable by increased activity or gnawing and is strikingly manifest against carbohydrate in carbohydrate vs. protein choice experiments.

Feeding in previously food deprived rats is decreased by the 5-HT agonists RU 24969, 1-(3-chlorophenyl)piperazine (mCPP) and 1-[3-(trifluoromethyl) phenyl]piperazine (TFMPP). Effects of antagonists suggest that RU 24969-induced hypophagia depends on $5\text{-}HT_{1B}$ receptors only while mCPP and TFMPP induce hypophagia at $5\text{-}HT_{1C}$ sites, though this effect also requires $5\text{-}HT_{1B}$ receptors for its expression. Responsible sites occur in the paraventricular nucleus of the hypothalamus as infusing either RU 24969 or TFMPP therein causes hypophagia. On systemic injection, the hypophagic drugs are particularly active in female rats, an effect of conceivable relevance to human anorexic illness.

REFERENCES

Asarch, K.E., Ransom, R.W., and Shih, J.S., 1985, $5\text{-}HT_{1A}$ and $5\text{-}HT_{1B}$ heterogeneity, Life Sci., 36:1265-1273.
Baxter, M.J., Miller, A.A., and Soroko, F.E., 1970, The effect of Cyroheptadine on food consumption in the fasted rat, Br. J. Pharmacol., 39:229-230.
Bedard, P., and Pycock, C.J., 1977, Wet-dog shake behavior in the rat: a possible quantitative model of central 5-hydroxytryptamine activity, Neuropharmacology, 16:663-670.
Bendotti, C., and Samanin, R., 1986, 8-Hydroxy-2-(di-n-propylamine)tetralin-(8-OH-DPAT) elicits eating in free feeding rats by acting on central serotonin neurons, Eur. J. Pharmacol., 121:147-180.
Bendotti, C., and Samanin, R., 1987, The role of putative $5\text{-}HT_{1A}$ and $5\text{-}HT_{1B}$ receptors in the control of feeding in rats, Life Sci., 41:635-642.
Blundell, J.E., 1977, Is there a role for serotonin (5-hydroxytryptamine) in feeding? Int. J. Obesity, 1:15-42.
Blundell, J.E., and Hill, A.J., 1987, Influence of tryptophan on appetite and food selection in main, in: "Amino Acids in Health and Disease: New Perspectives", Kaufman, S., ed., Alan R. Liss, New York, pp. 403-419.
Booth, D.A., 1987, Central dietary "feedback onto nutrient selection": not even a scientific hypothesis, Appetite, 8:195-201.
Cangiano, C., Cascino, A., Ceci, F., Muscaritoli, M., Rossi Fanelli, F., Menichetti, E.T., and Mulieri, M., 1986, Plasma and CSF tryptophan in patients with cancer anorexia, in: "Progress in Tryptophan and Serotonin Research", Schlossberger, H.G., Kochen, W., Linzen, B., and Steinhart, H., eds., de Gruyter, Berlin, pp. 269-272.
Chaouloff, F., Serrurrier, B., Mérino, D., Laude, D., and Elghozi, J.L., 1988, Feeding responses to a high dose of 8-OH-DPAT in young and adult rats: influence of food texture, Eur. J. Pharmacol., 151:267-273.
Clineschmidt, B.V., Flataker, L.M., Faison, E., and Holmes, R., 1979, An in vivo model for investigating α_1 and α_2 receptors in the CNS: studies with mianserin, Arch. Int. Pharmacodyn., 242:59-76.
Closse, A., 1983, [^{3}H]Mesulergine, a selective ligand for serotonin-2-receptors, Life Sci., 32:2485-2495.
Coil, J.D., Hankins, W.G., Jenden, A.J., and Garcia, J., 1978, The attenuation of a specific cue to consequence association by antiemetic agents, Psychopharmacology, 56:21-25.
Conn, P.J., and Sanders-Bush, E., 1987, Relative efficacies of piperazines at the phosphoinositide hydrolysis-linked serotonergic ($5\text{-}HT_2$ and $5\text{-}HT_{1C}$) receptors, J. Pharmacol. Exp. Ther., 242:552-557.
Costall, B., and Naylor, R.J., 1974, Stereotyped and circling behavior induced by dopaminergic agonists after lesions of the midbrain raphe nuclei, Eur. J. Pharmacol., 29:206-222.
Curzon, G., 1988, Feeding, stress exercise and the supply of tryptophan to the brain, in: "Amino Acid Availability and Brain Function in Health and Disease", Huether, G., ed., Springer, Berlin, pp. 39-59.

Donohoe, T.P., Kennett, G.A., and Curzon, G., 1987, Immobilization stress-induced anorexia is not due to gastric ulceration, **Life Sci.**, 40:467-472.

Dourish, C.T., Clark, M.L., and Iversen, S.D., 1988, 8-OH-DPAT elicits feeding and not chewing: evidence from liquid diet studies and a diet choice test, **Psychopharmacology**, 95:185-188.

Dourish, C.T., Hutson, P.H., and Curzon, G., 1985, Low doses of the putative serotonin agonist 8-hydroxy-2-(di-n-propylamino)tetralin (8-OH-DPAT) elicit feeding in the rat, **Psychopharmacology**, 86:197-204.

Dourish, C.T., Hutson, P.H., and Curzon, G., 1986a, Parachlorophenylalanine prevents feeding induced by the serotonin agonist 8-hydroxy-2-(di-n-propylamino) tetralin (8-OH-DPAT), **Psychopharmacology**, 89:467-471.

Dourish, C.T., Hutson, P.H., Kennett, G.A., and Curzon, G., 1986b, 8-OH-DPAT induced hyperphagia: its neural basis and possible therapeutic relevance, **Appetite**, Suppl., 7:127-140.

Fernstrom, J.D., 1987, Food-induced changes in brain serotonin syntheses: is there a relationship to appetite for specific macronutrients, **Appetite**, 8:163-182.

Fernstrom, J.D., Effects of the diet and other metabolic phenomena on brain tryptophan uptake and serotonin synthesis, this volume.

Fernstrom, J.D., and Wurtman, R.J., 1971, Brain serotonin content: physiological dependence on plasma tryptophan levels, **Science**, 173:149-152.

Fernstrom, J.D., and Wurtman, R.J., 1972, Brain serotonin content physiological regulation by plasma neutral amino acids, **Science**, 178:414-416.

Ghosh, M.N., and Parvathy, S., 1973, The effect of cyproheptadine on water and food intake and on body weight in the fasted adult and weanling rat, **Br. J. Pharmacol.**, 48:328-329.

Gilbert, F., and Dourish, C.T., 1987, Effects of the novel anxiolytics gepirone, buspirone and ipsapirone on free feeding and on feeding induced by 8-OH-DPAT, **Psychopharmacology**, 93:349-352.

Gillman, P.K., Bartlett, J.R., Bridges, P.K., Hunt, A., Patel, A.J., Kantamaneni, B.D., and Curzon, G., 1981, Indolic substances in plasma, cerebrospinal fluid and frontal cortex of human subjects infused with saline or tryptophan, **J. Neurochem.**, 37:410-417.

Green, A.R., Guy, A.P., and Gardner, C.R., 1984, The behavioral effects of RU24969, a suggested 5-HT receptor agonist in rodents and the effect on the behavior of treatment with antidepressants, **Neuropharmacology**, 23: 655-661.

Haleem, D.J., 1988, Serotonergic functions in rat brain: sex-related differences and responses to stress, **Ph.D. Thesis**, University of London.

Haleem, D.J., Kennett, G., and Curzon, G., 1988, Adaptation of female rats to stress: Shift to male pattern by inhibition of corticosterone synthesis, **Brain Res.**, 458:339-347.

Hewson, G., Leighton, G.E., Hill, R.G., and Hughes, J., 1988, Ketanserin antagonizes the anorectic effect of DL-fenfluramine in the rat, **Eur. J. Pharmacol.**, 145:227-230.

Hoyer, D., 1988, Functional correlates of serotonin 5-HT$_1$ recognition sites, J. Receptor Res., 8:59-81.

Hrboticky, N., Leiter, L.A., and Anderson, G.H., 1985, Effects of L-tryptophan on short term food intake in lean men, **Nutr. Res.**, 5:595-607.

Hutson, P.H., Dourish, C.T., and Curzon, G., 1986, Neurochemical and behavioral evidence for mediation of the hyperphagic action of 8-OH-DPAT by 5-HT cell body autoreceptors, **Eur. J. Pharmacol.**, 129:347-352.

Hutson, P.H., Donohoe, T.P., and Curzon, G., 1987, Neurochemical and behavioral evidence for an agonist action of 1-[2-(4-aminophenyl)-4-(3 trifluoromethylpenyl) piperazine at central 5-HT receptors, **Eur. J. Pharmacol.**, 138:215-223.

Hutson, P.H., Donohoe, T.P., and Curzon, G., 1988a, Infusion of the 5-hydroxytryptamine agonists RU 24969 and TFMPP into the paraventricular nucleus of the hypothalamus causes hypophagia, **Psychopharmacology**, 95:550-552.

Hutson, P.H., Dourish, C.T., and Curzon, G., 1988b, Evidence that the hyperphagic response to 8-OH-DPAT is mediated by 5-HT$_{1A}$ receptors, Eur. J. Pharmacol., 150:361-366.

Hutson, P.H., Sarna, G.S., O'Connell, M.T., and Curzon, G., 1989, Hippocampal 5-HT synthesis and release in vivo is decreased by infusion of 8-OH-DPAT into the nucleus raphe dorsalis, Neurosci. Lett., 100:276-280.

Invernizzi, R., Cotecchia, S., De Blasi, A., Mennini, T., Pataccini, R., and Samanin, R., 1981, Effects of m-chlorophenylpiperazine on receptor binding and brain metabolism of monoamines in rats, Neurochem. Int., 3:239-244.

Kennett, G.A., Chaouloff, F., Marcou, M., and Curzon, G., 1986, Female rats are more vulnerable than males in an animal model of depression: the possible role of serotonin, Brain Res., 382:416-421.

Kennett, G.A., and Curzon, G., 1988a, Evidence that mCPP may have behavioral effects mediated by 5-HT$_{1C}$ receptors, Br. J. Pharmacol., 94:137-147.

Kennett, G.A., and Curzon, G., 1988b, Evidence that hypophagia induced by mCPP and TFMPP requires 5-HT$_{1C}$ and 5-HT$_{1B}$ receptors: hypophagia induced by RU 24969 only requires 5-HT$_{1B}$ receptors, Psychopharmacology, 96:93-100.

Kennett, G.A., and Curzon, G., 1988c, The antiemetic drug trimethobenzamide prevents hypophagia due to acetyl salicylate but not to 5-HT$_{1B}$ or 5-HT$_{1C}$ agonists, Psychopharmacology, 96:101-103.

Kennett, G.A., Dourish, C.T., and Curzon, G., 1987, 5-HT$_{1B}$ agonists induce anorexia at a postsynaptic site, Eur. J. Pharmacol., 141:429-435.

Knott, P.J., and Curzon, G., 1972, Free tryptophan in plasma and brain tryptophan metabolism, Nature, 239:452-453.

Leysen, J.E., Awouters, F., Kennis, L., Laduron, P.M., Vandenberk, J., and Janssen, P.A.J., 1981, Receptor binding profile of R41 468, a novel antagonist of 5-HT$_2$ receptors, Life Sci., 28:1015-1022.

Li, E.T.S., and Anderson, G.H., 1984, 5-Hydroxytryptamine: a modulator of food composition but not quantity?, Life Sci., 34:2453-2460.

Menon, M.K., Clark, W.G., and Aures, D., 1971, Effect of Thiazol-4-xylmethoxyamine, a new inhibitor of histamine synthesis on brain histamine, monoamine levels and behavior, Life Sci., 10:1097-1109.

Middlemiss, D.N., 1984, 8-Hydroxy-2-(di-n-propylamino) tetralin is devoid of activity at the 5-hydroxytryptamine autoreceptor in rat brain. Implications for the proposed link between the autoreceptor and the [^{3}H]5-HT recognition site, Naunyn-Schmiedeberg's Arch. Pharmacol., 327:18-22.

Middlemiss, D.N., 1987, Lack of effect of the putative 5-HT$_{1A}$ receptor agonist, 8-OH-DPAT on 5-HT release in vitro, in: "Brain Receptors: Behavioral and Neurochemical Pharmacology", Dourish, C.T., Ahlenius, S., and Hutson, P.H., eds., Ellis Horwood, Chichester, pp. 82-93.

Montgomery, A.M.J., Willner, P., and Muscat, R., 1988, Behavioral specificity of 8-OH-DPAT induced feeding, Psychopharmacology, 94:110-114.

Moore, B.D., and Deutsch, J.R., 1985, An antiemetic is antidotal to the satiety effect of cholecystokinin, Nature, 315:321-322.

Peters, J.C., and Harper, A.E., 1987, A skeptical view of the role of central serotonin in the selection and intake of protein, Appetite, 8:206-210.

Rowland, N.E., 1986, Effect of continuous infusions of dexfenfluramine on food intake, body weight and brain amines in rats, Life Sci., 39:2581-2586.

Rowland, N.E., and Carlton, J., 1986, Neurobiology of an anorectic drug: fenfluramine, Progr. Neurobiol., 27:13-62.

Samanin, R., Mennini, T., Bendotti, C., Barone, D., Caccia, S., and Garattini, S., 1989, Evidence that central 5-HT$_2$ receptors do not play an important role in the anorectic activity of D-fenfluramine in the rat, Neuropharmacology, 28:465-469.

Samanin, R., Mennini, T., Ferraris, A., Bendotti, C., Borsini, F., and Garattini, S., 1979, m-Chlorophenylpiperazine: a central agonist causing powerful anorexia in rats, Naunyn-Schmiedeberg's Arch. Pharmacol., 308:159-163.

Sarna, G.S., Kantamaneni, B.D., and Curzon, G., 1984, Variables influencing the effect of a meal on brain tryptophan, J. Neurochem., 44:1575-1580.

Schechter, L.E., and Simansky, K.J., 1988, 1-(2,5-Dimethoxy-4-iodophenyl)-
2-aminopropane (DOI) exerts an anorexic action that is blocked by 5-HT$_2$
antagonists in rats, **Psychopharmacology**, 94:342-346.

Schwartz, D.H., McClane, S., Hernandez, L., and Hoebel, B.G., 1989, Feeding
increases extracellular serotonin in the lateral hypothalamus of the rat
as measured by microdialysis, **Brain Res.**, 479:349-354.

Shor-Posner, G., Grinker, J.A., Marmeson, C., Brown, O., and Leibowitz, S.F.,
1986, Hypothalamic serotonin in the control of meal patterns and macronutrient selection, **Brain. Res. Bull.**, 17:663-671.

Sills, M.A., Wolfe, B.B., and Frazer, A., 1984, Determination of selective
and non-selective compounds for the 5-HT$_{1A}$ and 5-HT$_{1B}$ receptor subtypes in
rat frontal cortex, **J. Pharmacol. Exp. Ther.**, 231:480-487.

Smith, T.M., and Suckow, R.F., 1985, Trazodone and m-chlorophenyl-piperazine:
concentration in brain and receptor activity in regions of the brain associated with anxiety, **Neuropharmacology**, 24:1067-1071.

Tricklebank, M.D., Middlemiss, D.N., and Neill, J., 1986, Pharmacological
analysis of the behavioral and thermoregulatory effects of the putative
5-HT$_1$ receptor agonist RU 24969 in the rat, **Neuropharmacology**, 25:877-886.

Verge, D., Duval, G., Patey, A., Gozlan, S., El Mestikaway, S., and Hamon,
M., 1985, Presynaptic 5-HT autoreceptors on serotonergic cell bodies
and/or dendrites but not terminals are of the 5-HT$_{1A}$ subtype, **Eur. J.
Pharmacol.**, 113:463-464.

Weinberger, S.B., Knapp, S., and Mandell, A.J., 1978, Failure of tryptophan
load-induced increases in brain serotonin to alter food intake in the rat,
Life Sci., 22:1595-1602.

Weissman-Nanopoulos, D., Mach, E., Magre, J., Demassey, Y., and Pujol, J.F.,
1985, Evidence for the localization of 5-HT$_{1A}$ binding sites on serotonin
containing neurons in the raphe dorsalis and raphe centralis nuclei of the
rat brain, **Neurochem. Int.**, 7:1061-1072.

White, P.J., Cybulski, K.A., Primus, R., Johnson, D.F., Collier, G.H., and
Wagner, G.L., 1988, Changes in macronutrient selection as a function of
dietary tryptophan, **Physiol. Behav.**, 43:73-77.

Wurtman, J.J., and Wurtman, R.J., 1979, Drugs that enhance central serotonergic transmission diminish elective carbohydrate consumption by rats,
Life Sci., 24:895-904.

Yap, C.Y., and Taylor, D.A., 1983, Involvement of 5-HT$_2$ receptors in the wet-
dog shake behavior induced by 5-hydroxytryptophan in the rat, **Neuropharmacology**, 22:801-804.

Young, S.N., Tourjman, S.V., Teff, K.L., Pihl, R.O., and Anderson, G.H.,
1988, The effect of lowering tryptophan on food selection in normal males,
Pharmacol. Biochem. Behav., 31:149-152.

THE ROLE OF SEROTONIN (5-HT) IN FEEDING RESPONSES TO AMINO ACIDS

D.W. Gietzen, V.A. Hammer, J.L. Beverly, and Q.R. Rogers

Departments of Physiological Sciences
and Psychiatry and Food Intake Laboratory
University of California
Davis, California 95616
USA

INTRODUCTION

A model of rapid onset anorexia has emerged from nutritional, biochemical and neurophysiological studies over the past 30 years, in which rats are fed diets containing imbalanced proportions of amino acids (Harper et al., 1970; Rogers and Leung, 1977; Gietzen et al., 1986a). The neurochemical changes engendered by ingestion of these diets are now beginning to be elucidated. In this model, the limiting amino acid is decreased in several brain areas, and both the noradrenergic and serotonergic systems appear to be activated as the animals begin to decrease their intake of the imbalanced diet. The studies presented here suggest a role for serotonin (5-HT), via a specific receptor subtype, the $5-HT_3$ receptor, in the initial anorectic response of rats to imbalanced amino acid diets.

The imbalanced amino acid model

Diets containing disproportionate amounts of amino acids are associated with a marked depression of growth and food intake (Willcock and Hopkins, 1906; Harper et al., 1970; Rogers and Leung, 1977). This consistent and robust anorectic response is seen in rats fed diets that differ by as little as 0.4% of the diet as an essential amino acid, illustrating the sensitivity of the rat to small changes in the amino acid composition of its internal milieu. These responses to amino acid imbalanced diets do not require olfactory (Leung et al., 1972), vagal, or gastric input (Stickney et al., 1976), nor an intact adrenal system (Hammer et al., 1990). An increase in circulating adrenal corticoids, from diet-induced stress, probably does not cause the depression in imbalanced diet intake (Leung et al., 1968; Hammer et al., 1990). Rather, several lines of evidence support the idea that the brain is the site of control in animals fed an amino acid imbalanced diet. For example, the food intake depression of rats (Leung and Rogers, 1969) and cockerels (Tobin and Boorman, 1979) fed amino acid imbalanced diets is prevented if a small quantity of the most limiting amino acid is infused into the carotid artery. Infusion of the limiting amino acid into the portal or jugular vein requires a much greater quantity than that given into the carotid artery to prevent the food intake depression (Leung and Rogers, 1969). Also, specific neuroanatomical lesions can abolish or attenuate the responses (Leung and Rogers, 1987), and injection of the specific limiting amino acid into a brain

area that has been shown to be essential for the response ameliorates the
feeding depression (Beverly et al., 1988). These findings suggest that the
brain plays a crucial role in the food intake responses of animals ingesting
amino acid imbalanced or devoid diets.

Serotonin and amino acid imbalance

Increased activity of brain 5-HT systems has been shown to have an inhibitory effect on feeding (suggested by Joyce and Morovsky, 1964; reviewed
in Blundell, 1984a). Since the response of rats to dietary amino acid imbalance is a reduction in food intake, an involvement of the serotonin system
might be expected. Therefore, in connection with a survey of neurotransmitter concentrations, measurements of 5-HT and its metabolite were made in several brain areas of rats fed an imbalanced diet.

Method for monoamine determinations: The ratio of the metabolite,
5-hydroxyindoleacetic acid, to 5-HT (5-HIAA/5-HT) was determined in brains of
rats after acute exposure to either the isoleucine imbalanced, basal or corrected diet (Leung and Rogers, 1969; Gietzen et al., 1986b, 1987). Six animals per group were adapted to the basal diet for at least 2 weeks and allowed to eat only during the 12 hr dark period for 4 days prior to the experiment to assure initiation of the first meal promptly at the beginning of the
feeding period. The rats were decapitated 3.5 hrs after initiation of feeding. 5-HT and its primary metabolite, 5-HIAA, were determined by high performance liquid chromatography with electrochemical detection (HPLC-EC)
(Gietzen et al., 1986b) in microdissected brain areas.

Results: The ratio of 5-HIAA/5-HT, a putative measure of 5-HT turnover,
especially in acute tests (Fuller, 1985), was found to be increased in 3 of
the 14 microdissected brain slices studied after 3.5 hrs of feeding. In the
imbalanced diet group, the ratio of 5-HIAA/5-HT was increased to 155% of control levels in the raphe nuclei, to 178% in the locus ceruleus and to 140% in
the hippocampus, by comparison with the basal diet group. Thus, an increase
in 5-HT activity was suggested in at least 3 brain areas of rats fed an imbalanced amino acid diet (Fig. 1). No significant changes were seen in the
remaining 11 areas.

Conclusion: The brain areas with increased 5-HT activity have also been
shown to be important for 5-HT function. The raphe is the site of the 5-HT
cell bodies, the hippocampus receives much 5-HT input, and the locus ceruleus
is the site of the noradrenergic cell bodies and may serve as a target for

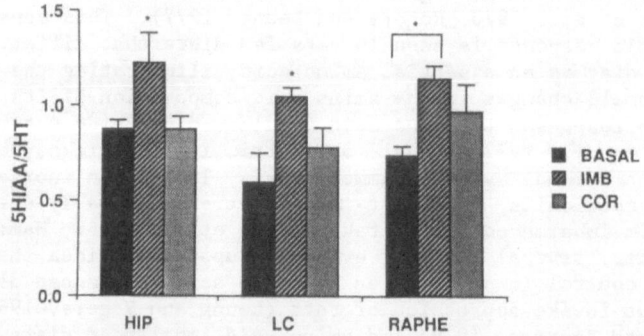

Fig. 1. Ratios of 5-HIAA/5-HT in 3 brain areas of rats after feeding
isoleucine basal, imbalanced or corrected diets for 3.5 hr.
HIP: hippocampus, LC: locus ceruleus, RAPHE: raphe nuclei,
*significant difference from basal diet group; $P < 0.05$
(least-significant differences (LSD) test after significant
overall ANOVA, within a brain area).

5-HT modulation of noradrenergic function (see below). From these observations, we formulated the hypothesis that the final common pathway in the anorectic response caused by imbalanced amino acid diets includes a serotonergic component.

Precursor loading

We first tested our hypothesis by asking the question: would provision of the 5-HT precursor tryptophan (TRP) affect the feeding response?

Methods: To test the effects of precursor loading, the amino acid precursor of 5-HT, L-TRP (Sigma), was given in two ways. Animals were adapted to the basal diet for 10 days prior to the experiment. In the first trial, L-TRP was injected i.p. in a single dose of 100 mg/kg, in a saline suspension, 30 min before the initiation of the feeding test with the isoleucine imbalanced diet, at the beginning of the dark period. This dose has been reported to increase 5-HT concentrations in the brain between 30 min and 2 hrs after injection (Peters et al., 1984). Food intake was measured for the first 24 hrs in 3 groups: TRP injected with basal or imbalanced diet feeding, and saline injected with imbalanced diet feeding. In the second experiment, an additional 1% TRP each was added to the isoleucine basal diet and the isoleucine imbalanced diet (plus-TRP diets). A 10-day adaptation period on the basal diet also preceded this precursor loading test. The total dietary concentration of TRP in each of the plus-TRP diets was 1.1%. In this experiment, there were four groups: basal diet without TRP, basal-plus-TRP, imbalanced diet without TRP and imbalanced-plus-TRP. Again, the food intake was measured for the first 24 hrs.

Results: The TRP injection did not change the initial response of the animals to the isoleucine imbalanced diet. The saline and TRP injected animals ate 9.75 ± 1.4 and 9.5 ± 1.6 g of the imbalanced diet, respectively, in 24 hr ($P > 0.05$). However, when 1% of TRP was added to the diet, the additional TRP in the imbalanced diet resulted in a severe exacerbation of the depression in food intake (F: 3,19 = 5.09; $P < 0.01$), while addition of 1% TRP did not alter basal diet intake (Fig. 2).

Conclusion: The exacerbation of the anorexia seen when TRP was added to

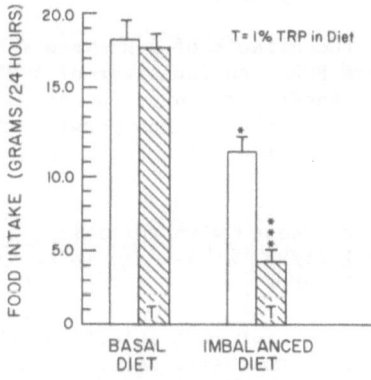

Fig. 2. Effects of addition of 1% tryptophan (TRP) to basal and imbalanced diets on food intake. Bars marked "T" represent food intake for rats given added TRP in their diets. *P < 0.05, ***P < 0.001 by LSD after ANOVA as described in Fig. 1. Vertical bars indicate the mean and SEM. (From Gietzen et al., 1987, with permission of The American Physiological Society).

the diet could be due to an increase in the dietary imbalance, but addition of TRP did not affect basal diet intake at all. An alternative suggestion is that the TRP provided an additional source of 5-HT, which was then responsible for the increased anorexia. These results are consistent with our hypothesis that increased 5-HT activity, caused by amino acid imbalance, plays a role in the anorectic response. We then elected to use a standard pharmacological approach in further tests of our hypothesis.

GENERAL METHODS FOR PHARMACOLOGICAL EXPERIMENTS

Animals: Male Sprague-Dawley rats (Bantin and Kingman, Lafayette, CA) weighed 140-280 g at the beginning of the experiments. They were housed individually in an artificially illuminated laboratory (22 ± 2°C) with light onset at 24:00 and a 12:12 h light:dark cycle. Daily food intake and body weight were measured and spillage collected. Basal (low protein) diets, with L-amino acids as the protein source, were used prior to feeding of the amino acid imbalanced diets and to provide baseline food intake data. Rats were then switched to the experimental diets. For the pharmacological trials, animals were divided into groups, using 6-7 animals/ group, usually in a 2 x 2 factorial design with 2 drug conditions (drug and vehicle) and 2 diet conditions (imbalanced and basal diets). Unless noted otherwise, i.p. injections were given at 11:00 hrs, just prior to the onset of the dark cycle, at noon, 12:00 hr. The diets were the same as reported previously (Leung and Rogers, 1969; Gietzen et al., 1986b, 1987).

Drugs: 8-Hydroxy-dipropylaminotetralin (8-OH-DPAT, Research Biochemicals, Inc.), para-chlorophenylalanine (PCPA, Sigma), quipazine, pirenpirone and ICS 205-930 (a gift of Sandoz Pharmaceuticals Corporation, East Hanover, NJ) were dissolved in 0.9% NaCl. 5,7-Dihydroxytryptamine (DHT, Sigma) was dissolved in saline with 0.1% ascorbic acid. MDL 72,222 (a gift of Merrell Dow Research Institute, Cincinnati, OH) was dissolved in propylene glycol, and metergoline was dissolved in 5% ascorbic acid. Quipazine, pirenpirone and metergoline were a gift of Dr. J. Miller, Stanford University. For each experiment, control injections of the vehicle were given to parallel groups of animals.

Site of action: pre- or post-synaptic?

To evaluate whether the effects of 5-HT were mediated at a pre- or a postsynaptic site, we gave PCPA, an inhibitor of 5-HT synthesis (Koe and Weismann, 1966), prior to feeding the experimental diets. This approach was taken, since responses to treatments which persist after depletion of 5-HT from presynaptic terminals provide evidence for postsynaptic action of the treatment.

Method: Rats were pretreated with three i.p. injections of PCPA, each 100 mg/kg in a volume of 1 ml/kg, at 3-4 day intervals for a total dose of 300 mg/kg. Control animals were given 1 ml/kg of 0.9% NaCl at the same intervals. After food intake on the basal diet had returned to preinjection levels, 3 to 4 days after the last PCPA injection, the animals were offered either the imbalanced diet or the basal diet. Food intake was measured at 24 hr. Terminal measurements of the 5-HT concentration were made by HPLC/EC in hemibrains including cerebellum, 8 days or 1 month after the last PCPA injection.

Results: On the first day of feeding the imbalanced diet, the PCPA treated animals ate significantly less than controls. When given the isoleucine imbalanced diet, saline injected animals reduced their 24 hr food intake by 18%. In contrast, the PCPA treated animals had a 60% reduction from their intake of the basal diet (overall ANOVA — F: 3,20 = 9.594; P < 0.001; Table

1). Thus, PCPA pretreatment resulted in a significant exacerbation of the depression in food intake of rats on the first day of their exposure to the isoleucine imbalanced diet, and had no effect on intake of the basal diet.

Conclusion: This result may be taken as evidence that the increased 5-HT activity induced by the imbalanced diet is translated into an anorectic response via a postsynaptic receptor. In addition, since the response was exaggerated after PCPA, it appears that the exaggeration of the anorexia is the result of postsynaptic supersensitivity.

Site of action: peripheral or central?

The experiments just described were done with peripheral injections. The next question was: would central injection of the serotonergic cytotoxin, DHT, replicate the PCPA result?

Method: To achieve a longer lasting depletion of cerebral 5-HT, groups of 12 rats were treated with intracerebroventricular (i.cv.) injections of DHT (a specific 5-HT cytotoxin (Baumgarten and Lachenmeyer, 1977) which has been shown to decrease 5-HT levels over several months). Prior to DHT or vehicle injection, the animals were given desmethylimipramine (DMI, Merrell Dow, Cincinnati, OH; 25 mg/kg, i.p.) to block the uptake of DHT into catecholamine neurons. Thirty min later, the rats were anesthetized with metophane for a 10 μl injection of DHT (200 μg/rat in saline + 0.1% ascorbic acid) into the cisterna magna. Needle placement in the cisternal space was verified by withdrawal of clear cerebrospinal fluid, prior to injection. Controls were given DMI i.p. and 10 μl of the vehicle intracisternally. All animals were maintained on the basal diet until food intake of the DHT groups had again returned to pre-injection levels. The animals were then offered the isoleucine imbalanced diet. Parallel groups were given the basal diet throughout. Terminal measurements of 5-HT concentrations in the brains of the animals were made by HPLC-EC.

Results: Pretreatment with intracerebral DHT also exacerbated the food intake depression of rats on the first day of exposure to the isoleucine imbalanced diet, compared to the group given vehicle injections and the imbalanced diet (F: 3,20 = 16.947; P < 0.001), as had peripheral injection of PCPA. The vehicle pretreated group's 24 hr intake of the imbalanced diet was reduced by 15%, while the DHT pretreated group's intake of the isoleucine imbalanced diet during that period was further reduced by 52% (Table 1). Terminal 5-HT concentrations were also decreased in the DHT treated group compared with those in vehicle injected animals (F: 3,20 = 3.88; P = 0.02), as they were 8 days after the last PCPA injection (see above).

Conclusion: These results suggest that not only was the 5-HT acting at a postsynaptic site, but a central locus of action was supported.

Which receptor mediates the anorectic response?

The question remained: which receptor, among the several 5-HT receptor subtypes, is the specific site of action in this animal model of acute onset anorexia? Several different 5-HT receptor sub-types have been reported (Peroutka, 1988). The basic categories of the 5-HT receptor types are listed as 1, 2 and 3. The 1s have been subdivided into at least 4 subtypes, 1A-1D; the 2s are of high and low affinity and the 3s may also be subdivided into at least 3 categories. The function and location of the $5-HT_1$ and $5-HT_2$ receptors that underlie the general depression in feeding seen with increased 5-HT activity (as for example with fluoxetine injection) have been discussed at length. Some workers favor the $5-HT_2$ (Massi and Marini, 1987; Schechter and Simansky, 1988), some the $5-HT_{1B}$ (Kennett et al., 1987) hypothesis. We herewith present evidence that the $5-HT_3$ receptor mediates 5-HT induced anorexia, at least in rats fed amino acid imbalanced diets. In the studies described below, we used agonists and antagonists of various 5-HT systems to evaluate the role of the several 5-HT receptors in this model. The working hypothesis for these studies was: if the anorectic effect of 5-HT in the imbalanced diet model occurs via a specific receptor which is stimulated by increased 5-HT activity from the imbalanced diet, inhibition of the receptor should increase intake of the imbalanced diet. Conversely, stimulation of the receptor should decrease imbalanced diet intake. This would establish a requirement for the specific receptor that underlies the effect of 5-HT in the anorectic response.

The 5-HT$_1$ receptor system

It has been suggested that 5-HT inhibition is mediated via the 5-HT$_1$ receptor (Peroutka et al., 1981). The 5-HT$_{1A}$ receptor has been shown to be inhibitory to 5-HT function at several sites (Peroutka, 1988). The 5-HT autoreceptor appears to be of the 5-HT$_1$ class but a subtype has not been clearly established (Engel et al., 1986); both pre- and postsynaptic sites have been proposed for both the 1A and 1B subtypes (Hutson et al., 1987). Changes in food intake have been postulated to be autoreceptor-mediated (Chaouloff and Jean Renaud, 1987). This is consistent with work describing increased feeding with the 5-HT$_{1A}$ agonist 8-OH-DPAT at low doses, whereas decreased feeding and increased 5-HT-induced behavior occur at higher doses (Dourish et al., 1986; Hutson et al., 1987). In addition, Cooper's (1987) findings suggest that feeding can be induced in a satiated animal by an inhibitory effect on serotonergic activity, mediated by agonists acting selectively at 5-HT$_{1A}$ autoreceptors in the raphe nuclei.

Method for studies of the 5-HT$_{1A}$ receptor: The 5-HT agonist, 8-OH-DPAT (500 μg/kg, s.c.), was given 30 min before introduction of the isoleucine imbalanced diet at the beginning of the dark cycle, after the usual basal diet-adaptation period. This is the dose that has been reported to have the maximum stimulatory effect on feeding in rats (Dourish et al., 1986). Six of the 8-OH-DPAT treated rats were given the imbalanced diet, and six the basal diet. Saline injected rats were given the imbalanced diet. In our experience, food intake is not increased during the first 24 hrs after a single s.c. injection of saline. Therefore, for basal comparisons, we used the mean of 3 days food intake just prior to the experiment. The drug conditions were: 8-OH-DPAT and saline. The diet conditions were basal and imbalanced diets. Food intake was measured for the first 24 hrs of feeding the imbalanced diet.

Results: 8-OH-DPAT attenuated the reduced feeding response of rats to the isoleucine imbalanced diet. When compared with their basal intake, the animals given 8-OH-DPAT injections and fed the imbalanced diet only decreased their intake by 7%, while the saline treated animals reduced their intake by 18%, i.e. similar to the non-injected controls in previous experiments. The reduction in imbalanced diet intake after 8-OH-DPAT was not significant, and it is clear that the exacerbation of the feeding response previously seen with the postsynaptic 5-HT agonist, or with chronic depletion of 5-HT, was not seen with this 1A agonist. Rather, the depression in intake of the imbalanced diet was attenuated after treatment with 8-OH-DPAT (Table 1).

Conclusion: Activation of the 5-HT$_{1A}$ receptor by the specific 1A agonist, 8-OH-DPAT, blocked the anorectic response to amino acid imbalance, suggesting that a generalized decrease in 5-HT activity could ameliorate the feeding depression caused by an amino acid imbalanced diet. The 8-OH-DPAT was given at a low dose to reduce neural transmission in the 5-HT system by stimulation of the autoreceptor, which should inhibit 5-HT release. This treatment was selected to decrease activity in the 5-HT system because of its reported specificity for the autoreceptor at the dose used, and because it has been shown to increase food intake in the rat at 2 and 4 hrs after administration. This timepoint corresponds to the time of the initial depression in feeding during the acute phase (Gietzen et al., 1986a,b). It should be noted that, while 8-OH-DPAT has been shown to induce hyperphagia in some feeding models, it did not increase 24 hr intake of the basal diet in our paradigm. Hyperphagia alone, such as that seen with VMH lesions, does not abolish the animal's anorectic response to amino acid imbalance (Leung and Rogers, 1970).

Next, in order to test the effect of a postsynaptic 5-HT agonist, we used quipazine, an agonist at several sites.

Method for quipazine experiments - studies of an agonist at the 1B and 2 sites, as well as a specific ligand at the type 3 sites: Naive rats were either pretreated with DHT, the DHT vehicle, 0.1% ascorbic acid or untreated. Three weeks later, after adaptation to the basal diet, they were injected with quipazine (5 mg/kg, i.p.), in a treatment simulating acute replacement or administration of 5-HT, one hr before being fed the isoleucine imbalanced diet. In this 3 x 2 experiment, the drug conditions were: quipazine, DHT and saline. After the single acute injection at 11:00 hr, the animals were offered either the imbalanced or the basal diet at noon, and 24 hr food intake measurements were made.

Results: Although treatment with quipazine did not significantly reduce the 24 hr food intake of rats given the basal diet, it did exacerbate the food intake depression of animals fed the isoleucine imbalanced diet by 57%. In addition, after a combination of both DHT pretreatment and acute injection of quipazine, there was a further exacerbation of the initial feeding depression in rats fed the imbalanced diet, to a 70% reduction of 24 hr intake of the imbalanced diet, a quantity of food less than that eaten by rats treated with either DHT or quipazine alone (overall ANOVA - F: 3,18 - 9.473; P < 0.001; Table 1).

Conclusions: We have seen an effect of general inhibition of 5-HT activity with the $5HT_{1A}$ subtype which restores feeding, and an exacerbation after treatments which can cause postsynaptic supersensitivity, or after treatment with quipazine, which includes in its profile the role of an agonist at the 1B site. There is other evidence that action of the $5\text{-}HT_{1B}$ receptor is at a postsynaptic as well as a presynaptic site, and that increased availability of 5-HT at postsynaptic receptors causes feeding inhibition in the rat. Several studies have differentiated locomotor effects from these anorectic effects of the 1B agonists, and have shown persistence of the anorectic effects after raphe lesions or PCPA treatment, suggesting action at a postsynaptic site (Garattini et al., 1986; Kennett et al., 1987).

Since quipazine, which is approximately equipotent at the $5\text{-}HT_{1B}$ and $5\text{-}HT_2$ receptor sites (Leysen, 1985), exacerbated intake of the amino acid imbalanced diet, and this effect was even more pronounced after treatment with DHT, the effect of quipazine, on intake of the imbalanced diet, is most likely to be mediated by a postsynaptic site. We did not see head shakes or other features of the 5-HT syndrome at the dose used (5 mg/kg, i.p.). However, at a dose of 10 mg/kg of quipazine there were elements of the 5-HT syndrome that were blocked by metergoline (1 mg/kg, i.p.).

In sum, with regard to the type 1 receptor, the 1A agonist restored feeding with the imbalanced diet. Since, in the untreated animal, an imbalanced diet causes reduction in feeding, the 1A site cannot be the primary locus of effect, and if the type 1 receptor subtype subserves the response, it must be via a subtype other than the 1A. Second, the receptor must have a postsynaptic locus, perhaps at the 1B site.

Table 1. Effects of serotonergic treatments on intake of a mild
isoleucine imbalanced diet

		Food intake response	
Drug	Dose	Basal diet	Imbalanced diet
		(g/24 hr)	
PCPA	300 mg/kg, i.p.	15.7 ± 0.9	6.3 ± 0.9[*]
Saline	1 ml/kg, i.p.	15.9 ± 1.3	13.0 ± 1.6
DHT	200 µg/rat, i.cv.	18.2 ± 0.5	8.8 ± 1.1[*]
Saline	10 µl/rat, i.cv.	18.3 ± 2.3	11.7 ± 1.0[*]
(+0.1% ascorbic acid)			
Quipazine	5 mg/kg, i.p.	18.9 ± 0.7	8.1 ± 0.6[*]
Saline	1 ml/kg, i.p.	18.3 ± 2.3	16.2 ± 1.0
DHT + quipazine	200 µg/rat + 5 mg/kg	16.5 ± 2.4	5.4 ± 0.4[*]
Saline + saline	10 µl/rat + 1 ml/kg	18.3 ± 2.3	15.1 ± 2.4
8-OH-DPAT	500 µg/kg, s.c.	17.6 ± 1.4	16.3 ± 0.8
Saline	1 ml/kg, s.c.	ND	14.4 ± 1.8
Pirenpirone	50 µg/kg, i.p.	12.6 ± 0.7	4.0 ± 0.4[*]
Saline	1 ml/kg, i.p.	11.1 ± 2.3	9.3 ± 1.8
ICS 205-930	9 mg/kg, i.p.	16.3 ± 0.7	16.2 ± 0.8
Saline	1 ml/kg, i.p.	15.4 ± 0.7	11.6 ± 1.1[*]
MDL 72,222	2 mg/kg, i.p.	14.9 ± 0.6	15.3 ± 0.5
Propylene glycol	1 ml/kg, i.p.	16.1 ± 0.8	13.5 ± 0.6[*]
		(g/100g body weight/24 hr)	
Metergoline	1 mg/kg, i.p.	7.3 ± 0.6	6.0 ± 0.6[*]
Saline	1 ml/kg, i.p.	6.9 ± 0.6	5.3 ± 0.6[*]
(+5% ascorbic acid)			

Abbreviations are the same as those used in the text. Values are the
mean ± SEM. [*]Significantly different from basal diet intake, P ≤
0.05 by LSD after ANOVA as in Fig. 1. ND: not determined.

The type 2 site

In contrast to the 5-HT$_1$ sites, 5-HT$_2$ receptors are a homogeneous population (Peroutka, 1988) and appear to mediate 5-HT excitation. These excitatory actions may reflect a facilitation of excitatory influences on other
substances (Peroutka et al., 1981). Subchronic treatment with the selective
5-HT$_2$ antagonist ritanserin produced very small and transient increases in
food intake (Massi and Marini, 1987). Ritanserin also attenuated the anorectic effect of exogenous 5-HT administered i.p., but was completely inactive on the anorectic effect of 5-HT injected into the paraventricular nucleus. The 5-HT agonist 1-(2,5-dimethoxy-4-iodophenyl)-2-aminopropane (DOI)
inhibited feeding. This anorectic effect was blocked by centrally acting

5-HT$_2$ antagonists, but not by a peripherally-acting 5-HT$_2$ antagonist, suggesting a central effect (Schechter and Simansky, 1988).

The effect of quipazine in depressing intake in rats fed an isoleucine imbalanced diet, described above, could have been mediated by the 5-HT$_{1B}$ site. However, quipazine also acts at the type 2 receptor (Leysen, 1985) and has been reported to act as a ligand at a type 3 site as well (Peroutka and Hamik, 1988). Therefore, we used pirenpirone, a specific 5-HT$_2$ antagonist (Colpaert and Janssen, 1983) to examine effects at the type 2 site.

Method: Rats were assigned to 4 groups, in a 2 x 2 design with the 2 drug conditions: pirenpirone at a dose of 50 μg/kg body weight or saline, and the basal and imbalanced diet conditions. Food intake was measured for 24 hr.

Results: Contrary to what would be expected if the 5-HT$_2$ receptor were involved in our feeding paradigm, specific antagonism of the 5-HT$_2$ receptor subtype exacerbated the feeding depression to the mild isoleucine imbalanced diet. Food intake after pirenpirone was similar to that seen in the DHT plus quipazine treated animals, in which we had assumed the response to be due to agonist enhancement of postsynaptic supersensitivity. This was not a generalized anorexia as might be seen with an excessive dose, since intake of the basal diet was not affected (Table 1).

Conclusion: This may be considered a paradoxical response, but Lakoski and Aghajanian saw just such an effect with iontophoresis of a 5-HT$_2$ antagonist on a 5-HT-inhibited neuron, and observed a similar paradoxical response. They suggested that the 5-HT$_2$ receptors may modulate an inhibitory response of 5-HT. Thus, the action of the type 2 antagonist may be to release other 5-HT receptors for full inhibitory activity (Lakoski and Aghajanian, 1985). These results suggest that, since we have seen an exacerbation of the anorectic response with both an antagonist and an agonist of the 5-HT$_2$ receptor subtype, the type 2 receptor is unlikely to be the specific receptor that mediates the 5-HT induced anorexia in this model.

The type 3 site

The 5-HT$_3$ receptor has been shown to be synonymous with what was first described as the 5-HT 'M' receptor (Bradley et al., 1986). The 5-HT$_3$ receptor has been extensively characterized in the periphery and appears to mediate the excitatory effects of 5-HT there (Richardson and Engel, 1986). There is evidence for at least three different types of 5-HT$_3$ receptors in the periphery. At these sites, 5-HT appears to modulate the action of other neurotransmitters (Richardson and Engel, 1986). 5-HT$_3$ receptors have also been found centrally (Kilpatrick et al., 1987; Peroutka and Hamik, 1988), although the function of the 5-HT$_3$ receptors in the CNS has not yet been elucidated.

The availability of potent and selective 5-HT$_3$ antagonists, like ICS 205-930 and MDL 72,222, has resulted in a better understanding of the function of 5-HT$_3$ receptors. One of the proposed roles of the 5-HT$_3$ receptor is mediation of gastric emptying, either via an action within the gut (Costall et al., 1988) or centrally (Costall et al., 1985). Also, chemotherapeutic agents may cause emesis by enhancing 5-HT function. Blockade of the 5-HT$_3$ receptor with ICS 205-930, MDL 72,222 and other compounds have been shown to antagonize cisplatin-induced emesis in ferrets (Costall et al., 1987). It appears that the action of 5-HT on 5-HT$_3$ receptors located on afferent nerve pathways from the viscera to the area postrema may evoke vomiting, and afford a peripheral site of action for the antagonist drugs. However, emesis can also be induced by the injection of cisplatin directly into the brain ventricular system and a central site of action is possible (Costall, 1988). 5-HT$_3$ antagonists have also been tested in models of experimental or provoked anxiety. These data indicate that antagonism of 5-HT$_3$ function can reduce anxiety in animal models (Costall et al., 1988). There is evidence that manipulations of the 5-HT$_3$ receptor may also moderate limbic dopamine over-activity (Costall, 1987), and it is known that dopamine has important influences on feeding. Thus, several lines of evidence suggest that the 5-HT$_3$ receptor subtype could also be involved in feeding.

In a preliminary experiment with ICS 205-930, we noticed an attenuation of the anorectic response to an imbalanced diet. However, food intake of the ICS 205-930 treated, basal diet group appeared to be slightly depressed in the first three hrs, suggesting a possible toxic effect of the drug even at the 15 mg/kg dose, which was the lowest effective dose listed in the Sandoz Investigators Brochure. Therefore, an experiment was designed to determine the minimum drug dose necessary to achieve attenuation of the food intake depression and to minimize any possible toxic effect of the drug.

Methods: ICS 205-930 (2 mg/ml) was injected at doses of 0, 6, 9 or 12 mg/kg body weight. The 0 mg/kg group received an injection volume equivalent to that of the 12 mg/kg group. Each group received the mild isoleucine imbalanced diet at the onset of the dark cycle. Food intake was measured during the first day.

A second experiment was designed to test whether ICS 205-930 would also attenuate the food intake depression seen with a severe isoleucine imbalanced diet, which is more anorexigenic than the mild isoleucine imbalanced diet. This experiment was set up in a 2 x 3 factorial design. Rats were injected with either saline or ICS 205-930 (9 mg/kg) and were fed either the basal, the mild isoleucine imbalanced or the severe isoleucine imbalanced diet. Food intake was measured at 24 hr.

An amino acid imbalanced diet limiting in threonine was used in the next experiment to determine if the effects of ICS 205-930 are specific to the diets having isoleucine as the limiting amino acid. Saline or ICS 205-930 (9 mg/kg) was injected into rats prefed a threonine basal diet. The rats were then presented with either the basal or the threonine imbalanced diet, and food intake was measured at 24 hr.

MDL 72,222 is also a potent antagonist specific for the 5-HT$_3$ receptor (Costall et al., 1988). This drug was used to determine if feeding behavior with an amino acid imbalanced diet would be altered similarly with a different 5-HT$_3$ antagonist. Rats were reused from the severe isoleucine experiment (see above), and assigned to groups such that an animal previously treated with drug or imbalanced diet would not receive a similar treatment in this experiment. Only two of the rats previously receiving the severe isoleucine imbalanced diet were reused. Again, the mild isoleucine imbalanced diet was used. MDL 72,222 (2 mg/kg) or the vehicle, propylene glycol, was injected at 1 ml/kg.

Results: ICS 205-930, a 5-HT$_3$ receptor antagonist, prevented the initial depression in food intake normally seen with a mild isoleucine imbalanced diet, while, as usual, the imbalanced diet caused a significant depression in food intake in the vehicle injected group (F: 4,21 = 3.62; P < 0.05; Table 1).

In the dose-response study, the 9 mg/kg dose was found to be most effective in reversing the food intake depression with dietary amino acid imbalance. By the end of the first day, imbalanced diet intake of rats given 9 mg/kg ICS 205-930 was 30% higher than that of the vehicle treated group, compared with 6 and 19% increases in food intake with the 6 mg/kg and 12 mg/kg doses, respectively.

ICS 205-930 (9 mg/kg) was also given to rats fed a more severe isoleucine imbalanced diet. Comparison of results from the two imbalanced diets showed that the depression in food intake was greater and became apparent more quickly with the severe isoleucine imbalance than with the mild isoleucine imbalanced diet, as has been observed previously (Gietzen et al., 1986b). ICS 205-930 did not alter intake of the basal diet, but at 24 hr it had increased intake of the mild imbalanced diet and severe unbalanced diet by 39 and 98%, respectively (Fig. 3). Thus, the drug had an even greater effect in restoring normal food intake when rats were fed the severe imbalanced diet.

When a threonine imbalanced diet was tested in place of the isoleucine imbalanced diets, the animals given ICS 205-930 again responded with increased intake of the imbalanced diet (F: 4,23 = 13.40; P < 0.01). ICS 205-930 caused intake of the threonine imbalanced diet to increase by 54%. This increase was greater than the 39% increase seen with the mild isoleucine imbalanced diet, but less than the 98% increase seen with the severe isoleucine imbalanced diet.

MDL 72,222, another 5-HT$_3$ antagonist, also attenuated the food intake depression seen with the mild isoleucine imbalanced diet. The food intakes were not significantly different, since intake of the imbalanced diet only tended to be depressed in the vehicle + imbalanced diet group. However, there was no anorexia seen with the imbalanced diet in the drug treated group. Thus, the effect of this second type 3 antagonist, MDL 72,222, was similar to that seen with ICS 205-930.

Conclusions: The dramatic increases in intake of the imbalanced diets seen with the potent and specific 5-HT$_3$ antagonists ICS 205-930 and MDL 72,222 have provided evidence that the 5-HT$_3$ receptor is involved in the feeding depression seen with amino acid imbalance. There are few treatments which will increase food intake but many which will inhibit feeding - some by merely interfering with the natural expression of behavior (Blundell, 1984b). This also appears to hold true for the consumption of amino acid imbalanced diets. In addition, treatments which are able to influence intake of the mild isoleucine imbalanced diet may be ineffective when tested with a severe isoleucine imbalanced diet (Leung and Rogers, 1970; Gietzen, unpublished observations). In our experiments, ICS 205-930 increased intake of the more severely imbalanced diets to 83 and 85% of control for the severe isoleucine imbalanced diet and the threonine imbalanced diet, respectively. The ability of 5-HT$_3$ antagonists to increase intake of imbalanced diets suggests a specific role for this receptor in the feeding depression associated with these diets. Thus, although the 5-HT$_1$ and 5-HT$_2$ receptors have not been conclusively ruled out, the involvement of the 5-HT$_3$ receptor in the food intake depression seems clearly demonstrated.

The anti-anorectic action of ICS 205-930 appears not to be specific for a particular amino acid imbalance, since the drug increased intake of the threonine imbalanced diet as well as the isoleucine diets. It is interesting to note that ICS 205-930 also increased intake of the threonine basal diet by about 13%. There was also a slight, but non-significant, increase in intake of the isoleucine basal diet in one experiment. This does not necessarily indicate an ability of this drug to increase intake of an adequate diet, as these basal diets may also be aversive, being low in protein and slightly limiting in one amino acid. It is possible that ICS 205-930 had a similar effect on the basal diet as it did with the imbalanced diet. This is consistent with data which suggest that the drug effect is more robust with the more severely imbalanced diets.

The mechanism by which the 5-HT$_3$ receptor is involved in this response remains speculative. Until recently, the 5-HT$_3$ receptor was thought only to

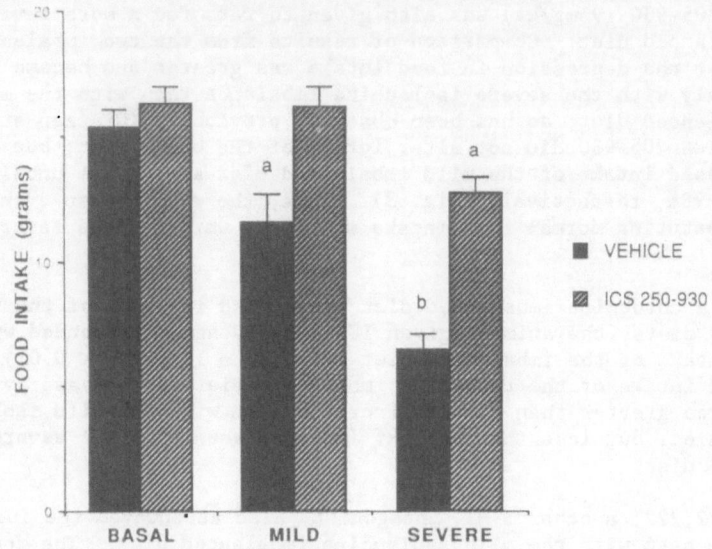

Fig. 3. Effects of ICS 205-930 on 24 hr food intake of isoleucine basal, mild imbalanced or severly imbalanced diet, as indicated on the abcissa. Black bars signify vehicle treated group means ± SEM, hatched bars signify drug treated groups. Superscript letters indicate significance: a: P < 0.05 vs. basal, b: P < 0.01 vs. basal, by multiple ANOVA with Fisher's LS means test posthoc.

exist in the periphery, but now has also been shown to be widely distributed centrally (Kilpatrick et al., 1987; Peroutka and Hamik, 1988). A possible central action of the 5-HT$_3$ antagonists in this paradigm is more consistent with what is known about the mechanism of the feeding depression associated with amino acid imbalance (Gietzen et al., 1986a; Leung and Rogers, 1987), although, as noted above, several of the peripheral actions of 5-HT$_3$ antagonists may affect food intake.

Studies with a general 5-HT antagonist

We selected metergoline, an antagonist at the 5-HT$_1$, 5-HT$_2$ and dopamine, but not the 5-HT$_3$, receptor sites (Dourish et al., 1987), to determine whether attenuation of the food intake depression was specific to the blockade of the 5-HT$_3$ receptor.

Method: Metergoline, in a dose of 1 mg/kg, was injected in a volume of 1 ml/kg in a 5% ascorbic acid solution in water. This dose of metergoline, in our hands, was able to block the 5-HT syndrome induced by 10 mg/kg of quipazine. Food intake was recorded during the first 24 hr after initiation of eating at the onset of the dark cycle.

Results: By 24 hr, the effect of the imbalanced diet was significant (overall F: 3,20 = 4.01; P = 0.02; for effect of diet, F: 11.59; P < 0.01). Intake of the vehicle + imbalanced diet group tended to be decreased relative to the vehicle basal group (P = 0.07). Food intake of the metergoline + imbalanced diet group did not differ from that of the vehicle + imbalanced diet group, and was significantly less than that of either of the basal diet groups.

Conclusion: In contrast to the restoration of feeding that was seen with blockade of the $5-HT_3$ receptor subtype, blockade of the $5-HT_{1A}$, $5-HT_{1B}$, $5-HT_{1C}$ and dopamine receptors (Dourish et al., 1987) had no effect on intake of the amino acid imbalanced diet. This result supports the specificity of the $5-HT_3$ receptor subtype in the activity of 5-HT that is involved in the anorectic response to amino acid imbalanced diets.

GENERAL DISCUSSION

After our initial observation of increased 5-HT activity in the brains of rats fed imbalanced amino acid diets, we have studied the role of 5-HT in the feeding depression seen with these diets. We have concluded that the action of 5-HT in this anorectic response may be central, and located at a postsynaptic site.

In the pharmacological trials, 8-OH-DPAT, an agonist at the $5-HT_{1A}$ receptor, ameliorated the anorectic response, and quipazine, an agonist, equipotent at the $5-HT_{1B}$ and $5-HT_2$ sites (Leysen, 1985), exacerbated the food intake depression seen with the mild isoleucine imbalanced diet, as did pirenpirone, the type 2 antagonist. We can rule out the 1A site as mediating the anorexia, since in the untreated animal the response is to decrease intake of the imbalanced diet. Thus, our requirements, outlined above for the specific receptor underlying the response, are not met. Likewise, since the specific type 2 antagonist, pirenpirone, exacerbated the anorexia, we can eliminate the type 2 site as mediating the response to imbalanced diets. If quipazine was acting at the 1B site, its action in furthering the anorexia suggests that the 1B site does fit the requirement, but blockade of the 1A, 1B, and 1C sites by metergoline resulted in no change in the response. Thus, it is unlikely that the type 1 receptors mediate the anorectic response under study here. The effects of the antagonists at the $5-HT_3$ receptor subtype do meet the requirements for a specific receptor, however, since they restored feeding in the rats fed the imbalanced amino acid diets. Quipazine has also been reported to be a specific ligand at the type 3 receptor in the brain (Peroutka and Hamik, 1988). If quipazine is acting as an agonist in rats fed imbalanced amino acid diets, then the results with quipazine still fit the requirements for a specific receptor listed above and are consistent with our results with the type 3 antagonists.

A possible mechanism of the $5-HT_3$ receptor in the amino acid imbalanced diet model is through mediation of dopamine activity. Dopaminergic drugs have been found to affect food intake (Smith and Schneider, 1988) and there is evidence that manipulations of the $5-HT_3$ receptor sub-type may moderate limbic dopamine overactivity (Coscall et al., 1988). The results with metergoline, in which dopamine receptors were blocked along with the $5-HT_1$ and $5-HT_2$ receptors, suggest that direct action of dopamine, alone, is not important since there was no change in the response to the imbalance. Indirect action of dopamine mediated by the $5-HT_3$ receptor would be consistent with the present data, however, and we do have preliminary data from experiments with dopaminergic agents that support this suggestion. We conclude that the results of the studies presented here support a specific role for $5-HT_3$ receptors in mediating the role of 5-HT in this model of anorexia.

The regulation of food intake undoubtedly involves a host of interactive mechanisms. The use of the imbalanced diet model of rapid onset anorexia may help to elucidate the roles of the various neurotransmitters in feeding. Indeed, the use of feeding models, such as the amino acid imbalanced diet used in these studies, may allow us to identify pieces in the very complex physiological puzzle of how animals control their food intake.

REFERENCES

Baumgarten, H.G., and Lachenmayer, L., 1977, Chemically induced degeneration of the indoleamine-containing nerve terminals in rat brain, **Brain Res.**, 38:228-232.

Beverly, J.L., Gietzen, D.W., Leung, P.M.B., and Rogers, Q.R., 1988, Threonine (Thr) injection into the prepyriform cortex increases the intake of thr imbalanced diet, Soc. Neurosci. Abstr., 14:215.6.

Blundell, J.E., 1984a, Serotonin and appetite, **Neuropharmacology**, 23:1537-1551.

Blundell, J.E., 1984b, Systems and interactions, an approach to the pharmacology of eating and hunger, in: **"Eating and its Disorders"**, Stunkard A.J., and Stellar, E., eds., Raven Press, New York, pp. 39-65.

Bradley, P.B., Engel, G., Feniuk, W., Fozard, J.R., Humphrey, P.P.A., Middlemiss, D.N., Mylecharane, E.J., Richardson, B.P., and Saxena, P.R., 1986, Proposals for the classification and nomenclature of functional receptors for 5-hydroxytryptamine, **Neuropharmacology**, 25:563-576.

Chaouloff, F., and Jean Renaud, B., 1987, 5-HT$_{1A}$ and alpha$_2$ adrenergic receptors mediate the hyperglycemic and hypoinsulinemic effects of 8-hydroxy-2-(di-n-propylamino)tetralin in the conscious rat, J. **Pharmacol. Exp. Therap.**, 243:1159-1166.

Colpaert, F.C., and Janssen, P.A.J., 1983, The head-twitch response to intraperitoneal injection of 5-hydroxytryptophan in the rat: antagonist effects of purported 5-hydroxytryptamine antagonists and of pirenpirone, an LSD antagonist, **Neuropharmacology**, 22:993-1000.

Cooper, S.J., 1987, Palatability-induced food consumption is stimulated by 8-hydroxy-2-di-n-propylaminotetralin (8-OH-DPAT) in: **"Brain 5-HT$_{1A}$ Receptors"**, Dourish, C.T., Ahlenius, S., and Hutson, P.H., eds., Ellis Horwood Ltd., Chichester, England, pp. 233-242.

Costall, B., Domeney, A.M., Naylor, R.J., and Tattersall, J.D., 1987, Emesis induced by cisplatin in the ferret as a model for the detection of antiemetic drugs, **Neuropharmacology**, 26:1321-1326.

Costall, B., Gunning, S.J., and Naylor, R.J., 1985, An analysis of the hypothalamic sites at which substituted benzamide drugs act to facilitate gastric emptying in the guinea pig, **Neuropharmacology**, 24:869-875.

Costall, B., Naylor, R.J., and Tyers, M.B., 1988, Recent advances in the neuropharmacology of 5-HT$_3$ agonists and antagonists, **Rev. Neurosci.**, 2:41-65.

Dourish, C.T., Ahlenius, S., and Hutson, P.H., 1987, Classification and summary tables, in: **"Brain 5-HT$_{1A}$ Receptors, Behavior and Neurochemical Pharmacology"**, Dourish, C.T., Ahlenius, S., and Hutson, P.H., eds., Ellis Horwood Ltd., Chichester, England, pp. 11-12.

Dourish, C.T., Hutson, P.H., Kennett, G.A., and Curzon, G., 1986, 8-OH-DPAT-induced hyperphagia: its neural basis and possible therapeutic relevance, **Appetite**, Suppl., 7:127-140.

Engel, G., Gothert, M.K., Hoyer, D., Schlicker, E., and Hillenbrand, K., 1986, Identity of inhibitory presynaptic 5-hydroxytryptamine (5-HT) autoreceptors in the rat brain cortex with 5-HT$_{1B}$ binding sites, **Naunyn-Schmiedeberg's Arch. Pharmacol.**, 357:1-7.

Fuller, R.W., 1985, Drugs altering serotonin synthesis and metabolism. in: **"Neuropharmacology of Serotonin"**, Green, A.R., ed., Oxford University Press, New York, pp. 1-20.

Garattini, S., Mennini, T., Bendotti, C., Invernizzi, R., and Samanin, R., 1986, Neurochemical mechanism of action of drugs which modify feeding via the serotonergic system, **Appetite**, Suppl., 7:15-38.

Gietzen, D.W., Leung, P.M.B., Castonguay, T.W., Hartman, W.J., and Rogers, Q.R., 1986a, Time course of food intake and plasma and brain amino acid concentrations in rats fed amino acid-imbalanced or deficient diets, in: **"Interaction of the Chemical Senses with Nutrition"**, Kare, M.R., and Brand, J.G., eds., Academic Press, New York, pp. 415-456.

Gietzen, D.W., Leung, P.M.B., and Rogers, Q.R., 1986b, Norepinephrine and amino acids in prepyriform cortex of rats fed amino acid-imbalanced diets, **Physiol. Behav.**, 36:1071-1080.

Gietzen, D.W., Rogers, Q.R., Leung, P.M.B., Semon, B., and Piechota, T., 1987, Serotonin and feeding responses of rats to amino acid imbalance: the initial phase, **Am. J. Phsyiol.**, 253:R763-R771.

Hammer, V.A., Gietzen, D.W., Beverly, J.L., Sworts, V.D., and Rogers, Q.R., 1990, The role of adrenal hormones in the initial feeding response and adaptation of rats to amino acid imbalance, **J. Nutr.**, in press.

Harper, A.E., Benevenga, N.J., and Wohlheuter, R.M., 1970, Effects of ingestion of disportionate amounts of amino acids, **Physiol. Rev.**, 50:428-558.

Hutson, P.H., Dourish, C.T., and Curzon, G., 1987, 8-Hydroxy-2-(di-n-propyl-amino)tetralin (8-OH-DPAT)-induced hyperphagia: neurochemical and pharmacological evidence for an involvement of 5-hydroxytryptamine somatodendritic autoreceptors, in: **"Brain 5-HT$_{1A}$ Receptors"**, Dourish, C.T., Ahlenius, S., and Hutson, P.H., eds., Ellis Horwood Ltd., Chichester, England, pp. 211-232.

Joyce, D., and Mrosovsky, N., 1964, Eating, drinking and activity in rats following 5-hydroxytryptophan (5-HTP) administration, **Psychopharmacologia**, 5:417-423.

Kennett, G.A., Dourish, C.T., and Curzon, G., 1987, 5HT$_{1B}$ agonists induce anorexia at a postsynaptic site, **Eur. J. Pharmacol.**, 141:429-435.

Kilpatrick, G.J., Jones, B.J., and Tyers, M.B., 1987, Identification and distribution of 5-HT$_3$ receptors in rat brain using radioligand binding, **Nature**, 330:746-748.

Koe, B.K., and Weissman, A., 1966, p-Chlorophenylalanine: a specific depletor of brain serotonin, **J. Pharmacol. Exp. Therap.**, 154:499-516.

Lakoski, J.M., and Aghajanian, G.K., 1985, Effects of ketanserin on neuronal responses to serotonin in the prefrontal cortex, lateral geniculate and dorsal raphe nucleus, **Neuropharmacology**, 24:265-273.

Leung, P.M.B., Larson, D.M., and Rogers, Q.R., 1972, Food intake and preference of olfactory bulbectomized rats fed amino acid imbalanced or deficient diets, **Physiol. Behav.**, 9:553-557.

Leung, P.M.B., and Rogers, Q.R., 1969, Food intake: regulation by plasma amino acid pattern, **Life Sci.**, 8:1-9.

Leung, P.M.B., and Rogers, Q.R., 1970, Effect of amino acid imbalance and deficiency on food intake of rats with hypothalamic lesions, **Nutr. Rep. Intl.**, 1:1-10.

Leung, P.M.B., and Rogers, Q.R., 1987, The effect of amino acids and protein on dietary choice, in: **"Umami: A Basic Taste"**, Kawamura, Y., and Kare, M.R., eds., Marcel Dekker, New York, pp. 565-610.

Leung, P.M.B., Rogers, Q.R., and Harper, A.E., 1968, Effect of cortisol on growth, food intake, dietary preferences and plasma amino acid pattern of rats fed amino acid imbalanced diets, **J. Nutr.**, 96:139-151.

Leysen, J.E., 1985, Characterization of serotonin receptor binding sites, in: **"Neuropharmacology of Serotonin"**, Green, A.R., ed., Oxford University Press, New York, pp. 76-116.

Massi, M., and Marini, S., 1987, Effect of the 5HT$_2$ antagonist ritanserin on food intake and on 5HT-induced anorexia in the rat, **Pharmacol. Biochem. Behav.**, 26:333-340.

Peroutka, S.J., 1988, 5-Hydroxytryptamine receptor subtypes: molecular, biochemical and physiological characterization, **Trends Neurosci.**, 11:496-500.

Peroutka, S.J., and Hamik, A., 1988, [^{3}H] Quipazine labels 5-HT$_3$ recognition sites in rat cortical membranes, **Eur. J. Pharmacol.**, 148:297-299.

Peroutka, S.J., Lebovitz, R.M., and Snyder, S.H., 1981, Two distinct central serotonin receptors with different physiological functions, **Science**, 212:827-829.

Peters, J.C., Bellissimo, D.B., and Harper, A.E., 1984, L-Tryptophan injection fails to alter nutrient selection by rats, **Physiol. Behav.**, 32:253-259.

Richardson, B.P., and Engel, C., 1986, The pharmacology and function of 5-HT$_3$ receptors, **Trends Neurosci.**, 7:424-428.

Rogers, Q.R., and Leung, P.M.B., 1977, The control of food intake: when and how are amino acids involved, in: "**The Chemical Senses and Nutrition**", Kare, M.R., and Maller, O., eds., Academic Press, New York, pp. 213-249.

Schechter, L.E., and Simansky, K.J., 1988, 1-(2,5-dimethoxy-4-iodophenyl)-2-aminopropane (DOI) exerts an anorectic action that is blocked by 5-HT$_2$ antagonists in rats, **Psychopharmacology**, 94:342-346.

Smith, G.P., and Schneider, L.H., 1988, Relationships between mesolimbic dopamine function and eating behavior, **Annal. N.Y. Acad. Sci.**, 537:254-261.

Stickney, G.D., Leung, P.M.B., Rogers, Q.R., Lepkovsky, S., and Schmidt, P., 1976, The effect of total gastrectomy on free feeding patterns in rats, **Fed. Proc.**, 35:Abstr. 520.

Tobin, G., and Boorman, K.N., 1979, Carotid or jugular amino acid infusion and food intake in the cockerel, **Br. J. Nutr.**, 41:157-162.

Willcock, E.G., and Hopkins, F.G., 1906, The importance of individual amino acids in metabolism; observations on the effect of adding tryptophane to a dietary in which zein is the sole nitrogenous constituent, **J. Physiol.**, 35:88-102.

SESSION VII

CLINICAL ASPECTS

CARDIOVASCULAR AND ENDOCRINE PROPERTIES OF L-TRYPTOPHAN IN COMBINATION WITH

VARIOUS DIETS

H. Lehnert and J. Beyer

IIIrd Medical Clinic
Dept. of Endocrinology and Metabolism
University of Mainz
Germany

INTRODUCTION

Brain serotonin neurons are intimately involved in a number of relevant physiological functions such as cardiovascular regulation, neuroendocrine output from the anterior pituitary (e.g. ACTH, prolactin), regulation of behavior (e.g. agression, sleep, locomotor and sexual behavior), mood or appetite control (Fernstrom, 1983; Lehnert et al., 1987; Spring et al., 1987; Wurtman, 1987). The synthesis of brain serotonin is dependent on the availability of the large neutral amino acid L-tryptophan that is hydroxylated to 5-L-hydroxytryptophan and subsequently decarboxylated to yield serotonin. The rate-limiting enzyme tryptophan hydroxylase has a Michaelis constant of approximately $2-3 \times 10^{-5}$ M with tetrahydrobiopterin used as a cofactor (Tong and Kaufmann, 1975) and thus approximates normal brain tryptophan concentrations of about $1-5 \times 10^{-5}$ M. Therefore, the enzyme is not saturated under normal circumstances and an increased availability of brain tryptophan will lead to an enhanced synthesis of brain serotonin (Fernstrom and Wurtman, 1971; Fernstrom, 1983).

The availability of tryptophan to the central nervous system is further enhanced, when the amino acid is ingested together with a high-carbohydrate diet. This is based on the observation, that the carbohydrate-induced insulin secretion readily promotes the transport of large neutral amino acids into striated muscle tissue that compete with tryptophan for transport across the blood-brain barrier. Thus, as an index of tryptophan availability for brain serotonin synthesis a ratio of tryptophan to amino acids with comparable affinity for the cerebrovascular transport sites is calculated (i.e. tyrosine, valine, phenylalanine, leucine and isoleucine). This ratio will be referred to as amino acid ratio. Tryptophan can be bound to serum albumin and is thus not subject to transport into the striated muscle. The albumin binding of tryptophan can be further enhanced by the insulin-induced dissociation of non-esterified fatty acids from albumin (Madras et al., 1974). Since both free and albumin-bound tryptophan can cross the blood brain barrier due to the high affinity of the transport molecule, the capacity of tryptophan to bind to albumin is thus important for prevention of striated muscle uptake and subsequent increase of the amino acid ratio and transport across the blood brain barrier. These mechanisms explain the clinically most relevant dependency of brain serotonin synthesis on the availability of tryptophan, the amino acid ratio and a priori on the composition of a previous meal.

Unlike brain serotonin, the precursor dependency of catecholamine synthesis in the brain is coupled to the firing rate of the tyrosine hydroxylase (TOH) containing neurons. A large number of studies has clearly demonstrated, that a supplementation of L-tyrosine does not augment the synthesis of catecholamines under basal conditions, while an enhanced neuronal activity will increase synthesis and release of dopamine and noradrenaline (Lehnert et al., 1984; Milner and Wurtman 1986; Wurtman, 1987). Physiological examples are increases in noradrenaline metabolites following tyrosine administration in experimental stress (Reinstein et al., 1984) or due to pharmacological interventions designed to enhance TOH activity (Roth et al., 1975; Sved and Fernstrom, 1982). The biochemical basis for this phenomenon is the depolarization-induced phosphorylation of TOH, which leads to a conformational change of the enzyme and renders it no longer dependent on the availability of its cofactor tetrahydrobiopterin, but rather on the availability of the substrate, i.e. tyrosine (Roth et al., 1975). The phosphorylated form of TOH also appears to be much less sensitive to end-product inhibition. As mentioned above, tyrosine supplementation does not affect brain catecholamine synthesis under basal conditions (in the quiescent state). The main reason is the saturation of TOH by about 80%, the Michaelis constant being 25-50 μM (Morgenroth et al., 1976).

The state-dependency of transmitter production from their precursors thus constitutes a major difference between brain serotonin and catecholamine synthesis, a finding that bears most significant implications for the design of experimental and clinical studies. A major focus of our group has been the investigation of physiological effects of precursor amino acids, e.g. on the cardiovascular and endocrine system as well as on animal and human behavior. We will thus briefly summarize here findings with respect to blood pressure behavior in different groups of hypertensive patients and some endocrine changes, namely effects on anterior pituitary hormone secretion.

ANTIHYPERTENSIVE PROPERTIES OF L-TRYPTOPHAN IN HUMANS

In previous studies, we were able to demonstrate that the administration of the immediate brain serotonin precursor 5-hydroxytryptophan in conjunction with a monoamine oxidase inhibitor and the selective peripheral L-amino acid decarboxylase inhibitor carbidopa to cats increased cerebrospinal fluid concentrations of serotonin and its major metabolite 5-HIAA and led to a decrease of sympathetic nervous system activity. Ventricular fibrillation threshold was elevated, blood pressure and heart rate were decreased and efferent sympathetic activity was suppressed in the normal and ischemic heart (Lehnert et al., 1983, 1987; Raeder et al., 1987). The neuroanatomical basis for this effect is the dense innervation of sympathetic preganglionic neurons by bulbospinal serotoninergic neurons, originating in the raphe nuclei (Bowker et al., 1979).

We then decided to investigate the cardiovascular, endocrine and behavioral effects of L-tryptophan in conjunction with two different diets (high-protein vs. high-carbohydrate diet) in humans (Lehnert et al., 1988, 1989; Lehnert et al., submitted). In this study, 42 essential hypertensive patients were enrolled in a placebo-controlled, randomized and double-blind trial. After a wash-out phase, they received either tryptophan (50 mg/kg) or placebo for three weeks at 8 a.m.: in the first week, they ingested it with a standard meal, in the second and third week either with a high-protein (28 g CHO, 75 g protein, 10 g fat) or high-carbohydrate diet (75 g CHO, 7 g protein, 14 g fat) in an intra-group cross-over design (Fig. 1). Blood pressure was taken four times daily under standardized conditions, and blood samples were collected 90 minutes following intake of tryptophan for subsequent anal-

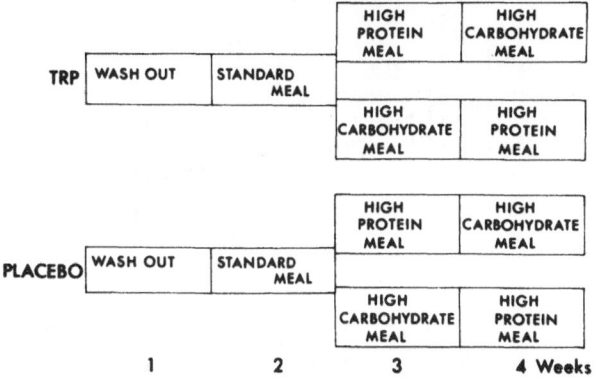

Fig. 1. Schematic representation of the study design: effect of
tryptophan and macronutrients on blood pressure behavior
in essential hypertensives (Lehnert et al., 1988).

ysis of plasma amino acids, plasma renin activity, serum aldosterone and cortisol.

At the end of the four week trial, blood pressure was significantly lower in the tryptophan group when compared with placebo: 140.3 ± 11.8 / 86.2 ±
7.2 vs. 147.5 ± 13.8 / 93.0 ± 7.7 mm Hg (P < 0.01) (Fig. 2). The subgroup
analysis in the tryptophan-treated patients revealed lower blood pressure
levels in those patients who received a carbohydrate-rich meal first as compared with those receiving a protein-rich meal first (Fig. 3). This latter
difference did not reach statistical significance. Serum concentrations of
tryptophan were increased significantly with no differences between the subgroups receiving the amino acid. Also not surprisingly, there was a marked
elevation in the plasma amino acid ratio in the groups receiving L-tryptophan, that was further enhanced when the patients ingested tryptophan together with a high-carbohydrate meal. Thus, the amino acid ratio was 1.5 ± 0.54
in the tryptophan plus carbohydrate group vs. 0.66 ± 0.29 in tryptophan plus
protein group. Following a protein meal alone, the amino acid ratio was
found to be 0.12 ± 0.1, following a high-carbohydrate meal 0.25 ± 0.1 (P <
0.05). Furthermore, a significant inverse correlation was observed between
the height of the ratio and blood pressure levels at almost all measurements
taken. This and the observed blood pressure lowering effects of carbohydrates in the fourth week suggests to us that carbohydrate intake may augment
the tryptophan-mediated drop in blood pressure. The co-administration of
proteins and tryptophan led to a lower amino acid ratio when compared with
carbohydrates, that was nevertheless significantly higher than the ratio under control conditions. The protein content of the high-carbohydrate meal
was approximately 6% and reduced the ratio by about 50%. Thus, an attenuation of tryptophan's effect on the amino acid ratio was clearly observed.

The results of this study are in accordance with a number of animal
studies demonstrating depressor effects of brain serotoninergic neurons and
in particular of L-tryptophan (Kuhn et al., 1980; Echizen and Freed, 1982;
Sved et al., 1982; Fregly et al., 1987) and one human study that investigated
the acute effects of tryptophan on blood pressure in hypertensives (Feltkamp
et al., 1984). The study on the antihypertensive effects of L-tryptophan in
spontaneously hypertensive rats (Sved et al., 1982) clearly suggested that L-
tryptophan acted by enhancing the serotonin release from the presynaptic neuron, since tryptophan's effects could be enhanced by co-administration of the
presynaptic serotonin reuptake-blocker fluoxitene. It should be noted
though, that the involvement of brain serotonin in cardiovascular regulation

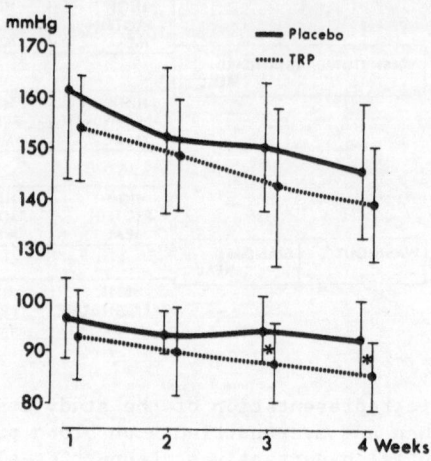

Fig. 2. Effects of L-tryptophan on blood pressure in essential hypertensives (Lehnert et al., 1988).

is a highly controversial subject, and a number of studies point to a centrally mediated pressor effect, that in general appears to be species- and
site-dependent. For example, injection of serotonin into the lateral ventricles or stimulation of the dorsal and median raphe nuclei have been reported
to elevate blood pressure (Smits et al., 1978; Wolf et al., 1981). Taken
together, our animal and human data suggest that the administration of the
serotonin precursor acts through a decrease in sympathetic nervous system
activity to lower blood pressure. In addition, it is suggested that the
amino acid ratio is of potential predictive relevance.

DIFFERENTIAL EFFECTS OF TRYPTOPHAN AND TYROSINE ON BLOOD PRESSURE

In a number of studies, the antihypertensive properties of L-tyrosine
were investigated in the animal model with unequivocal results. Thus, a
dose-dependent lowering of arterial blood pressure with concurrent increases
of the norepinephrine metabolite MHPG sulfate in brain stem of spontaneously
hypertensive rats (SHRs) was found following intraperitoneal injection of L-

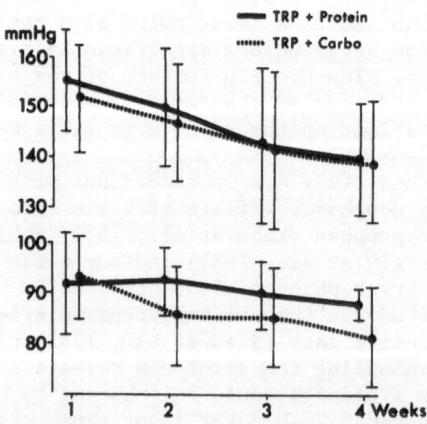

Fig. 3. Effects of different diets within the tryptophan group on blood
pressure in essential hypertensives (Lehnert et al., 1988).

tyrosine (Sved et al., 1979). Similarly, a decrease in blood pressure was found following both acute and chronic infusion of L-tyrosine directly into the lateral ventricles in SHRs (Bresnahan et al., 1980; Yamori et al., 1980). In the same animal model a tyrosine-enriched diet was found to attenuate the development of hypertension (Bossy et al., 1983). Vice versa, the administration of tyrosine to rats in hemorrhagic shock led to a significant increase of blood pressure, an effect likely mediated by enhanced peripheral secretion of catecholamines as suggested by increased adrenomedullary and spinal cord sympathetic neuron firing (Conlay et al., 1981). These data thus apparently confirm that the blood pressure affecting properties of tyrosine are closely related to the state of the animal, i.e. to the starting blood pressure values. While tyrosine administration has little or no effect in normotensive rats, it lowers blood pressure in hypertensive rats presumably by augmenting norepinephrine synthesis and release in relevant brain stem areas, and increases blood pressure in experimentally hypotensive rats. We have further investigated the anatomical correlates of the antihypertensive effect of tyrosine in SHRs and Wistar-Kyoto rats by monitoring the concentrations of MHPH-sulfate as an index for norepinephrine turnover. In SHRs a significant increase of MHPG-sulfate was found in hypothalamus, amygdala, locus coeruleus and remaining brain stem, with highest increases in locus coeruleus and amygdala (Lehnert et al., 1985).

So far, only sparse data are available in human essential hypertension; one controlled study was performed for two weeks in patients with mild essential hypertension and no effects of tyrosine were observed on either supine or standing blood pressure, heart rate and plasma norepinephrine levels, indicating that the administration of tyrosine has no beneficial effect on blood pressure (Sole et al., 1985). We have recently concluded a controlled trial (3 x 1 g of tyrosine) for a period of 10 weeks in a cross-over design and obtained data that suggest a similar interpretation (Walger et al., submitted).

These observations also clearly suggest that the experimental design and in particular the state of the organism is crucial to the outcome and implications of such experiments. Along these lines we have decided to investigate the effects of both tyrosine and tryptophan in a well defined subgroup of hypertensive patients, namely young borderline hypertensives. Borderline hypertensives represent a subgroup of patients with typical cardiovascular features, namely elevated systolic and normal to mildly elevated diastolic blood pressure levels with a high risk to develop stable hypertension. They are defined as having systolic blood pressure values between 140 and 160 mm Hg and/or diastolic blood pressure values between 90 and 95 mm Hg. In addition, strong hemodynamic reactions are observed during various stressful tasks such as mental challenge or cold pressure test (Folkow, 1987). The rationale for this study was to investigate whether a stress-induced increase in blood pressure is sensitive to precursor amino acid application.

A group of 15 male students that were previously identified as suffering from borderline hypertension were studied under three different, randomly applied conditions. Therefore, they received either 5 g of tyrosine, tryptophan or placebo and were exposed to a stressful task consisting of mental arithmetic for 15 minutes under continuous noise distraction of 90 dB delivered over headphones. While the baseline blood pressure levels were similar for all three conditions, the stress-induced increase in systolic blood pressure was significantly attenuated in the tyrosine group with 155.6 ± 10.8 mm Hg when compared with tryptophan (163.0 ± 11.9 mm Hg) and placebo (164.0 ± 10.5) (Fig. 4). Moreover, sub-group analysis revealed an inverse relation between pre- and intra-stress blood pressure levels in the tyrosine group; in other words, the higher the starting blood pressure, the more pronounced was tyrosine's antihypertensive effect (Diebschlag et al., 1989). Thus, while we did not observe an effect of tryptophan on blood pressure behavior in a

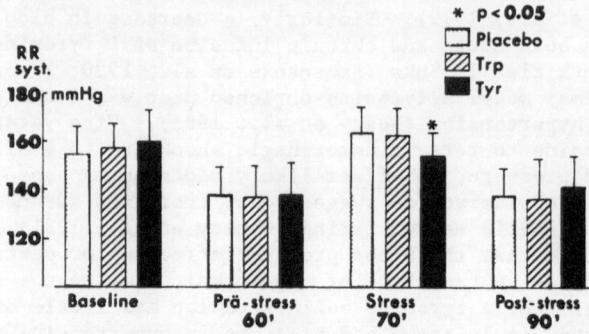

Fig. 4. Effects of tyrosine and tryptophan on stress-induced systolic
blood pressure changes in borderline hypertensives.

stressful task, tyrosine appears to be beneficial in stress-dependent as
opposed to everyday blood pressure behavior.

ENDOCRINE EFFECTS OF L-TRYPTOPHAN AND MACRONUTRIENT INTAKE IN HUMANS

A number of hormonal effects are associated with the administration of
L-tryptophan and little doubt exists that the intravenous application of L-
tryptophan stimulates growth hormone and prolactin secretion in human subjects (Cowen et al., 1985; Winokur et al., 1986). On the other hand, no consistent effects were obtained following oral intake and in general little, if
any, effect on the secretion of these hormones is observed. The effects on
ACTH release are even more inconsistent, with stimulatory (Modlinger et al.,
1980; Lewis and Sherman, 1984) and inhibitory (Woolf and Lee, 1977) modes of
action described. The most likely explanation for the apparent discrepancy
in these studies appeared to be the different time points following tryptophan intake when samples were taken for hormonal analysis and thus different
plasma concentrations. Thus, tryptophan could be stimulatory for the hypo-
thalamic-pituitary-adrenal-(HPA-)axis in humans without excluding the possibility of an early inhibition of the activity of the HPA-axis shortly after
administration of L-tryptophan. In favor of a stimulatory role of the brain
serotonergic system is the finding that the spontaneous release of CRF from
incubated hypothalami was stimulated by serotonin in a dose-dependent fashion
and that this effect was antagonized by methysergide (Buckingham and Hodges,
1977; Holmes et al., 1982). The postsynaptic receptor involved appears to be
of the $5-HT_2$ type. In addition, central serotonergic neurons appear to stimulate aldosterone secretion both in vivo and in vitro, and serotonin antagonists have been shown to decrease serum aldosterone concentrations in patients with idiopathic hyperaldosteronism (Shenker et al., 1985).

In the study in essential hypertensive patients, we did not find data
robust enough to suggest a relation between tryptophan intake and cortisol or
aldosterone secretion. Admittedly though, blood samples were only taken once
90 minutes following nutrient intake and hormone secretion was thus not studied continuously. The data did suggest a negative correlation between the
amino acid ratio and plasma renin activity with renin activity being even
significantly lower following intake of tryptophan and carbohydrates as compared to tryptophan and proteins. Since renin secretion may be enhanced by
activation of beta-adrenoceptors, lowering of plasma renin activity might indicate a reduction in sympathetic nervous tone achieved by the intervention.

Similarly, tryptophan intake had no major effect on serum cortisol concentrations, while independent of tryptophan intake the carbohydrate diet clearly appeared to lower serum cortisol levels (placebo group: 10.6 ± 5.9 in the high-carbohydrate group vs. 13.1 ± 1.8 μg/dl in the protein group; tryptophan group: 10.2 ± 4.4 in the high carbohydrate group vs. 14.7 ± 2.5 μg/dl in the high-protein group; $P < 0.05$, respectively). As mentioned above, this finding has to be interpreted very cautiously with respect to episodic hormone secretion, but is nevertheless in accordance with findings from a recent study that investigated the effects of different diets on testosterone and cortisol concentrations in normal men. Consistently lower concentrations of cortisol were found during the high-carbohydrate diet than during the high-protein diet with parallel changes in the level of the cortisol binding globulin (Anderson et al., 1987). While there is no ready explanation for this finding, it appears possible that alterations in hormone binding globulins represent the intrahepatic effects of dietary factors on synthesis and degradation of liver-derived factors.

In a different study, we administered oral tryptophan and tyrosine (5 g each) to normal men and did not find a significant effect of tryptophan administration on prolactin and TSH secretion, while there was a minor increase in growth hormone levels. Similarly, tryptophan did not affect the TRH- or metoclopramide-(MCP) stimulated prolactin secretion. Very consistently though, we observed a significant increase of basal levels of prolactin and TSH following intake of L-tryptophan with differential effects on the stimulated levels of these hormones. When stimulated with TRH and MCP following tyrosine administration, the prolactin surge was significantly blunted when compared with tryptophan and placebo. A ceiling effect of basal prolactin levels on stimulated levels could be excluded. On the other hand, the TRH-stimulated TSH secretion was further enhanced by tyrosine administration. Since prolactin regulates its own secretion by stimulation of the tuberoinfundibular dopaminergic system, tyrosine may act to amplify the inhibitory effects of dopamine (Lehnert et al., 1988). This again confirms the importance of neuronal activation for tyrosine's mode of action. As demonstrated recently, tyrosine's effects on anterior pituitary hormone secretion appear to be mediated through central alpha-1 adrenoceptors (Al-Damluji, 1988). As opposed to the enhanced prolactin secretion in normal men, in patients bearing a prolactinoma the stimulatory effect of tyrosine on prolactin secretion was completely lost and not different from tryptophan or placebo administration. This observation may be due to a toxic effect of chronic hyperprolactinemia on dopamine neurons (Sarkar et al., 1984).

CONCLUSION

In summary, relevant effects of the precursor amino acids L-tryptophan and L-tyrosine on physiological functions in humans can undoubtedly be observed. The different influence on their transmitter products, that have been found experimentally dependent on the state of the organism, are reflected in the various cardiovascular and endocrine responses described. Thus, the effects of tryptophan on blood pressure behavior in hypertensive patients appears to be a tonic one, while tyrosine rather exerts a phasic action in situations where neuronal activation occurs and presumably brain catecholaminergic neurons fire more rapidly.

Whether these effects are clinically relevant, is too early to tell. Clearly though, the effects of miconutrients such as precursor amino acids on physiological systems and their potential amplification by a concurrent diet have to be considered for diagnostic (i.e. probing a potential deficiency of a neurotransmitter) and therapeutic approaches.

REFERENCES

Al-Damluji, S., 1988, Adrenergic mechanisms in the control of corticotrophin secretion, J. Endocrinol., 119:5-14.

Anderson, K.E., Rosner, W., Khan, M., New, M., Pang, S., Wissel, P., and Kappas, A., 1987, Diet-hormone interactions: protein/carbohydrate ratio alters reciprocally the plasma levels of testosterone and cortisol and their binding globulins in man, Life Sci., 40:1761-1768.

Bossy, J., Guidox, R., Milon, H., and Werzner, H.P., 1983, Development of hypertension in spontaneously hypertensive rats fed l-tyrosine supplemented diets, Zeitschr. Ernährwiss., 22:45-49.

Bowker, R.M., Steinbusch, H.W., and Coulter, J.D., 1979, Serotonergic and peptidergic projections to the spinal cord demonstrated by combined retrograde HRP histochemical and immunocytochemical staining method, Brain Res., 211:412-417.

Bresnahan, M.R., Hatzinikolaou, P., Brunner, H.R., and Gavras, H., 1980, Effects of tyrosine infusion in normotensive and hypertensive rats, Am. J. Physiol., 239:H206-211.

Buckingham, J.C., and Hodges, J.R., 1977, Production of corticotrophin releasing hormone by the isolated hypothalamus of the rat, J. Physiol., 272:469-479.

Conlay, L.A., Maher, T., and Wurtman, R.J., 1981, Tyrosine increases blood pressure in hypotensive rats, Science, 212:559-560.

Cowen, P.J., Gadhvi, H., Gosden, B., and Kolakowska, T., 1985, Responses of prolactin and growth hormone to L-tryptophan infusion: effects in normal subjects and schizophrenic patients receiving neuroleptics, Psychopharmacology, 86:164-169.

Diebschlag, U., Lehnert, H., Reche, A., Warnecke, W., Hellhammer, D., and Beyer, J., 1989, Effects of the precursor amino acids L-tyrosine and L-tryptophan on stress-induced blood pressure increases in borderline hypertensives, Acta Endocrinol., 120, Suppl. 1:257.

Echizen, K., and Freed, C.R., 1982, Long-term infusion of 5-L-hydroxytryptophan increases brain serotonin turnover and decreases blood pressure in normotensive rats, J. Pharmacol. Exp. Ther., 220:579-584.

Feltkamp, H., Meurer, K.A., and Godehardt, E., 1984, Tryptophan-induced lowering of blood pressure and changes of serotonin uptake by platelets in patients with essential hypertension, Klin. Wschr., 62:1115-1119.

Fernstrom, J.D., 1983, Role of precursor availability in control of monoamine biosynthesis in the brain, Physiol. Rev., 63:484-545.

Fernstrom, J.D., and Wurtman, R.J., 1971, Brain serotonin content: physiological dependence on plasma tryptophan levels, Science, 173:149-152.

Folkow, B., 1987, Psychosocial and central nervous influences in primary hypertension, Circul. Monogr., 6, 76:10-19.

Fregly, M.J., Lockley, O.E., van der Voort, J., Sumners, C., and Henley, W.N., 1987, Chronic dietary administration of tryptophan prevents the development of deoxycorticosterone acetate salt induced hypertension in rats, Can. J. Physiol. Pharmacol., 65:753-764.

Holmes, M.C., Renzo, G.D., Becford, U., Gillham, B., and Jones, M.T., 1982, Role of serotonin in the control of secretion of CRF, J. Endocrinol., 93:151-160.

Kuhn, D.M., Wolf, W.A., and Lovenberg, W., 1980, Review of the role of the central serotonergic neuronal system in blood pressure regulation, Hypertension, 2:243-255.

Lehnert, H., Beyer, J., Cloer, E., Gutberlet, I., and Hellhammer, D., Effects of L-tryptophan and various diets on behavioral functions in essential hypertensives, Neuropsychobiology, 21:84-89.

Lehnert, H., Beyer, J., Czernik, C., Schneider, K.P., Schrezenmeir, J., and Krause, U., 1988, Control of prolactin secretion by the dopamine precursor amino acid L-tyrosine, Akt. Ernähr., 14:41-43.

Lehnert, H., Beyer, J., Heismann, I., Siekermann, H., Schmidt, H., Ullrich, K., and Vetter, H., 1988, Einfluss von L-Tryptophan und kohlenhydratreicher Diät auf das Blutdruckverhalten essentieller Hypertoniker, Akt. Ernähr., 13:1-5.

Lehnert, H., Beyer, J., Hellhammer, D., Gutberlet, I., Ullrich, K., and Vetter, H., Effects of L-tryptophan and macronutrient intake on blood pressure, plasma amino acids, and endocrine parameters in essential hypertensives, submitted.

Lehnert, H., Lombardi, F., Raeder, E.A., Lorenzo, A.V., Verrier, R.L., Lown, B., and Wurtman, R.J., 1987, Increased release of brain serotonin reduces vulnerability to ventricular fibrillation in the cat, J. Cardiovasc. Pharmacol., 10:389-397.

Lehnert, H., Lombardi, F., Verrier, R.L., Lown, B., and Wurtman, R.J., 1983, Suppression of cardio-cardiac reflexes by increasing central serotonergic neurotransmission, J. Am. Coll. Cardiol., 1:606.

Lehnert, H., Maher, T., Yokogoshi, H., and R.J. Wurtman, 1985, Effects of tyrosine on blood pressure and turnover of norepinephrine in different brain areas of the spontaneously hypertensive rat, J. Hypertens., 3,412.

Lehnert, H., Reinstein, D.K., Strowbridge, B.W., and Wurtman, R.J., 1984, Neurochemical and biochemical consequences of an acute uncontrollable stress: effects of dietary tyrosine, Brain Res., 303:215-223.

Lewis, D.A., and Sherman, B.M., 1984, Serotonergic stimulation of adrenocorticotropin secretion in man, J. Clin. Endocrinol. Metabol., 58:458-462.

Madras, B.K., Cohen, E.L., Messing, R., Munro, H.N., and Wurtman, R.J., 1974, Relevance of serum free tryptophan to tissue tryptophan concentrations, Metabolism, 23:1107-1116.

Milner, J.D., and Wurtman, R.J., 1986, Catecholamine synthesis: physiological coupling to precursor supply, Biochem. Pharmacol., 35:875-881.

Modlinger, R.S., Schonmuller, J.M., and Arora, S.P., 1980, Adrenocorticotropin release by tryptophan in man, J. Clin. Endocrinol. Metabol., 50:360-363.

Morgenroth, V.A., Walters, J.R., and Roth, R.H., 1976, Dopaminergic neurons - alterations in the kinetic properties of tyrosine hydroxylase after cessation of impulse flow, Biochem. Pharmacol., 25:655-661.

Raeder, E.A., Berger, R., Kenet, R., Kiely, J.P., Lehnert, H., Cohen, R.J., and Lown, B., 1987, Assessment of autonomic cardiac control by power spectrum of heart rate fluctuations, J. Appl. Cardiol., 2:283-300.

Reinstein, D.K., Lehnert, H., Scott, N.A., and Wurtman, R.J., 1984, Tyrosine prevents behavioral and neurochemical correlates of acute stress in rats, Life Sci., 34:2225-2231.

Roth, R.H., Morgenroth, J.D., and Salzman, P.M., 1975, Tyrosine hydroxylase: allosteric activation induced by stimulation of activated noradrenergic neurons, Naunyn-Schmiedeb. Arch. Pharmacol., 289:327-334.

Sarkar, D.K., Gottschall, P.E., and Meites, J., 1984, Decline of tuberoinfundibular dopaminergic function resulting from chronic hyperprolactinemia in rats, Endocrinology, 115:1269-1274.

Shenker, Y., Gross, M.B., and Grekin, R.J., 1985, Central serotonergic stimulation of aldosterone secretion, J. Clin. Invest., 76:1485-1490.

Smits, J.F., van Essen, H., and Stuyker-Boudier, H.A.J., 1978, Serotonin-mediated cardiovascular responses to electrical stimulation of the raphe nuclei in the rat, Life Sci., 23:173-178.

Sole, M.J., Benedict, C.R., Myers, M.G., Leenen, F.H.H., and Anderson, G.H., 1985, Chronic dietary tyrosine supplements do not affect mild essential hypertension, Hypertension, 7:593-596.

Spring, B., Chiodo, J., and Bowen, D.J., 1987, Carbohydrates, tryptophan and behavior: a methodological review, Psychol. Bull., 102:234-256.

Sved, A., and Fernstrom, J.D., 1982, Tyrosine availability and dopamine synthesis in the striatum: studies with gamma-butyrolactone, Life Sci., 29:743-748.

Sved, A.F., Fernstrom, J.D., and Wurtman, R.J., 1979, Tyrosine administration reduces blood pressure and enhances brain norepinephrine release in spontaneously hypertensive rats, **Proc. Nat. Acad. Sci. USA**, 76:3511-3514.

Sved, A.F., van Itallie, C.M., and Fernstrom, J.D., 1982, Studies on the antihypertensive action of L-tryptophan, **J. Pharmacol. Exp. Ther.**, 221:329-333.

Tong, J.H., and Kaufmann, S., 1975, Tryptophan hydroxylase. Purification and some properties of the enzyme from rabbit hindbrain, **J. Biol. Chem.**, 250:4152-4158.

Walger, P., Lehnert, H., Baumgart, P., Rahn, K.H., and Vetter, H., Chronic effects of L-tyrosine on circadian blood pressure in borderline hypertensives, submitted.

Winokur, A., Lindberg, N.O., Lucki, I., Philipps, J., and Amsterdam, J.W., 1986, Hormonal and behavioral effects associated with intravenous L-tryptophan administration, **Psychopharmacology**, 88:213-219.

Wolf, W.A., Kuhn, D.M., and Lovenberg, W., 1981, Pressor effects of dorsal raphe stimulation and intrahypothalamic serotonin in the spontaneously hypertensive rat, **Brain Res.**, 208:192-197.

Woolf, P.D., and Lee, L., 1977, Effect of the serotonin precursor tryptophan on pituitary hormone secretion, **J. Clin. Endocrinol. Metabol.**, 45:123.

Wurtman, R.J., 1987, Nutrients affecting brain composition and behavior, **Integr. Psychiatry** 5: 226-238.

Yamori, Y., Fujiwara, M., Horie, K., and Lovenberg, W., 1980, The hypotensive effect of centrally administered tyrosine, **Eur. J. Pharmacol.**, 68:201-204.

ACUTE EFFECTS OF MEALS ON BRAIN TRYPTOPHAN AND SEROTONIN IN HUMANS

S.N. Young

Dept. of Psychiatry
McGill University
Montreal, Quebec H3A 1A1
Canada

INTRODUCTION

Under normal circumstances, tryptophan hydroxylase, the rate-limiting enzyme on the pathway from tryptophan to serotonin, is about half saturated with tryptophan. This is true for both rats (Grahame-Smith, 1971) and humans (Young and Gauthier, 1981). As a result, alterations in brain tryptophan can influence serotonin synthesis. Brain tryptophan is in some circumstances altered by food intake, leading to the rather surprising conclusion that a neurotransmitter, serotonin, which is involved in the control of a number of important mental functions, can be influenced by the diet. As discussed below, in the rat, protein and carbohydrate meals have opposite effects on brain serotonin. There is also evidence that serotonin can influence macronutrient selection, leading to the suggestion that serotonin may be a part of a system regulating dietary intake (Wurtman and Wurtman, 1986). Thus, a carbohydrate meal would raise brain serotonin which would inhibit subsequent carbohydrate selection, thereby ensuring that protein and carbohydrate intakes would stay within certain limits over the long run.

The ideas discussed above are derived primarily from work on rodents. The purpose of this article is to examine critically the extent to which they apply to humans.

BIOCHEMICAL EFFECTS OF MEALS IN RODENTS

To enter the brain, tryptophan must compete with other large neutral amino acids for transport across the blood-brain barrier by a common carrier system (Oldendorf and Szabo, 1976). Therefore it is the plasma ratio of tryptophan to the sum of its competitors (the plasma tryptophan ratio) which becomes the determining factor of tryptophan entry into the brain. The dietary macronutrients protein and carbohydrate affect the ratio in opposite ways. Protein, which contains a high concentration of competitors relative to tryptophan, tends to lower the plasma tryptophan ratio. As a result, rat brain tryptophan (Glaeser et al., 1983) and serotonin (Teff and Young, 1988) may decline after a protein meal. On the other hand, carbohydrate raises brain serotonin. Insulin, released after carbohydrate ingestion, causes uptake of the branched-chain amino acids into muscle. The concentration in plasma of the competitors is lowered and the tryptophan ratio increases, allowing more tryptophan to enter the brain. As a result, brain serotonin may increase (Fernstrom and Wurtman, 1971). Work on the rat suggests that serotonin may be involved in the control of macronutrient selection (Anderson, 1979) and, indeed, pharmacological experiments suggest that potentiation of serotonin function will decrease carbohydrate intake selectively (Wurtman and Wurtman, 1977). These results have led to the hypothesis that serotonin is part of a homeostatic system regulating food intake, in which the relative amount of protein or carbohydrate in a meal influences brain serotonin, thereby altering subsequent macronutrient selection and maintaining macronutrient intake within certain limits (Wurtman and Wurtman, 1986). This hypothesis is appealing because it explains why serotonin should be vulnerable to dietary uptake. However, the assumptions that this mechanism operates under normal physiological circumstances in the rat, and that it also operates in humans, are not, at the moment, supported by convincing evidence.

THE EFFECT OF MEALS ON PLASMA AMINO ACIDS IN HUMANS

Measurement of the plasma amino acid ratio after dietary manipulation in humans has been employed as an indicator of potential changes in brain tryptophan and serotonin (Fernstrom et al., 1979; Ashley et al., 1982; Ashley et al., 1985; Lieberman et al., 1986). These studies all found that protein-containing meals lowered the plasma tryptophan ratio or that carbohydrate meals raised the ratio. However, the magnitude of the changes were smaller than those seen in the rat. Ashley et al. (1985) have argued that any change in the plasma tryptophan ratio will cause a smaller change in brain tryptophan and a still smaller change in brain serotonin. From a consideration of the kinetics of the blood-brain barrier transport system, and of tryptophan hydroxylase, they concluded that a 50-100% increase or a 30-50% decrease in the plasma tryptophan ratio would be needed to cause a significant alteration in brain serotonin. As the changes in the plasma tryptophan ratio in humans after meal ingestion are smaller than these limits, they suggest that human brain serotonin may be less vulnerable than rat brain serotonin to manipulation by dietary intake.

In the rat, quite small amounts of protein are capable of antagonizing the carbohydrate-induced rise in the plasma tryptophan ratio (Yokogoshi and Wurtman, 1986). We have studied whether this is also true of humans (Teff et al., 1989a). Young adult males had a blood sample taken and then ingested a breakfast consisting of a chocolate pudding which contained carbohydrate with 0,4,8 or 12% protein. The puddings were isocaloric (400 kcal) and contained no fat. Two hours later a second blood sample was taken. While the 0% protein (i.e. pure carbohydrate) breakfast caused a modest (25%) but significant (P < 0.05) rise in the plasma tryptophan ratio, the changes seen with the 4% (+8%), 8% (-6%) and 12% (-8%) protein breakfasts were all not significant.

There are few foods in which the protein content is less than 4% of the carbohydrate content. This is true for some, but not all, fruits. Thus, most normal meals which would be eaten by humans, even those regarded as high carbohydrate meals, would fail to raise the plasma tryptophan ratio. To test this, subjects were given a high carbohydrate breakfast consisting of a danish pastry and a cup of coffee. This meal contained 342 kcal and the protein content was 11% of the carbohydrate content, and provided 5.7% of the calories. The meal, traditionally regarded as a carbohydrate breakfast, resulted in a 2% decline in the plasma tryptophan ratio.

The experiments described above suggest that, in humans, meals may lower the plasma tryptophan ratio if they have a sufficiently high protein content, but will not raise the ratio unless they are almost pure carbohydrate. Thus, alterations in the plasma tryptophan ratio might be part of a mechanism that regulates the intake of specific carbohydrate foods, but it is unlikely to regulate overall protein and carbohydrate intake. This makes sense in evolutionary terms. For example, some types of fruit such as mangoes are highly preferred by certain species of monkeys. A mechanism would be needed to stop these monkeys eating nothing but mangoes during mango season, as this would result in inadequate protein intake. Our results suggest that only a meal that was almost pure carbohydrate would increase the plasma tryptophan ratio and therefore have the potential to raise brain serotonin, and decrease subsequent carbohydrate selection.

THE EFFECT OF MEALS ON BRAIN SEROTONIN IN HUMANS

The use of the plasma tryptophan ratio as an indicator of the availability of tryptophan to the human brain is valid only if the carrier system which transports amino acids across the blood-brain barrier has similar properties in humans and rats. While there is some evidence supporting this idea (Pardridge and Choi, 1986), a more direct index of human brain tryptophan and serotonin after ingestion of meals is needed.

Two studies have looked at the effect of meals on the levels of tryptophan and the serotonin metabolite, 5-hydroxyindoleacetic acid (5HIAA), in human lumbar cerebrospinal fluid (CSF). Perez-Cruet et al. (1974) fed neurologic patients a balanced lunch containing 25 g protein, 25 g fat, 55 g carbohydrate and 545 kcal. CSF and blood were taken before and four hours after the meal. When the free (i.e. non-albumin bound) plasma tryptophan concentration was used to calculate the plasma tryptophan ratio, the meals caused a 50% decline in the plasma tryptophan ratio. However, when the total plasma tryptophan concentration was used there was a 1% increase in the plasma tryptophan ratio. This is somewhat surprising as other studies, mentioned in the section above, found declines in the plasma tryptophan ratio after protein-containing meals when the total plasma tryptophan was used to calculate the plasma tryptophan ratio. Studies on the uptake of tryptophan from blood to brain indicate that some but not all of the albumin-bound tryptophan is available for transport into the brain (Etienne et al., 1976; Yuwiler et al., 1977). Therefore the effective plasma tryptophan ratio should be somewhere in between the ratios calculated using total plasma tryptophan and free plasma tryptophan. The decrease in availability of tryptophan after the meals was large enough to influence the brain as the meals caused a significant 20% decline in the CSF tryptophan and a significant decline in CSF 5HIAA which ranged from 20 to 35%. In subjects who did not eat the meal there was no change in CSF indoles.

We have studied patients suffering from normal pressure hydrocephalus who underwent three lumbar punctures in the space of a week for diagnostic purposes (Teff et al., 1989b). Three hours before each lumbar puncture the subjects were given one of three treatments, 500 ml water, 100 g carbohydrate in 500 ml water (385 kcal), or 45 g of protein, 12 g carbohydrate and 3 g fat in the form of a chocolate pudding (210 kcal). The carbohydrate meal contained more calories than the protein meal because humans normally ingest more carbohydrate than protein. The protein meal resulted in a 25% decline in the plasma tryptophan ratio (calculated using the total plasma tryptophan concentration) while the carbohydrate meal increased it 47%. As might be expected with these relatively small changes, neither meal caused a significant alteration in CSF tryptophan or 5HIAA.

Our results (Teff et al., 1989b) do not, at first sight, agree with those of Perez-Cruet et al. (1974). However, there are important differences in the design of the two studies. In our study the patients fasted overnight before their meal, while in the study of Perez-Cruet et al. subjects who remained fasting had not eaten for almost 24 hours by the time of the second lumbar puncture. In addition to this longer fast, the calorie intake was higher. Therefore, while meals may alter brain serotonin after longer fasts and larger meals, they are unlikely to have important influences under normal circumstances. Indeed, in the study of Perez-Cruet et al. (1974) some of the change in CSF 5HIAA may have been due to factors other than an alteration in tryptophan availability. CSF tryptophan declined by 20%. In humans, tryptophan hydroxylase is normally about half saturated with tryptophan. Consideration of enzyme kinetics indicates that under these circumstances a 20% decline in tryptophan availability would lead to a 11% decrease in the rate of serotonin synthesis. However, the decline in CSF 5HIAA was larger than this.

The overall conclusion from the two studies on the effect of meals on CSF indices of tryptophan metabolism in humans is that, in most circumstances, alterations in the plasma tryptophan ratio after ingestion of meals will not have any important effect on brain serotonin metabolism.

MENTAL EFFECTS OF MEALS IN HUMANS

The acute effects of meals on mood and behavior in humans have been studied to some extent and some of the results have been interpreted in terms of serotonin-mediated mechanisms. Thus, Spring et al. (1983) compared the effects of carbohydrate and high protein meals on mood. The carbohydrate meals had a greater sedative effect than the protein meals. This was seen as increased drowsiness in the females and increased calmness in the males. As pointed out by Spring et al. (1983) these data are consistent with the carbohydrate meal increasing brain tryptophan levels and brain serotonin function. Indeed, tryptophan, when given in purified form, can have both a sedative and hypnotic effect. Thus, it decreases sleep latency and has been used with some success in the treatment of mania (Young, 1986). Although the sedation seen after a carbohydrate meal is the type of effect that might occur if the carbohydrate meal raised brain tryptophan and serotonin, the evidence linking increased tryptophan to this effect is merely circumstantial. Craig (1986) has also studied the decline in performance that can occur after lunch, a phenomenon that may be mediated by the sedative effect of the meal. He has shown that extroverts seem more susceptible to a food-induced decrement in performance than introverts, and a heavy but nutritionally balanced three course lunch (1000 kcals) caused a decline in performance while a light nutritionally balanced 300 kcal lunch did not. Craig (1986) has also pointed out that studies in which lunch ended a long fast tended not to demonstrate a post-lunch decline in performance. None of these results discussed by Craig (1986) are explained by food-mediated changes in brain tryptophan and serotonin.

We have carried out two experiments looking at the effects of meals on possible alterations in serotonin function. The measure we used was macronutrient selection. This was chosen because of the hypothesis, mentioned in the first section of this article, that food-mediated alterations in serotonin result in subsequent changes of macronutrient selection, thereby insuring that intakes of protein and carbohydrate remain within certain limits (Wurtman and Wurtman, 1986).

In the first experiment we used artificial meals comprised of amino acids (Young et al., 1988). In a cross-over design, normal male subjects received a nutritionally balanced (B) mixture containing all the essential amino acids and a tryptophan deficient (T-) mixture which was the same except that tryptophan was omitted. The B mixture, like any meal of protein, caused a small decline in the plasma tryptophan ratio and served as the control treatment. However, the T- mixture caused a large decline in plasma tryptophan and the plasma tryptophan ratio. This is because the T- mixture, like any protein meal, induces synthesis of labile protein stores. The tryptophan that is incorporated into this protein comes from the free tryptophan in blood and tissues, resulting in a dramatic decline in plasma tryptophan (Gessa et al., 1974). In our experiment, five hours after ingestion of the amino acid mixtures, when plasma tryptophan was at its lowest after ingestion of the T- mixture, the plasma tryptophan ratio when the subjects received the T-mixture was only 16% of the ratio when subjects received the B mixture. Thus, the effect of the T- mixture on the plasma tryptophan ratio was qualitatively similar to that of a protein meal, but the effect was several fold greater.

Five hours after the ingestion of the amino acid mixtures the subjects were allowed to select lunch from a buffet. Macronutrient selection was determined. Relative to the B group, the T- group showed a modest (14%) but significant (P < 0.05) decline in selection of protein, with no significant alteration in selection of carbohydrate or total calories. These results are consistent with the animal data of Anderson (1979) which suggest serotonin may influence protein selection, rather than those of Wurtman and Wurtman (1977) which indicate an effect on carbohydrate intake. These theories are not necessarily incompatible as serotonin may be involved primarily with the relative intakes of protein and carbohydrate. Thus, carbohydrate intake would tend to decline as protein intake increases and vice versa. However, the important aspect of our results is not that a decline in tryptophan availability diminished protein intake rather than increased carbohydrate selection. It is that an alteration of the plasma tryptophan ratio that was several times larger than would ever occur with a real meal had only a small effect on macronutrient selection. Thus, it is unlikely that changes in the plasma tryptophan ratio which occur with real meals would be large enough to alter macronutrient selection. We tested this in a second experiment (Teff et al., 1989a).

The design of the second experiment was similar to the first except that the treatments were protein or carbohydrate meals, instead of amino acid mixtures, and the interval between this initial meal and lunch, when macronutrient selection was measured, was three hours, instead of the five hour interval used in the experiment with amino acid mixtures. As might be expected from the discussion in previous paragraph, protein and carbohydrate premeals had no significant effect on overall macronutrient selection. The carbohydrate premeal did decrease selection of one of the food items given at lunch. This was the apple, the only relatively pure carbohydrate food available, possibly indicating that this was a real finding and not just a chance occurrence. If so, then there may be a mechanism whereby carbohydrate meals decrease subsequent selection of pure carbohydrate foods, as suggested earlier in this article. However, our results do show that there is no general mechanism by which the alterations in the plasma tryptophan ratio, which occur after a meal, regulate subsequent macronutrient intake.

CONCLUSIONS

The idea that dietary intake alters tryptophan availability and brain serotonin synthesis, and that the alterations in serotonin influence subsequent dietary intake to ensure that intakes of protein and carbohydrate stay within certain limits, is an appealing one. It explains why serotonin should be vulnerable to dietary influences and suggests a mechanism for regulation of macronutrient selection. There is evidence consistent with this idea. For example (i) in the rat, protein and carbohydrate meals can have opposite effects on brain serotonin (ii) pharmacological manipulations of serotonin function in rodents can result in altered macronutrient selection. However, there is ample evidence that in humans this system does not operate under normal circumstances. As detailed above the data on humans shows that (i) alterations in the plasma tryptophan ratio after meals are small (ii) moderate sized meals taken in the absence of an unusually long fast do not seem to alter CSF tryptophan or 5HIAA (iii) protein or carbohydrate meals failed to alter subsequent macronutrient selection in one experiment. Of course, these data were collected under specific experimental circumstances and can not necessarily be generalized to other situations. Nonetheless they do suggest that food-mediated alterations in brain serotonin are not a general phenomenon in humans and are not a normal physiological mechanism for regulating macronutrient selection. It also seems unlikely that other effects of food on brain function, such as the sedative effect of carbohydrate meals in humans, are mediated by alterations in brain serotonin. It remains to be seen whether extremes of dietary intake will result in altered brain serotonin in humans and whether such changes may play a role in various forms of pathology. The fact that extreme tryptophan depletion in normal human subjects failed to increase carbohydrate selection argues against the idea that low serotonin is in any simple way involved in the carbohydrate craving that can occur in conditions such as bulimia and seasonal affective disorder (Rosenthal and Hefferman, 1986). However, control of food intake is complex and probably involves several interacting systems. The idea that serotonin is one of the systems which is involved in the control of food intake in physiological or pathological situations is a useful working hypothesis. It remains to be seen whether serotonin plays any role in the regulation of food intake in humans and, if it does have some effect, whether it is a major or a minor one.

ACKNOWLEDGEMENTS

Work in the author's laboratory is supported by the Medical Research Council of Canada.

REFERENCES

Anderson, G.H., 1979, Control of protein and energy intake: role of plasma amino acids and brain neurotransmitters, Can. J. Physiol. Pharmacol., 57:1043-1057.

Ashley, D.V.M., Barclay, D.V., Chauffard, F.A., Moennoz, D., and Leathwood, P.D., 1982, Plasma amino acid responses in humans to evening meals of differing nutritional composition, Am. J. Clin. Nutr., 36:143-153.

Ashley, D.V.M., Liardon, R., and Leathwood, P.E., 1985, Breakfast meal composition influences plasma tryptophan to large neutral amino acid ratios of healthy lean young men, J. Neural Transm., 63:271-283.

Craig, A., 1986, Acute effects of meals on perpetual and cognitive efficiency, Nutr. Rev., 44:163-171.

Etienne, P., Young, S.N., and Sourkes, T.L., 1976, Inhibition by albumin of tryptophan uptake by rat brain, Nature, 262:144-145.

Fernstrom, J.D., and Wurtman, R.J., 1971, Brain serotonin content: increase following ingestion of carbohydrate diet, Science, 174:1023-1025.

Fernstrom, J.D., Wurtman, R.J., Hammarstrom-Wiklund, B., Rand, W.M., Munro, H.N., and Davidson, C.S., 1979, Diurnal variations in plasma concentrations of tryptophan, tyrosine, and other neutral amino acids: effect of dietary protein intake, Am. J. Clin. Nutr., 32:1912-1922.

Gessa, G.L., Biggio, G., Fadda, F., Corsini, G.V., and Tagliamonte, A., 1974, Effect of the oral administration of tryptophan-free amino acid mixtures on serum tryptophan, brain tryptophan and serotonin metabolism, J. Neurochem., 22:869-870.

Glaeser, B.S., Maher, T.J., and Wurtman, R.J., 1983, Changes in brain levels of acidic, basic and neutral amino acids after consumption of single meals containing various proportions of proteins, J. Neurochem., 41:1016-1021.

Grahame-Smith, D.G., 1971, Studies in vivo on the relationship between tryptophan, brain 5-HT synthesis and hyperactivity in rats treated with a monoamine oxidase inhibitor and L-tryptophan, J. Neurochem., 18:1053-1066.

Lieberman, H.R., Spring, B.J., and Garfield, G.S., 1986, The behavioral effects of food constituents: strategies used in studies of amino acids, protein, carbohydrate and caffeine, Nutr. Rev., Suppl., 44:61-70.

Oldendorf, W.H., and Szabo, J., 1976, Amino acid assignment to one of three blood-brain barrier amino acid carriers, Am. J. Physiol., 230:94-98.

Pardridge, W.M., and Choi, T., 1986, Amino acid transport at the human blood-brain barrier, Fed. Proc., 45:2073-2078.

Perez-Cruet, J., Chase, T.N., and Murphy, D.L., 1974, Dietary regulation of brain tryptophan metabolism by plasma ratio of free tryptophan and neutral amino acids in humans, Nature, 248:693-695.

Rosenthal, N.E., and Hefferman, M.M., 1986, Bulimia, carbohydrate craving, and depression: a central connection?, in: "Nutrition and the Brain, Volume 7, Food Constituents Affecting Normal and Abnormal Behaviors", Wurtman, R.J., and Wurtman, J.J., eds., Raven Press, New York, pp. 139-166.

Spring, B., Maller, O., Wurtman, J., Digman, L., and Cozolino, L., 1983, Effects of protein and carbohydrate meals on mood and performance: interactions with sex and age, J. Psychiatr. Res., 17:155-167.

Teff, K.L., and Young, S.N., 1988, Effects of carbohydrate and protein administration of rat tryptophan and 5-hydroxytryptamine: differential effects on the brain, intestine, pineal, and pancreas, Can. J. Physiol. Pharmacol., 66:683-688.

Teff, K.L., Young, S.N., and Blundell, J.E., 1989a, The effect of protein or carbohydrate breakfasts on subsequent plasma amino acid levels, satiety and nutrient selection in normal males, Pharmacol. Biochem. Behav., 34:829-837.

Teff, K.L., Young, S.N., Marchand, L., and Botez, M.I., 1989b, Acute effect of protein or carbohydrate breakfasts on human cerebrospinal fluid monoamine precursor and metabolite levels, J. Neurochem., 52:235-241.

Wurtman, J.J., and Wurtman, R.J., 1977, Fenfluramine and fluoxetine spare protein consumption while suppressing calorie intake by rats, Science, 198:1178-1180.

Wurtman, R.J., and Wurtman, J.J., 1986, Carbohydrate craving, obesity and brain serotonin, Appetite, Suppl., 7:99-103.

Yokogoshi, H., and Wurtman, R.J. 1986, Meal composition and plasma amino acid ratios: effect of various proteins or carbohydrates, and of various protein concentrations, Metabolism, 35:837-842.

Young, S.N., 1986, The clinical psychopharmacology of tryptophan, in: "Nutrition and the Brain, Volume 7, Food Constituents Affecting Normal and Abnormal Behaviors", Wurtman, R.J., and Wurtman, J.J., eds., Raven Press, New York, pp. 49-88.

Young, S.N., and Gauthier, S., 1981, Effect of tryptophan administration on tryptophan, 5-hydroxyindoleacetic acid and indoleacetic acid in human lumbar and cisternal cerebrospinal fluid, J. Neurol. Neurosurg. Psych., 44:323-328.

Young, S.N., Tourjman, S.V., Teff, K.L., Pihl, R.O., and Anderson, G.H., 1988, The effect of lowering plasma tryptophan on food selection in normal males, Pharmacol. Biochem. Behav., 31:149-152.

Yuwiler, A., Oldendorf, S.M., Geller, E., and Braun, L., 1977, Effect of albumin binding and amino acid competition on tryptophan uptake into brain, J. Neurochem., 28:1015-1023.

IMPLICATIONS OF INTERFERON-INDUCED TRYPTOPHAN CATABOLISM IN CANCER, AUTO-IMMUNE DISEASES AND AIDS

R.R. Brown[1], Y. Ozaki[1], S.P. Datta[3], E.C. Borden[2],
P.M. Sondel[1], and D.G. Malone[2]

[1]Departments of Human Oncology and [2]Medicine
University of Wisconsin Medical School
Madison, Wisconsin 53792

[3]University of Wisconsin-Parkside
Kenosha, Wisconsin 53141
USA

SUMMARY

Tryptophan (Trp) is an indispensable amino acid required for biosynthesis of proteins, serotonin and niacin. Indoleamine 2,3-dioxygenase (IDO) is induced by infections, viruses, lipopolysaccharides, or interferons (IFNs) and this results in significant catabolism of Trp along the kynurenine (Kyn) pathway. Intracellular growth of Toxoplasma gondii and Chlamydia psittaci in human fibroblasts in vitro is inhibited by IFN-gamma and this inhibition is negated by extra Trp in the medium. Similarly, growth of a number of human cell lines in vitro is inhibited by IFN-gamma and addition of extra Trp restores growth. Thus, in some in vitro systems, antiproliferative effects of IFN-gamma are mediated by induced depletion of Trp. We find that cancer patients given Type I or Type II IFNs can induce IDO which results in decreased serum Trp levels (20-50% of pretreatment) and increased urinary metabolites of the Kyn pathway (5 to 500 fold of pretreatment). We speculate that in vivo antineoplastic effects of IFNs and clinical side effects are mediated, at least in part, by a general or localized depletion of Trp.

In view of reported increases of IFNs in autoimmune diseases and our earlier findings of elevated urinary Trp metabolites in autoimmune diseases, it seems likely that systemic or local depletion of Trp occurs in autoimmune diseases and may relate to degeneration, wasting and other symptoms in such diseases. We find high levels of IDO in cells isolated from synovia of arthritic joints.

IFNs are also elevated in human immunodeficiency virus (HIV) patients and increasing IFN levels are associated with a worsening prognosis. We propose that IDO is induced chronically by HIV infection, is further increased by opportunistic infections, and that this chronic loss of Trp initiates mechanisms responsible for the cachexia, dementia, diarrhea and possibly immunosuppression of AIDS patients. In these symptoms, AIDS resembles classical pellagra due to dietary deficiency of Trp and niacin. In preliminary studies, others report low levels of Trp and serotonin, and elevated levels of Kyn and quinolinic acid in AIDS patients. The implications of these data in cancer, autoimmune diseases and AIDS are discussed.

INTRODUCTION

Tryptophan (Trp) is an essential (indispensible) amino acid required for biosynthesis of proteins, serotonin (a neurotransmitter), the pineal hormone melatonin, and a major source of the vitamin niacin (Bender, 1987). About 175 mg/day of Trp is required for women and 250 mg/day for men, with a typical western diet providing approximately 1 g/day. Intake of less than the required amount promptly results in a negative nitrogen balance and significant reductions in Trp metabolite levels in blood and urine (Vivian et al. 1966). Because of the Trp released from breakdown of body protein during Trp deprivation, it is probably not possible to decrease serum Trp below a certain level as long as body protein pools are available for breakdown. Metabolism of Trp to the niacin pathway and to carbon dioxide is initiated in liver by the classical Trp dioxygenase (TDO, tryptophan pyrrolase, tryptophan peroxidase oxidase) well studied by the early pioneers in Trp metabolism, Yoshiro and Yahito Kotake, Luigi Musajo, and W. Eugene Knox. The excellent work in Osamu Hayaishi's group later showed the presence of another distinct non-hepatic enzyme able to form Kyn from Trp (Yoshida et al., 1981; Hayaishi et al., 1984; Yasui et al., 1986; Takikawa et al., 1986) which was named indoleamine 2,3-dioxygenase (IDO) since it also showed slight, but measurable activity against certain other indoles. Most exciting were reports that IDO was highly induced by stimulation of the immune system with bacterial endotoxins (Yoshida and Hayaishi, 1978; Yoshida et al., 1981; Carlin et al., 1989a) by virus infection (Yoshida et al., 1979), or by interferons (IFNs) (Yoshida et al., 1981; Yasui et al., 1986; Byrne et al., 1986b; Ozaki et al., 1987, 1988; Werner et al., 1987; Takikawa et al., 1988). Independently, it was shown that the inhibition of growth of Toxoplasma gondii in human cells in culture by interferon-gamma was the result of depletion of Trp from the medium by the induced IDO activity. The resulting Trp metabolites were not toxic, and repletion of media Trp negated the IFN inhibition (Pfefferkorn, 1984; Pfefferkorn et al., 1986). Similar mechanisms of IFN-gamma inhibition of proliferation have been shown for Chlamydia psittaci (Byrne et al., 1986a), for a number of human cell lines in culture (de la Maza and Peterson, 1988; Ozaki et al., 1988; Takikawa et al., 1988; Carlin et al., 1989c), and for rejection of a transplantable mouse tumor in an allogeneic host (Yoshida et al., 1988). Thus, in some but not all cell culture systems, the antiproliferative effect of interferon-gamma results from depletion of Trp by induced IDO. Deficiency of Trp can result in antitumor effects in vitro (Wooley et al., 1974) and in vivo (Roberts et al., 1979).

Interferon, immune stimulation, and tryptophan metabolism in humans

Advances in recombinant DNA technology provided a variety of interferons in larger quantities and in high states of purity so that clinical trials of these agents began. To assess in vivo effects of IFNs in humans, we studied serum Trp and urine Kyn levels in patients given various interferons. Both type I and type II IFNs induced significant decreases in serum Trp levels accompanied by marked increases in urinary Kyn excretion (Brown et al., 1986, 1987; Byrne et al., 1986b; Datta et al., 1987a; Carlin et al., 1989c). Neopterin and biopterin, metabolites of GTP which are increased in a variety of immune stimulated conditions (Wachter et al., 1983; Huber and Troppmair, 1985; Werner et al., 1985; Reibnegger et al., 1986), were also increased in patients by interferon dosages (Brown et al., 1987; Datta et al., 1987b) (Fig. 1).

Interferon-gamma resulted in dose-dependent increases in urine Kyn (Fig. 2), and these were inversely related to serum Trp levels. Type I interferons (IFN-βser and IFN-alpha2a) also caused significant decreases in serum Trp.

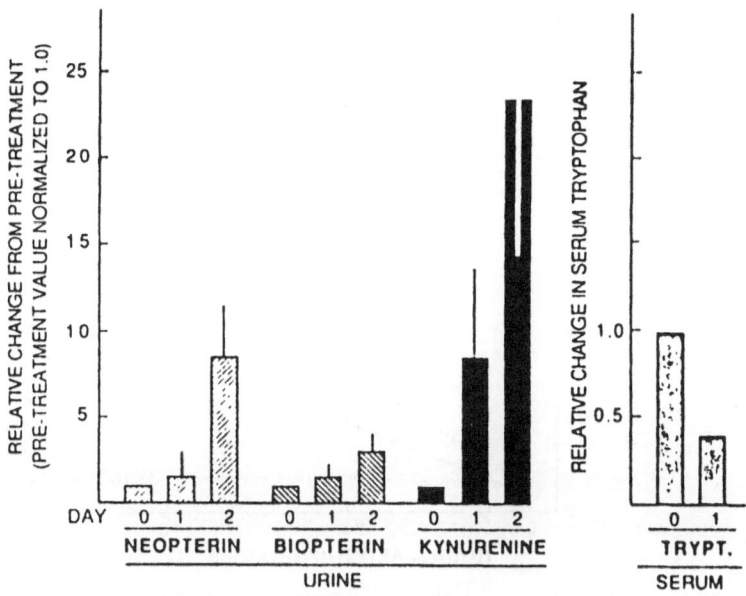

Fig. 1. Relative levels of neopterin, biopterin and Kyn in urine and Trp in
serum in 5 cancer patients before (day 0) and after a single i.v.
dose of recombinant human interferon-gamma (300 x 10^6 units/m^2).
Error bars are standard deviations.

Losses of serum Trp appeared as early as 2-4 hours after an intravenous dose
of IFN-gamma, at a time long before induction of IDO was measurable in various in vitro systems. These data suggest that other mechanisms may cause
the initial drop in serum Trp. It is indeed reported that IFN may enhance
Trp uptake into blood mononuclear cells at such early times (Finochiaro et
al., 1989).

    Decreases in serum Trp, associated with corresponding increases in urin-
ary Kyn and neopterin, also occurred after administration of interleukin-2
(IL-2) to patients (Brown et al., 1986, 1989; Datta et al., 1987a) (Fig. 3).
IL-2 causes induction of IDO in cultures of blood mononuclear cells comprised
of lymphocytes and monocytes, but not in isolated monocytes (Werner et al.,
1987; Carlin et al., 1989b,c; Ozaki et al., this volume). Supernatants from
lymphocytes cultured in IL-2 were able to induce IDO in monocyte/macrophage
preparations and this induction was prevented by anti-IFN-gamma antibodies.
Thus, the induction of IDO by IL-2 is most likely mediated by the well-documented stimulation of IFN-gamma release from T-lymphocytes stimulated with
IL-2. Tumor necrosis factor (TNF, cachectin), a product of macrophages stimulated by IFN-gamma, did not induce IDO in vitro nor did it cause significant
changes in serum Trp or urine Kyn at maximum tolerated doses (Datta et al.,
1987a).

Immune stimulation, endogenous interferons and altered tryptophan metabolism

    As shown above, IFNs administered exogenously to humans result in de-
creased serum Trp levels. The production of endogenous IFNs in patients in
response to stimulation of immunological processes by HIV infections (Eyster
et al., 1983), endotoxin-producing bacterial infections, tissue transplant
rejection, or autoimmune diseases (Talal, 1985; Cleveland et al., 1988),
would also be expected to alter Trp metabolism. Although there is some disa-
greement in the literature regarding levels and types of IFNs in such dis-

427

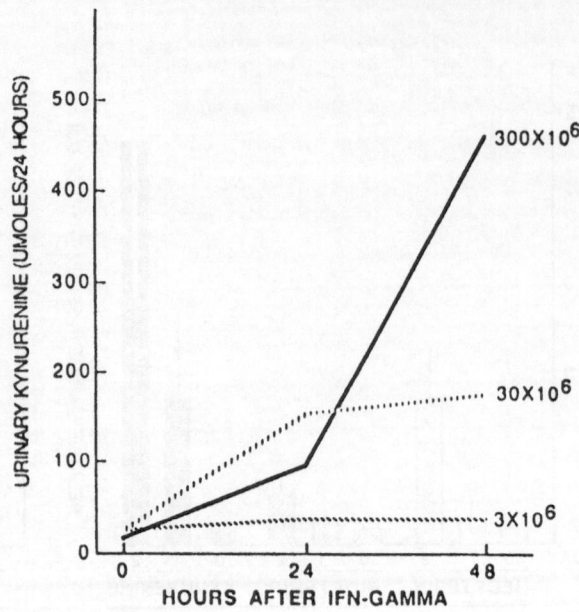

Fig. 2. Urinary kyn levels in patients give single intravenous injections of interferon-gamma in doses of 3, 30, or 300 x 10^6 units/m^2 at time 0.

eases, there is also indirect evidence to indicate production of IFNs even though circulating levels may not be detected. Neopterin elevations are a good index of IFN-gamma production and neopterin is elevated in these diseases (Wachter et al., 1983; Goebel et al., 1985; Huber and Troppmair, 1985; Werner et al., 1985). Additionally, we have shown that subcutaneous administration of IFN-ß results in decreased serum Trp and elevation of several cell markers and of neopterin, even though measurable blood levels of IFN were not detected (Goldstein et al., 1989). Serum $\beta2$-microglobulin is also elevated in various infections, autoimmune diseases and AIDS (Zolla-Pazner et al., 1984; Goebel et al., 1985; Burkes et al., 1986) and in IFN treatment and is correlated with neopterin and Kyn production (Datta et al., 1987b). Further, earlier studies reported enhanced excretion of Trp metabolites in autoimmune diseases (Price et al., 1957; Flinn et al., 1964; Hansotia et al., 1969) and in experimental infections (Rapoport and Biesel, 1971). We have recently found that cells isolated from synovia of patients with rheumatoid arthritis have elevated levels of IDO which suggests a possible role for local depletion of Trp at such sites of inflammation (Ozaki et al., unpublished data). Alterations of tryptophan metabolism have been suggested in multiple sclerosis and degenerative disease (Monaco et al., 1979).

Tryptophan metabolism in AIDS

Elevated levels of an acid-labile IFN are reported in AIDS and increasing levels are believed to indicate a worsening prognosis (DeStefano et al., 1982; Eyster et al., 1983; Green and Spruance, 1984; Fuchs and Wachter, 1988). Elevated IFN-gamma levels are reported in patients with HIV-1 infections (Fuchs and Wachter, 1988; Murray et al., 1988). Such elevations would predict that neopterin levels would be increased, and that Trp and serotonin levels would be decreased in AIDS patients. These predictions have been confirmed with the reports that serum Trp is low in AIDS patients (Werner et al., 1988; Larsson et al. 1989; Heyes et al., this volume), that blood serotonin is also very low (Launay et al. 1988; Larsson et al. 1989) and that

428

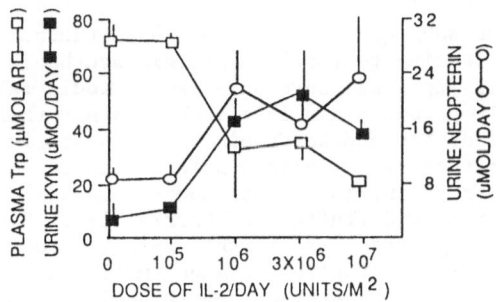

Fig. 3.   Levels of plasma Trp and urinary Kyn and neopterin after 3 daily
          doses of IL-2 given i.v. by bolus or continuous infusion (Brown et
          al., 1989).

neopterin levels are elevated (Wachter et al., 1983; Werner et al., 1988;
Melmed et al., 1989).

     Dementia is a common finding in AIDS patients (Navia et al., 1986; Grant
et al., 1987) and similar neuropsychiatric changes are seen in IL-2 and IFN-
treated cancer patients (McDonald et al., 1978; Denicoff et al., 1987), and
in various autoimmune diseases. Potential reasons for such dementias would
be lowered serotonin levels (Launay et al., 1988; Larsson et al. 1989) and
elevated brain quinolinic acid levels (Moroni et al., 1986; Heyes et al.,
1989 and this volume) resulting from Trp metabolism by IFN-induced IDO activity. Dermatologic problems are also manifest in AIDS patients (Laurence,
1987) and are associated with altered Trp metabolism in other conditions
(Sternberg et al., 1980).

Possible role for tryptophan catabolism in the pathogenesis of AIDS

     Consideration of the observations cited above suggests the hypothesis
that many of the clinical problems in AIDS are the result of a chronic deficiency of Trp produced by a long-term elevation of IDO induced by infection
with HIV. The scheme shown in Fig. 4 summarizes this hypothesis and predicts
changes associated with a chronic and worsening depletion of Trp. Thus, with

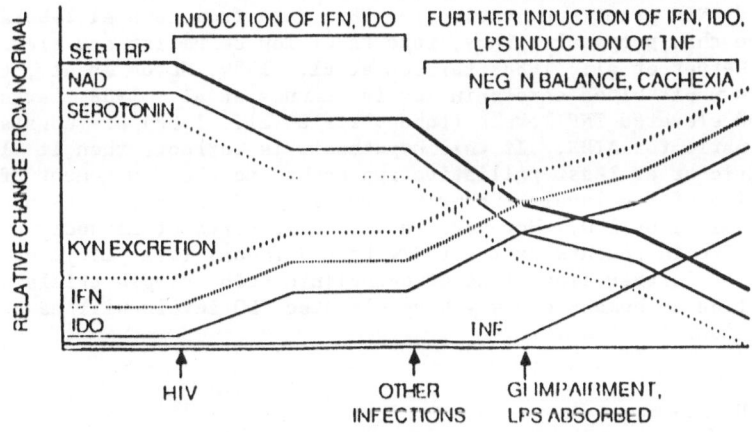

Fig. 4.   Predicted changes in Trp, nicotinamide coenzymes (NAD), serotonin,
          Kyn and tumor necrosis factor (TNF) production and IDO levels re-
          sulting from HIV and other infections in AIDS patients.

HIV-1 infection and subsequent infections (opportunistic or endemic), Trp
levels would fall leading to impaired protein synthesis and negative nitrogen
balance. With this Trp loss, serotonin levels would fall, as would NAD levels if peripherally produced Kyn is excreted rather than processed to NAD and
niacin. With loss of Trp and serotonin in brain and increase in other neuroactive metabolites such as quinolinic acid, dementias may develop. With intestinal loss of serotonin, intestinal mucosal repair, bowel integrity and
function would be impaired (Tutton and Barkla, 1987; Hasegawa et al., 1988)
which could lead to diarrhea, and further malnutrition. We suggest that the
opportunistic intestinal infections common in AIDS may be a result of rather
than the cause of the AIDS diarrhea. Additionally, with a damaged intestinal
mucosa, absorption of LPS (bacterial endotoxins) may occur and lead to increased production of tumor necrosis factor (TNF)/cachectin (Beutler, 1988)
and enhanced IDO induction, further enhancing malnutrition and wasting.

Possible secondary effects of induced tryptophan catabolism

IDO-mediated catabolism of Trp can be reasonably expected to influence
protein synthesis, and production of serotonin and niacin. Additionally, a
number of other secondary or indirect effects may result. Niacinamide is a
putative angiogenic factor (Kull et al., 1987). Trp or one of its pyridine
metabolites was shown to be necessary for IFN-gamma mediated induction of
tumor cell cytotoxicity in macrophages (Leyko and Varesio, 1989). Serotonin
may influence release of prolactin from the pituitary (MacIndoe and Turkington, 1973; Ferrari et al., 1978; Spampinato et al., 1979), and a recent report indicates that prolactin is important for macrophage activation and T-
lymphocyte function in mice (Bernton et al., 1988). Perhaps related to these
effects is the report that serotonin may also influence TNF-mediated antitumor activity (Manda et al., 1988). At an immunological level, serotonin
reduces IFN-gamma induced expression of HLA-Ia surface antigens on mouse macrophages (Sternberg et al., 1986), competes with muramyl peptides for receptor binding (Karnovsky, 1986), and influences immune response (Jackson et
al., 1985). Serotonin acts as a growth factor for intestinal mucosal cells
(Tutton and Barkla, 1987) and blockage of serotonin synthesis by inhibition
of biopterin synthesis leads to intestinal necrosis in mice which is relieved
by administration of tetrahydrobiopterin, the cofactor for hydroxylation of
Trp (Hasegawa et al., 1988).

Fig. 4 also suggests the kinds of measurements needed to confirm this
hypothesis. Indeed, since proposing this hypothesis at a workshop on Pteridine Metabolism at St. Christoph, Austria, Feb. 27 - Mar. 5, 1988, and in an
(unfunded) NIH grant application in 1988, data from several laboratories have
supported this proposal. Thus, reports of low serum Trp and elevated Kyn
levels (Werner et al., 1988; Larsson et al., 1989; Heyes et al., this volume), low whole blood serotonin levels (Launay et al., 1988; Larsson et al.,
1989) and elevated TNF levels (Lahdevirta et al., 1988) are consistent with
this scenario for AIDS. If this hypothesis is correct, then it also suggests
therapeutic or at least palliative approaches to the management of AIDS, i.e.
restoration of Trp levels either by Trp supplementation, or by prevention of
Trp catabolism by IDO. However, in view of reports of markedly enhanced toxicity of Trp in rodents pretreated with LPS (Moon, 1971; Lloyd et al., 1983),
and risk of further elevations of quinolinic acid, simple trials of Trp supplementation in humans who may have elevated IDO levels will have to proceed
with caution.

ACKNOWLEDGEMENTS

These studies have been supported in part by grants from Glaxo Inc.,
Triton Biosciences and by NIH Contract BRMP-NO1-CM47669, NIH grants CA-32685
and RR-03186, and American Cancer Soc. Grant CH-237.

REFERENCES

Bender, D.A., 1987, The relative importance of dietary tryptophan and pre-formed nicotinic acid and nicotinamide as precursors of nicotinamide nucleotide coenzymes, in: "Progress in Tryptophan and Serotonin Research 1986", Bender, D.A., Joseph, M.H., Kochen, W., and Steinhart, H.W., eds., de Gruyter, Berlin, pp. 159-164.

Bernton, E.W., Meltzer, M.S., and Holaday, J.W., 1988, Suppression of macrophage activation and T-lymphocyte function in hypoprolactinemic mice, Science, 239:401-404.

Beutler, B., 1988, The presence of cachectin/tumor necrosis factor in human disease states, Am. J. Med., 85:287-288.

Brown, R.R., Borden, E.C., Sondel, P.M., Byrne, G.I., Lee, C.M., Nunnink, J.C., Schiller, J.H., and Lehman, L.K., 1986, Effects of interferons and interleukin-2 on tryptophan metabolism in humans, J. Cell Biochem. Suppl., 10C:230.

Brown, R.R., Borden, E.C., Sondel, P.M., and Lee, C.M., 1987, Effects of interferons and interleukin-2 on tryptophan metabolism in humans, in: "Progress in Tryptophan and Serotonin Research 1986", Bender, D.A., Joseph, M.H., Kochen, W., and Steinhart, H.W., eds., de Gruyter, Berlin, pp. 19-26.

Brown, R.R., Lee, C.M., Kohler, P.C., Hank, J.A., Storer, B.E., and Sondel, P.M., 1989, Altered tryptophan metabolism in cancer patients treated with recombinant interleukin-2, Cancer Res., 49:4941-4944.

Burkes, R.L., Sherrod, A.E., Stewart, M.L., Gill, P.S., Aguilar, S., Taylor, C.R., Krailo, M.D., and Levine, A.M., 1986, Serum beta-2 microglobulin levels in homosexual men with AIDS and with persistent, generalized lymphadenopathy, Cancer, 57:2190-2192.

Byrne, G.I., Lehmann, L.K., Kirschbaum, J.G., Borden, E.C., Lee, C.M., and Brown, R.R., 1986b, Induction of tryptophan degradation in vitro and in vivo: a gamma interferon-stimulated activity, J. Interferon Res., 6:389-396.

Byrne, G.I., Lehmann, L.K., and Landry, G.J., 1986a, Induction of tryptophan catabolism is the mechanism for gamma-interferon-mediated inhibition of intracellular Chlamydia psittaci replication in T24 cells, Infect. Immun., 53:344-351.

Carlin J.M., Borden, E.C., and Byrne, G.I., 1989b, Enhancement of indoleamine 2,3-dioxygenase activity in cancer patients receiving interferon-βser, J. Interferon Res., 9:167-173.

Carlin, J.M., Borden, E.C., Sondel, P.M., and Byrne, G.I., 1989a, Interferon-induced indoleamine 2,3-dioxygenase activity in human mononuclear phagocytes, J. Leukocyte Biol., 45:29-34.

Carlin, J.M., Ozaki, Y., Byrne, G.I., Brown, R.R., and Borden, E.C., 1989c, Interferons and indoleamine 2,3-dioxygenase: role in antimicrobial and antitumor effects, Experientia, 45:535-541.

Cleveland, M.G., Annable, C.R., and Klimpel, G.R., 1988, In vivo and in vitro production of IFN-β and IFN-γ during graft vs. host disease, J. Immunol., 141:3349-3356.

Datta, S.P., Brown, R.R., Borden, E.C., Sondel, P.M., and Trump, D.L., 1987a, Interferon and interleukin-2 induced changes in tryptophan and neopterin metabolism: possible markers for biologically effective doses, Proc. Am. Assoc. Cancer Res., 28:338.

Datta, S.P., Brown, R.R., Schiller, J.H., Storer, B.E., and Borden, E.C., 1987b, Neopterin (NP) levels in cancer patients treated with human recombinant interferon (IFN). Abstract, Annual Meeting of the International Soc. for Interferon Research, Washington, D.C., November 2-6.

de la Maza, L.M., and Peterson, E.M., 1988, Dependence on the in vitro antiproliferative activity of recombinant human g-interferon on the concentration of tryptophan in culture media, Cancer Res., 48:346-350.

Denicoff, K.D., Rubinow, D.R., Papa, M.Z., Simpson, C., Seipp, C.A., Lotze, M.T., Chang, A.E., Rosenstein, D., and Rosenberg, S.A., 1987, The neuropsychiatric effects of treatment with interleukin-2 and lymphokine-activated killer cells, **Ann. Internal. Med.**, 107:293-300.

DeStefano, E., Friedman, R.M., Friedman-Kien, A.E., et al., 1982, Acid-labile human leukocyte interferon in homosexual men with Kaposi's sarcoma and lymphadenopathy, J. Infect. Dis., 146:451-455.

Eyster, M.E., Goedert, J.J., Poon, M-C., and Preble, O.T., 1983, Acid-labile alpha interferon. A possible preclinical marker for the acquired immunodeficiency syndrome in hemophilia, **N. Engl. J. Med.**, 309:583-586.

Ferrari, C., Caldara, R., Romussi, M., Rampini, P., Telloli, P., Zaatar, S., and Curtarelli, G., 1978, Prolactin suppression by serotonin antagonists in man: further evidence for serotoninergic control of prolactin secretion, **Neuroendocrinology**, 25:319-328.

Finocchiaro, L.M.E., Arzt, E.S., Fernandez-Castelo, S., Criscuolo, M., Finkielman, S., and Nahmod, V.E., 1988, Serotonin and melatonin synthesis in peripheral blood mononuclear cells: stimulation by interferon-gamma as part of an immunomodulatory pathway, **J. Interferon Res.**, 8:705-716.

Flinn, J.H., Price, J.M., Yess, N., and Brown, R.R., 1964, Excretion of tryptophan metabolites by patients with rheumatoid arthritis, **Arthr. Rheumatol.**, 7:201-210.

Fuchs, D., and Wachter, H., 1988, Inflammatory joint disease and HIV infection, **Brit. Med. J.**, 297:422-423.

Goebel, F.D., Erfle, V., Piechowiak, H., Hien, P., Schloz, R., and Hehlmann, R., 1985, The relations of HTLV-III antibodies to neopterin and beta-2-microglobulin in the serum of patients with AIDS or persons at risk, in: **"Biochemical and Clinical Aspects of Pteridines"**, Vol. 4, Wachter, H., Curtius, H.C., and Pfleiderer, W., eds., de Gruyter, Berlin, pp. 319-333.

Goldstein, D., Sielaff, K.M., Storer, B.E., Brown, R.R., Datta, S.P., Witt, P.L., Teitelbaum, A.P., Smalley, R.V., and Borden, E.C., 1989, Human biological response modification by interferon in the absence of measurable serum concentrations: a comparative trial of subcutaneous and intravenous interferon-βser, **J. Nat. Cancer Inst.**, 81:1061-1068.

Grant, I., Atkinson, J.H., Hesselink, J.R., Kennedy, C.J., Richman, D.D., Spector, S.A., and McCutchan, J.A., 1987, Evidence for early central nervous system involvement in the acquired immunodeficiency syndrome (AIDS) and other human immunodeficiency virus (HIV) infections, **Ann. Int. Med.**, 107:828-836.

Green, J.A., and Spruance, S.L., 1984, Acid-labile alpha interferon, **New Engl. J. Med.**, 310:922-923.

Hansotia, P., Peters, H., Bennet, M., and Brown, R.R., 1969, Chelation therapy in Wegner's granulomatosis treatment with EDTA, **Ann. Otol. Rhinol. Laryngol.**, 78:388-402.

Hasegawa, H., Kobayashi, T., and Ichiyama, A., 1988, 2,4-Diamino-6-hydroxy-pyrimidine (DAHP) induces intestinal disorder in mice: an animal model of serotonin deficiency by inhibition of tetrahydrobiopterin (THBP) synthesis, **Biol. Chem. Hoppe-Seyler**, 369:532.

Hayaishi, O., Yoshida, R., Takikawa, O., Yasui, H., 1984, Indoleamine dioxygenase - a possible biological function, in: **"Progress in Tryptophan and Serotonin Research"**, Schlossberger, H.G., Kochen, W., Linzen, B., and Steinhart, H., eds., de Gruyter, Berlin, pp. 33-42.

Heyes, M.P., Rubinow, D., Lane, C., and Markey, S.P., 1989, Cerebrospinal fluid quinolinic acid concentrations are increased in acquired immune deficiency syndrome, **Ann. Neurol.**, 26:275-277.

Huber, C., Troppmair, J., Fuchs, D., Hausen, A., Lang, D., Niederwieser, D., Reibnegger, G., Swetly, P., Wachter, H., and Margreiter, R., 1985, Immune response-associated production of neopterin-release from macrophages under control of lymphokines, **Transplant Proc.**, 17:582-585.

Jackson, J.C., Cross, R.J., Walker, R.F., Markesberry, W.R., Brooks, W.H., and Roszman, T.L., 1985, Influence of serotonin on the immune response, **Immunology**, 54:505-512.

Karnovsky, M.L., 1986, Muramyl peptides in mammalian tissues and their effects at the cellular level, Fed. Proc., 45:2556-2560.

Kull, F.C., Jr., Brent, D.A., Parikh, I., and Cuatrecasas, P., 1987, Chemical identification of a tumor-derived angiogenic factor, Science, 236: 843-845.

Lahdevirta, J., Maury, C.P.J., Teppo, A-M., and Repo, H., 1988, Elevated levels of circulating cachetin/tumor necrosis factor in patients with acquired immunodeficiency syndrome, Am. J. Med., 85:289-291.

Larsson, M., Hagberg, L., Norkrans, G., and Forsman, A., 1989, Indoleamine deficiency in blood and cerebrospinal fluid from patients with human immunodeficiency virus infection, J. Neurosci. Res., 23:441-446.

Launay, J-M., Copel, L., Callebert, J., Corvaia, N., Lepage, E., Bricare, F., Saal, F., and Peries, J., 1988, Decreased whole blood 5-hydroxytryptamine (serotonin) in AIDS patients, J. Acquired Immune Defic. Synd., 1:324-325.

Laurence, J., 1987, Dermatologic manifestations of HIV infection, Infections in Med., Jul/Aug, 241-251.

Leyko, M.A., and Varesio, L., 1989, Role for tryptophan and its metabolites in the activation of macrophage tumor cytotoxicity, FASEB J., 3:A822.

Lloyd, P., Stribling, D., and Pogson, C.I., 1983, Endotoxin and tryptophan-induced hypoglycaemia in rats, Biochem. Pharmacol., 31:3571-3576.

MacIndoe, J.H., and Turkington, R.W., 1973, Stimulation of human prolactin secretion by intravenous infusion of L-tryptophan, J. Clin. Invest., 52:1972-1978.

Manda, T., Nishigaki, F., Mor, J., and Shimomura, K., 1988, Important role of serotonin in the antitumor effects of recombinant tumor necrosis factor-a in mice, Cancer Res., 48:4250-4255.

McDonald, E.M., Mann, A.H., and Thomas, H.C., 1978, Interferons as mediators of psychiatric morbidity, Lancet, 2:1175-1178.

Melmed, R.N., Taylor, J.M.G., Detels, R., Bozorgmehri, M., and Fahey, J.L., 1989, Serum neopterin changes in HIV-infected subjects: indicator of significant pathology, CD4 T cell changes, and the development of AIDS, J. Acquired Immune Defic., 2:70-76.

Monaco, F., Fumero, S., Mondino, A., and Mutani, R., 1979, Plasma and cerebrospinal fluid tryptophan in multiple sclerosis and degenerative dis-diseases, J. Neurol. Neurosurg. Psych., 42:640-641.

Moon, R.J., 1971, Tryptophan oxygenase and tryptophan metabolism in endo-toxin-poisoned and allopurinol-treated mice, Biochim. Biophys. Acta, 230:324-348.

Moroni, F., Lombardi, G., Carla, V., Lal, S., Etienne, P., and Nair, N.P.V., 1986, Increase in the content of quinolinic acid in cerebrospinal fluid and frontal cortex of patients with hepatic failure, J. Neurochem., 47:1667-1671.

Murray, H.W., DePamphilis, J., Schooley, R.T., and Hirsch, M.S., 1988, Circulating interferon gamma in AIDS patients treated with interleukin-2, New Engl. J. Med., 318:1538-1539.

Navia, B.A., Cho, E-S., Petito, C.K., and Price, R.W., 1986, The AIDS dementia complex: II. Neuropathology, Ann. Neurol., 19:525-535.

Ozaki, Y., Edelstein, M.P., and Duch, D.S., 1987, The action of interferon and antiinflammatory agents on induction of indoleamine 2,3-dioxygenase in human peripheral blood monocytes, Biochem. Biophys. Res. Commun., 144: 1147-1153.

Ozaki, Y., Edelstein, M.P., and Duch, D.S., 1988, Induction of indoleamine 2,3-dioxygenase: a mechanism of the antitumor activity of interferon-gamma, Proc. Natl. Acad. Sci. USA, 85:1242-1246.

Pfefferkorn, E.R., 1984, Interferon-gamma blocks the growth of Toxoplasma gondii in human fibroblasts by inducing the host cells to degrade tryptophan, Proc. Natl. Acad. Sci. USA, 81:908-912.

Pfefferkorn, E.R., Eckel, M., and Rebhun, S., 1986, Interferon-gamma suppresses the growth of Toxoplasma gondii in human fibroblasts through starvation for tryptophan, Mol. Biochem. Parasitol., 20:215-224.

Price, J.M., Brown, R.R., Rukavina, J.G., Mendelson, C., and Johnson, S.A.M., 1957, Scleroderma (Acrosclerosis). II. Tryptophan metabolism before and during treatment by chelation (EDTA), J. Invest. Dermatol., 29:289-298.

Rapoport, M.I., and Beisel, W.R., 1971, Studies of tryptophan metabolism in experimental animals and man during infectious illness, Am. J. Clin. Nutr., 24:807-814.

Reibnegger, G., Bollbach, R., Fuchs, D., Hausen, A., Judmaier, G., Prior, C., Rotthauwe, H.W., Werner, E.R., and Wachter, H., 1986, A simple index relating clinical activity in Crohn's disease with T cell activation: hematocrit, frequency of liquid stools and urinary neopterin as parameters, Immunobiology, 173:1-11.

Roberts, J., Schmid, F.A., and Rosenfeld, H.J., 1979, Biologic and antineoplastic effects of enzyme-mediated in vivo depletion of L-glutamine, L-tryptophan, and L-histidine, Cancer Treatment Rep., 63:1045-1054.

Spampinato, S., Locatelli, V., Cocchi, D., Vicentini, L., Bajusz, S., Ferri, S., and Müller, E.E., 1979, Involvement of brain serotonin in the prolactin-releasing effect of opioid peptides, Endocrinology, 105:163-170.

Sternberg, E.M., Trial, J., and Parker, C.W., 1986, Effect of serotonin on murine macrophages: suppression of Ia expression by serotonin and its reversal by $5\text{-}HT_2$ serotonergic receptor antagonists, J. Immunol., 137:276-282.

Sternberg, E.M., Van Woert, M.H., Young, S.N., Magnussen, I., Baker, H., Gauthier, S., and Osterland, C.K., 1980, Development of a scleroderma-like illness during therapy with L-5-hydroxytryptophan and carbidopa, New Engl. J. Med., 303:782-787.

Takikawa, O., Kuroiwa, T., Yamazaki, F., and Kido, R., 1988, Mechanism of interferon-gamma action; characterization of indoleamine 2,3-dioxygenase in cultured human cells induced by interferon-gamma and evaluation of the enzyme-mediated tryptophan degradation in its anticellular activity, J. Biol. Chem., 263:2041-2048.

Takikawa, O., Yoshida, R., Kido, R., and Hayaishi, O., 1986, Tryptophan degradation in mice initiated by indoleamine 2,3-dioxygenase, J. Biol. Chem., 261:3648-3653.

Talal, N., 1985, Interleukins, interferon and rheumatic disease, Clin. Rheum. Dis., 11:633-644.

Tutton, P.J.M., and Barkla, D.H., 1987, Biogenic amines as regulators of the proliferative activity of normal and neoplastic intestinal epithelial cells, Anticancer Res., 7:1-12.

Vivian, V.M., Brown, R.R., Price, J.M., and Reynolds, M.S., 1966, Some aspects of tryptophan and niacin metabolism in young women consuming a low tryptophan diet supplemented with niacin, J. Nutr., 88:93-99.

Wachter, H., Fuchs, D., Hausen, A., Huber, C., Knosp, O., Reibnegger, G., and Spira, T., 1983, Elevated urinary neopterin levels in patient's with the acquired immunodeficiency syndrome (AIDS), Hoppe Seyler's Z. Physiol. Chem., 364:1345-1346.

Werner, E.R., Bitterlich, G., Fuchs, D., Hausen, A., Reibnegger, G., Szabo, G., Dierich, M.P. and Wachter, H., 1987, Human macrophages degrade tryptophan upon induction by interferon-gamma, Life Sci., 41:273-280.

Werner, E.R., Fuchs, D., Hausen, A., Jaeger, H., Reibnegger, G., Werner-Felmayer, G., Dierich, M.P., and Wachter, H., 1988, Tryptophan degradation in patients infected by human immunodeficiency virus, Biol. Chem. Hoppe-Seyler, 369:337-340.

Werner, E.R., Fuchs, D., Hausen, A., Lutz, H., Reibnegger, G., and Wachter, H., 1985, Interferon-gamma-induced in vitro-excretion of neopterin and 3-hydroxyanthranilic acid by human macrophages, in: "Biochemical and Clinical Aspects of Pteridines", Vol. 4, Wachter, H., Curtius H.C., and Pfleiderer, W., eds., de Gruyter, Berlin, pp. 473-406.

Wooley, P.V., III, Dion, R.L. and Bono, V.H., Jr., 1974, Effects of tryptophan deprivation on Ll210 cells in culture, Cancer Res., 34:1010-1014.

Yasui, H., Takai, K., Yoshida, Y., and Hayaishi, O., 1986, Interferon enhances tryptophan metabolism by inducing pulmonary indoleamine 2,3-dioxy-
genase: its possible occurrence in cancer patients, Proc. Natl. Acad. Sci.
USA, 83:6622-6626.

Yoshida, R., and Hayaishi, O., 1978, Induction of pulmonary indoleamine 2,3-
dioxygenase by intraperitoneal injection of bacterial lipopolysaccharide,
Proc. Natl. Acad. Sci. USA, 75:3998-4001.

Yoshida, R., Imanishi, J., Oku, T., Kishida, T., and Hayaishi, O., 1981, Induction of pulmonary indoleamine 2,3-dioxygenase by interferon, Proc.
Natl. Acad. Sci. USA, 78:129-132.

Yoshida, R., Park, S.W., Yasui, H., and Hayaishi, O., 1988, Tryptophan degradation in transplanted tumor cells undergoing rejection, J. Immunol., 141:
2819-2823.

Yoshida, R., Urade, M., Tokuda, M., and Hayaishi, O., 1979, Induction of indoleamine 2,3-dioxygenase in mouse lung during virus infection, Proc.
Natl. Acad. Sci. USA, 76:4084-4088.

Zolla-Pazner, S., William, D., El-Sadr, W., Marmor, M., and Stahl, R., 1984,
Quantitation of β-2-microglobulin and other immune characteristics in a
prospective study of men at risk for acquired immune deficiency syndrome,
J. Am. Med. Assoc., 251:2951-2955.

INDUCTION OF INDOLEAMINE 2,3-DIOXYGENASE IN TUMOR CELLS TRANSPLANTED INTO

ALLOGENEIC MOUSE: INTERFERON-γ IS THE INDUCER

O. Takikawa, A. Habara-Ohkubo, and R. Yoshida

Department of Cell Biology
Osaka Bioscience Institute
Osaka 565
Japan

INTRODUCTION

The major catabolic pathway of tryptophan in mammals is the kynurenine pathway. In adult young man in fact about 90% of dietary tryptophan is metabolized along this pathway (Fig. 1) (Wolf, 1974). Although it has long been generally accepted that "hepatic" L-tryptophan 2,3-dioxygenase (tryptophan pyrrolase) is the sole enzyme that catalyzes the first step of this pathway, recent studies have revealed that another enzyme called indoleamine 2,3-dioxygenase (IDO), which catalyzes the same reaction as "hepatic" enzyme, does occur in various "extra-hepatic" organs of mammals [mouse (Yoshida et al., 1980; Yoshida et al., 1981), rat (Cook et al., 1980), rabbit (Hayaishi et al., 1975 and human (Yamazaki et al., 1985)].

What is the physiological role of IDO? An involvement of IDO in the self-defence mechanism against some pathogens was shown by the findings that IDO in mice was dramatically (up to 120-fold) induced during viral infection (Yoshida et al., 1979) or endotoxin shock (Yoshida and Hayaishi, 1978) or by the addition of interferon (IFN) to mouse lung slices (Yoshida et al., 1981). During the endotoxin shock the catabolism of tryptophan was markedly enhanced, as shown by several-fold increase both in the plasma level of kynur-

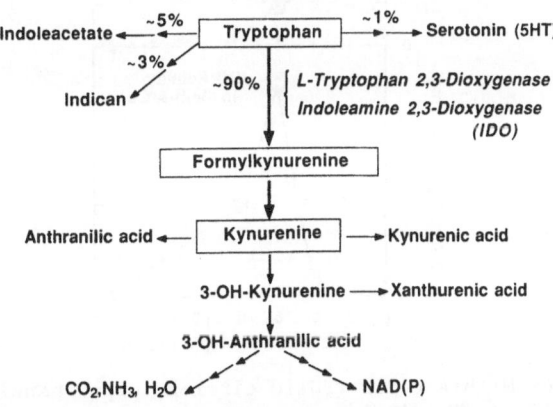

Fig. 1. Metabolic pathways of tryptophan in adult man.

enine and in the urinary level of xanthurenic acid, a metabolite of kynurenine (Takikawa et al., 1986). Recently, it has been demonstrated in some
human cultured cells that tryptophan depletion or starvation caused by induction of IDO was the mechanism by which IFN-γ blocked the replication of intracellular parasites such as <u>Toxoplasma gondii</u> (Pfefferkorn, 1984), <u>Chla-</u>
<u>mydia psittaci</u> (Byrne et al., 1986), and <u>Chlamydia trachomatis</u> (Shemer et
al., 1987). More importantly, such tryptophan deprivation mechanism was also
shown to be operative in the <u>in vitro</u> antiproliferative action of IFN-γ
against several human tumor cells (de la Maza and Peterson, 1988; Ozaki et
al., 1988; Takikawa et al., 1988).

These <u>in vitro</u> studies with tumor cells and the fact that the IDO level
was significantly higher in the human lung tissue bearing tumor than that in
the normal tissue (Yasui et al., 1986), enable us postulate that tryptophan
depletion caused by IDO induction may be an <u>in vivo</u> anti-tumor mechanism.
Being consistent with this idea, we have recently found that a marked induction of IDO occurred in tumor cells (Meth-A cells) undergoing rejection from
mice (C57BL/6) (Yoshida et al., 1988). The IDO induction was restricted to
the tumor cells, while there was no change in the IDO activity of host cells
infiltrated into the transplantation loci. Furthermore, we demonstrated that
the induction was mediated by membrane-permeable factor(s) released from the
infiltrated host cells through the interaction between the host cells and the
tumor cells. In this study, we have characterized the <u>in vivo</u> factor(s), and
found that the factor is IFN-γ.

RESULTS AND DISCUSSION

Induction of IDO in tumor cells undergoing rejection

As reported previously (Yoshida et al., 1988), when Meth-A cells (3 x
10^6 cells/mouse) were given i.p. to allogeneic mice (C57BL/6), the tumor
cells grew about 2-fold 4 days after the transplantation, but on the 6th day
the growth was suppressed with host cell infiltration into the transplanta-

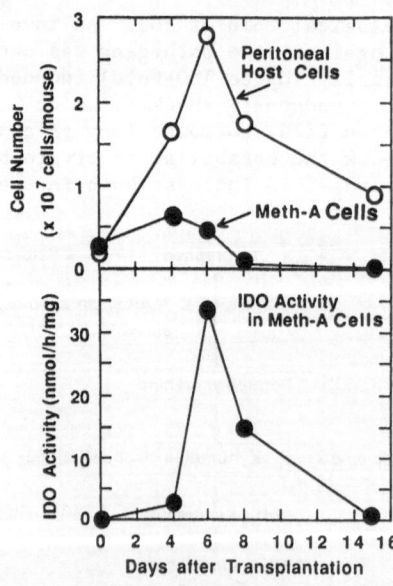

Fig. 2. Growth of Meth-A cells (solid circles, upper panel), infiltration of
host cells (open circles, upper panel), and induction of IDO in the
tumor cells (lower panel).

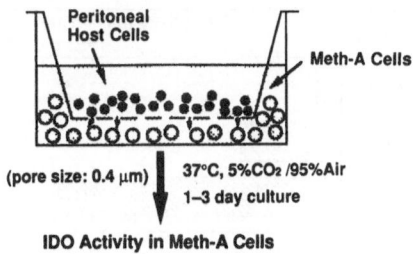

Peritoneal Host Cells → Meth-A Cells

(pore size: 0.4 μm) | 37°C, 5%CO₂ /95%Air
1–3 day culture

IDO Activity in Meth-A Cells

Fig. 3.   In vitro assay of the IDO induction factor(s) with a Transwell™
culture dish (Costar 3412). The volumes of upper and lower wells
are 1.5 and 2.6 ml, respectively.

tion loci, and on the 8th day the tumor cells were almost rejected from the
peritoneal cavity (Fig. 2, upper panel). The induction of IDO in the tumor
cells was observed around the 6th day, when the tumor cells were strongly
rejected (Fig. 2, lower panel). Such induction, however, was not observed
when Meth-A cells were transplanted into syngeneic mice (Balb/c) (data not
shown). Thus, IDO was induced when the tumor cells were in the process of
being rejected from allogeneic mice, and this suggests that the tryptophan
depletion caused by the induction of IDO is one of the mechanisms of tumor
rejection.

An in vitro assay of IDO induction factor(s)

To assay the in vivo IDO inducer(s), we used a 35 mm special culture
dish (Transwell™), which consisted of two wells divided vertically with a
membrane (0.4 μm pore) as illustrated in Fig. 3. Host cells (2-8 x 10^6
cells/ml) that infiltrated into the transplantation loci were cultured in the
upper well, and untreated Meth-A cells (1 x 10^5 cells/ml) in the lower well.
If the host cells in the upper well released soluble factor(s), it would
reach the lower well by diffusion and induce IDO in the tumor cells. As expected, we detected membrane-permeable factor(s) that induced IDO in the tum-
or cells. Fig. 4 shows the time course of the IDO induction and the effect
of density of the host cells on the enzyme induction.

IDO induction factor(s) in conditioned medium

If the membrane-permeable factor was stable, the factor would be accumulated in the culture medium. To examine this possibility, the culture supernatants (conditioned medium) were added to untreated fresh Meth-A cells, and
the cells were incubated for further 2 days. As indicated in Fig. 5, the

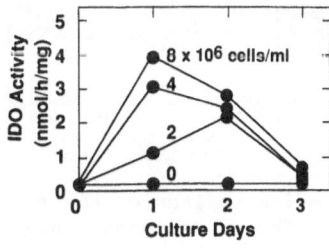

Fig. 4.   The time course of the IDO induction in Meth-A cells and the effect
of density of the host cells on the induction with a Transwell™
culture dish. Each number represents the density of the host cells
in the upper well.

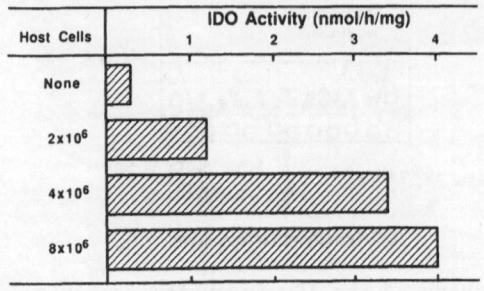

Host Cells	IDO Activity (nmol/h/mg)

Fig. 5. The IDO induction in Meth-A cells by the factors in the conditioned
medium. The conditioned medium after 1-day culture with a Trans-
well™ culture dish was prepared by centrifugation at 400 x g for 10
min.

conditioned medium was found to contain the active component(s) associated
with the induction of IDO. The ability of the supernatants to induce IDO in
the tumor cells was almost comparable to that obtained with the host cells in
the Transwell™ (Fig. 4). The active factor(s) in the supernatants could be
stored at 4°C for at least 2 weeks without loss of activity, indicating that
the factor(s) is fairly stable in the culture medium.

Neutralization of factor(s) with antibody to IFN-γ

Viruses and LPS are known to be inducers of IFNs. The incubation of
mouse and human lung slices with IFNs resulted in an induction of IDO (Yoshida et al., 1979; Yasui et al., 1986). In addition, IFN-γ induced IDO in
various cultured human cells (Pfefferkorn, 1984; Byrne et al., 1986; Pfefferkorn et al., 1986; Carlin et al., 1987; Ozaki et al., 1987; Shemer et al.,
1987; Werner et al., 1987a,b; de la Maza and Peterson, 1988; Ozaki et al.,
1988; Rubin et al., 1988; Takikawa et al., 1988; Werner et al., 1988; Murray
et al., 1989). Therefore, we examined whether the inducer activity in the
culture supernatants was due to IFNs. Fig. 6 illustrates that an antibody to
IFN-γ completely neutralized the inducer activity in the culture medium,
while an antibody to IFN-α/β did not affect the activity. The concentration
of IFN-γ in the culture supernatants was found to be 2-3 U/ml based on the
neutralization curve (Fig. 6). At this concentration, recombinant IFN-γ was
able to induce IDO in Meth-A cells to almost the same extent as the inducer
in the culture medium. Natural IFN-α or recombinant IFN-β was completely

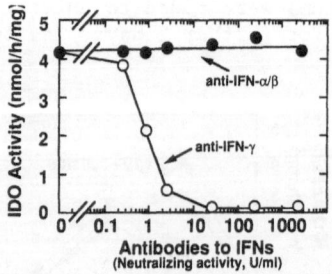

Fig. 6. Effects of antibodies to IFNs on the factor-mediated IDO induction
in Meth-A cells. The conditioned medium, which was obtained after
1-day culture of host cells at 4 x 10^6/ml was incubated with
various indicated concentrations of antibodies to IFNs for 2 h at
37°C. The remaining inducer activity in the medium was determined
by the incubation of Meth-A cells in the antibody-treated medium
for further 2 days.

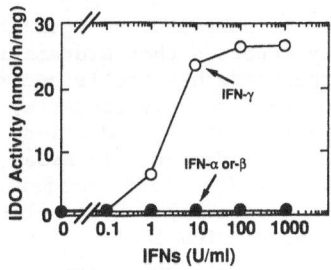

Fig. 7. IDO induction in Meth-A cells by IFNs. The IDO activity was determined after 2-day culture with various concentrations of recombinant or natural IFNs.

inactive for the IDO induction (Fig. 7). These results indicated that the
factor responsible for the induction of IDO in the culture medium was IFN-γ
but not IFN-α or IFN-β.

Molecular size of IFN-γ released during tumor rejection

To determine the molecular weight (MW) of IFN-γ released from the host
cells during the rejection of the tumor cells, we subjected IFN-γ in the culture medium to gel permeation chromatography. Fig. 8 illustrates that IFN-γ
was eluted as a single entity with MW of about 37,000 on Superose 12 by FPLC.
The MW was close to the apparent MW (about 38,000) for mitogen-induced IFN-γ
produced by splenocytes (Wietzerbin et al., 1978), but was different from
those (28,000 and 56,000) obtained after the injection of BCG into mice
(Stefanos et al., 1982).

SUMMARY

Tryptophan depletion observed during induction of indoleamine 2,3-dioxy-
genase (IDO) in cultured cells has been suggested to involve a mechanism

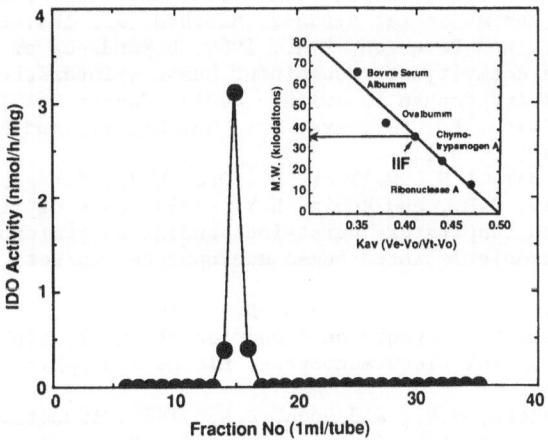

Fig. 8. Elution profile of IFN-γ released from the host cells on Superose
12. The conditioned medium, which was obtained after 1-day culture
of host cells at 4 x 10^6/ml in a Transwell™ culture dish, was concentrated 30-fold by a Centriprep-10™ membrane (Amicon) and applied on the column. The mobile phase was phosphate buffered saline
(pH 7.4) and the flow rate was 1 ml/min.

identical to that employed in self-defense against inhaled microorganisms and tumor growth. We recently reported that a dramatic induction of IDO occurred in i.p. transplanted tumor (Meth-A) cells undergoing rejection from allogeneic mice (C57BL/6), and that soluble factor(s) released from infiltrated host cells was responsible for the IDO induction. Here we report on the characterization of the soluble factor. To assay the factor, we used a 35 mm special culture dish (Transwell™), which consisted of two wells divided vertically with a membrane (0.4 μm pore). Host cells (mainly lymphocytes) that infiltrated into the transplantation loci were cultured in the upper well, and untreated Meth-A cells in the lower well. With this in vitro system, the membrane-permeable factor, released by the host cells (upper well), induced IDO in the tumor cells (lower well). The culture superna tants, obtained by centrifuging the culture media from the upper and lower wells, contained the IDO inducer. The inducer activity was completely neutralized by the addition of antibody against interferon-γ (IFN-γ) but not by antibody against INF-α/β. The concentration of INF-γ in the medium after 1-day culture with a Transwell™ culture dish was found to be 2-3 U/ml based on the neutralization curve with the antibody. At this concentration, recombinant IFN-γ induced IDO in Meth-A cells to the same extent as the inducer in the culture medium. These observations indicate that the in vivo factor for IDO induction in the allografted tumor cells is IFN-γ.

REFERENCES

Byrne, G.I., Lehmann, L.K., Kirschbaum, J.G., Bordon, E.C., Lee, C.M., and Brown, R.R., 1986, Induction of tryptophan degradation in vitro and in vivo: a γ-interferon-stimulated activity, J. Interferon Res., 6:389-396.

Byrne, G.I., Lehmann, L.K., and Landry, G.J., 1986, Induction of tryptophan catabolism is the mechanism for gamma-interferon-mediated inhibition of intracellular chlamydia psittaci replication in T24 cells, Infect. Immun., 53:347-351.

Carlin, J.M., Bordon, E.C., Sondel, P.M., and Byrne, G.I., 1987, Biologic response modifier-induced indoleamine 2,3-dioxygenase activity in human peripheral blood mononuclear cell cultures, J. Immunol., 139:2414-2418.

Cook, J.S., Pogson, C.I., and Smith, S.A., 1980, Indoleamine 2,3-dioxygenase, a new rapid, sensitive, radiometric assay and its application to the study of the enzyme in rat tissues, Biochem. J., 189:461-466.

de la Maza, L.M., and Peterson, E.M., 1988, Dependence of the in vitro antiproliferative activity of recombinant human γ-interferon on the concentration of tryptophan in culture media, Cancer Res., 48:346-350.

Hayaishi, O., Hirata, F., Fujiwara, M., Ohnishi, T., and Nukiwa, T., 1975, Proc. FEBS Meet., 10:131-141.

Murray, H.W., Sudol, A.S., Wellner, D., Oca, M.J., Granger, A.M., Libby, D.M., Rothemel, C.D., and Rubin, B.Y., 1989, Role of tryptophan degradation in respiratory burst-independent an timicrobial activity of gamma interferon-stimulated human macrophages, Infect. Immun., 57:845-849.

Ozaki, Y., Edelstein, M.P., and Duch, D.S., 1987, The actions of interferon and antiinflammatory agents on induction of indoleamine 2,3-dioxygenase in human peripheral blood monocytes, Biochem. Biophys. Res. Commun., 144:1147-1153.

Ozaki, Y., Edelstein, M.R., and Duch, D.S., 1988, Induction of indoleamine 2,3-dioxygenase: a mechanism of the antitumor activity of interferon-γ, Proc. Natl. Acad. Sci. USA., 85:1242-1246.

Pfefferkorn, E.R., 1984, Interferon γ blocks the growth of toxoplasma gondii in human fibroblasts by inducing the host cells to degrade tryptophan, Proc. Natl. Acad. Sci. USA, 81:908-912.

Pfefferkorn, E.R., Rebhun, S., and Eckel, M., 1986, Characterization of an indoleamine 2,3-dioxygenase induced by gamma-interferon in cultured human fibroblasts, J. Interferon Res., 6:267-279.

Rubin, B.Y., Anderson, S.L., Hellermann, G.R., Richardson, N.K., Lunn, R.M., and Valinsky, J.E., 1988, The development of antibody to the interferon-induced indoleamine 2,3-dioxygenase and the study of the regulation of its synthesis, J. Interferon Res., 8:691-702.

Shemer, Y., Kol, R., and Sarov, I., 1987, Tryptophan reversal of recombinant human gamma-interferon, inhibition of chlamydia trachomatis growth, Curr. Microbiol., 16:9-13.

Stefanos, S., Wietzerbin, J., Huygen, K., and Falcoff, E., 1982, Chromatographic behavior, purification, and properties of mouse BCG-induced gamma interferon, J. Interferon Res., 2:447-456.

Takikawa, O., Kuroiwa, T., Yamazaki, F., and Kido, R., 1988, Mechanism of interferon-γ action: characterization of indoleamine 2,3-dioxygenase in cultured human cell induced by interferon-γ and evaluation of the enzyme-mediated tryptophan degradation in its anticellular activity, J. Biol. Chem., 263:2041-2048.

Takikawa, O., Yoshida, R., Kido, R., and Hayaishi, O., 1986, Tryptophan degradation in mice initiated by indoleamine 2,3-dioxygenase, J. Biol. Chem., 261:3648-3653.

Werner, E.R., Bitterlich, G., Fuchs, D., Hausen, A., Reibnegger, G., Szabo, G., Dierich, M.P., and Wachter, H., 1987, Human macrophages degrade tryptophan upon induction by interferon-gamma, Life Sci., 41:273-280.

Werner, E.R., Kauffmann, M.H., Fuchs, D., Hausen, A., Reibnegger, G., Schweiger, M., and Wachter, H., 1987, Interferon-γ-induced degradation of tryptophan by human cells in vitro, Biol. Chem. Hoppe-Seyler, 368:1407-1412.

Werner, E.R., Werner-Felmayer, G. Fuchs, D., Hausen, A., Reibnegger, G., and Wachter, H., 1988, Influence of interferon-gamma and extracellular tryptophan on indoleamine 2,3-dioxygenase activity in T24 cells as determined by a non-radiometric assay, Biochem. J., 256:537-541.

Wietzerbin, J., Stefanos, S., Falcoff, R., Lucero, M., Catinot, L., Falcoff, E., 1978, Immune interferon induced by phytohemagglutinin in nude spleen cells, Infect. Immun., 21:966-972.

Wolf, H., 1974, Studies on tryptophan metabolism in man, Scand. J. Clin. Lab. Invest., 33:Suppl. 136.

Yamazaki, F., Kuroiwa, T., Takikawa, O., and Kido, R., 1985, Human indoleamine 2,3-dioxygenase, its tissue distribution, and characterization of the placental enzyme, Biochem. J., 230:635-638.

Yasui, H., Takai, K., Yoshida, R., and Hayaishi, O., 1986, Interferon enhances tryptophan metabolism by inducing pulmonary indoleamine 2,3-dioxygenase: its possible occurrence in cancer patients, Proc. Natl. Acad. Sci. USA, 83:6622-6626.

Yoshida, R., and Hayaishi, O., 1978, Induction of pulmonary indoleamine 2,3-dioxygenase by intraperitoneal injection of bacterial lipopolysaccharide, Proc. Natl. Acad. Sci. USA, 75:3998-4000.

Yoshida, R., Imanishi, J., Oku, T., Kishida, T., and Hayaishi, O., 1981, Induction of pulmonary indoleamine 2,3-dioxygenase by interferon, Proc. Natl. Acad. Sci. USA, 78:129-132.

Yoshida, R., Nukiwa, T., Watanabe, Y., Fujiwara, M., Hirata, F., and Hayaishi, O., 1980, Regulation of indoleamine 2,3-dioxygenase in the small intestine and the epididymis of mice, Arch. Biochem. Biophys., 203:343-351.

Yoshida, R., Park, S.W., Yasui, H., and Takikawa, O., 1988, Tryptophan degradation in transplanted tumor cells under going rejection, J. Immunol., 141:2819-2823.

Yoshida, R., Urade, Y., Nakata, K., Watanabe, Y., and Hayaishi, O., 1981, Specific induction of indoleamine 2,3-dioxygenase by bacterial lipopolysaccharide in the mouse lung, Arch. Biochem. Biophys., 212:629-637.

Yoshida, R., Urade, Y., Tokuda, M., and Hayaishi, O., 1979, Induction of indoleamine 2,3-dioxygenase in mouse lung during virus infection, Proc. Natl. Acad. Sci. USA, 76:4084-4086.

SHORT COMMUNICATIONS

PLASMA FREE 5HT, PLASMA 5HIAA AND WHOLE BLOOD 5HT IN THE RAT

J. Ortiz, P. Celada, E. Martinez, and F. Artigas

Dept. of Neurochemistry
CSIC
08034 Barcelona
Spain

INTRODUCTION

Previous results from this and other laboratories (Anderson et al., 1987) showed that a free pool of serotonin (5HT) occurs in human plasma and that it can be pharmacologically differentiated after several treatments including clomipramine (Sarrias et al., 1987), lithium (Artigas et al., 1989) or carbidopa (Ortiz et al., 1988). Also, free 5HT in plasma and platelet 5HT undergo independent seasonal changes (Sarrias et al., 1989). These data suggest that, when working with human blood, it is possible to have only negligible or no contribution from platelets to free plasma 5HT. As a consequence, it is of interest to study how both 5HT pools (free and stored) are affected by pathologies in which 5HT is thought to be involved and to examine their change after drug treatment. Since 5HT uptake inhibitors (acting in CNS and in platelets) are widely used as antidepressant drugs, such a peripheral model could be useful in the study of their "in vivo" mode of action in patients and in experimental animals. In the latter, CNS changes after drug treatment can also be studied, enabling a direct comparison of CNS and peripheral effects. For that purpose, we have explored the possibility of using rat blood to study the plasma free and platelet pools of 5HT.

EXPERIMENTAL

Male albino Wistar rats (200-250 g) were used. Animals were anesthetized with i.p. pentobarbital (40 mg/kg) and the carotid or tail arteries exposed and cannulated with a 6-7 cm long polythene tubing filled with 0.5% EDTA (PORTEX, UK, i.d.: 0.58 mm, o.d.: 0.96 mm). Blood was collected using 1.5 ml plastic tubes containing 40 μl of 25% K_2EDTA. Tubes were immediately mixed by inversion. The first tube collected (500 μl of blood) was used for the analysis of whole blood 5HT. Plasma was obtained after two centrifugations (of blood and platelet-poor-plasma; 15 and 30 min respectively, 1000 x g) (Ortiz et al., 1988). Plasma free 5HT was routinely analyzed in the 2nd tube (see Results). Plasma tryptophan (TP), plasma free 5HT, plasma 5HIAA and whole blood 5HT were analyzed as previously described (Artigas et al., 1985; Ortiz et al., 1988).

Table 1. Differences between tail artery or carotid artery cannulation

Group	Plasma free 5HT		Plasma 5HIAA		Whole blood 5HT	
	Tail	Carotid	Tail	Carotid	Tail	Carotid
1	25.6 ± 33[a]	3.3 ± 1.5	28.3 ± 4.9	28.6 ± 5.4	1784 ± 354	2058 ± 380
2	9.2 ± 7.1[a]	2.9 ± 1.3	40.5 ± 17.0	36.5 ± 8.6	2676 ± 649[a]	2174 ± 469
3	—	4.3 ± 3.0	—	29.5 ± 3.7	—	2195 ± 506

Results are expressed as ng/ml of plasma or whole blood and are the mean ±
SD (N = 8-10 animals per group). Total plasma TP was also analyzed but no
tail-carotid differences were observed (mean values between 13 and 17 µg/ml)
[a]Significantly different (P < 0.05, Wilcoxon rank t-test) compared to the
corresponding carotid artery.

RESULTS

    In a first experiment, three different groups of 8-10 animals each were
examined according to the following scheme: Group 1: The animals had their
tail artery cannulated and immediately their carotid artery cannulated.
Group 2: There was a time interval of 4 hours between cannula implantation
in the tail artery and blood extraction through tail and carotid arteries.
Group 3: Animals had only their carotid artery cannulated. Our purpose was
to evaluate the possible use of the tail artery to obtain repeated sampling
from the same animal, avoiding the use of anesthesia needed for cannulation
of the carotid artery. This would also provide more powerful experimental
designs, since parallel samples could be obtained.

    As shown in Table 1, the plasma free 5HT levels obtained by cannulation
of the tail artery were considerably higher than the respective carotid values. Moreover, some results from tail artery were extremely high, showing

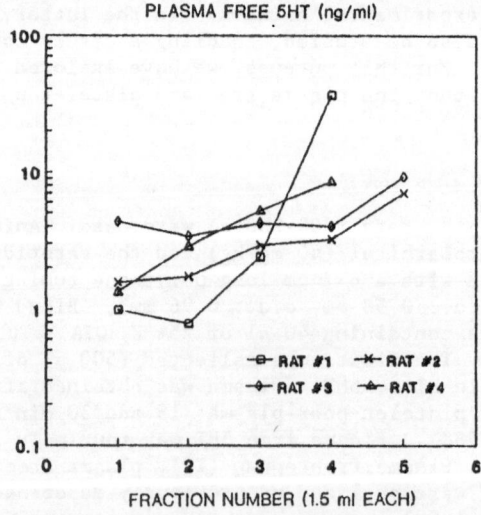

Fig. 1. Individual plasma 5HT values obtained from successive fractions of
         blood extracted from 4 animals through the carotid artery. Volume
         of each fraction: 1.5 ml.

that this procedure is prone to give artifactually high plasma 5HT values. Also, the passage of blood through the cannula was faster when the carotid artery was cannulated (usually, the blood used for plasma 5HT analysis was obtained within 10 sec after the start of bleeding through the carotid artery). This may explain why erratic high values were obtained through the tail. No other metabolite examined showed tail-carotid differences except whole blood 5HT in group #3 in which a significant 23% increase was observed in blood obtained through the tail. This difference is difficult to explain since it was not seen in group #1 and may be related to the implantation of the cannula for 4 hrs before bleeding. It is interesting to note that plasma 5HIAA did not exhibit the differences found for free 5HT, suggesting that tail cannulation may be a suitable method for the analysis of 5HIAA in rat plasma. This would permit repeated sampling through implanted cannulae. However, plasma free 5HT should be analyzed from blood obtained through the carotid artery. Otherwise, abnormally high values are likely to be obtained due to platelet release during cannulation.

In another experiment, 4 rats were cannulated through the carotid artery plasma 5HT was analyzed in successive tubes. The results are shown in Fig. 1. Plasma free 5HT levels increased with the volume of blood extracted. The explanation for such a phenomenon is not clear, but might be related to an activation of 5HT release from platelets or enterochromaffin cells (E.C.) as a physiological response to bleeding. Also, since the speed of passage of the blood through the cannula decreased with time, such an increase could also be due to a non-physiological release of 5HT from platelets in the cannula.

Fig. 2 shows the statistical distribution of the concentrations of plasma free 5HT, plasma 5HIAA and whole blood 5HT in control rats from 8 to 11 different experiments, obtained in a time span of 10 months. Interestingly,

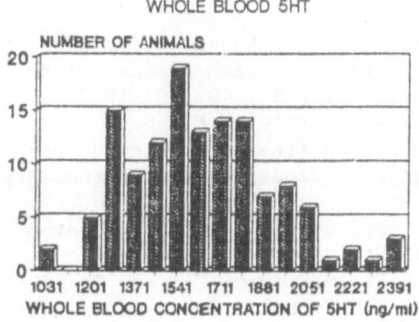

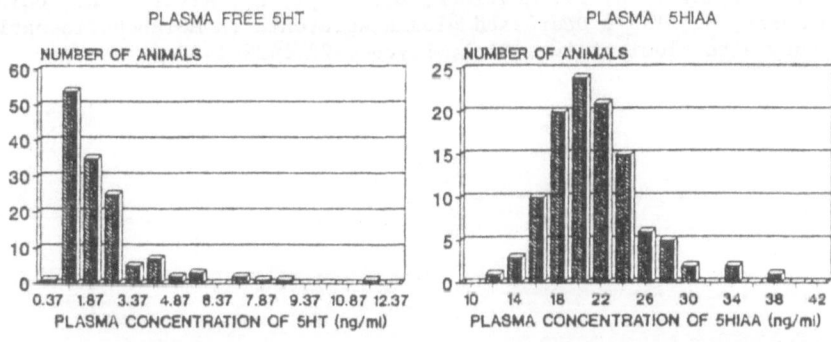

Fig. 2. Statistical distributions corresponding to control values observed in 8-11 different experiments of 8-10 animals each.

plasma free 5HT showed a skewed distribution, similar to that previously reported in humans (Ortiz et al., 1988). These values have been obtained by cannulation of the carotid artery, plasma 5HT and 5HIAA being analyzed in the second tube.

These results illustrate the absolute requirement for using rat blood obtained by cannulation of the carotid artery to avoid erratic values of plasma 5HT. Also, extraction of large volumes of blood leads to higher values. Accordingly, plasma free 5HT should be analyzed in the second tube (1.5 ml of blood) obtained by cannulation of the carotid artery and collected in EDTA-containing tubes. Using these precautions, reproducible values of 5HT (low nM range) are routinely found in platelet-free plasma. That concentration is very similar to that found in human plasma, while whole blood 5HT is ten times higher in rats. This seems to support the non-artifactual origin of rat plasma 5HT: if plasma 5HT were a contribution from platelets, the values found in rats would probably be higher than those in humans. Also, the treatment of rats with drugs known to affect the 5HT uptake process (clomipramine, tianeptine) elicit changes in both free and stored pools of 5HT consistent with their mode of action (Ortiz et al., submitted).

Taken together, these results support the use of platelet-free plasma 5HT as a measure of the actual value of free 5HT in rat blood, provided that appropriate sampling procedures are used.

ACKNOWLEDGEMENTS

This work was supported by FIS grant No. 89/0387 (Fondo de Investigaciones Sanitarias de la Seguridad Social).

REFERENCES

Anderson, G.M., Feibel, F.C., and Cohen, D.J., 1987, Determination of serotonin in whole blood, platelet-rich-plasma, platelet-poor-plasma and plasma ultrafiltrate, Life Sci., 40:1063-1070.
Artigas, F., Sarrias, M.J., Martinez, E., Gelpi, E., Alvarez, E., and Udina, D., 1989, Increased plasma free serotonin but unchanged platelet serotonin in bipolar patients treated with lithium, Psychopharmacology, 99: 328-332.
Ortiz, J., Artigas, F., and Gelpi, E., 1988, Serotonergic status in human blood, Life Sci., 43:983-990.
Sarrias, M.J., Artigas, F., Martinez, E., and Gelpi, E., 1989, Seasonal changes of plasma serotonin and related parameters: correlations with environmental measures, Biol. Psych., 26:695-706.
Sarrias, M.J., Artigas, F., Martinez, E., Gelpi, E., Alvarez, E., Udina, C., and Casas, M., 1987, Decreased plasma serotonin in melancholic patients: a study with clomipramine, Biol. Psych., 22:1429-1438.

EFFECTS OF CLOMIPRAMINE ON EXTRACELLULAR SEROTONIN IN THE RAT FRONTAL CORTEX

A. Adell and F. Artigas

Department of Neurochemistry
CSIC
08034 Barcelona
Spain

INTRODUCTION

The tricyclic antidepressant clomipramine (CIM) has been shown to inhibit the uptake of serotonin (5-hydroxytryptamine, 5-HT) into rat brain synaptosomes as well as into human platelets. The acute administration of CIM decreases 5-HT turnover (Carlsson and Lindqvist, 1978; Marco and Meek, 1979). This finding agrees with electrophysiological studies showing a reduction in the firing rate of 5-HT neurons after a single dose of CIM (Gallager and Aghajanian, 1975; Dresse and Scuvée-Moreau, 1979). The increased synaptic availability caused by the inhibition of 5-HT uptake possibly activates a feedback mechanism mediated by presynaptic and/or somatodendritic autoreceptors. This may lead to a decrease in the activity of 5-HT neurons thereby lowering its release.

The present work is aimed at studying the effects of CIM administered through a dialysis probe or intraperitoneally (i.p.) on the extracellular concentration of 5-HT and its main metabolite 5-hydroxyindoleacetic acid (5-HIAA) in the frontal cortex of the rat.

METHODS

Male Wistar rats (300-350 g) were used. Concentric dialysis probes were implanted under sodium pentobarbitone anesthesia (60 mg/kg, i.p.). They were made up of 25 gauge stainless steel tubing and fitted with vitrous silica tubing. Cuprophan hollow fibres (Enka AG, Wuppertal, FRG) were used for dialysis membranes. Animals were allowed to recover for approximately 18 hrs following implantation. Artificial CSF (mM: NaCl, 125; KCl, 2.5; MgCl$_2$, 1.18, CaCl$_2$, 1.26) was perfused at 0.25 μl/min. Successive 20 min (5 μl) dialysis samples were collected and analyzed by HPLC with amperometric detection at + 0.65 V.

When delivered through the probe, CIM was dissolved in artificial CSF. After approximately 3 hrs (8 dialysis samples) to obtain basal values for dialysate 5-HT, the perfusion fluid was switched to artificial CSF that contained CIM, and samples were collected for approximately 6 hrs. Following this period (16 dialysis samples), CIM was withdrawn, and artificial CSF was perfused for 2 more hrs (6 dialysis samples).

Kynurenine and Serotonin Pathways
Edited by R. Schwarcz et al., Plenum Press, New York, 1991

451

When administered i.p., CIM was dissolved in 0.9% NaCl and injected 1 or 2 hrs after the onset of perfusion.

Results are expressed as the area under the curve (AUC) computed from selected dialysate fractions (i.e. pretreatment, once the effect of CIM was fully established (maximal effect) and after withdrawal of the drug). In the study in which CIM was injected i.p., AUC was calculated from fractions collected before the treatment, and over 1 and 2 hrs after drug administration.

The effects of CIM were assessed by one-way ANOVA followed by Newman-Keuls test. Differences were considered significant at $P < 0.05$.

RESULTS AND DISCUSSION

The recovery of 5-HT through the dialysis membrane was dependent on the flow rate of the perfusion. Thus, the highest percentage recovery (66%) was obtained at the lowest rate (0.25 μl/min). The basal value found for dialysate 5-HT was 0.006 pmol/5 μl. The administration of CIM through the probe caused a dose-dependent increase in dialysate 5-HT. At 5 μM, the increase in extracellular 5-HT was 200% over basal values whereas at 10 μM a 300% increment was found. This finding is consistent with an in vivo inhibition of 5-HT uptake. The higher 5-HT dialysate levels remained significantly increased 2 hrs after withdrawal of the drug.

The effect of the drug started one hour after the onset of its perfusion and the increase in dialysate 5-HT was found to be maximal three hrs after the onset of CIM administration. This delayed effect of CIM could be due to pharmacokinetic reasons (e.g. saturation of non-neuronal stores) or other mechanisms involved in the interaction of the drug with the recognition site of the 5-HT uptake system.

When CIM was administered i.p., neither 10 nor 20 mg/kg of the drug caused any significant effect on dialysate 5-HT. Although this finding was unexpected, the lack of effect of this route of administration could be due to several reasons. An uneven distribution of CIM across the brain after i.p. administration may lead to a lesser accumulation of the drug in frontal cortex. Thus, it has been shown that CIM accumulates preferentially and for longer time in septum and hypothalamus, whilst the level of the drug in other

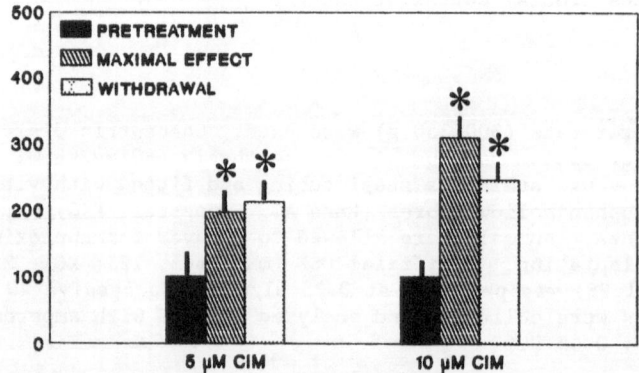

Fig. 1. Effect of different doses of CIM administered through the probe on dialysate 5-HT. Results (AUC) are expressed as the mean + SEM of 5 animals per dose. Bars represent the AUC obtained over 3 hrs before CIM administration (pretreatment), 3 hrs after the onset (maximal effect) or 2 hrs after the withdrawal of the drug. *P < 0.05 (Newman-Keuls test).

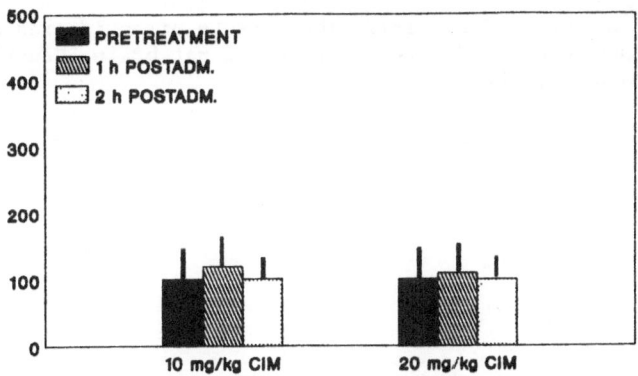

Fig. 2.   Effect of 10 mg/kg (N = 4) or 20 mg/kg (N = 5) CIM, i.p., on
dialysate 5-HT.   Results are expressed as the mean ± SEM.

brain regions is substantially lower (Friedman and Cooper, 1983).   On the
other hand, it has also been shown that CIM decreases the firing rate of 5-HT
neurons whose cell bodies are located in the raphe nuclei (Gallager and
Aghajanian 1975; Dresse and Scuvée-Moreau, 1979).   We hypothesize that after
i.p. administration, CIM distributes throughout the brain and can act on 5-HT
cell bodies in the raphe nuclei thereby decreasing their firing rate.   This
effect could lead to a decreased release of 5-HT from nerve endings.   Thus,
the action of CIM on the raphe nuclei could oppose the effect of the drug in
the frontal cortex.

In summary, the present study shows that CIM dose-dependently increases
dialysate 5-HT in frontal cortex but only when the drug is administered
through the probe.   In contrast, this effect was not seen following i.p. administration of the drug, at least at the doses studied.

ACKNOWLEDGEMENTS

The skillful technical assistance and dedication of Miss A. Carceller is
greatly acknowledged.   Thanks are also given to Ciba-Geigy for kindly providing us with clomipramine.   This work was supported by FIS grant No.
8910387 (Fondo de Investigaciones Sanitarias de la Seguridad Social).

REFERENCES

Carlsson, A., and Lindqvist, M., 1978, Effects of antidepressant agents on
    the synthesis of brain monoamines, J. Neural Transm., 43:73-91.
Dresse, A.E., and Scuvée-Moreau, J., 1979, Action of antidepressants on the
    spontaneous firing of locus coeruleus and dorsal raphe neurons of the
    rat, in: "Catecholamines: Basic and Clinical Frontiers", Usdin, E.,
    Kopin, I.J., and Barchas, J., eds., Pergamon, New York, pp. 640-642.
Friedman, E., and Cooper, T.B., 1983, Pharmacokinetics of chlorimipramine and
    its demethylated metabolite in blood and brain regions of rats treated
    acutely and chronically with chlorimipramine, J. Pharmacol. Exp. Ther.,
    225:387-390.
Gallager, D.W., and Aghajanian, G.K., 1975, Effects of chlorimipramine and
    lysergic acid diethylamide on efflux of precursor-formed ^{3}H-serotonin:
    correlations with serotonergic impulse flow, J. Pharmacol. Exp Ther.,
    193:785-795.

Marco, E.J., and Meek, J.L., 1979, The effects of antidepressants on
serotonin turnover in discrete regions of rat brain, Naunyn-Schmiedeberg's
Arch. Pharmacol., 306:75-79.

TRYPTOPHAN METABOLISM IN SENILE CATARACT PATIENTS ASSESSED BY HIGH-PERFORMANCE LIQUID CHROMATOGRAPHY

A. Elderfield and R. Truscott

Chemistry Department
University of Wollongong
Wollongong NSW
Australia

INTRODUCTION

In 1975, it was suggested that a relationship may exist between the development of senile nuclear cataract and the metabolism of tryptophan (Ogino and Ichihara, 1957). Two metabolites of tryptophan, anthranilic acid and 5-hydroxyanthranilic acid, were isolated from the urine of patients with cataractous lenses. The oxidation product of 5-hydroxyanthranilic acid was demonstrated to initiate cataract formation. In 1977, anthranilic acid was iso-

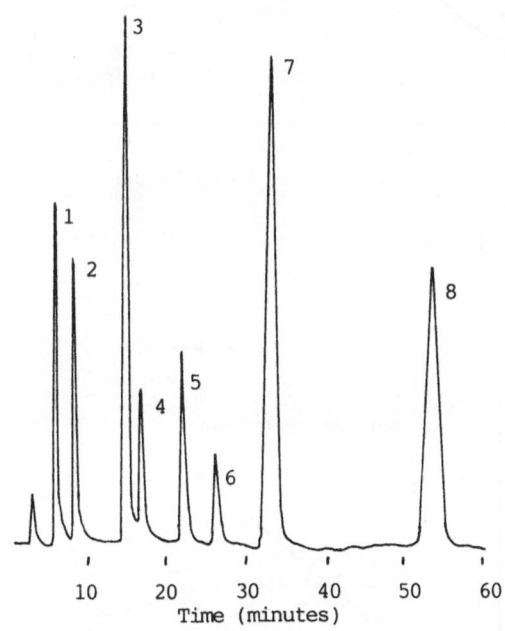

Fig. 1. HPLC profile of tryptophan and its metabolites.
1: 5-hydroxyanthranilic acid; 2: 3-hydroxykynurenine;
3: 5-hydroxytryptophan; 4: kynurenine; 5: 3-hydroxy-
anthranilic acid; 6: xanthurenic acid; 7: tryptophan;
8: anthranilic acid.

lated from the proteolytic digest of cataractous lens protein, but not from normal lenses (Truscott et al., 1977). The authors suggested that the anthranilic acid was derived from protein-bound tryptophan. In the 1980's, some controversy arose as to the status of plasma tryptophan in senile cataract patients (Cotlier and Sharma, 1980; Allegri and Angi, 1981; Chadwick et al., 1981; Cotlier and Sharma, 1981; Allegri et al., 1984; Augusteyn et al., 1987), but the findings were consistent with the idea that cataract patients may metabolize tryptophan differently after a meal or a tryptophan load.

We have now used a sensitive reverse-phase HPLC technique developed in our laboratory (Elderfield et al., 1989) to thoroughly investigate the response of cataract patients and normal controls to an oral loading dose of tryptophan.

MATERIALS AND METHODS

Cataract patients were selected at random from patients presenting for surgical removal of the cataractous lens. They fasted for 14 hours prior to removal of the lens. After 8 hours fasting, the patients were asked to give a urine specimen, and 10 ml blood were removed; subsequently, they were given a loading dose of 50 mg/kg L-tryptophan with 250 ml water.

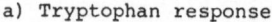

a) Tryptophan response

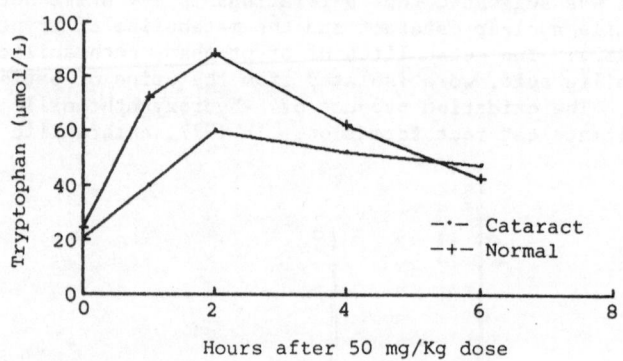

b) Cataract patient

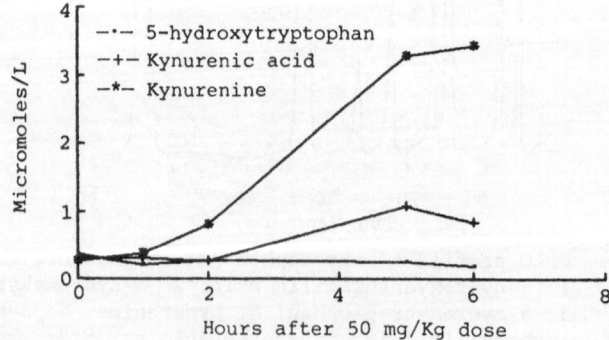

Fig. 2. Tryptophan load.

Blood was taken 1,2,4 and 6 hours post-tryptophan. Urine was collected when passed. Aqueous humor was withdrawn immediately prior to removal of the lens. The lens was then removed, photographed (for classification) and frozen until analysis. Blood samples were allowed to clot and then centrifuged. Proteins were precipitated with 10% cold trichloroacetic acid, and the clear supernatant was analyzed directly by HPLC.

The lens was homogenized in acetate buffer and proteins precipitated with cold 10% trichloroacetic acid. The clear supernatant was applied directly to the HPLC column. Aqueous humor samples were injected directly onto the HPLC column.

HPLC with dual electrochemical and programmable fluorescence detection was performed under the conditions previously described (Elderfield et al., 1989). Standardization was accomplished by means of an external aqueous standard consisting of tryptophan and eight of its metabolites, namely 5-hydroxyanthranilic acid, xanthurenic acid, 3-hydroxykynurenine, 5-hydroxytryptophan, kynurenine, 3-hydroxyanthranilic acid, kynurenic acid and anthranilic acid.

RESULTS

The HPLC system is capable of detecting tryptophan metabolites between 6 and 200 picomoles. A typical separation is shown in Fig. 1. Fig. 2 shows the response of tryptophan and some of its metabolites to a loading dose of L-tryptophan given to a patient with senile nuclear cataract. In individuals tested so far, serum tryptophan levels peaked at approximately 2 hours, leading to a slow increase in kynurenine and kynurenic acid which both peaked at 4-6 hours (Fig. 2). Thus far we have not been able to detect anthranilic acid or 5-hydroxyanthranilic acid in the serum of either controls or cataract patients.

REFERENCES

Allegri, G., and Angi, M., 1981, Free and total serum tryptophan in senile cataract, Lancet 2:157-158.
Allegri, G., Angi, M., Costa, G., and Bettero, A., 1984, Tryptophan and kynurenine in senile cataract, in: "Progess in Tryptophan and Serotonin Research", Bender, D., ed., de Gruyter, Berlin, pp. 469-472.
Augusteyn, R., Maclean, H., Boyd, A., and Merkrebs, J., 1987, Serum tryptophan and cataract, Aust. N. Z. J. Ophthalmol., 64:708-710.
Chadwick, C., Phipps, D., and Powell, C., 1981, Serum tryptophan and cataract, Lancet, 2:583.
Cotlier, E., and Sharma, Y., 1980, Plasma tryptophan and senile cataract, Lancet, 1:607.
Cotlier, E., and Sharma, Y., 1981, Senile cataract and plasma tryptophan, Lancet, 2:534.
Elderfield, A., Truscott, R., Gan, I., and Schier, G., 1989, A method for the separation of tryptophan metabolites by rp-HPLC with amperometric and fluorescence detection, J. Chromat., 4945:71-80.
Ogino, S., and Ichihara, T., 1957, Biochemical studies on cataract: V. Biochemical genesis of senile cataract, Amer. J. Ophthalmol., 43:754-764.
Truscott, R., Faull, K., and Augusteyn, R., 1977, The identification of anthranilic acid in proteolytic digests of cataractous lens proteins, Ophthalmol. Res., 9:263-268.

THE EFFECT OF ALUMINUM ON RAT BRAIN AND LIVER QUINOLINIC ACID CONCENTRATIONS

J.H. Connick[1], G. Lombardi[2], F. Moroni[2], E. Hall[3], A. Taylor[3], and T.W. Stone[1]

[1]Dept. Pharmacology
Univ. of Glasgow
UK

[2]Dept. Pharmacology
Univ. of Florence
Italy

[3]Robens Institute
Univ. of Surrey
UK

INTRODUCTION

Aluminum has been shown in a number of studies to cause intoxication and cell death in a variety of animal species. There is also considerable evidence that aluminum plays some role in the pathogenesis of Alzheimer's disease and in encephalopathy suffered by patients undergoing hemodialysis. Fatal dialysis encephalopathy in patients undergoing haemodialysis was first described in 1972 by Alfrey et al. and was subsequently shown to be caused by the accumulation of aluminum in the grey matter (Alfrey et al., 1976). The aluminum is derived from aluminum containing antacids and from the dialysate itself.

A suggested involvement of aluminum derived from cooking utensils and domestic drinking water has also recently been the subject of more general interest.

The precise mechanism underlying the toxic action of aluminum is unknown but an inhibitory effect of the metal upon a number of important enzymes and enzyme activities in the central nervous system (CNS) has been well reported. These include choline acetyltransferase, acetylcholinesterase, ADP ribosylation, cytosolic and mitochondrial glycolysis and monoamine oxidase activity (Yales et al., 1980; McLachlan et al., 1983; Lai and Blass, 1984; Sharp and Rosenberry, 1985; Tsuzuki and Marquis, 1985). It may also increase the permeability of the blood-brain barrier (Banks and Kastin, 1983).

When extracts of brain tissue from a variety of species including rat, hamster and human are exposed to levels of aluminum similar to those found in the brains of patients who had died of aluminum encephalopathy, the activity of another enzyme of central importance in brain metabolism, dihydropteridine reductase, is reduced by 40% (Leeming and Blair, 1979). This enzyme is essential for the regeneration of tetrahydrobiopterin which is an essential co-factor in the activity of three aromatic amino acid hydroxylases: phenylala-nine, tyrosine and tryptophan hydroxylase. These enzymes are critical to the functioning of the CNS, being responsible for the synthesis of DOPA, nor-adrenaline and 5-hydroxytryptamine respectively. Our interest was aroused by the possibility that aluminum might inhibit the activity of tryptophan hy-droxylase by decreasing the concentration of tetrahydrobiopterin in the brain. Tryptophan levels in the brain would then increase, causing the sub-strate induction of tryptophan pyrrolase, the rate limiting step at the start of the other main catabolic pathway for tryptophan, the kynurenine pathway.

This pathway has become of considerable interest recently following the discovery of Perkins and Stone in 1982, that one component of the kynurenine pathway, quinolinic acid, was able to excite neurones in the mammalian CNS. This interest was compounded in 1983, when Schwarcz and his colleagues demon-strated that quinolinic acid could also produce neurodegeneration upon focal injection into the rat hippocampus and striatum.

Thus, any increase in the amount of tryptophan passing down the kynuren-ine pathway might tend to increase the concentration of quinolinic acid in the brain, which could in turn lead to excitotoxic cell damage. This could explain some aspects of the pathology referred to above.

TRYPTOPHAN PYRROLASE

The rate limiting step in the conversion of tryptophan to kynurenine is tryptophan pyrrolase (tryptophan dioxygenase). This enzyme was the first en-zyme found to be adaptively controlled by its own substrate (Knox and Mehler, 1951). The substrate induction of this enzyme has now been studied inten-sively and it has been found that tryptophan allows the apoenzyme (which is inactive) to become more fully saturated with its essential iron porphyrin cofactor. Following aluminum treatment, the activity of this enzyme would be expected to increase in response to the increased level of tryptophan, which would normally be metabolized to serotonin.

In order to test this hypothesis, we treated rats with aluminum. Male Wistar rats which had been starved for 24 h before treatment were injected with 150 mg/kg (i.p.) $Al_2(SO_4)_3.16 H_2O$. Control rats were injected with sa-line. After 1 h, the animals were killed and the liver removed and homoge-nized prior to the analysis of tryptophan pyrrolase activity.

Aluminum treatment resulted in a 92% increase in total tryptophan pyr-rolase activity, whilst the holoenzyme was also increased by 150% compared with control animals. A simple increase in free tryptophan levels would be expected to increase the activity of tryptophan pyrrolase by substrate in-duction. This was indeed the case, in addition to an increase in the holo-enzyme, indicating some other potentiating factor, possibly stress.

QUINOLINIC ACID

Since the amount of tryptophan catabolized via tryptophan pyrrolase was increased, we investigated the effect of aluminum on the concentration of quinolinic acid in the brain and liver of rats 1 h after injection.

Rats were killed and the cerebral cortex and livers immediately removed, weighed and frozen at -80°C. The tissue was then homogenized in alkaline ethanol together with the internal standard, centrifuged and analyzed as pre-viously described (Moroni et al., 1984). Briefly, the tissue was loaded onto an anion exchange resin and interfering compounds eluted with successive

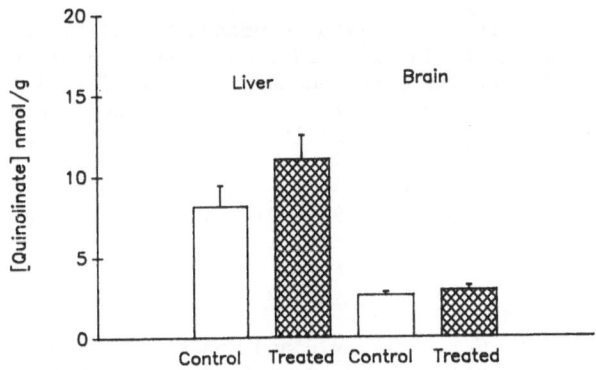

Fig. 1. A histogram of the amount of quinolinic acid detected in both the
liver and cerebral cortex of rats treated with either saline (control) or aluminum. The columns indicate the mean ± SEM from
four animals.

washing with water, 0.1 M NaOH and 0.1 M formic acid. Quinolinic acid was
then eluted with 5 M formic acid, dried in a vacuum centrifuge before derivatization with hexafluoroisopropanol and quantification using gas chromato-
graphy/mass spectrometry. The results of these experiments are given in Fig.
1.

The liver and cerebral cortex and control rats contained 8.1 ± 1.3 nmol/
g and 2.6 ± 0.2 nmol/g quinolinic acid, respectively (N = 4). This increased
to 11.0 ± 1.5 nmol/g and 2.9 ± 0.3 nmol/g, respectively, in aluminum treated
animals (N = 4), but neither of these changes was statistically significant.

Our hypothesis that an increased flux of tryptophan via the kynurenine
pathway is stimulated by aluminum administration does not appear to be supported by these results. Neither rat brain or liver quinolinate concentrations are changed by an acute dose of aluminum. It is possible that a chronic treatment might allow more time for an increase in quinolinate levels.
This is the subject of a current investigation.

In addition, aluminum has been shown to increase the permeability of the
blood-brain barrier (Banks and Kastin, 1983). Thus, it is possible that the
gradient between the periphery and the CNS for quinolinate might cause an increased exposure of the CNS to quinolinic acid, which may contribute to the
encephalopathy associated with aluminum intoxication.

ACKNOWLEDGEMENT

Supported by Action Research for the Crippled Child.

REFERENCES

Alfrey, A.C., Mishell, J.M., and Burks, J., 1972, Syndrome of dyspraxia and
multifocal seizures associated with chronic hemodialysis, Trans. Am.
Soc. Artif. Intern. Organs, 18:257-260.
Alfrey, A.C., LeGendre, G.R., and Kaehny, W.D., 1976, The dialysis encephalopathy syndrome: possible aluminum intoxication, New Engl. J. Med., 294:
184-188.

Banks, W.A., and Kastin, A.J., 1983, Aluminum increases permeability of the blood-brain barrier to labelled DSIP and β-endorphin: possible implications for senile and dialysis dementia, Lancet, 2:1227-1229.

Knox, W.E., and Mehler, A.H., 1951, The adaptive increase of the tryptophan peroxidase-oxidase system of the liver, Science, 113:237-238.

Lai, J.C.K., and Blass, J.P., 1984, Inhibition of brain glycolysis by aluminum, J. Neurochem., 42:438-446.

Leeming, R.J., and Blair, J.A., 1979, Dialysis dementia, aluminum, and tetrahydrobiopterin metabolism, Lancet, 1:556.

McLachan, C., Dam T-V., Farnell, B.J., and Lewis P.N., 1983, Aluminum inhibition of ADP-ribosylation in vivo and in vitro, Neurobehav. Toxicol. Teratol., 5:645-650.

Moroni, F., Lombardi, G., Carlà, V., and Moneti, G., 1984, The excitotoxin quinolinic acid is present and unevenly distributed in the rat brain, Brain Res., 295:352-355.

Perkins, M.N., and Stone, T.W., 1982, An iontophoretic investigation of the actions of convulsant kynurenines and their interactions with the endogenous excitant quinolinic acid, Brain Res., 247:184-187.

Schwarcz, R., Whetsell, W.O., and Mangano, R.M., 1983, Quinolinic acid: an endogenous metabolite that produces axon sparing lesions in rat brain, Science, 219:316-318.

Sharp, T.R., and Rosenberry, T.L., 1985, Ionic strength dependence of acetlycholinesterase activity of Al^{3+}, Biophys. Chem., 21:261.

Tsuzuki, Y., and Marquis, J.K., 1985, Investigations of the interaction of aluminum with bovine plasma monoamine oxidase, Bull. Environ. Contam. Toxicol., 34:451-455.

Yates, C.M., Simpson, J., Russell, D., and Gordon, A., 1980, Cholinergic enzymes in neurofibrillary degeneration produced by aluminum, Brain Res., 197:269-274.

KYNURENINE AND LIPID METABOLISM

V. Rudzite and E. Jurika

Latvian Research Institute of Cardiology
Riga
USSR

INTRODUCTION

Previous observations showed an increased total cholesterol and tri-
glycerol content and decreased phospholipid and lecithin content in blood
serum of patients with elevated kynurenine accumulation in blood serum after
tryptophan loading, i.e. in the case of pyridoxal-5-phosphate (P-5-P) deficiency in the organism (Rudzite et al., 1988). The increase of saturated and
the decrease of unsaturated fatty acid content were typical for serum phospholipids obtained from these patients. The present work reports further
studies of lipid metabolism and phospholipid metabolism, especially in experimentally induced P-5-P deficiency.

MATERIALS AND METHODS

P-5-P deficiency was induced in white Wistar male rats by isoniazide
(SERVA) in an oral dose of 20 mg/kg of body weight once a day for 3 months.
The deficiency was monitored by the observation of P-5-P concentrations in
blood serum, liver, kidney and heart tissue (Serfontein et al., 1985), as
well as of kynurenine levels in blood serum (Joseph and Risby, 1975). 20
experimental and 10 control (intact) rats were used for these experiments.

Lipids were extracted from blood serum, liver, kidney and heart tissue
by the method of Folch and others (Folch et al., 1957). Cholesterol was determined by the method of Engelhart and Smirnowa (1950). Phospholipids were
extracted from total lipids by the method of Kates (1978) and measured by the
method of Urbach and Raabe (Homolka, 1961). Gas-chromatography was used for
the determination of fatty acids in phospholipids. Methylesters and fatty
acids (SERVA) were used as standards.

The incorporation of fatty acids into liver phospholipids was monitored
by the method of Nilsson (1970). The influence of P-5-P and kynurenine on
the incorporation of fatty acids in phospholipids was investigated by adding
P-5-P (1 mg/g tissue wet weight) or L-kynurenine (10^{-3} mg/g tissue wet
weight) to all samples. The statistical significance between the values
obtained from control and experimental animals was calculated by the method
of Weber (1957).

RESULTS AND DISCUSSION

Fig. 1 shows the changes in P-5-P, cholesterol and phospholipid concentration in blood serum, liver, kidney and heart tissue in isoniazide-treated rats. Fig. 2 compares the results obtained from isoniazide-treated rats and control rats with regard to the fatty acid content in phospholipids. Isoniazide decreased P-5-P concentration significantly both in blood serum and in different peripheral tissues. The changes in phospholipids as well as in the fatty acid content of phospholipids observed in isoniazide-treated rats corresponded to data obtained in patients with kynurenine accumulation in blood serum on an empty stomach (from 9.9 μmol/l in controls to 15.0 μmol/l in isoniazide-treated rats). The influence of kynurenine and P-5-P on the process of fatty acid incorporation into phospholipids was investigated next. Kynurenine addition (Fig. 3) caused a decreased incorporation of linoleic, arachidonic and stearic acid into liver tissue phospholipids. A very high oleic acid incorporation in liver tissue phospholipids was observed after L-kynurenine addition to samples from both intact and experimental rats. P-5-P addition (Fig. 4) was followed by the normalization of fatty acid incorporation into liver tissue phospholipids from isoniazide-treated rats but was ineffective in tissue from intact rats.

CONCLUSION

These findings allow us to conclude that kynurenine impairs lipid metabolism due to a failure in phospholipid biosynthesis and fatty acid incorporation in phospholipids.

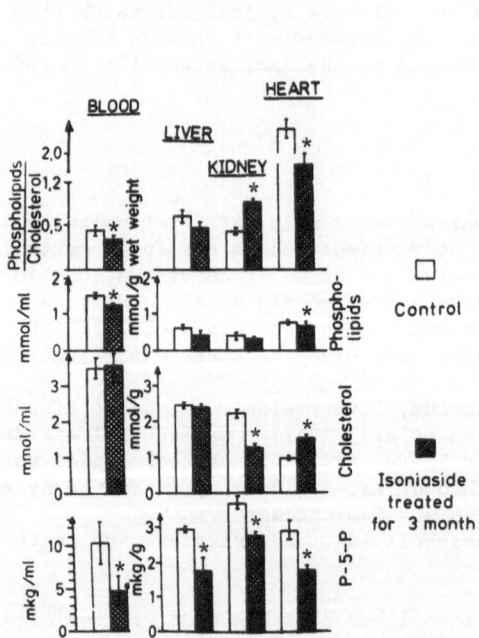

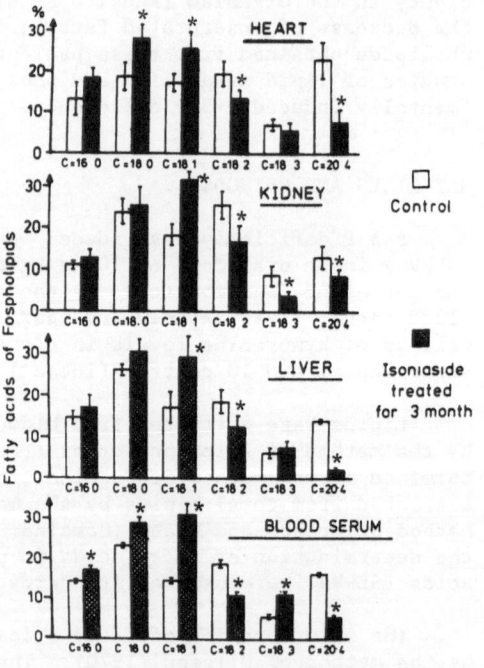

Fig. 1. P-5-P, cholesterol and and phospholipid content of blood serum, liver, kidney and heart tissue of control and isoniazide treated rats.
*P < 0.005 (t-test).

Fig. 2. Fatty acid content in phospholipids of blood serum, liver, kidney and heart tissue of control and isoniazide-treated rats.
*P < 0.005 (t-test).

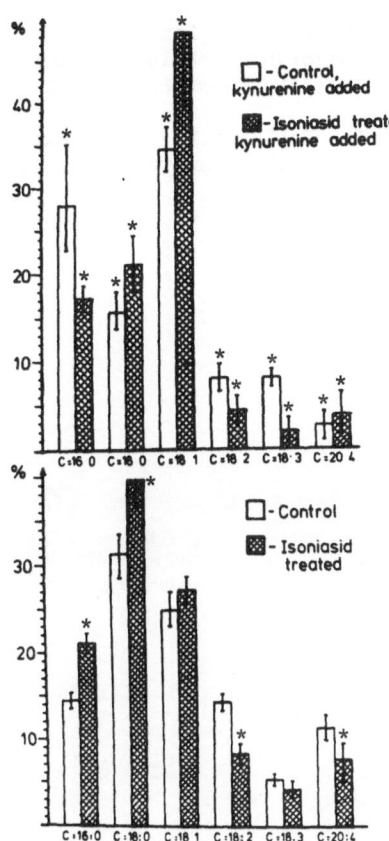

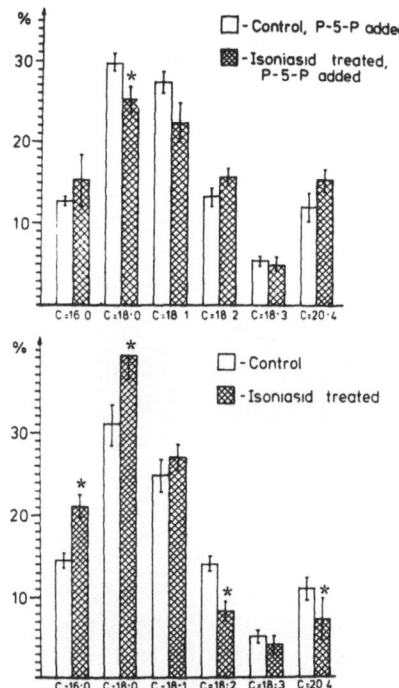

Fig. 3. Influence of
L-kynurenine on fatty
acid incorporation in
liver tissue phospho-
lipids of control and
isoniazide-treated
rats.
*P < 0.05 (t-test).

Fig. 4. Influence of P-5-P
acid incorporation on
liver tissue phospholi-
pids of control and
isoniazide-treated
rats.
*P < 0.05 (t-test).

REFERENCES

Folch, J., Lees M., and Stanley, G.H., 1957, A simple method for isolation
   and purification of total lipids from animal tissues. J. Biol. Chem.,
   226:497-509.
Homolka, J., 1961, Bestimmung des anorganischen Phosphors nach Urbach-Raabe
   (modifiziert), in: "Chemische Diagnostik im Kindesalter mit Bevorzugung
   der Mikrometrischen Blutanalysen", Homolka, J., ed., VEB Verlad Volk und
   Gesundheit, Berlin, pp. 354-355.
Joseph M.H., and Risby, D., 1975, The determination of kynurenine in plasma,
   Clin. Chim. Acta, 63:197-204.
Kates, M., 1978, Technique of lipidology, Mir, Moscow, pp. 201-203.
Nilsson, S., 1970, Synthesis and secretion of bilary phospholipids in man,
   Acta Chirurg. Scand., Suppl., 405:1-38.
Predtechensky, V.E., Borowskaya, V.M., and Margolona, L.T., 1950, "Handbook
   of Clinical Laboratory Investigations", Megiz, Moscow, USSR.

Rudzite, V., Jirgensons, J., Jurika, E., Sileniece, G., Zirne R., and Jirgensons, S., 1988, Besonderheiten der Nikotinsäurebildung bei koronarer Herzkrankheit unter Berücksichtigung von Herzrhythmusstörungen, Z. ges. inn. Med., 43:60-65.

Serfontein, W.J., Ubink, J.B., DeViller, L.S., Rapley C.H., and Becker, P.J., 1985, Plasma pyridoxal-5-phosphate level as risk index for coronary artery disease, Atherosclerosis, 55:357-361.

Weber, E., 1957, "Grundriss der biologischen Statistik für Naturwissenschaftler, Landwirte und Mediziner", VEB Gustav Fischer Verlag, Jena.

CONTENT OF TRYPTOPHAN IN HUMAN HAIR

G. Allegri, C. Costa, M. Biasiolo, and R. Arban

Department of Pharmaceutical Sciences
Padova University
35131 Padova
Italy

INTRODUCTION

Previous papers have shown that tryptophan is involved in the process of melanogenesis (Musajo et al., 1966; Musajo et al., 1969; De Antoni et al., 1970; Allegri et al., 1972; De Antoni et al., 1974; Costa et al., 1975). In fact, tryptophan and also kynurenines and indolic hydroxylated metabolites form melanins by the action of tyrosinase from Sepia officinalis and of polyphenoloxydase from potato and from Psalliota campestris mushroom (Musajo et al., 1966; Musajo et al., 1969; De Antoni et al., 1970). The study of the precursors which form during the process of melanization of tryptophan revealed that one of the substances which give rise to the formation of black pigments is 3-hydroxykynurenine (Musajo et al., 1966; Musajo et al., 1969).

Within the general framework of our extensive research on the origin and structure of melanins, and considering that Fitzpatrick et al. (1958) had suggested that red hair pigments were derived from tryptophan, the present work investigated the content of tryptophan and its metabolites "via kynurenine" in human hair in an attempt to explore possible differences associated with its pigmentary variations.

EXPERIMENTAL

Materials

Hair samples, without chemical treatment, were obtained from 200 healthy subjects, divided by gender (100 males aged from 2 to 80 years and 100 females aged from 2 to 92 years), age groups and hair color (fair, red, brown, black, grizzled and white).

Methods

The hair specimens were washed consecutively in non-ionic detergent and deionized water, then defatted by diethyl ether for 5 hours, dried overnight at 105°C, cut and pulverized using the Mikro dismembrator II (Braun). The powder was extracted with distilled water (100 ml/g hair) at 50°C for 12 hours under constant stirring, filtered, lyophilized and then solubilized with 1 ml 50% ethanol. Tryptophan determination was carried out by HPLC ac-

Kynurenine and Serotonin Pathways
Edited by R. Schwarcz *et al.*, Plenum Press, New York, 1991

Table 1. Mean values (μg/g hair ± SE) of
tryptophan in hair of 200 subjects
in relation to gender

Number of subjects	Tryptophan (μg/g hair)
Total	50.8 ± 2.2
100 Males	59.4 ± 3.4
100 Females	42.2 ± 2.7

cording to Costa et al. (1987).   The metabolites of the kynurenine pathway
were determined according to the method of Coppini et al. (1959).

RESULTS

Neither kynurenines nor other tryptophan metabolites were detected in
normal human hair.

Table 1 reports the mean values of tryptophan in the hair of 200 sub-

Table 2. Mean values (μg/g hair ± SE) of tryptophan in hair
of 200 subjects divided by age

Age	Total	Males	Females
2-5	66.2 ± 7.6 (22)	69.9 ± 9.8 (16)	56.3 ± 2.7 (6)
6-12	39.2 ± 3.2 (46)	43.1 ± 4.7 (28)	33.1 ± 3.7 (18)
13-19	33.3 ± 3.9 (26)	42.5 ± 5.3 (15)	20.7 ± 3.2 (11)
20-40	44.8 ± 4.0 (60)	65.8 ± 7.3 (25)	29.8 ± 2.4 (35)
41-60	63.8 ± 5.6 (26)	66.3 ± 5.7 (9)	62.5 ± 8.2 (17)
61-80	88.2 ± 8.6 (13)	105.2 ± 12.6 (7)	68.4 ± 4.3 (6)
> 80	78.1 ± 10.4 (7)	—	78.1 ± 10.4 (7)

The number of subjects is given in parentheses.

jects (50.8 ± 2.2 μg/g dry hair) and the mean concentrations in relation to gender: 59.4 ± 3.4 μg/g hair in 100 males and 42.2 ± 2.7 μg/g hair in 100 females. Values were found to be significantly higher (P < 0.001) in males than in females.

Table 2 shows the mean tryptophan concentrations in hair of the same subjects divided by age. Between the ages of 2 to 5, tryptophan levels in hair were relatively high in comparison with the age groups up to 40. Then values increased again, reaching the highest levels at age 61 and beyond. Female subjects always had lower concentrations of tryptophan.

Table 3 reports the variations in the content of hair tryptophan in relation to hair color. Tryptophan concentrations increased from fair to dark and from dark to white hair in both males and females. However, in females values were always markedly lower than in males.

DISCUSSION

Barnicot (1959), qualitatively analyzing aqueous extracts of human hair by paper chromatography, observed that tryptophan could be detected in some specimens but not in others, and that its presence did not appear to be related to color.

In this study, we always found high amounts of tryptophan in all specimens of human hair, examined over a wide range of colors. Gender influences the content of tryptophan, as males show significantly higher values of hair tryptophan than females. Moreover, the concentrations of this amino acid in hair appear to be correlated with age and differences in hair pigmentation in both sexes. Thus, tryptophan levels are high between the ages of 2 and 5 years, probably due to the influence of food. They then decrease markedly to rise again and reach their highest values around the age of 80 and beyond, when tryptophan accumulates in the hair during the process of keratinization. The highest levels of this amino acid are found in grizzled and white hair.

Table 3. Mean levels (μg/g hair ± SE) of tryptophan in hair of 200 subjects in relation to hair color

Color	Total	Males	Females
Fair	35.7 ± 2.8 (44)	41.9 ± 3.3 (22)	29.5 ± 4.2 (22)
Red	36.3 (3)	——	36.3 (3)
Brown	45.8 ± 3.0 (99)	55.5 ± 4.8 (54)	34.2 ± 2.6 (45)
Black	57.9 ± 6.7 (22)	74.9 ± 9.9 (10)	43.8 ± 7.5 (12)
Grizzled	87.0 ± 6.2 (23)	93.7 ± 8.4 (13)	78.3 ± 9.6 (10)
White	75.4 ± 8.3 (9)	——	75.4 ± 8.3 (9)

The number of subjects is given in parentheses.

In support of our previous results, demonstrating that tryptophan is involved in melanogenesis, it can be observed that, even in human hair, this amino acid is related to hair pigmentation, as its concentration increases from fair to black in all age groups (Musajo et al., 1966, 1969; De Antoni et al., 1970, 1974; Allegri et al., 1972; Costa et al., 1975).

REFERENCES

Allegri, G., De Antoni, A., Baccichetti, F., and Costa, C., 1972, Incorporation of D,L-tryptophan-benzene ring ^{14}C(U) and D,L-tryptophan-methylene-^{14}C into the melanin of mouse Harding-Passey melanoma, Gazz. Chim. Ital., 102:426-430.

Barnicot, N.A., 1959, Paper chromatography of human hair follicles and hair extracts, Br. J. Derm., 71:303-308.

Coppini, D., Benassi, C.A., and Montorsi, M., 1959, Quantitative determination of tryptophan metabolites (via kynurenine) in biologic fluids, Clin. Chem., 5:391-401.

Costa, C., Allegri, G., and De Antoni, A., 1975, Studies on melanogenesis of tryptophan in Harding-Passey mouse melanoma, Acta Vitaminol. Enzymol., 29: 223-226.

Costa, C., Bettero, A., and Allegri, G., 1987, Rapid measurement of tryptophan, 5-hydroxytryptophan, serotonin and 5-hydroxyindoleacetic acid in human CSF by selective fluorescence and HPLC, Gior. It. Chim. Clin., 12: 307-312.

De Antoni, A., Allegri, G., and Costa, C., 1970, Azione di sistemi enzimatici ad attivita polifenolossidasica sul triptofano e suoi derivati metabolici, Gazz. Chim. Ital., 100:1039-1049.

De Antoni, A., Allegri, G., Costa, C., and Bordin, F., 1974, Melanogenesis from tryptophan. Biogenetic experiments with Harding-Passey mouse melanoma, Experientia, 30:600-604.

Fitzpatrick, T.B., Brunet, P., and Kukita, A., 1958, Nature of hair pigment, in: "The Biology of Hair Growth", Montagna, W., and Ellis, R.A., eds., Academic Press, New York, pp. 255-303.

Musajo, L., Benassi, C.A., Allegri, G., Levorato, E., and De Antoni, A., 1966, Azione della tirosinasi sul triptofano e su suoi derivati metabolici, Chim. e Ind., 48:1221.

Musajo, L., De Antoni, A., Allegri, G., Levorato, E., and Benassi, C.A., 1969, Azione della tirosinasi sul triptofano e su suoi derivati metabolici, in: "Chimica del Farmaco e dei Prodotti Biologicamente Attivi del CNR", Roma, pp. 283-285.

DETECTION OF A NEW INTERMEDIATE PRODUCT OF INDOLEALKYLAMINE METABOLISM IN THE BRAIN

H. Rommelspacher, R. Susilo, and M. Schühle

Department of Neuropsychopharmacology
Free University
1000 Berlin 19
Germany

INTRODUCTION

Recently, we have isolated a new biogenic aldehyde adduct following the incubation of tryptamine with brain homogenate from rat, bovine and pig. The compound could be identified as (4R)-2-(3'-indolyl-methyl)-1,3-thiazolidine-

Table 1. Effect of Thiazolidinecarboxylic Acid Derivatives on Various Neuronal Mechanisms

1 mM	R_1	R_2	^{3}H-Flunitra-zepam receptor binding % activity	^{3}H-GABA$_A$	^{3}H-GABA uptake
ITCA	(indolyl-CH$_2$-)	-H	83.2$\pm$5 (7)		84.4$\pm$7 (3)
ITCA-Methylester	- " -	-CH$_3$	69.5$\pm$7 (5)		
C-TCA	COOH - CH$_2$ - CH$_2$ -	-H	no changes	no changes	no changes
PTCA	C$_6$H$_5$ - CH$_2$ -	-H			
MTCA	CH$_3$ -	-H			
MTCA-Methylester	CH$_3$ -	-CH$_3$			
TCA	H -	-H			

4-carboxylic acid (ITCA), a condensation product of indoleacetaldehyde (IAAL) with L-cysteine (Susilo et al., 1987, 1988). Similar experiments were conducted with 5-hydroxytryptamine (5-HT).

RESULTS

The formation of the thiazolidine derivative occurs spontaneously. Only the next step requires an enzyme, carbon-sulfur lyase (C-S-lyase, EC 4.4.1), which is a cytosolic protein. Indoleacetic acid (IAA) and L-cysteine are the products of the reaction (Fig. 1). The portion of ITCA compared with IAA depends on the concentration of tryptamine: the lower the concentration, the higher the portion of ITCA in an in vitro assay with rat brain homogenate. Thus, it can be predicted that the formation of ITCA predominates at the low physiological concentrations of tryptamine in the brain (ca. 0.5 ng/g). At higher concentrations, a certain amount of the IAAL may be oxidized directly by aldehyde dehydrogenase (EC 1.2.1.3). An analogous reaction was observed with 5-HT as substrate and brain homogenate (Susilo et al., 1989). The concentration of 5'-HITCA was higher than that of ITCA during the observation period (4 h). This finding suggests that ITCA is a better substrate of C-S-lyase than 5'-HITCA.

The biological importance of this new pathway of indoles might be a control mechanism of the chemically active aldehydes by rapid formation of a biologically and chemically inactive product. This has been examined in various assay systems. In rat brain synaptosomal mitochondria, the enzyme MAO (EC 1.4.3.4) was inhibited by ITCA with an IC_{50} of 8.0×10^{-4} M. At a con-

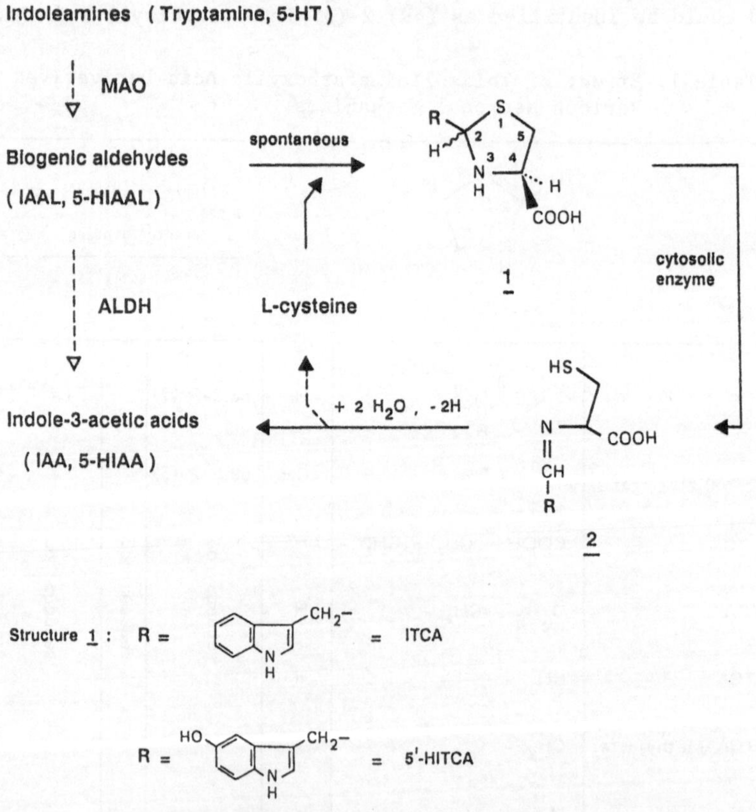

Fig. 1. Synthesis and catabolism of thiazolidine carboxylic acid derivatives (from Susilo et al., 1989).

centration of 0.5 mM, the activity of MAO-A was reduced by 70.4 ± 0.5% and that of MAO-B by 14.6 ± 0.6% (method: Krajl, 1965). Benzodiazepine and GABA$_A$-binding as well as high affinity GABA uptake was only slightly or not at all inhibited by a number of thiazolidinecarboxylic acid derivatives (Table 1) at 1 mM. Angiotensin-converting enzyme activity (method: Cushman and Cheung, 1971) was not affected by 100 μM concentrations of ITCA, C-TCA, MTCA, TCA and PTCA.

CONCLUSION

Indoleamines (tryptamine, 5-HT) are inactivated by MAO to the chemically active indoleacetaldehydes. Under physiological conditions, the aldehydes react with L-cysteine to thiazolidinecarboxylic acid derivatives (e.g. ITCA and 5'-HITCA). A cytosolic enzyme (C-S-lyase) catalyzes the formation of indoleacetic acid and L-cysteine.

REFERENCES

Cushman, D.W., and Cheung, H.S., 1971, Spectrophotometric assay and proper-ties of the angiotensin-converting enzyme of rabbit lung, Biochem. Pharmacol., 20:1637-1648.

Krajl, M., 1965, A rapid mcirofluorimetric determination of monoamine oxi-dase, Biochem. Pharmacol., 14:1683-1685.

Susilo, R., Höfle, G., and Rommelspacher, H., 1987, Degradation of trypta-mine in pig brain: identification of a new condensation product, Biochem. Biophys. Res. Comm., 148:1045-1052.

Susilo, R., Rommelspacher, H., and Höfle, G., 1989, Formation of thiazol-idine-4-carboxylic acid represents a main metabolic pathway of 5-hydroxy-tryptamine in rat brain, J. Neurochem., 52:1793-1800.

THE EXCRETION OF N^1-METHYL-2-PYRIDONE-5-CARBOXYLIC ACID AND RELATED COMPOUNDS

IN HUMAN SUBJECTS AFTER ORAL ADMINISTRATION OF NICOTINIC ACID, TRIGONALLINE

AND N^1-METHYL-2-PYRIDONE-5-CARBOXYLIC ACID

S. Yuyama and T. Suzuki

Department of Nutrition and Biochemistry
The Institute of Public Health
Tokyo 108
Japan

INTRODUCTION

During the development of a procedure for the determination of trigonel-line (Tg) in human urine (Yuyama and Suzuki, 1987), an organic acid with a similar ultra-violet spectrum was isolated. The compound was found to be N^1-methyl-2-pyridone-5-carboxylic acid (Tg-2Py)(Lindenblad et al., 1956), which Holman et al. (1950) were unable to detect in normal human urine. In the present study, Tg-2Py was determined by HPTLC. The urinary excretion of Tg-2Py, Tg and N^1-methylnicotinamide (MNA) were determined before and after the oral administration of nicotinic acid (NiA), Tg and Tg-2Py.

METHODS

Subjects and diet. Nine female volunteers, aged 20-21 years, who had no history of serious renal, gastrointestinal or metabolic disease, were used in the studies. They were kept at an ordinary diet for 7 days before and 3 days during the experimental period. Urine samples were collected in bottles containing 2 ml acetic acid. After collecting one 24 hr urine sample, groups of three subjects received 50 mg (406 μmole) of NiA, 50 mg (288 μmole) of Tg or 50 mg (326 μmole) of Tg-2Py, respectively. All samples were stored at -20°C until analysis.

Extraction of Tg-2Py from urine. One-two per cent of the 24 hr urine were diluted to 50 ml with water. The extraction procedure of Tg-2Py is shown in Scheme 1. The extraction and determination of Tg and MNA from human urine has been described previously (Yuyama and Suzuki, 1987).

Determination of Tg-2Py. 5-20 μl portions of the Tg-2Py extract were applied in a small band at the origin of a HPTLC cellulose plate using a Linomat III sample applicator. Plates were developed in n-butyl alcohol:water:acetic acid (60:20:3, by volume), and were then scanned in the reflectance mode at 254 nm using an HPTLC/TLC Scanner. Tg-2Py was quantified by comparing the peak areas with authentic Tg-2Py standard. The recovery of internal standards of Tg-2Py added to urine was 96 ± 2 (SD)%.

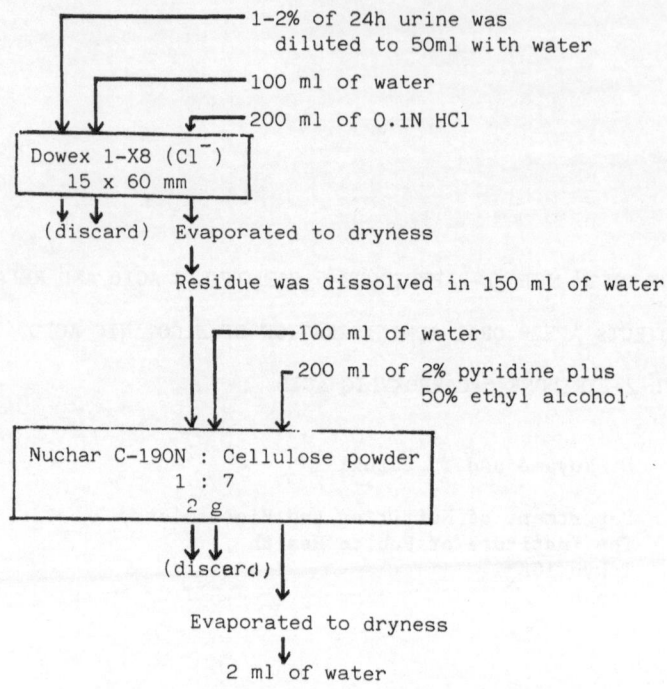

```
                    ┌──── 1-2% of 24h urine was
                    │         diluted to 50ml with water
                    │    ┌── 100 ml of water
                    │    │ ┌─ 200 ml of 0.1N HCl
                    ▼    ▼ ▼
         ┌──────────────────────────┐
         │ Dowex 1-X8 (Cl⁻)         │
         │    15 x 60 mm            │
         └──────────────────────────┘
                │            │
           (discard)    Evaporated to dryness

              Residue was dissolved in 150 ml of water

                         ┌──── 100 ml of water
                         │  ┌─ 200 ml of 2% pyridine plus
                         │  │      50% ethyl alcohol
                    ▼    ▼  ▼
         ┌──────────────────────────────────────┐
         │ Nuchar C-190N : Cellulose powder      │
         │          1 : 7                        │
         │          2 g                          │
         └──────────────────────────────────────┘
                    │
                (discard)
                    ▼
              Evaporated to dryness
                    ▼
              2 ml of water
```

Scheme 1.   Extraction of Tg-2Py from urine.

RESULTS

Excretion of Tg-2Py, Tg and MNA by 9 healthy subjects on the ordinary diet. The daily excretion of Tg-2Py, Tg and MNA in the urine is shown in Table 1. The excretion of Tg-2Py in this group ranged between 28.9 and 41.6 μmole with an average of 36.0 μmole. The excretion of Tg and MNA in the urine ranged between 20.6 and 28.3 μmole and 19.8 and 26.5 μmole, respectively.

Excretion of MNA, Tg and Tg-2Py in 3 healthy subjects after oral administration of 50 mg (406 μmole) of NiA. Table 2 shows the results of studies on the relationship of MNA, Tg and Tg-2Py to the metabolic fate of the administered NiA. The excretion of MNA, Tg and Tg-2Py significantly increased on day 1. Values ranging between 48.2 and 53.7 μmole/day were obtained for Tg-2Py. The excretion of Tg-2Py returned almost to the value of the ordinary diet group after 2 days. The actual conversion of NiA into Tg-2Py was about 4.0% of the dose, while NiA conversion into MNA and Tg was 5.2 and 13.0% of the dose, respectively. Thus, NiA is a source of MNA, Tg and Tg-2Py.

Excretion of MNA, Tg and Tg-2Py in 3 healthy subjects after oral administration of 50 mg (288 μmole) of Tg. The data describing the excretion of MNA, Tg and Tg-2Py by 3 healthy subjects after the administration of Tg are summarized in Table 3. The daily excretion of Tg-2Py in this group for 1 day after the administration of Tg increased on the first day and decreased on the second day. The excretion of Tg-2Py in the Tg group returned almost to the value of the ordinary diet after 2 days. About 8.9% of the dose of Tg was excreted in the urine as Tg-2Py within 24 hr. The daily excretion of Tg in this group was between 80.2 and 89.5 μmole, and about 21% of Tg was recov-

Table 1. Urinary excretion of N^1-methyl-2-pyridone-5-carboxylic acid, trigonelline and N^1-methylnicotinamide by 9 healthy subjects on the ordinary diet

	N^1-Methyl-2-pyridone-5-carboxylic acid (Tg-2Py)	Trigonelline (Tg)	N^1-Methylnicotin-amide (MNA)
		(μmole/day)	
Subjects (9)	28.9 - 41.6	20.6 - 28.3	19.8 - 26.5
Average	36.0 ± 4.7	25.1 ± 2.6	24.0 ± 2.8

ered in the 24 hr urine. No increase of MNA excretion in the urine was observed.

Excretion of MNA, Tg and Tg-2Py in 3 healthy subjects after oral administration of 50 mg (326 μmole) of Tg-2Py. When Tg-2Py was administered, the daily excretion of Tg-2Py ranged between 198.2 and 206.1 μmole (Table 4). The excretion of Tg-2Py returned almost to the value of the ordinary diet after 2 days. About 50% of the dose of Tg-2Py was recovered in the urine within 24 hr as unchanged Tg-2Py. No increase of MNA and Tg excretion in the urine was observed.

DISCUSSION

The daily excretion of Tg-2Py in 9 healthy subjects given the ordinary diet was 36.0 μmole. The range of values for Tg-2Py in this group of subjects is of the same order of magnitude as that found by Lindenblad et al. (1956) in 3 male adults (3-6 mg, 19.6 - 39.1 μmole). The present value is the first report on the excretion of Tg-2Py in Japanese human subjects. The excretion of Tg in 6 healthy subjects given the ordinary diet was 25.1 μmole and the level was almost identical to that of MNA excretion.

Table 2. Urinary excretion of various metabolites of nicotinic acid before and after oral administration of 50 mg (406 μmole) of nicotinic acid

Derivatives of nicotinic acid	N	Urinary excretion (μmole per day)		% of dose excreted in the urine
		Before	After	
MNA	3	22.2	75.0	13.0
Tg	3	22.8	44.0	5.2
Tg-2Py	3	34.9	51.1	4.0

Figures are expressed as the average of 3 subjects.

Table 3. Urinary excretion of various metabolites of nicotinic acid before and after the oral administration of 50 mg (288 μmole) of trigonelline

Derivatives of nicotinic acid	N	Urinary excretion (μmole per day)		% of dose excreted in the urine
		Before	After	
MNA	3	23.6	25.2	—
Tg	3	24.8	84.9	20.9
Tg-2Py	3	35.9	61.7	8.9

Figures are expressed as the average of 3 subjects.

After the administration of NiA, the actual conversion of NiA into MNA, Tg and Tg-2Py was 13.0, 5.2 and 4.0% of the dose, respectively. The conversion of NiA into Tg was described in a previous report (Yuyuma and Suzuki, 1987). As far as we know, this is the first report on the conversion of NiA into Tg-2Py via Tg. The conversion of NiA into MNA has been reported by several investigators.

After the administration of Tg, the excretion of Tg-2Py and Tg increased, the actual conversion of Tg into Tg-2Py was about 9.0% of the dose, and only 21% of the dose was recovered in the 24 hr-urine as unchanged Tg. No increase of MNA excretion in the urine was observed. Therefore Tg is converted into Tg-2Py and is thus a source of Tg-2Py.

When Tg-2Py was administered, the recovery of Tg-2Py in the urine was about 50% of the dose. Although Lindenblad et al. (1956) and Holman et al. (1950) showed that about 22% of the dose could be accounted for in the urine, it is not evident that Tg-2Py is an end product. In fact, further metabolism of this compound in the body is likely.

Table 4. Urinary excretion of various metabolites of nicotinic acid before and after the oral administration of 50 mg (326 μmole) of N^1-methyl-2-pyridone-5-carboxylic acid

Derivatives of nicotinic acid	N	Urinary excretion (μmole per day)		% of dose excreted in the urine
		Before	After	
MNA	3	24.7	25.6	—
Tg	3	27.6	28.9	—
Tg-2Py	3	37.1	201.6	50.4

Figures are expressed as the average of 3 subjects.

The combined results show that the body is capable of converting NiA into MNA via NAD coenzymes. NiA can be converted to Tg and Tg-2Py, and Tg to Tg-2Py. These changes do not appear to be reversible, and there is no evidence that any other inter-conversion occurs. It is interesting to note that NiA is methylated directly by the body, and that the methylated compound (Tg) can be oxidized to a pyridone.

REFERENCES

Holman, W.I.M., and Lange, D.J., 1950, Metabolism of nicotinic acid and related compounds by humans, Nature, 165:604-605.

Lindenblad, G.E., Kaihara, M., and Price, J.M., 1956, The occurrence of N-methyl-2-pyridone-5-carboxylic acid and its glycine conjugate in normal human urine, J. Biol. Chem., 219:893-901.

Yuyama, S., and Suzuki, T., 1987, Urinary excretion of trigonelline and N^1-methylnicotinamide by human beings and rats following oral administration of nicotinic acid or trigonelline, in: "Progress in Tryptophan and Serotonin Research", Bender, D.A., Joseph, M.H., Kochen, W., and Steinhart H., eds., de Gruyter, Berlin, pp. 197-202.

STABLE ISOTOPE-LABELED TRYPTOPHAN AS A PRECURSOR FOR STUDYING THE DISPOSITION

OF QUINOLINIC ACID IN RABBITS

R.L. Boni[1], M.P. Heyes[1], J.D. Bacher[2], J.A. Yergey[3], X.-D. Ji[1],
F.P. Abramson[1,4], and S.P. Markey[1]

[1]Laboratory of Clinical Science
NIMH
Bethesda, Maryland 20892

[2]Veterinary Resources Branch
NIH
Bethesda, Maryland 20892

[3]Laboratory of Clinical Studies
NIAAA
Bethesda, Maryland 20892

[4]Dept. of Pharmacology
George Washington University Medical Center
Washington, District of Columbia 20037
USA

INTRODUCTION

A major pathway for metabolism of the essential amino acid tryptophan
via the kynurenine pathway leads to the production of quinolinic acid as a
precursor for nicotinamide containing nucleotides (Nishizuka and Hayaishi,
1963). Quinolinic acid, an agonist at the NMDA receptor, is also a neurotoxin and convulsant (Stone and Connick, 1985). It has been postulated to be
involved in several neurodegenerative diseases (Schwarcz et al., 1984). Recently, elevated concentrations of quinolinic acid have been found in the
cerebrospinal fluid of patients with acquired immune deficiency syndrome
(Heyes et al., 1989). Little is known of the kinetics of the in vivo disposition of quinolinic acid (i.e. rates of formation, distribution and elimination). The use of stable isotope-labeled precursors offers many advantages in the study of the disposition of endogenous compounds (Wolfe, 1984).
We have begun investigations into the use of stable isotope-labeled tryptophan as a precursor for the in vivo production of labeled quinolinic acid in
rabbits with a view of developing methods applicable for studying the disposition of quinolinic acid in man.

METHODS

Deuterated tryptophan, $(2',4',5',6',7')-[^2H]_5$-L-tryptophan (MSD Isotopes), at doses of 50, 25 or 10 mg/kg, was administered into the lateral ear
vein of male New Zealand white rabbits weighing approximately 3 kg. Serial

blood samples (approximately 0.5 ml), drawn via an indwelling catheter in an external jugular vein, were taken prior to the administration of the deuter-ated tryptophan and at 10, 20, 30, 45, 60, 75, 90, 120, 180, 240 and 360 minutes after administration. The blood was placed in heparinized test tubes and stored on ice. After centrifugation, the plasma was separated. Approxi-mately 100 μl cerebrospinal fluid (CSF) were drawn at the same time inter-vals via a chronic catheter implanted in the cisterna magna (Glue et al., 1988). All plasma and CSF samples were stored at -80°C until analysis.

Endogenous (TRP) and labeled tryptophan (TRP-D5) plasma concentrations were determined by gas chromatography-electron ionization mass spectrometry. Briefly, after the addition of the isotopomer internal standard, $(3,3)-[^2H]_2$-D,L-tryptophan, plasma samples (20-100 μl) were extracted with methanol and the methanol extracts dried in vacuo. The tryptophans were first methylated with trimethylsilyldiazomethane and then N-acylated with heptafluorobutyryl-imidazole. Samples were assayed on a Hewlett-Packard 5790 MSD.

The concentrations of endogenous and labeled quinolinic acid were deter-mined by gas chromatography-negative chemical ionization mass spectrometry. Internal standard, $[^{18}O]_4$-quinolinic acid, was added to 30 μl of plasma or CSF, the samples freeze-dried and derivatized with hexa-fluoroisopropanol (Heyes and Markey, 1988).

The concentration-time dependence of deuterated tryptophan and deuter-ated quinolinic acid in plasma was analyzed pharmacokinetically. Initial es-timates of the rate constants involved were obtained manually. Final esti-mates of these pharmacokinetic parameters were obtained by least-squares non-linear regression analysis with MLAB (Knott and Reece, 1972).

RESULTS AND DISCUSSION

Deuterium-labeled tryptophan was measured in rabbit plasma at concentra-tions ranging from 412 nmol/l to 421 μmol/l. The administered tryptophan was present exclusively as TRP-D5. No significant TRP-D4 or TRP-D3 was detected, indicating no loss or randomization of label either biologically or chemical-ly. Hayashi et al. (1986) administered 10 mg/kg $(3,3)-[^2H]_2$-L-tryptophan orally to a human subject and also found no loss or scrambling of label.

The concentration of TRP-D5 in plasma declined rapidly in a mono-expo-nential manner. ,The calculated biological half-lives (0.693/elimination rate constant) for rabbits administered TRP-D5 were 29.5, 23.3 and 31.8 min (50 mg/kg), 34.4 and 37.1 min (25 mg/kg), and 29.8 and 47.0 min (10 mg/kg). At the two higher doses, there was some initial nonlinearity at concentrations greater than 100 μmol/l, suggesting saturation of tryptophan elimination.

Selectively-labeled quinolinic acid (QUIN-D3) was detected in the plas-ma. No loss or randomization of the deuterium atoms to produce QUIN-D2 or QUIN-D1 was detected. Plasma QUIN-D3 concentrations ranged from 1.3 to 416.4 nmol/l. The concentration of QUIN-D3 in the plasma appeared to follow a bi-exponential curve. Plasma QUIN-D3 was measurable at 10 min. (first sampling time) and maximal between 1 and 2 hours. Fig. 1 illustrates QUIN-D3 concen-tration-time curves obtained in a single rabbit.

The estimated disposition rate parameters showed considerable variabil-ity among individual rabbits especially at the larger doses of TRP-D5. The apparent formation half-lives calculated from the hybrid rate constant, which describes the appearance of QUIN-D3 in the plasma, were 91.4, 59.3 and 43.6 min (50 mg/kg), 43.6 and 44.6 min (25 mg/kg), and 29.8 and 32.9 min (10 mg/kg). The calculated biological half-lives for the QUIN-D3 formed were 2.2, 2.4 and 4.2 h (50 mg/kg), 4.5 and 1.1 h (25 mg/kg), and 0.6 and 1.5 h (10

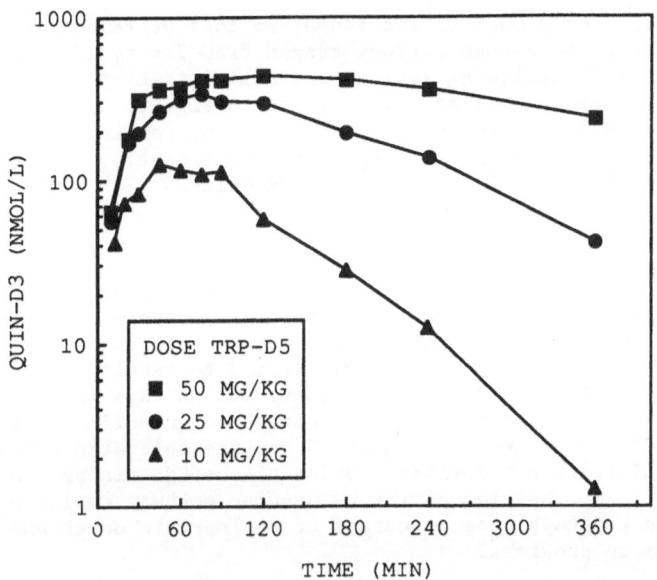

Fig. 1.   Concentration-time curves for QUIN-D3 in the plasma
          following the intravenous administration of 50, 25
          and 10 mg/kg TRP-D5 to a single rabbit.

mg/kg). The inter-individual variability, particularly at the higher doses
may have been due in part to difficulties in obtaining accurate computer fits
to the data because of the shallow slopes associated with the elimination
phase.   Saturation of one or more QUIN-D3 elimination processes is suggested
by increased biological half-lives after the higher doses.

QUIN-D3 was detected in CSF following intravenous administration of

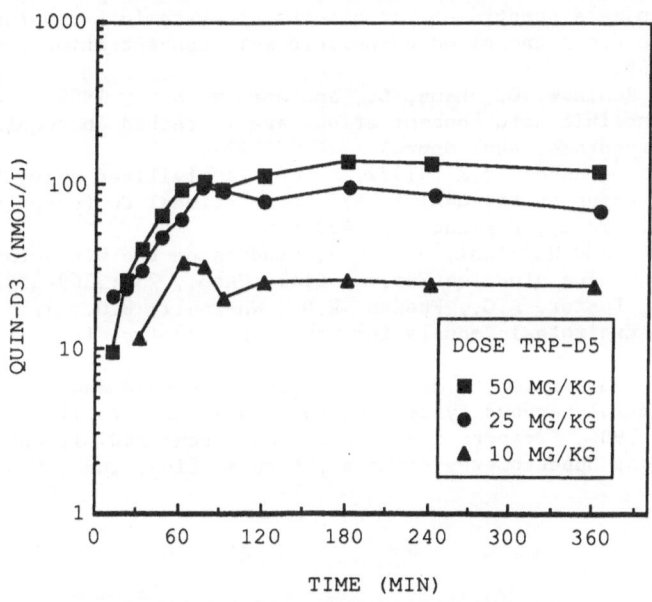

Fig. 2.   Concentration-time curves for QUIN-D3 in the CSF
          following the intravenous administration of 50, 25
          and 10 mg/kg TRP-D5 to a single rabbit.

TRP-D5. Again, only QUIN-D3 was found; no loss or randomization of label was detected. Measured concentrations ranged from 2.6 to 134.1 nmol/l. The CSF levels of QUIN-D3 following intravenous administration of 50, 25 or 10 mg/kg TRP-D5 to the same rabbit are illustrated in Fig. 2. The concentration of QUIN-D3 reached a plateau between 1-3 hours and remained elevated, although the plasma concentrations had decreased significantly, indicating a longer half-life for CSF QUIN-D3 than for plasma QUIN-D3.

SUMMARY

    The utility of stable isotope-labeled tryptophan as a precursor for studying the disposition of quinolinic acid was investigated. TRP-D5 at doses of 50, 25 or 10 mg/kg was administered to rabbits. Blood and CSF samples were taken for up to 6 hours. There was no loss of deuterium from the tryptophan and the specifically tri-deuterated quinolinic acid measured in plasma and CSF. CSF levels of QUIN-D3 remained elevated 6 hours following TRP-D5 administration. Further studies of the CNS disposition of quinolinic acid and other metabolites of the kynurenine pathway employing stable isotope-labeled tryptophan as precursor at appropriate doses and with extended sampling are in progress.

REFERENCES

Glue, P., Bacher, J.D., and Nutt, D.J., 1988, A technique for chronic catheterization of the cisterna magna in rabbits, **Lab. Anim. Sci.**, 38:740-742.

Hayashi, T., Shimamura, M., Matsuda, F., Minatogawa, Y., and Naruse, H., 1986, Sensitive determination of deuterated and non-deuterated tryptophan, tryptamine and serotonin by combined capillary gas chromatography and negative ion chemical ionization mass spectrometry, **J. Chrom. Sci.**, 383: 259-269.

Heyes, M.P., and Markey, S.P., 1988, Quantification of quinolinic acid in rat brain, whole blood, and plasma by gas chromatography and negative chemical ionization mass spectrometry: effects of systemic L-tryptophan administration on brain and blood quinolinic acid concentrations, **Anal. Biochem.**, 174:349-359.

Heyes, M.P., Rubinow, D., Lane, C., and Markey, S.P., 1989, Cerebrospinal fluid quinolinic acid concentrations are increased in acquired immune deficiency syndrome, Ann. Neurol., 26:275-277.

Knott, G.D., and Reece, D.K., 1972, MLAB: A civilized curve-fitting system, in: "Proceeding of the ONLINE 1972 International Conference", Vol. 1, Brunel University, England, pp. 497-526.

Nizhizuka, Y., and Hayaishi, O., 1963, Studies on the biosynthesis of nicotinamide adenine dinucleotide, J. Biol. Chem., 238:3369-3377.

Schwarcz, R., Foster, A.C., French, E.D., Whetsell, W.O., Jr., and Köhler, C., 1984, Excitotoxic models for neurodegenerative disorders, **Life Sci.**, 35:19-32.

Stone, T.W., and Connick, J.H., 1985, Quinolinic acid and other kynurenines in the central nervous system, **Neuroscience**, 15:597-617.

Wolfe, R.R., 1984, "**Tracers in Metabolic Research: Radioisotope and Stable Isotope/Mass Spectrometry Methods**", Alan R. Liss, Inc., New York.

EVIDENCE FOR THE PREFERENTIAL PRODUCTION OF 3-HYDROXYANTHRANILIC ACID FROM

ANTHRANILIC ACID IN THE RAT BRAIN

H. Baran and R. Schwarcz

Maryland Psychiatric Research Center
Baltimore, Maryland 21228
USA

INTRODUCTION

The excitotoxic brain metabolite quinolinic acid (QUIN) has been hypothetically linked to the pathogenesis of neurodegenerative and seizure disorders. Thus, intracerebral QUIN injections in rats produce patterns of nerve cell loss in the hippocampus and striatum which are remarkably similar to those seen in temporal lobe epilepsy and Huntington's disease, respectively. More recently (Heyes et al., this volume), QUIN has also been speculatively related to the pathogenesis of viral disorders such as AIDS.

As detailed elsewhere in this volume (cf. chapter by Schwarcz and Du), elements of the kynurenine pathway, namely the enzymes immediately responsible for the production and degradation of QUIN [3-hydroxyanthranilic acid oxygenase (3HAO) and quinolinic acid phosphoribosyltransferase], have been described and characterized in rat and human brain tissue. However, since the activities of kynurenine 3-hydroxylase and kynureninase in the rat brain are very low, there is very poor incorporation of either tryptophan or kynurenine into QUIN in vivo (Speciale et al., 1989). We therefore set out to explore the existence of an alternative biosynthetic avenue for brain QUIN. We report here data from our investigation of the relative potencies of 3-hydroxykynurenine (3HK) and anthranilic acid (ANA) as bioprecursors of 3-hydroxyanthranilic acid (3HANA) in the rat brain. For this purpose, rat brain slices were used, and a novel analytical technique for the measurement of small amounts of 3HANA was developed.

METHODS AND RESULTS

The column separation/extraction method for the determination of 3HANA in tissue is shown schematically in Fig. 1. After lyophilization at the end of the procedure depicted in the Figure, the samples were resuspended in 60 μl 0.05 M perchloric acid, and an appropriate aliquot subjected to HPLC with coulometric detection. HPLC was performed using a 8 cm C_{18} reversed-phase column (3 μm spherical particles) and a mobile phase consisting of 0.1 M sodium phosphate, 0.75 mM octanesulfonic acid and 7% methanol, pH 3.0. 3HANA was eluted at a flow rate of 0.5 ml/min and detected at + 0.40 V. The retention time was routinely between 9.5 and 11 minutes.

One mm thick brain slices were obtained from male Sprague-Dawley rats

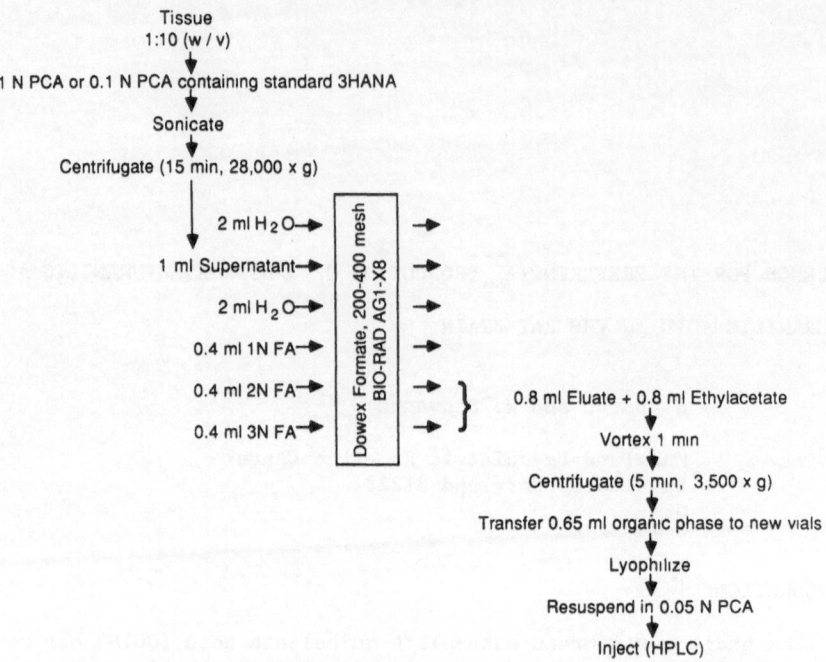

Fig. 1.   Schematic representation of the purification
procedure for 3HANA.

(approximately 200 g) by standard procedures and incubated in culture wells
(10 slices/well) containing 1 ml oxygenated Krebs-Ringer buffer, pH 7.4, and
10 μM of the potent 3HAO inhibitor 4-chloro-3HANA (4-Cl-3HANA; cf. Walsh et
al., this volume). The tissues were preincubated for 10 min at 37°C up to
120 min in the presence or absence of 3HK or ANA, two putative bioprecursors
of 3HANA. Following incubation, the wells were immediately placed on ice,
and the slices rapidly separated from the incubation medium. Slices were
then sonicated (1:10, w/v) in 0.1 M perchloric acid, the resulting homogenate
was centrifuged (15 min, 28,000 x g), and the 3HANA content of the resulting
supernatant determined as described above.

     In pilot experiments, de novo synthesis of 3HANA in brain slices was

Table 1.   De novo production of 3-hydroxyanthranilic acid in rat brain slices

	Anthranilic acid		3-Hydroxykynurenine	
	—	+ 4-Cl-3HANA	—	+ 4-Cl-3HANA
5 min	108.3 ± 8.3	89.0 ± 9.3	< 1.0	9.0 ± 0.3
120 min	11.3 ± 1.0	187.5 ± 11.2	2.1 ± 0.2	51.6 ± 7.9

Slices of rat cortex were incubated with 1 mM ANA or 1 mM 3HK in the presence
or absence of 10 μM 4-Cl-3HANA as described previously (Baran and Schwarcz,
1990). Data represent newly produced 3HANA in excess of endogenously synthesized 3HANA. Data are expressed in fmol/mg tissue and are the mean ± SEM of
3-6 independent determinations.

verified by degrading the compound enzymatically using a highly purified preparation of rat liver 3HAO (Okuno et al., 1987; Baran and Schwarcz, 1990).

As shown in Table 1, 3HANA was synthesized from both ANA and 3HK. In the presence of 4-Cl-3HANA, the production of 3HANA from both bioprecursors could be readily assessed after 5 or 120 min of incubation. Lack of 3HAO inhibition resulted in substantially lower yields of 3HANA, particularly at the 120 min timepoint, presumably due to further metabolism to QUIN (Foster et al., 1986). Under all experimental conditions, ANA proved to be a better precursor of 3HANA than 3HK. The preferential effect of ANA was particularly noticable at the short incubation time of 5 min. In fact, the production of 3HANA from ANA occurred so rapidly that the presence of 4-Cl-3HANA did not result in further increases, indicating that the speed of anthranilate hydroxylation alone (rather than 3HAO activity) determined the extent of 3HANA production under this experimental condition.

DISCUSSION

The present results indicate that brain-specific mechanisms are in place to avert potentially catastrophic consequences of a tryptophan overload to the brain. It appears that the kynurenine pathway, which in the periphery is almost exclusively responsible for tryptophan catabolism, is very ineffective in the (rat) brain. Thus, while both kynurenine-3-hydroxylase (Uemura and Hirai, this volume) and kynureninase (Okuno and Kido, this volume) are clearly present in brain tissue, their activities appear to be lower than that of anthranilic acid hydroxylase (Schwarcz et al., 1989). In peripheral organs, the situation is reversed and anthranilate hydroxylase clearly plays only a minor role (Ueda et al., 1978; Baran and Schwarcz, 1990). The brain therefore has the ability to synthesize 3HANA - and hence the N-methyl-D-aspartate (NMDA) receptor agonist QUIN - quite independent from tryptophan. This alternative biosynthetic route may be of particular importance when rapid mobilization of brain QUIN for NMDA receptor activation is required during brain development or in the mediation of memory processes (Watkins and Collingridge, 1989). Given this hypothetical scenario, it is mandatory to examine the presence of ANA in brain tissue, to investigate its penetration through the blood-brain barrier and, most importantly, to explore the characteristics of brain anthranilic acid hydroxylase. Such studies are currently in progress in our laboratory.

ACKNOWLEDGEMENTS

Supported by USPHS grants NS 16102 and NS 28236 and a fellowship from the Max Kade Foundation (to H.B.).

REFERENCES

Baran, H., and Schwarcz, R., 1990, Presence of 3-hydroxyanthranilic acid in rat tissues and evidence for its production from anthranilic acid in the brain, J. Neurochem., 55:738-744.

Foster, A.C., White, R.J., and Schwarcz, R., 1986, Synthesis of quinolinic acid by 3-hydroxyanthranilic acid oxygenase in rat brain tissue in vitro, J. Neurochem., 47:23-30.

Okuno, E., Köhler, C., and Schwarcz, R., 1987, Rat 3-hydroxyanthranilic acid oxygenase: purification from the liver and immunocytochemical localization in the brain, J. Neurochem., 49:771-780.

Schwarcz, R., Baran, H., and Okuno, E., 1989, Anthranilate hydroxylase in rat brain, Soc. Neurosci. Abstr., 15:328.6.

Speciale, C., Ungerstedt, U., and Schwarcz, R., 1989, Production of extracellular quinolinic acid in the striatum studied by microdialysis in unanesthetized rats, Neurosci. Lett., 104:345-350.
Ueda, T., Otsuka, H., Goda, K., Ishiguro, I., Naito, J., and Kotake, Y.,
  1978, The metabolism of [carobxyl-^{14}C]anthranilic acid. I. The incorporation of radioactivity into NAD$^+$ and NADP$^+$, J. Biochem., 84:687-696.
Watkins, J.C., and Collingridge, G.L., eds., 1989, "The NMDA Receptor", IRL,
  Oxford, 242 pp.

CHRONOBIOLOGICAL EFFECTS OF L-TRYPTOPHAN IN HUMANS: INFLUENCE ON MELATONIN

SECRETION

T. Sielaff, L. Demisch, P. Gebhart, A. Blumhofer, A. Khazai,
and B. Lemmer[1]

Dept. of Psychiatry and [1]Dept. of Pharmacology
Hospital of the Goethe-University
6000 Frankfurt a. Main
Germany

INTRODUCTION

Melatonin synthesis in the pineal organ results from acetylation of serotonin (5-HT) by arylalkylamine-N-acetyltransferase (NAT) and subsequent conversion of N-acetylserotonin to melatonin by hydroxyindole-O-methyltransferase. A substantial amount of data derived from animal experiments indicates that the circadian rhythm of melatonin synthesis with an about tenfold increase in the dark phase is closely associated with the oscillation in the activity of pineal NAT, the rate-limiting enzyme of melatonin production. The major biological event controlling NAT and thus melatonin synthesis is the β-adrenoreceptor-linked increase in pineal cAMP production. However, the precise mechanism by which the noradrenergic (NE) input from sympathetic nerve endings controls the circadian melatonin rhythm is a matter of debate (circadian changes in NE input, β-adrenoreceptor density, uncoupling of the receptor after agonist interaction, synergism with α_1-adrenoreceptors, etc.; for summary, see Klein et al., 1987; Cardinali et al., 1987; Gonzalez-Brito and Reiter, 1987). In addition to the predominant role of mechanisms controlling NAT activity, many data are compatible with the suggestion that the availability of serotonin may be involved in regulating melatonin synthesis (Wurtman and Ozaki, 1978; Chan and Ebadi, 1981; for summary, see Ebadi, 1984). Recently, Demisch et al. (1986, 1987) reported that the administration of the selective 5-HT reuptake inhibitor fluvoxamine at night led to significant increases of nocturnal melatonin concentrations in the plasma of healthy volunteers and a delayed decline of plasma melatonin in early morning hours. This observation led to the hypothesis that the availability of serotonin as a substrate of NAT may be of relevance in the process of terminating the synthesis of melatonin. This study was designed to examine this hypothesis by elucidating the influence of L-tryptophan on melatonin plasma levels in healthy subjects.

SUBJECTS AND METHODS

6 healthy subjects participated in this study (3 women and 3 men, aged between 21 and 32 years). They gave informed consent and had not taken any drugs during the last 6 weeks. Placebo or 50 mg/kg L-tryptophan (obtained from Fesenius AG, Oberursel, FRG as a 1.5% solution) was infused within 30

minutes by means of an infusion pump (Volumed, Fresenius, FRG) started at 8:00 a.m., 10:00 a.m. or 12:00 p.m. The subjects had fasted at least two hours before starting the experiments. 10 ml blood samples were obtained at intervals of 15 minutes by means of an indwelling catheter up to 4 hours following the end of infusion. Melatonin concentrations were measured in aliquots of heparinized plasma samples (stored at -20°C until analysis) by using a highly specific melatonin antibody obtained from J. Arendt and a direct plasma melatonin radioimmunoassay method (Fraser et al., 1983). Plasma tryptophan concentrations were measured by HPLC with fluorimetric detection.

RESULTS and DISCUSSION

Surprisingly, the infusion of 50 mg/kg L-tryptophan was followed by increases of melatonin concentrations in the plasma of all subjects and at all time points investigated. Peak concentrations were measured within 60 minutes after starting the infusion (range of peak levels: 8:00 - 9:00 a.m.: 40 - 180 pg/ml, mean: 124 ± 59 pg/ml; 10:00 - 11:00 a.m.: 50 - 240 pg/ml, mean: 98 ± 76 pg/ml; 12:00 - 1:00 p.m: 60 - 120 pg/ml, mean: 95 ± 25 pg/ml). Analysis of variance for dependent groups did not show a significant effect of time with respect to peak levels of melatonin secretion induced by tryptophan whereas a significantly higher amount of melatonin (area under the curve) was found when 50 mg/kg tryptophan was infused at 8:00 a.m. as compared to 10:00 a.m. and 12:00 p.m.

In order to rule out the possibility that the effect of 50 mg/kg tryptophan on melatonin synthesis/secretion was due to increases in the sympathetic noradrenergic input to pinealocytes, the subjects were pretreated with 100 mg of the β-adrenoreceptor blocker atenolol (at 6:00 a.m.), and 50 mg/kg L-tryptophan was infused at 8:00 a.m. There was a stimulation of melatonin production/secretion in all subjects. Peak concentration and duration of increases of plasma melatonin increases were slightly but significantly smaller compared to stimulation of melatonin following 50 mg/kg tryptophan at 8:00 a.m. without β-adrenoreceptor blockade.

In accordance with published data, tryptophan peak concentrations were found to range between 0.2 and 1.0 mg/ml plasma. A number of methodological experiments were carried out in order to exclude the possibility of measuring erroneous tryptophan instead of melatonin due to the cross-reactivity of the melatonin antibody.

In conclusion, the results indicate that stimulation of the synthesis and/or secretion of melatonin can be induced in humans by a load with 50 mg/kg L-tryptophan. This effect is apparently not mediated by pineal β-adrenoreceptor activity. Therefore it is suggested that substrate availability by active 5-HT re-uptake mechanisms into pinealocytic vesicle-compartments is involved in the termination of melatonin production in early morning hours in humans.

REFERENCES

Cardinali, D.P., Vacas, M.I., and Rosenstein, R.E., 1987, Pineal: a multi-effector organ, in: "Fundamentals and Clinics in Pineal Research", Trentini, G.P., De Gaetani, C., and Pevet, P., eds., Raven Press, New York, pp. 137-150.
Chan, A., and Ebadi, M., 1981, Reciprocal relationship between the concentration of serotonin and the activity of serotonin N-acetyltransferase in rat pineal glands in culture, Endocrin. Res. Commun., 8:25-44.

Demisch, K., Demisch, L., Bochnik, H.J., Nickelsen, T., Althoff, P.H., Schöffling, K., and Rieth, R., 1986, Melatonin and cortisol increase after
fluvoxamine, Br. J. Clin. Pharmacol., 22:620-622.

Demisch, K., Demisch, L., Nickelsen, T., and Rieth, R., 1987, The influence
of acute and subchronic administration of various antidepressants on early
morning melatonin plasma levels in healthy subjects: increases following
fluvoxamine, J. Neural Transm., 68:257-270.

Ebadi, M., 1984, Regulation of the synthesis of melatonin and its significance to neuroendocrinology, in: "The Pineal Gland", Reiter, R.J., ed.,
Raven, New York, pp. 1-38.

Fraser, S., Cowen, P., Franklin, M., Franey, C., and Arendt, J., 1983, Direct
radioimmunoassay for melatonin in plasma, Clin. Chem., 28:396-397.

Gonzalez-Brito, A., Reiter, R.J., 1987, Mammalian pineal beta-adrenergic receptors: their regulation and physiological significance, in: "Fundamentals and Clinics in Pineal Research", Trentini, G.P., De Gaetani, C., and
Pevet, P., eds., Raven Press, New York, pp. 121-124.

Klein, D.C., Weller, A.L., Sudgen, A.L., Sudgen, D., Vanecek, J., Chik, C.L.,
and Ho, A.K., 1987, Integration of multiple receptor mechanisms regulating pineal cyclic nucleotides, in: "Fundamentals and Clinics in Pineal
Research", Trentini, G.P., De Gaetani, C., and Pevet, P., eds., Raven,
New York, pp. 111-120.

Wurtman, R.J., and Ozaki, Y., 1978, Physiological control of melatonin synthesis and secretion: mechanisms generating rhythms in melatonin, methoxytryptophol, and arginine vasotocin levels and effects on the pineal of
endogenous catecholamines, the estrous cycles, and environmental lighting,
J. Neural Transm., Suppl., 13:59-70.

CIRCADIAN PATTERNS OF SALIVARY MELATONIN AND URINARY 6-SULFATOXY-MELATONIN

BEFORE AND AFTER A 9 HOUR TIME-SHIFT

T. Nickelsen[1], A. Samel, H. Maass, M. Vejvoda, H. Wegmann,
and K. Schöffling[1]

DFLVLR Institute of Aerospace Medicine
Köln

[1]Department of Endocrinology
University Hospital
Frankfurt/Main
Germany

INTRODUCTION

During the past decade, many laboratories have replaced melatonin (MT) determination in serum by the measurement of the hormone in saliva, or its main metabolite 6-sulfatoxymelatonin (MTS) in urine. Advantages of these methods include non-invasive sampling techniques which have proven extremely beneficial during longitudinal studies in humans; in addition, determination of urinary MTS excretion allows to calculate more closely the total amount of MT production. Several groups have reported high correlations between serum and salivary MT on the one hand (Vakkuri 1985; Miles et al., 1987; Nowak et al., 1987), and between serum MT and urinary MTS on the other (Markey et al., 1985; Bojkowski et al., 1987). However, all these studies were done under baseline conditions, i.e. with subjects living in a regular 24 h-time pattern.

In the present study, we report the relationship between circadian salivary MT and urinary MTS in eight subjects studied longitudinally before and after a 9 hour time-shift. These results are part of a more comprehensive study on the effects of exogenous MT on jet lag whose results will be reported elsewhere in detail.

MATERIALS AND METHODS

This double-blind, cross-over experiment consisted of two parts lying two months apart. Except for the medication, the trials were identical, each one lasting 15 days. Eight healthy male volunteers (age 25.8 ± 4.0 years, range 20 to 32) were housed in the isolation unit of the DFVLR Institute of Aerospace Medicine. They were submitted to a strict time schedule with meals at 0900, 1245 and 1830 h and a dark period from 2300 until 0800 h. On day 7 at 0200 h, laboratory time was phase advanced by 9 hours, and subjects were awakened; this resulted in a partial sleep deprivation in the night of the time shift. With respect to the new laboratory time, subjects continued to live on the same time schedule as before. In a cross-over design, each sub-

ject received a daily oral dose of either 5 mg MT or placebo from day 5 to
11; the drug was ingested at 1800 h before the time shift, and at 2300 h after the time shift. Urine was collected at 3-hourly intervals throughout the
study at 0800, 1100, 1400, 1700, 2000, 2300, 0200, and 0500 h. Saliva was
sampled at the same hours, but only on days 3,4, and 12 to 15. During the
dark phase, subjects were awakened and sampling was performed under dim red
light. None of the volunteers experienced serious difficulties in returning
to sleep afterwards.

Salivary MT was determined using a modification of the direct radioimmunoassay described by Fraser et al. (1983). The standard curve was set up
in pooled saliva which had been collected by laboratory staff members and
other healthy volunteers after exposure to bright sunlight for at least two
hours. Each individual saliva sample contributing to the pool was checked
against the previous pool; only samples containing MT concentrations below
the lower limit of detection were used for subsequent pooling. Assay sensitivity, recovery and precision were similar to those reported by Fraser for
the serum assay. MTS was measured using a commercially available kit (IBL,
Hamburg, West Germany).

Statistical methods included determination of the mean, standard deviation, Student's t-test and linear regression if groups of values were normally distributed. Correspondingly, median, standard error of the median, Wilcoxon and Spearman tests were used where parameters failed to show normal
distribution.

RESULTS

The mean salivary and urinary concentrations of MT and MTS, respectively, on days 3,4, and 12 to 15 of the melatonin and the placebo trial are presented in Fig. 1. On days 3 and 4, the salivary melatonin rhythm exhibited a
significant rise in both groups between 2300 and 0200 h. After the time
shift and discontinuation of drug intake, the salivary MT rhythm on the group
level appeared to be reestablished, although the mean nighttime rise failed
to reach statistical significance in 3 out of 8 nights. In contrast, the
circadian rhythmicity of urinary MTS excretion, which had been clearly present on days 3 and 4, was considerably dampened after MT and almost abolished
after placebo. None of the 8 nights observed after the time shift exhibited
a significant rise in urinary MTS. Correlations between the means of both
parameters are presented in Table 1. (Since in the MT trial MTS levels remained elevated until 0800 h on day 13 due to the preceding drug intake,
these data were excluded from statistical calculation.)

Under baseline conditions, 10 out of 16 individual urinary and salivary
profiles correlated significantly (62.5%), whereas after time shift this per-

Table 1. Correlations between mean salivary MT and urinary MTS
concentrations in 8 volunteers before (days 3,4) and
after (days 12 to 15) a 9 hour time-shift on day 7 and
daily ingestion of 5 mg MT or placebo from day 5 to 11

| | Before time shift | | After time shift | |
	r	P	r	P
MT	0.75	0.001	0.31	n.s.
Placebo	0.72	0.001	0.20	n.s.

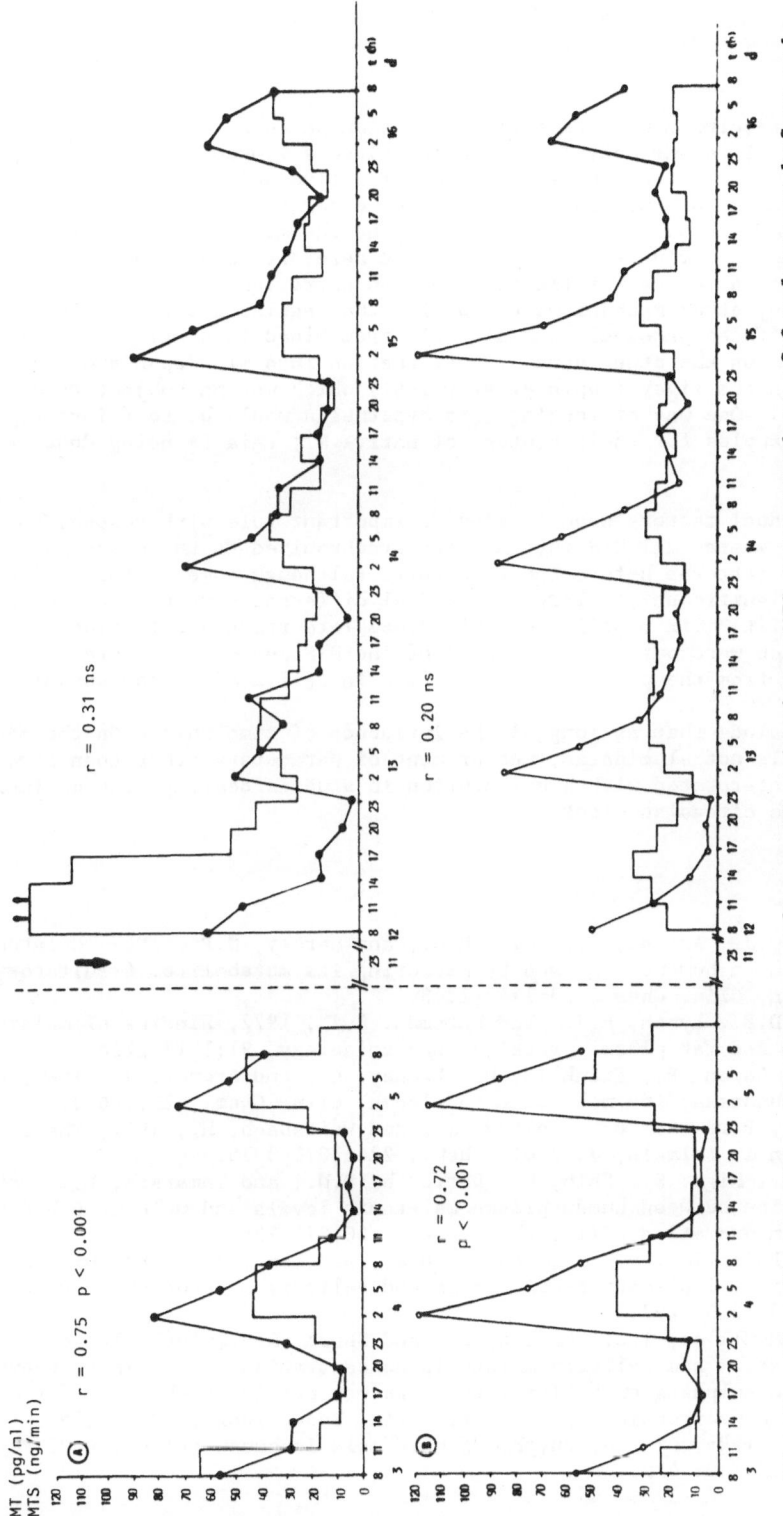

MT (pg/nl)
MTS (ng/min)

r = 0.75 p < 0.001

r = 0.72
p < 0.001

r = 0.31 ns

r = 0.20 ns

Fig. 1. Circadian profiles of the mean salivary MT and urinary MTS concentrations of 8 volunteers before and after a 9 hr-phase advance, accompanied by a daily intake of 5 mg MT or placebo for one week. h: laboratory clocktime, d: day, *significant rise as compared to preceeding value; the arrow indicates the time of the last drug intake.

centage shrank to 50 (4 out of 8) after MT, and to 25 (2 out of 8) after placebo.

CONCLUSIONS

These results demonstrate that circadian patterns of MT and its metabolite MTS in different body fluids, being closely linked to one another under usual circumstances, may drift apart after a time shift. Although the underlying mechanism for this phenomenon is not clear at present, we believe that a delay in MT metabolism in the liver may be the main reason. Although serum was not sampled during the study described here, it is hypothesized that salivary MT content closely reflects serum concentrations regardless of time shifts or any other mechanisms disrupting the regular pattern of secretion, since MT diffuses passively and directly from blood into saliva (Cardinali et al., 1972). On the other hand, transformation into MTS depends on metabolic processes in the liver (Kopin et al., 1961) which may be subject to circadian rhythmicity. One way of testing this hypothesis would be to reinvestigate the urine samples for their content of native MT; this is being done at present.

Individual factors seem to play an important role with respect to MT metabolism; wheras all MTS rhythms were synchronized to the corresponding salivary MT patterns before the time shift (although some of them did not correlate significantly), large individual differences were observed after the time shift. Six profiles exhibited complete resynchronization, but the remaining ten were out of phase, some of the MTS peaks being more than 12 hours apart from the preceeding MT maximum in saliva (data not shown).

We conclude that as long as the influence of time shifts on the metabolism of MT is not elucidated, measurement of parameters other than serum MT should be interpreted with great caution in studies dealing with manipulations of the circadian clock.

REFERENCES

Bojkowski, C.J., Arendt, J., Shih, M.C., and Markey, S.P., 1987, Melatonin secretion in humans assessed by measuring its metabolite, 6-sulfatoxy-melatonin, Clin. Chem., 33:1343-1348.
Cardinali, D.P., Lynch, H.J., and Wurtman, R.J., 1972, Binding of melatonin to human and rat plasma proteins, Endocrinology, 91:1213-1218.
Fraser, S., Cowen, P., Franklin, M., Franey, C., and Arendt, J., 1983, Direct radioimmunoassay for melatonin in plasma, Clin. Chem., 29:396-397.
Kopin, I.J., Pare, C.M.B., Axelrod, J., and Weissbach, H., 1961, The fate of melatonin in animals, J. Biol. Chem., 236:3072-3075.
Markey, S.P., Higa, S., Shih, M., Danforth, D.N., and Tamarkin, L., 1985, The correlation between human plasma melatonin levels and urinary 6-hydroxy-melatonin excretion, Clin. Chim. Acta, 150:221-225.
Miles, A., Philbrick, D.R.S., Thomas, D.R., and Grey, J., 1987, Diagnostic and clinical implications of plasma and salivary melatonin assay, Clin. Chem., 33:1295-1297.
Nowak, R., McMillen, I.C., Redman, J., and Short, R.V., 1987, The correlation between serum and salivary melatonin concentrations and urinary 6-hydroxy-melatonin sulphate excretion rates: two noninvasive techniques for monitoring human circadian rhythmicity, Clin. Endocrinol., 27:445-452.
Vakkuri, O., 1985, Diurnal rhythm of melatonin in human saliva, Acta Physiol. Scand., 124:409-412.

TRYPTOPHAN METABOLISM IN MICE INFECTED WITH SCHISTOSOMA MANSONI

E.N.M. Njagi and D.A. Bender

Department of Biochemistry
University College London
London WC1E 6BT
UK

Biharzia due to infection with Schistosoma mansoni is a major and increasing problem in tropical countries, associated especially with irrigation schemes, which provide a habitat for the water snail which is the alternate host of the parasite. Infection results from bathing or other contact with water containing infected snails, which release large numbers of cercariae, the infectious form which penetrates human skin.

A number of studies have demonstrated abnormalities of tryptophan metabolism in patients with bilharzia. These have been interpreted as indicating vitamin B_6 deficiency, although they are not wholly corrected by the administration of vitamin B_6 supplements (Abdel-Tawab et al., 1966, 1969; Abdel-Daim et al., 1971).

The effect of infection on tryptophan metabolism has been assessed in both intact mice and in isolated hepatocytes. There was a very marked effect on tryptophan metabolism in vivo, but not in hepatocytes. This was unrelated to vitamin B_6 nutritional status. It seems that parasites or their eggs avidly take up tryptophan from the host's circulation.

METHODS

Male mice weighing 20 g were infected by application of cercariae of S. mansoni to shaved skin. 7 weeks later, they received a tracer dose of [^{14}C] tryptophan by intraperitoneal injection. Exhaled $^{14}CO_2$ was collected for 2 h. Urine was collected for the measurement of the basal excretion of xanthurenic and kynurenic acids, kynurenine, 3-hydroxykynurenine, methyl nicotinamide and methyl pyridone carboxamide.

Isolated hepatocytes were prepared by perfusion with collagenase and were incubated with tryptophan over the range of 0.6 - 6.8 mmol/l. Incubations were performed with both [^{14}C] and non-radioactive tryptophan. The total metabolism of tryptophan was determined from the sum of radioactivity in CO_2 and non-aromatic metabolites. After incubation with non-radioactive tryptophan, the formation of xanthurenic acid, kynurenine, 3-hydroxykynurenine, NAD(P), niacin and methyl nicotinamide was determined.

Kynurenine and Serotonin Pathways
Edited by R. Schwarcz *et al.*, Plenum Press, New York, 1991

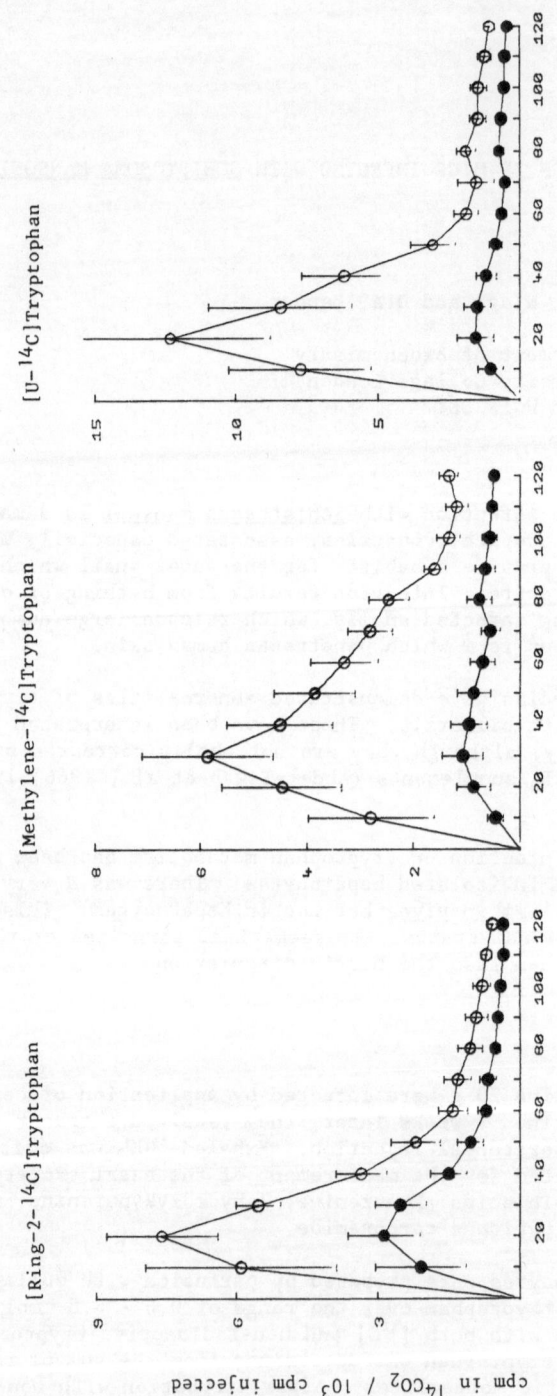

Fig. 1. Production of $^{14}CO_2$ from $[^{14}C]$tryptophan over 2 h after intraperitoneal injection. ○: Controls, ●: mice infected with S. mansoni.

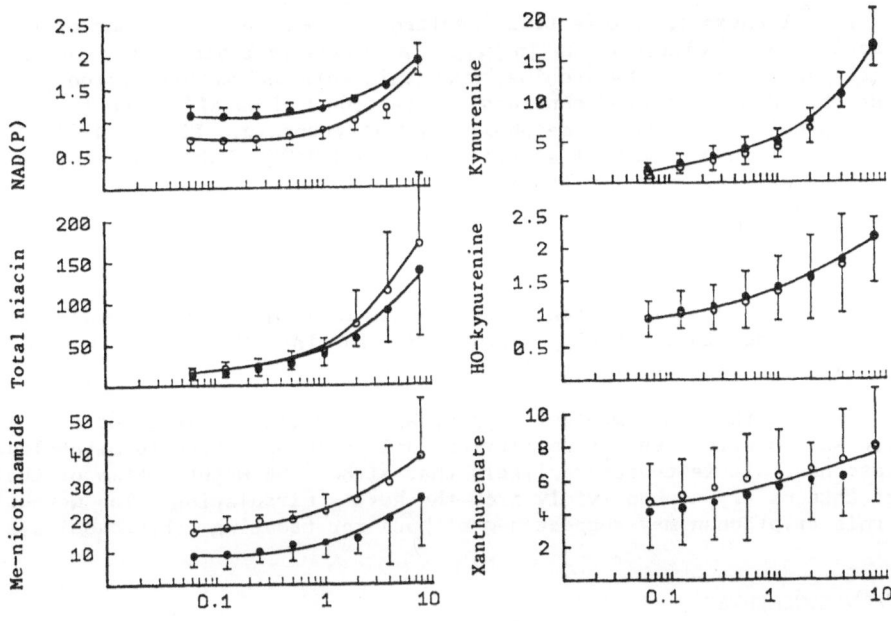

Tryptophan, mmol/l in incubation

Fig. 2. Metabolism of tryptophan in isolated hepatocytes. Points (means ± SEM), show nmol product formed/mg dry weight of cells for 5 animals in each group. O: Control mice, ●: mice infected with S. mansoni.

RESULTS

On dissection, the infected animals showed considerable accumulation of parasite eggs. The livers were grey and granular, and a large number of eggs were retained on the filter used to strain the cells. There was no significant difference in the recovery of cells from the two groups, and no significant depletion of liver or red cell vitamin B_6. Plasma tryptophan was slightly lower in infected animals.

Table 1. 24 h basal urinary excretion of tryptophan metabolites by mice infected with S. mansoni and control mice

	Control		Infected	
	Mean	SEM	Mean	SEM
Kynurenine (μmol)	2.13	0.49	1.21	0.21**
3-Hydroxykynurenine (nmol)	704	104.9	779	111.5
Kynurenic acid (μmol)	49	7.7	36	7.8*
Xanthurenic acid (μmol)	89	18.5	77	21.8*
N-methyl nicotinamide (nmol)	356	145.7	511	297.6
Methyl pyridone carboxamide (nmol)	1095	394	224	97.4**
5-Hydroxytryptamine (μmol)	14	1.7	13	3.5
5-Hydroxyindoleacetic acid (μmol)	6.6	1.34	5.3	0.76

Significance of differences (t-test): *0.1 > P > 0.05, **0.05 > P > 0.001. Data are the average of 5 animals in each group.

Fig. 1 shows that infection resulted in a marked impairment of the metabolism of [^{14}C]tryptophan _in vivo_. As shown in Table 1, the infected animals excreted less kynurenine, kynurenic acid and methyl pyridone carboxamide than did uninfected controls. Infection had no effect on the ability of isolated heaptocytes to metabolise [^{14}C]tryptophan. Fig. 2 shows that cells from infected animals synthesized less NAD(P), but more methyl nicotinamide and niacin, from tryptophan.

DISCUSSION

The effects of infection on tryptophan metabolism are very different from those seen in vitamin B_6 deficiency. The infected animals showed no significant depletion of vitamin B_6.

Despite the impairment of tryptophan metabolism in infected animals, there was no effect on the ability of isolated hepatocytes to metabolize tryptophan. It is therefore likely that either the mature worms or their eggs take up tryptophan avidly from the host's circulation. The metabolism of this tryptophan may suggest new methods for treating schistosomiasis.

ACKNOWLEDGEMENTS

ENMN is supported by a British Council Technical Development Training Award. We are grateful to Dr. S.R. Smithers of the National Institute of Medical Research, London, UK, for providing cercariae.

REFERENCES

Abdel-Daim, M.H., Konbar,A.A., Kelada, F.S., and Moustafa, M.H., 1971, Studies on the functional capacity of the tryptophan-niacin pathway in bilharzial children from rural areas, Trans. Roy. Soc. Trop. Med. Hyg., 65:668-671.
Abdel-Tawab, G.A., Kelada, F.S., Kelada, N.L., Abdel-Daim, M.H. and Makhgrun, N., 1966, Studies on the aetiology of bilharzial carcinoma of the urinary bladder: V. excretion of tryptophan metabolites in urine, Int. J. Cancer, 1:377-382.
Abdel-Tawab, G.A., Kholief, T.S., El-Sewedy, S.M., Abbassy, A.S., Zeitoun, M.M., and Hassanein, E.A., 1969, Studies on tryptophan metabolism in bilharzial hepatic fibrosis and non-bilharzial hepatic cirrhosis in childhood, Acta Paed. Scand., 58:54-58.

INHIBITION OF TRYPTOPHAN → NIACIN METABOLISM BY DIETARY LEUCINE AND BY

LEUCINE AND 2-OXO-ISOCAPROATE IN VITRO

D.A. Bender

Department of Biochemistry
University College London
London WC1E 6BT
UK

INTRODUCTION

A relative excess of leucine in dietary proteins has been implicated in the etiology of the tryptophan-niacin deficiency disease pellagra in those parts of India where the dietary staple is jowar (Sorghum vulgare) (Gopalan and Srikantia, 1960).

Previous studies have shown that feeding rats on a low tryptophan/high leucine diet results in depletion of the nicotinamide nucleotide coenzymes NAD and NADP, if they are wholly or partly dependent on de novo synthesis of NAD(P) from tryptophan, but not when the diet provides a minimally adequate intake of preformed niacin. Leucine-fed rats showed a significant reduction in blood NAD(P) and urinary excretion of methyl nicotinamide (Magboul and Bender, 1983). Leucine inhibits kynureninase, and its oxo-acid, 2-oxo-iso-caproate, inhibits kynurenine hydroxylase; both leucine and 2-oxo-isocap-roate increase the activity of tryptophan dioxygenase in liver homogenates (Magboul and Bender, 1983). Rats fed the high leucine diet excreted more kynurenine and formed less $^{14}CO_2$ from [methylene-^{14}C] tryptophan than did control animals (Bender, 1983).

These results suggest that a dietary excess of leucine inhibits the oxidative metabolism of tryptophan, and hence reduces NAD(P) synthesis. However, studies of the metabolic flux of [^{14}C]tryptophan in hepatocytes isolated from rats fed the same high leucine/low tryptophan diet are at variance with these observations. In this system, leucine has no effect on tryptophan metabolism beyond what would be expected from competition between leucine and tryptophan for uptake (Salter et al., 1985).

There is thus an obvious discrepancy between the results of studies in intact animals, which suggest that the inhibition of kynureninase and kynurenine hydroxylase by leucine and its oxo-acid result in a reduced capacity to synthesize NAD(P) from tryptophan (Bender, 1983; Magboul and Bender, 1983), and studies in isolated hepatocytes, which suggest that inhibition of these two enzymes, both of which have negligible Control Coefficients for metabolic flux of [^{14}C]tryptophan, does not affect the ability to metabolize tryptophan (Salter et al., 1985). The present study was designed in order to investigate the effects of leucine and 2-oxo-isocaproate on tryptophan metabolism in isolated hepatocytes.

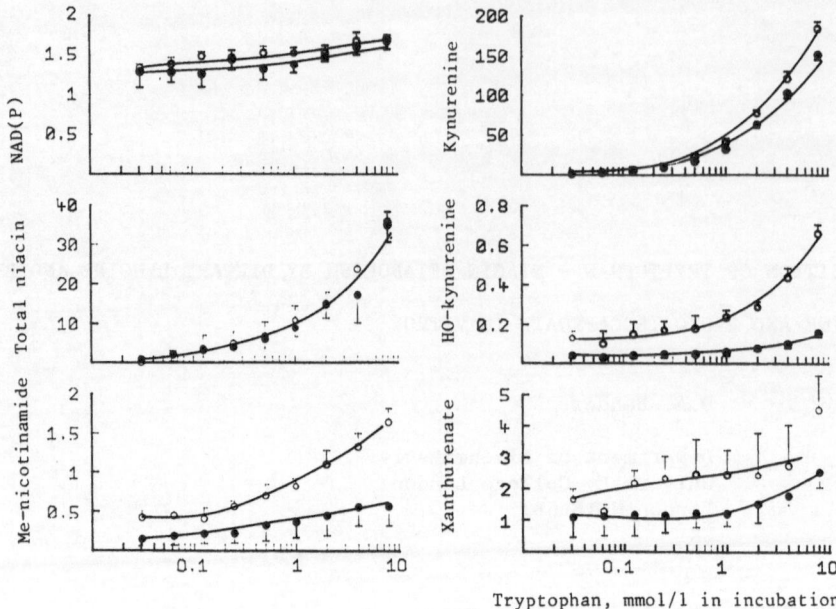

Tryptophan, mmol/l in incubation

Fig. 1. Metabolism of tryptophan in hepatocytes isolated from rats fed a low
tryptophan diet, or with an additional 15 g leucine/kg diet. Points
show nmol product formed/mg dry weight of cells (mean ± SEM of 5
animals in each group). ○: Low tryptophan control diet, ●: high
leucine diet.

METHODS

Rats were maintained from weaning on a diet based on maize, gelatine and
casein (Magboul and Bender, 1983). It provided 1030 mg tryptophan/kg, and no
preformed niacin other than that present in the maize meal, most of which is
biologically unavailable. In the high leucine diet, 15 g leucine was added
per kg diet, at the expense of gelatine. After 7-8 weeks, hepatocytes were
isolated by perfusion of the liver with collagenase.

The ability of the cells to metabolize tryptophan was assessed by the
formation of NAD(P), niacin (nicotinic acid + nicotinamide), methyl nicotinamide, kynurenine, 3-hydroxykynurenine and xanthurenic acid after incubation
with tryptophan over a range of concentrations from 0.32 to 8 mmol/l. Cells
from animals fed the control diet were also incubated with 8 mmol/l leucine
and 2-oxo-isocaproate.

RESULTS

As shown in Fig. 1, hepatocytes from rats fed the high leucine diet
showed impaired synthesis of both NAD(P) and N-methyl nicotinamide from tryptophan, and reduced formation of kynurenine, 3-hydroxykynurenine and xanthurenic acid.

As shown in Fig. 2, the addition of leucine to the incubation medium resulted in reduced synthesis of niacin from tryptophan, and increased accumulation of 3-hydroxykynurenine. As shown in Fig. 3, the addition of 2-oxo-
isocaproate resulted in increased synthesis of NAD(P), niacin and 3-hydroxy-
kynurenine.

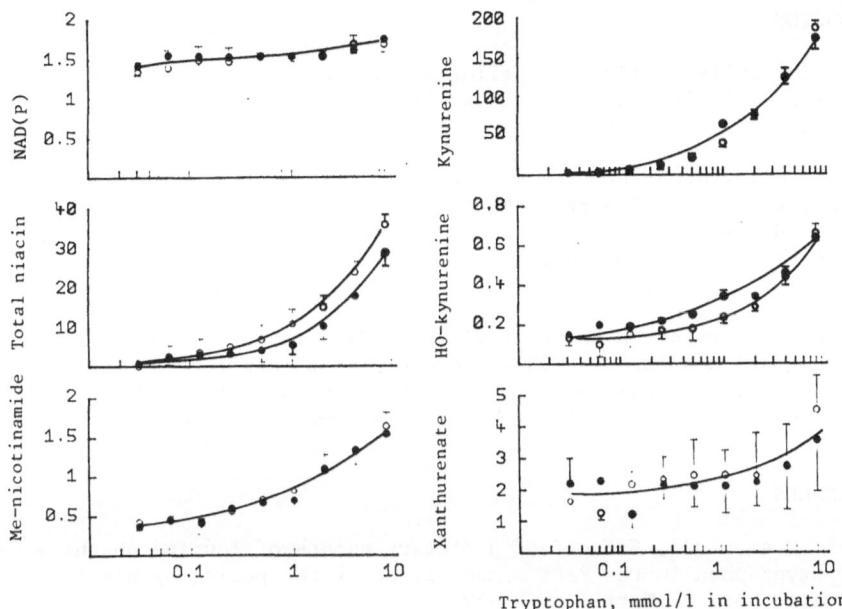

Fig. 2. Effect of adding 8 mmol/l leucine on the metabolism of tryptophan in hepatocytes isolated from rats fed the low tryptophan control diet. Points show nmol product formed/mg dry weight of cells
(mean ± SEM of 5 animals in each group). ○: Control; ●: +leucine.

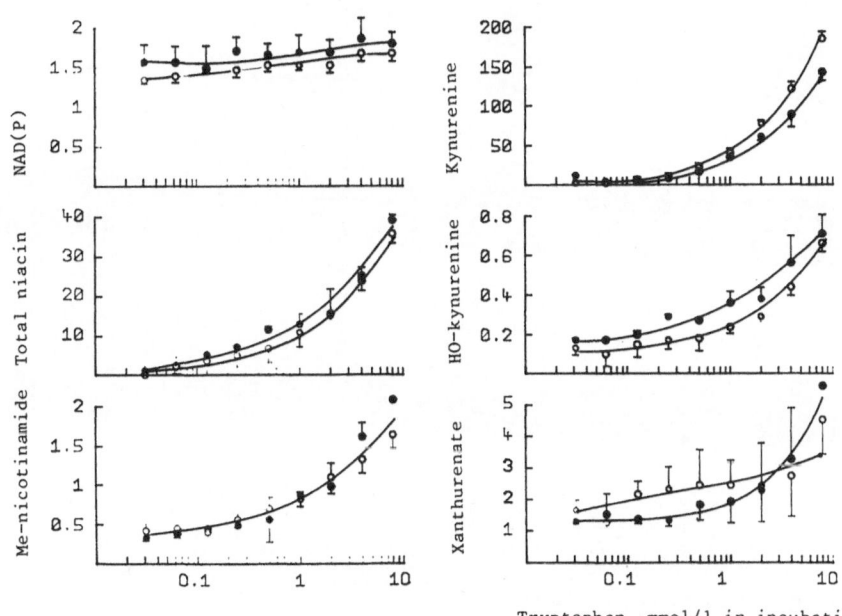

Fig. 3. Effect of adding 8 mmol/l 2-oxo-isocaproate on the metabolism of
tryptophan in hepatocytes isolated from rats fed the low tryptophan
control diet. Points show nmol product formed /mg dry weight of
cells (mean ± SEM of 5 animals in each group). ○: Control; ●:
+ 2-oxo-isocaproate.

DISCUSSION

The results confirm the pellagragenic action of leucine in the diet. Cells from animals fed the high leucine diet had an impaired ability both to synthesize niacin metabolites and to metabolize tryptophan.

The differences between the effects of feeding a dietary excess of leucine and adding leucine or 2-oxo-isocaproate to the incubation medium suggest that the dietary excess has a more complex effect on enzymes of tryptophan-niacin metabolism than the simple inhibition seen with liver homogenates (Magboul and Bender, 1983).

The differences between previous studies in intact animals (Bender, 1983; Magboul and Bender, 1983) and the present studies in hepatocytes suggest that extra-hepatic metabolism of tryptophan, initiated by indoleamine dioxygenase, may be more important than has hitherto been believed.

REFERENCES

Bender, D.A., 1983, Effects of a dietary excess of leucine on the metabolism of tryptophan in the rat: a mechanism for the pellagragenic action of leucine, Br. J. Nutr., 50:25-32.

Gopalan, C., and Srikantia, S.G., 1960, Leucine and pellagra, Lancet, 1:954-957.

Magboul, B.I., and Bender, D.A., 1983, Effects of a dietary excess of leucine on the synthesis of nicotinamide nucleotides in the rat, Br. J. Nutr., 49:321-329.

Salter, M., Bender, D.A., and Pogson, C.I., 1985, Leucine and tryptophan metabolism in rats, Biochem. J., 225:277-281.

INDUCTION OF INDOLEAMINE 2,3-DIOXYGENASE IN HUMAN CELLS IN VITRO

G. Werner-Felmayer, E.R. Werner, D. Fuchs, A. Hausen,
G. Reibnegger, and H. Wachter

Institute of Medical Chemistry and Biochemistry
University of Innsbruck
6020 Innsbruck
Austria

INTRODUCTION

Human macrophages (MO) release neopterin and degrade tryptophan by in-doleamine 2,3-dioxygenase (IDO), when stimulated with interferon (IFN)-gamma in vitro (Werner et al., 1987a). Also, in a panel of non-immune cells, in-duction of IDO by IFN-gamma is strictly correlated to pteridine synthesis, which results in intracellularly increased levels of biopterin and/or neo-pterin (for details see Werner et al., 1989 and this volume). Out of the peripheral blood mononuclear cell fraction, MO are the cell population re-sponsible for formation of neopterin and for IDO activity (Huber et al., 1984; Werner et al., 1987a). Lymphocytes do not release neopterin and are not inducible for IDO.

In vitro, IFN-gamma is the only inducer of IDO in a panel of human cell lines (Takikawa et al., 1988). In vivo, however, treatment of patients with IFN-alpha, IFN-beta or IFN-gamma leads to enhanced excretion of neopterin and tryptophan metabolites (Datta et al., 1987). Further, IDO is induced in mice treated with bacterial lipopolysaccharide (LPS), tumor necrosis factor alpha (TNF), IFN-alpha and IFN-gamma (Bianchi et al., 1988). In peripheral blood mononuclear cells (PBMC), all three IFN species induce IDO (Carlin et al., 1987).

We studied the effect of IFNs in PBMC, MO, human dermal fibroblasts (FB) and a panel of tumor cell lines concerning IDO induction. Further, the capa-city of IFNs, TNF and LPS to induce neopterin release and IDO in PBMC and MO was assessed.

METHODS

Tryptophan and metabolites were detected in supernatants by HPLC accord-ing to Werner et al. (1987a). The pattern of tryptophan metabolites produced by individual cells is described in Werner-Felmayer et al. (1989). IDO acti-vity is given as the sum of detectable tryptophan metabolites per mg of total cell protein. Neopterin was determined in supernatants by solid phase ex-traction and on-line elution HPLC (Werner et al., 1987b). Cell culture con-ditions and preparation of MO from buffy coats of peripheral blood of healthy

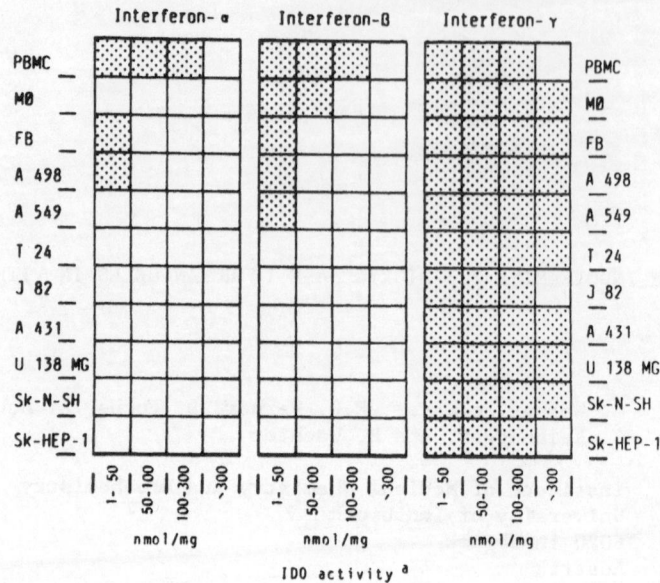

Fig. 1. Confluent monolayers, 3 x 10^6/ml of PBMC and 5 x 10^5/ml of MO were treated with 10^3 U/ml of IFNs. (a) IDO activity is expressed as the sum of tryptophan metabolites detectable by HPLC in supernatants after 48 or 72 (PBMC, MO) hours per mg of total cell protein.

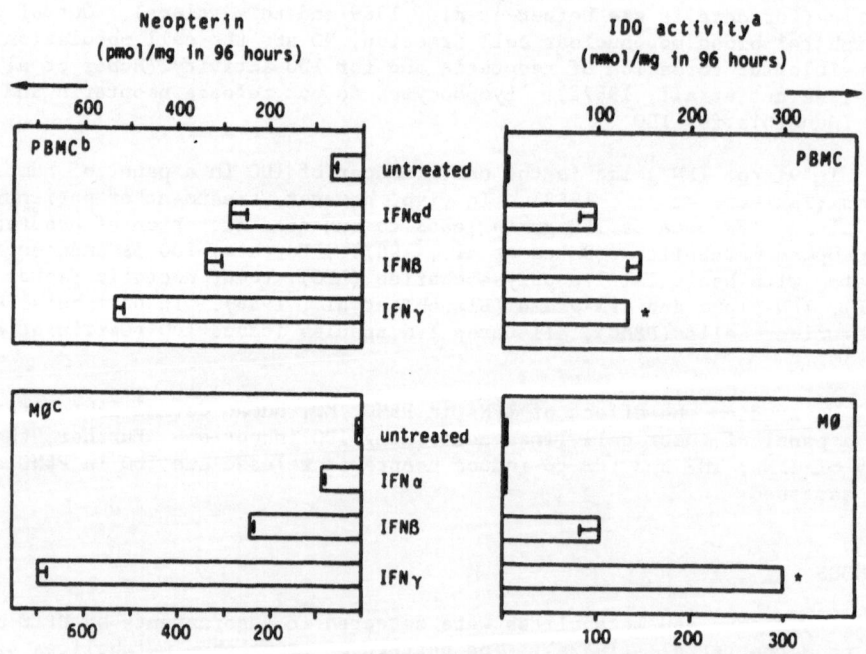

Fig. 2. Values are the mean ± SEM of triplicate cultures from 3 donors (N = 9. (a) IDO activity is expressed as the sum of tryptophan metabolites detected by HPLC in supernatants per mg of total cell protein; (b) 3 x 10^6/ml; (c) 5 x 10^5/ml; (d) 10^3 U/ml. *available tryptophan was totally degraded.

donors by two-step density gradient centrifugation (Ficoll-Paque, Percoll) and plastic adherence is described in Werner-Felmayer et al. (1989). PBMC were obtained by centrifugation over Ficoll-Paque.

RESULTS

1. Potential of IFNs to induce IDO. Only in PBMC, all three IFN species had a pronounced IDO activating potential (Fig. 1). In all other tested cell types comprising MO, FB and the human tumor cell lines A 498 (kidney), A 549 (lung), T 24 (bladder), J 82 (bladder), A 431 (epidermis), U 138 MG (glioblastoma), SK-N-SH (neuroblastoma) and SK-HEP-1 (hepatoma), IFN-gamma was the most active inducer of IDO.

2. Comparison of IFN effects in PBMC with those in MO. In PBMC, all three IFNs induced IDO and neopterin to a comparable degree (Fig. 2). However, in highly purified MO (> 99% non-specific esterase-positive cells on day 3 of culture), IFN-gamma was by far the best inducer of IDO and neopterin. We therefore conclude that the pronounced action of IFN-alpha and IFA-beta found in PBMC is supported by lymphocyte factors.

3. Effects of LPS. LPS induced IDO and pteridine synthesis in PBMC, MO and FB. Data obtained for IDO are shown in Fig. 3. In presence of lymphocytes (= PBMC), LPS acts via induction of endogenous IFN-gamma. This was demonstrated by inhibition of the effect by dexamethasone (Dex) and by detection of 82 ± 32 U/ml (N = 5) of IFN-gamma in supernatants of LPS-treated PBMC by ELISA, which dropped < 2 U/ml in the presence of Dex.

In contrast, Dex enhanced the LPS induced IDO activity in MO. In FB,

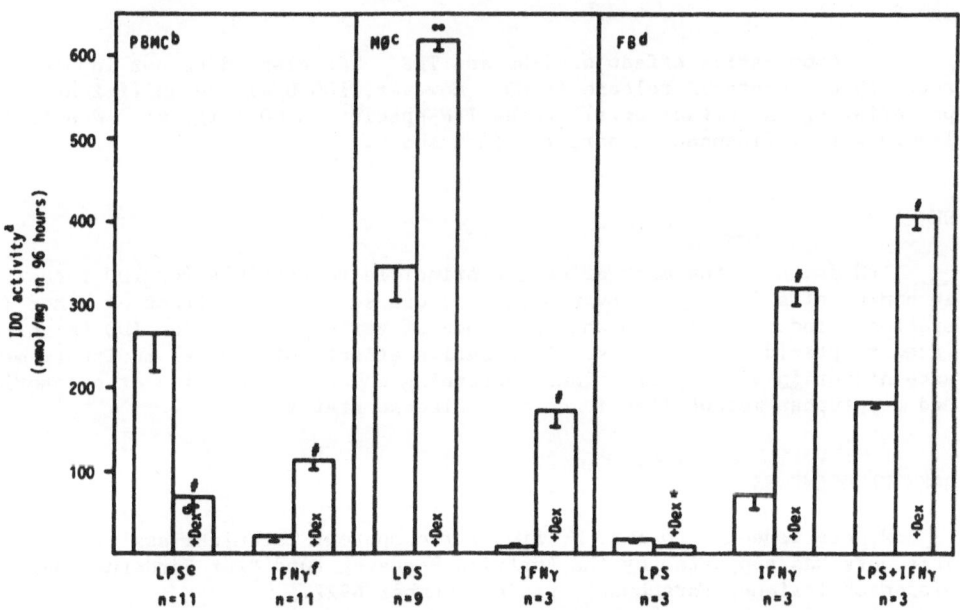

Fig. 3. (a) IDO activity (mean ± SD) is expressed as the sum of tryptophan metabolites detectable by HPLC in supernatants; (b) 3×10^6/ml; (c) 5×10^5/ml; (d) 10^5/ml plated and grown to confluency; (e) LPS from E.coli (10 μg/ml); (f) IFN-gamma (10 U/ml); (g) 50 nM Dex, added one hour before other stimuli. Significantly different from Dex-free cultures according to Student's t-test: *P < 0.02, **P < 0.001, #P < 0.0001.

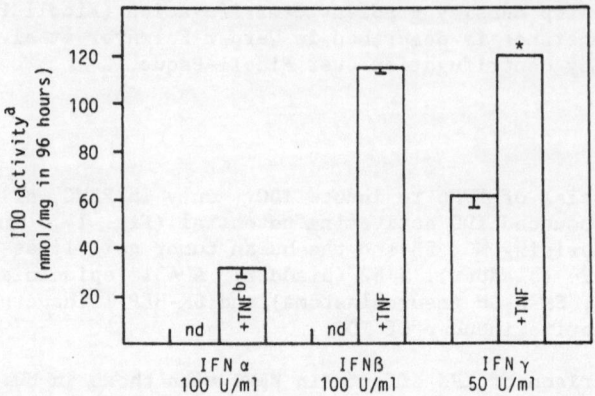

Fig. 4. (a) IDO activity (mean ± SD) is expressed as the sum of tryptophan
metabolites detected by HPLC in supernatants of MO (5 x 10^5/ml) (N
= 3); (b) 100 U/ml; nd: not detectable. *available tryptophan was
totally degraded.

the LPS effect was significantly decreased by Dex, but the effect of LPS in
combination with IFN-gamma or with IFN-gamma alone was drastically increased
by Dex. The phenomena of superinduction of IFN-gamma-activated IDO was described previously by Ozaki et al. (1987). Neopterin release to the supernatant and intracellular pteridine synthesis in the case of FB were influenced
in parallel (not shown).

   Thus, in the presence of lymphocytes, LPS acts via induction of IFN-gamma. Purified MO react upon LPS like FB in presence of small amounts of IFN-
gamma.

   4. Cooperative effect of IFNs and TNF. TNF alone does not induce IDO
activity or neopterin release in NO. However, 100 U/ml are sufficient for
potentiating the effect of all three IFN-species in MO (Fig. 4). Neopterin
levels were influenced in parallel (not shown).

SUMMARY

   IFN-gamma is the most effective principle responsible for IDO induction
in human cells in vitro. Lymphocyte factors support the effect of other IFN
species. Induction of IDO and influence of various factors on IDO is correlated to pteridine synthesis. Cooperative effects of several mediators may
account for in vivo observations concerning enhanced excretion of neopterin
and tryptophan metabolites in certain disease states.

ACKNOWLEDGEMENTS

   We are indebted to Ing. M. Broz for competent technical assistance.
This work was supported by the Austrian Research Fund "Zur Förderung der
wissenschaftlichen Forschung", project number 6922.

REFERENCES

Bianchi, M., Bertini, R., and Ghezzi, P., 1988, Induction of indoleamine
   2,3-dioxygenase by interferon in mice: a study with different recombinant
   interferons and various cytokines, Biochem. Biophys. Res. Commun., 152:
   237-242.

Carlin, J.M., Borden, E.C., Sondel, P.M., and Byrne, G.I., 1987, Biologic response modifier-induced indoleamine 2,3-dioxygenase activity in human peripheral blood mononuclear cell cultures, J. Immunol., 139:2414-2418.

Datta, S.P., Brown, R.R., Borden, E.C., Sondel, P.M., and Trump, D.L., 1987, Interferon and interleukin 2 induced changes of tryptophan and neopterin metabolism: possible markers for biologically effective doses, Proc. Am. Assoc. Cancer Res., 28:338.

Huber, C., Batchelor, J.R., Fuchs, D., Hausen, A., Lang, A., Niederwieser, D., Reibnegger, G., Swetly, P., Troppmair, J., and Wachter, H., 1984, Immune response-associated production of neopterin, J. Exp. Med., 160:310-316.

Ozaki, Y., Edelstein, M.P., and Duch, D.S., 1987, The actions of interferon and anti-inflammatory agents on induction of indoleamine 2,3-dioxygenase in human peripheral blood monocytes, Biochem. Biophys. Res. Commun., 155: 1147-1153.

Takikawa, O., Kuroiwa, T., Yamazaki, F., and Kido, R., 1988, Mechanism of interferon-gamma action, characterization of indoleamine 2,3-dioxygenase in cultured human cells induced by interferon-gamma and evaluation of the enzyme-mediated tryptophan degradation in its anticellular activity, J. Biol. Chem., 263:2041-2048.

Werner, E.R., Bitterlich, G., Fuchs, D., Hausen, A., Reibnegger, G., Szabo, G., Dierich, M.P., and Wachter, H., 1987a, Human macrophages degrade tryptophan upon induction by IFN-gamma, Life Sci., 41:273-280.

Werner, E.R., Fuchs, D., Hausen, A., Reibnegger, G., and Wachter, H., 1987b, Simultaneous determination of neopterin and creatinine in serum with solid-phase extraction and on-line elution liquid chromatography, Clin. Chem., 33:2028-2033.

Werner, E.R., Werner-Felmayer, G., Fuchs, D., Hausen, A., Reibnegger, G., and Wachter, H., 1989, Parallel induction of tetrahydrobiopterin biosynthesis and indoleamine 2,3-dioxygenase activity in human cells and cell lines by interferon-gamma, Biochem. J., 262:861-866.

Werner, E.R., Werner-Felmayer, G., Fuchs, D., Hausen, A., Reibnegger, G., and Wachter, H., Relationships between pteridine synthesis and tryptophan degradation, this volume.

Werner-Felmayer, G., Werner, E.R., Fuchs, D., Hausen, A., Reibnegger, G., and Wachter, H., 1989, Characteristics of interferon induced tryptophan metabolism in human cells in vitro, Biochim. Biophys. Acta, 1012:140-147.

DIFFERENTIAL EFFECTS OF ESTRADIOL AND HYDROCORTISONE ON LIVER TRYPTOPHAN

PYRROLASE IN RATS OF VARIOUS AGES

S.K. Patnaik

Department of Pharmaceutics
Institute of Technology
Banaras Hindu University
Varanasi 221005
India

INTRODUCTION

Changes in the levels of certain regulatory enzymes and their induc-
ibility by several hormones might account for the deterioration of physiological efficiency during aging in animals. Such changes may be brought
about by the alterations in the template activity of their corresponding
genes (Kanungo, 1975). Tryptophan pyrrolase (TP; L-tryptophan-oxygen 2,3-
oxidoreductase, E.C. 1.13.11.11) is a liver cytosolic enzyme that catalyzes
the first and rate-limiting step in the degradation of tryptophan (Knox and
Mehler, 1950), resulting in the synthesis of nicotinamide cofactors (Nishizuka and Hayaishi, 1963). The present paper describes the activity and induction pattern of liver TP by estradiol and hydrocortisone in rats of different ages to elucidate molecular mechanisms of aging.

MATERIALS AND METHODS

Immature (6-weeks), young-adult (13-weeks), adult (33-weeks) and senescent (85-weeks) albino rats (female rats for estradiol treatment and male
rats for hydrocortisone treatment) of the Wistar strain, kept under controlled conditions of temperature and light, were used. They were fed with a
freshly prepared standard diet. Tap water was supplied ad libitum.

Pilot experiments were undetaken to examine the time and dose dependence
of this enzyme on estradiol and hydrocortisone in rats of various ages. Rats
of all ages were divided into seven sets (N = 4-5 each). Set I served as a
control. Rats in sets II, III, and IV were bilaterally ovariectomized (Ov)
and kept under laboratory conditions for 21 days. On the 22nd day, sets III
and IV were injected with estradiol (200 μg/kg) and set IV rats was given
actinomycin D (100 μg/kg) 1 hour prior to the estradiol injection. Further,
sets V, VI, and VII were bilaterally adrenalectomized (Ad) and kept under
laboratory conditions for 10 days. On the 11th day, sets VI and VII were injected with hydrocortisone (30 mg/kg) and set VII was given actinomycin D
(100 μg/kg) 1 hour prior to the hydrocortisone injection. Sets II and V were
given 1.0 ml of the vehicle solution and served as controls for the induction
studies in Ov and Ad rats, respectively. All injections were given intraperitoneally at a fixed time of the day and the rats were killed after 4 hours

(for estradiol treatment) and on the 4th day (for hydrocortisone treatment) of the hormone injections. Their livers were removed, washed in normal saline, and blotted dry.

A 25% (w/v) homogenate of the tissue was prepared in 0.02 M potassium phosphate buffer, pH 7.0, containing 0.14 M KCl. The homogenate was centrifuged at 14,500 x g for 1 hour and the supernatant was used for the enzyme assay according to the method of Knox et al. (1970). An endogenous reaction occurred, which yielded about 0.03 absorbance units per 10 min. This value was substracted from the actual measurements of enzyme activity. Enzyme activity was calculated by using an extinction coefficient of 4,530 (Knox and Mehler, 1950). One unit of enzyme activity was defined by the formation of 1.0 μmol kynurenine/h/g wet weight. Protein content in the soluble fraction of the liver was measured (Lowry et al., 1951) and enzyme activity was expressed as units per mg protein.

RESULTS AND DISCUSSION

Several changes in physiological and biochemical functions are known to occur during various phases of the life span of an organism (Kanungo, 1980). Since enzymes are responsible for specific functions, the initiation, duration, and termination of various phases such as differentiation, development, reproductive maturity, and senescence may depend on alterations in the levels of several enzymes or their isozymes. These alterations can be due to either temporary or permanent "switching on" and/or "off" of the gene(s) responsible for their synthesis (Patnaik, 1979; Kanungo, 1975).

The specific activity of liver TP activity does not change until adulthood but decreases significantly thereafter (Tables 1 and 2). The higher activity of TP in immature, young-adult, and adult rats as compared to the senescent rat may support the active growth phases of these animals. Furthermore, changes in TP activity may also influence the normal behavior of the animal by regulating the synthesis of neurotransmitters in the brain (Sharma and Patnaik, 1983). On the other hand, the decrease in TP activity in senescent rats may reflect the cessation of growth and cell proliferation following the attainment of reproductive maturity. Further, the low level of this

Table 1. Effects of ovariectomy (Ov), estradiol (Est) and actinomycin D (A) on the activity of tryptophan pyrrolase in the liver of rats of various ages

Treatment	6-weeks	13-weeks	33-weeks	85-weeks
Control	36.2 ± 2.8	37.7 ± 2.1	41.3 ± 2.2	20.5 ± 1.3
Ov	24.3 ± 1.5*	16.6 ± 0.5**	20.7 ± 1.3**	9.7 ± 0.9*
Ov + Est	44.2 ± 2.4*	52.4 ± 2.1**	40.0 ± 0.4**	15.1 ± 1.0*
Ov + A + Est	26.4 ± 2.7*	20.9 ± 1.5**	20.0 ± 0.8**	10.0 ± 1.0*

Data were collected from 4-5 rats of each age group and are the mean ± SD. Enzyme activity is expressed in units x 10^3/mg protein. *P < 0.01, **P < 0.001 (Student's t-test).

Table 2. Effects of adrenalectomy (Ad), hydrocortisone (HC), and actinomycin D (A) on the activity of tryptophan pyrrolase in the liver
of rats of various ages

Treatment	6-weeks	13-weeks	33-weeks	85-weeks
Control	33.2 ± 1.2	35.3 ± 2.0	36.5 ± 2.9	18.6 ± 1.1
Ad	$17.1 \pm 1.1^{**}$	$16.9 \pm 0.9^{**}$	$18.8 \pm 0.9^{**}$	$12.9 \pm 0.7^{**}$
Ad + HC	$91.2 \pm 5.4^{**}$	$116.5 \pm 13.5^{**}$	$86.2 \pm 4.3^{**}$	$43.7 \pm 2.4^{**}$
Ad + A + HC	$26.6 \pm 1.3^{**}$	$24.1 \pm 2.1^{**}$	$22.2 \pm 1.9^{**}$	$15.7 \pm 1.9^{**}$

Data were collected from 4-5 rats of each age group and are the mean $\pm$ SD.
Enzyme activity is expressed in units x 10^2/mg protein. $^{**}P < 0.001$
(Student's t-test).

enzyme may cause behavioral disorders due to an increase in the level of
5-hydroxytryptamine in the senescent rat.

It is well known that removal of hormone-secreting organs from an animal
causes a change in the levels of many enzymes/isozymes in different tissues
(Sharma and Patnaik, 1983, 1984, 1985; Patnaik et al., 1987a,b; Patnaik,
1989; Patnaik and Patnaik, 1989). Our present observations show that the activity of liver TP decreases significantly following Ov and increases after
the administration of 17-beta-estradiol in rats of all ages (Table 1). It is
noteworthy that the decrease following Ov and the increase after estradiol
treatment are highest in the mature rat. Moreover, the degree of induction
of this enzyme by estradiol is about 2.6, 2.3, and 3.8 fold higher than those
of the immature, adult, and senescent rats, respectively. This induction,
however, is significantly inhibited by actinomycin D. The degree of inhibition of the enzyme following actinomycin D treatment is also proportional to
the degree of induction by estradiol in different age groups.

Ad significantly decreases the activity of liver TP at all ages (Table
2), and the administration of hydrocortisone to Ad rats increases TP activity
significantly in the liver of rats of all ages. However, the degree of decrease or increase, respectively, in enzyme activity is much higher in the
young-adult rat. These studies show that adrenal steroids play a major role
in the regulation of TP activity. Furthermore, the degree of induction of
the enzyme by hydrocortisone in young-adult rats is 1.4, 1.7, and 2.8-fold
higher than those of the immature, adult, and senescent rats, respectively
(Table 2). This clearly shows that induction of TP by hydrocortisone in
young-adult rat is more pronounced due to the higher transcriptional modulation (i.e. higher de novo synthesis) by this hormone in this phase of the
animal's life.

The hormone-mediated induction of TP is actinomycin D-sensitive, which
clearly indicates that estradiol and hydrocortisone induce the enzyme by
stimulating transcription of mRNA(s) responsible for its synthesis. The decrease in the degree of induction of TP by estradiol and hydrocortisone during the phases of growth, adulthood and senescence could be due to the gradual loss in the level or decrease in the number of the receptors for these

hormones (Singer et al., 1973; Roth and Adelman, 1974; Kanungo et al., 1975; Roth and Adelman, 1975).

On the basis of the present findings, it may be concluded that the endogenous levels of TP are dependent on different types of physiological controls. The responsiveness of this enzyme to estradiol and hydrocortisone depends on developmental stages and undergoes specific alterations at different phases of the life span. Such alterations may be due to regulatory changes in the template activities of the corresponding gene(s) which are brought about by various factors such as hormones according to a specific genetic program (Kanungo, 1975; Patnaik, 1979).

REFERENCES

Kanungo, M.S., 1975, A model for aging, J. Theor. Biol., 53: 253-261.
Kanungo, M.S., Patnaik, S.K., and Koul, O., 1975, Decrease in 17-beta-estradiol receptor in brain of aging rats, Nature, 253:366-367.
Knox, W.E., and Mehler, A.H., 1950, The conversion of tryptophan to kynurenine in liver: the coupled tryptophan peroxidase system forming formyl-kynurenine, J. Biol. Chem., 187:419-430.
Knox, W.E., Yip, A., and Reshef, L., 1970, L-tryptophan 2,3-di-oxygenase (tryptophan pyrrolase) (rat liver), Meth. Enzymol., 17a:415-421.
Lowry, O.H., Rosebrough, N.J., Farr, A.L., and Randall, R.J., 1951, Protein estimation with Folin-Ciocalteau reagent, J. Biol. Chem., 193:265-275.
Nishizuka Y., and Hayaishi, O., 1963, Studies on the biosynthesis of nicotinamide adenine dinucleotide: enzymatic synthesis of niacin ribonucleotides from 3-hydroxyanthranilic acid in mammalian tissues, J. Biol. Chem., 238:3369-3377.
Patnaik, S.K., 1979, Changes in the sub-types of soluble alanine aminotransferase in the liver of rats during development and aging, Cell Biol. Int. Rep., 7:607-614.
Patnaik, S.K., 1989, Evidence for estradiol induced differential expression of genes for brain RNA polymerase during developmental stages of the rat, Biochem. Int., 18:721-728.
Patnaik, S.K., and Patnaik, R., 1989, Tissue-specific differential induction of arginase and ornithine aminotransferase by hydrocortisone during various developmental stages of the rat, Biochem. Int., 18:709-719.
Patnaik, R., Sarangi, S.K., and Patnaik, S.K., 1987a, Tissue-specific differential induction of ornithine aminotransferase by estradiol in rats of various ages, Biochem. Int., 14:843-850.
Patnaik, S.K., Sharma, R., and Patnaik, R., 1987b, Differential effects of hydrocortisone on aspartate aminotransferase isozymes of the liver of rats during growth, development, and senescence, Biochem. Int., 15:611-617.
Roth, G.S., and Adelman, R.C., 1974, Age-dependent regulation of mammalian DNA synthesis and cell division in vivo by glucocorticoids, Exp. Gerontol., 9:27-31.
Roth, G.S., and Adelman, R.C., 1975, Age-related changes in hormone binding by target cells and tissues, possible role in altered adaptive responsiveness, Exp. Gerontol., 10:1-11.
Sharma, R., and Patnaik, S.K., 1983, Induction of phosphoenolpyruvate carboxykinase by hydrocortisone in rat liver and brain as a function of age, Biochem. Int., 7:535-540.
Sharma, R., and Patnaik, S.K., 1984, Age-related response of citrate synthase to hydrocortisone in the liver and brain of male rats, Experientia, 40:97-98.
Sharma, R., and Patnaik, S.K., 1985, Age-dependent regulation of aspartate aminotransferase isozymes by hydrocortisone in the brain of male rats, Mol. Physiol., 7:195-200.
Singer, S., Ito, H., and Litwack, G., 1973, ^{3}H-Cortisol binding by young and old human liver cytosol protein in vitro, Int. J. Biochem., 4:569-573.

INDOLE-3-PYRUVIC ACID AS A DIRECT PRECURSOR OF KYNURENIC ACID

V. Politi, M.V. Lavaggi, G. Di Stazio, and A. Margonelli

Polifarma Research Centre
00155 Roma
Italy

INTRODUCTION

Kynurenic acid (KYNA) is found in large amounts after tryptophan load, because tryptophan-2,3-dioxygenase (EC 1.13.11.11), present in the liver, opens its indole ring, while kynurenine transaminase (EC 2.6.1.7) is able to catalyze the transformation of kynurenine to KYNA. In the last few years, it was demonstrated that KYNA is an important endogenous antagonist of excitatory amino acid receptors (Ganong et al., 1983; Stone and Burton, 1988) and that it reduces cerebral ischemic effects when administered at high dosages in vivo (Germano et al., 1987). Unfortunately, tryptophan is not a good precursor of cerebral KYNA, because it simultaneously increases kynurenine and quinolinic acid, two well-known pro-convulsant agents (Lapin, 1983; Foster et al., 1984). On the other hand, indole-3-pyruvic acid (IPA), the keto-analog of tryptophan, increases KYNA content in several rat organs after i.p. injection (Russi et al., 1989). Inside the brain, conversion to KYNA appears more effective for IPA than for tryptophan, suggesting a different metabolic pathway. Experiments were therefore performed in order to find the new pathway leading from IPA to KYNA.

RESULTS

IPA is effectively transformed to KYNA in the absence of enzymatic systems, provided that oxygen is present in the incubation mixture. For this reason, care must be taken to avoid interferences. Determination of KYNA was performed as soon as possible after incubation at 37°C and acidification. Mixtures were applied to disposable sulphonic acid columns (Baker 7090-3) and KYNA was eluted with methanol:10% ammonia (13:2). Using HPLC with fluorimetric detection (330-410 nm), the recovery of KYNA was > 95%.

IPA is transformed in various organ homogenates, but also in systems deprived of any enzyme (Table 1). This means that the transformation is very easy and depends on specific chemical structures: keto-enol tautomerism seems to be the main operating factor, as only the "enol" form, which originates in slightly acidic medium and is easily identified from UV spectra, is able to give rise to substantial KYNA production. Furthermore, the chemical transformation needs a radical attack from reactive oxygen species (mainly: ·OH), since KYNA is found in far larger amounts in the presence of free radical forming systems. It can be concluded, therefore, that IPA acts as a radical scavenger, probably interfering with pathological changes activated by oxygen

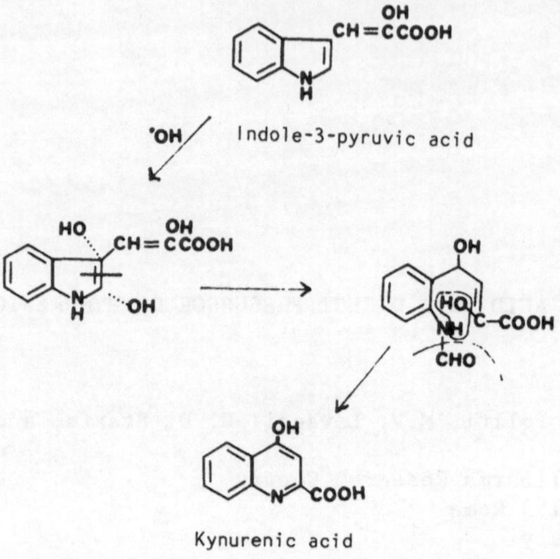

Fig. 1. Proposed transformation of IPA to kynurenic acid after a radical attack.

free radicals, along the pathway leading to KYNA.

The evidence is shown in Table 2. IPA effectively scavenges free radicals in abiotic systems (luminol luminescence) as well as in the presence of cellular structures (malondialdehyde formation).

Fig. 1 illustrates a proposed mechanism through which the attack by oxygen free radicals can trigger the chemical transformation of IPA to KYNA. The "enol" conformation of IPA seems to be the preferred target of radicals because of the presence of two conjugated double bonds in the carbon frame. In fact, if the attack is towards the indole ring, the molecule can be partly rearranged to form KYNA. If, however, the attack is on the pyruvate moiety, glyoxylic acid is the main detectable reaction product (data not shown).

Fig. 2 emphasizes the fact that KYNA is formed from tryptophan by means

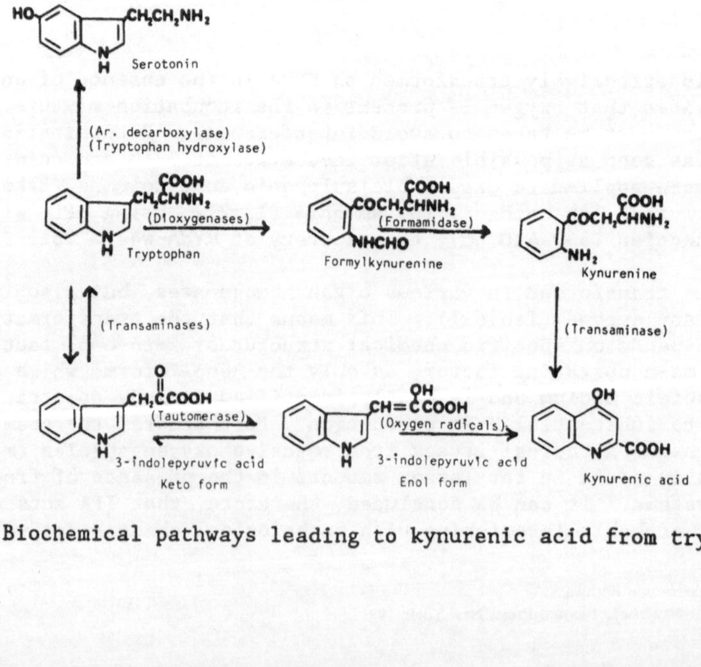

Fig. 2. Biochemical pathways leading to kynurenic acid from tryptophan.

Table 1. Amount of KYNA formed after incubation with IPA

System	KYNA
1. Rat kidney homogenate, pH 7.4	110 ± 14 ng
2. Rat liver homogenate, pH 7.4	32 ± 5 ng
3. Rat brain homogenate, pH 7.4	27 ± 6 ng
4. Rat brain homog. plus free radicals	184 ± 22 ng
5. (Tryptophan in brain homogenate)	0
6. IPA keto form	6 ± 2 ng
7. IPA keto form plus free radicals	12 ± 5 ng
8. IPA enol form (in methanol)	24 ± 5 ng
9. IPA enol (in methanol) + free radicals	251 ± 38 ng
10. IPA enol form (in acidic medium)	390 ± 52 ng

Incubations were performed at 37°C for 30 min, using 0.5 mg
of IPA. Systems 6 to 10 were free of enzymes. Free radicals
were obtained with ascorbate/Fe/hydrogen peroxide. Each
value is the mean of 6 experiments.

of two different biochemical pathways: the first should be mainly effective
in peripheral organs, and is activated by enzymes like tryptophan-2,3-dioxy-
genase or indoleamine-2,3-dioxygenase. The second one could be much more
relevant inside the brain, and should be related to physiological mechanisms
regulating free radicals and excitatory amino acid effects. In this case,
transamination of tryptophan and the possible presence of tautomerases (a
class of poorly described enzymes, able to shift keto to enol form of aromatic ketoacids inside the tissues) are the triggering factors.

In conclusion, the newly discovered pathway of KYNA formation suggests
that, besides its role as cerebral serotonin precursor, tryptophan can be a
neuronal protecting agent when it is transaminated to IPA. Furthermore, many
important pharmacological effects observed after tryptophan administration
(sleep, sedation, analgesia, etc.) could be related to IPA transformation to
KYNA.

Table 2. Radical scavenging effects of IPA

	IC_{50}
A. Inhibition of chemiluminescence	
IPA keto form	2.5×10^{-7} M
IPA enol form	8.8×10^{-8} M
B. Inhibition of malondialdehyde (MDA)	
IPA keto form	$> 10^{-3}$ M
IPA enol form	1×10^{-5} M

Luminol chemiluminescence was determined by the method of Rao
et al. (BBRC, 150:39, 1988). MDA formation was assessed in
brain homogenates.

REFERENCES

Foster, A.C., Vezzani, A., French, E.D., and Schwarcz, R., 1984, Kynurenic acid blocks neurotoxicity and seizures induced in rats by the related brain metabolite quinolinic acid, Neurosci. Lett., 48:273-278.

Ganong, A.H., Lanthorn, T.H., and Cotman, C.W., 1983, Kynurenic acid inhibits synaptic and acidic amino acid-induced responses in the rat hippocampus and spinal cord, Brain Res., 273:170-174.

Germano, I.M., Pitts, L.H., Meldrum, B.S., Bartkowski, H.M., and Simon, R.P., 1987, Kynurenate inhibition of cell excitation decreases strock size and deficits, Ann. Neurol., 22:730-734.

Lapin, I.P., 1983, Antagonism of kynurenine-induced seizures by picolinic, kynurenic and xanthurenic acids, J. Neural Transm., 56:177-185.

Russi, P., Carlà, V., Moroni, F., 1989, Indolepyruvic acid administration increases the brain content of kynurenic acid. Is this a new avenue to modulate excitatory amino acid receptors in vivo? Biochem. Pharmacol., 38:2405-2409.

Stone, T.W., and Burton, N.R., 1988, NMDA receptors and ligands in the vertebrate CNS, Progr. Neurobiol., 30:333-368.

TRYPTOPHAN METABOLISM IN HEALTHY SUBJECTS: INFLUENCE OF PYRIDOXINE AFTER

SINGLE OR REPEATED ADMINISTRATIONS

L. Demisch and P. Kaczmarczyk

Department of Psychiatry
Hospital of the Goethe-University
6000 Frankfurt/Main
Germany

INTRODUCTION

Common wisdom presently suggests that the neurobiological basis for
therapies of sleep disorders with oral loading doses of L-tryptophan (TRP) is
the increase in synthesis, storage and/or release of serotonin (5-HT) in synaptosomal compartments from TRP transported from plasma to central neurons.
Quantitatively, the conversion of TRP to 5-HT and/or biologically active indole derivatives in brain structures is very small compared to tryptophan
oxidation in the periphery through both dioxygenases (indole 2,3-dioxygenase
in lung, intestine and other organs and liver tryptophan 2,3-dioxygenase).
Therefore, the elucidation of strategies leading to an increased transport of
the administered TRP dose into the brain and thus utilization within neurons
is of considerable importance.

Beside the influences of the large neutral amino acids, kynurenine (Kyn)
has been shown to inhibit the transport of TRP from plasma into the brain.
Kyn has been found to accumulate in plasma of healthy volunteers following
longer-term treatment with loading doses of TRP (Green and Aronson, 1983).
This may be caused by depletion of pyridoxine (vitamin B-6), since several of
the degradative enzymes along the kynurenine-nicotinamide pathway of tryptophan oxidation are pyridoxine dependent. In support of this hypothesis,
Green and Aronson (1983) found that when TRP (100 mg/kg) was given with 0.5
mg/kg pyridoxine, the basal Kyn concentration before the last load following
one week of treatment was significantly lower, as was the peak following the
load. In addition, Bender and Totoe (1984) reported that administration of
vitamin B-6 (10 mg/kg) to rats led to an increase in both the concentration
of TRP in the brain and the transport of peripherally administered tritiated
TRP. There was a considerable reduction of hepatic tryptophan 2,3-dioxy-
genase and reduced excretions of N-methyl-nicotinamide and methyl pyridone,
indicating inhibition of TRP metabolism through the kynurenine-nicotinamide
pathway.

The present study was designed to elucidate this effect of pyridoxine on
the transport and utilization of 5-HT from TRP in healthy volunteers. Since
the studies of Green and Aronson (1983) were performed with high doses of TRP
(100 mg/kg per day), the dependency of dose and time course of application of
TRP with and without a 80 mg pyridoxine hydrochloride supplement was examined.

Kynurenine and Serotonin Pathways
Edited by R. Schwarcz *et al.*, Plenum Press, New York, 1991

SUBJECTS AND METHODS

The group consisted of 6 healthy females and 4 males, aged between 22 and 45 years (mean: 35 years). All subjects were drug-free and gave their informed consent. 80 mg pyridoxine hydrochloride (B-6-Vicotrat, Heyl, FRG) and 0.5 g, 1.0 g or 2.0 g L-tryptophan (given as 500 mg tablets, Kalma™, Fresenius, FRG) were administered as single doses or in conjunction (Kalma™ together with pyridoxine; TRP/B-6) at 22.00 hrs. The subjects fasted for 2 hrs before the treatment. B-6, TRP or TRP/B-6 were given in the above described doses for 7 days with one week intervals (wash-out). Night (20.00 - 08.00 hrs) and day (08.00 - 20.00 hrs) urine collections were made on day 1 and day 7 of the respective medication intervals. Urine samples were collected in bottles containing 10 ml 1 N HCl and 5 ml 10% sodium-metabisulfite. Aliquots of the urine collections were stored at -20°C until the assay. 5-Hydroxyindoleacetic acid (5-HIAA) was measured using a radioimmunological method and a commercially available kit (DDV, Marburg, FRG). Kyn was determined according to the method of Joseph and Risby (1975). The data were analyzed using the SPSS-8 (Statistical Package for the Social Sciences) (Hull and Nie, 1981).

RESULTS AND DISCUSSION

A significant increase in the urinary 5-HIAA excretion at night was measured following 7 days of treatment with 80 mg/night pyridoxine and also after a single and repeated administration of 2 g TRP together with 80 mg vitamin B-6 (paired t-test, $P < 0.05$). A single dose of 2 g TRP/night led to a small and not significant increase in urinary 5-HIAA. The urinary 5-HIAA excretion rate decreased significantly after 7 days of administration when the subjects were given 2 g TRP. This was not the case when TRP was administered together with pyridoxine (Fig. 1). Comparable increases in urinary 5-HIAA excretion were also measured following a single or repeated administration of 1 g and 0.5 g TRP in conjunction with 80 mg pyridoxine. Analysis of variance for dependent groups (ANOVA) showed neither a significant effect of the TRP dose nor the time course of application.

Table 1. Influence of a single and repeated pyridoxine and tryptophan administration on urinary Kyn excretion

	Kynurenine (mg/night)			
	B-6 (80 mg)	TRP (2 g)	TRP (1 g)	TRP (0.5 g)
Single dose	2.9 ± 2.0	6.7 ± 2.5	4.0 ± 1.9	3.2 ± 1.4
Subchronic	3.2 ± 1.2	5.4 ± 1.4	4.0 ± 1.3	3.0 ± 1.3
Tryptophan + Vitamin B-6				
Single dose		6.7 ± 3.2	4.0 ± 1.3	3.0 ± 1.3
Subchronic		6.6 ± 2.5	3.1 ± 0.4	3.5 ± 1.4

Collection period: 20.00 - 08.00 hrs; data are expressed as the mean ± SD. Control Kyn concentration: 2.5 ± 0.6 mg/night.

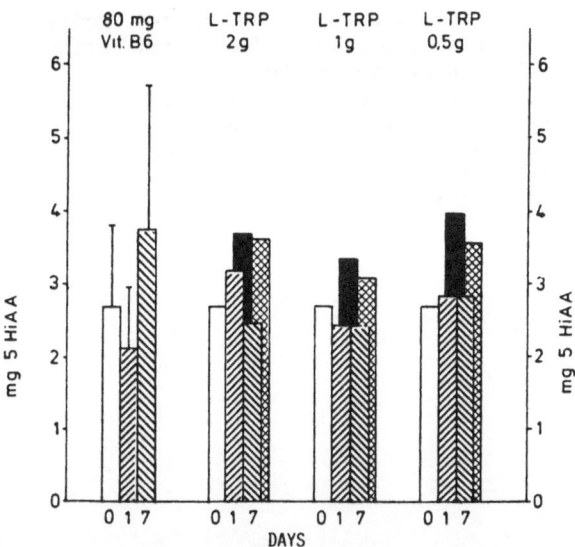

Fig. 1. Influence of pyridoxine (80 mg) and L-tryptophan on the nocturnal
urinary excretion of 5-HIAA (collection period: 20.00 - 08.00 hrs).
□ : basal value before treatment; ▨ : single dose; ▨ : after subchronic treatment (7 days); ■ : together with 80 mg B-6 and single
dose of TRP; ▨ : together with 80 mg B-6 and after subchronic
treatment (7 days).

A dose-dependent increase in urinary Kyn excretions at night was measured following a single or subchronic administration of TRP or TRP together
with B-6. Pyridoxine and subchronic administration of TRP did not lead to
significant changes in urinary Kyn (Table 1).

In conclusion, the major result of this study is the significant increase in urinary 5-HIAA excretion and thus the synthesis of 5-HT after subchronic administrations of 80 mg of vitamin B-6 or TRP together with B-6.
This elevation may be due to an increase of TRP transport into serotonergic
neurons induced by pyrodoxine in accordance with the findings of Bender and
Totoe (1984) in rats. It appears unlikely that the increase is caused by a
reduction in plasma Kyn levels since no effect of pyridoxine on urinary Kyn
excretion was found. Møller (1981) found that Kyn clearance varied with Kyn
plasma levels. A significant positive correlation between plasma Kyn and the
corresponding urinary Kyn excretion was determined above 30 to 40 nmol Kyn/ml
plasma. Therefore, it is conceivable that pyridoxine led to reduced Kyn levels in the plasma of the subjects which was not reflected in urinary Kyn excretion rates.

REFERENCES

Bender, D.A., and Totoe, L., 1984, High doses of vitamin B_6 in the rat are
associated with inhibition of hepatic tryptophan metabolism and increased
uptake of tryptophan into the brain, J. Neurochem., 43:733-736.
Green, A., and Aronson, J.K., 1983, The pharmakokinetics of oral L-tryptophan; effects of dose and concomitant pyridoxine, allopurinol or nicotinamide administration, in: "Management of Depression with Monoamine Precursors", Van Praag, M.H., and Mendlewicz, J., eds., Adv. Biol. Psychiat.,
Vol. 10, Karger, Basel, pp. 67-81.

Hull, C.H., and Nie, N.H., 1981, SPSS Update 7-9, McGraw-Hill, New York.
    Joseph, M.H., and Risby, D., 1975, The determination of kynurenine in
    plasma, Clin. Chim. Acta, 63:197-204.
Møller, S.E., 1981, Pharmacokinetics of tryptophan, renal handling of kynure-
    nine and the effect of nicotinamide on its appearance in plasma and urine
    following L-tryptophan loading in healthy subjects, Eur. J. Clin.
    Pharmacol., 21:137-142.

ON KYNURENINASE ACTIVITY

Y. Shibata, F. Takeuchi, R. Tsubouchi, M. Haneda, T. Ohta,
M. Nakatsuka, Y. Nisimoto, T. Tamai, S. Nomura, H. Fujimoto,
M. Sakata, T. Maesaki, and K. Goda[1]

Department of Biochemistry
Aichi Medical University Nagakute
Aichi 480-11

[1]Kobe-Gakuin University
Nisi-ku, Kobe-City
Hyogo 673
Japan

INTRODUCTION

In 1952, Yahito Kotake et al. reported that xanthurenic acid (XA) might be a diabetogenic agent in vitamin B_6 (B_6)-deficient rats. From that time, many studies have been performed in B_6-deficient rats and in diabetic (DM) patients. The results are as follows:

1) The pyridoxal phosphate/pyridoxal ratio in plasma decreased in DM patients compared with the control group [DM group: 0.54 (N = 9); control group: 3.84 (N = 8)], suggesting that pyridoxal kinase activity may be inhibited in DM patients (Hattori et al., 1984).

2) In vitro, pyridoxal kinase activity was inhibited by the addition of XA (to 8%), and pyridoxine 5-phosphate oxidase was activated to 225% by the addition of 3-hydroxykynurenine (Takeuchi et al., 1985).

3) Serum Ca values decreased and serum Mg values slightly increased in B_6-deficient rats and DM patients. Other authors reported that serum Mg values decreased in some DM patients (Shibata et al., 1989; Teraki and Uchiumi, 1989; Yamatani and Okada, 1989).

4) Administration of excess methionine resulted in a B_6-deficient condition (as determined by XA excretion after a tryptophan load), but doses of excess taurine or oyster extract (which contains much taurine) did not (Shibata et al., 1986, 1989).

5) In Mg-deficient rats, kynureninase activity decreased [control group: 114.0 ± 1.0 nmol/h/g liver (N = 5); Mg-deficient group: 95.1 ± 8.8 nmol/h/g liver (N = 7) (Shibata et al., 1988)].

We have now examined the regulation of XA metabolism related to 3-hydroxykynurenine.

Kynurenine and Serotonin Pathways
Edited by R. Schwarcz *et al.*, Plenum Press, New York, 1991

Tryptophan
↓
Kynurenine
↓
3-Hydroxyanthranilic acid ← 3-Hydroxykynurenine → Xanthurenic acid
↑ ↑
Kynureninase   Kynurenine aminotransferase

EXPERIMENTAL METHODS

Kynureninase (EC 3.7.1.3) was purified 600-fold from the liver cytosol of male Wistar Rats as previously reported (Takeuchi et al., 1980).

RESULTS

1. a) Experiments using the holoenzyme of kynureninase: DTNB (5,5'-dithio-bis-(2-nitrobenzoic acid)) did not inactivate the enzyme.

   b) Experiments using the apoenzyme of kynureninase: Kynureninase activity was inactivated by DTNB treatment of the apokynureninase at several concentrations (Fig. 1).

2. Experiment using the apoenzyme of kynureninase: Pyridoxal phosphate protected the apoenzyme. Inactivation of apokynureninase by DTNB was reversed by the addition of S-2-(benzothiazolyl)-L-cysteine, a substrate of cysteine-conjugate-β-lyase (4.4.1.13).

3. Effect of DTT (1,4-dithiothreitol) addition on apokynureninase activity: Kynureninase was activated by the addition of DTT, especially with 3-hydroxykynurenine as a substrate (Table 1).

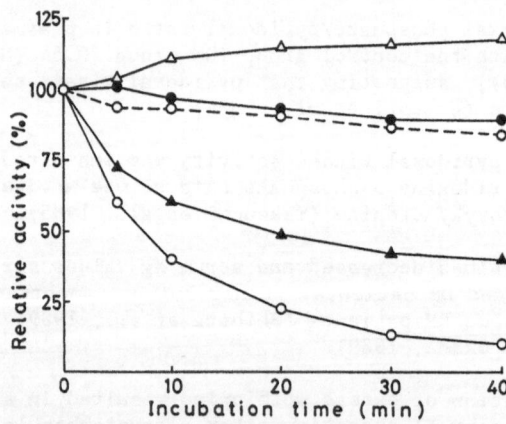

Fig. 1. Effect of pyridoxal phosphate (PLP) and S-2-(benzothiazolyl)-L-cysteine on the inactivation of apoenzyme with 10 μM DTNB. Apoenzyme was dialyzed against 50 mM Tris/HCl containing 1 mM EDTA, pH 8.0, and incubated at 37°C and pH 8.0 with 25 mM Tris/HCl containing 0.5 mM EDTA, 10 μM DTNB and one or none of the following: 7 μM PLP ($\triangle$), 1 μM PLP ($\bullet$), 50 μM S-2-(benzothiazolyl)-L-cysteine ($\blacktriangle$) or none ($\bigcirc$). The broken line shows the incubation without DTNB (control). Residual activity was determined with the standard assay.

Table 1. Effect of DTT and pyridoxal phosphate on kynureninase apoenzyme activity

Treatment	Relative activity (%)	
	Kynurenine	3-Hydroxykynurenine
Standard assay	100	100
1 mM DTT incubation	183	220

Phenylhydrazine-treated kynureninase was dialyzed against 50 mM Tris/HCl containing 0.1 mM dithiothreitol (DTT), pH 8.0. 1 mM DTT incubation: after incubation of the apoenzyme with 90 mM Hepes/KOH containing 1 mM DTT, pH 8.3, the activity was determined by standard assay. Kynurenine and 3-hydroxykynurenine, respectively, were used as substrates.

4. Effect of α-keto acids on kynureninase activity: Kynureninase activity was inhibited by p-hydroxyphenylpyruvate, a metabolite of phenylalanine or tyrosine, but was not inhibited by tyrosine itself (Table 2).

SUMMARY

1) In Mg-deficient rats, kynureninase activity is decreased.
2) p-Hydroxyphenylpyruvate inhibits kynureninase activity.
3) -SH groups in the apoenzyme of kynureninase play a very important role in the enzymatic reaction.
4) 3-Hydroxykynurenine may be a very important regulative metabolite in the 3-hydroxykynurenine → xanthurenic acid pathway.

Table 2. Effect of metabolites on phenylalanine and tyrosine on kynureninase activity

Addition to assay mixture	Relative activity (%)			
	Kynurenine		3-Hydroxykynurenine	
	1 mM	5 mM	1 mM	5 mM
None	100		100	
Tyrosine	98		95	
p-Hydroxyphenylpyruvate	22	0.2	23	0.7
Phenylalanine	100	102	97	92
Phenylpyruvate	74	27	85	44

REFERENCES

Hattori, M., Kotake, Y., Kotake, Y., Otsuka, H., and Shibata, Y., 1984, Studies on the urinary excretion of xanthurenic acid in diabetes, in: "Progress in Tryptophan and Serotonin Research", Schlossberger, H.G., Kochen, W., Linzen, B., and Steinhart, H., eds., de Gruyter, Berlin, pp. 347-354.

Shibata, Y., Hattori, M., Takeuchi, F., Otsuka, H., Tsubouchi, R., Ugata, M., Shiraishi, S., Ohta, T., Sugino, N., Sotokawa, Y., Takahashi, R., Kotake Y., and Kotake, Y., 1986, Influence of feeding oysters on the metabolism of serotonin in alloxan-diabetic rats, in: "Progress in Tryptophan and Serotonin Research", Bender, D.A., Joseph, M.H., Kochen, W., and Steinhart, H., eds., de Gruyter, Berlin, pp. 87-90.

Shibata, Y., Ohta, T., and Nakatsuka, M., 1989, Relation between tryptophan and calcium-magnesium metabolism, Trace Metal Metab., 17:109-112.

Shibata, Y., Ohta, T., Nakatsuka, M., Tsubouchi, R., Takeuchi, F., Sakata, M., Okinaka, O., and Nakasa, T., 1989, Relation between tryptophan metabolism and methionine metabolism, 43rd Annual Meeting of Japanese Society of Nutrition and Food Science, 12th May, Tokyo, Abstracts, 193.

Shibata, Y., Tsubouchi, R., Haneda, M., Ohta, T., Nakatsuka, M., Nomura, J., and Tanaka, T., 1988, Tryptophan metabolism in magnesium deficient rats, Fifth International Magnesium Symposium, 8th, August, Kyoto, Abstract, 150.

Takeuchi, F., Otsuka, H., and Shibata, Y., 1980, Purification and properties of kynureninase from rat liver, J. Biochem., 88:987-994.

Takeuchi, F., Tsubouchi, R., and Shibata, Y., 1985, Effect of tryptophan metabolites on the activities of rat liver pyridoxal kinase and pyridoxamine 5-phosphate oxidase in vivo, Biochem. J., 277:537-544.

Teraki, Y., and Uchiumi, A., 1989, Calcium, magnesium, phosphate metabolism in experimental diabetes mellitus in the rat, Trace Metal Metab., 17:103-107.

Yamatani M., and Okada, T., 1989, Comparative study of urinary excretion values of zinc and magnesium in each renal function, Trace Metal Metab., 17:113-116.

RAT SKIN TRYPTOPHAN 2,3-DIOXYGENASE

J. Naito, I. Ishiguro, Y. Nagamura, and H. Ogawa

Fujita-Gakuen Health University
Toyoake
Aichi 470-11
Japan

INTRODUCTION

The fact that a considerable amount of kynurenine (Kyn) is species-specifically accumulated in rat fur was discovered in 1960 by Ishiguro and co-workers (Hotta et al., 1960). The cause of this phenomenon has been exhaustively investigated in our laboratory, but no adequate explanation has been obtained so far.

In the course of the present study, we have developed a more accurate and sensitive method for the determination of tryptophan 2,3-dioxygenase (TPO) (Naito et al., 1987). By using this method, we have found an enzyme in rat skin which catalyzes the reaction from L-tryptophan (Trp) to Kyn, presumably via formyl-Kyn. Some features of the enzyme have been described recently (Naito et al., 1989).

METHODS

Rat skin homogenate or its supernatant fraction, prepared with 0.14 M KCl by using Polytron homogenization, was used as the enzyme source. Enzyme activity was assayed in the medium (1.0 ml) containing 50 mM potassium phosphate buffer, pH 7.0, 4 mM Trp, 5 mM sodium ascorbate, 15 μg hematin, 0.1 mg catalase and the enzyme solution. After incubation (37°C, 1h), the reaction was terminated by heating at pH 4.9, and the Kyn produced during the reaction was determined by HPLC with electrochemical detection following the enzymatic conversion of Kyn to 3-hydroxy-Kyn (Naito et al., 1987).

RESULTS

Physiological differences between skin and hepatic TPO were observed. In contrast to hepatic TPO activity, which remained rather constant throughout the rats' life (unless animals were treated with Trp or cortisone), the activity of skin TPO changed dramatically according to the growth cycle of the fur. As shown in Fig. 1, skin TPO in 5-weeks-old rats began to increase in coincidence with the period in which the new fur began to grow older. After reaching its highest value, the activity of the skin enzyme decreased suddenly, and its extremely low activity continued at least for another 8

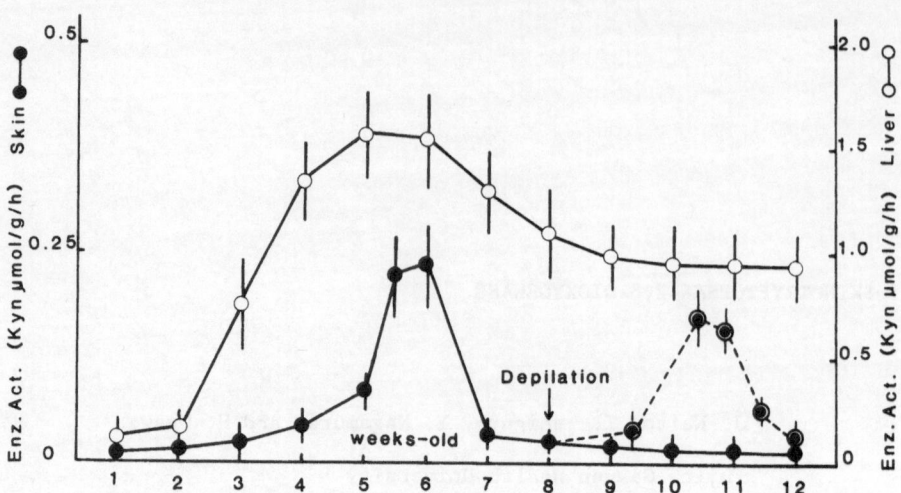

Fig. 1. Changes in tryptophan dioxygenase activity in rat skin and liver
homogenate with age.

weeks and was not accompanied by further apparent fur growth. However, 2
weeks after treatment of 9-weeks-old rats whose hair on part of the dorsal
posterior region had been eliminated, the enzymatic activity of the bared
skin increased together with fur growth as described above. The origin of
rat fur kynurenine was examined as follows: Furs from the dorsal posterior
region of rats (4-8 weeks old) were plucked 1 hr after injection of a small
amount of ^{3}H-trp. Radioactive substances were then extracted with 0.01 M HCl
at 80°C for 15 min, or the fur was solubilized by mixing with "Soluene-350"

Table 1. Radioactivity in rat fur one hour after intravenous
injection of ^{3}H-L-tryptophan

	Age (weeks)	Extraction[a]	No extraction[b]
		(dpm/10 mg fur)	
Rat	4	12 ± 5	104 ± 20
	5	164 ± 22	125 ± 13
	6	1,248 ± 145	882 ± 98
	7	23 ± 6	20 ± 6
	8	18 ± 7	19 ± 5
Guinea pig	6	204 ± 40	140 ± 16

1 h after the i.v. injection of ^{3}H-Trp (5 μCi), dissolved in 0.2 ml
Hanks' balanced salt solution containing 10^{-3} M Trp per 100 g body
weight, 10 mg of the fur from the dorsal posterior region were
plucked. [a]The furs were heated at 80°C for 15 min with 0.01 M HCl
in order to extract acid-soluble materials from the furs. [b]Each
fur which remained unextracted was solubilized for 48 h with 1.0 ml
"Soluene 350" at room temperature. Radioactivity is presented as
the mean ± SE of four rats.

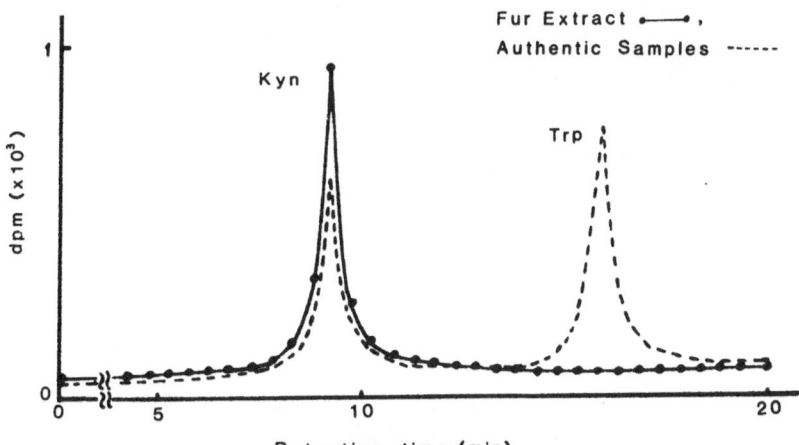

Fig. 2. Radiochromatography of the fur extract obtained from 6 weeks-
old rats injected with ^{3}H-L-trytophan. By using a fraction-
collector connected to a HPLC apparatus, the radioactivity of
each fraction was determined after decolorization of the
samples with hydrogen peroxide.

at 20°C for 24 hrs. Results obtained from the determination of radioactivity
in these samples revealed that the highest amount of isotope was seen in the
extract as well as the non-extracted sample prepared from the fur of 6 weeks
old rats (Table 1). All other samples showed extremely low levels of activity. Analysis of the radioactive substances in the fur extract of 6 weeks-
old rats revealed that most of the radioactivity was ^{3}H-L-Kyn (Fig. 2).

CONCLUSION

TPO is present in rat skin and produces Kyn, a general constituent of rat
fur. The enzymatic activity is highest at times of vigorous fur growth and
disappears when the length of the fur shows no further increase.

REFERENCES

Hotta, K., Ishiguro, I., Naito, J., and Kuzuya, H., 1960, Studies on kynurenine in rat fur, **Seikagaku**, 32:28-31.
Naito, J., Ishiguro, I., Murazumi, T., and Morimoto, M., 1987, Determination
of kynurenine in serum by high-performance liquid chromatography after
enzymatic conversion to 3-hydroxykynurenine, **Anal. Biochem.**, 161:16.
Naito, J., Ishiguro, I., Nagamura, Y., and Ogawa, H., 1989, Tryptophan 2,3-
dioxygenase activity in rat skin, **Arch. Biochem. Biophys.**, 270:236-241.

KYNURENINE 3-MONOOXYGENASE ACTIVITY OF RAT BRAIN MITOCHONDRIA DETERMINED

BY HIGH PERFORMANCE LIQUID CHROMATOGRAPHY WITH ELECTROCHEMICAL DETECTION

T. Uemura and K. Hirai

Neurochemistry Department
Psychiatric Research Institute of Tokyo
Tokyo 156
Japan

INTRODUCTION

Among the enzymes of the kynurenine pathway in brain, the presence of the following five enzymes has been reported: IDO (Gál, 1974), kynurenine 3-monooxygenase (Battie and Verity, 1981), kynureninase (Kawai et al., 1988), 3-OHAADO (Foster et al., 1986) and QPRT (Foster et al., 1985)[*]. The reported specific activity of rat brain mitochondrial kynurenine 3-monooxygenase is rather higher than that of liver. We have now reexamined this enzyme activity in rat brain mitochondria by using a new, highly sensitive assay method (HPLC-ECD). The enzyme activity of rat brain mitochondria is lower than that of liver mitochondria.

METHODS

Rat brain mitochondria were isolated by a Ficoll gradient method (Booth and Clark, 1978) and dialyzed overnight against buffered 0.32 M sucrose (pH 7.6) containing 1 mM DTT, 0.1 mM EDTA-Na_2 and 5 μM FAD. The reaction mixture contained, in a final volume of 1.2 ml, 0.1 M Tris-acetate (pH 8.1), 0.01 M KCl, 5 mM $MgCl_2$, 10 μM FAD (only for brain), 1.5 mM NADPH, 5 mM G6P, 3.5 U G6PDH, a suitable amount of mitochondrial protein and 0.3 mM L-KYN. After stopping the reaction by adding 0.05 ml of PCA, 1 ml of ethanol, 0.1 ml of 0.2 M EDTA-Na_2 and 50 ng of α-MD as an internal standard, the mixture was centrifuged at 15,000 rpm for 10 min, and the supernatant was passed through a Millipore filter (0.22 μm, GV). 20 μl of the eluate was directly injected into an HPLC column (ODS, 5 μm, 4.6 x 250 mm) equipped with a BAS PM-60 pump and a TL-5A Glassy Carbon electrode whose potential was set at 0.5 V versus the Ag/AgCl electrode. The quantity of 3-OHKYN formed was calculated by comparing a known standard composed of mitochondrial protein, 500 pmol of 3-OHKYN and 50 ng of α-MD.

Fig. 1 shows the HPLC elution pattern. 3-OHKYN and α-MD were eluted at

---

[*]IDO: indoleamine-2,3-dioxygenase; 3-OHAADO: 3-hydroxyanthranilic acid dioxygenase; QPRT: quinolinic acid phosphoribosyltransferase; 3-OHKYN: 3-hydroxykynurenine; KYN: L-kynurenine; α-MD: α-methyldopa; DTT: dithiothreitol; G6P: glucose-6-phosphate; G6PDH: glucose-6-phosphate dehydrogenase; PCA: perchloric acid.

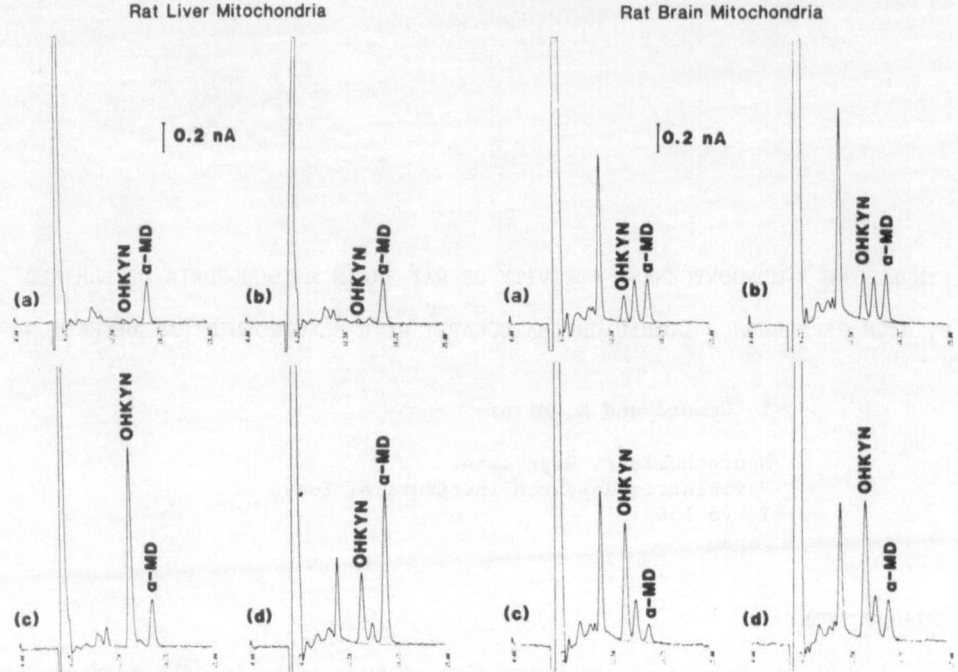

Fig. 1. HPLC elution profiles obtained from the reaction mixture for the assay of kynurenine 3-monooxygenase.

14.5 and 17.4 min, respectively. (a) and (b) show the results with boiled mitochondria and without NADPH, respectively, and these served as blanks. (c) and (d) show a complete system and a known standard, respectively.

RESULTS

A significant amount of 3-OHKYN was formed non-enzymatically (a,b) in the

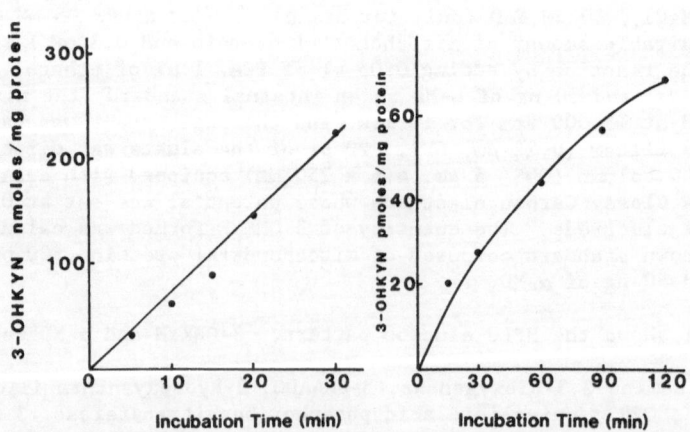

Fig. 2. The rate of 3-hydroxykynurenine formation using rat liver (left) or brain (right) mitochondria as enzyme source at 37°C.

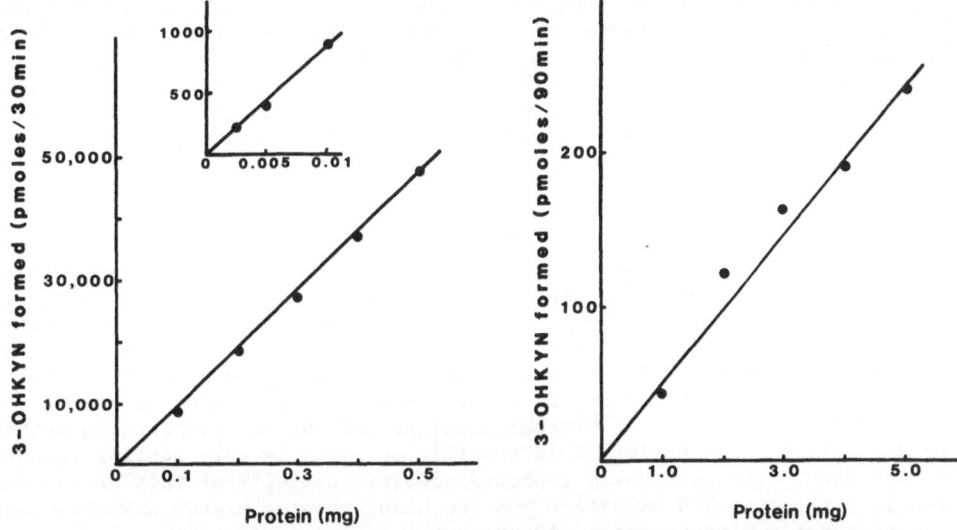

Fig. 3. Kynurenine 3-monooxygenase activity in rat liver (left) or
brain (right) mitochondria as a function of mitochondrial
protein.

case of brain. The true amount of 3-OHKYN formed was calculated by sub-
tracting (a) or (b) from (c). The enzyme reaction proceeded linearly at 37°C
for 30 and 90 min for liver and brain, respectively (Fig. 2). The amount of
3-OHKYN formed showed linearity with the amount of mitochondrial protein up
to 0.5 and 5 mg for liver and brain, respectively. We could measure the activity by using only 2.5 μg of liver mitochondrial protein (Fig. 3).

    Fig. 4 shows Lineweaver-Burk plots illustrating the effect of L-KYN in
the presence of the NADPH generating system on the rate of 3-OHKYN formation

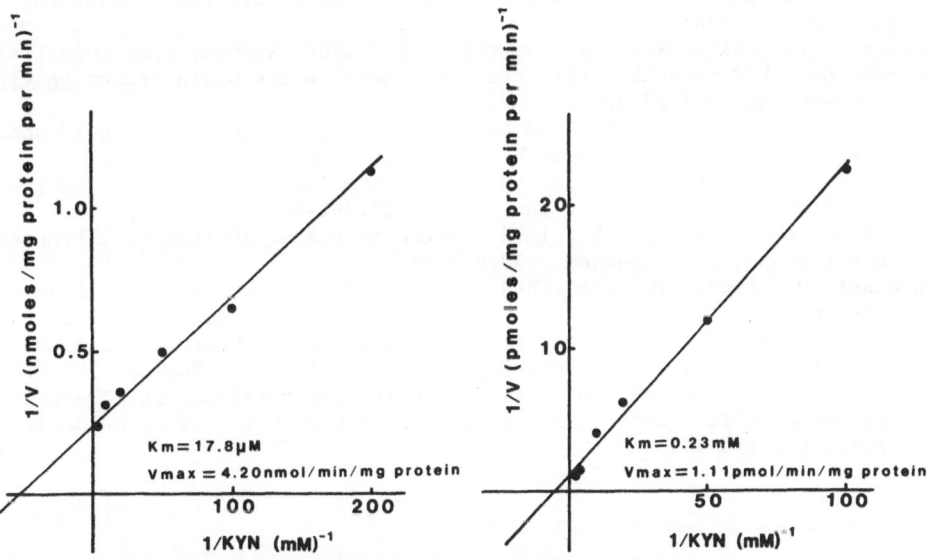

Fig. 4. Lineweaver-Burk plots illustrating the effect of the concentration
of L-kynurenine on the rate of 3-hydroxykynurenine formation. Liver
mitochondria (left), brain mitochondria (right).

by liver (left) and brain (right) mitochondria. The K_m values of L-KYN were 17.8 and 230 μM for the liver and brain enzyme, respectively. The V_{max} values were 4,200 and 1 pmol min^{-1} mg protein^{-1} for the liver and brain enzyme, respectively. Both K_m and V_{max} values for the liver enzyme agree well with previously reported values (Hayaishi, 1962; Okamoto, 1970). The cerebral specific activity was about 1/4,000 of the hepatic specific activity.

DISCUSSION

Concerning the kynurenine 3-monooxygenase activity in rat brain mito-chondria, there are only two reports up to the present, one based on in vivo studies (Gàl and Sherman, 1978), and the other in vitro (Battie and Verity, 1981). In vivo, the rate of synthesis of [^{14}C]3-OHKYN from [^{14}C]tryptophan was 0.47 nmol g^{-1} hr^{-1}. Although this does not correspond directly to the monooxygenase activity, it may be converted to 0.78 pmol min^{-1} mg protein^{-1} on the basis that the yield of mitochondrial protein by our preparation method was about 10 mg/g of brain. This specific activity is very similar to our present data. The previously reported specific activity of this enzyme meas-ured in vitro by NADPH oxidation was too high, showing higher activity than that of hepatic mitochondria. At present, we cannot offer an explanation for those data.

Our preliminary results indicate the activation of this enzyme activity by some electron acceptors, such as menadione in the presence of Fe-EDTA com-plex. Whether this activation occurs at the level of the purified brain mi-tochondrial outer membrane through the outer membrane electron transport sys-tem(s) or directly at the level of this flavin monooxygenase remains to be elucidated.

REFERENCES

Battie, C., and Verity, M.A., 1981, Presence of kynurenine hydroxylase in developing rat brain, J. Neurochem., 36:1308-1310.
Booth, R.F.G., and Clark, J.B., 1978, A rapid method for the preparation of relatively pure metabolically competent synaptosomes from rat brain, Biochem. J., 176:365-370.
Foster, A.C., White, R.J., and Schwarcz, R., 1986, Synthesis of quinolinic acid by 3-hydroxyanthranilic acid oxygenase in rat brain tissue in vitro, J. Neurochem., 47:23-30.
Foster, A.C., Zinkand, W.C., and Schwarcz, R., 1985, Quinolinic acid phos-phoribosyltransferase in rat brain, J. Neurochem., 44:446-454.
Gàl, E.M., 1974, Cerebral tryptophan-2,3-dioxygenase (pyrrolase) and its induction in rat brain, J. Neurochem., 22:861-863.
Gàl, E.M., and Sherman, A.D., 1978, Synthesis and metabolism of L-kynurenine in rat brain, J. Neurochem., 30:607-613.
Hayaishi, O., 1962, Kynurenine hydroxylase, in: "Methods in Enzymology V", Colowick, S.P., and Kaplan, N.O., eds., Academic Press, pp. 807-809.
Kawai, J., Okuno, E., and Kido, R., 1988, Organ distribution of rat kynuren-inase and changes of its activity during development, Enzyme, 39:181-189.
Okamoto, H., 1970, Kynurenine 3-monooxygenase (hydroxylase) in: "Methods in Enzymology XVII", Part A, Tabor, H., and Tabor, C.W., eds., Academic Press, pp. 460-463.

INDOLIC AND KYNURENINE PATHWAY METABOLITES OF TRYPTOPHAN IN RAT BRAIN:

EFFECT OF PRECURSOR AVAILABILITY ON IN VIVO RELEASE

M.J. During, M.P. Heyes[1], A. Freese[2], K.J. Swartz[2],
W.R. Matson[3], S.P. Markey[1], J.B. Martin[2], and R.H. Roth

Departments of Pharmacology and Psychiatry
Yale University School of Medicine
New Haven, Connecticut 06510

[1]Section on Analytical Biochemistry
NIMH
Bethesda, Maryland 20892

[2]Department of Neurology
Mass. General Hospital and Harvard Medical School
Boston, Massachusetts 02114

[3]ESA, Inc.
Bedford, Massachusetts 01730
USA

INTRODUCTION

Changes in tryptophan metabolism which influence central nervous system function are generally considered to be mediated through changes in serotonin (5HT) neurotransmission. It has been known for nearly 30 years that tryptophan loading or other interventions which increase brain tryptophan availability increase brain tissue levels of 5HT (Green et al., 1962; Wang et al., 1962) and that such an increase appears secondary to increased synthesis by increasing substrate availability as tryptophan hydroxylase is unsaturated in vivo (Carlsson and Lindquist, 1978). Over the last decade, there has been increasing interest in another route of tryptophan metabolism - the kynurenine pathway. Several indoleamine-2,3 dioxygenase metabolites are neuroactive and have been implicated in the pathogenesis of a number of neurological disorders. The kynurenine pathway metabolites which are perhaps of most interest are the excitatory amino acid (EAA) receptor ligands, quinolinic acid (QUIN) and kynurenic acid (KYNA). QUIN is an agonist at the N-methyl-D-aspartate (NMDA) receptor (Stone and Perkins, 1981) and a potent neurotoxin (Schwarcz et al., 1983). The enzymes leading to its synthesis from tryptophan have all been identified in the mammalian brain, and it is therefore likely it is formed de novo in the brain. However, the presence of QUIN in brain was not determined until sensitive gas chromatography/mass spectrometry (GCMS) assays were developed (Wolfensberger et al., 1983; Heyes and Markey, 1988). KYNA is a broad spectrum glutamate receptor antagonist and protects against QUIN (and other excitotoxin)-induced neurotoxicity. KYNA has also recently been shown to·exist within the brain (Moroni et al., 1988). Both QUIN and KYNA tissue levels are increased following tryptophan loading, however such a change in tissue levels does not necessarily reflect a change in their concentrations in the active compartment, the extracellular fluid (ECF). Within the ECF, QUIN and KYNA have access to EAA receptors and may therefore modulate glutamate (and aspartate) neurotransmission. We were therefore interested in determining whether QUIN and KYNA existed within the ECF and whether their concentrations would be influenced by alterations in precursor availability.

Using the technique of microdialysis, which enables on-line monitoring of the ECF in vivo, we followed the time course of indole and kynurenine pathway metabolites following acute intraperitoneal (i.p.) loading of tryptophan.

METHODS

Male Sprague-Dawley rats (250 - 350 g) were housed in a animal facility with ad libitum access to Agway 3000 (Syracuse, NY) rat chow. On the morning (0800-0900) of the day of experimentation, rats were anesthetized with chloralose/urethane (50/500 mg kg^{-1}, i.p.) with small, additional doses administered as necessary to maintain stable levels of anesthesia. Rats were placed in a Kopf stereotaxic frame and kept at 37°C using a homeostatically controlled heating pad. Dialysis probes were implanted into left and right striata at coordinates AP: +0.5, L: 2.5, V: -7.0 (Paxinos and Watson, 1982). Correct probe placement was verified by post-mortem sectioning and light microscopic visualization. The probes were of concentric design (300 μm o.d.) and had a 4 mm length of exposed membrane. The diffusion surface spanned the entire dorsoventral coordinates of the striatum. The probes were perfused with artificial ECF (Na$^+$ 125 mM, K$^+$ 2.8 mM, Mg^{2+} 1.0 mM, Ca^{2+} 1.2 mM, ascorbate 100 μM, phosphate 2 mM, pH 7.4) at a flow rate of 0.6 μl/min. Probes were calibrated in vitro by placing them in standard solutions (10^{-6} M and 10^{-7} M) and determining the relative concentration recovered.

The left striatal dialysates were assayed for QUIN using negative ionization GCMS as previously described (Heyes and Markey, 1988) and the right dialysates were analyzed for tryptophan, 5-HT and its acidic metabolite, 5-hydroxyindole acetic acid (5-HIAA) using HPLC with electrochemical detection. The HPLC system used a 15 cm x 2.1 mm 3 μm C$_{18}$ narrow bore column with a low volume (< 0.1 μl) electrochemical cell with an applied voltage of 0.95 V using a BAS LC3 potentiostat (Bioanalytical Systems, West Lafayette, IN). The mobile phase used to achieve separation was a 0.06 M sodium phosphate buffer, pH 2.1, containing 0.2 mM EDTA, heptane sulfonic acid 0.5 mM and methanol 25% (v/v). Dialysate samples were injected directly onto the HPLC column without any need for an internal standard or sample preparation. Chromatograms were completed within 25 minutes. Detection limits with a signal to noise ratio exceeding 2 to 1 were 5 fmoles for 5-HT and 5-HIAA, and 200 fmoles per sample for tryptophan. Groups of rats received saline or tryptophan in doses of 12.5, 50 and 100 mg/kg i.p. An additional group of rats received tryptophan 50 mg/kg with co-administration of large neutral amino acids (LNAAs) 100 mg/kg (leucine, isoleucine and valine; 1.5:1.0:0.8 weight ratio) i.p. All amino acids (Sigma, St. Louis, MO) were administered as a saline solution in a volume of 5 ml/kg. The amino acids were dissolved in 1 M NaOH, diluted close to final volume with 0.9% saline, then adjusted to pH 7.4 using 6 M HCl, with saline added to the final volume. Treatments were administered following three 30 minute baseline samples, collected at approximately 120 to 210 minutes following probe implantation. Perfusion continued for three hours after treatments, with dialysates taken at 30 minute intervals. Samples for QUIN assay were immediately frozen on dry ice and stored at -70°C until assay. The dialysates for tryptophan, 5-HIAA and 5-HT measurement were directly injected onto the HPLC system.

As an additional control to ensure that dialysate QUIN represented de novo brain synthesis and ECF spillover, and not disruption of the blood-brain barrier, a group of rats were cannulated with intrafemoral catheters prior to dialysis probe implantation. These rats underwent simultaneous dialysis and plasma measurements following the intravenous administration of QUIN (2.5 mmol/kg). Plasma QUIN levels increased by one-hundred fold, whereas dialysate QUIN concentrations remained unchanged (During et al., 1989).

A further group of animals received subchronic tryptophan by i.p. administration (100 mg/kg) or 0.9% saline three times daily for the ten days prior to dialysis. Two rats from each group underwent dialysis and were administered an acute 100 mg/kg load of tryptophan.

RESULTS

Dialysate tryptophan levels remained stable in the saline treated animals but increased from 5 ± 1 fold (mean ± SEM) in the 50 mg/kg group to 8 ± 3 fold in the 100 mg/kg group. QUIN increased markedly by 3 ± 1 fold in animals treated with 12.5 mg/kg and by 97 ± 14 fold in the 100 mg/kg group (Fig. 1).

The levels of QUIN peaked at 150 minutes in the 100 mg/kg animals and at 30 minutes in the 12.5 mg/kg animals. The rate of increase in QUIN levels over the first 30 minutes, however, appeared to be identical, suggesting saturation of the enzyme even at the lowest dose of tryptophan and obtainment of maximal enzymatic rates (V_{max}). Tryptophan levels peaked between 30 and 60 minutes, 5-HT levels slightly increased and peaked at 90 minutes, and 5-HIAA levels had a delayed increase reaching a plateau at 150 to 180 minutes.

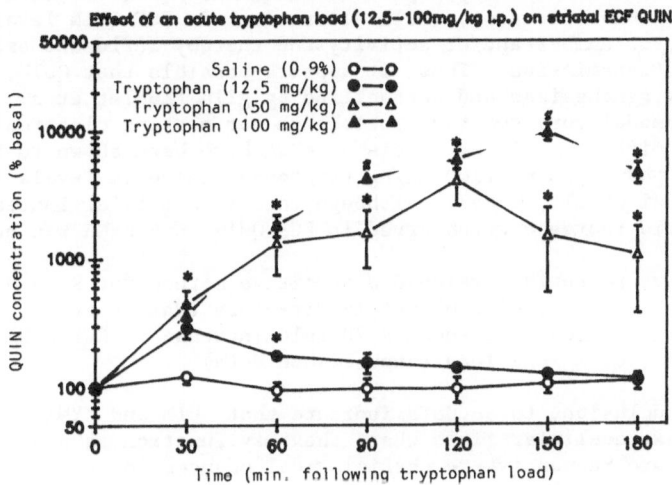

Fig. 1. Rats were anesthetized and microdialysis probes were implanted into left and right striata. 120 minutes following probe implantation, 3 consecutive 30 minute samples were collected to obtain a baseline measurement. Tryptophan (12.5, 50 and 100 mg/ kg, all groups: N = 5) or 0.9% saline (N = 5) was then administered i.p., and dialysis continued for a further 180 minutes. The left striatal dialysates were immediately frozen and subsequently assayed for QUIN using GC/MS. Data represent dialysate QUIN concentrations as a percentage of baseline (45 ± 9 nM) with vertical bars representing the SEM. *Significantly elevated above saline-treated control values (P < 0.05; unpaired Student's t-test).

The peak QUIN increase following tryptophan (50 mg/kg, i.p.) in the rats who received coadministration of LNAAs was decreased from 41 ± 15 fold to 6.5 ± 0.4 fold of baseline QUIN concentrations. In the rats who received 10 days of subchronic tryptophan administration, the peak QUIN following an acute tryptophan load was attenuated from a 26 fold increase above baseline values in the saline pretreated animals to a 6.6 fold increase above baseline concentrations in the tryptophan pretreated animals.

DISCUSSION

These data suggest that changes in brain tryptophan availability have a marked influence on levels of QUIN in striatal ECF. In addition, this relationship exists even with doses of tryptophan which may result in alterations in plasma levels well within the physiological range. This coupling between acute tryptophan loading and striatal ECF QUIN is of much greater magnitude than the relationship which exists between brain tryptophan availability and ECF 5-HT. However, the ECF levels of 5-HT obtained by microdialysis may not accurately reflect synaptic release, and are perhaps more appropriately considered as an indicator of synaptic spillover, as high affinity uptake of the transmitter may attenuate the changes in release as monitored by microdialysis.

Levels of QUIN of 10 μM have been shown to be toxic in organotypic neuronal cell cultures (Whetsell, 1984), and injection of as little as 12 nmoles directly into the striatum causes neuronal death at the site of infusion (Schwarcz et al., 1983). Peak estimated ECF levels of QUIN after the 100 mg/kg dose exceeded 10 μM; these levels may prove to be neurotoxic. The functional significance of the QUIN elevations observed following the administration of lower doses of tryptophan are unknown. However, as a specific agonist of the NMDA receptor, small alterations in ECF QUIN levels may modulate NMDA receptor number and/or activity and thereby influence excitatory amino acid neurotransmission. Thus, it appears possible that QUIN, a compound predominantly synthesized and catabolized by glia (Köhler et al., 1988) may act as a neuromodulatory substance, although not meeting classical criteria for neurotransmitters. Tissue levels of KYNA have been shown to increase following tryptophan loading (100 mg/kg tryptophan increase levels less than 2 fold; Moroni et al., 1988). Although this is a trivial increase compared to the 100 fold increase we observed in ECF QUIN, ECF KYNA was not measured.

We have recently developed a sensitive method for KYNA determination using a series array of coulometric detectors (Matson et al., 1987) and in preliminary studies have shown a 20 fold increase in ECF KYNA following a tryptophan (100 mg/kg) load (unpublished data).

In conclusion, these data indicate that QUIN and KYNA are present in rat striatal extracellular fluid where they may function as neuromodulators and that they are responsive to physiological changes in precursor availability.

REFERENCES

Carlsson, A., and Lindquist, M., 1978, Dependence of 5-HT and catecholamine synthesis on concentrations of precursor amino acids in rat brain, Naunyn-Schmiedeberg's Arch. Pharmacol., 303:157-165.
During, M.J., Heyes, M.P., Freese, A., Markey, S.P., Martin, J.B., and Roth, R.H., 1989, Quinolinic acid concentrations in striatal extracellular fluid reach potentially neurotoxic levels following systemic L-tryptophan loading, Brain Res., 476:384-387.

Green, H., Greenburg, S.M., Erickson, R.W., Sawyer, J.L., and Ellizon, T., 1962, Effect of dietary phenylalanine and tryptophan upon rat brain amine levels, J. Pharmacol. Exp. Ther., 136:174-178.

Heyes, M.P., and Markey, S.P., 1988, Quantification of quinolinic acid in rat brain, whole blood and plasma by gas chromatography and negative chemical ionization mass spectrometry: effects of systemic L-tryptophan administration on brain and blood quinolinic acid concentrations, Anal. Biochem., 174:349-359.

Köhler, C., Eriksson, L.G., Flood, P.R., Hardie, H.A., Okuno, E., and Schwarcz, R., 1988, Quinolinic acid metabolism in the rat brain. Immunohistochemical identification of 3-hydroxyanthranilic acid oxygenase and quinolinic acid phosphoribosyltransferase in the hippocampal region, J. Neurosci., 8:975-987.

Matson, W.P., Gamache, P.G., Beal, M.F., and Bird, E.D., 1987, EC array sensor concepts and data, Life Sci., 41:905-908.

Moroni, F., Russi, P., Lombardi, G., Beni, M., and Carla, V., 1988, Presence of kynurenic acid in the mammalian brain, J. Neurochem., 51:177-180.

Paxinos, G., and Watson, C., 1982, "The Rat Brain in Stereotaxic Coordinates", Academic Press, NY.

Schwarcz, R., Whetsell, W.O., and Mangano, R.M., 1983, Quinolinic acid: an endogenous metabolite that produces axon-sparing lesions in rat brain, Science, 219:316-318.

Stone, T.W., and Perkins, M.N., 1981, Quinolinic acid: a potent endogenous excitant of amino acid receptors in CNS, Eur. J. Pharmacol., 72:411-412.

Wang, H.L., Harwalker, V.H., and Waisman, H.A., 1962, Effect of dietary phenylalanine and tryptophan on brain serotonin, Arch. Biochem. Biophys., 97:181-184.

Whetsell, W.O., 1984, The use of organotypic tissue culture for the study of amino acid neurotoxicity and its antagonism in the mammalian CNS, Clin. Neuropharmacol., 8:248-250.

Wolfensberger, M., Amsler, U., Cuénod, M., Foster, A.C., Whetsell, W.O., and Schwarcz, R., 1983, Identification of quinolinic acid in rat and human brain tissue, Neurosci. Lett., 41:247-251.

KINETIC PROPERTIES OF HUMAN LIVER TRYPTOPHAN PYRROLASE

Y. Minatogawa, I.S.L. Matsui, and R. Kido

Department of Biochemistry
Wakayama Medical College
Wakayama 640
Japan

ABSTRACT

Human liver tryptophan pyrrolase (TPO) activity exhibited substrate level regulation. TPO showed biphasic activity to tryptophan when low ascorbate was used as an activator. The high affinity form (K_m for tryptophan: 0.05 mM) was promoted by low ascorbate and low tryptophan. The low affinity form (K_m for tryptophan: 0.4 mM) was induced by high concentrations of tryptophan or ascorbate. Both high and low affinity forms showed the same affinity to oxygen. The high affinity form was also induced by pyrroloquinoline quinone, but this effect was decreased by catalase, suggesting the participation of H_2O_2.

INTRODUCTION

Tryptophan pyrrolase (tryptophan 2,3-dioxygenase; EC 1.13.11.11) (TPO) essentially requires L-tryptophan and molecular oxygen as substrates and some reducing agent, such as L-ascorbate or Methylene Blue (Watanabe et al., 1980). However, cysteine and glutathione inhibit the reaction (Tanaka and Knox, 1959). The activation and inactivation mechanisms of these reducing agents are not clear. Partially purified human liver TPO was studied with regard to its kinetic properties using L-tryptophan, oxygen and L-ascorbate. The effect of PQQ (pyrroloquinoline quinone), a putative co-factor of oxidoreductases on the kinetics was also examined.

MATERIAL AND METHODS

Pyrroloquinoline quinone (PQQ) was purchased from Kanto Chemicals Co., Ltd., Tokyo, Japan. Other reagent quality chemicals were purchased from Sigma or Wako Chemicals, Osaka, Japan.

Enyzme preparation

Frozen human liver was homogenized in 9 vol. of 5 mM potassium phosphate buffer, pH 7.5, containing 1 mM L-tryptophan (full speed, Polytron, Kinematica GmbH, Luzern, Switzerland). The homogenate was centrifuged at 10,000 x g for 30 min at 4°C. The supernatant was heated at 60°C for 1 min, and then

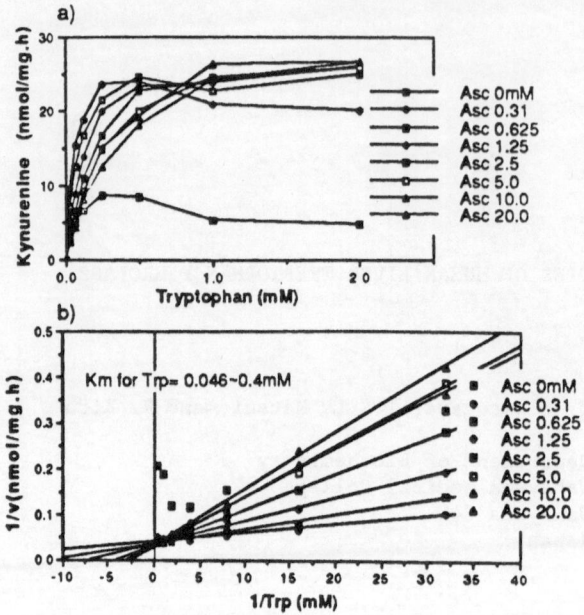

Fig. 1. Tryptophan saturation curves of TPO with or without ascorbate, and Lineweaver-Burk plots for tryptophan.

centrifuged for 30 min at 10,000 x g at 4°C. The resultant supernatant was fractionated by solid ammonium sulfate. The protein which was precipitated between 50 to 90% saturation of ammonium sulfate was collected and dialysed against 50 mM potassium phosphate buffer, pH 7.5, overnight in a cold room. This preparation was used for studies of tryptophan pyrrolase.

Tpo assay

Tryptophan pyrrolase activity was assayed in the presence of L-trypto-phan, L-ascorbate and 50 mM potassium phosphate buffer, pH 7.5, and enzyme preparation in 1 ml reaction mixture at 37°C for 30 min. The reaction was started by the addition of L-tryptophan and stopped by 0.2 ml of 30% tri-chloroacetate. The blank had L-tryptophan added after the trichloroacetate. The reaction mixture was then heated in a boiling bath for 5 min and depro-teinized by centrifugation. Newly formed kynurenine was determined using p-dimethylaminobenzaldehyde as follows: 0.7 ml of the clear supernatant was taken in a reaction tube and the same volume of 2% of p-dimethylaminobenzal-

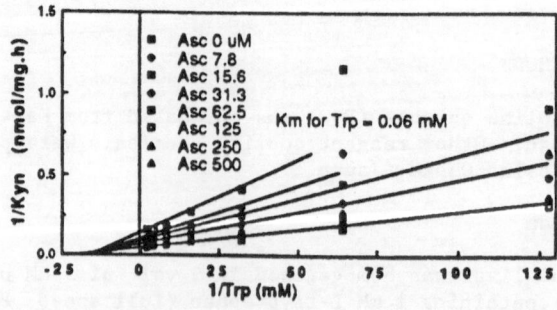

Fig. 2. Lineweaver-Burk plots of TPO at low tryptophan and low ascorbate concentrations.

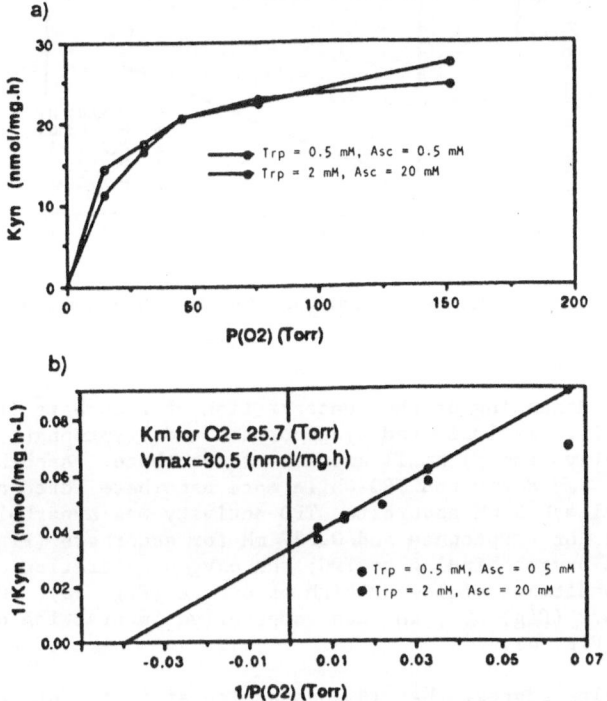

Fig. 3. Oxygen saturation curves of TPO at low concentrations of tryptophan (0.5 mM) and ascorbate (0.5 mM), and high concentrations of tryptophan (2 mM) and ascorbate (20 mM).

dehyde (dissolved in acetic acid) was added. The ΔOD was determined at 480 nm.

RESULTS AND DISCUSSION

The relationship between L-tryptophan and L-ascorbate with regard to TPO is shown in Fig. 1. Low concentrations of ascorbate showed biphasic reaction curves. A high affinity phase to tryptophan was observed at low concentrations of tryptophan, and low affinity condition at high concentrations of tryptophan. On the other hand, only the low affinity situation was observed at high concentrations of ascorbate. The K_m value of tryptophan varied from

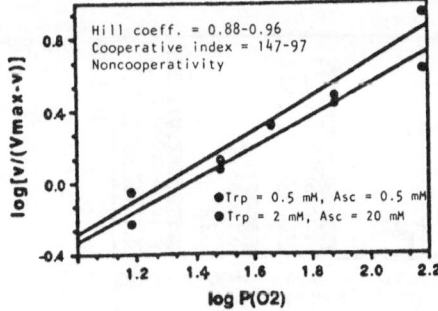

Fig. 4. Hill plots of oxygen and TPO activity under the low and high concentration conditions for tryptophan and ascorbate as described in Fig. 3.

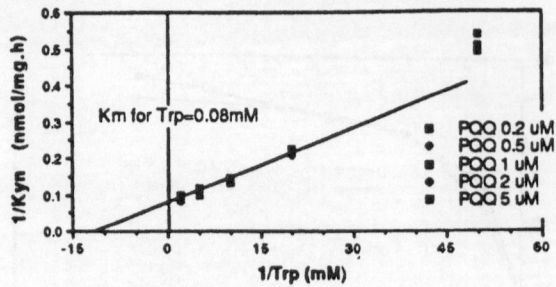

Fig. 5. Lineweaver-Burk plots of TPO in the presence of PQQ.

0.05 to 0.4 mM depending on the concentration of ascorbate (Fig. 1b). Thus, TPO was competitively inhibited by ascorbate and tryptophan. However, TPO did not get activation by small amounts of ascorbate. Ascorbate concentrations up to 0.3 mM activated TPO while more ascorbate acted as a competitive inhibitor. Below 0.5 mM ascorbate, TPO activity was hyperbolic, with K_m values of 0.06 mM for tryptophan and 0.027 mM for ascorbate (Fig. 2). Ascorbate did not affect oxygen binding to TPO: the oxygen saturation curve was hyperbolic under conditions of low or high ascorbate (Fig. 3a). Oxygen had a K_m value of 25 torr (Fig. 3b), and non-cooperative interaction of oxygen and TPO was observed (Fig. 4).

These results suggest that TPO has a high affinity form and a low affinity form with regard to binding tryptophan. Tryptophan is not only a substrate of TPO, but also a negative allosteric effector of the enzyme. Tryptophan may bind to TPO at a regulatory site and convert it to a low affinity form at high concentrations of tryptophan.

A similar or identical conversion may also result from a high concentration of ascorbate. Ascorbate reduces Fe^{+++} to Fe^{++} of TPO (Tanaka and Knox, 1959). This process is essential to TPO activation. Low ascorbate cooperates with low tryptophan to induce the high affinity form of TPO, but high ascorbate changes TPO to a low affinity form even if the tryptophan concentration is low. The high and low affinity forms of TPO with regard to tryptophan have the same affinity for oxygen. Oxygen saturation of TPO was hyperbolic and non-cooperative, like with myoglobin.

PQQ had an activating effect on TPO without ascorbate (Fig. 5). PQQ-activated TPO had a K_m of 0.08 mM for tryptophan. This value indicates that TPO is in a high affinity form to tryptophan. The effect of PQQ was largely

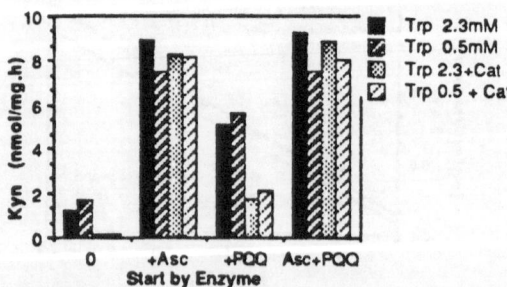

Fig. 6. TPO activity in the presence or absence of ascorbate, PQQ and both. Reaction mixture contains high (2.3 mM) or low (0.5 mM) concentrations of tryptophan with or without catalase.

abolished by catalase, suggesting the participation of H_2O_2 (Fig. 6). However, stimulation of TPO by ascorbate was not significantly changed by catalase (Fig. 6). The high affinity form may be induced by low ascorbate or H_2O_2, and this form may be transformed to the low affinity form by high ascorbate or tryptophan. TPO is a multi-hormonal inducible enzyme in liver (Nakamura et al., 1987), and it is stabilized by tryptophan itself _in vivo_ (Schimke et al. 1965). Additionally, the human TPO activity may also be finely regulated both by its own substrate, tryptophan, and its own activators, namely reducing agents.

REFERENCES

Nakamura, T., Niimi, S., Nawa, K., Noda, C., Ichihara, A., Takagi, Y., Anai, M., and Sakaki, Y., 1987, Multihormonal regulation of transcription of the tryptophan 2,3-dioxygenase gene in primary cultures of adult rat hepatocytes with special reference to the presence of transcriptional protein mediating the action of glucocorticoids, J. Biol. Chem., 262:727-733.

Schimke, R.T., Sweeney, E.W., and Berlin, C.M., 1965, The role of synthesis and degradation in the control of rat liver tryptophan pyrrolase, J. Biol. Chem., 240:322-330.

Tanaka, T., and Knox, W.E., 1959, The nature and mechanism of tryptophan pyrrolase (peroxidase-oxidase) reaction of Pseudomonas and of rat liver, J. Biol. Chem., 234:1162-1170.

Watanabe, Y., Fujiwara, M., Yoshida, R., and Hayaishi, O., 1980, Stereospecificity of hepatic L-tryptophan 2,3-dioxygenase, Biochem. J., 189: 393-405.

INTERFERON TYPE I AND II ANTAGONISM: A NOVEL REGULATORY MECHANISM OF

INDOLEAMINE DIOXYGENASE INDUCTION IN HUMAN PERIPHERAL BLOOD MONOCYTES AND

PERITONEAL MACROPHAGES

Y. Ozaki, E.C. Borden, R.V. Smalley, and R.R. Brown

University of Wisconsin Clinical Cancer Center
Madison, Wisconsin 53792
USA

INTRODUCTION

Antiproliferative effects of interferon (IFN)-γ in vitro on some tumor cells (de la Maza and Peterson, 1988; Ozaki et al., 1988a; Takikawa et al., 1988), Chlamydia psittaci (Byrne et al., 1986) and Toxoplasma gondii (Pfefferkorn, 1984) are mediated by deprivation of tryptophan (TRP) which results when the cytokine induces indoleamine 2,3-dioxygenase (IDO) activity in the cancer or host cells. Among the human peripheral blood mononuclear cells as well as granulocytes, IFN-mediated induction of IDO is localized in peripheral blood monocytes (PBM) or PBM-derived macrophages (DM; Ozaki et al., 1987a; Werner et al., 1987; Rubin et al., 1988; Carlin et al., 1989), however the mechanism of control of extra-hepatic TRP metabolism by the IFNs has not been fully elucidated in these cells or in peritoneal macrophages (PM). The occurrence of IFN type I antagonism against IFN type II-elicited release of H_2O_2, expression of Ia antigen as well as mannosyl/fucosyl receptors, high-affinity binding of type II IFN, and tumor cell killing by macrophages has been reported (Ling et al., 1985; Ezekowitz et al., 1986; Garotta et al., 1986; Inaba et al., 1986; Pace 1987; Yoshida et al., 1988). However, the down-regulation of IFN-γ-mediated induction of IDO by IFN-α, IFN-β and 12-O-tetradecanoylphorbol-13 acetate (TPA) has never been documented. To this end, we report clear differences between PBMs, DMs and PMs in response to IFNs as well as to a potent tumor promoter, TPA. Further, the occurrence of a novel regulatory mechanism of TRP metabolism through modulation of type II IFN-induced IDO activity by type I IFNs or by TPA, in monocytes and macrophages is demonstrated in this report.

MATERIALS AND METHODS

Cell culture: Peripheral blood mononuclear cells (PBL) were isolated from heparinized (16 U/ml) whole blood of normal volunteers and were fractionated using Ficoll-Paque (Pharmacia, Piscataway, NJ) gradient centrifugation (Boyum, 1968; Ozaki, et al., 1987a). The PBL layer at the gradient interface after the centrifugation at 400 x g for 30 min was removed and diluted 1/5 (v:v) with PBS (phosphate buffered saline) lacking calcium and magnesium. The cell suspension was centrifuged at 250 x g for 10 min. The cell pellet was washed twice more by resuspending in PBS and by centrifugation.

Kynurenine and Serotonin Pathways
Edited by R. Schwarcz et al., Plenum Press, New York, 1991

PBLs were resuspended in culture medium consisting of RPMI 1640 (Hazelton Research Products, Inc., Lenexa, KS) supplemented with 2 mM L-glutamine (Sigma Chemical Co., St. Louis, MO), heat-inactivated 10% fetal bovine serum (Hy-Clone Lab, Inc., Logan, UT), 50 μg/ml of gentamicin sulfate (GIBCO Laboratories, Grand Island, NY) and 20 mM Hepes buffer (GIBCO). A PBM-enriched fraction was prepared by allowing cells to adhere to the plastic tissue culture plates (Costar, Cambridge, MA) for 1 hour at 37°C in an atmosphere of 5% CO_2 in air. Nonadherent cells were removed by vigorous washing of the plates with warm PBS. Adherent cells with more than 80% monocytes morphology (van Furth et al., 1979) were incubated for 2-10 days with fresh medium in presence of absence of near optimum IDO inducing concentrations of human recombinant (Hr) IFN-α 2a (1,800 IU/ml; Hoffman-LaRoche Inc., Nutley, NJ), IFN-βser (1,800 IU/ml; Triton Bio-science Inc., San Francisco, CA), IFN-γ(1,000 IU/ml; Biogen Research Co., Cambridge, MA), or optimum IDO suppressing doses of TPA (5-10 nM in final concentration of 0.1% dimethyl sulfoxide; Sigma) and their vehicles. PMs were obtained from the ascitic fluids of patients with advanced ovarian carcinoma. Peritoneal effusions were centrifuged at 400 x g for 30 min. The cell pellets were suspended in PBS and fractionated on a Ficoll-Paque gradient. PMs were similarly separated from lymphocytes and tumor cells on the basis of adherence to the plastic tissue culture plates (Costar, Cluster[6], 35 mm) at 37°C for 45 min. Adherent cells (1-2 x 10[6]) with more than 95% macrophage morphology were subjected to the various cytokine treatments. Adherent PBM, DM and PM were incubated in 0.2% EDTA in PBS at 4°C for 15 min and were harvested using a cell scraper (Coster). Cell viability, determined by exclusion of 0.2% trypan blue, was greater than 95%.

IDO assay: IDO activity was determined using slight modifications of the method described previously (Ozaki et al., 1986, 1987b). The harvested cell suspension was centrifuged at 260 x g for 15 min, the resultant cell pellet was lysed in 0.1 ml of 5 mM EDTA in 50 mM potassium phosphate buffer (KPB; pH 7.2) and centrifuged at 12,000 x g for 15 min. Aliquots of 5 to 50 μl of this supernatant fraction were subjected to the Bradford protein assay (BIO-RAD, Richmond, CA) or IDO assay. The standard IDO assay mixture consisted of 50 mM KPB, the enzyme preparation, 40 μM L-[ring-2-¹⁴C]-tryptophan (1480 Bq;

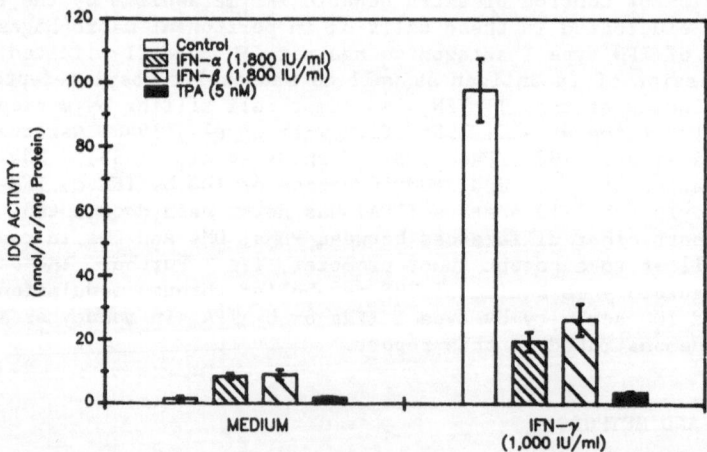

Fig. 1. Modulation of HrIFN-γ-mediated induction of IDO activity by HrIFN-α2a, HrIFN-βser and phorbol ester (TPA) in human peripheral blood monocytes (PBMs). Adherent PBMs (2 x 10[6] cells) were incubated in 2 ml of the culture medium containing IFN-α(1,800 IU/ml), -β(1,800IU/ml), -γ(1,000 IU/ml), TPA (5 nM) and their vehicles (PBS and DMSO) alone or type I IFN plus type II IFN, concurrently, for 48 hours. Adherent PBMs were harvested and assayed for IDO as described above. Data are the mean ± SD of triplicate samples.

Research Product International Co., Chicago, IL), catalase (4,200 units; Sigma), 10 mM sodium formate, oxygen at ambient tension (20.9%), 25 μM methylene blue, 10 mM ascorbic acid and 5 mM EDTA in total volume of 0.1 ml. The reaction mixture was incubated at 37°C for 15-30 min. The reaction was terminated by the addition of 0.1 ml of 6% perchloric acid (PCA) and complete hydrolysis of [^{14}C]-N-formylkynurenine was accomplished at 37°C for 30 min. The labeled formate was separated from its percursors by adding 1 ml of 10% (w:v) charcoal, equilibrated with 1 M formic acid in 3% PCA, to the tube. The tubes were vortexed for 5 sec and centrifuged at 3,000 x g for 5 min. The radioactivity of the supernatant fluids (0.6 ml) was determined by liquid scintillation spectrometry.

RESULTS

Human recombinant (Hr) IFNs induced IDO in PBM, DM and PM in vitro. In PBM and DM, but not in PM, this increase in IDO activity by HrIFN-γ (1,000 IU/ml) was antagonized by HrIFN-α2a (1,800 IU/ml), HrIFN-βser(1,800 IU/ml), or by TPA (5-10 nM) treatment. The most potent inducer of IDO activity in PBMs was HrIFN-γ resulting in enzyme activity 100-fold higher than control levels (P < 0.001, two-tailed Student's t-test) and more than 10-fold higher than levels induced by HrIFN-α2a or HrIFN-βser alone (P < 0.001; Fig. 1). Thus, type I IFNs were less potent inducers of IDO than type II IFN. Simultaneous incubation of PBMs with optimal IDO-inducing concentrations of type I and II IFNs limited IDO induction to 20-25% (P < 0.005) of the level observed with type II IFN alone (Fig. 1). However, the type I IFNs were ineffective when added 24 hour (but not 2 hour) after IFN-γ treatment. Interestingly, exposure of PBM to TPA (which induces monocytoid differentiation in HL60 cells) completely abolished enzyme induction. These antagonistic effects of type I IFNs or TPA were independent of PKC as indicated by a lack of effect by the PKC inhibitor 1-(5 isoquinolinyl-sulfonyl)-2-methylpiperazine (H7; Fig. 2) or by retinoic acid (data not shown). Serotonin, a neurotransmitter, shown to reduce IFN-γ induced expression of Ia antigen in murine macrophages (Sternberg et al., 1986), did not affect IDO induction by HrIFN-γ (data not shown). Type I IFNs were more effective in inducing IDO in PMs than in PBMsor DMs. Antagonistic actions of type I IFNs against IDO induction by type II IFN were more pronounced in PBM than in macrophages of DM and PM

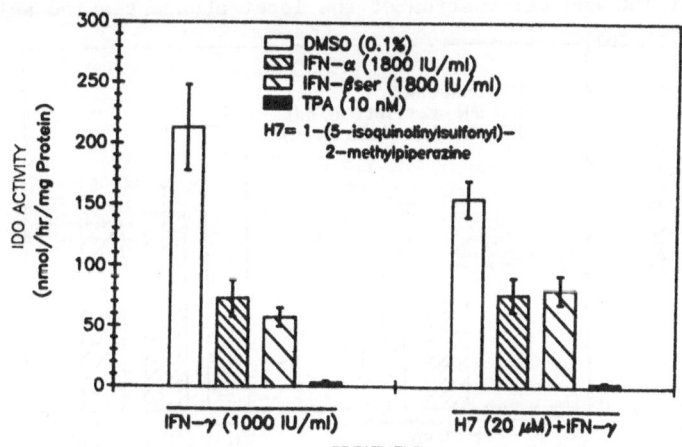

Fig. 2. Effect of PKC inhibitor H7 pretreatment on modulation of HrIFN-γ-mediated induction of IDO activity by HrIFN-α2a, IFN-βser and TPA in human PBMs. Adherent PBMs (2 x 10^6 cells) were pretreated with either HrIFN-γ alone or H7 plus IFN-γ for 2 hours; HrIFN type I was added to the culture medium (2 ml) and incubation was continued for 48 hours.

origin (Figs. 1,3 and 4). Interestingly, TPA, a potent tumor promoter, an enhancer of IFN-γ-elicited IDO induction in THP-1 monocytic leukemia cells (Ozaki et al., 1988b) and inducer of natural killer cells adherence (Argov et al., 1985), abolished type II IFN induction of IDO activity in DM but did not affect IDO induction in PM (Fig. 4). Thus, these data show that DM and PM are quite different entities with regard to their responses to these factors.

DISCUSSION

These observations clearly demonstrate that among IFNs, type II IFN is the strongest IDO inducer and type I IFNs are weak IDO inducers. However, in PBMs, type I IFNs are strong suppressors of type II-induced IDO. Conversely, type I IFNs are effective IDO inducers in PM and are less effective suppressors or modulators of the type II action in PMs and DMs. Thus, the maturation stages of monocytes/ macrophages and/or the state of the cells (primed and activated or down-regulated) are important factors influencing responsiveness of mononuclear phagocytes towards IFNs. For example, simple adherence of monocytes to the plastic culture plate itself increases the number of low affinity IFN-γ receptors 25-fold and alters the saturability from 1.2 nM to 40-50 nM within 2-3 hours of incubation (Finbloom and Wahl, 1989).

Although TRP and kynurenine metabolites have been reported to be necessary for the IFN-γ-mediated activation of macrophage tumor cell killing (Leyko and Varesio, 1989), the exact role of IDO in the immune system has been unclear. The sensitivity of PM to IFN-βser for IDO induction in vitro (Fig. 3) reinforces the argument that the increased extra-hepatic TRP metabolism and kynurenine formation in vivo in patients given HrIFN-γ (Brown et al., 1987) or IFN-βser (Datta et al., 1987) may stem from the contribution of mature tissue macrophages (PM) rather than from that of pluripotent monocytes (PBM). Type I IFNs are products of monocytes/macrophages and induced by type II IFN (Johnson and Torres, 1985; Gessani et al., 1988). We have found that type I IFNs also suppress IFN type II induction of HLA class II antigens in monocytes (data not shown). Therefore, the modulation of IFN-γ action by type I IFNs may represent a general type of feedback control, not only of IDO induction but of other IFN-γ induced proteins such as murine or human HLA class II (Ia) antigens, or Fe receptors. It is conceivable that the overexpression of IDO and elimination of the least abundant amino acid, L-TRP,

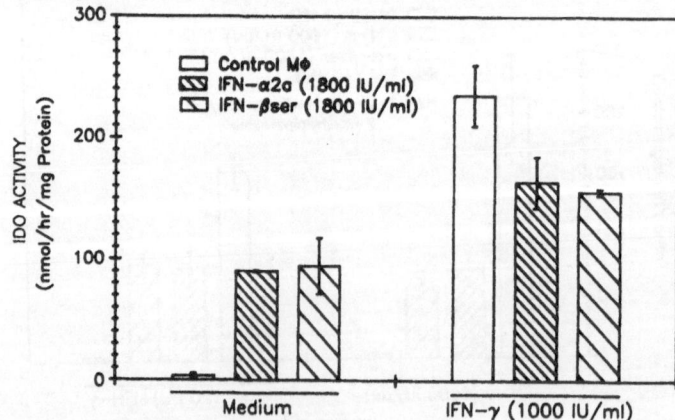

Fig. 3.   Effect of HrIFN-α2a and IFN-βser (alone or in combination with HrIFN-γ) on the induction of IDO activity in human peritoneal macrophages (PM) isolated from peritoneal effusions. Adherent PMs (2 x 10^6 cells) were incubated in the presence or absence of the indicated IFNs, in 2 ml of the culture medium for 48 hours before assay.

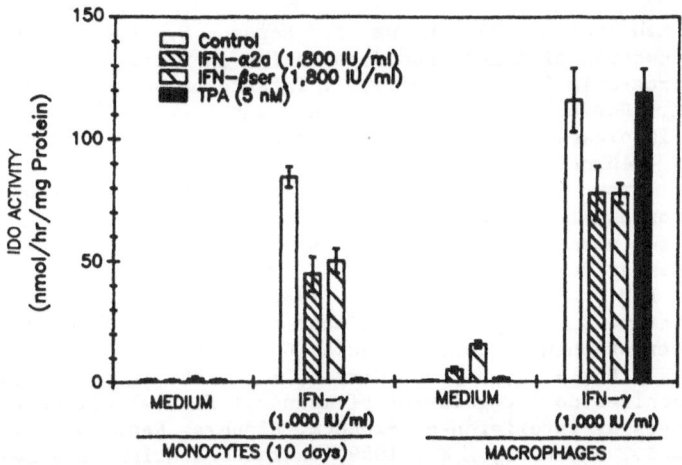

Fig. 4. Effects of HrIFN-α2a, HrIFN-βser and TPA on HrIFN-γ-mediated induction of IDO activity in PBM-derived macrophages (DM: monocytes [10 days]) and peritoneal macrophages (PM: macrophages). Adherent PBMs (2 x 10⁶ cells) were allowed to mature for 10 days in plastic culture plates, and the cells were further incubated in the presence or absence of IFNs and TPA in the culture medium (2 ml) for 48 hours. Fresh adherent PMs (2 x 10⁶ cells) were similarly treated with IFNs and TPA in the same studies.

may be more harmful than beneficial to the integrity of macrophages. Inhibitory action of type II IFN on proliferation of PBMs induced by colony stimulating factors (Hoover et al., 1989) might also be mediated by TRP deprivation (Ozaki et al., 1988a). These observations further suggest that the regulation of their own actions by IFNs constitute an immunomodulatory action of cytokines. In addition to the uniqueness of mammalian dioxygenase in using flavins, tetrahydrobiopterin and superoxide anion for TRP oxidative-decyclization (Ozaki et al., 1986, 1987b), control of IDO induction and regulatory mechanisms for TRP metabolism by substances (IFNs) other than pathway substrates suggest a novel control system in amino acid metabolism. These data further suggest likely mechanisms for interaction of nutritional, immunological and neurochemical systems.

ACKNOWLEDGEMENTS

We thank Ms. D. Helgeson, Ms. M. Malone, and Mr. R. Wagner for their excellent technical support. These studies are supported in part by a grant from Glaxo, Inc.

REFERENCES

Argov, S., Hebdon, M., Cuatrecasas, P., and Koren, H.S., 1985, Phorbol ester-induced lymphocyte adherence: selective action on NK cells, J. Immunol., 134:2215-2222.
Boyum, A., 1968, Isolation of mononuclear cells and granulocytes from human blood, Scand. J. Clin. Lab. Invest., Suppl. 97, 21:77-89.
Brown, R.R., Borden, E.C., Sondel, P.M., and Lee, C.M., 1987, Effects of interferons and interleukin-2 on tryptophan metabolism in humans, in: "Progress in Tryptophan and Serotonin Research", Bender, D.A., Joseph, M.H., Kochen, W., and Steinhart, H., eds., de Gruyter, Berlin, pp. 19-26.

Byrne, G., Lehmann, L., Kirschbaum, J., Borden, E., Lee, C., and Brown, R., 1986, Induction of tryptophan degradation in vitro and in vivo: a γinterferon-stimulated activity, J. Interferon Res., 6:389-396.

Carlin, J.M., Borden, E.C., Sondel, P.M., and Byrne, G.I., 1989, Interferon-induced indoleamine 2,3-dioxygenase activity in human mononuclear phagocytes, J. Leukocyte Biol., 45:29-34.

Datta, S.P., Brown, R.R., Borden, E.C., Sondel, P.M., and Trump, D.L., 1987, Interferon and interleukin-2 induced changes in tryptophan and neopterin metabolism: possible markers for biologically effective doses, Proc. Am. Assoc. Cancer Res., 28:338.

de la Maza, L.M., and Peterson, E.M., 1988, Dependence of the in vitro antiproliferative activity of recombinant human γ-interferon on the concentration of tryptophan in culture media, Cancer Res., 48:346-350.

Ezekowitz, R.A.B., Hill, M., and Gordon, S., 1986, Interferon α/β selectively antagonizes down-regulation of mannosyl-fucosyl receptors on activated macrophages by interferon-γ, Biochem. Biophys. Res. Commun., 136:737-744.

Finbloom, D.S., and Wahl, L.M., 1989, Characterization of a novel low affinity receptor for IFN-γ on adherent human monocytes by radioligand binding studies and chemical cross-linking, J. Immunol., 142:2314-2320.

Garotta, G., Talmadge, K.W., Pink, R.L., Dewald, B., and Baggiolini, M., 1986, Functional antagonism between type I and type II interferons on human macrophages, Biochem. Biophys. Res. Commun., 140:948-955.

Gessani, S., Baglioni, C., Puddu, P., Di Marzio, P., and Belardelli, F., 1988, Bacterial lipopolysaccharide and IFN-γ induce transcription of IFN-β mRNA and IFN secretion in murine macrophages, J. Interferon Res., Suppl. 1, 8:S50.

Hoover, D.L., Gendelman, H., Vargo, M., and Metzer, M.S., 1989, IL4 and IFN-gamma inhibit replication of human peripheral blood monocytes, FASEB J., 3:A822.

Inaba, K., Kitaura, M., Kato, T., Watanabe, Y., Kawade, Y., and Muramatsu, S., 1986, Contrasting effect of α/β- and γ-interferons on expression of macrophage Ia antigens, J. Exp. Med., 163:1030-1037.

Johnson, H.M., and Torres, B.A., 1985, Mechanism of calcium ionophore A23187-induced priming of bone marrow-derived macrophages for tumor cell killing: Relationship to priming by interferon, Proc. Natl. Acad. Sci. USA, 82: 5959-5962.

Leyko, M.A., and Varesio, L., 1989, Role for tryptophan and its metabolites in the activation of macrophage tumor cytotoxicity, FASEB J., 3:A822.

Ling, P.D., Warren, M.K., and Vogel, S.N., 1985, Antagonistic effect of interferon-β on the interferon-γ-induced expression of Ia antigen in murine macrophages, J. Immunol., 135:1857-1863.

Ozaki, Y., Edelstein, M., and Duch, D., 1987a, The actions of interferon and antiinflammatory agents on induction of indoleamine 2,3-dioxygenase in human peripheral blood monocytes, Biochem. Biophys. Res. Commun., 144: 1147-1153.

Ozaki, Y., Edelstein, M., and Duch, D., 1988a, Induction of indoleamine 2,3-dioxygenase: a mechanism of the antitumor activity of interferon-γ, Proc. Natl. Acad. Sci. USA, 85:1242-1246.

Ozaki, Y., Edelstein, M., and Duch, D., 1988b, Induction of indoleamine 2,3-dioxygenase by interferon-γ in tumor cells: potentiation by phorbol ester, J. Interferon Res., Suppl. 1, 8:S104.

Ozaki, Y., Nichol, C.A., and Duch, D., 1987b, Utilization of dihydroflavin mononucleotide and superoxide anion for the decyclization of L-tryptophan by murine epididymal indoleamine 2,3-dioxygenase, Arch. Biochem. Biophys., 257:207-216.

Ozaki, Y., Reinhard, J.F., and Nichol, C.A., 1986, Cofactor activity of dihydroflavin mononucleotide and tetrahydrobiopterin for murine epididymal indoleamine 2,3-dioxygenase, Biochem. Biophys. Res. Commun., 137:1106-1111.

Pace, J.L., Mackay, R.J., and Hayes, M.P., 1987, Suppressive effect of inter-feron-β on development of tumoricidal activity in mouse macrophages, J. Leukocyte Biol., 41:257-262.

Pfefferkorn, E., 1984, Interferon-γ blocks the growth of Toxoplasma gondii in human fibroblasts by inducing host cell to degrade tryptophan, Proc. Natl. Acad. Sci. USA, 81:908-912.

Rubin, R.Y., Anderson, S.L., Hellermann, G.R., Richardson, N.K., Lunn, R.M., and Valinsky, J.E., 1988, The development of antibody to the interferon-induced indoleamine 2,3-dioxygenase and the study of the regulation of its synthesis, J. Interferon Res., 8:691-702.

Sternberg, E.M., Trial, J., and Parker, C.W., 1986, Effect of serotonin on murine macrophages: suppression of Ia expression by serotonin and its reversal by $5HT_2$ serotoninergic receptor antagonists, J. Immunol., 137: 276-282.

Takikawa, O., Kuroiwa, T., Yamazaki, F., and Kido, R., 1988, Mechanism of interferon-γ action: characterization of indoleamine 2,3-dioxygenase in cultured human cells induced by interferon-γ and evaluation of the enzyme-mediated tryptophan degradation in its anticellular activity, J. Biol. Chem., 263:2041-2048.

van Furth, R., Raeburn, J.A., and van Zwet, T.L., 1979, Characteristics of human mononuclear phagocytes, Blood, 54:485-500.

Werner, E.R., Bitterlich, G., Fuchs, D., Hausen, A., Reibnegger, G., Szabo, G., Dierich, P.M., and Wachter, H., 1987, Human macrophages degrade tryptophan upon induction by interferon-γ, Life Sci., 41:273-280.

Yoshida, R., Murray, H.W., and Nathan, C.F., 1988, Agonist and antagonist effects of interferon α and β on activation of human macrophages: two classes of interferon γ receptors and blockade of the high-affinity sites by interferon α or β, J. Exp. Med., 167:1171-1185.

QUINOLINIC ACID CONCENTRATIONS ARE INCREASED IN CEREBROSPINAL FLUID OF RHESUS

MACAQUES (Macaca Mulatta) NATURALLY INFECTED WITH SIMIAN RETROVIRUS TYPE

M.P. Heyes[1], M. Gravell[2], M. April[3], D. Blackmore[4],
W.T. London[5], J.A. Yergey[6], and S.P. Markey[1]

[1]Section on Analytical Biochemistry
Laboratory of Clinical Science
NIMH

[2]Laboratory of Central Nervous System Studies
NINDS

[3]NIH Animal Centre
Poolesville, Maryland 20837

[4]Comparative Medicine Branch
NIEMS
Research Triangle Park, North Carolina 27709

[5]Georgetown University
Washington, District of Columbia 20057

[6]Section of Analytical Chemistry
Laboratory of Clinical Studies
NIAAA
Bethesda, Maryland 20892
USA

INTRODUCTION

Infection of macaques in captivity with simian retrovirus type-D (SRV-D) is associated with reductions in T4 cell counts, immunosuppression, neoplasms and opportunistic infections (Arthur et al., 1986; King, 1986). This disease is referred to as simian acquired immunodeficiency syndrome (SAIDS). Infection of macaques with SRV-D currently occurs naturally in the primate colony at the NIH. The symptoms of SAIDS closely resemble the clinical features of patients with acquired immune deficiency syndrome (AIDS) which is caused by infection with the human immunodeficiency virus (HIV).

We have found that the concentration of quinolinic acid (QUIN) is increased in the cerebrospinal fluid (CSF) of patients with AIDS (Heyes et al., 1989) and in the cerebral cortex of mice following systemic administration of endotoxin (Heyes et al., 1988). QUIN is an agonist of N-methyl-D-aspartate (NMDA) excitatory amino acid receptors (Perkins and Stone, 1983) when injected directly into the brains of experimental animals. Therefore, we postulated that an increase in QUIN concentrations in the brain may be involved in the neuropathology of AIDS (Heyes et al., 1989).

Kynurenine and Serotonin Pathways
Edited by R. Schwarcz et al., Plenum Press, New York, 1991

In the present study, we have measured the concentrations of QUIN in the CSF of macaques infected with SRV-D or having antibodies to SRV-D compared to non-infected controls. None of the SRV-D positive macaques had overt symptoms of SAIDS and none had evidence of opportunistic infections during the course of the study.

MATERIALS AND METHODS

Nine non-infected control macaques and 13 SRV-D positive macaques were studied at the NIH Animal Center in Poolesville, MD. Macaques were housed in individual standard laboratory primate cages. Macaques were fed monkey chow, supplementary fresh fruit and vegetables. Water was provided ad libitum. The SRV-D positive macaques had been identified as either virus positive or antibody positive to SRV-D at least 18 months prior to this study.

Two samples of cisternal CSF collected 4 to 10 weeks apart from each SRV-D positive macaque and one sample from each control macaque were collected after ketamine anesthesia. QUIN was quantified by electron capture negative chemical ionization mass spectrometry (Heyes and Markey, 1988a,b).

Results were analyzed by analysis of variance and paired and unpaired 't'-tests. Regression analysis was done using the method of least squares. A P value of < 0.05 was considered significant.

RESULTS

The concentration of QUIN in the CSF of control macaques was 20.9 ± 2.1 nmol/l. In the first sample of CSF from the SRV-D positive macaques, QUIN concentration was 72.2 ± 19.4 nmol/l (P < 0.01 compared to control macaques) and in the second follow-up sample the concentration was 99.7 ± 27.9 nmol/l (P < 0.01). There was a significant correlation between CSF QUIN concentrations at the two time points examined (r = 0.74, P < 0.005).

DISCUSSION

The results show that the concentration of QUIN in the CSF of SRV-D positive macaques is increased and maintained at either elevated or normal level within the same macaque. QUIN is formed by the catabolism of L-TRP through the kynurenine pathway (Bender, 1982), although the degree to which CSF QUIN concentrations are dependant on synthesis from L-TRP within the brain is unknown. Changes in the activity of QUIN precursors within brain could account for increased CSF QUIN concentrations in the SRV-D positive macaques.

In a model of acute infection, endotoxin administered to mice increases the activity of indoleamine-2,3 dioxygenase (IDO) in brain and other extra-hepatic organs (Takikawa et al., 1986; Yoshida et al., 1986) and is associated with increased concentrations of QUIN in brain (Heyes et al., 1988). It is conceivable that the increase in CSF QUIN concentrations in the SRV-D positive macaques reflects increased IDO activity within the brain. The increased CSF QUIN may also reflect infiltration into the brain by macrophages which increase L-TRP catabolism through the kynurenine pathway in response to interferon (Werner et al., 1987). It is likely that CSF QUIN concentrations would be further increased in both SRV-D positive macaques and AIDS patients at times of opportunistic infections (Heyes et al., 1988). In a preliminary study, Shaskan (1987) reported that the activity of 3-hydroxyanthranilate-3,4-dioxygenase in glioblastoma C6 cells in vitro was increased following acute infection with herpes simplex virus. This enzyme converts the kynurenine pathway metabolite 3-hydroxyanthranilic acid to QUIN. Therefore, it is

556

possible that the increased concentration of QUIN SRV-D positive macaques reflects increased 3-hydroxyanthranilate-3,4-dioxygenase activity.

Increased CSF QUIN concentrations in SRV-D positive macaques most likely reflect increased concentrations of QUIN in the extracellular fluid (ECF) compartment of brain. QUIN in the ECF would have access to NMDA receptors which mediate the excitatory effects of QUIN. If extracellular fluid QUIN concentrations in SRV-D positive macaques approximate the CSF concentrations, the maintained levels of QUIN in some macaques are in the range of QUIN concentrations toxic to neurons in vitro over prolonged periods (Whetsell and Schwarcz, 1989). Although brain atrophy and neuronal cell loss have been reported in extreme cases of AIDS, particularly in demented patients (Navia et al., 1986a,b), brain atrophy and neurodegeneration have not been reported in SAIDS caused by SRV-D infection. NMDA receptors are localized in cerebral cortex, basal ganglia and hippocampus. These receptors are involved in long term potentiation, neuronal plasticity, learning and memory (Morris et al., 1986). Therefore, increased concentrations of QUIN in the extracellular space could induce neuronal dysfunction by alterations in the degree of NMDA receptor activity and neuronal electrical activity. If increased CSF QUIN concentrations do have neuropathologic consequences, it may be manifested as subtle functional disturbances detectable only by quantitative learning, memory and motor tasks.

We conclude that SRV-D positive macaques offer a model to investigate the mechanism involved in the increases in QUIN in the CSF of AIDS patients. SRV-D positive macaques also offer a model to investigate the consequences of increased concentrations of QUIN in the brain. It is likely that other infectious diseases are also associated with increased concentrations of QUIN in the brain. Further studies of the role of QUIN as a neuropathogen in infectious diseases are warranted.

ACKNOWLEDGEMENT

We appreciate the careful technical assistance of B.J. Quearry, R. Sherman, D. Mackenzie, C. Belcher, C. Hatcherson and M. Der, and the useful discussions with B.J. Brew and R. Price (Memorial Sloan-Kettering Cancer Center, NY).

REFERENCES

Arthur, L.O., Gilden, R.V., Marx, P.A., and Bardner, M.B., 1986, Simian acquired immunodeficiency syndrome, Prog. Allergy, 37:332-352.

Bender, D.A., 1982, Biochemistry of tryptophan in health and disease, Mol. Asp. Med., 6:101-197.

Heyes, M.P., Kim, P., and Markey, S.P., 1988, Systemic lipopolysaccharide and pokeweed mitogen increase quinolinic acid concentrations in mouse cerebral cortex, J. Neurochem., 51:1946-1948.

Heyes, M.P., and Markey, S.P., 1988a, Quantification of quinolinic acid in rat brain, whole blood and plasma by gas chromatography and negative chemical ionization mass spectrometry: effects of systemic L-tryptophan administration on brain and blood quinolinic acid concentrations, Anal. Biochem., 174:349-359.

Heyes, M.P., and Markey, S.P., 1988b, (^{18}O)-quinolinic acid: Its esterification without back exchange for use as internal standard in the quantification of brain and CSF quinolinic acid, Biomed. Environ. Mass Spectrom., 15:291-293.

Heyes, M.P., Rubinow, D., Lane, C., and Markey, S.P., 1989, Cerebrospinal fluid quinolinic acid concentrations are increased in acquired immune deficiency syndrome, Ann. Neurol., 26:275-277.

King, N.W., 1986, Simian models of acquired immunodeficiency syndrome (AIDS): a review, Vet. Pathol., 23:345-353.

Morris, R.G.M., Anderson, E., Lynch, G.S., and Baudry, M., 1986, Selective impairment of learning and blockade of long-term potentiation by the N-methyl-D-aspartate receptor antagonist, AP5, Nature, 319:774-776.

Navia, B.A., Cho, E.-S, Petito, C.K., and Price, R.W., 1986a, The AIDS dementia complex: II. Neuropathology, Ann. Neurol., 19:525-535.

Navia, B.A., Jordan, B.D., and Price, R.W., 1986b, The AIDS dementia complex: I. Ann. Neurol., 19:517-524.

Perkins, M.N., and Stone, T.W., 1983, Pharmacology and regional variations of quinolinic acid-evoked excitations in rat central nervous system, J. Pharmacol. Exp. Ther., 226:551-557.

Shaskan, E.G., 1987, Increased 3-hydroxyanthranilic acid oxidase and decreased picolinic carboxylase activities in C6-glioblastoma cells, following serum deprivation or herpes virus infection, Soc. Neurosci. Abstr., 13:138.8.

Takikawa, O., Yoshida, R., Kido, R., and Hayaishi, O., 1986, Tryptophan degradation in mice initiated by indoleamine-2,3-dioxygenase, J. Biol. Chem., 261:3648-3653.

Werner, E.R., Bitterlich, G. Fuchs, D., Hausen, A., Reibnegger, G., Szabo, G., Dierich, M.P., and Wachter, H., 1987, Human macrophages degrade tryptophan upon induction by interferon-gamma, Life Sci., 41:273-280.

Whetsell, W.O., and Schwarcz, R., 1989, Prolonged exposure to submicromolar concentrations of quinolinic acid causes excitotoxic damage in organotypic cultures of rat corticostriatal system, Neurosci. Lett., 97:271-275.

Yoshida, R., Oku, T., Imanishi, J., Kishida, T., and Hayaishi, O., 1986, Interferon: a mediator of indoleamine-2,3-dioxygenase induction by lipopolysaccharide, Poly (I), Poly (C) and pokeweek mitogen in mouse lung, Arch. Biophys., 249:596-604.

INCREASED L-TRYPTOPHAN, 5-HYDROXYINDOLEACETIC ACID, 3-HYDROXYKYNURENINE AND QUINOLINIC ACID CONCENTRATIONS IN CEREBRAL CORTEX FOLLOWING SYSTEMIC ENDOTOXIN ADMINISTRATION

M.P. Heyes and B.J. Quearry

Section on Analytical Biochemistry
Laboratory of Clinical Science, NIMH
Bethesda, Maryland 20892
USA

INTRODUCTION

Systemic infections and administration of lipopolysaccharide (LPS) are known to increase L-tryptophan (L-TRP) release from skeletal muscle, accelerate systemic L-TRP catabolism through the kynurenine pathway (Rapoport et al., 1970; Takikawa et al., 1986), and increase the excretion of L-kynurenine and xanthurenic acid (Rapoport et al., 1970; Takikawa et al., 1986). This increase in kynurenine pathway metabolites may, in part, be secondary to the induction of indoleamine-2,3-dioxygenase (IDO), the first enzyme in the kynurenine pathway in extrahepatic tissues (Takikawa et al., 1986; Yoshida et al., 1986). Although there are numerous reports on the effects of infection and LPS on systemic L-TRP metabolism, the effects of immunologic stimulation on brain L-TRP metabolism have received little attention.

Increases in brain 5-hydroxytryptamine concentration (5-HT; Endo, 1983) and increases in cerebral cortex L-TRP and quinolinic acid (QUIN) concentrations (Heyes et al., 1988) have been reported in mice following systemic LPS administration. In addition, elevations in the concentration of QUIN in the cerebrospinal fluid of patients infected chronically with the human immunodeficiency virus have been demonstrated (Heyes et al., 1989). It is possible that changes in the concentrations of these L-TRP and kynurenine pathway metabolites during immunologic activation may have behavioral and neurologic consequences given the excitotoxic effects of QUIN and 3-hydroxykynurenine (3-HKYN) (Lapin 1982; Perkins and Stone 1983; Schwarcz et al., 1983) and the behavioral effects of brain serotonergic systems.

To further investigate the effects of systemically administered LPS on brain L-TRP, 5-HT, 5-hydroxyindoleacetic acid (5-HIAA), 3-HKYN and QUIN, C57BL6 mice were given an intraperitoneal injection of LPS from Salmonella abortus equii. Samples of serum and cerebral cortex were collected 24 h later.

METHODS

Seven mice were given an intraperitoneal (i.p.) injection of sterile saline as control and eight mice received an i.p. injection of LPS (10 μg/mouse). Mice were decapitated 24 h after the injection of LPS or saline. Trunk blood was collected and serum was isolated by centrifugation. The brain was quickly removed and chilled on a glass plate over ice. The cerebellum and brain stem were removed and the cerebral cortex peeled away from the sub-cortical structures. One cerebral cortex (80-110 mg) was placed in tared 1.5 ml polypropylene tubes and immediately frozen on dry ice. Samples were analyzed for their concentration of L-TRP, 5-HT and 5-HIAA by HPLC; QUIN was determined by gas chromatography/mass spectrometry (Heyes and Markey, 1988a,b) and 3-HYKN by HPLC (Heyes and Quearry, 1988).

Results were analyzed by one-way analysis of variance and Dunnett's 't'-test. Correlation coefficients were calculated by the method of least squares. Results are expressed as the mean ± one SEM, and tissue concentrations expressed as moles per wet weight. A P value of < 0.05 was considered significant.

RESULTS

The results (Table 1) show significant increases in the concentrations of L-TRP, 5-HIAA, QUIN and 3-HKYN in the cerebral cortex of LPS treated mice. No changes in the concentrations of 5-HT were observed. In plasma, L-TRP concentrations were reduced from 74.0 ± 6.7 nmol/ml to 57.7 ± 6.6 nmol/ml by LPS.

DISCUSSION

The results demonstrate that systemic administration of LPS from Salmonella abortus equii increases the concentrations of metabolites of the kynurenine and serotonin pathways in brain. The mechanisms responsible for these changes are likely complex and may be dependent on several interacting responses.

The reduction in plasma L-TRP concentration may, in part, reflect increased catabolism of L-TRP through the kynurenine pathway due to activation of IDO in extrahepatic tissues (Takikawa et al., 1986; Yoshida et al., 1986). Total L-TRP metabolism in extrahepatic tissues may be facilitated by decreases in hepatic tryptophan-2,3-dioxygenase (TDO) activity following LPS (Takikawa et al., 1986). Small increases in IDO activity in brain have been reported following systemic LPS administration (Takikawa et al., 1986). Therefore, it is possible that the LPS-induced increases in cortical 3-HKYN and QUIN concentrations reflect increased metabolism of L-TRP through the kynurenine pathway in brain. However, the activity of kynurenine-3-hydroxylase in brain is low (cf. Okuno and Kido, this meeting).

Table 1. L-TRP, 5-HT, 5-HIAA, QUIN and 3-HKYN in cerebral cortex of mice 24 h after treatment with sterile saline (control) or LPS (10 μg)

	L-TRP (nmol/g)	5-HT (nmol/g)	5-HIAA (nmol/g)	QUIN (pmol/g)	3-HKYN (pmol/g)
Saline	14.4 ± 0.9	4.61 ± 0.39	1.69 ± 0.17	137 ± 10	54.6 ± 5.8
LPS	20.1 ± 1.0[*]	4.46 ± 0.32	2.83 ± 0.14[***]	660 ± 147[**]	221.8 ± 43.0

[*]$P < 0.05$, [**]$P < 0.005$, [***]$P < 0.0001$ as compared to saline.

Increased cortical L-TRP concentration, at a time when plasma L-TRP concentration is reduced, may result from a disproportionately larger decrease in the plasma concentrations of other amino acids in plasma which compete with L-TRP for transport across the blood-brain barrier (Gál and Sherman, 1980; Fernstrom, 1983). A similar mechanism may also contribute to increased 3-HKYN in brain following LPS. Further studies are required to identify the source of QUIN in brain in response to LPS.

Increases in the concentration of 5-HIAA in cerebral cortex imply increased serotonergic activity. Intracerebral serotonergic systems have been implicated in a variety of behaviors including sleep, satiety and mood, all of which are disturbed during infection. In addition, neuronal effects of kynurenine pathway metabolites, including 3-HKYN and QUIN, have been demonstrated in experimental animals, including increased neuronal activity (Perkins and Stone, 1983), seizures (Lapin, 1982) and neurodegeneration (Schwarcz et al., 1983). Therefore, our results raise the possibility that changes in the concentrations of indoleamines and kynurenine pathway metabolites occur during infectious diseases and activation of the immune system, and that behavioral and neuropathologic responses result. The most dramatic increases in CSF QUIN concentrations we have observed occur in patients infected with the human immunodeficiency virus and AIDS patients with opportunistic infections. Consequently, we have postulated that QUIN and other kynurenine pathway metabolites may be involved in the neuropathology of AIDS (Heyes et al., 1989 and this volume).

REFERENCES

Endo, Y., 1983, Lipopolysaccharide and concanavalin A induce variations of serotonin levels in mouse tissues, Eur. J. Pharmacol., 91:493-499.
Fernstrom, J.D., 1983, Role of precursor availability in control of monoamine biosynthesis in brain, Physiol. Rev., 63:484-546.
Gál, E.M., and Sherman, A.D., 1980, L-kynurenine: its synthesis and possible regulatory function in brain, Neurochem. Res., 5:223-239.
Heyes, M.P., Kim, P., and Markey, S.P., 1988, Systemic lipopolysaccharide and pokeweek mitogen increase quinolinic acid concentrations in mouse cerebral cortex, J. Neurochem., 51:1946-1948.
Heyes, M.P., and Markey, S.P., 1988a, [^{18}O]-Quinolinic acid: its esterification without back exchange for use as internal standard in the quantification of brain and CSF quinolinic acid, Biomed. Environ. Mass Spectrom., 15:291-293.
Heyes, M.P., and Markey, S.P., 1988b, Quantification of quinolinic acid in rat brain, whole blood and serum by gas chromatography and negative chemical ionization mass spectrometry: effects of systemic L-tryptophan administration on brain and blood quinolinic acid concentrations, Anal. Biochem., 174:349-359.
Heyes, M.P., and Quearry, B.J., 1988, Quantification of 3-hydroxykynurenine in brain by high performance liquid chromatography and electrochemical detection, J. Chromat., 428:340-344.
Heyes, M.P., Rubinow, D., Lane, C., and Markey, S.P., 1989, Cerebrospinal fluid quinolinic acid concentrations are increased in acquired immune deficiency syndrome, Ann. Neurol., 26:275-277.
Lapin, I.P., 1982, Convulsant action of intracerebroventricularly administered L-kynurenine sulphate, quinolinic acid and other derivatives of succinic acid, and effects of amino acids: structure-activity relationships, Neuropharmacology, 21:1227-1233.
Perkins, M.N., and Stone, T.W., 1983, Pharmacology and regional variations of quinolinic acid-evoked excitations in rat central nervous system, J. Pharmacol. Exp. Ther., 226:551-557.
Rapoport, M.I., Beisel, W.R., and Hornick, R.B., 1970, Tryptophan metabolism during infectious illness in man, J. Infect. Dis., 122:159-169.

Schwarcz, R., Whetsell, W.O., and Mangano, R.M., 1983, Quinolinic acid: an endogenous metabolite that produces axon sparing lesions in rat brain, Science, 219:316-318.

Takikawa, O., Yoshida, R., Kido, R., and Hayaishi, O., 1986, Tryptophan degradation in mice initiated by indoleamine-2,3-dioxygenase, J. Biol. Chem., 261:3648-3653.

Yoshida, R., Oku, T., Imanishi, J., Kishida, T., and Hayaishi, O., 1986, Interferon: a mediator of indoleamine 2,3-dioxygenase induction by lipopolysaccharide, Poly(I), Poly(C), and pokeweed mitogen in mouse lung, Arch. Biophys., 249:596-604.

NON-COMPETITIVE INHIBITION OF 3-HYDROXYANTHRANILATE-3,4-DIOXYGENASE BY

4-CHLORO-3-HYDROXYANTHRANILIC ACID IN WHOLE BRAIN OF RAT

X.-D. Ji, M. Nishimura, and M.P. Heyes

Section on Analytical Biochemistry
Laboratory of Mental Health
Bethesda, Maryland 20892
USA

INTRODUCTION

The catabolism of L-tryptophan through the kynurenine pathway has long been recognized as the predominant route for systemic L-tryptophan metabolism and synthesis of nicotinamide containing nucleotides in mammalian species (see Bender et al., 1982). The potential importance of this pathway in brain is highlighted by the fact that many intermediates and products of the kynurenine pathway, including quinolinic acid (QUIN), are neuroactive (Lapin, 1982; Perkins and Stone, 1983; Schwarcz et al., 1983).

QUIN is produced by a non-enzymatic reaction from 2-amino-3-carboxymuconic acid semialdehyde, which itself is formed from 3-hydroxyanthranilic acid (3-HAA) by an oxidation reaction catalyzed by 3-hydroxyanthranilate-3,4-dioxygenase (3-HAD; Foster et al., 1986). This enzyme appears to be localized within glial cells in brain (Okuno et al., 1987). We have found that the synthesis of QUIN in rat brain in vivo from 3-HAA is attentuated by a co-administration of 4-chloro-3-hydroxyanthranilate (Cl-HAA). This exogenous compound was first recognized as an inhibitor of 3-HAD by Parli et al. (1980), who demonstrated that systemic administration of Cl-HAA to rodents resulted in a marked increase in the exretion of 3-hydroxyanthranilate (3-HAA) in the urine. In vitro studies using Lineweaver-Burke plot analysis of hepatic 3-HAD activity indicated that Cl-HAA was a non-competitive inhibitor of 3-HAD.

In the present study, the inhibitory effects of Cl-HAA on the conversion of 3-HAA to QUIN in aqueous extracts of whole brain were investigated and the Lineweaver-Burke plot characteristics of the reaction were studied.

MATERIALS AND METHODS

Cl-HAA was a gift from Dr. C. J. Parli (Lilly Research Laboratories, Indianapolis, IN) and $[^{14}C]$-3-HAA was a gift from Dr. E. Shaskan (University of Connecticut, Farmington, CT). 3-HAD activity was quantified by the method of Okuno et al. (1987) except that ascorbate was added to the incubation buffer. The reaction was terminated by the addition of 1 ml of 1 mol/l HCl containing 150 pmol/ml of $[^{18}O]$-QUIN as internal standard. QUIN was quantified

Kynurenine and Serotonin Pathways
Edited by R. Schwarcz et al., Plenum Press, New York, 1991

by electron capture negative chemical ionization mass spectrometry (Heyes and Markey, 1988a,b).

Lineweaver-Burke plots were generated by regression analysis using the method of least squares.

RESULTS

Incubation of 5 mg samples of whole brain with 325 pmol of 3-HAA resulted in the generation of a peak on GC/MS which had the same time of elution as authentic QUIN when monitored at m/z 467 and m/z 316. When [^{14}C]-3-HAA was substituted for 3-HAA, no ion current was found at m/z 467, other than endogenous QUIN as quantified in blanks. However, large ion currents were detected at m/z 469 and 318, corresponding to the synthesis of [^{14}C]-QUIN, confirmed by comparison to authentic [^{14}C]-QUIN.

Small ion current peaks at m/z 467 corresponding to endogenous QUIN were found in boiled blanks. In un-boiled blanks incubated without exogenous 3-HAA, background QUIN levels were larger than the boiled blanks, indicating synthesis of QUIN from endogenous substrate. However, the amounts of QUIN in blanks were substantially smaller than the amounts of QUIN synthesized from exogenous 3-HAA in the assay.

Lineweaver-Burke plot analysis. Cl-HAA inhibited whole brain 3-HAD activity in a concentration dependent manner. The inhibition produced Lineweaver-Burke plots consistent with non-competitive inhibition.

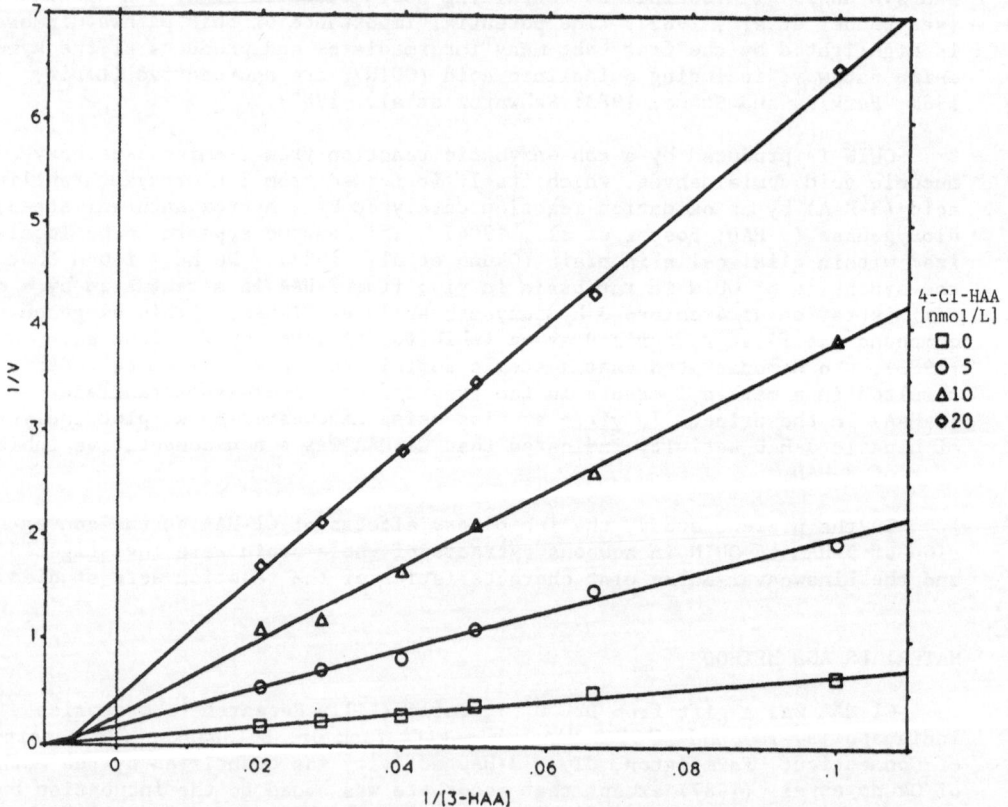

Fig. 1. Lineweaver-Burke plots of 4-chloro-3-hydroxyanthranilate
inhibition of 3-hydroxyanthranilate-3,4-dioxygenase.

DISCUSSION

The present results confirm our previous observation that Cl-HAA inhibits the synthesis of QUIN from 3-HAA (and [^{14}C]-3-HAA) in rat brain (Heyes et al., 1988a). In our in vitro studies, the values obtained for K_m (1 to 4 μmol/l) and V_{max} (77.8 to 186.7 pmol QUIN/h/mg) are very similar to those reported by Foster et al. (1986; K_m = 3.5 ± 0.5 μmol/l; V_{max} 73.7 ± 9.5 pmol QUIN/h/mg). The fact that QUIN formation from 3-HAA is precursor sensitive (Heyes et al., 1988) indicates that the rate of QUIN synthesis in brain can be influenced by the concentration of 3-HAA.

The Lineweaver-Burke plot analysis of the inhibitory effects of Cl-HAA on regional brain 3-HAD activity is consistent with a non-competitive form of inhibition. This conclusion is consistent with the observations of Parli et al. (1980), who showed that in liver Cl-HAA was also a non-competitive inhibitor of 3-HAD, and of Okuno et al. (1987), who found evidence that brain and liver 3-HAD may be the same protein.

ACKNOWLEDGEMENT

We appreciate the careful technical assistance of J.A. Yergey and M. Der.

REFERENCES

Bender, D.A., 1982, Biochemistry of tryptophan in health and disease, Mol. Aspects Med., 6:101-197.

Foster, A.C., White, R.J., and Schwarcz, R., 1986, Synthesis of quinolinic acid by 3-hydroxyanthranilic acid oxygenase in rat brain tissue in vitro, J. Neurochem., 47:23-30.

Heyes, M.P., Hutto, B., and Markey, S.P., 1988, 4-Chloro-3-hydroxyanthranilic acid inhibits 3-hydroxyanthranilic acid oxidase in brain, Neurochem. Int., 13:405.

Heyes, M.P., and Markey, S.P., 1988a, [^{18}O]Quinolinic acid: its esterification without back exchange for use as internal standard in the quantification of brain and CSF quinolinic acid, Biomed. Environ. Mass Spectrom., 15:291-293.

Heyes, M.P., and Markey, S.P., 1988b, Quantification of quinolinic acid in rat brain, whole blood and serum by gas chromatography and negative chemical ionization mass spectrometry: effects of systemic L-tryptophan administration on brain and blood quinolinic acid concentrations, Anal. Biochem., 174:349-359.

Lapin, I.P., 1982, Convulsant action of intracerebroventricularly administered L-kynurenine sulphate, quinolinic acid and other derivatives of succinic acid, and effects of amino acids: structure-activity relationships, Neuropharmacology, 21:1227-1233.

Okuno, E., Köhler, C., and Schwarcz, R., 1987, Rat 3-hydroxyanthranilic acid oxygenase: purification from the liver and immunocytochemical localization in the brain. J. Neurochem., 49:771-780.

Parli, C.J., Krieter, P., and Schmidt, B., 1980, Metabolism of 6-chlorotryptophan to 4-chloro-3-hydroxyanthranilic acid: a potent inhibitor of 3-hydroxyanthranilic acid oxidase, Arch. Biochem. Biophys., 203:161-166.

Perkins, M.N., and Stone, T.W., 1983, Pharmacology and regional variations of quinolinic acid evoked excitations in rat central nervous system, J. Pharmacol. Exp. Ther., 226:551-557.

Schwarcz, R., Whetsell, W.O., and Mangano, R.M., 1983, Quinolinic acid: an endogenous metabolite that produces axon-sparing lesions in rat brain, Science, 219:316-318.

KYNURENINE-PYRUVATE AMINOTRANSFERASE IN RAT KIDNEY AND BRAIN

T. Ishikawa, E. Okuno, M. Tsujimoto, M. Nakamura, and R. Kido

Wakayama Medical College
Dept. of Biochemistry
Wakayama 640
Japan

INTRODUCTION

Kynurenic acid, the product of kynurenine aminotransferase, has increasingly attracted attention because of its physiological effects in the brain, such as blockade of the neurotoxicity induced by quinolinic acid in rat (Foster et al., 1984) or decrease of the stroke size and deficits in human (Germano et al., 1987). Also, it can protect neonatal rats from hypoxia-induced brain edema (Simon et al., 1986).

It has been reported that rat brain contains two kinds of kynurenine aminotransferases. One is kynurenine-pyruvate aminotransferase (Minatogawa et al., 1974) and another is kynurenine-2-oxoglutarate aminotransferase, which is identical with mitochondrial aspartate aminotransferase (Noguchi et al., 1975). Kynurenine-2-oxo-glutarate aminotransferase in rat brain has a very high K_m value for kynurenine (unpublished data), and the activity is strongly inhibited by the addition of aspartate or aromatic amino acids (Noguchi et al., 1975). In rat liver and kidney, there is another kynurenine-2-oxo-glutarate aminotransferase, which is identical with L-α-aminoadipate aminotransferase (Tobes and Mason, 1975, 1977). Kynurenine-pyruvate aminotransferase in kidney is also identical with histidine-pyruvate (glutamate-phenylpyruvate) aminotransferase (Noguchi and Kido, 1976). The kynurenic acid concentration in rat brain, (14 pmol/g wet weight) has been measured (Carlà et al., 1988). In spite of the physiological importance of kynurenic acid in brain, the K_m value of kynurenine-pyruvate aminotransferase for kynurenine was reported as 14 mM (Minatogawa et al., 1974) or 1.4 mM (Noguchi et al., 1975). This difference might be caused by impurity of enzyme.

We describe here the properties of pure kynurenine-pyruvate aminotransferase and the identification of kidney and brain enzymes.

METHODS

Purification of kynurenine-pyruvate aminotransferase

The purification procedure is shown in Table 1. Rat kidney (250 g) was homogenized with 5 vol. (w/v) of 0.15 M KCl containing 50 mM pyridoxal 5'-phosphate and 10 mM 2-mercaptoethanol in a teflon/glass homogenizer. The

homogenate was centrifuged at 10,000 x g for 10 min, and the pellet was discarded. After heat treatment (60°C, 1 min) and ammonium sulfate (AS) fractionation from 40% to 70% saturation, the enzyme fraction (pellet) was dissolved in 5 mM potassium phosphate buffer, pH 7.5 (buffer A). All buffers
contained 50 mM pyridoxal phosphate and 10 mM 2-mercaptoethanol unless state
otherwise. The solution was heat-treated for a second time (65°C, 1 min) an
centrifuged. The supernatant was desalted by gel filtration (Sephadex G-25
column equilibrated with buffer A). The sample was then applied to a DEAE-
Sepharose column (2.5 x 15 cm) and eluted by a linear gradient of 5 - 250 mM
potassium phosphate buffer without pyridoxal phosphate (500 ml each). The
active fractions were concentrated by AS, desalted and applied onto a hydro-
xyapatite column (2.5 x 7 cm) equilibrated with buffer A. The enzyme was
eluted by a linear gradient of buffer A and 300 mM potassium buffer, pH 7.5
(200 ml each). AS (20%) was added to the active fraction, and the fraction
was applied to a phenyl-Sepharose column equilibrated with a 20% saturated A:
solution. The enzyme was eluted by a linear gradient of 20% saturated AS
solution and buffer A (100 ml each).

The active fraction was concentrated by AS, resuspended in buffer A and
applied to a Sephacryl S-200 column (2.5 x 100 cm) equilibrated with 100 mM
potassium phosphate buffer. The enzyme was eluted by the same buffer. The
purified enzyme was obtained after concentration using a collodion bag in
vacuo.

Preparation of antibody

Anti-rat kidney kynurenine-pyruvate aminotransferase antibodies were
obtained using 3 mg of purified enzyme and "adjuvant" injected into a white
New Zealand rabbit. The antibody from rabbit serum was purified (Okuno et
al., 1987) using AS fractionation, DEAE-Sephacel and protein A-Sepharose.

Table 1. Purification of kynurenine-pyruvate aminotransferase from rat
kidney

	Vol. (ml)	Protein (mg/ml)	Activity (μmol/h/ml)	Specific activity (μmol/h/mg)	Purification (fold)	Recovery (%)
Super-natant	1150	27.2	1.44	0.053	1	100
Heat	940	14.8	1.17	0.079	1.2	66
AS	66	47.1	16.13	0.34	6.4	64
2nd heat	50	46.4	15.41	0.33	6.2	47
DEAE Sepharose	57	7.8	13.43	1.73	32.6	43
Hydroxy-apatite	6.2	12.1	98.28	8.12	153	37
Phe-Sepharose	4.4	10.5	118.1	11.25	212	31
Sephacryl S-200	2.1	9.3	172.4	18.54	350	22

Immunoblot

The immunoblot was performed using anti-rat kidney kynurenine-pyruvate aminotransferase, purified rat kidney and brain kynurenine-pyruvate amino-transferases (Towbin et al., 1979) using a Bio-Rad immuno-blot assay kit.

All other methods have been described previously (Okuno et al., 1980).

RESULTS

Table 1 shows that a 350-fold purification of kynurenine-pyruvate amino-transferase, with a yield of 22%, was obtained after the final purification step. On polyacrylamide gel electrophoresis, the purified enzyme migrated toward the anode as a single protein band (mobility 0.43), which coincided with kynurenine-pyruvate aminotransferase, glutamine aminotransferase and phenylalanine-pyruvate aminotransferase activities. The result of polyacrylamide gel electrophoresis of the purified enzyme with sodium dodecyl sulfate yielded a single band, corresponding to a molecular weight of 48,000.

Characteristics of kidney and brain enzyme

The influence of pH on kynurenine-pyruvate aminotransferase activity of kidney enzyme was investigated over the range pH 6.5 - 9.2 and revealed pH optima of 8.0 - 8.3. The brain enzyme showed the same profile of pH optima (data not shown).

The Michaelis constants for the kynurenine aminotransferase activities were measured. Absolute K_m values for kynurenine-pyruvate aminotransferase were 2.8 mM for kynurenine and 3.8 mM for pyruvate, respectively.

Substrate specificity

The purified enzyme showed broad specificity for 2-oxo acids with kynurenine (Table 2). The specificity for various L-amino acids, using phenylpyruvate or pyruvate on kidney and brain kynurenine-pyruvate amino-transferase, is shown in Table 3.

Table 2. Relative activities of transamination between L-kynurenine and various 2-oxo acids of rat kidney and brain enzymes

2-Oxo acid	Kidney KPT activity	Brain KPT activity
	%	%
Pyruvate	100	100
2-Oxo n-valerate	117	120
2-Oxo n-caproate	98	96
2-Oxo iso-caproate	95	94
Oxaloacetate	91	72
2-Oxobutyrate	84	94
2-Oxo-4-methylthiobutyrate	65	71
Glyoxylate	27	35
2-Oxoadipate	11	4
Phenylpyruvate	14	14
2-Oxo iso-valerate	4	3
2-Oxoglutarate	4	2

Table 3. Substrate specificity of rat kidney and brain kynurenine aminotransferase for various L-amino acids with phenyl-pyruvate and pyruvate

Amino acid	Phenylpyruvate		Pyruvate	
	Kidney	(Brain)	Kidney	(Brain)
Met	169	(158)		
Gln	100	(100)		
Leu	8.1	(---)		
Arg	6.0	(---)		
Asn	2.2	(---)		
Orn	1.5	(---)		
Asp	1.5	(---)		
Val	0.8	(---)		
Thr	0.5	(---)		
Ala	N.D.	(---)		
Ser	N.D.	(---)		
Glu	N.D.	(---)		
Ile	N.D.	(---)		
Cys	N.D.	(---)		
Phe	---	(---)	1.2	(5.4)
Tyr	---	(---)	5.9	(---)
His	---	(---)	5.6	(11.5)
Trp	---	(---)	1.1	(---)
Kyn	0.8	(---)	5.8	(6.0)
3-OHkyn	---	(---)	N.D.	(---)

Inhibition

Carbonyl reagents (1 mM) were used as inhibitors: Hydrazine, semicarbazide and hydroxylamine inhibited the enzyme activity 10 - 20%. Potassium cyanide inhibited about half of the activity. p-Chloromercuribenzoate (PCMB) completely blocked the activity, and addition of 5 mM mercaptoethanol restored about 85% of the activity (Table 4).

Immunological tests

The purified kidney enzyme was injected into a rabbit to obtain anti-rat kidney kynurenine-pyruvate aminotransferase. The purified antibodies were used for immunoprecipitation tests. After mixing kidney or brain enzyme with the IgG fraction, Protein A was added and centrifuged. The activities in the supernatant and precipitate were measured (Fig. 1). Fig. 2 shows the results of Western blotting on kidney and brain kynurenine aminotransferases using anti-kidney antibodies.

CONCLUSION

Kynurenine-aminotransferase from rat kidney was purified to homogeneity, as judged by polyacrylamide gel electrophoresis.

Identity of the enzyme from rat kidney and brain was proven by biochemical, physico-chemical and immunological techniques.

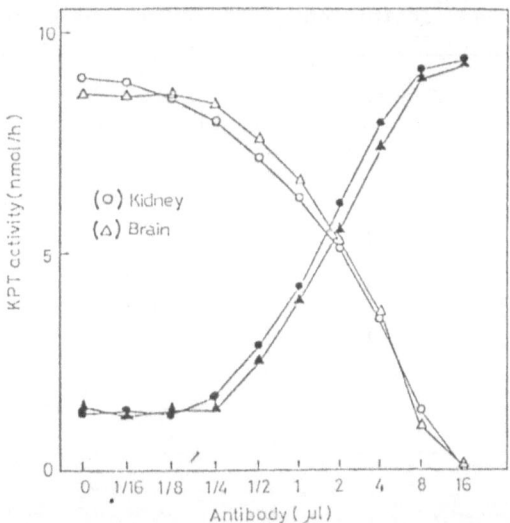

Fig. 1. Immunoprecipitation of rat kidney and brain kynurenine amino·
transferase by anti-kidney kynurenine aminotransferase.

The optimum pH of purified enzyme was obtained between 8.0 and 8.3 for
both kidney and brain enzymes. Kinetic analysis revealed absolute K_m
values of 2.8 mM and 3.8 mM for the substrates, kynurenine and pyruvate,
respectively. The order of effectiveness of 2-oxo acids was 2-oxo n-vale-
rate > pyruvate > 2-oxo n-caproate > 2-oxo iso-caproate > oxalo-acetate >
2-oxo butyrate > 2-oxo 4-methylthiobutyrate, with kynurenine as the amino
donor.

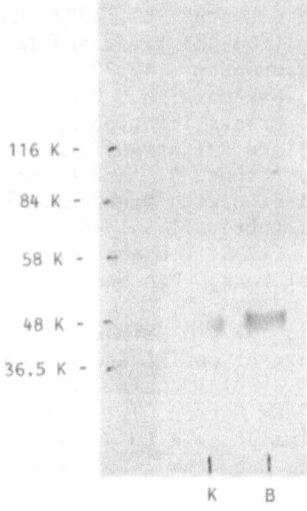

Fig. 2. Western blotting of rat kidney and brain kynurenine aminotransferase
using anti-kidney enzyme.

Table 4. Inhibition of KPT by various compounds

	Concentration (mM)	Kidney KPT activity	Brain KPT activity
		%	%
None	1	100	100
Hydrazine	1	84.3	86.6
Semicarbazide	1	87.9	91.6
Hydroxylamine	1	80.9	77.7
KCN	1	43.8	48.0
PCMB	1	0	0
PCMB + mercaptoethanol	1 5	85.4	87.2

REFERENCES

Carlà, V., Lombardi, G., Beni, M., Russi, P., Moneti, G., and Moroni, F., 1988, Identification and measurement of kynurenic acid in the rat brain and other organs, Anal. Biochem., 169:89-94.

Foster, A.C., Vezzani, A., French, E.D., and Schwarcz, R., 1984, Kynurenic acid blocks neurotoxicity and seizures induced in rats by the related brain metabolite quinolinic acid, Neurosci. Lett., 48:273-278.

Germano, I.M., Pitts, L.H., Meldrum, B.S., Bartkowski, H.M., and Simon, R.P., 1987, Kynurenate inhibition of cell excitation decreases stroke size and deficits, Ann. Neurol., 22:730-734.

Minatogawa, Y., Noguchi, T., and Kido, R., 1974, Kynurenine pyruvate trans-aminase in rat brain, J. Neurochem., 23:271-272.

Noguchi, T., Minatogawa, Y., Okuno, E., Nakatani, M., Morimoto, M., and Kido, R., 1975, Purification and characterization of kynurenine-2-oxoglutarate aminotransferase from the liver, brain and small intestine of rats, Biochem. J., 151:399-406.

Noguchi, T., and Kido, R., 1976, Identity of kynurenine:pyruvate aminotrans-ferase with histidine:pyruvate aminotransferase, Hoppe-Seyler's Z. Physiol. Chem., 357:649-656.

Okuno, E., Köhler, C., and Schwarcz, R., 1987, Rat 3-hydroxyanthranilic acid oxygenase: purification from the liver and immunocytochemical localiza-tion in the brain, J. Neurochem., 49:771-780.

Okuno, E., Minatogawa, Y., Nakamura, M., Kamoda, N., Nakanishi, J., Makino, M., and Kido, R., 1980, Crystallization and characterization of human liver kynurenine-glyoxylate aminotransferase, Biochem. J., 189:581-590.

Simon, R.P., Young, R.S.K., Stont, S., and Cheng, S., 1986, Inhibition of excitatory neurotransmission with kynurenate reduces brain edema in neo-natal anoxia, Neurosci. Lett., 71:361-364.

Tobes, M.C., and Mason, M., 1975, L-kynurenine aminotransferase and L-α-aminoadipate aminotransferase, Biochem. Biophys. Res. Commun., 62:390-397.

Tobes, M.C., and Mason, M., 1977, α-Aminoadipate aminotransferase and kynurenine aminotransferase: purification, characterization and further evidence for identity, J. Biol. Chem., 252:4591-4599.

Towbin, H., Staehlin, T., and Gordon, J., 1979, Electrophoretic transfer of proteins from polyacrylamide gels to nitrocellulose sects: procedure and some applications, Proc. Nat. Acad. Sci. USA, 76:4350-4354.

ENZYMATIC SYNTHESIS OF PAPILIOCHROME II

Y. Umebachi and T. Yokoyama

Department of Biology
Faculty of Science
Kanazawa University
Kanazawa 920
Japan

INTRODUCTION

Papiliochrome II is the pale yellow pigment in the wings of Papilio butterflies. The structure has been reported to be N^{ar}-[α-(3-amino-propionyl-
aminomethyl)-3,4-dihydroxybenzyl]-L-kynurenine, which consists of one molecule each of L-kynurenine and N-β-alanyldopamine (NBAD) (Rembold and Umebachi, 1984). There are two kinds of isomers with different optical activity, called IIA and IIb. Both isomers readily decompose to L-kynurenine and
N-β-alanylnorepinephrine (NBANE) by being heated in 10^{-2} N HCl at 100°C for 2
hr. The latter compound decomposes further to β-alanine and ($\pm$)-norepine-
phrine by being heated in 1 N HCl at 100°C for 2 hr (Umebachi and Yamashita,
1977; Umebachi, 1985). The structure of this pigment is interesting in that
the aromatic amino group of kynurenine is bonded to the β-carbon of the side
chain of catecholamine.

Changes in the levels of kynurenine, β-alanine, dopamine and NBAD in the
haemolymph and wings during the pupal stage of Papilio xuthus have been investigated in detail. It has been suggested that free kynurenine and free
NBAD enter the wings (or scale cells) through haemolymph and there combine
with each other to form Papiliochrome II (Ishizaki and Umebachi, 1988).

In 1987, Yago et al. reported that Papiliochrome IIa and IIb could be
enzymatically synthesized from kynurenine and NBAD by phenoloxidase which was
isolated from the collaterial glands of praying mantis (Yago et al., 1987).
We have investigated this enzyme reaction with tyrosinase and confirmed that
the IIa and IIb are indeed synthesized by the enzyme.

METHODS

The reaction mixture contained 0.3 mol L-kynurenine, 0.3 mol NBAD, and
150 units mushroom tyrosinase in 5 ml phosphate buffer (pH 6.0 - 6.6). In
some experiments, N-acetyldopamine (NADA) was used instead of NBAD. Incubation was conducted at 25°C for 5-120 min. Kynurenine, NADA and tyrosinase
were obtained from Sigma Chemical Co. NBAD was chemically synthesized from
N-t-BOC-β-alanine-p-nitrophenylester and dopamine in our laboratory.

Kynurenine and Serotonin Pathways
Edited by R. Schwarcz et al., Plenum Press, New York, 1991

Thin-layer chromatography (TLC) was performed on cellulose plate (Merck No. 5716, 20 x 20 cm). The solvent for one-dimensional TLC was a mixture of n-butanol, water and acetic acid (BAW)(120:50:30). For two-dimensional TLC, the first solvent was 70% methanol, and the second, BAW. After the development, the phosphomolybdic acid-ammonia test for phenols or the ninhydrin reaction was performed. Amino acid analysis, based on o-phthalaldehyde method, was carried out with a Shimazu LC-4 HPLC amino acid analyzer.

RESULTS

Absorption spectra of reaction mixture

An aliquot of the reaction mixture was taken at different intervals of time, and the absorption spectrum was taken. The results are given in Fig. 1a (kyn. + NBAD + tyrosinase) and Fig. 1b (kyn. + NADA + tyrosinase). The figures show that the absorption peak of kynurenine at 360 nm shifts to about 370 nm. A similar shift was not seen in the incubations of (1) kynurenine plus tyrosinase, (2) NBAD (or NADA) plus tyrosinase, and (3) kynurenine, NBAD (or NADA) plus heat-pretreated tyrosinase, nor in the case of addition of kynurenine after the NBAD (or NADA) plus tyrosinase incubation. These facts indicate that the enzyme reaction is involved in the shift.

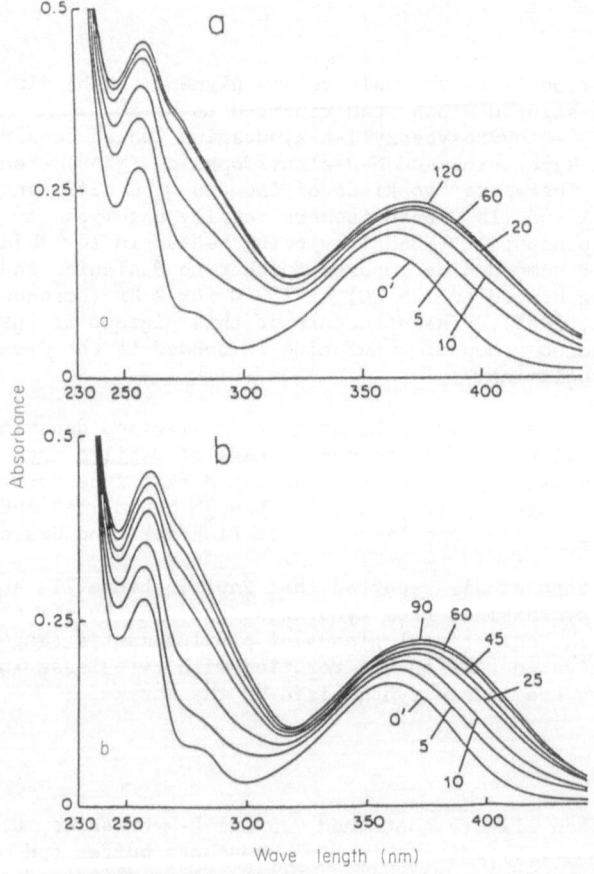

Fig. 1. Absorption spectra of the reaction mixture.
(a) kynurenine, NBAD, and tyrosinase. (b) kynurenine, NADA, and tyrosinase.

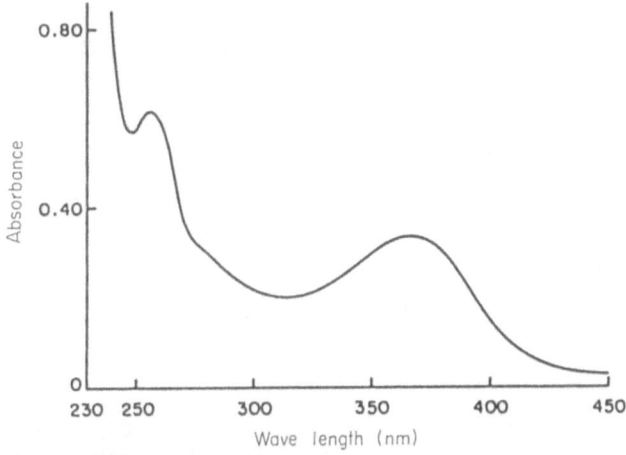

Fig. 2. TLC of the reaction-products. (a) and (b): see Fig. 1.
K: kynurenine.

Reaction products

     After a 2 hr incubation, the reaction mixture was submitted to one-
dimensional TLC.   The chromatograms are given in Figs. 2a and 2b.   In the
incubation of kynurenine, NBAD and tyrosinase (Fig. 2a), the main products
were IIa and IIb, which had pale yellow fluorescence and were positive in
both the phosphomolybdic acid-ammonia test and the ninhydrin reaction.   In
the case of kynurenine, NADA, and tyrosinase (Fig. 2b), the main product was
"D", which also had pale yellow fluorescence and was positive in the phenol
test but negative in the ninhydrin reaction.   The D component could be
separated to two spots in two-dimensional TLC.

     The IIa and IIb products in Fig. 2a were scraped, respectively, and ex-
tracted with 70% ethanol.   After lyophilization, the sample was dissolved in
phosphate buffer (pH 6.5) and an absorption spectrum was taken.   The spectrum, given in Fig. 3, was very similar to that of natural Papiliochrome II
(Umebachi, 1985).

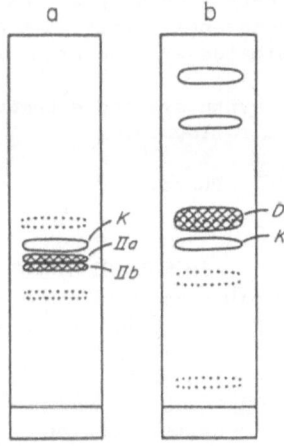

Fig. 3.   Absorption spectrum of the synthesized Papiliochrome II.

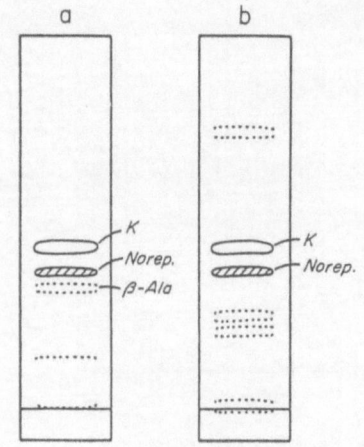

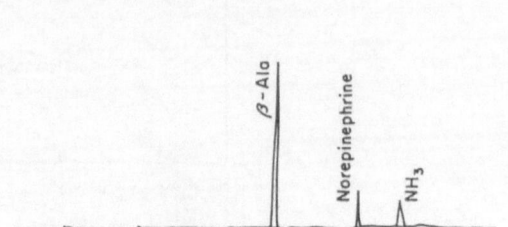

Fig. 4. Decomposition of the reaction products. (a) and (b), see Fig. 1.

Fig. 5. Amino acid analysis of the degradation products from the synthesized Papilio-chrome II.

Decomposition of the reaction products

The IIa and IIb spots of Fig. 2a and the D product of Fig. 2b were scraped and extracted with 70% ethanol. The extract was lyophilyzed and hy-drolyzed in 1 N HCl at 106°C for 4 hr. The hydrolysate was submitted to TLC and amino acid analysis. The results indicated the presence of kynurenine, β-alanine, and norepinephrine in the case of IIa and IIb (Fig. 4a and Fig. 5). On the other hand, in the case of "D", kynurenine and norepinephrine were found (Fig. 4b).

DISCUSSION AND CONCLUSION

It was confirmed that the products of Papiliochrome-type are formed from L-kynurenine and NBAD (or NADA) by phenoloxidase. In other words, the aroma-tic amino group of kynurenine is bonded to the β-carbon of the side chain of NBAD (or NADA). This fact is interesting in that the link between aromatic amine and catecholamine is catalyzed by phenoloxidase. The reaction mechan-ism may be explained by Sugumaran's theory that quinone methide is formed from catecholamine by phenoloxidase (Sugumaran, 1988).

Finally, the enzymatic synthesis and decomposition described in the pre-sent paper are summarized as follows:

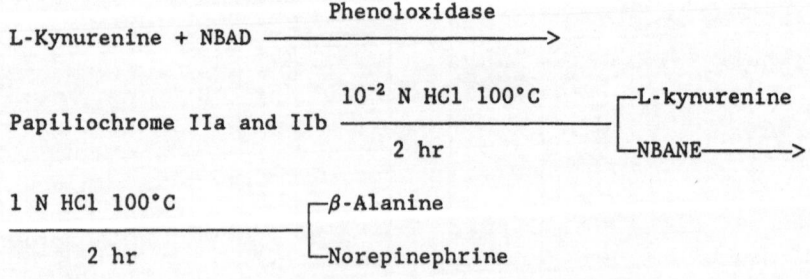

REFERENCES

Ishizaki, Y., and Umebachi, Y., 1988, Level changes of β-alanine, dopamine, and N-β-alanyldopamine during the pupal stage of Papilio xuthus (Lepidoptera: Papilionidae), Comp. Biochem. Physiol., 90C:83-87.

Rembold, H., and Umebachi, Y., 1984, The structure of Papiliochrome II, the yellow wing pigment of the papilionid butterflies, in: "Progress in Tryptophan and Serotonin Research", Schlossberger, H.G., Kochen, W., Linzen, B., and Steinhart, H., eds., de Gruyter, Berlin, pp. 743-746.

Sugumaran, M., 1988, Molecular mechanism for cuticular sclerotization, in: "Advances in Insect Physiology", Vol. 21, Evans, P.D., and Wigglesworth, V.B., eds., Academic Press, New York, pp. 179-231.

Umebachi, Y., 1985, Papiliochrome, a new pigment group of butterfly, Zool. Sci., 2:163-174.

Umebachi, Y., and Yamashita, H., 1977, Beta-alanine as a constituent of the dopamine derivative, SN-1 of Papilio xuthus, Comp. Biochem. Physiol., 56B:5-8.

Yago, M., Sato, H., and Kawasaki, H., 1987, Synthesis of Papiliochrome II by phenoloxidase from praying mantis, Zool. Sci., 4:1010.

4-HALO-3-HYDROXYANTHRANILATES ARE POTENT INHIBITORS OF 3-HYDROXYANTHRANILATE

OXYGENASE IN THE RAT BRAIN IN VITRO AND IN VIVO

J.L. Walsh, W.P. Todd[1], B.K. Carpenter[1], and R. Schwarcz

Maryland Psychiatric Research Center
Baltimore, Maryland 21228

[1]Cornell University
Dept. Chemistry
Baker Laboratory
Ithaca, New York 14853
USA

INTRODUCTION

   3-Hydroxyanthranilic acid oxygenase (3HAO; E.C. 1.13.11.6) is the enzyme responsible for the metabolism of 3-hydroxyanthranilic acid (3HANA) to
α-amino-β-carboxymuconic acid-ω-semialdehyde, which in turn spontaneously rearranges to form the endogenous excitotoxin quinolinic acid (QUIN). Since
increased levels of QUIN can trigger the structural deterioration of nerve
cells, it has been implicated in the etiology of neurodegenerative diseases
(Schwarcz et al., 1984). It follows that specific inhibitors of 3HAO may be
of therapeutic value in diseases which can be traced to an overabundance of
QUIN. 4-Chloro-3-hydroxyanthranilic acid (4-Cl-3HANA), originally described
by Parli et al. (1980), is a very effective blocker of 3HAO, and therefore
can serve as a lead compound in the development of novel 3HAO inhibitors.
The present study was designed to compare the inhibitory potency of 4-Cl-
3HANA with that of 4-Br- and 4-F-3HANA in cell free homogenates of rat brain,
and to evaluate the effectiveness of 3HAO inhibition in vivo by microdialysis using the most potent compound discovered so far (4-Br-3HANA).

MATERIALS AND METHODS

Materials

   4-F-, 4-Cl- and 4-Br-3HANA were synthesized as described previously
(Todd et al., 1989). All other chemicals were obtained from commercial
sources.

Tissue preparation and in vitro assay

   Male Sprague-Dawley rats (approx. 250 g) were killed by decapitation,
their brains were rapidly removed, and their forebrains dissected on ice.
The tissue was sonicated in four volumes (w/v) of distilled water, and the
homogenate centrifuged at 50,000 x g for 20 min at 4°C. Measurement of 3HAO
activity was performed on the supernate in the presence or absence of enzyme

inhibitors as described by Foster et al. (1986b). The assay mixture contained 20 μl of tissue homogenate, 0.3 mM $FeSO_4$, 38 mM HEPES, pH 6.0, and 5 μM [1-carboxy-^{14}C]3HANA, in a final volume of 195 μl. Inhibitors were added to the incubation mixture immediately prior to the addition of ^{14}C-3HANA. Boiled tissue was used to obtain blank values.

In vivo microdialysis

Microdialysis was performed in the hippocampus of unanesthetized animals as described in detail previously (Speciale et al., 1989). Briefly, a guide cannula (outer diameter: 0.65 mm) was implanted under chloral hydrate anesthesia over the dorsal hippocampus, protruding < 0.1 mm into the brain. On the next day, a microdialysis probe with a 2 mm dialysis section (CMA 10, Carnegie Medicin, Stockholm, Sweden) was inserted into the hippocampus through the guide. Using a microdialysis pump, Krebs-Ringer buffer, pH 7.3, containing 500 μM 3HANA was perfused through the probe for 4 hours at a speed of 1 μl/min. In some animals, the perfusion solution was then switched to a solution containing, in addition, 500 μM 4-Br-3HANA; the perfusion was continued for another 4 hours.

QUIN was measured in the dialysate using the radioenzymatic method of Foster et al. (1986a).

RESULTS and DISCUSSION

As shown in Fig. 1, the inhibitors demonstrated the following order of potency: 4-Br-3HANA > 4-Cl-3HANA > 4-F-3HANA. The IC_{50} values for all three compounds were found to be in the nanomolar range. Thus, measurement of the production of QUIN, which is not the immediate reaction product of 3HAO, yielded virtually identical data as the spectrophotometric determination of α-amino-β-carboxymuconic-ω-semialdehyde production in the presence of the inhibitors (Walsh et al., 1989). Kinetic analysis, performed during the first seconds of the enzymatic reaction, demonstrated that the tightly binding 3HAO inhibitors block 3HAO competitively, and subsequent further in vitro

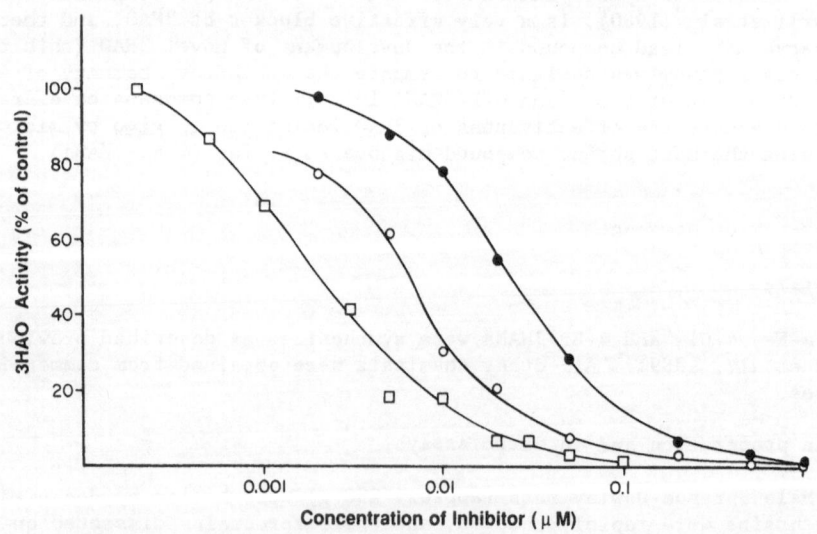

Fig. 1. Inhibitory potency of 3HAO blockers in rat brain homogenate. Assays using 4-Br-3HANA (□), 4-Cl-3HANA (○) or 4-F-3HANA (●) were performed as described in the text.

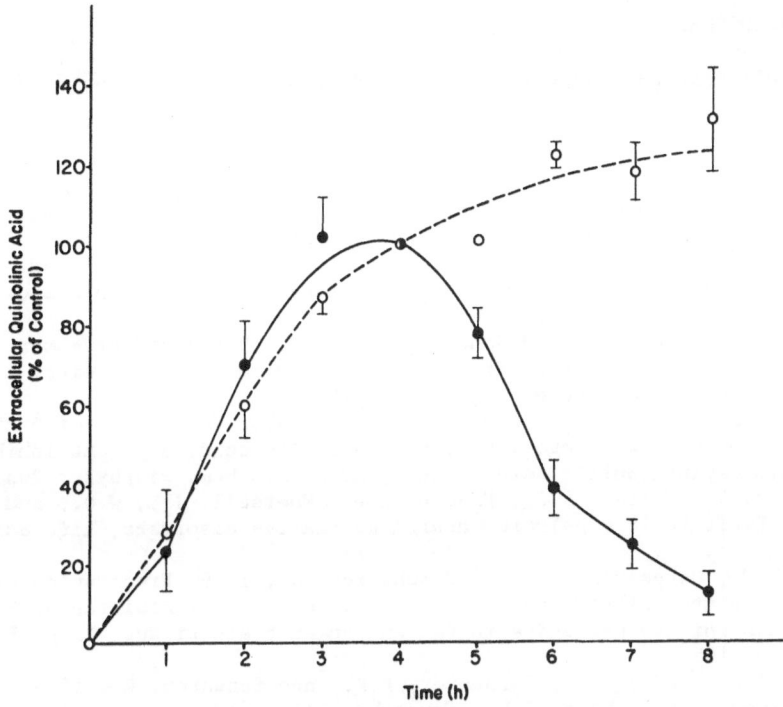

Fig. 2. Inhibition of 3HAO by 4-Br-3HANA. The experiments using 3HANA
alone (○; N = 6) or 3HANA + 4-Br-3HANA (●; N = 9) were performed as described in the text. The data are expressed as a
percentage of extracellular QUIN levels reached after 4 hours
perfusion with 3HANA alone and are the mean ± SEM.

tests showed that all three 4-halo-3HANAs are <u>reversible</u> inhibitors of enzyme
activity (Walsh et al., 1989).

Microdialysis experiments, which so far have been conducted with 4-Br-
3HANA only, clearly demonstrated the ability of the compound to decrease the
concentration of QUIN in an <u>in vivo</u> situation as well. As compared to the <u>in
vitro</u> potency of the inhibitor, however, a surprisingly high concentration of
the drug was needed to suppress brain 3HAO activity totally (Fig. 2). The
discrepancy between <u>in vitro</u> and <u>in vivo</u> potency of the drug may be related
to poor penetration of 4-Br-3HANA through the dialysis membrane into the
brain and/or limited uptake into astrocytes, which harbor 3HAO in the brain
(Okuno et al., 1987). Current experiments are designed to examine the relative potency of the three established 3HAO inhibitors <u>in vivo</u>, to investigate
their penetration through the blood-brain barrier and to determine the re-
versibility of 3HAO inhibition after discontinuation of drug administration
<u>in vivo</u>.

Regardless of their precise mechanism of action, 4-halo-3HANAs are promising research tools for the elucidation of QUIN metabolism and function. In
particular, the drugs described here can be expected to serve as lead compounds for the development of 3HAO inhibitors with improved pharmacokinetic
properties and, in appropriately radiolabelled form, may constitute probes
for brain PET studies. Derivatives of 4-halo-3HANAs may also become of
therapeutic value in diseases which are caused by hyperphysiological quantities of QUIN (see Introduction).

ACKNOWLEDGEMENTS

This work was supported by USPHS grants NS 16102 and NS 28236.

REFERENCES

Foster, A.C., Okuno, E., Brougher, D.S., and Schwarcz, R., 1986a, A radioenzymatic assay for quinolinic acid, Anal. Biochem., 158:98-103.
Foster, A.C., White, R.J., and Schwarcz, R., 1986b, Synthesis of quinolinic
acid by 3-hydroxyanthranilic acid oxygenase in rat brain tissue in vitro,
J. Neurochem., 47:23-30.
Okuno, E., Köhler, C., and Schwarcz, R., 1987, Rat 3-hydroxyanthranilic acid
oxygenase: purification from the liver and immunocytochemical localization
in the brain, J. Neurochem., 49:771-780.
Parli, C.J., Krieter, P., and Schmedt, B., 1980. Metabolism of 6-chloro-
tryptophan to 4-chloro-3-hydroxyanthranilic acid: a potent inhibitor of
3-hydroxyanthranilic acid oxidase, Arch. Biochem. Biophys., 203:161-166.
Schwarcz, R., Foster, A.C., French, E.D., Whetsell, Jr., W.O., and Köhler,
C., Excitotoxic models for neurodegenerative disorders, Life Sci., 35:19-
32.
Todd, W.P., Carpenter, B.K., and Schwarcz, R., 1989, Preparation of 4-halo-3-
hydroxyanthranilates and demonstration of their inhibition of 3-hydro-
xyanthranilate oxygenase in rat and human brain tissue. Prep. Biochem.,
19:155-165.
Walsh, J.L., Todd, W.P., Carpenter, B.K., and Schwarcz, R., 1989, 4-Chloro-,
4-fluoro- and 4-bromo-3-hydroxyanthranilic acids are potent competitive
inhibitors of 3-hydroxyanthranilic acid oxygenase, Soc. Neurosci. Abstr.,
15:328.11.

ON THE DISPOSITION OF DE NOVO SYNTHESIZED QUINOLINIC ACID IN RAT BRAIN

TISSUE

C. Speciale and R. Schwarcz

Maryland Psychiatric Research Center
Baltimore, Maryland 21228
USA

INTRODUCTION

Because of the putative role of the cerebral metabolite and specific N-methyl-D-aspartate receptor agonist quinolinic acid (QUIN) in physiological and pathological processes in the brain (Schwarcz et al., 1984; Stone and Burton, 1989; Lapin et al., this volume), the mechanisms by which QUIN function is regulated in the normal and diseased brain have become the object of investigative interest. The presence of 3-hydroxyanthranilic acid oxygenase (3HAO), QUIN's immediate biosynthetic enzyme, in brain tissue (Foster et al., 1986b) indicated the possibility to assess the dynamics of QUIN disposition in whole cell preparations (tissue slices) which had been exposed to the substrate of 3HAO, 3-hydroxyanthranilic acid (3HANA). Preliminary experiments showed that QUIN can indeed be recovered from both tissue and incubation medium after the administration of 3HANA to slices from rat liver or brain in a physiological milieu (Speciale et al., 1989). Assay parameters have now been scrutinized in detail and optimal conditions for the examination of QUIN function in vitro have been established (Speciale et al., submitted for publication). In the present report, we describe the use of this experimental system for the initial assessment of mechanisms of QUIN release in normal and lesioned rat brain.

MATERIALS AND METHODS

One mm thick slices (1.5-2.0 mg/slice) from cerebral cortex or striatum were obtained from male Sprague-Dawley rats (200-250 g). They were randomly transferred to culture wells (10 slices/well) containing 450 μl oxygenated Krebs-Ringer buffer (KRB; NaCl 118 mM, KCl 4.8 mM, CaCl$_2$ 1.8 mM, MgSO$_4$ 1.2 mM, NaHPO$_4$ 12.9 mM, Na$_2$HPO$_4$ 3 mM, glucose 5 mM, pH 7.4). After preincubation (10 min, 37°C), 30 μM 3HANA was added to the wells and the incubation was continued for 30 min. The incubation was stopped by rapid manual separation of the tissue from the incubation medium and QUIN determined in solubilized tissue or in medium according to the radioenzymatic method of Foster et al. (1986a).

QUIN efflux from tissue slices was examined in separate experiments. To this end, slices were exposed to 64 μM 3HANA for 60 min as described above, and the tissue QUIN content measured in one (control) set of slices. QUIN efflux was assessed in parallel in separate groups following transfer of the

Table 1. Newly produced QUIN in striatal slices: effects of ibotenate-
lesions

	Tissue	Medium
	(pmoles/well)	
Control	84.3 ± 11.5	18.0 ± 4.0
Ibotenate-lesioned	502.4 ± 35.4[**]	182.2 ± 10.0[**]

Rats were injected intrastriatally with ibotenic acid (40 µg/2 µl) or
saline (2 µl) and sacrified seven days later. Striatal slices were
prepared and incubated 30 min, 37°C) with 30 µM 3HANA as described in
the text. Data are the mean ± SEM of 4 separate animals. [**]P < 0.01
as compared to controls (Student's t-test).

slices into 500 µl fresh KRB and further incubation for 30, 60 or 90 min in
the presence of the selective 3HAO inhibitor 4-Cl-3-hydroxyanthranilic acid
(10 µM; Todd et al., 1989; kindly provided by Dr. B.K. Carpenter, Cornell
University, Ithaca, NY) to prevent further QUIN synthesis from trapped 3HANA
during the efflux period. The incubation was terminated by manual removal of
the tissue, and QUIN content determined in both tissue and medium as described above.

Stereotaxic ibotenate (40 µg/2 µl) injections into the rat striatum were
performed as described by Schwarcz et al. (1979).

RESULTS

As shown in Table 1, newly synthesized QUIN could be readily detected in
both tissue and incubation medium after exposure of striatal slices to 3HANA.
Notably, substantially more QUIN was recovered from the tissue under these
assay conditions. Striatal slices obtained 7 days after an intrastriatal
ibotenate injection showed not only a significant increase in the total
amount of QUIN produced but also revealed a redistribution of QUIN between
the intra- and extracellular compartment. Thus, relatively more QUIN was recovered in the incubation medium when lesioned slices were used.

Efflux experiments, performed with cerebral cortical slices, provided
more direct evidence for the brain's ability to retain QUIN. As depicted in
Fig. 1, QUIN entered the extracellular compartment only rather slowly, causing a proportional slow decline in intracellular QUIN content over the 90 min
incubation period.

DISCUSSION

The present data indicate that brain-specific mechanisms exist which
control the release of QUIN into the extracellular space. Thus, in contrast
to the liver (Speciale et al., 1989), brain tissue has the ability to retain
newly synthesized QUIN, possibly by storage in intracellular organelles or
binding to selective cellular macromolecules. The precise nature of the
mechanisms which are in place in the brain to store QUIN and to control its
release is of obvious relevance for QUIN function. In particular, it will be
necessary to examine the dependency of QUIN efflux on the ionic milieu, energy status and pH of the microenvironment of astrocytes, which in the rat

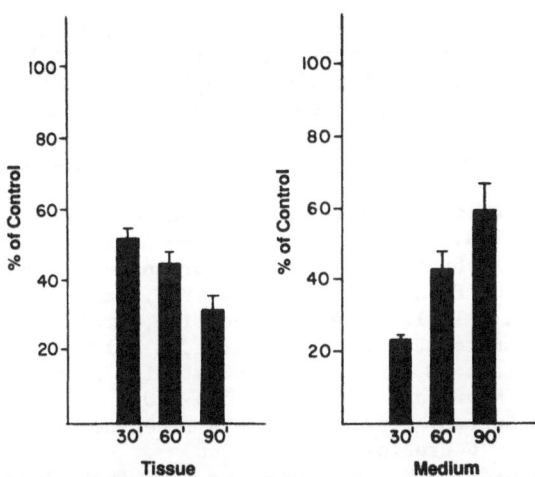

Fig. 1. QUIN content of tissue and medium compartments 30, 60 and 90
         min after placement of preloaded slices into fresh incuba-
         tion medium. Experimental details of this QUIN efflux study
         are given in the text. Data are expressed as a percentage
         of the QUIN tissue content of slices exposed to 64 μM 3HANA
         for one hour at 37°C and are the mean ± SEM of 4 separate
         experiments.

brain harbor 3HAO almost exclusively (Köhler et al., 1988). It will also be
interesting to investigate the existence of feedback mechanisms which may
influence the nature or speed of QUIN release. For example, the blockade of
QUIN-sensitive neuronal N-methyl-D-aspartate receptors [e.g. by synthetic
agents such as MK-801 (Wong et al., 1986)] may increase QUIN liberation into
the extracellular space.

The results obtained with neuron-depleted astrogliotic tissue indicate
that the disposition of brain QUIN may be altered substantially under pathological conditions. While the overall increase in de novo QUIN production in
the ibotenate-lesioned striatum is likely due to the large increase in 3HAO
activity (Schwarcz et al., 1989), the relatively larger proportion of QUIN
found in the extracellular compartment of lesioned tissue was unexpected. It
remains to be seen if changes in the QUIN disposition in pathological tissues
also occur in vivo and if evidence can be found for their possible relevance
in disease processes which are initiated or propagated by QUIN.

ACKNOWLEDGEMENTS

The technical assistance of Mrs. D. Clary is gratefully acknowledged.
This work was supported by USPHS grants NS 28236 and MH 44211.

REFERENCES

Foster, A.C., Okuno, E., Brougher, D.S., and Schwarcz, R., 1986a, A radio-
    enzymatic assay for quinolinic acid, Anal. Biochem., 158:98-103.
Foster, A.C., White, R.J., and Schwarcz, R., 1986b, Synthesis of quinolinic
    acid by 3-hydroxyanthranilic acid oxygenase in rat brain tissue in vitro,
    J. Neurochem., 47:23-30.

Köhler, C., Eriksson, L.G., Okuno, E., and Schwarcz, R., 1988, Localization of quinolinic acid metabolizing enzymes in the rat brain. Immunohisto-chemical studies using antibodies to 3-hydroxyanthranilic acid oxygenase and quinolinic acid phosphoribosyltransferase, Neuroscience 27:49-76.

Schwarcz, R., Foster, A.C., French, E.D., Whetsell Jr., W.O., and Köhler, C., 1984, Excitotoxic models for neurodegenerative disorders, Life Sci., 35:19-32.

Schwarcz, R., Hökfelt, T., Fuxe, K., Jonsson, G., Goldstein, M., and Terenius, L., 1979, Ibotenic acid induced neuronal degeneration: a morpho-logical and neurochemical study, Exp. Brain Res., 37:199-216.

Schwarcz, R., Okuno, E.,and White, R.J., 1989, Basal ganglia lesions in the rat: effects on quinolinic acid metabolism, Brain Res., 490:103-109.

Speciale, C., Ungerstedt, U., and Schwarcz, R., 1989, Production of extracel-lular quinolinic acid in the striatum studied by microdialysis in unanes-thetized rats, Neurosci. Lett., 104:345-350.

Stone, T.W., and Burton, N.R., 1988, NMDA receptors and ligands in the verte-brate CNS, Prog. Neurobiol., 10:333-368.

Todd, W.P., Carpenter, B.K., and Schwarcz, R., 1989, Preparation of 4-halo-3-hydroxyanthranilates and demonstration of their inhibition of 3-hydroxy-anthranilate oxygenase in rat and human brain tissue, Prep. Biochem., 19:155-165.

Wong, E.H., Kemp, J.A., Priestley, T., Knight, A.R., Woodruff, G.N., and Iversen, L.L., 1986, The novel anticonvulsant MK 801 is a potent N-methyl-D-aspartate antagonist, Proc. Nat. Acad. Sci. USA, 83:7104-7108.

BRAIN-SPECIFIC CONTROL OF KYNURENIC ACID PRODUCTION BY DEPOLARIZING AGENTS

J.B.P. Gramsbergen, W.A. Turski, and R. Schwarcz

Maryland Psychiatric Research Center
Baltimore, Maryland 21228
USA

INTRODUCTION

Kynurenic acid (KYNA), a prominent tryptophan metabolite found in high concentration in urine, has recently attracted the attention of neuroscientists studying the function of excitatory amino acids in the brain. KYNA, which is present in the mammalian brain (Carlà et al., 1988; Turski et al., 1988; cf. Moroni et al., this volume), is a neuroinhibitory compound which exerts its effect by interacting with all three established ionotropic excitatory amino acid receptors (Foster and Fagg, 1984). In experimental systems, KYNA has also been shown to act as a neuroprotectant and anticonvulsant (Foster et al., 1984; Germano et al., 1987). Because of the inferred possible role of KYNA in human neuropsychiatric disorders, it therefore became of interest to examine brain mechanisms which control the function of endogenous KYNA.

Previous studies conducted on rat brain slices had documented that KYNA can be produced in vitro from its immediate bioprecursor L-kynurenine (KYN). KYNA synthesis takes place preferentially in astrocytes and is dependent on the ionic composition of the incubation medium (Turski et al., 1989). The experiments also demonstrated that newly synthesized KYNA is rapidly liberated into the extracellular compartment where it can be readily measured using HPLC with spectrophotometric detection. We have now examined in greater detail the mechanisms governing the production of extracellular KYNA in slices from rat brain and liver.

MATERIALS AND METHODS

Male Sprague-Dawley rats (200-250 g) were used for all experiments. The animals were killed by decapitation, and their brains and livers rapidly removed on ice. Tissue slices were prepared, preincubated for 10 min, and then incubated with 50 μM KYN in Krebs-Ringer buffer as detailed by Turski et al. (1989). After a 2 hour incubation at 37°C, tissue slices were removed, and KYNA determined in the incubation medium following its separation from KYN using a Dowex 50W (H$^+$-form) ion exchange resin (Turski et al., 1989). Addition of veratridine or tetrodotoxin (TTX), and changes in the ionic composition of the Krebs-Ringer buffer were made prior to the 10 min pre-incubation period. Substitution of high K$^+$ and Mg^{2+} was made at the expense of Na$^+$ to maintain osmolarity. Protein was measured according to Lowry et al. (1951).

RESULTS

Under depolarizing conditions, the production of KYNA in brain (striatum) was significantly impeded. Thus, 50 mM K^+ and 50 μM veratridine resulted in 41.5% and 79% decreases in extracellular KYNA as compared to control slices incubated in the presence of KYN alone. Under the same assay conditions, no effect of the two depolarizing agents was observed in liver slices (Table 1).

A second set of experiments was designed to explore in more detail the mechanism(s) by which high K^+ and veratridine cause decreases in KYNA release in the brain. Using cerebral cortical slices, the K^+ effect was shown to be completely blocked by the omission of Ca^{2+} and the presence of 20 mM Mg^{2+} in the incubation medium. Under these conditions, veratridine still caused a dramatic reduction in KYNA output. In contrast, inclusion of 1 μM TTX in the incubation medium completely antagonized the action of veratridine without affecting that of 50 mM K^+ (Table 2).

DISCUSSION

The present data confirm and extend the notion that neuronal depolarization can significantly decrease the production of KYNA recovered in the extracellular compartment. Not only a high K^+ concentration (cf. Turski et al., 1989) but also, to an even greater extent, veratridine was found to be extremely potent in this respect. The effect was clearly brain-specific since both depolarizing agents failed to affect KYNA production in liver slices.

Since 50 mM K^+ is known to rather indiscriminately cause Ca^{2+}-dependent neuronal transmitter release, the next set of experiments was designed to examine the effect of Ca^{2+}-deletion. Under these conditions, and in the presence of 20 mM Mg^{2+}, high K^+ was found to be inactive, indicating that a K^+-releasable endogenous compound of neuronal origin is capable of depressing KYNA production. The effect of veratridine, too, was apparently caused by the primary effect of the drug on neurons, which in turn release (an) as yet unidentified substance(s) to influence KYNA release. In accordance with prevalent views, the effect of the Na^+-channel activator veratridine was blocked by TTX but was unaffected by the deletion of Ca^{2+} from the incubation medium. Regardless of their precise chemical identity, it therefore appears that neuron-derived factors, released upon depolarization, are capable of dramatical-

Table 1. Effect of depolarizing agents on KYNA production in striatal and liver slices

	Striatum	% of Control	Liver	% of Control
Control	47.9 ± 5.3	100.0	85.5 ± 9.3	100.0
50 mM K^+	28.0 ± 2.4**	58.5	72.6 ± 8.9	84.9
50 μM Veratridine	10.1 ± 1.2**	21.0	79.5 ± 13.2	93.0

Slices were prepared and incubated with 50 μM KYN as described by Turski et al. (1989). Data are expressed as pmol KYNA/2 h/mg protein and are the mean ± SEM of 6-12 individual wells. **$P < 0.01$ as compared to controls (Student's t-test).

Table 2. Effect of tetrodotoxin or 0 Ca^{2+}/20 mM Mg^{2+} on depolarization-induced reduction of KYNA production

	TTX (1 μM)	0 Ca^{2+}/20 mM Mg^{2+}
Control	128.0 ± 4.1** (24)	104.0 ± 4.2 (11)
50 mM K^+	52.3 ± 1.6** (29)	101.4 ± 3.0 (12)
5 μM Veratridine	103.5 ± 4.1 (6)	33.2 ± 1.5** (12)

Rat cerebral cortical slices were prepared and incubated with 50 μM KYN as described by Turski et al. (1989). Data are expressed as a percentage of control values (incubation in regular Krebs-Ringer buffer; 171.2 ± 3.9 pmol KYNA/well) and are the mean ± SEM of the number of individual wells indicated in parentheses. **$P < 0.01$ as compared to controls (Student's t-test).

ly affecting the function of KYNA, which in the rat brain is known to be preferentially produced in and liberated from astrocytes (Turski et al., 1989; Schwarcz and Du, this volume). The nature of this neuron-glia interaction is somewhat counterintuitive since one would assume that neuronal depolarization should result in an _increase_ in KYNA output - and therefore enhance neuronal inhibition. In contrast, the present results raise the intriguing possibility that KYNA could serve an _amplification_ function in the brain by removing its tonic inhibition of excitatory amino acid receptors upon increased neuronal firing. The validity of this scenario, and its possible implications for the function and dysfunction of central excitatory amino acid-related neurotransmission, are currently under investigation in our laboratory.

ACKNOWLEDGEMENTS

This work was supported by USPHS grants NS 16102 and NS 28236.

REFERENCES

Carlà, V., Lombardi, G., Bensi, M., Russi, P., Moneti, G., and Moroni, F., 1988, Identification and measurement of kynurenic acid in the rat brain and other organs, Anal. Biochem., 169:89-94.

Foster, A.C., Fagg, G.E., 1984, Acidic amino acid binding sites in mammalian neuronal membranes: their characteristics and relationship to synaptic receptors, Brain Res. Rev., 7:103-164.

Foster, A.C., Vezzani, A., French, E.D., and Schwarcz, R., 1984, Kynurenic acid blocks neurotoxicity and seizures induced in rats by the related brain metabolite quinolinic acid, Neurosci. Lett., 48:273-278.

Germano, I.M., Pitts, L.H., Meldrum, B.S., Bartkowski, H.M., and Simon, R.P., 1987, Kynurenate inhibition of cell excitation decreases stroke size and deficits, Ann. Neurol., 22:730-734.

Lowry, O.H., Rosebrough, N.J., Farr, A.L., and Randall, R.J., 1951, Protein measurement with the Folin phenol reagent, J. Biol. Chem., 193:265-275.

Turski, W.A., Gramsbergen, J.B.P., Traitler, H., and Schwarcz, R., 1989, Rat brain slices produce and liberate kynurenic acid upon exposure to L-kynurenine, J. Neurochem., 52:1629-1636.

Turski, W.A., Nakamura, M., Todd, W.P., Carpenter, B.K., Whetsell, W.O., and Schwarcz, R., 1988, Identification and quantification of kynurenic acid in human brain tissue, **Brain Res.**, 454:164-169.

EFFECTS OF 5-HYDROXYTRYPTOPHAN ON EATING BEHAVIOR AND ADHERENCE TO DIETARY

PRESCRIPTIONS IN OBESE ADULT SUBJECTS

C. Cangiano[1], F. Ceci[2], M. Cairella[3], A. Cascino[1], M. Del Ben[3],
A. Laviano[1], M. Muscaritoli[1], and F. Rossi-Fanelli[1]

[1]Lab. Clin. Nutrition
3rd Dept. Internal Medicine
Rome

[2]Dept. Human Biopathology

[3]Serv. Dietologia
Ist. Terapia Medica Sistematica
Univ. "La Sapienza"
Rome
Italy

INTRODUCTION

Changes in plasma amino acid levels by affecting the availability of neurotransmitter amino acid precursors within the brain may influence eating behavior (Wurtman et al., 1981; Fernstrom, 1983). Among the different neurotransmitter systems, a number of theoretical and experimental data support the role played by serotonin in the regulation of eating habits (Lytle, 1977; Li et al., 1983). Rather recently, R.J. Wurtman et al. (1981) have shown that the serotonergic system plays an important role also in the selection of macronutrients, especially in obese people consuming preferentially carbohydrate-rich foods. Moreover, a "normalization" of such aberrant behavior was obtained by pharmacologically enhancing brain serotonin synthesis (J.J. Wurtman et al., 1981). A significant reduction in food intake has been reported

Table 1. Prevalence of anorexia and related symptoms during 5-HTP and placebo treatments

| | 5-HTP | | Placebo | |
	period I	period II	period I	period II
Taste alteration	2/7	1/7	0/7	0/7
Smell alteration	2/7	1/7	0/7	0/7
Meat aversion	3/7	0/7	0/7	1/7
Early satiety	7/7	6/7	2/7	2/7
Nausea	5/7	0/7	1/7	0/7
Anorexia	7/7	6/7	2/7	3/7

by Blundell and Leshem (1975) in hyperphagic obese rats by administration of
tion of 5-hydroxytryptophan (5-HTP). Previous experiences from this laboratory in a group of obese adult female subjects have shown that the oral administration of 5-HTP caused the onset of typical anorexia-related symptoms,
decreased food intake and promoted weight loss in the absence of any dietary
restriction (Ceci et al., 1989). The aim of the present study was to confirm
these data in a longer study period and to verify whether the adherence to
dietary prescriptions could be improved by oral administration of 5-HTP.

MATERIALS AND METHODS

Fourteen hyperphagic adult female patients, aged 24 to 51 years (mean
age: 38 yrs), with body mass index ranging between 30 and 40 were studied.
Patients were randomly assigned to receive either 5-HTP or placebo (300 mg,
30 min before each meal). The twelve weeks overall study period for each
group of patients was subdivided into two consecutive six week periods. During the first period, patients received no dietary prescriptions, whereas
during the second period a 1200 Kcal diet was prescribed. Patients were seen
every two weeks to evaluate body weight, diet diary, self evaluation of appetite and satiety, and the presence of anorexia. Anorexia was evaluated by
means of a questionnaire from which the presence of meat aversion, taste or
smell alterations, early satiety, nausea and/or vomiting was recorded. Patients reporting one or more of the above mentioned symptoms were considered
"anorectic". To verify patient compliance to treatment, the 24 hr urinary
excretion of 5-hydroxy-indoleacetic acid (5-HIAA) was assayed according to
Udenfriend et al. (1955).

RESULTS AND DISCUSSION

The prevalence of anorexia was higher in the group of patients receiving
5-HTP than in the placebo group (Table 1). This was true during both periods
of observation (i.e. with and without diet prescription). From patients' records of the questionnaire, early satiety was responsible for the reduced
food intake in the majority of the cases. Interestingly, nausea, which was
present in about 70% of the cases treated with 5-HTP during the first six
weeks, disappeared during the second study period, suggesting that this symp-
tom may be partly considered as a temporary side effect of 5-HTP administration.

Table 2. Body weight and urinary excretion of 5-HIAA
during 5-HTP and placebo treatments

	5-HTP	Placebo
Weight (kg)		
Baseline	104.3 ± 6.4	94.4 ± 5.8
Period I	102.7 ± 6.3	93.9 ± 5.9
Period II	99.6 ± 5.8	93.2 ± 5.9
Urinary 5-HIAA (mg/day)		
Baseline	8.8 ± 1.3	7.3 ± 0.5
Period I	505 ± 97	7.5 ± 0.7
Period II	496 ± 90	6.9 ± 0.8

Data are the mean ± SEM.

In the 5-HTP group the presence of anorexia was correlated during the first study period with a weight loss similar to that previously observed (Ceci et al., 1989); moreover, during the second study period patients almost doubled their weight loss, whereas no reduction of body weight was observed in the control group (cf. Table 2). These data suggest that a better adherence to the dietary prescription was obtained in the 5-HTP group. The significant difference in 5-HIAA urinary excretion between the two groups of patients provided evidence for patients' compliance to the treatment.

The results obtained in this study confirm the key role played by the serotonergic system in the regulation of eating behavior and are clearly in favor of the usefulness of 5-HTP administration to obtain a good adherence of obese people to dietary prescriptions.

REFERENCES

Blundell, J.E., and Leshem, M.B., 1975, The effect of 5-hydroxytryptophan on food intake and on the anorexic action of amphetamine and fenfluramine, J. Pharm. Pharmacol., 27:31-37.
Ceci, F., Cangiano, C., Cairella, M., Cascino, A., Del Ben, M., Muscaritoli, M., Sibilia, L., and Rossi-Fanelli, F., 1989, The effects of oral 5-hydroxytryptophan administration on feeding behavior in obese adult female subjects, J. Neural Transm., 76:109-117.
Fernstrom, J.D., 1983, Role of precursor availability in control of monoamine biosynthesis in brain. Physiol. Rev., 63:484-546.
Li, E.T.S., and Anderson, G.H., 1983, Amino acids in the regulation of food intake, Nutr. Abstr. Rev. Clin. Nutr., 53:169-181.
Lytle, L.D., 1977, Control of eating behavior, in: "Nutrition and Brain", Raven Press, New York, pp. 2-145.
Udenfriend, S., Titus, E., and Welssabach, H., 1955, The identification of 5-hydroxy-3-indoleacetic acid in normal urine and a method for its assay, J. Biol. Chem., 216:499-505.
Wurtman, J.J., Wurtman, R.J., Growdon, J.H., Henry, P., Lipscomb, M.A., and Zeisel, S.H., 1981, Carbohydrate craving in obese people: suppression by treatments affecting serotoninergic neurotransmission, Int. J. Eating Dis., 1:2-15.
Wurtman, R.J., Hefti, F., and Melamed, E., 1981, Precursor control of neurotransmitter synthesis, Pharmacol. Rev., 32:315-331.

THE INVOLVEMENT OF BULBOSPINAL SEROTONERGIC NEURONS IN THE EFFECT OF

5-HYDROXYTRYPTOPHAN ON ALPHA AND GAMMA MOTONEURONS

N.R. Myslinski

Department of Physiology
University of Maryland School of Medicine
Baltimore, Maryland 21201
USA

INTRODUCTION

As indicated by the work of other investigators, there exists a system
of serotonin (5-HT)-containing neurons which originate in the nucleus raphe
and terminate near motoneurons in the spinal cord (Carlsson et al., 1964;
Dahlström and Fuxe, 1965; Anderson and Holgerson, 1966). It has also been
shown that the i.v. administration of the 5-HT precursor, 5-hydroxytryptophan
(5-HTP), leads to the facilitation of both alpha and gamma motoneurons (Myslinski and Anderson, 1978). The general belief is that 5-HTP produces this
effect via the bulbospinal 5-HT pathway which converts it to 5-HT and subsequently releases it to activate the postsynaptic receptor sites (Anderson,
1972). If 5-HTP does produce this effect via the 5-HT neurons, the concurrent use of a 5-HT reuptake inhibitor should potentiate its effects; pretreatment with a 5-HT synthesis inhibitor should depress its effects; and the
electrical stimulation of the bulbospinal 5-HT system should produce effects
similar to that of 5-HTP. We found all three cases to be true, which argues
that 5-HTP works via the 5-HT pathway to facilitate alpha and gamma motoneurons.

METHODS AND MATERIALS

Twenty-eight adult cats of both sexes were used. The methods, including
those used to discriminate alpha and gamma motoneurons, were the same as
those detailed previously (Myslinski and Anderson, 1978). Spinal cats were
used except in the raphe stimulation experiments where decerebrate cats were
used. To stimulate 5-HT-containing cell bodies in the medulla, concentric
stainless steel electrodes were stereotaxically placed in the nucleus raphe
magnus along the midline 4 mm below the surface of the fourth ventricle
(Horsley-Clark coordinates: P: 7.1, L: 0.0, V: 7.5). Electrode placement was
verified histologically. For evoked activity, 30 msec trains of 0.5 msec
square waves at 300 Hz were delivered to the brainstem electrodes. The alpha
and gamma motoneuron spike activity recorded from an L7 or S1 ventral root
was used to generate a post-stimulus interspike interval histogram. The following drugs were administered i.v.: DL-5-HTP and L-tryptophan (Nutritional
Biochemicals), cinanserin HCl (Squibb), methysergide bimaleate (Sandoz), imipramine (Ciba), and gallamine triethiodide (Davis and Geck). Doses refer to
the salt. Parachlorophenylalanine (PCPA) (Calbiochem) was administered in-

traperitoneally as a suspension made with pectin. It was given in two 300 mg/kg doses, 48 and 24 hrs before the experiment.

RESULTS

5-HTP and imipramine

These experiments were carried out to observe if imipramine, which preferentially inhibits 5-HT reuptake (Carlsson et al., 1969), would potentiate the actions of 5-HTP. Imipramine was injected slowly over a period of 10 minutes in order to minimize any increases in blood pressure. Administered alone, 5 mg/kg of imipramine produced no significant change in alpha activity but decreased gamma activity by an average of 30%. Ten mg/kg of 5-HTP alone produced no significant change in alpha activity, but increased gamma activity by 50%. This dose was used since it produced a submaximal effect; 75 mg/kg produces the maximal effect (Myslinski and Anderson, 1978). When imipramine was administered 60 minutes after 10 mg/kg of 5-HTP, it initiated alpha activity within 10 minutes. This activity reached its maximum within 15 minutes. Gamma activity was increased more than 3-fold with maximal facilitation occurring at 10 minutes. Subsequent administration of 4 mg/kg of the 5-HT antagonist, cinanserin, 30 minutes after imipramine injection produced a return to the control baseline within 15 min for both types of fibers.

Experiments were also performed using 100 mg/kg of the precursor, tryptophan (TRY), instead of 5-HTP. Previous studies indicated that it has effects similar to 5-HTP on motoneurons, and presumably works by a similar mechanism (Myslinski and Anderson, 1978). Our results were qualitatively the same with TRY as with 5-HTP. The alpha fibers were activated and gamma activity increased 2-fold when imipramine was injected.

5-HTP and PCPA

The effects of 5-HTP were examined in cats pretreated with the TRY hydroxylase inhibitor, PCPA. Previous work has shown that PCPA paradoxically blocks the effects of 5-HTP, but only when 5-HT fibers are intact (Taber and Anderson, 1973). Whereas in the non-pretreated animal 5-HTP (75 mg/kg) always initiated alpha activity and markedly increased gamma activity (Myslinski and Anderson, 1978), in the PCPA-pretreated cats 5-HTP initiated alpha activity in only one third of the cats and increased gamma activity 17 ± 5%.

Nucleus raphe stimulation

If the effects of 5-HTP are mediated by 5-HT neurons, they should be mimicked by electrical stimulation of the serotonergic pathway. Raphe stimulation evoked potentials with 2 different latencies: a short latency, and a long latency of 60 to 80 msec. These potentials are similar to the VRP-1 and VRP-2 of Proudfit and Anderson (1974). Unlike the short latency response, the long latency response was susceptible to inhibition by cinanserin (4 mg/kg) and methysergide (2 mg/kg). This indicated that the long latency potentials were produced by the 5-HT system. These potentials were of both alpha and gamma types. The alpha fibers tended to respond with a single spike. The gamma fibers responded with 2 or more discharges, and usually responded to a lower stimulus intensity than the alpha fibers.

DISCUSSION

Since 5-HT itself does not cross the blood-brain barrier, 5-HTP has long been administered to experimental animals to activate 5-HT receptor sites indirectly via CNS conversion to 5-HT, and to potentiate or mimic the actions of 5-HT pathways. This method was used to determine the effects of the 5-HT system on alpha and gamma motoneuron activity in the spinal cord (Myslinski and Anderson, 1978). Since L-aromatic amino acid decarboxylase converts 5-HTP to 5-HT, there remains the possibility that 5-HTP is working via non-5-HT terminals to produce its effect on alpha and gamma motoneurons. However, this mechanism is less likely since the administration of TRY (which is converted to 5-HT only in 5-HT neurons) duplicates the effects of 5-HTP. Moreover, the 5-HT antagonists, cinanserin and methysergide, reverse the effects of 5-HTP (Myslinski and Anderson, 1978). The data presented here provide further evidence to support the contention that 5-HTP does work through 5-HT terminals to produce its effects.

Since imipramine is a preferential inhibitor of 5-HT reuptake, it should potentiate the effects of 5-HTP if they are mediated via 5-HT. Our data support this contention. The potentiation that we saw was not due to a direct effect of imipramine on 5-HT receptors since imipramine itself did not facilitate motoneuron activity. Thus, in the spinal cat when there is no release of 5-HT, imipramine has no effect on motoneuronal activity. When precursors are introduced, imipramine enhances motoneuronal activity. Since imipramine is not totally specific for 5-HT neurons, this evidence is only suggestive of their role in 5-HTP's effects. Since the completion of this study, other compounds have been claimed to be more specific. Further studies will have to determine their effects.

PCPA is a potent inhibitor of TRY hydroxylase and leads to a profound reduction in brain 5-HT levels (Taber and Anderson, 1973). It has been shown that PCPA inhibits the effects of 5-HTP (Taber and Anderson, 1973). This is explained by the fact that, unlike in intact animals, the 5-HT synaptic vesicles of PCPA-pretreated animals are empty. The sequestering of newly formed 5-HT by these empty vesicles minimizes the amount overflowing into the synaptic cleft, and therefore prevents the effect. Since PCPA prevented the facilitation of alpha and gamma activity in our study, it implicates the 5-HT pathway in this effect of 5-HTP. PCPA's actions, however, are not totally specific. It can also inhibit amino acid transport and phenylalanine hydroxylase, and reduce brain catecholamines (Koe and Weissman, 1968). The catecholamine decline, however, is less than 15%, while 5-HT reduction may reach 97% (Taber and Anderson, 1973).

The previous experiments used pharmacological methods to determine the effects of the descending 5-HT pathway on spinal motoneurons. Duplication of those effects by electrically activating the same pathway is necessary to verify these effects. It was found that stimulation of the caudal raphe nucleus evoked long latency potentials that consisted of both alpha and gamma motoneuron spikes. The conduction velocity of the descending fibers that activated both types of motoneurons was approximately 3 m/sec. This condition velocity is compatible with the calculated velocity of the bulbospinal 5-HT fibers which are 1-2 μm in diameter (Dahlström and Fuxe, 1965). The difficulty with this approach stems from the fact that even the most discrete pulse of current will indiscriminately activate any cell bodies or axons in its vicinity. These potentials, therefore, may be the result of a delayed activation of a non-5-HT multisynaptic pathway, such as the reticulospinal tract. However, two 5-HT antagonists, cinanserin and methysergide, reversibly blocked the evoked potentials, suggesting that the 5-HT pathway mediates the effect.

These data and those of others indicate that the bulbospinal 5-HT pathway has a nonspecific excitatory action on both alpha and gamma motoneurons. Alpha-gamma coactivation is common in the CNS, since it maintains the sensitivity of the muscle spindles even in a shortened muscle. Previous data indicate that this is also a non-specific effect on both extensor and flexor motoneurons (Myslinski and Anderson, 1978). This motor effect, in conjunction with the inhibitory effect of the bulbospinal 5-HT system on nociceptive primary afferents (Fields et al., 1977) provides a means for holding a limb rigid and reducing nociception at the same time. This suggests a mechanism for the overriding of the withdrawal reflex by higher centers.

REFERENCES

Anderson, E.G., 1972, Bulbospinal serotonin-containing neurons and motor control, Fed. Proc., 31:107-112.

Anderson, E.G., and Holgerson, L.O., 1966, The distribution of 5-hydroxy-tryptamine and norepinephrine in the cat spinal cord, J. Neurochem., 13:479-485.

Carlsson, A., Corrodi, H., Fuxe, K., and Hökfelt, 1969, The effect of anti-depressant drugs in the depletion of intraneuronal brain 5-HT stores caused by 4-methyl alpha-ethyl metatyramine, Eur. J. Pharmacol., 5:357-363.

Carlsson, A., Falck, B., Fuxe, K., and Hillarp, N.A., 1964, Cellular localization of monoamines in the spinal cord, Acta Physiol. Scand., 60:112-119.

Dahlström, A., and Fuxe, K., 1965, Evidence for the existence of monoamine neurons in the central nervous system. II. Experimentally induced changes in the intraneuronal amine levels of bulbospinal neuron systems, Acta Physiol. Scand., Suppl. 247, 64:1-36.

Fields, H.L., Basbaum, A.L., Clanton, C.H., and Anderson S.D., 1977, Nucleus raphe magnus inhibition of spinal cord dorsal horn neurons, Brain Res., 126:441-453.

Koe, B.K., and Weissman, A., 1968, p-Chlorophenylalanine: a specific depletor of brain 5-HT, Adv. Pharmacol., 6B:27-29.

Myslinski, N.R., and Anderson, E.G., 1978, The effect of precursors on alpha and gamma motoneuron activity. J. Pharmacol. Exp. Therap., 204:19-26.

Proudfit, H.K., and Anderson, E.G., 1974, New long latency bulbospinal evoked potentials blocked by serotonin antagonists, Brain Res., 65:542-546.

Taber, C.A., and Anderson, E.G., 1973, Paradoxical blockade by p-chloro-phenylalanine of 5-hydroxytryptophan facilitatory actions on spinal reflexes, J. Pharmacol. Exp. Therap., 187:229-238.

QUINOLINIC ACID-INDUCED DECREASES IN ACETYLCHOLINE RELEASE IN THE RAT NUCLEUS

BASALIS MAGNOCELLULARIS

R.H. Metcalf, K. Wainwright, and R.J. Boegman

Department of Pharmacology and Toxicology
Queen's University
Kingston, Ontario K7L 3N6
Canada

INTRODUCTION

The tryptophan metabolite quinolinic acid (Quin) is a potent agonist at N-methyl-D-aspartate (NMDA) receptors on central cholinergic neurons in the rat nucleus basalis magnocellularis (nbM; Stone and Connick, 1985). Quin injected into the nbM produces neuronal death and axon-sparing lesions of this cortically projecting pathway (Schwarcz and Köhler, 1983; Stone and Connick, 1985). The effects of such lesions on cortical cholinergic parameters have indicated that this pathway provides the major cholinergic input to the rat frontoparietal cortex. Both morphological and electrophysiological evidence suggests that the cholinergic neurons located in the nbM also provide local cholinergic input to this area of the brain in the form of axon collaterals (Lamour et al., 1986; Semba et al., 1987). Recently, we have shown that the release of acetylcholine (ACh) in the nbM can be stimulated by potassium depolarization and in addition requires calcium for optimal release of ACh (Metcalf and Boegman, 1987). In the present study, we used Quin to produce lesions of the nbM, then determined the effect of these lesions on the release of endogenous ACh from tissue slices prepared from the nbM.

MATERIALS AND METHODS

Under halothane anesthesia, male Sprague-Dawley rats (275-325 g) received a 1 μl infusion of Quin (120 nmoles/μl dissolved in 0.9% saline, pH adjusted to 7.4) into the right nbM. Coordinates were 0.8 mm posterior to bregma, 2.6 mm lateral to the midline and 8.0 mm ventral to the surface of the skull. The nbM was removed 24 hr post-injection, and 300 μm slices prepared using a McIlwain tissue chopper. Tissue slices of the nbM were dispersed in 3.5 ml of cold Krebs buffer (KRB), then transferred to static release vessels containing 3.5 ml of prewarmed (37°C), oxygenated KRB. After pre-incubation, KRB was removed and replaced by prewarmed KRB (3.5 ml) containing 20 μM physostigmine. Samples were collected at 10 min intervals. At T = 20 min, KRB was replaced by KRB containing 35 mM potassium. At T = 30 min, high potassium KRB was replaced by normal KRB containing physostigmine. At the end of the experiment, tissue slices were collected and placed in a pre-weighed Eppendorf tube, centrifuged to produce a tissue pellet, the supernatant discarded, and the tube weighed. ACh was extracted from the collection samples into dichloromethane using dipicrylamine. ACh was assayed by the gas chromatography-mass spectrometry method reported by Jenden et al. (1968) using a GC column of 10% Pennwalt 223, 4% KOH on 80/100 GasChrom R. For localization of cholinergic cell bodies in the nbM, sections were processed for acetylcholinesterase (AChE) staining using the method of Karnovsky and Roots (1964). Animals were anesthetized with ketamine and perfused through the ascending aorta with 0.9% saline followed by 4% paraformaldehyde in 0.1 M phosphate buffer, pH 7.4. The fixed brains were placed in 30% sucrose at 4°C for at least 18 hours. Transverse 45 μm sections were cut on a freezing microtome and collected in 0.9% saline. In separate experiments, choline acetyltransferase (ChAT) activity present in nbM homogenates (300 μl per 8 ng tissue wet weight, containing 10 mM EDTA, pH 7.4, and 0.5% Triton X-100) was measured according to the method of Fonnum (1975).

RESULTS

ChAT activity in the nbM prepared from both left and right nbM did not show any differences (72.7 ± 13.1 vs. 73.2 ± 9.6 nmoles/mg protein/hr). Injection of saline into the right nbM produced a slight non-significant decrease in ChAT activity. However, ChAT activity decreased to 62.9 ± 10.1% of that seen in the contralateral nbM 24 hr after Quin infusion. Fig. 1a illustrates the AChE staining of nbM cholinergic cell bodies. Neurons were large, with two or more dendritic processes. AChE-positive neurons of the nbM showed marked changes 24 hr after Quin (Fig. 1b). A decrease in the number and intensity of AChE-positive neurons was evident. Neurons that did stain displayed an altered morphology. The cell bodies were more rounded, apparently swollen, and lacked dendritic staining.

The release of ACh from tissue slices of the nbM is presented in Fig. 2. A 15-fold increase in 35 mM potassium-induced ACh release from the contralateral control nbM was observed over spontaneous release of ACh. By the fifth collection fraction, ACh release had returned to spontaneous levels. ACh release in the ipsilateral nbM 24 hr after Quin injection was significantly lower than control. Spontaneous release in the two pre-stimulation fractions was approximately 35% of that seen in the contralateral control. Upon potas-

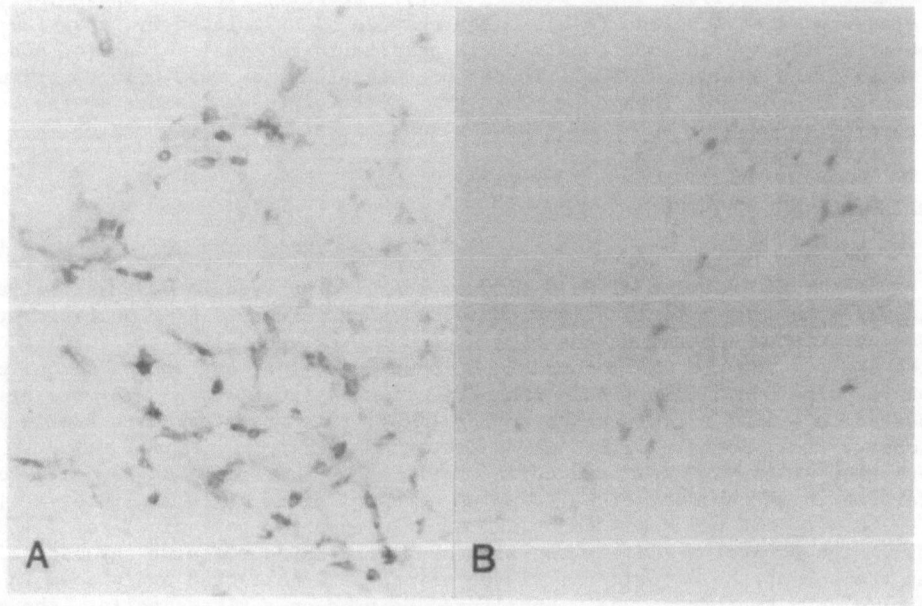

Fig. 1. AChE positive neurons in the nbM. A) Control; B) 120 nmoles Quin 24 hr post infusion. Magnification: x128.

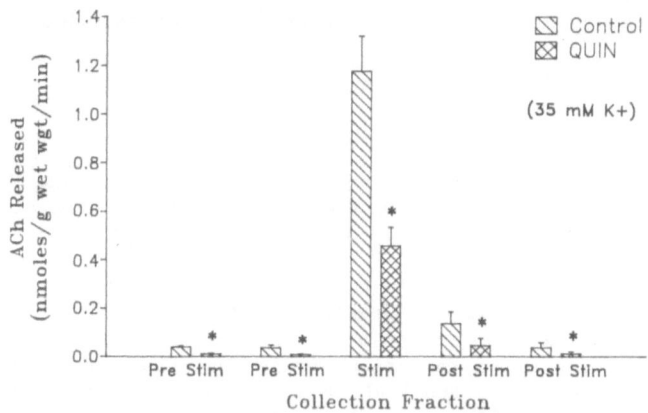

Fig. 2. ACh release in the nbM of 24 hr Quin-lesioned and contralateral
controls. *Significantly different from control (paired Student's
t-test; P ≤ 0.01, N = 4).

sium stimulation, ACh release increased to only 39.6 ± 8.2% of that seen in
the contralateral control. This increased release also returned to spontaneous levels by the end of the experiment. The potassium-induced ACh release
in the Quin injected nbM was approximately 25-fold higher than spontaneous
release, both of which were lower than that observed in the normal contralateral control tissue.

DISCUSSION

     The destruction of cholinergic neurons in the rat nbM by Quin has been
used to determine the origin of ACh release in this rat brain region. The
AChE-positive neurons seen in this study are similar in morphology to AChE
positive neurons observed previously (Fibiger, 1982) and to ChAT immunoreactive neurons in the nbM (Armstrong et al., 1983). A decrease in the number
and intensity of AChE-positive cells was observed 24 hr after Quin infusion.
Previously, we have shown that 24 hr after infusing Quin into the nbM, cortical high affinity choline uptake was decreased (Metcalf et al., 1987), suggesting that the ability of this pathway to function properly had been impaired. The decreased staining of cholinergic neurons seen in this study is
consistent with this hypothesis. Previous morphologic analysis of the cholinergic projection neurons suggests that they give off local collaterals
which display bouton-like swelling (Semba et al., 1987). The use of Quin to
destroy these projections selectively provides further support for the idea
that local cholinergic input to the nbM exists.

     The role of locally released ACh is at present unknown. A number of
other neurotransmitter systems, including glutamate, neuropeptide Y, the
enkephalins, neurotensin and somatostatin, are known to impinge on nbM neurons (Zaborszky and Braun, 1988). Locally released ACh from axon collaterals
may provide a means of regulating the activity of transmitter systems that
project to the nbM. Since iontophoretic application of cholinergic agonists,
including ACh, in vivo results in increased excitation of cortically projecting nbM neurons (Lamour et al., 1986), the ACh released locally in the nbM
may also modulate the activity of the cortically projecting cholinergic neurons themselves. In conclusion, destruction of nbM cholinergic neurons by
infusion of Quin into this region has demonstrated the local release of ACh
from cholinergic projection neurons.

Supported by Medical Research Council and Ontario Mental Health Foundation.

REFERENCES

Armstrong, D.M., Saper, C.B., Levey, A.I., Wainer, H., and Terry, R.D., 1983, Distribution of cholinergic neurons in rat brain, demonstrated by the immunocytochemical localization of choline acetyltransferase, J. Comp. Neurol., 216:53-68.

Fonnum, F., 1975, A rapid radiochemical method for the determination of choline acetyltransferase, J. Neurochem., 24:407-409.

Fibiger, H.C., 1982, The organization of some projections of cholinergic neurons of the mammalian forebrain, Brain Res. Rev., 4:327-388.

Jenden, D.J., Hanin, I., and Lamb, S.I., 1968, Gas chromatographic microestimation of acetylcholine and related compounds, Anal. Chem., 40:125-128.

Karnovsky, M.J., and Roots, L., 1964, A "direct-colouring" thiocholine method for cholinesterases, J. Histochem. Cytochem., 12:219-221.

Lamour, Y., Dutar, P., Rascol, O., and Jobert, A., 1986, Basal forebrain neurons projecting to the rat frontoparietal cortex: electrophysiological and pharmacological properties, Brain Res., 362:122-131.

Metcalf, R.H., and Boegman, R.J., 1987, The release of endogenous acetylcholine in tissue slices prepared from the rat nucleus basalis, Soc. Neurosci. Abstr., 13:326.4.

Metcalf, R.H., Boegman, R.J., Quirion, R., Riopelle, R.J., and Ludwin, S.K., 1987, Effect of quinolinic acid in the nucleus basalis magnocellularis on cortical high affinity choline uptake, J. Neurochem., 49:639-644.

Schwarcz, R., and Köhler, C., 1983, Differential vulnerability of central neurons of the rat to quinolinic acid, Neurosci. Lett., 38:85-90.

Semba, K., Reiner, P.B., McGeer, E.G., and Fibiger, H.C., 1987, Morphology of cortically projecting basal forebrain neurons in the rat as revealed by intracellular iontophoresis of horseradish peroxidase, Neuroscience, 20:637-651.

Stone, T.W., and Connick, J.H., 1985, Quinolinic acid and other kynurenines in the central nervous system, Neuroscience, 15:597-617.

Zaborszky, L., and Braun, A., 1988, Peptidergic afferents to forebrain cholinergic neurons, Soc. Neurosci. Abstr., 14:365.6.

THE ROLE OF TRANSAMINATIONS IN THE PHARMACOLOGICAL EFFECTS OF INDOLE-3-

PYRUVIC ACID

V. Politi, M.V. Lavaggi, G. De Luca, and A. Gorini

Polifarma Res. Centre
00155 Roma
Italy

INTRODUCTION

Indole-3-pyruvic acid (IPA), the keto-analog of tryptophan, is very much diffused in plants, where it can act as a phytohormone. In the past, it has been used as a substitute for tryptophan, with the aim to stimulate growth in rats (Jackson, 1930) or in chickens (Grigoriev and Truzhnikova, 1971). More recently, it has been shown that IPA decreases kynurenine formation by inhibiting tryptophan-2,3-dioxygenase (TPO; EC 1.13.11.11) (Lavaggi et al., 1987) and increases serotonin turnover in rat brains, inducing remarkable sedation and analgesia (Bacciottini et al., 1987). Because IPA can interfere with aromatic aminoacid metabolism, acting as substrate or inhibitor of transaminases, the present study was designed to find possible relationships between transaminases present in mammalian bodies and pharmacological effects observed after IPA administration.

RESULTS

Transamination of IPA is easily performed in several organ homogenates, and substantial amounts of tryptophan were detected even after a few minutes of incubation (data not shown). The same pattern of transamination appears activated in vivo, as high levels of tryptophan were found when IPA was administered to rats by injection or by the oral route. Fig. 1 shows that, after a single administration of IPA to rats (100 mg/kg "per os"), tryptophan increases very much in the brain and, apparently as a consequence of this, also serotonin and 5-OH-indoleacetic acid are synthesized inside the cerebral tissue. These results are in good agreement with those of Bacciottini et al. (1987) and suggest that the increased serotonin turnover in the brain is the explanation for the analgesic and sedative effects of IPA already described.

On the other hand, as an aromatic ketoacid, IPA could act on other aminotransferases, reducing their normal activity. One of the most interesting of such enzymes is tyrosine aminotransferase (TAT; EC 2.6.1.5), which is involved in the degradation of catecholamine precursor(s). Like TPO, TAT is stimulated by glucocorticoids and several other physiopathological stimuli (Groenwald et al., 1984). By reducing tyrosine available to the brain, TAT can act as an indirect hypertensive agent.

When tested on TAT, IPA shows a remarkable inhibiting effect in vitro,

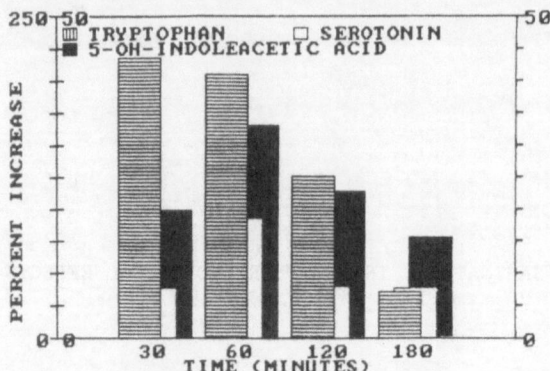

Fig. 1. IPA was administered "per os" at 100 mg/kg. Indoles were determined
by means of HPLC on all rat brain tissues. Bars express mean values
from six experiments.

as well as in vivo, in unstimulated and glucocorticoid-stimulated animals.
Fig. 2 illustrates a U-shaped curve, obtained measuring TAT activity in rats
stimulated by corticosteroids (hydrocortisone; 50 mg/kg, i.p.) and treated
with different amounts of IPA. The ketoacid is able to interfere with transaminases also when administered in very low amounts, and this can have a correlate in its pharmacological effects.

     In order to assess if an antihypertensive effect could derive from ac-
tivity on TAT, IPA was administered to Spontaneously Hypertensive Rats (SHR),
showing a clear dose-effect reduction in blood pressure after few days of
treatment (unpublished results). Normotensive control rats, however, did not
respond to the same treatment, suggesting that an abnormal metabolism of
cerebral catecholamines could be an important factor in the development of
hypertension in SHR. Furthermore, when IPA was administered for several
weeks to growing young SHR, no hypertension developed, in comparison with
age- and strain-matched untreated control rats.

     It is therefore possible to hypothesize that in the biochemical mechan-
isms of pathologic stress at least two factors might play a fundamental role:
on one side, the increased synthesis of TPO shifts tryptophan metabolism to
kynurenines, decreasing serotonin turnover in the brain, thus triggering the
"alert" system in the CNS. On the other side, TAT synthesis is able to induce indirectly systemic blood pressure increase. If this hypothesis is cor-

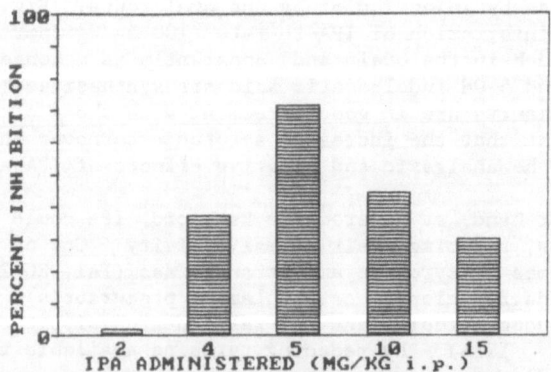

Fig. 2. IPA was administered to rats treated with glucocorticoids.
TAT activity was determined in liver.

rect, IPA could be seen as a major antagonist of pathologic stresses, due to its ability to simultaneously inhibit TPO and TAT. Finally, because some unstimulated aromatic aminotransferases showed a diurnal rhythm, with lowest levels in the morning and a 300% increase during the evening (Chia et al., 1984), it is possible that IPA's pharmacological effects are even more evident if the compound is administered to animals during specific hours of the day.

REFERENCES

Bacciottini, L., Pellegrini-Giampietro, D.E., Bongianni, F., De Luca, G., Beni, M., Politi, V., and Moroni, F., 1987, Biochemical and behavioral studies on indole-pyruvic acid: a keto-analogue of tryptophan, Pharmacol. Res. Comm., 19:803-817.

Chia, Y.C., Smith, G.W., and Lees, G.J., 1984, Differences in properties between aromatic amino acid: aromatic keto acid aminotransferases and aromatic amino acid: α-ketoglutarate aminotransferases, Life Sci., 34:2443-2452.

Grigoriev, N.G., and Truzhnikova, T.M., 1971, Formation of tryptophan from 3-indolylpyruvic acid in the organism of chicks, Sel'skokhoz. Biol., 6:97-101.

Groenwald, J.V., Terblanche, S.E., and Oelofsen, W., 1984, Tyrosine aminotransferase: characteristics and properties, Int. J. Biochem., 16:1-18.

Jackson, R.W., 1930, Indole derivatives in connection with a diet deficient in tryptophan, J. Biol. Chem., 84:1-21.

Lavaggi, M.V., Politi, V., De Luca, G., and Gorini, A., 1987, Effects of indole-3-pyruvic acid on tryptophan pyrrolase activity in vitro and in vivo, in: "Progress in Tryptophan and Serotonin Research", Bender, D.A., Joseph, M.H., Kochen, W., and Steinhart, H., eds., de Gruyter, Berlin, pp. 51-54.

ENDOCRINE RESPONSE OF PROLACTIN, CORTISOL, AND GROWTH HORMONE TO LOW DOSE

INTRAVENOUS L-TRYPTOPHAN IN HEALTHY SUBJECTS DURING DAY AND NIGHT

G. Hajak, A. Rodenbeck, J. Blanke, W. Wuttke[1], and E. Rüther

Department of Psychiatry
University of Göttingen
3400 Göttingen

[1]Clin. and Exp. Endocrinology
University of Göttingen
3400 Göttingen
Germany

INTRODUCTION

Neuroendocrine responses to intravenous (i.v.) administered L-trypto-phan (LTP) seem to be an index of brain serotonin (5-HT) function (Meites and Sonntag, 1981; Meltzer et al., 1982; Cowen, 1987, 1988). 5-HT seriously af-fects the regulation of the sleep wake cycle (Jouvet, 1984). Pituitary gland hormones such as growth hormone (GH), prolactin (PRL), and cortisol, show nycthemeral rhythmicity. Sleep as a part of a circadian rhythm has a masking effect on these hormones: maximal on GH, but minimal on cortisol (Parker et al., 1980; Clarenbach and Ries, 1985).

Resulting interactions between i.v. administration LTP, endocrine activ-ity, and sleep, have as yet only been shown for high doses (> 4 g) of LTP (McIndoe and Turkington, 1973; Woolf and Lee, 1977; Charney et al., 1982; Meltzer et al., 1984; Heninger et al., 1984; Cowen et al., 1985; Anderson and Cowen 1986; Glue et al., 1986; Cowen and Charig, 1987; Price et al., 1989). Our study was aimed to investigate dose-effect relations of low dose (1.0 g, 3.0 g, 5.0 g) i.v. administered LTP on neuroendocrine functions during day-time and during sleep.

SUBJECTS AND METHODS

20 healthy, male subjects (age 24-33, mean 27 ± 3) were randomly as-signed to the daytime (N = 10) and the overnight (n = 10) study. Every sub-ject performed four trials, with weekly intervals between the tests which were conducted from 8 a.m. to 2 p.m. and 10 p.m. to 7 a.m. respectively.

The subjects received a standardized xanthine and tyramine free and pro-tein and carbohydrate restricted diet from 24 hours prior to until the end of each test. During this period the subjects also refrained from alcohol. Within 40 minute periods the subjected received double-blind i.v. infusions of 0.0 (placebo), 1.0, 3.0, and 5.0 g LTP diluted in 500 ml of 0.9% NaCl so-lution. Blood was drawn from an antebrachial vein by a catheter every 15, 20 and 30 minutes, respectively, and collected into EDTA treated tubes. During the tests the following procedures were performed: continuous sleep recording with EEG, EOG, and submental EMG, Multiple-Sleep-Latency-Test (daytime-study), and self-rating with the CIPS-EWL-60-S-scale (Janke and Debus, 1978), the CIPS-SFA-scale and CIPS-SFB-scale (Görtelmeyer, 1986) for evaluation of mood changes and subjective sleep quality.

Biochemical methods: Plasma was rapidly separated and stored at -30°C before assay. PRL (Medgenix), GH (Biermann), and cortisol (Pharmacia) were assayed by a double antibody radioimmuno-assay system.

Data analysis: Within dose differences were analyzed by Wilcoxon matched pairs signed-rank test.

RESULTS

Daytime

1. PRL showed an increase of plasma concentrations following 1.0, 3.0, and 5.0 g LTP. No effect was seen with the placebo dose. The 3.0 and 5.0 g LTP dose responses differed significantly ($P < 0.05$) from the 1.0 g and placebo dose responses.

The maximal PRL response was obtained immediately after the infusions had been finished. The decreases of PRL levels were dose-dependently delayed to those of controls.

2. The above mentioned PRL responses were interindividually variable. 5 out of 10 subjects showed significant effects with 3.0 and 5.0 g of LTP (responders), while the other five subjects showed only insignificant or no effects at all (non-responders).

Nighttime

3. During sleep no significant effect of LTP on PRL was observed. In contrast to the daytime study, no initial increases in plasma PRL levels were noted. However, similar to the daytime results, there seems to be a tendency of clustering with respect to responders and non-responders, indicated by an elevation of PRL some hours after infusion in some subjects. However, onset and amount of this response could not yet be correlated to the respective LTP doses.

Daytime and nighttime

4. Neither cortisol nor GH levels revealed any significant effect suggesting that these hormones did not respond consistently to low-dose LTP.

CONCLUSIONS

The results of the present investigation indicate that i.v. LTP shows a neuroendocrine response, operationalized by increasing PRL plasma concentrations. This effect already shows up with low doses. There might be an important difference in the neuroendocrine response of subjects on LTP infusions (responders vs. non-responders). This could explain some controversial results of former precursor studies and contribute to the explanation of ambiguous therapeutical effects of LTP in patients with insomnia.

This study was supported by Fresenius AG, Homburg, Germany.

REFERENCES

Anderson, M., and Cowen, P.J., 1986, Clomipramine enhances prolactin and
   growth hormone responses to l-tryptophan, Psychopharmacology, 89:131-133.
Charney, D.S., Heninger, G.R., Reinhard, J.T., Sternberg, D.E., and Hafstead,
   K.M., 1982, The effect of intravenous l-tryptophan on prolactin and growth
   hormone and mood in healthy subjects, Psychopharmacology, 77:217-222.
Clarenbach, P., and Ries, F., 1985, Endocrinological and pharmacological aspects of slow-wave sleep in man, in: "Sleep 84", Koella, W.P., Rüther, E.,
   and Schulz, H., eds., Gustav Fischer, Stuttgart, New York, pp. 169-170.
Cowen, P.J., 1987, Psychotropic drugs and human 5-HT neuroendocrinology,
   Trends Pharmacol. Sci., 8:105-108.
Cowen, P.J., 1988, Neuroendocrine responses to tryptophan as an index of
   brain serotonin function, in: "Amino Acid Availability and Brain Function
   in Health and Disease", Hüther, G., ed., Springer, Berlin, Heidelberg, pp.
   285-290.
Cowen, P.J., and Charig, E.M., 1987, Neuroendocrine responses to intravenous
   tryptophan in major depression, Arch. Gen. Psych., 44:958-966.
Cowen, P.J., Gadhvi, H., Gosden, B., and Kolakowska, T., 1985, Response of
   prolactin and growth hormone to l-tryptophan infusion: effects in normal
   subjects and schizophrenic patients receiving neuroleptics, Psychopharmacology, 86:164-169.
Glue, P.W., Cowen, P.J., Nutt, D.J., Kolakowska, T., and Grahame-Smith, D.G.,
   1986, The effect of lithium on 5-HT mediated neuroendocrine responses and
   platelet 5-HT receptors, Psychopharmacology, 90:398-402.
Görtelmeyer, R., 1986, Schlaffragebogen A und B Selbstbeurteilungsskala, in:
   "Internationale Skalen für Psychiatrie", Collegium Internationale Psychi-
   atriae Scalarum, ed., Beltz, Weinheim.
Heninger, G.R., Charney, D.S., and Sternberg, D.E., 1984, Serotonergic func-
   tion in depression: prolactin response to intravenous tryptophan in de-
   pressed patients and healthy subjects, Arch. Gen. Psych., 41:398-402.
Janke, W., and Debus, G., 1978, "Die Eigenschaftswörterliste", Hofgrefe,
   Göttingen.
Jouvet, M., 1984, Indolamines and sleep-inducing factors, in: "Experimental
   Brain Research, Suppl. 8", Borbely, A., and Valatx, J.L., eds., Springer,
   Berlin, Heidelberg, pp. 81-94.
McIndoe, J.H., and Turkington, R.W., 1973, Stimulation of human prolactin
   secretion by intravenous infusion of l-tryptophan, J. Clin. Invest., 52:
   1972-1978.
Meites, J., and Sonntag, W.E., 1981, Hypothalamic, hypophysiotropic hormones
   and neurotransmitter regulation: current views, Ann. Rev. Pharmacol.
   Toxicol., 21:295-322.
Meltzer, H.Y., Lowy, M.T., Koyama, T., Robertson, A., Goodnick, P., and Jack-
   man, H.L., 1984, Stimulation of adrenal and pituitary hormone secretion
   by serotonergic agents in the affective disorders, Clin. Neuropharmacol.,
   Suppl. 1, 7:154-155.
Meltzer, H.Y., Wiita, B., Tricou, B.J., Simonovic, M., Fang, V.S., and Manov,
   G., 1982, Effect of serotonin precursors and serotonin agonists on plasma
   hormone levels, in: "Serotonin in Biological Psychiatry", Ho, B.T., et
   al., eds., Raven Press, New York, pp. 117-139.
Parker, D.C., Rossman, L.G., Kripke, D.F., Hershman, J.M., Gibson, W., Davis,
   C., Wilson, K., and Pekary, E., 1980, Endocrine rhythms across sleep wake
   cycles in normal young men under basal conditions, in: "Physiology in
   Sleep", Orem, J., and Barnes, C.D., eds., Academic Press, New York, pp.
   145-179.
Price, L.H., Ricaurte, G.A., Krystal, J.H., and Heninger, G.R., 1989, Neuroendocrine and mood responses to intravenous l-tryptophan in 3,4-methyl-
   enedioxymethamphetamine (MDMA) users, Arch. Gen. Psych., 46:20-22.
Woolf, P.D., and Lee, L., 1977, Effect of serotonin precursor, tryptophan, on
   pituitary hormone secretion, J. Clin. Invest., 52:1972-1978.

SPECIFIC BINDING OF L-TRYPTOPHAN TO SERUM ALBUMIN AND ITS FUNCTION IN VIVO

E. Sasaki, K. Saito, Y. Ohta, I. Ishiguro, Y. Nagamura,
R. Shinohara, H. Takahashi, and O. Tagaya[1]

Fujita-Gakuen Health University
Toyoake
Aichi 470-11

[1]Eisai Co. Ltd.
Kawashima
Gifu 483
Japan

INTRODUCTION

Many kinds of organic anions such as long-fatty acids, bilirubin, bile acids and hormones, bind to albumin in the plasma. L-Tryptophan (Trp) also binds to albumin in the plasma, and this property is quite different from those of other amino acids (Macmenamy and Oncley, 1958). Recent studies on the transport of organic anions into liver cells using perfused rat liver have suggested that ligands tightly bound to albumin rather than unbound ligands are transferred into liver cells readily, and that hepatic uptake of ligands is primarily mediated by the direct interactions between albumin-ligand complexes and the surface of liver cells (Weisigar et al., 1981).

Previously, we have demonstrated that bovine serum albumin plays a facilitating role in Trp uptake into isolated rat hepatocytes by interacting directly with the plasma membrane (Sasaki et al., 1987). However, the detailed role of serum albumin in hepatic Trp uptake in vivo remains unclear. We therefore investigated the role of serum albumin in Trp transport into liver using Nagase analbuminemic rats (NAR), which are the mutant strain established from Sprague-Dawley rats and are generally lacking in plasma albumin (Nagase et al., 1979).

MATERIALS AND METHODS

Normal male Sprague-Dawley rats were obtained from Clea Japan Inc., and male NAR were bred in our laboratory. Animals weighing 150-200 g were used throughout. Free serum Trp was separated by centrifugation through an ultra-filtration membrane Ultrafree C3TK (Millipore Co., Ltd.) as described previously (Sasaki et al., 1987). Concentrations of serum L-leucine (Leu) and L-phenylalanine (Phe) were measured by a Hitachi L-8500 amino acid analyzer. Pharmacokinetic parameters of Trp clearance were calculated by a non-linear iterative least square method. Rat hepatocytes were isolated by the collagenase perfusion method of Seglen (1976). Uptake of ^{3}H-Trp was determined by the method of Fehlmann et al. (1979) with a slight modification. Trp-2,3-dioxygenase (TPO) activity was assayed in fresh liver homogenates in the presence or absence of hematin (2 μmol) according to the method of Knox et al. (1970). Results are expressed as means ± SD. Student's t-test was used to determine a significant difference between NAR and normal rats.

RESULTS AND DISCUSSION

Trp is the only amino acid that exists mostly in an albumin-bound form in the plasma (Macmenamy and Oncley, 1958). Recently, a strain of rats (NAR) which genetically lack plasma albumin was established from Sprague-Dawley rats (Nagase et al., 1979). Although NAR have little plasma albumin, the concentration of total serum protein is known to be similar to that in normal rats (Nagase et al., 1979). Total serum Trp concentrations were first determined in NAR and normal rats fasted for 24 hours using blood samples obtained from the right jugular vein under pentobarbital anesthesia. The concentration of total serum Trp in NAR was 24.1 ± 4.7 μM and was significantly lower than in normal rats. In addition, most of the free Trp in NAR was slightly higher than in controls. Next, we examined the change in the serum concentrations in NAR and normal rats fasted for 24 hours after the injection of Trp (100 μmol/kg) into the right jugular vein (under pentobarbital anesthesia). Blood was drawn from the left jugular vein. As shown in Fig. 1, increased serum Trp levels observed after the Trp injection decreased rapidly in both NAR and normal rats, but the concentration of serum Trp in NAR was significantly lower than that in the control 30 seconds after Trp administration. The clearance of Leu or Phe, which are known to be unable to bind to albumin in the plasma, from the bloodstream was further examined in NAR and normal rats administered with each amino acid at the same dose as Trp (100 μmol/kg). Unlike in the case of Trp, the clearance rates of Leu and Phe

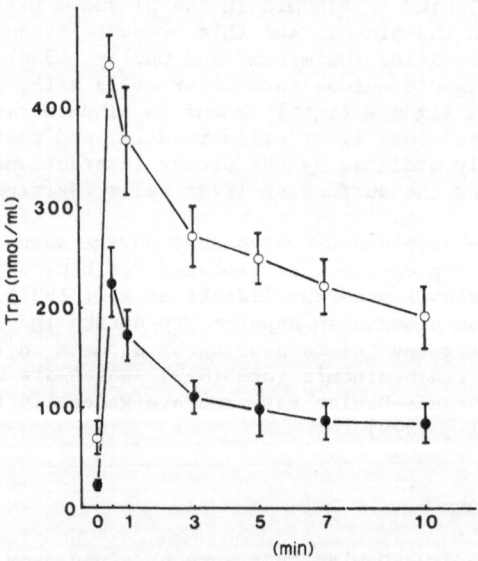

Fig. 1. Changes in serum Trp concentrations in NAR and normal rats after intravenous Trp administration. NAR and normal rats were fasted for 24 hours. Under pentobarbital (40 mg/kg) anesthesia, Trp (100 μmol/kg) was injected into the right jugular vein. After Trp administration, blood samples were drawn from the left jugular vein at the indicated times and serum Trp concentrations were determined by HPLC as described in the text. ●: NAR, ○: normal rats. Results are expressed as the mean ± SD (N = 5-11).

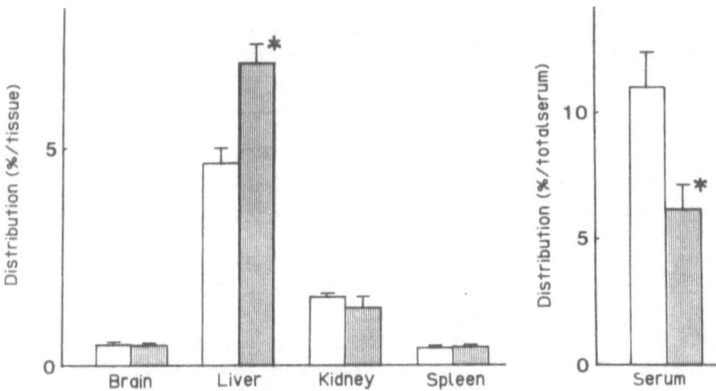

Fig. 2.   Distribution of radioactivity in tissues and sera of NAR and normal rats 10 min after the intravenous administration of ^{3}H-Trp.
Tissues and sera were taken from NAR and normal rats 10 min after
the intravenous administration of ^{3}H-Trp (100 μmol/kg).   The distribution of radioactivity originating from administered ^{3}H-Trp
was estimated from the radioactivity in the tissues and sera, expressed as a percentage of the total radioactivity administered.
Hatched bars: NAR, open bars: normal rats.   Results are the mean ±
SD (N = 5-8).   *P < 0.001.

in NAR were not different from those in controls (data not shown).   When
pharmacokinetic parameters of administered Trp were calculated from the disappearance curves of the amino acid from the bloodstream in NAR and normal
rats, the constant of Trp disappearance in NAR was larger than that in normal
rats at the first phase, but the disappearance constants of both groups were
similar at the second phase (data not shown).   The total distribution volume
of administered Trp in NAR was larger than that in normal rats (data not
shown).

The distribution of radioactivity in various tissues and serum was examined in NAR and normal rats 10 min after the intravenous administration of
^{3}H-Trp.   As shown in Fig. 2, the distribution of radioactivity in the liver
of NAR was significantly higher than that of the controls, but there was no
difference in the distribution of radioactivity between other tissues of both
groups.   Trp uptake into hepatocytes was further examined in NAR and control
rats which were not fasted, using isolated hepatocytes.   When the concentration of Trp used as substrate was 10 or 100 μM, the velocity of Trp uptake
into hepatocytes obtained from NAR in either cases was similar to that from
the control (data not shown).   When kinetic parameters for the Trp uptakes
were estimated from Lineweaver-Burk plots, there was no difference in K_m and
V_{max} values between NAR and normal rats (data not shown).   In addition, both
apo- and holoenzyme activities of liver TPO were not different between NAR
and normal rats (data not shown).   These results indicate that the catabolism
of Trp in the liver of NAR is similar to that of normal rats.   One can consider that in NAR, the rapid disappearance of administered Trp from the
bloodstream and the appearance of the amino acid in the liver are due to the
high concentration of free serum Trp which is caused by a low binding capacity of serum proteins to Trp.

In conclusion, the present results suggest that serum albumin allows the
concentration of total plasma Trp to remain high by its specific binding
ability to Trp.   In addition, free Trp in serum may regulate the transport of
Trp into hepatocytes since in NAR, lacking in plasma albumin, Trp was taken
up by the liver more rapidly than by other tissues.   It had been suggested

previously that free Trp may be an important factor in the regulation of Trp uptake by the liver (Smith and Pogson, 1980).

REFERENCES

Fehlmann, M., Cam, A.L., and Freychet, P., 1979, Insulin and glucagon stimulation of amino acid transport in isolated rat hepatocytes, J. Biol. Chem., 254:10431-10437.

Knox, W.E., Yip, A., and Reshelf, L., 1970, L-Tryptophan 2,3-dioxygenase. in: "Methods in Enzymology 17A", Tabor, H., and Tabor, C.W., eds., Academic Press, New York, London, pp. 415-421.

Macmenamy, R.H., and Oncley, J.L., 1958, The specific binding of L-tryptophan to serum albumin, J. Biol. Chem. 233:1436-1447.

Nagase, S., Shimamura, K., and Shumiya, S.,1979, Albumin deficient rat mutant, Science, 205:590-591.

Sasaki, E., Saito, K., Ohta, Y., Nagamura, Y., Shinohara, R., and Ishiguro, I., 1987, Facilitation of L-tryptophan uptake into rat hepatocytes by albumin. in: "Progress in Tryptophan and Serotonin Research", Bender, D.A., Joseph, M.H., Kochen, W., and Steinhart, H., eds., de Gruyter, Berlin, pp. 377-380.

Seglen, E.O., 1976, Preparation of isolated rat liver cells, Methods Cell Biol., 13:29-83.

Smith, S.A., and Pogson, C.I.,1980, The metabolism of L-tryptophan by isolated rat liver cells, Biochem. J., 186:977-986.

Weisigar, R., Golan, J., and Ockner, R., 1981, Receptor for albumin on the liver cell surface may mediate uptake of fatty acids and other albumin bound substances, Science, 211:1048-1051.

EFFECTS OF L-TRYPTOPHAN ON THE BRAINSTEM INDOLE AND CATECHOLAMINE METABOLITES

IN SPONTANEOUSLY HYPERTENSIVE RATS

D. Ghosh[1], B.O. Anyanwu[1], J. Guilford[2], and L.L. Henderson[1]

[1]Department of Biology
Texas Southern University

[2]College of Pharmacy
Houston, Texas 77004
USA

INTRODUCTION

L-Tryptophan has several beneficial effect in mammals including humans. The amino acid is known to alleviate insomnia (Hartman and Greenwald, 1984), and it does act as a natural tranquilizer as well as an analgesic (Lytle et al., 1975; Hosobuchi et al., 1980a,b). Two recent reports (Sved et al., 1982; Wolf and Kuhn, 1984) and our current laboratory observations (this paper) indicate that L-tryptophan elicits pronounced antihypertensive effects in spontaneously hypertensive rats (SHRs). Wistar-Kyoto rats (WKYs) were utilized as normotensive counterparts of SHRs in our experiments.

MATERIAL AND METHODS

All rats (both SHRs and WKYs obtained from Charles River Laboratories, Wilmington, MA) were male and ~3 months old. The animals were acclimatized for at least one week in our laboratory; during this period, the rats were further conditioned (exposed) to all the steps attendant with regular blood pressure monitoring by the Udenfriend tail-cuff method (Udenfriend, 1976). Each single brainstem sample was extracted in 0.05 M perchloric acid (= 16.7% homogenate). High pressure liquid chromatography/ electrochemical detection (HPLC/ECD) was performed on supernatants obtained from crude homogenates centrifuged at 15000 x g for 15 minutes (Kissinger, 1984; Brubaker, 1988). Norepinephrine (NE), dopamine (DA), 3,4-dihydroxyphenylacetic acid (DOPAC), homovanillic acid (HVA), 5-hydroxytryptamine (5-HT) and 5-hydroxyindole-3-acetic acid (5-HIAA) were assayed.

RESULTS AND DISCUSSION

(i) The mean control brainstem NE concentration of WKYs was 392 ± 32 ng/g (= 64%; N = 6) compared to 611 ± 30 ng/g (= 100%; N = 6) in SHRs. Thus, the spontaneously hypertensive rats showed 36% more of NE in the brainstem than their normotensive WKY counterparts. The two means were significantly different from each other ($P < 0.01$). Thus, it seems likely that elevated NE steady state level and its activity in the acceleratory cardiovascular center in the brainstem of SHRs would result in increased sympathetic outflow causing high blood pressure in these rats.

(ii) The 3-hour effect of L-tryptophan (300 mg/kg, intraperitoneal) in SHRs resulted in the brainstem NE value of 423 ± 23 ng/g (N = 6), a significant (31%) reduction (P < 0.01) compared to SHR controls (611 ± 30 ng/g); the reduced level of NE caused by L-tryptophan treatment was close to that of the normotensive WKY NE concentration. The systolic blood pressure of the L-tryptophan treated SHRs dropped by an average of 25 mm Hg from 175 ± 2 mm Hg to 150 ± 4 mm Hg (P < 0.01). Thus, the reduced level of NE caused by L-tryptophan treatment of the SHRs could be a key factor in the reduction of blood pressure in these rats.

(iii) Although DA in the brainstem apparently does not exert any effect on blood pressure, this biogenic amine is of significance as the immediate precursor of NE. The DA levels (in ng/g) in the brainstem were as follows:

(a) Wistar-Kyoto Normotensive rats (Controls):
86 ± 12 = 70% (N = 6)

(b) Spontaneously Hypertensive rats (Controls):
123 ± 26 = 100% (N = 6)

(c) Spontaneously Hypertensive rats (L-tryptophan treated; 300 mg/kg, 3 hours): 63 ± 10 = 51% (N = 6).

(iv) The control brainstem 5-hydroxytryptamine level in SHRs (252 ± 18 ng/g = 100%; N = 6) was somewhat higher than the corresponding steady-state concentration in WKYs (186 ± 14 ng/g = 74%), the difference between these two means being non-significant. The results do seem to indicate that higher level of 5-HT per se in SHRs compared to WKYs does not elicit any reduction of blood pressure in SHRs.

L-Tryptophan treatment caused a 37% increase in 5-HT (399 ± 52 ng/g) in the brainstem of the SHRs over the corresponding controls (252 ± 18 ng/g; P < 0.05). Such an increase is expected because the brainstem is rich in clusters of serotonergic neurons (the raphe nuclei). The raphe nuclei are capable of synthesizing significantly more of 5-HT than most other discrete areas of the brain. Reduction in blood pressure in SHRs could be due to simultaneous depression of NE and increase in 5-HT activities in the brainstem of these rats; however, infusion with 5-hydroxytryptophan has been demonstrated to increase serotonin turnover and to reduce blood pressure in normotensive male Sprague-Dawley rats (Freed et al., 1986).

Table 1. Brainstem concentrations of DOPAC and HVA

		DOPAC	HVA
a.	WKY (Control)	48 ± 8 (80%)	70 ± 8 (100%)
b.	SHR (Control)	60 ± 9 (100%)	70 ± 12 (100%)
c.	SHR (L-tryptophan) treated	79 ± 3 (123%)	61 ± 6 (88%)

Values are expressed in ng/g (N = 6).

(v) 5-HIAA is produced through the enzymatic conversion of 5-HT. 5-HIAA is an important index in the cerebral turnover of 5-HT. Control brainstem values of 5-HIAA in WKYs and SHRs were 342.5 ± 20.7 ng/g and 416 ± 20 ng/g, respectively (P < 0.01). L-Tryptophan treatment in SHRs caused a 3.2-fold increase in the brainstem 5-HIAA (1319 ± 68 ng/g) over the SHR controls (P < 0.001).

(vi) Brain DOPAC and HVA are produced from DA through the action of monoamine oxidase and, subsequently, aldehyde dehydrogenase (DOPAC) or catecholamine-0-methyltransferase (HVA).

The difference in DOPAC levels between WKY and SKR controls was non-significant; also, L-tryptophan-treated SHRs showed a non-significant 14% elevation of DOPAC compared to SHR controls. HVA levels did not exhibit any significant variation in WKYs and SHRs.

This work was supported by NIH grant RR 03045.

REFERENCES

Brubaker, A., 1988, personal communication, Bioanalytical Systems, Inc., West Lafayette, Indiana.
Freed, C.R., Echizen, H., and Bhaskaran, D., 1986, Serotonin metabolism and blood pressure regulation: insights form brain tissue assays and in vivo electrochemical recording, in: "Monitoring Neurotransmitter Release during Behavior", Joseph, M.H., Fillenz, M., MacDonald, I.A., and Marsden, C.A., eds., Ellis Horwood Ltd., Chichester, England, pp. 133-143.
Hartman, E., and Greenwald, D., 1984, Tryptophan and human sleep, in: "Progress in Tryptophan and Serotonin Research", Schlossberger, H.G., Kochen, W., Linzen, B., and Steinhart, H., eds., de Gruyter, Berlin, pp. 297-304.
Hosobuchi, Y., Lamb, S., and Bascom, D., 1980a, Tryptophan loading may reverse tolerance to opiate analgesics in humans, Pain, 9:161-169.
Hosobuchi, Y., Rossier, J., and Bloom, F.E., 1980b, Oral loading with L-tryptophan may augment the simultaneous release of ACTH and beta-endorphin that accompanies periaqueductal stimulation in humans, in: "Neural Peptides and Neuronal Communication", Costa, E., and Trabucchi, M., eds., Raven Press, New York, pp. 563-570.
Kissinger, P.T., 1984, Liquid chromatography/electrochemistry determination of biogenic amines: improved performance using low dead volume multiple electrode transducers, in: "Progress in Tryptophan and Serotonin Research", Schlossberger, H.G., Kochen, W., Linzen, B., and Steinhart, H., eds., de Gruyter, Berlin, pp. 45-43.
Lytle, L.D., Messing, R.B., Fisher, L., and Phebus, L., 1975, Effects of long-term consumption on brain serotonin and the response to electric shock, Science, 19:692-694.
Sved, A.S., Van Itallie, C.M., and Fernstrom, J.D., 1982, Studies on the antihypertensive action of L-tryptophan, J. Pharmacol. Exp. Ther., 221: 329-333.
Udenfriend, S., 1976, Committee on care and use of spontaneously hypertensive rats (SHRs), ILAR News, 19:G1-G20.
Wolf, W.A., and Kohn, D.M., 1984, Antihypertensive effects of L-tryptophan are not mediated by brain serotonin, Brain Res., 295:356-359.

EFFECTS OF TRYPTOPHAN ON THE DEVELOPMENT OF DEOXYCORTICOSTERONE ACETATE

(DOCA)-INDUCED HYPERTENSION IN RATS

M.J. Fregly

Department of Physiology
University of Florida
College of Medicine
Gainesville, Florida 32610
USA

INTRODUCTION

For some years, we have been interested in the role of neutral amino acids in the development and maintenance of hypertension. Our initial studies
with tyrosine suggested that chronic dietary treatment with this essential
amino acid could provide modest protection against the development of renal
hypertension in rats (Lockley et al., 1985). However, it was considerably
less effective in protecting against the development of deoxycorticosterone
(DOCA)-induced hypertension (Henley et al., 1986).

The experiments described here were carried out to assess the effect of
another neutral amino acid, L-tryptophan, on the development of DOCA-salt-induced hypertension in rats. A number of investigators has administered tryptophan acutely to spontaneously hypertensive rats (SHR) and assessed its effect on blood pressure (Sved et al., 1982; Wolf and Kuhn, 1984). There appears to be general agreement that acute doses in excess of 50 mg of trypto-
phan/kg, i.p., reduce the blood pressure of SHR. Within 4 hours after treatment, blood pressure returned to control level. Normotensive animals do not
respond in the same fashion and may actually show an increase in blood pressure. The possibility that the reduction in blood pressure of hypertensive
rats accompanying acute treatment with tryptophan occurs by way of an increased synthesis and release of serotonin is controversial. The weight of
the evidence does not appear either to support or negate the possibility.

Studies to assess the effect of chronic administration of tryptophan on
the development and maintenance of hypertension in rats, or any other animals, do not appear to have been conducted. Therefore, this was a major objective of the experiments reported here. In addition, an assessment was
made of potential mechanisms for the antihypertensive effect of tryptophan.

METHODS AND RESULTS

Two separate experiments were carried out. The methods and techniques
used have been described elsewhere (Fregly and Fater, 1986; Fregly et al.,
1987b). Statistical analyses were carried out by means of a one-way ANOVA,

using the pooled variance and t-test to assess differences between individual means.

Experiment 1: Effect of chronic dietary treatment with tryptophan on development of DOCA-induced hypertension

Thirty-two female Sprague-Dawley rats (180-225 g) were used. They were divided into four equal groups and allowed Purina Laboratory Chow and tap water ad libitum. During a two-week control period, systolic blood pressure and body weight were measured three times. All rats were then unilaterally nephrectomized while anesthetized with pentobarbital. At this time, 3 of the 4 groups were implanted s.c. with two 50-mm Silastic tubes filled with crystalline DOCA as described elsewhere (Fregly et al., 1987b). These tubes released 800 µg DOCA/kg/day. Two of the three DOCA-treated groups were administered 2.5 and 5.0% L-tryptophan mixed thoroughly into their food. In addition, all rats were allowed only isotonic saline to drink. The fourth group was uninephrectomized and given isotonic saline, but was treated with neither DOCA nor tryptophan.

The systolic blood pressure for the four groups of rats measured weekly during the course of the experiment is shown in Fig. 1. Beginning about the second week after unilateral nephrectomy, blood pressure of all rats receiving DOCA and tryptophan was reduced significantly below that of the group given DOCA alone for the duration of the experiment. Neither blood pressure nor body weight of the groups given the combined treatment differed significantly from those of the untreated controls.

During the fifth week after nephrectomy, the rats were caged individually and intake of isotonic saline and food, as well as urine output, were measured daily for 5 days. Treatment with DOCA increased significantly the daily intake of isotonic saline while administration of tryptophan to DOCA-treated rats reduced intake of saline in a dose-related fashion (Fig. 2). Output of urine paralleled intake of isotonic saline for each of the four groups. Intake of food was slightly, but not significantly, reduced in the two groups receiving tryptophan. Calculation of the daily intake of tryptophan added to the diet of the two treated groups revealed that 1.38 and 2.75 g tryptophan/kg of body weight, respectively, were ingested.

During the thirteenth week, the rats were killed and the heart removed, drained of blood, blotted and weighed on a torsion balance. Ratio of heart/body weight of the DOCA-treated group [360 ± 15 (SEM) mg/100 g body weight] was significantly (P < 0.05) greater than control (308 ± 9). Heart weights of the two tryptophan-treated groups were between these and did not differ significantly from either (2.5% group: 333 ± 11; 5.0% group: 336 ± 19).

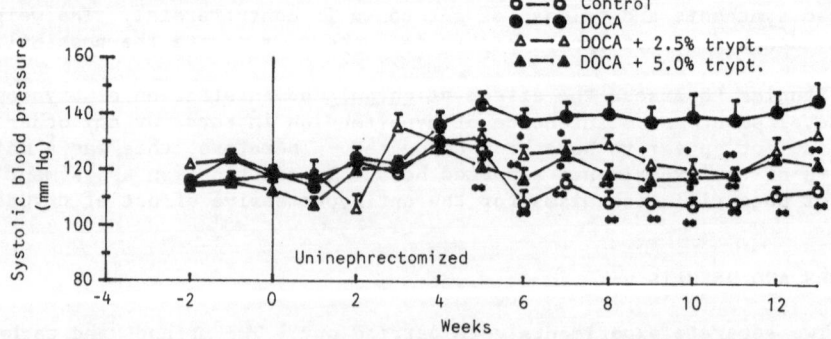

Fig. 1. Effect of chronic dietary treatment with tryptophan on the development of DOCA-salt-induced hypertension. Symbols are the mean ± SEM. *P < 0.05; **P < 0.01 compared to DOCA-treated group.

Experiment 2: Effect of chronic dietary treatment with tryptophan on brain catecholamines and on the binding of angiotensin II (Ang II) to diencephalic tissue of DOCA-treated rats

This experiment was carried out in a fashion identical to that of Experiment I. At the end of this experiment (week 15), the rats were killed by decapitation and their brains were rapidly removed from the skull. A detailed description of the membrane binding assay for Ang II in diencephalic membranes has been published (Wilson et al., 1986).

Chronic treatment with tryptophan returned the increased receptor binding of Ang II in the diencephalon of the brains of DOCA-treated rats toward control levels (Fig. 3).

For analysis of catecholamines in brain, the mesencephalon was removed, weighed, and homogenized as described earlier (Fregly et al., 1987b). The filtrate was assayed for norepinephrine, dopamine, serotonin (5-HT), and 5-hydroxyindole acetic acid (5-HIAA) by HPLC. All values were standardized to a unit of wet tissue weight.

The content of 5-HT in the mesencephalon did not increase after chronic treatment with tryptophan. However, 5-HIAA did show a significant ($P < 0.01$) increase (DOCA: 915 ± 21; DOCA + 2.5%: 1310 ± 73; DOCA + 5.0%: 1693 ± 110 ng/g tissue). This suggests that the turnover (5-HIAA/5-HT) of 5-HT was increased by treatment with dietary tryptophan. The content of norepinephrine and dopamine in the mesencephalic portion of the brain was unaffected by treatment with tryptophan.

Chronic treatment with DOCA induced a significant hypertrophy of the heart while treatment with tryptophan attenuated the hypertrophy (control: 301 ± 12; DOCA: 384 ± 23; DOCA + 2.5%: 347 ± 16; DOCA + 5.0%: 310 ± 16 mg/100 g body weight). The heart weight of the DOCA-treated group was significantly ($P < 0.01$) greater than that of the control group. The heart weight of the group treated with 5.0% tryptophan was significantly ($P < 0.01$) less than that of the DOCA-treated group.

DISCUSSION

The data presented here demonstrate that chronic dietary administration of the neutral amino acid, L-tryptophan, can attenuate the development of DOCA-induced hypertension and cardiac hypertrophy in rats without significantly affecting either their food intake or body weight.

An earlier study from this laboratory showed that chronic treatment with DOCA increased the number (B_{max}), but not the affinity (K_d) of the specific binding of Ang II to its receptors in the diencephalon (HTS) of the rat (Wilson et al., 1986). Chronic administration of either tryptophan or 5HTP to DOCA-treated rats returned to control level the specific binding of Ang II (Fregly et al., 1987a,b). We have postulated that an increase in the number of Ang II receptors in the diencephalon of DOCA-treated rats contributes to the development of hypertension, and thus, the antihypertensive effect of tryptophan may be associated with a reduction in specific binding of Ang II in the HTS.

These studies also showed that chronic administration of tryptophan reduced the spontaneous intake of a 0.15 M NaCl solution by DOCA-treated rats. Since ingestion of NaCl solution is an important cofactor in the development of this type of hypertension, reduction in blood pressure by tryptophan may be related in some fashion to the reduction in intake of NaCl solution. However, tryptophan has been shown to manifest antihypertensive effects in other

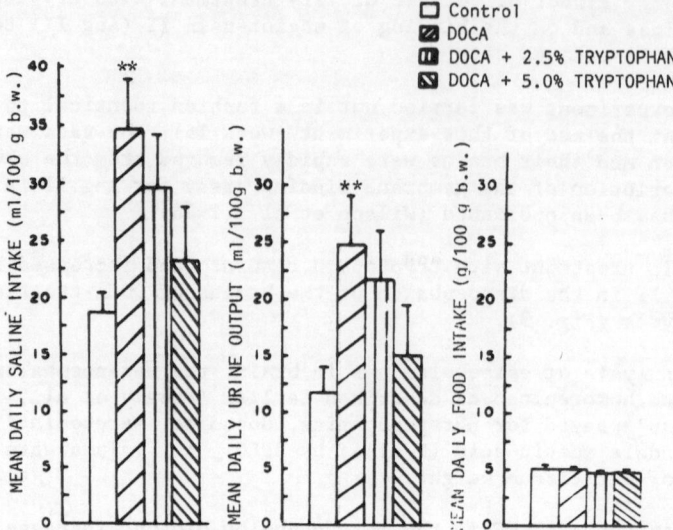

Fig. 2. Mean (± SEM) daily intake of isotonic saline (A), output of urine
(B) and intake of food (C) are shown for the groups designated in
the figure. **P < 0.01 compared to the control group.

models of hypertension, e.g. renal and SHR, that do not require ingestion of
isotonic saline as a cofactor (Fregly et al., 1988, 1989).

The increased production and turnover of serotonin may be involved in
the antihypertensive effects of tryptophan as suggested by the fact that
chronic administration of 5-hydroxytryptophan to DOCA-treated rats prevented
the development of hypertension, reduced intake of a NaCl solution and returned to control level the increased binding of Ang II (Fregly et al.,
1987a). Thus, a number of potential mechanisms present themselves as explanations for the antihypertensive effects of tryptophan. Additional studies
will be required to determine which of these, or others, is most important.

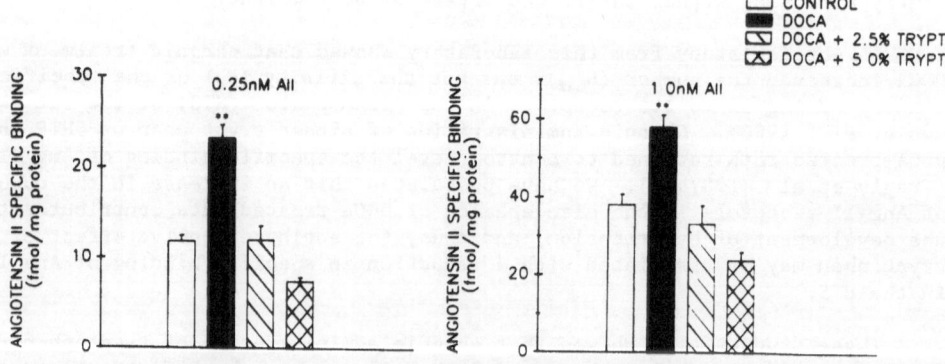

Fig. 3. Specific binding of Ang II to membranes from the diencephalic portion of brain from experimental rats. Measurements were made at
0.25 (A) and 1.00 nM [^{125}I]-Ang II. Data are the mean ± SEM. **P
< 0.01 compared to the control group. T: tryptophan.

REFERENCES

Fregly, M.J., and Fater, D.C., 1986, Prevention of DOCA-induced hypertension by chronic treatment with tryptophan, Clin. Exp. Pharmacol. Physiol., 13: 767-776.

Fregly, M.J., Lockley, O.E., and Cade, J.R., 1988, Effect of chronic dietary treatment with L-tryptophan on the development of renal hypertension in rats, Pharmacology, 36:91-100.

Fregly, M.J., Lockley, O.E., and Sumners, C., 1987a, Chronic treatment with L-5-hydroxytryptophan prevents the development of DOCA-salt-induced hypertension in rats, J. Hypertens., 5:621-628.

Fregly, M.J., Lockley, O.E., van der Voort, J., Sumners C., and Henley, W.N., 1987b, Chronic dietary administration of tryptophan prevents development of DOCA-salt-induced hypertension in rats, Can. J. Physiol. Pharmacol., 65:753-764.

Fregly, M.J., Sumners, C., and Cade, J.R., 1989, Effect of chronic dietary treatment with L-tryptophan on the maintenance of hypertension in spontaneously hypertensive rats, Can. J. Physiol. Pharmacol., 67:656-662.

Henley, W.N., Fregly, M.J., Wilson, K.M., and Hathaway, S., 1986, Physiologic responses to chronic dietary tyrosine supplementation in DOCA-salt-treated rats, Pharmacology, 33:334-347.

Lockley, O.E., Fregly, M.J., and Fater, D.C., 1985, Effect of L-p-tyrosine on the development of renal hypertension in rats, Pharmacology, 31:132-149.

Sved, A.F., van Itallie, C.M., and Fernstrom, J.D., 1982, Studies on the antihypertensive action of L-tryptophan, J. Pharmacol. Exp. Therap., 221:329-333.

Wilson, K.M., Sumners, C., Hathaway, S., and Fregly, M.J., 1986, Mineralocorticoids modulate central angiotensin II receptors in rats, Brain Res., 382:87-96.

Wolf, W.A., and Kuhn, D.M., 1984, Effects of L-tryptophan on blood pressure in normotensive and hypertensive rats, J. Pharmacol. Exp. Therap., 230: 324-329.

CYTOTOXICITY OF 3-HYDROXYKYNURENINE: IMPLICATIONS FOR CNS DAMAGE IN NEONATAL VITAMIN B-6 DEFICIENCY

C.L. Eastman and T.R. Guilarte

Department of Environmental Health Sciences
Johns Hopkins University School of Hygiene and Public Health
Baltimore, Maryland 21205
USA

INTRODUCTION

Pyridoxal phosphate, the biologically active form of vitamin B-6, plays a central role in the metabolism of amino acids and biogenic amines. A deficiency of this vitamin in human infants or in neonates of various species results in a deficiency syndrom characterized by marked neurological impairment (Coursin, 1969). The neurological deficits associated with neonatal vitamin B-6 deficiency include ataxia, irritability, tremor and seizures (Coursin, 1969). CNS histopathological changes resulting from neonatal vitamin B-6 deficiency have not been extensively studied. However, several investigators have reported neuropathological alterations suggestive of impaired development, neuronal loss or damage, or premature aging of CNS neurons (Morre et al., 1978; Root and Longenecker, 1983; Wasynczuk et al., 1983).

Despite increasing knowledge about the role of vitamin B-6 in general brain metabolism, the specific mechanism(s) by which the deficiency state affects the development and function of the CNS are still not well understood. Recent studies in this and other laboratories have implicated the tryptophan metabolite, 3-hydroxykynurenine (3HK), as having a possible role in the sei-

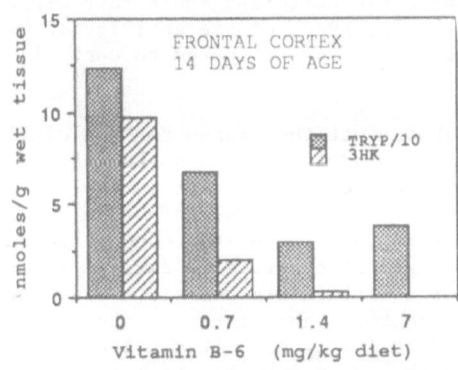

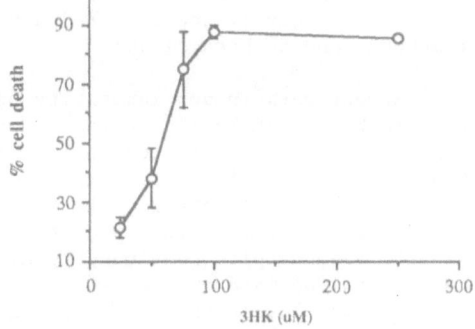

Fig. 1. Levels of tryptophan (1/10 x measured values) and 3HK as a function of maternal dietary vitamin B-6.

Fig. 2. Dose-response relationship for 3HK-induced cell death. Cultures were exposed to 3HK for 24 hours.

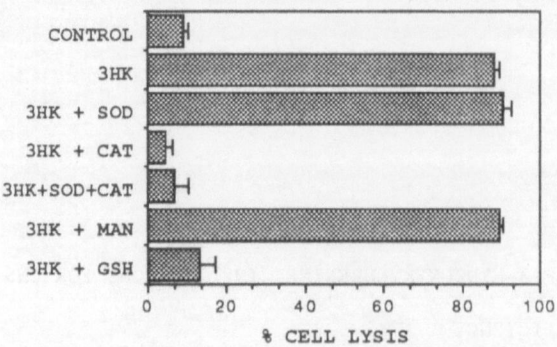

Fig. 3.   Effect of antioxidants on 3HK-induced cell death.   Cultures were
          exposed to 200 μM 3HK for 24 hours.

zure disorder associated with neonatal vitamin B-6 deficiency (Guilarte et
al., 1988).   In addition, our data indicate that 3HK may have neurotoxic
properties which may contribute to CNS damage resulting from inadequate dietary intake of vitamin B-6.

ELEVATION OF BRAIN 3HK IN NEONATAL VITAMIN B-6 DEFICIENCY

     Vitamin B-6 plays an important role in the metabolism of tryptophan
through the kynurenine pathway.   3HK is known to be metabolized to 3-hydroxy-
anthranilate (3HA) by the B-6 dependent enzyme, kynureninase.   Previous studies have shown that a dietary deprivation of vitamin B-6 in rodents produces
a significant reduction in the activity of tissue kynureninase (Takeuchi and
Shibata, 1984).   This enzyme is exquisitely sensitive to vitamin B-6 deficiency, and the administration of a loading dose of tryptophan results in the
excretion of elevated levels of 3HK and other kynurenine metabolites in the
urine (Henderson and Hulse, 1978).

     We have previously reported that increased concentrations of 3HK are
present in the brains of vitamin B-6 deficient neonatal rats but not in B-6
deficient adult rats (Guilarte and Wagner, 1987).   Elevated 3HK levels were
measured without the administration of exogenous tryptophan.   This was attributed to the blockade of the metabolism of 3HK to 3HA as a result of cofactor depletion for the enzyme, kynureninase.   However, this explanation
could not account for the fact that vitamin B-6 deficient adult rats did not
show increased brain 3HK despite the fact that kynureninase activity is known
to be significantly lower in B-6 depleted adult rats compared to controls
(Takeuchi and Shibata, 1984).

     Recent work in our laboratory has shown that the reason for the differ-
ence in brain 3HK levels between neonatal and adult rats is an increased concentration of brain tryptophan (TRYP) in the neonatal but not in the adult B-
6 deficient rat.   The data in Figure 1 show that the levels of brain TRYP and
3HK in the frontal cortex of 14 day old rats increase as the concentration of
vitamin B-6 decreases in the diet.   This was not observed in rats at 28 or 56
days of age, despite continued feeding of a vitamin B-6 deficient diet, or in
adult rats deprived of vitamin B-6 for 2 months.   Therefore, it appears that
the elevated levels of brain 3HK measured in 14 day old vitamin B-6 deficient
rat brain are the result of increased uptake and metabolism of TRYP in the
brain coupled to the decrease in kynureninase activity due to cofactor depletion.

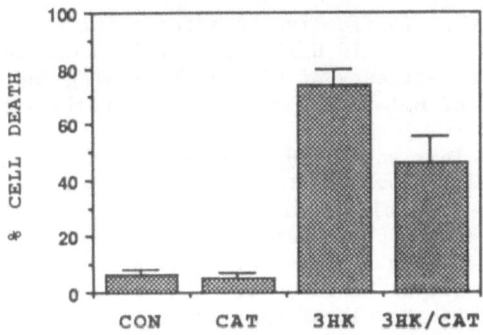

Fig. 4. Effect of post-exposure application of CAT on 3HK toxicity. Cell
lysis was assayed 24 hrs after a 2 hr exposure to 500 μM 3HK.

IN VITRO TOXICITY OF 3HK

A series of reports have shown that several kynurenine pathway metabolites, including 3HK, induce convulsions upon i.c.v. administration in rodents (Lapin, 1978, 1981; Pinelli et al., 1984). One of these, quinolinic
acid, has excitotoxic properties and has been postulated as having a possible
etiological role in neurodegenerative disorders (Schwarcz et al., 1984).
While the neurotoxicity of quinolinic acid has been well characterized, the
neurotoxic properties of other convulsant kynurenines have not previously
been evaluated. Since CNS levels of 3HK are elevated dramatically as a result of neonatal vitamin B-6 deficiency, the neurotoxicity of this metabolite
is of particular interest.

We have investigated the toxic properties of 3HK in a mouse neuroblastoma x embryonic rat retinal neuron hybrid cell line (N18-RE-105). 3HK was
the most toxic among several kynurenine metabolites tested, and its in vitro
toxicity was evident at concentrations comparable to those previously measured in CNS tissue from vitamin B-6 deficient rat pups (Guilarte and Wagner,
1987). The dose-response curve (Fig. 2) for 3HK-induced cell lysis is quite
steep and concentrations above 100 μM were toxic to virtually all cultured
cells within 24 hours (Eastman and Guilarte, 1989).

Exposure to other kynurenine metabolites (500 μM for 24 hrs) was either
non-toxic (L-kynurenine and quinolinate) or much less toxic (xanthurenate) to
N18-RE-105 cells than 3HK (Eastman and Guilarte, 1989). The failure of quinolinate to kill these cells is explained by the lack, in N18-RE-105 cells, of
the excitatory amino acid receptor subtype at which quinolinate is thought to
act (Malouf et al., 1984). The lack of toxicity of L-kynurenine, a metabolic
precursor and close structural analog of 3HK, suggests that the presence of a
phenolic function may be an important determinant of toxicity. This suggestion is supported by the observed toxicity of xanthurenate. A possible mechanistic basis for the importance of a phenolic function is that it may serve
to permit oxidation of 3HK or xanthurenate to quinoid products. The oxidation of 3HK to yield H_2O_2 and the corresponding quinoneimine has been described previously (Tomoda et al., 1986).

A role for oxidative stress in 3HK toxicity is supported by the ability
of antioxidant treatments to prevent toxic cell death (Fig. 3). The protective effects of superoxide dismutase (SOD), catalase (CAT), glutathione
(GSH), and the hydroxyl radical scavenger, mannitol (MAN), have been assessed
in cells exposed to 200 μM 3HK (Eastman and Guilarte, 1989). None of the
antioxidants, alone, had any effect on baseline cell lysis, and MAN (10 μM)

627

and SOD (150 U/ml) were ineffective in attenuating toxicity. However, 3HK toxicity was abolished when 10 U/ml CAT or 2.5 mM GSH were present in the incubation medium. The efficacy of CAT in abolishing 3HK toxicity indicates that the generation of H_2O_2 is a critical step in the toxic process.

It is not yet clear whether H_2O_2 is generated within affected cells or in the culture medium. However, the presence of an intracellular pool of H_2O_2 is indicated by the ability of CAT applied after termination of a short 3HK exposure to attenuate cell lysis. Such delayed toxicity is evident when cell death is assayed both immediately after a 2 hr exposure to 500 μM 3HK and after an additional 22 hr incubation in fresh culture medium (Eastman and Guilarte, 1989). While levels of cell lysis measured immediately upon termination of exposure do not differ from control levels, substantial cell death is observed in 3HK-exposed cultures harvested 24 hrs after the start of the exposure period. When CAT was added to the post-exposure culture medium, the extent of 3HK-induced cell lysis was significantly diminished (Fig. 4). Thus, the H_2O_2 produced in, or residing in, a compartment other than the culture medium must be involved in delayed cell lysis.

DISCUSSION

Inadequate vitamin B-6 nutrition in the neonatal period results in the development of neurological signs including ataxia, irritability, tremor, seizures, and CNS neuropathology. Neonatal vitamin B-6 deficiency also results in a dramatic increase of brain 3HK which occurs approximately coincidently with the onset of neurological signs in neonatal rats. Such an elevation of 3HK is not seen in adult rats which do not experience similar neurological impairment as a result of prolonged vitamin B-6 deprivation.

While the convulsant properties of 3HK and related compounds have been described by others, the possible neurotoxic properties of 3HK have not been studied previously. We have shown that 3HK, at concentrations in the range of those previously measured in CNS tissue from vitamin B-6 deficient rat pups, is toxic to a neuronal cell line. Our data also demonstrate that the generation of H_2O_2, probably as a result of the oxidation of 3HK, is a critical step in the toxic process. If the in vitro dose-response parameters described here reflect the in vivo vulnerability of neuronal elements to 3HK, the dramatic increase in brain 3HK levels may lead to oxidative stress and CNS damage in neonatal vitamin B-6 undernutrition.

REFERENCES

Beal, M.F., Kowall, N.W., Ellison, D.W., Mazurek, M.F., Swartz, K.W., and Martin, J.B., 1986, Replication of the neurochemical characteristics of Huntington's disease by quinolinic acid, Nature, 321:168-171.

Coursin, D.B., 1969, Vitamin B-6 and brain function in animals and man, Ann. N.Y. Acad. Sci., 166:7-15.

Eastman, C.L., and Guilarte, T.R., 1989, Cytotoxicity of 3-hydroxykynurenine in a neuronal hybrid cell line, Brain Res., 495:225-231.

Guilarte, T.R., Block, L.D., and Wagner, H.D., 1988, The putative endogenous convulsant 3-hydroxykynurenine decreases benzodiazepine receptor binding affinity: implications to seizures associated with neonatal vitamin B-6 deficiency, Pharmacol. Biochem. Behav., 30:665-668.

Guilarte, T.R., and Wagner, H.D., 1987, Increased concentrations of 3-hydroxykynurenine in vitamin B-6 deficient neonatal rat brain, J. Neurochem., 49:1918.

Henderson, L.M., and Hulse, J.D., 1978, Vitamin B6 relationship in tryptophan metabolism, in: "Human Vitamin B6 Requirements", Nat. Acad. Sci., Washington, D.C., pp. 21-36.

Lapin, I.P., 1978, Stimulant and convulsive effects of kynurenines injected into brain ventricles in mice, J. Neural Transm., 42:37-43.

Lapin, I.P., 1981, Kynurenines and seizures, Epilepsia, 22:257-265.

Malouf, A.T., Coyle, J.T., and Schnaar, R.L., 1984, Agonists and cations regulate the glutamic acid receptors on inact neuroblastoma hybrid cells, J. Biol. Chem., 259:12763-12768.

Morre, D.M., Kirksey, A., and Das, G.D., 1978, Effects of vitamin B-6 deficiency on the developing central nervous system of the rat. Gross measurements and cytoarchitectural measurements, J. Nutr., 108:1250-1259.

Pinelli, A., Ossi, C., Columbo, R., Tofanetti, O., and Spazzi, L., 1984, Experimental convulsions in rats induced by intraventricular administration of kynurenine and structurally related compounds, Neuropharmacology, 23:333-337.

Root, E.J., and Longenecker, J.B., 1983, Brain cell alterations suggesting premature aging induced by dietary deficiency of vitamin B-6 and/or copper, Am. J. Clin. Nutr., 37:540-553.

Schwarcz, R., Foster, A.C., French, E.D., Whetsell, W.O., and Köhler, C., 1984, Excitotoxic models for neurodegenerative diseases, Life Sci., 35: 19-32.

Takeuchi, F., and Shibata, Y., 1984, Kynurenine metabolism in vitamin B-6 deficient rat liver after tryptophan injection, Biochem. J., 220:693-699.

Tomoda, A., Shirasawa, E., and Yoneyama, Y., 1986, Reactions of oxy- and met-hemoglobin with tryptophan metabolites 3-hydroxykynurenine and 3-hydroxyanthranilic acid, Hemoglobin, 10:33-44.

Wasynczuk, A., Kirksey, A., and Morre, D.M., 1983, Effects of maternal vitamin B-6 deficiency on specific regions of developing rat brain: the extrapyramidal motor system, J. Nutr., 113:746-754.

PELLAGRA, MYCOTOXINS AND TRYPTOPHAN-NIACIN METABOLISM

R.B. Sashidhar, Y. Ramakrishna[1], and R.V. Bhat[1]

Department of Biochemistry
University College of Science
Osmania University
Hyderabad 500 007

[1]Food and Drug Toxicology Research Centre
National Institute of Nutrition
Hyderabad 500 007
India

INTRODUCTION

Pellagra has long been known to be a nutritional syndrome caused by niacin deficiency. Endemic pellagra has been traditionally associated with consumption of corn-based diet and is uncommon where rice or wheat is the staple
food. Pellagra, in an endemic form among sorghum (Sorghum vulgare) eaters,
was first described among poor agricultural laborers in Hyderabad, India
(Gopalan and Srikantia, 1960). Extensive studies from India have shown that
leucine toxicity, leucine-isoleucine imbalance and concomitant deficiency of
vitamin B_6 may be the causative factors in the pathogenesis of pellagra in
sorghum eaters (Srikantia, 1978). Several studies in pellagrins and in experimental animals have shown that there is a disturbance in tryptophan-niacin metabolism. Although rats do not show the characteristic dermal changes
of pellagra, they show a biochemical disturbance of the tryptophan-niacin
pathway.

Involvement of mycotoxins in the development of pellagra was postulated
in the recent past among sorghum and corn eaters (Schoental, 1980; Rao, 1983)
as both grains are susceptible to fungal contamination and mycotoxin elaboration. Sorghum is susceptible to Aspergillus sp. and Fusarium sp. infestation
and elaboration of aflatoxins and T-2 toxin respectively (Anon, 1979). The
biochemical effects of mycotoxins are more apparent when there is nutritional
stress and marginal intake of dietary proteins and vitamins.

In view of the above considerations, the possibility that mycotoxins may
be involved in the development of pellagra appears to be interesting. Studies were therefore undertaken to test the validity of the hypothesis. The
experimental investigations were two-pronged, i.e. assessed (i) the biochemical basis of mycotoxins, and (ii) the field contamination of mycotoxin in
sorghum-growing areas, and its relation to the prevalence of pellagra in a
sorghum-eating population.

Kynurenine and Serotonin Pathways
Edited by R. Schwarcz et al., Plenum Press, New York, 1991

MATERIALS AND METHODS

N^1methylnicotinamide (N^1MeN), NAD, NADP, nicotinic acid mononucleotide (NMN), quinolinic acid, niacin and T-2 toxin were obtained from Sigma Chemical Company, USA. Aflatoxin mixture (AFB_1; AFB_2; AFG_1; AFG_2; 44: 3.3:31.2: 3.2) was purchased from CSIR, Centre for Biochemicals, New Delhi. ^{14}C-Niacin (^{14}C-carboxyl; specific activity: 17.6 mCi/mmol) was obtained from BARC, Isotope Division, Bombay.

Biochemicals studies

All biochemical studies were carried out in male albino rats (Wistar/NIN), which were fed diets containing marginal protein (9%) and vitamin B-complex. This diet approximates the dietary intake of poor people in the rural areas of India. The effect of dietary aflatoxins (4 ppm and 10 ppm) and T-2 toxin (3 ppm) on tryptophan-niacin metabolism and nicotinamide nucleotide synthesis in liver were studied in two separate animal experiments. At the end of the experimental period, 72 hrs urine was collected in plastic containers. ^{14}C-Niacin, at a dose of 15 μCi/100 g body weight in 0.2 ml saline, was injected intraperitoneally to rats. After 2 hrs, the rats were sacrificed and their liver dissected out. ^{14}C-Niacin incorporation into liver nicotinamide nucleotides (NMN, NAD, NADP) were analyzed by paper chromatography. Total nicotinamide nucleotides in liver and N^1MeN in urine were estimated by a fluorimetric method. The major urinary metabolites of tryptophan-niacin metabolism, quinolinic acid and niacin, were estimated microbiologically (Sashidhar et al., 1988a,b, 1989).

Field studies

The sorghum samples were collected during the months of October and November (Kharif) 1985. Major sorghum growing districts of Andhra Pradesh in India, namely Mahbubnagar and Medak, were selected for the present study. The samples were collected in the Kharif (rainy) season when the production of sorghum is maximal. The purposive sampling method was adopted to collect the sorghum samples. In each of the districts, the taluks (administrative units) were ranked according to the total area under sorghum cultivation, and the first two or three taluks with the highest area under cultivation were selected in each district. From these taluks, three villages were selected for sample collection, based again on the highest area under cultivation. From each village, ten samples of sorghum were collected. The samples were either freshly harvested or stored grains. A total of 150 samples were collected randomly from 137 households. These samples included three varieties of sorghum, namely Red, Yellow and White.

The samples were selected to ensure representation of all categories of householders, namely agricultural laborers, small farmers, medium farmers and large farmers. Approximately 1 kg samples were collected, carefully packed in plastic bags and sealed securely. Adequate precautions were taken to avoid any contamination during transportation. Information on the harvesting procedure, time, quality of seeds, period of storage, method of storage, etc. was collected.

A diet survey was conducted to assess the levels of mycotoxin ingestion, if any, through staple food. The 24 h dietary recall method was chosen for the estimation of cereal intake. The dietary pattern for the past one year was also recorded. These assessments were based on questionnaires.

To assess the individual intake, a set of 12 standard aluminum vessels of known volume were used. The vessels were used to estimate the amount of raw food consumed by each of the family members.

Table 1. Concentration of urinary metabolites, total liver pyridine nucleotides and ^{14}C-niacin uptake and its incorporation into pyridine nucleotides in rats fed aflatoxins and T-2 toxin

	Total liver pyridine nucleotides (μg/g liver)	N^{1}MeN (μg)	Niacin (μg)	Quinolinic acid (μg)
		(per mg creatinine)		
Control[*]	698 ± 63	8.5 ± 3.0	2.9 ± 1.4	10.8 ± 3.6
Experimental (Aflatoxin-fed)[*]	586 ± 66[a]	5.1 ± 1.2[a]	4.0 ± 1.6	8.6 ± 3.1
Control[**]	540 ± 78	8.5 ± 1.4	7.5 ± 1.4	10.4 ± 1.6
Experimental (T-2 toxin-fed)[**]	553 ± 45	10.4 ± 2.1[a]	7.1 ± 1.7	8.7 ± 2.9

	% ^{14}C-Niacin incorporated into liver (per g liver)	% ^{14}C-Niacin incorporated into pyridine nucleotides (per g liver)		
		NMN	NAD	NADP
Control[*]	3.7 ± 0.9	0.03 ± 0.01	0.7 ± 0.3	0.2 ± 0.1
Experimental (Aflatoxin-fed)[*]	3.9 ± 1.0	0.04 ± 0.01	0.8 ± 0.3	0.3 ± 0.1
Control[**]	6.7 ± 1.4	0.09 ± 0.04	2.2 ± 0.7	0.6 ± 0.2
Experimental (T-2 toxin-fed)[**]	7.1 ± 1.9	0.11 ± 0.03	2.4 ± 0.9	0.7 ± 0.2

[*]N = 12; [**]N = 10; Data are the mean ± SD; [a]: $P < 0.05$ as compared to controls.

The method described by Booth (1971) was employed for isolating the fungi from the collected sorghum samples. From each one kg sorghum sample collected, twenty five lots were drawn and numbered. From these lots, ten subsamples (randomly drawn) were pooled (approx. 200 g). 100 g of pooled sample was powdered in the grinder to yield a fine powder. From the powdered sample, 25 g of flour were used for toxin extraction. Aflatoxin B_1 and T-2 toxin were estimated by non-competitive and competitive ELISA respectively (Sashidhar and Rao, 1988; Sashidhar et al., 1989).

RESULTS

There were no significant changes in body and liver weight between experimental and control groups fed aflatoxins and T-2 toxin, respectively. The concentration of total nicotinamide nucleotides in liver was significantly lower in rats receiving aflatoxins ($P < 0.05$). The urinary excretion of N^1MeN was significantly lower ($P < 0.05$). There was no significant change in the excretion of quinolinic acid and niacin in urine. No significant changes were observed in the uptake of ^{14}C-niacin by liver, and there was no significant difference in the rate of nicotinamide nucleotide (NMN, NAD, NADP) synthesis in liver (Table 1). In T-2 toxin fed rats, there was no significant change in the levels of nicotinamide nucleotides in liver. The uptake of ^{14}C-niacin and its incorporation into nicotinamide nucleotides were also not significantly altered. There was no significant difference in urinary excretion of niacin and quinolinic acid. The excretion of N^1MeN in urine was significantly higher ($P < 0.05$) in experimental animals (Table 1).

Field studies showed that 25% (37/150) of the sorghum samples were contaminated with various fungi, including Aspergillus sp. and Fusarium sp. Only two samples were positive for aflatoxin B_1 with a concentration of 16 ppb and 40 ppb (1.4%). T-2 toxin was not detected in any of the samples collected (Table 2). No cases of pellagra were detected during the survey, nor did the records of the local primary health centres reveal any cases of pellagra. The contribution of sorghum to the total cereal intake was found to be 45%.

DISCUSSION

The present study was undertaken with the hypothesis that pellagra in sorghum eaters may be a mycotoxicosis. The field study in two major sorghum growing areas of Andhra Pradesh failed, however, to reveal the presence of pellagra. Examination of the primary health centre records as well as discussions with the villagers did not reveal any information regarding the occurrence of pellagra. Earlier, no epidemiological studies had been conducted to establish the prevalence of pellagra in these sorghum eating areas and to find out its period prevalence.

The mycotoxin contamination of the sorghum samples was not high (1.4%). Therefore, the role of the mycotoxins, particularly aflatoxin and T-2 toxin, in the pathogenesis of pellagra could not be assessed in the field studies. It is therefore difficult to conclude, on the basis of field studies alone, whether absence of pellagra in the survey area was due to the absence of mycotoxin contamination or due to other factors. It is of interest to note here that sorghum was not the dietary staple in these two sorghum growing areas. Rice is presently the staple. It contributes more than 50% to the total cereal intake. Agricultural records show that the area under sorghum cultivation in the survey region has declined by about 11% from 19%. Simultaneously, there has been an increase of 18.6% in the area under paddy cultivation. This has been facilitated by advances in agricultural practices and better irrigation facilities. Thus, the absence of pellagra in this area is perhaps due to a change in dietary pattern of the population. Since there

Table 2. Fungal profile and percent fungal contamination of sorghum
          varieties

	Profile of contaminating fungi (% contamination)				
	A. flavus	A. niger	Fusarium sp.	Rhizopus sp.	Unidentified
Red	25	21	23	2	--
Yellow	31	14	15	8	32
White	11	9	21	12	19
Total	67	44	59	22	51

The number of samples analyzed were 25 (Red), 87 (Yellow) and 38 (White).
Of these, 7 (Red), 27 (Yellow) and 3 (White) samples were contaminated
(i.e. a total of 25% of the samples were contaminated by fungi).

are no other studies on the extent of mycotoxin contamination of sorghum in
the field, it is not possible to judge whether this has changed over the
years.

     None of the biochemical changes of tryptophan-niacin metabolism observed
in the present study are similar to that seen in pellagrins, namely increased
excretion of quinolinic acid in urine and impaired nicotinamide nucleotide
synthesis (Srikantia, 1978). The reduced concentration of total nicotinamide
nucleotides in liver of aflatoxin-fed rats can be explained by the shift in
the requirement of pyridine nucleotides as a donor of reducing equivalents
for biotransformation of aflatoxins, since liver is the target organ for
aflatoxin (Ueno, 1985). The increased excretion of N^1MeN appears to be a
non-specific effect of T-2 toxin on tryptophan-niacin metabolism.

     Although the findings of the field study do not permit any conclusion
regarding the role of mycotoxins in the development of pellagra, the laboratory experiments suggest that there is no specific effect of mycotoxins on
tryptophan-niacin metabolism, which is found to be disturbed in pellagrins.
They do not support Schoental's hypothesis (1980) that pellagra may be a manifestation of mycotoxicosis. It may be concluded that mycotoxins do not have
a significant role in the development of pellegra.

REFERENCES

Anon, 1979, Perspective of mycotoxins, FAO, Food and Nutrition paper 13, FAO,
    Un, Rome, pp. 1-167.
Booth, C., 1971, Introduction of general methods, in: "Methods in Microbio-
    logy", Vol. 4, Booth, C., ed., Academic Press, New York, pp. 4-47.
Gopalan, C., and Srikantia, S.G., 1960, Leucine and pellagra, Lancet, 1:954-
    957.
Rao, K.S.J., 1983, Pellagra in sorghum eaters: a mycotoxicosis? Eco. Food
    Nutr., 13:59-62.

Sashidhar, R.B., Ramakrishna, Y., and Bhat, R.V., 1989, Moulds and mycotoxins in sorghum stored in traditional containers in India, **J. St. Prod. Res.**, in press.

Sashidhar, R.B., and Rao, B.S.N., 1988, Non-competitive enzyme linked immuno-sorbent assay for detection of aflatoxin B_1, 26:984-989.

Sashidhar, R.B., Rao, K.S.J., and Rao, B.S.N., 1988a, Effect of dietary afla-toxins on tryptophan-niacin metabolism, **Nutr. Rep. Int.**, 37:515-521.

Sashidhar, R.B., Rao, K.S.J., and Rao, B.S.N., 1988b, Effect of dietary T-2 toxin on tryptophan-niacin metabolism, **Nutr. Rep. Int.**, 37:867-873.

Sashidhar, R.B., Rao, K.S.J., and Rao, B.S.N., 1989, Effect of dietary afla-toxins on nicotinamide nucleotide synthesis in liver, **Nutr. Rep. Int.**, 31:1037-1043.

Schoental, R., 1980, Mouldy grains and aetiology of pellagra: the role of toxic metabolites of _Fusarium_, **Biochem. Soc. Trans.**, 8:147-150.

Srikantia, S.G., 1978, Endemic pellagra among jowar eaters, **Ind. J. Med. Res. Suppl.**, 68:38-47.

Ueno, Y., 1985, The toxicology of mycotoxins, **CRC Rev. Toxicol.**, 14:99-132.

LEVELS OF NON-PROTEIC TRYPTOPHAN IN MILK AND SOYBEAN FORMULAS

M. Biasiolo, C. Costa, A. Bettero, and G. Allegri

Department of Pharmaceutical Sciences
Padova University
35131 Padova
Italy

INTRODUCTION

Although human milk is the main source of nutrition for infants, for many years artificial formulas based on cow's milk have been introduced into infant nutrition. It is known that human milk contains far less protein but many more free amino acids (Lönnerdal et al., 1976; Svanberg et al., 1977) than cow's milk (Macy et al., 1953; Fomon, 1974; Lee and Lorenz, 1978; George and Lebenthal, 1981). Studies on the content of free amino acids in milk have excluded tryptophan for methodological reason, and consequently little is known of the nutritional factors that may influence its availability in early life (Tricklebank et al., 1979; De Antoni et al., 1980). Tryptophan is in effect the only amino acid bound between 80 and 90% to plasma albumin. Only the small free fraction is able to cross the haematoencephalic barrier to enter the brain to form serotonin (McMenamy and Oncley, 1958).

In this work we investigated the differences in content of total non-proteic and free tryptophan in human milk, fresh cow's milk, soy "milk" from a commercial source, cow's milk formulas and soybean formulas used in infant nutrition.

EXPERIMENTAL

Materials

We analyzed: a) ten samples of human milk obtained from 10 mothers in the morning after they nursed their infants, during the first 30 days after delivery; b) ten samples of cow's milk obtained from 10 cows during the first 30 days post partum; c) ten randomized samples each of fresh cow's milk and soy "milk" from a commercial source; d) ten randomized samples each of soy-based formulas: ten of Isomil (Abbott, Netherlands) (dilution at 13% p/v) and ten of Humana Sinelac (Humana, Germany) (dilution at 13.2% p/v); e) twenty randomized samples of cow's milk adapted formulas, ten of Preaptamil (Milupa Colmar, France) (dilution 13.2% p/v), and ten samples of Nan (Nestlè, Vevey, Switzerland) (dilution at 13.2% p/v).

Kynurenine and Serotonin Pathways
Edited by R. Schwarcz *et al.*, Plenum Press, New York, 1991

Method

Tryptophan was measured by HPLC according to the method of Costa et al. (1987). The free form was obtained after ultrafiltration using an Amicon Model 12 with an XM-50 Diaflo Membrane (De Antoni et al., 1980).

RESULTS

Table 1 reports the mean levels (mg/l ± SD) of total non-proteic and free tryptophan in human and cow's milk samples obtained during the first month post partum. Colostrum contains much more of both forms of non-proteic tryptophan than mature human milk and cow's milk. Also, the percentage of free tryptophan is higher in human milk. However, it increases in both milks from the first days of lactation to 30 days after delivery.

Table 2 shows the mean values of total non-proteic and free tryptophan in commercial cow's milk. The values and percentage of free amino acid are similar to those obtained in cow's milk on the 30th day after delivery. As appears from Table 2, the levels of non-proteic tryptophan in cow's milk are much lower than those observed in soy "milk".

Table 3 reports the levels of both forms of non-proteic tryptophan in soy-based formulas and cow's milk-based formulas. Values are expressed as milligrams of amino acid per liter of solution, prepared according to the manufacturer's indications. The content of non-proteic tryptophan is much higher in soybean formulas than in cow's milk formulas. In any case, the values are significantly lower than those of colostrum.

Table 1. Mean levels (± SD) of total non-proteic and free tryptophan in human milk and cow's milk

Days after delivery	Human milk (10) Tryptophan (mg/l)			Cow milk (10) Tryptophan (mg/l)		
	Total	Free	% of Free	Total	Free	% of Free
1				1.7 ± 0.5	0.3 ± 0.1	19
2	10.9 ± 2.3	7.9 ± 1.2	73	1.4 ± 0.4	0.4 ± 0.0	27
3				1.3 ± 0.1	0.4 ± 0.0	29
4	5.8 ± 0.9	4.3 ± 1.2	74	1.1 ± 0.2	0.3 ± 0.1	28
5				1.0 ± 0.2	0.3 ± 0.1	27
6	2.0 ± 0.5	1.5 ± 0.4	75	1.0 ± 0.1	0.3 ± 0.1	29
10	1.6 ± 0.2	1.2 ± 0.2	76	0.7 ± 0.3	0.2 ± 0.1	31
20				0.6 ± 0.1	0.3 ± 0.1	42
30	1.3 ± 0.2	1.1 ± 0.2	84	0.5 ± 0.1	0.3 ± 0.1	60

The number of subjects is indicated in parentheses.

Table 2. Mean concentrations (± SD) of total non-proteic and free tryptophan in fresh cow's milk and soy "milk" from a commercial source

Fresh cow's milk (10)			Soy "milk" (10)		
Tryptophan (mg/l)			Tryptophan (mg/l)		
Total	Free	% of Free	Total	Free	% of Free
0.6 ± 0.1	0.3 ± 0.1	57	3.2 ± 0.3	1.7 ± 0.2	52

The number of samples is given in parentheses.

DISCUSSION

The high levels of the free fraction of tryptophan found in human milk in the first few days after birth seem to make available a quota for uptake by the brain. This observation is of particular interest considering the role of this amino acid in the control of synthesis of cerebral serotonin (Fernstrom and Wurtman, 1971).

The content of both forms of non-proteic tryptophan is higher in colostrum. It decreases rapidly after the first few days of lactation, and then more slowly in the subsequent days when it becomes mature milk.

The levels of non-proteic tryptophan are markedly higher in soy "milk" than in the cow's milk on the market. In the soy-based formulas, the levels of non-proteic tryptophan are also higher than cow's milk formulas.

The portion of non-proteic tryptophan in soy-based formulas may add to the lower proteic content of tryptophan in soybean protein and therefore contribute to protein synthesis. However, if total needs for protein synthesis can be met in infants fed with different formulas, research should aim at establishing the real significance of the high levels of non-proteic tryptoplan in colostrum and its role in the control of serotonin synthesis in the brain.

Table 3. Content of total non-proteic and free tryptophan (mean values ± SD) in soybean and cow's milk formulas for term infants

	Tryptophan (mg/l)		
	Total	Free	% of Free
Isomil (10)	3.4 ± 0.4	2.7 ± 0.2	81
Humana (10)	3.1 ± 0.4	1.5 ± 0.1	47
Preaptamil[*](10)	0.7 ± 0.1	0.4 ± 0.1	53
Nan[*] (10)	0.5 ± 0.1	0.3 ± 0.1	63

The number of samples is given in parentheses; [*]Cow's milk formula.

REFERENCES

Costa, C., Bettero, A., and Allegri, G., 1987, Rapid measurement of tryptophan, 5-hydroxytryptophan, serotonin and 5-hydroxyindoleacetic acid in
human CSF by selective fluorescence and HPLC, **Gior. Ital. Chim. Clin.**,
12:307-312.

De Antoni, A., Allegri, G., Costa, C., Vanzan, S., Bertolin, A., Carretti,
N., and Zanardo, V., 1980, Total and free tryptophan in serum of newborn
infants, **Acta Vitaminol. Enzymol.**, 2:17-20.

Fernstrom, J.D., and Wurtman, R.J., 1971, Brain serotonin content: physiological dependence on plasma tryptophan levels, **Science**, 173:149-159.

Fomon, S.J., 1974, **Infant Nutrition**, W.B. Saunders, Philadelphia, p. 575.

George, D.E., and Lebenthal, E., 1981, Human breast milk in comparison to
cow's milk, in: "**Textbook of Gastroenterology and Nutrition in Infancy**",
Lebenthal, E., ed., Raven Press, New York, pp. 295-320.

Lee, V.A., and Lorenz, K., 1978, The nutritional and physiological impact of
milk in human nutrition, **CRC Crit. Rev. Food Sci. Nutr.**, 11:41-116.

Lönnerdal, B., Forsum, E., and Hambraeus, L., 1976, A longitudinal study of
the protein, nitrogen, and lactose contents of human milk from Swedish
well-nourished mothers, **Am. J. Clin. Nutr.**, 29:1127-1133.

Macy, I.G., Kelly, H.J., and Sloan, R.E., 1953, The composition of milks, in:
"**National Academy of Science Research Council**", Publ. 254, Washington,
D.C., p. 70.

McMenamy, R.H., and Oncley, J.L., 1958, The specific binding of L-tryptophan
to serum albumin, **J. Biol. Chem.**, 233:1436-1447.

Svanberg, U., Gebre-Medhin, M., Ljungqvist, B., and Olsson, M., 1977, Breast
milk composition in Ethiopian and Swedish mothers. III. Amino acids and
other nitrogenous substances, **Am. J. Clin. Nutr.**, 30:499-507.

Tricklebank, M.D., Pickard, F.J., and De Souza, S.W., 1979, Free and bound
tryptophan in human plasma during the perinatal period, **Acta Paediatr.
Scand.**, 68:199-204.

CHARACTERIZATION OF L-TRYPTOPHAN TRANSPORT INTO LIVER IN SUCKLING RATS

K. Saito, Y. Nagamura, Y. Ohta, E. Sasaki, and I. Ishiguro

Department of Biochemistry
School of Medicine
Fujita-Gakuen Health University
Toyoake
Aichi 470-11
Japan

INTRODUCTION

In adult rats, about 10% of total plasma L-tryptophan (Trp) (ca. 100 μM) exists in a free form (not albumin-bound) under physiological conditions (Madras et al., 1974). We have demonstrated that Trp uptake into adult rat hepatocytes occurs mainly via a trypsin-sensitive high-affinity transport component (K_m = 14 μM) under physiological conditions (Saito et al., 1986). We have also shown that Trp uptake into adult rat hepatocytes via the high-affinity transport component is closely related to liver tryptophan 2,3-dioxygenase (TPO) activity (Saito et al., 1987). However, rat liver TPO activity is known to appear rapidly about two weeks after birth (Franz and Knox, 1967). In addition, the proportion of free Trp, which can be utilized readily in tissues, in the plasma of suckling rats has been shown to be high compared to that of young and adult rats (Sarna et al., 1970). Taking these findings into account, it is suggested that Trp transport into suckling rat liver is different from that into young and adult rat liver. In order to elucidate the characteristics of Trp transport into suckling rat liver, we have examined the changes of liver TPO activity and Trp levels not only in serum and liver but also brain and muscle during development, using male rats aged 10 to 42 days. We also compared the mode of Trp uptake into hepatocytes between suckling and adult rats using isolated hepatocytes.

MATERIALS AND METHODS

Male Wistar rats aged 10 to 42 days (Shizuoka animal laboratory, Hamamatsu) were used. They were maintained ad libitum on cube diet Oriental MF (Oriental Yeast Ltd., Tokyo) and water. Suckling rats were kept with their dams. All animals were killed between 10:00 and 11:00 h. Free serum Trp was separated by centrifugation through an ultrafiltration membrane Ultrafree C3TK (Millipore, Tokyo) as described previously (Saito et al., 1986). Total and free serum Trp were determined using high performance liquid chromatography (HPLC) with electrochemical detection (Saito et al., 1986). Serum non-esterified fatty acid (NEFA) and albumin were measured using commercial clinical test kits. Concentrations of Trp in liver, brain, and muscle were determined by the HPLC methods described above after extracting Trp from these tissues with perchloric acid. Total TPO activity was assayed in fresh liver homogenates in the presence of added hematin (2 μM) (Knox et al., 1970). Hepatocytes were isolated from rats aged 10 and 42 days by collagenase dissociation of the liver as described previously (Saito et al., 1986). Cell viability, estimated by trypan blue exclusion, was above 85%. Trp uptake was determined as described previously (Saito et al., 1986), and the rate was estimated by the amount of the ^{3}H-labeled amino acid taken up. Results were analyzed statistically by Student's t-test.

RESULTS AND DISCUSSION

We have demonstrated that Trp uptake into adult rat hepatocytes via a high affinity transport component, which is trypsin-sensitive and works mainly under physiological conditions, is closely related to liver TPO activity (Saito et al., 1987). However, liver TPO activity is known to appear rapidly about one week before weaning (21 days after birth) (Franz and Knox, 1967). In addition, no information is available on Trp transport into suckling rat livers. In order to characterize the transport of Trp into the liver of suckling rats, we therefore first examined developmental change of total TPO activity using rats aged 10 to 42 days. As reported previously (Franz and Knox, 1967), liver TPO activity increased rapidly before and after weaning and reached the level of adult rats 28 days after birth. This value was about 11-fold higher than that obtained at 10 days (data not shown). The concentration of total and free serum Trp was determined next since the proportion of free Trp, which can be taken up readily into liver, in the plasma of suckling rats has been shown to be higher than that of young and adult rats (Sarna et al., 1982). A little drop in total serum Trp level was found after weaning, whereas the level of free serum Trp decreased by 88% between 10 and 21 days after birth and remained constant thereafter (Fig. 1). Such a marked change of free serum Trp level with aging seemed to be due to the change in the level of serum NEFA rather than albumin because the concentration of NEFA, which is known to weaken the binding of Trp to albumin (Curzon et al., 1973), in the serum of 10-day-old rats was about three-fold higher than that of 21-day-old rats (ca. 100 μEq/l), whereas the concentration of serum albumin increased by 46% between both groups of rats (data not shown).

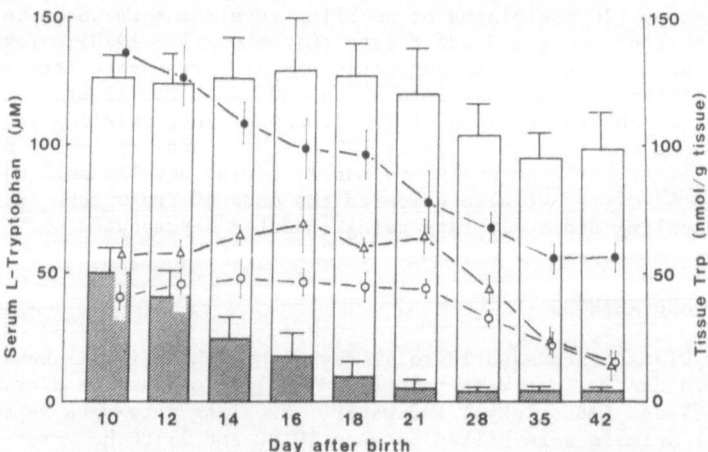

Fig. 1. Developmental changes in the levels of total and free serum Trp and liver, brain, and muscle Trp. Trp concentrations of serum and tissues were determined as described in the text. Total bars: total serum Trp; hatched bars: free serum Trp. ●: Liver Trp; ○: brain Trp; △: muscle Trp. Results are expressed as means ± SD (N = 6).

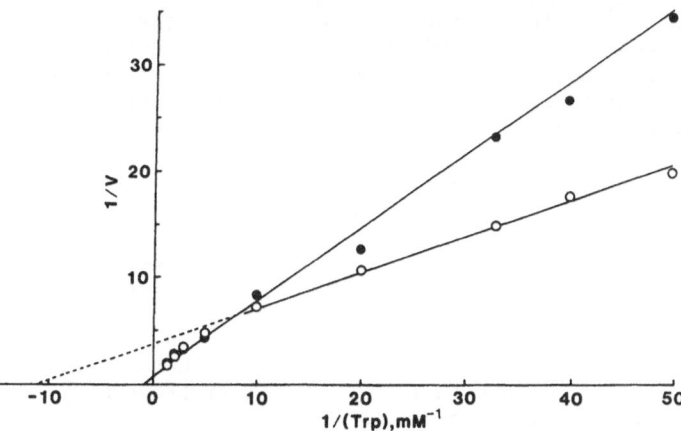

Fig. 2.   Lineweaver-Burk plots of kinetics of Trp uptake by isolated
hepatocytes of rats aged 10 and 42 days. The preparation of
isolated hepatocytes and the determination of Trp uptake were
carried out as described in the text. Uptake rates of Trp are
expressed as nmol of substrate taken up per 10^6 cells for the
initial first minute. ●: Measured rate of Trp uptake by isolated hepatocytes of rats aged 10 days; ○: measured rate of
Trp uptake by isolated hepatocytes of rats aged 42 days. Dot-
ted line is the extension of reciprocal plot kinetics of Trp
uptake at concentrations less than 0.15 mM. The data are the
average of six determinations.

Since it has been suggested that alterations in free Trp concentrations -
following changes in albumin binding - may be an important factor in
regulating Trp uptake by the liver (Smith and Pogson, 1980), the age-related
change of liver Trp level was compared with those of brain and muscle Trp
levels. As shown in Fig. 1, a rapid decrease of liver Trp level was found
between 10 and 21 days after birth. This change was very similar to that in
free serum Trp level. Liver Trp concentrations of rats aged 10 and 21 days
were 0.140 and 0.076 μmol/g liver (mean value), respectively. In contrast to
the finding in liver, a rapid change of Trp level with aging did not occur in
the brain and muscle (Fig. 1). The relationship between liver, brain, or
muscle Trp level and total or free serum Trp level at different ages was further examined. The concentration of liver Trp correlated well with that of
free serum Trp ($r = 0.924$, $P < 0.001$), but not with that of total serum Trp
($r = 0.481$, $P > 0.01$). In contrast, the concentration of brain or muscle Trp
correlated well with that of total serum Trp ($r = 0.734$, $P < 0.001$ for brain;
$r = 0.609$, $P < 0.001$ for muscle). Thus, the level of liver Trp was found to
be closely related to that of free serum Trp. We have demonstrated that in
adult rat hepatocytes, Trp uptake occurs via at least two saturable (one
high-affinity and one low-affinity) and one non-saturable component, and that
the high-affinity component, which is trypsin-sensitive, functions mainly as
a carrier in Trp transport into hepatocytes under physiological conditions
(Saito et al., 1986). The mode of Trp uptake into hepatocytes was compared
between rats aged 10 and 42 days using isolated hepatocytes. As shown in
Fig. 2, a Lineweaver-Burk plot of the results obtained from 42-day-old rats
yielded a linear curve which corresponded to the curve observed at concentrations above 0.15 mM in adult rats. Thus, the mode of Trp uptake into hepatocytes in suckling rats was partially different from that in adult rats: suckling rat liver was lacking in the high-affinity transport component for Trp
uptake which is present in adult rat liver.

In conclusion, the present results indicate that the system of Trp transsport into liver in suckling rats is different from that in young and adult rats, and that in suckling rats without liver Trp uptake via a high-affinity transport component, a marked increase in free serum Trp level contributes to an efficient transport of the amino acid into liver.

REFERENCES

Franz, J.M., and Knox, W.E., 1967, The effect of development and hydrocorti-
    sone on tryptophan oxygenase, formamidase, and tyrosine aminotransferase
    in the livers of young rats, Biochemistry, 6:3464-3471.
Curzon, G., Friedel, J., and Knott, P.J., 1973, The effect of fatty acids on
    the binding of tryptophan to plasma protein, Nature, 242:198-200.
Knox, W.E., Yip, A., and Reshef, L., 1970, L-tryptophan 2,3-dioxygenase,
    Methods Enzymol., 17A:415-421.
Madras, B.K., Cohen, E.L., Messing, R., Munro, H.N., and Wurtman, R.J., 1974,
    Relevance of free tryptophan in serum to tissue tryptophan concentrations,
    Metabolism, 23:1107-1116.
Saito, K., Sasaki, E., Ohta, Y., Nagamura, Y., and Ishiguro, I., 1986, Mode
    of L-tryptophan uptake into rat hepatocytes via trypsin-sensitive high-af-
    finity transport system, Biochem. Int., 13:873-883.
Saito, K., Sasaki, E., Ohta, Y., Nagamura, Y., and Ishiguro, I., 1987, The
    effect of tryptophan 2,3-dioxygenase on L-tryptophan transport into rat
    liver, Seikagaku, 59:732.
Sarna, G.S., Tricklebank, M.D., Kantamaneni, B.D., Hunt, A., Patel, A.J., and
    Curzon, G., 1982, Effect of age on variables influencing the supply of
    tryptophan to the brain, J. Neurochem., 39:1283-1290.
Smith, S.A., and Pogson, C.I., 1980, The metabolism of L-tryptophan by iso-
    lated rat liver cells, Biochem. J., 186:977-986.

EFFICIENCY OF EXOGENOUS QUINOLINIC ACID, A METABOLITE OF TRYPTOPHAN, AS A NIACIN PRECURSOR IN RATS

K. Shibata and K. Murata[1]

Department of Food Science and Nutrition
Teikoku Women's University
Moriguchi
Osaka 570

[1]Research Institute for Education
Teikoku Gakuen
Moriguchi
Osaka 570
Japan

INTRODUCTION

It is generally known that protein nutrition, especially tryptophan metabolites, niacin intake and amino acid balance are important for the prevention of pellagra. The pathway from tryptophan to NAD via kynurenine, 3-hydroxyanthranilic acid and quinolinic acid (QA) has been shown to be particularly important from the nutritional standpoint (Krehl et al., 1946; Henderson and Hirsch, 1949; Nishizuka and Hayaishi, 1963).

QA is known as a key intermediate of the tryptophan-NAD pathway. It is spontaneously produced from α-amino-β-carboxymuconate-ε-semialdehyde and transformed into nicotinic acid mononucleotide in the presence of 5-phosphoribosyl-1-pyrophosphate by quinolinate phosphoribosyltransferase. The bioavailability of QA was originally studied by Henderson (1949). He observed that the administration of large quantities of QA to rats fed a tryptophan imbalanced diet had a marked effect on their growth. It was tentatively concluded that QA was 1/20 to 1/100 as active as niacin. Therefore, QA given to mammals may not be efficiently utilized as niacin. Ijichi et al. (1966) found that QA hardly penetrated into liver cells compared with nicotinic acid (NA) and its amide. Hagino et al. (1968) reported a similar finding. Since our daily foods contain appreciable amounts of QA as well as niacin (Shibata et al., 1985), we have now explored the utilization of QA from a nutritional standpoint, using an _in vivo_ test in rats.

EXPERIMENTAL METHOD

Weanling male rats of the Sprague-Dawley strain were divided into 10 groups. One group received the basal diet shown in Table 1, which is tryptophan-limited and niacin-free. Four groups received NA at the levels of 0.5, 1.0, 2.0 and 3.0 mg per 100 g of the above basal diet. Another five groups were given the basal diet supplemented with QA at 4.1, 8.2, 12.2, 24.4 and

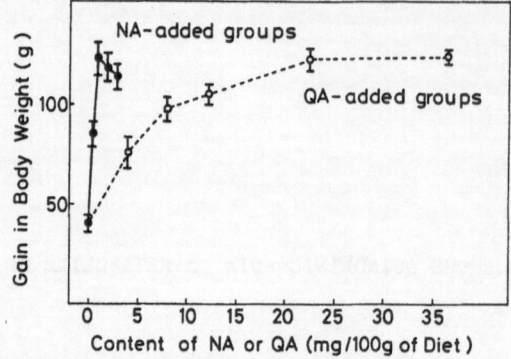

Fig. 1. Relative activities of NA and QA in promoting the growth of weanling rats. Values are the mean ± SEM.

36.7 mg per 100 g of the diet, respectively. Urinary niacin content was measured microbiologically using Lactobacillus plantarum (Iwai et al., 1967). Body weight and food intake were monitored every 2-3 days during the 31-day feeding. At the end of the experiment, rats were killed. Blood, liver, upper small intestine and brain were taken for NAD analysis. NAD was determined as described by Shibata and Murata (1986).

RESULTS AND DISCUSSION

The relationships between the body weight and the content of QA or NA are shown in Fig. 1. NAD levels in blood, liver, upper small intestine, and brain clearly demonstrated that exogenous QA was converted into NAD, but the efficiency was lower than that of NA. The relative activity of QA to NA was different with different indices (Table 2). From the nutritional standpoint, the gain in body weight could be the most important. Therefore, we propose that the efficiency of exogenous QA would be 1/18 of that of NA on a molar basis and 1/25 on a weight basis.

Table 1. Tryptophan-limited and niacin-free diet

Materials	g/kg diet
Vitamin free milk casein	90
Glycine	20
L-Threonine	0.78
L-Cystine	2
Sucrose	777.22
Corn oil	50
Mineral mixture[a]	50
Vitamin mixture[b]	10

[a] A.E. Harper, J. Nutr., 68:405 (1959). [b] Oriental's ratio (NA-free).

Table 2. Relative activity of QA to NA

Index	QA/NA (molar basis)
Gain in body weight	1/18
Food efficiency ratio	1/3
Blood NAD	1/6
Liver NAD	1/4.5
Upper small intestine NAD	1/9
Brain NAD	1/6
Urinary excretion of niacin	1/9

REFERENCES

Hagino, Y., Lan, S.J., Ng, C.Y., and Henderson, L.M., 1968, Metabolism of pyridinium precursors of pyridine nucleotides in perfused rat liver, J. Biol. Chem., 243:4980-4986.

Henderson, L.M., 1949, Quinolinic acid metabolism. II. Replacement of nicotinic acid for the growth of the rat and Neurospora, J. Biol. Chem., 181:677-685.

Henderson, L.M., and Hirsch, H.M., 1949, Quinolinic acid metabolism. I. Urinary excretion by the rat following tryptophan and 3-hydroxyanthranilic acid administration, J. Biol. Chem., 181:667-675.

Ijichi, H., Ichiyama, A., and Hayaishi, O., 1966, Studies on the biosynthesis of nicotinamide adenine dinucleotide. III. Comparative studies on nicotinic acid, nicotinamide and quinolinic acid as precursors of nicotinamide adenine dinucleotide, J. Biol. Chem., 241:3701-3707.

Iwai, K., Okinaka, O., and Yokomizo, H., 1967, A uniform medium for microbiological determination of the B-vitamins with various lactic acid bacteria, Vitamins, 35:387-394.

Krehl, W.A., Sarma, F.S., Teply, L.J., and Elvehjem, C.A., 1946, Factors affecting the dietary nicotinic acid and tryptophan requirement of the growing rat, J. Nutr., 31:85-106.

Nishizuka, Y., and Hayaishi, O., 1963, Studies on the biosynthesis of nicotinamide adenine dinucleotide. I. Enzymic synthesis of niacin ribonucleotide from 3-hydroxyanthranilic acid in mammalian tissues, J. Biol. Chem., 238:3369-3377.

Shibata, K., and Murata, K., 1986, Blood NAD as index of niacin nutrition, Nutr. Int., 2:177-181.

Shibata, K., Tanaka, K., Hayakawa, F., and Murata, K., 1985, Efficiency of exogenous quinolinic acid as niacin, and the quinolinic acid content in food, Bull. Teikoku-Gakuen, 11:1-8.

WHY DOES A DIFFERENCE IN GROWTH RATE BETWEEN RATS OF THE WISTAR AND SPRAGUE

DAWLEY STRAINS OCCUR WHEN A NIACIN-FREE AND TRYPTOPHAN-LIMITING DIET IS FED?

T. Hayakawa, K. Shibata[1], and K. Iwai[2]

Dept. of Nutrition
Tokyo Univ. of Agriculture
Tokyo 156

[1]Dept. of Food Science and Nutrition
Teikoku Women's Univ.
Osaka 570

[2]Faculty of Home Economics
Kobe Women's Univ.
Kobe 654
Japan

INTRODUCTION

In the course of our experiments on tryptophan (Trp)-NAD metabolism, we have found that the rats of the Sprague-Dawley (SD) strain showed good growth as compared to those of the Wistar strain (Shibata et al., 1980) when they were fed a niacin-free diet containing 8.0 % amino acid mixture simulating rice protein supplemented with some limiting amino acid except for Trp (Yoshida, 1971). We also found that the total amounts of Trp metabolites in liver, kidneys and urine were higher in the SD strain than in the Wistar strain (Shibata et al., 1982). These findings prompted us to compare enzyme activities involved in the Trp-NAD pathway in the two strains of rats fed a normal diet.

MATERIALS AND METHODS

Six-week-old rats of the SD and Wistar strains (Clea Japan Inc.) were housed individually in wire-bottomed stainless cages and were used after 8 days feeding of a commercial food (CE-2, Clea Japan Inc.) and water ad lib-itum in a conditioned room (temperature: 22 ± 2°C; light schedule: light: 06:00-18:00, dark: 18:00-06:00). On the last day, the rats (body weight expressed as mean ± SEM of six rats; SD strain: 214.2 ± 6.0 g, Wistar strain: 202.5 ± 7.8 g) were anesthetized with an intraperitoneal injection of sodium pentobarbital and their livers were excised. Each liver (SD, 8.65 ± 0.30 g; Wistar, 8.44 ± 0.19 g) was minced and homogenized in five volumes of cold 50 mM potassium phosphate buffer, pH 7.0, with a Yamato ultradisperser model LK-21. These homogenates were used as enzyme sources for measuring the activities of tryptophan oxygenase (EC 1.13.11.11), kynureninase (EC 3.7.1.3), quinolinate phosphoribosyltransferase (QPRTase, EC 2.4.2.19), nicotinate phosphoribosyltransferase (EC 2.4.2.11), nicotinamide mononucleotide adenylyltransferase (EC 2.7.7.1), NAD synthetase (EC 6.3.5.1), nicotinamide methyltransferase (EC 2.1.1.1) and kynurenine aminotransferase (EC 2.6.1.7). One portion of each homogenate was centrifuged at 10,000 x g for 20 min and the resulting supernatant was used for measuring the activities of 3-hydroxyanthranilic acid oxygenase (EC 1.13.11.6) and amino-β-carboxymuconate-semialdehyde decarboxylase (ACMSDase: EC 4.1.1.45). The methods used were as follows: tryptophan oxygenase: Knox et al. (1970); kynureninase: Takeuchi et al. (1980); 3-hydroxyanthranilic acid oxygenase: Decker et al. (1961); ACMSDase: Ichiyama et al. (1965); QPRTase: Iwai and Taguchi (1973); nicotinate phosphoribosyltransferase: Honjo (1966); nicotinamide mononucleotide adenylyltransferase: Kurokawa et al. (1966); NAD synthetase: Yu and Dietrich (1972); nicotinamide methyltransferase: Cantoni (1951); and kynurenine aminotransferase: Tobes and Mason (1975).

RESULTS AND DISCUSSION

The activities of the enzymes involved in Trp-NAD metabolism in the livers of SD and Wistar strains of rats are shown in Table 1. The activities of tryptophan oxygenase, 3-hydroxyanthranilic acid oxygenase and nicotinate phosphoribosyltransferase in the livers were significantly higher in the SD strain than in the Wistar strain. The activities of NAD synthetase, nicotinamide methyltransferase and kynurenine aminotransferase were almost the same between the two strains. The activity of ACMSDase tended to be lower (0.6 fold) in the SD strain than in the Wistar strain. The activity of QPRTase tended to be higher (1.4 fold) in the SD strain than in the Wistar strain. The lower activity of ACMSDase means that quinolinic acid formation from tryptophan is higher in the SD strain than in the Wistar strain. The higher activity of QPRTase means that phosphoribosylation of quinolinic acid, which

Table 1. Comparison of the enzyme activities in the tryptophan-NAD pathway between the livers of the Sprague Dawley and Wistar strains of rats

	Wistar	Sprague Dawley
	(μmol/hr/g wet weight)	
Tryptophan oxygenase	1.57 ± 0.08[a]	2.06 ± 0.15[b]
Kynureninase	0.324 ± 0.012[a]	0.259 ± 0.024[b]
3-Hydroxyanthranilic acid oxygenase	491 ± 37[a]	693 ± 28[b]
Aminocarboxymuconate-semialdehyde decarboxylase	1.42 ± 0.20	0.89 ± 0.15
Quinolinate phosphoribosyltransferase	0.267 ± 0.022	0.365 ± 0.043
Nicotinate phosphoribosyltransferase	0.049 ± 0.005[a]	0.104 ± 0.017[b]
Nicotinamide mononucleotide adenylyltransferase	5.50 ± 0.15[a]	6.57 ± 0.22[b]
NAD synthetase	0.88 ± 0.04	0.91 ± 0.07
Nicotinamide methyltransferase	0.724 ± 0.030	0.701 ± 0.014
Kynurenine aminotransferase	21.2 ± 1.7	26.4 ± 1.8

Values are the mean ± SEM of six rats; means in the same line not sharing a common superscript letter differ significantly at P < 0.05. Significant differences between means with the same or different variance by F-test were evaluated by Student's t-test and Welch's test, respectively.

is a rate-limiting step in the de novo biosynthetic pathway of NAD, is also
higher in the SD strain than in the Wistar strain. Generally speaking, the
lower activity of ACMSDase and the higher activity of QPRTase in the SD
strain showed that the Trp-NAD metabolism was more in favor of NAD formation
from Trp than in the Wistar strain. The higher production of the Trp metabolites and the higher growth rate of the SD strain than of the Wistar strain
(Shibata et al., 1980) under niacin-free and Trp-limiting conditions are due
to the higher activity of the Trp-NAD pathway in the SD strain. This could
be the reason why rats of the SD strain have a greater resistance to a niacin-free and Trp-limiting diet as compared to those of the Wistar strain.

REFERENCES

Cantoni, G.L., 1951, Methylation of nicotinamide with a soluble enzyme sys-
    tem from rat liver, J. Biol. Chem., 189:203-216.
Decker, R.H., Kang, H.H., Leach, F.R., and Henderson, L.M., 1961, Purifica-
    tion and properties of 3-hydroxyanthranilic acid oxidase, J. Biol. Chem.,
    236:3076-3082.
Honjo, T., 1971, Nicotinate phosphoribosyltransferase from baker's yeast, in:
    "Methods in Enzymology, Vol. 18B", McCormick, D.B., and Wright, L.D.,
    eds., Academic Press, New York, pp. 123-127.
Ichiyama, A., Nakamura, S., Kawai, H., Honjo, T., Nishizuka, Y., Hayaishi,
    O., and Senoh, S., 1965, Studies on the metabolism of the benzene ring of
    tryptophan in mammalian tissues; Enzymic formation of α-aminomuconic acid
    from 3-hydroxyanthranilic acid, J. Biol Chem., 240:740-749.
Iwai, K., and Taguchi, H., 1973, Distribution of quinolinate phosphoribosyl-
    transferase in animals, plants and microorganisms, J. Nutr. Sci. Vitamin-
    ol., 19:491-499.
Knox, W.E., Yip, A., and Reshef, L., 1970, L-tryptophan 2,3-dioxygenase
    (tryptophan pyrrolase) (rat liver), in: "Methods in Enzymology, Vol. 17A",
    Tabor, H., and Tabor, C.W., eds., Academic Press, New York, pp. 415-421.
Kurokawa, M., Kato, T., and Inamura, H., 1966, Unequal distribution of ATP:
    NMN adenylyltransferase activity among neuronal and glial cell nuclei,
    Proc. Jap. Acad., 42:1217-1222.
Shibata, K., Motooka, K., and Murata, K., 1980, Difference in growth rates
    between Wistar and Sprague Dawley strains of rats fed on the tryptophan-
    limiting diet, Bull. Teikoku-Gakuen, 6:1-6.
Shibata, K., Motooka, K., and Murata, K., 1982, The differences in growth and
    activity of the tryptophan-NAD pathway between Wistar and Sprague Dawley
    strains of rats fed on tryptophan-limited diets, J. Nutr. Sci. Vitaminol.,
    28:11-19.
Takeuchi, F., Otsuka, H., and Shibata, Y., 1980, Purification and properties
    of kynureninase from rat liver, J. Biochem., 88:987-994.
Tobes, M.C., and Mason, M., 1975, L-Kynurenine aminotransferase and L-α-
    aminoadipate aminotransferase: Evidence for identity, Biochem. Biophys.
    Res. Commun., 62:390-397.
Yoshida, A., 1971, Effect of amino acid supplementation to a rice diet on
    niacin requirement of rats, Agric. Biol. Chem., 35:1943-1949.
Yu, C.K., and Dietrich, L.S., 1972, Purification and properties of yeast
    nicotinamide adenine dinucleotide synthetase, J. Biol. Chem., 247:4794-
    4802.

PLASMA FREE 5HT AND PLATELET 5HT IN DEPRESSION: CASE-CONTROL STUDIES AND

THE EFFECT OF ANTIDEPRESSANT THERAPY

M.J. Sarrias, E. Martinez, P. Celada, C. Udina[1],
E. Alvarez[1], and F. Artigas

Department of Neurochemistry, CSIC
Jordi Girona, 18-26
08034 Barcelona

[1]Department of Psychiatry
Hospital de Sant Pau
Barcelona
Spain

INTRODUCTION

In recent years, a number of laboratories have reported on the existence of a non-platelet pool of 5HT in human blood, in the low nM concentration range (Artigas et al., 1985; Anderson et al., 1987). The physiological and pharmacological aspects of the platelet 5HT pool have been extensively studied but little is known about the free 5HT pool in blood. The 5HT found in human platelet-free-plasma (PFP) is thought to be representative of the actual concentration of free 5HT in blood. Results from this laboratory showed that 5HT in PFP and platelet 5HT behave separately after several pharmacological treatments, which provides evidence that 5HT in PFP does not derive artifactually from platelets. In particular, chronic treatment with clomipramine - a 5HT uptake inhibitor - elicits a drastic decrease of platelet 5HT (-90% of pre-treatment values) without affecting PFP 5HT (Sarrias et al., 1987). Furthermore, platelet 5HT does not experience marked seasonal changes (± 20%), while plasma 5HT exhibits a 6-fold variation throughout the year in man (Sarrias et al., 1989) and in a large number of healthy people of both sexes. Both are poorly correlated (Ortiz et al., 1988). All these data strongly suggest that most 5HT found in PFP is representative of the actual free 5HT in blood, provided that adequate precautions are taken during blood sampling and treatment. The 5HT in the bloodstream mainly originates in the enterochromaffin cells (E.C.) of the gut, from which it is secreted to either the portal circulation or the gut lumen in response to a variety of different stimuli (Ahlman and Dahlström, 1983; Schwörer et al., 1987). Two mechanisms inactivate the freely circulating 5HT which has strong vasoactive properties (for a review, see Vanhoutte, 1985): uptake (into endothelial cells and platelets) and MAO-A deamination, yielding 5-hydroxyindoleacetic acid (5HIAA). Accordingly, the balance between 5HIAA and 5HT in plasma should be influenced by treatments with MAO inhibitors (MAOI). Also, the ratio between 5HT in platelets and free 5HT in plasma provides a good peripheral index to examine in vivo the effects of drugs affecting 5HT uptake. Both kinds of drugs (MAOI and uptake inhibitors) are effective anti-depressants.

Thus, the measure of 5HT in platelets (or whole blood) and in PFP, together with plasma 5HIAA, provides a peripheral model to study the state of presynaptic components of 5HT neurons in psychiatric pathologies and the mechanism of action of antidepressant therapies. In this report we describe the levels of 5HT and 5HIAA found in different types of depressed patients and the effect of lithium and MAOI therapies on these measures.

METHODS

Patients and controls

Case-control study: Twenty patients with major depressive disorder (DSM-III) with melancholia were examined. All had been free of antidepressant drugs for several months or two weeks for benzodiazepines (before they relapsed). Some had never been treated before with antidepressant medication. They all showed marked diurnal mood changes. Seven patients with major depressive disorder without melancholia were also included in the study. Twenty-three controls were examined. They were matched to the patients for sex, age and body weight, and their blood sampling was done shortly after the patients, in order to avoid artefactual differences due to the seasonal rhythm exhibited by plasma 5HT and plasma 5HIAA (Sarrias et al., 1989). The melancholic patients included in this study were carefully selected according to the above criteria (long wash-out periods and specific symptoms) since 1984, and the results of a previous stage of the work have been published (Sarrias et al., 1987). The severity of depression was assessed by means of Hamilton (HDRS), Beck (BDI) and Minnesota Multiphasic Personality Inventory (MMPI) tests. The mean HDRS score of untreated patients was 25 at their entry in the study.

Lithium treatment: Twenty-two patients (13 male, 9 female) treated chronically with lithium were examined. Most of them (19) were bipolar patients (DSM-III) in an euthymic state at the moment of blood sampling. Fourteen controls (8 male, 6 female), age and weight matched to the patients, were simultaneously examined. Blood sampling was carried out for both controls and patients during the month of March, to avoid time-related changes of the parameters examined. The mean serum level of lithium was 0.75 mmol/l and the daily dosage was 1280 ± 380 mg (mean ± SD) of lithium carbonate.

Treatment with MAO inhibitors: Seven patients treated with therapeutic doses of either phenelzine, a non-specific MAO inhibitor, or brofaromine, a specific MAO-A inhibitor, were examined prior to treatment and after 6 weeks of daily treatment. Patients had been taking placebo for one week before the start of the treatment and had not been treated with tricyclic antidepressants for at least eight weeks before.

Sample treatment and analyses of 5HT, 5HIAA and tryptophan (TP)

Blood extraction of controls and patients was performed by venopuncture. The blood was collected into EDTA-containing tubes as described previously (Sarrias et al., 1987; Ortiz et al., 1988). All subjects fasted overnight and the blood was drawn between 9 and 11 a.m., before the morning dose in the case of patients. Blood treatment to obtain PFP was carried out as described (Ortiz et al., 1988). Aliquots of whole blood were taken for the analysis of platelet 5HT. In some samples, an intermediate step to obtain platelet-rich-plasma was included prior to sedimentation of the platelets. The compounds analyzed were: plasma free 5HT, plasma 5HIAA, platelet 5HT (or whole blood 5HT, expressed as ng 5HT/ml of blood) and plasma total tryptophan (TP). All compounds were analyzed by HPLC coupled to fluorimetric or electrochemical detectors, using methods previously described (Artigas et al., 1985; Sarrias et al., 1987; Ortiz et al., 1988).

RESULTS AND DISCUSSION

Initially, blood was collected in open plastic tubes containing EDTA (final concentration 1%). Subsequently, EDTA-containing vacuum tubes (Venoject) were used for their simplicity of use and higher standardization. Accordingly, an experiment was carried out in 8 male controls to compare the results of plasma free 5HT obtained with these two different tubes. Blood was obtained by two independent venopunctures (anterocubital vein) and collected in either kind of tubes (the blood corresponding to the open tubes was collected first in all cases). The concentration of plasma free 5HT was 2.12 ± 0.50 (SEM) in the open tubes and 2.14 ± 0.48 in Venoject tubes (n.s., paired Student's t-test). This result suggests that both kind of tubes are appropriate for the analysis of plasma 5HT.

Case-control study

Fig. 1 shows the concentration of the 5HT variables studied in controls and patients. Plasma free 5HT, platelet 5HT and plasma 5HIAA were significantly decreased in melancholic patients. Plasma free 5HT showed the most severe reduction (-60%, $P < 0.001$, Student's t-test). Plasma 5HIAA was only slightly reduced (-20%, $P < 0.05$), while platelet 5HT was decreased 30% with respect to control values ($P < 0.005$). Plasma TP was not different from controls. Given the marked reduction of both 5HT pools in blood (and also plasma 5HIAA), and the unchanged TP in plasma, these results suggest that the melancholic patients examined might have a reduced conversion of TP to 5HT when compared to the mean value of controls. This reduction does not seem to exist in non-melancholic major depressives. Although the number of patients examined was low (N = 7), we did not find any tendency towards lower values in this sample. The factors controlling the synthesis and release of 5HT from E.C. cells to the bloodstream are still poorly understood. Both cholinergic (Schwörer et al., 1987) and adrenergic (Ahlman and Dahlström, 1983) innervation of the gut seems to be involved, so that an abnormal functioning of either of these systems could account for a reduced 5HT secretion to the blood. A decreased activity of tryptophan hydroxylase in melancholic patients might also explain these results. The levels of plasma free 5HT and plasma 5HIAA in healthy people seem to be individually determined. Repeated sampling of 8 healthy individuals for 14 months yielded consistently ranked values, independent of the seasonal changes observed. In other words, individuals who were "high" or "low" in plasma free 5HT displayed such a characteristic at all times (Sarrias et al., 1989). This may be due to genetic or to environmental reasons, since it has been shown that platelet 5HT in pri-

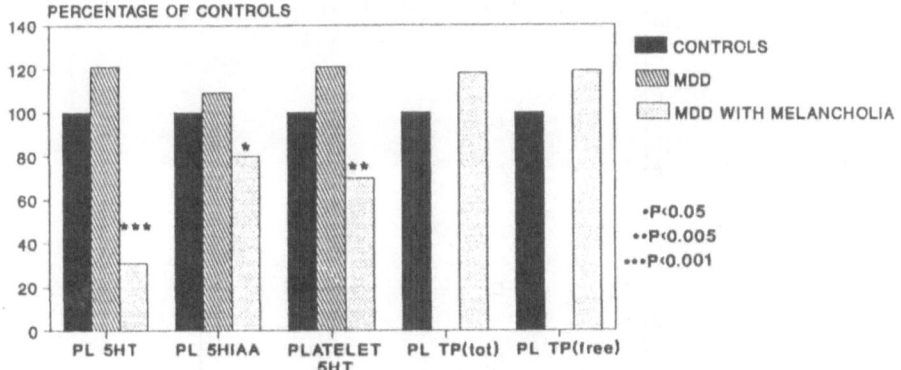

Fig. 1. Mean values of plasma free 5HT (PL 5HT), plasma 5HIAA (PL 5HIAA), platelet 5HT and plasma TP (PL TP) in major depressed patients and in controls. ·Statistical significances were calculated by Student's t-test.

mates is under the influence of social variables (Raleigh et al., 1984). In any case, the results suggest that "low 5HT" people are more prone to this particular subtype (melancholia) of depressive illness. We have no indication that these patients could exhibit a trend towards a decreased 5HT function in CNS, but other authors (Åsberg et al., 1984) have shown a reduction of 5HIAA in CSF of depressive melancholic patients. Accordingly, plasma free 5HT might become a valuable tool for the diagnosis and classification of depressed patients.

Effect of chronic treatment with lithium salts

Chronic treatment with lithium salts induces a marked increase in plasma free 5HT when compared to controls ($P < 0.01$). Most lithium treated patients displayed plasma free 5HT values well above the mean value of controls (Fig. 2). In contrast, whole blood 5HT was not changed in these patients (data not shown). Platelets take up 5HT from plasma, so that an increase in plasma 5HT should have led to a higher concentration of the 5HT in platelets (> 99% of whole blood 5HT). Since this was not observed, these in vivo results suggest that chronic treatment with lithium might interfere with the 5HT high affinity uptake process in platelets. The effect of treatment with lithium salts on the in vitro 5HT uptake by platelets has been examined with conflicting results. Both increases (Born et al., 1980; Meltzer et al., 1983) and decreases (Poirier et al., 1988) of ^{3}H-5HT uptake of humans treated repeatedly with lithium have been reported. Methodological factors and the effect of co-administered antidepressants may account for the controversy. According to our in vivo results, two different actions seem to be elicited by lithium: a) increased synthesis and/or release of 5HT by E.C. cells, as suggested by the higher levels of circulating free 5HT, b) impaired uptake by platelets, as shown by the different effects of lithium on both pools of blood 5HT. In any case, the treatment with lithium elicits a dramatic increase of the free pool of 5HT in plasma, indicating enhanced serotonergic function as suggested by different experimental approaches (Perez-Cruet et al., 1971; Müller-Oerlinghausen, 1985).

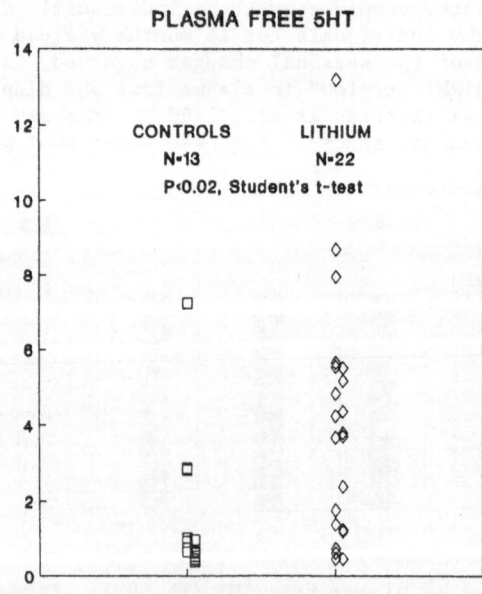

Fig. 2. Individual plasma free 5HT concentrations in 13 controls and in 22 patients treated chronically with lithium salts. Most patients on lithium had free 5HT levels above the median of controls.

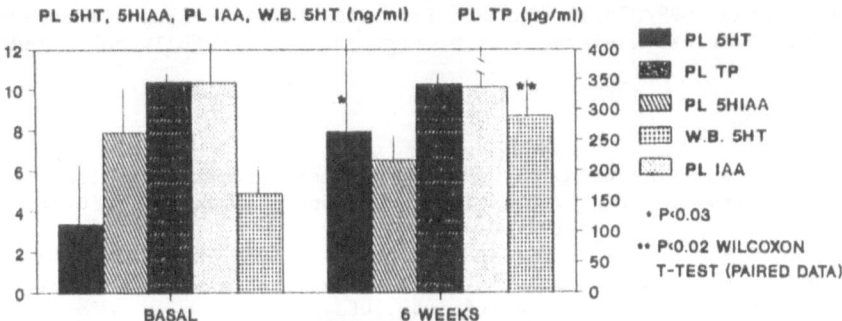

Fig. 3. Mean and deviation values of the variables indicated before and after treatment with MAO inhibitors for 6 weeks. Plasma free 5HT and whole blood 5HT were significantly higher after treatment, while plasma 5HIAA had a tendency to decrease. Abbreviations as in Fig. 1. PL IAA = Plasma indoleacetic acid.

Effects of MAOI treatment

Fig. 3 shows the preliminary results of the application of this peripheral model of serotonergic function to the study of MAO inhibition. Although only seven patients were studied, the results show clear-cut increases in both plasma free 5HT and platelet 5HT ($P < 0.03$ and $P < 0.02$, respectively, Wilcoxon t-test for paired data). There was a tendency towards decreased 5HIAA in plasma after repeated treatment but this was not statistically significant. Neither plasma TP nor plasma indoleacetic acid (IAA) showed any significant change. Plasma 5HT is taken up by endothelial tissues and converted to 5HIAA by MAO-A (Gillis and Pitt, 1982). MAO in liver is also able to catalyze the oxidative deamination of 5HT. Accordingly, the increased plasma 5HT after MAOI treatment is probably due to the partial blockade of MAO that would partly inhibit the degradative pathway of 5HT, leading to higher 5HT in the free pool. Also, since platelets take up 5HT from the surrounding plasma a higher platelet 5HT would be expected. It is interesting to note that the mean levels of plasma free 5HT after MAOI treatment increase more than two-fold, becoming higher than those of plasma 5HIAA.

These results indicate that MAOI treatments produce actual increases of either the free and stored pools of 5HT. Also unlike in vitro measurements of platelet MAO (which is of the B-type), in vivo measure of plasma 5HT and 5HIAA provides a real estimate of the actual efficacy of MAOI drugs in inhibiting 5HT degradation in patients. The variability of plasma 5HT is very high after MAOI treatment, some patients displaying large increases, some not changing their pre-treatment value. Since these results are part of an ongoing study on the clinical efficacy of two different MAO inhibitors, it will be interesting to relate the change of plasma free 5HT to the clinical improvement.

In summary, we have shown that a peripheral in vivo model may be useful in searching case-control differences of 5HT-mediated pathologies and to study the effect of antidepressant drugs on particular aspects (synthesis, uptake, MAO-A activity) of 5HT neurotransmission in psychiatric patients, using a simple experimental approach.

ACKNOWLEDGEMENTS

We thank the FIS (Fondo de Investigaciones Sanitarias) for financial support (grant #89/0387). Support from CIBA-GEIGY is also acknowledged. The skillful technical assistance of M. Figueras is gratefully acknowledged.

REFERENCES

Ahlman, H., and Dahlström, A., 1983, Vagal mechanisms controlling serotonin release from the gastrointestinal tract and pyloric motor function, J. Auton. Nerv. Sys., 9:119-140.

Anderson, G.M., Feibel, F.C., and Cohen, D.J., 1987, Determination of serotonin in whole blood, platelet-rich-plasma, platelet-poor-plasma and plasma ultrafiltrate, Life Sci., 40:1063-1070.

Artigas, F., Sarrias, M.J., Martinez, E., and Gelpi, E., 1985, Serotonin in body fluids: characterization of human plasma and cerebrospinal fluid pools by means of a new HPLC method, Life Sci., 37:441-447.

Åsberg, M., Bertilsson, L., Martensson, B., Scalia-Tomba, G.P., Thoren, P., and Traskman-Bendz, L., 1984, CSF monoamine metabolites in melancholia, Acta Psych. Scand., 69:201-219.

Born, G.V.R., Grignani, G., and Martin, L., 1980, Long-term effect of lithium on the uptake of 5-hydroxytryptamine by human platelets, Br. J. Clin. Pharmac., 9:321-325.

Gillis, C.N., and Pitt, B.R., 1982, The fate of circulating amines within the pulmonary circulation, Ann. Rev. Physiol., 44:269-281.

Meltzer, H.Y., Arora, R.C., and Goodnick, P., 1983, Effect of lithium carbonate on serotonin uptake in blood platelets of patients with affective disorders, J. Aff. Disord., 5:215-221.

Müller-Oerlinghausen, B., 1985, Lithium long-term treatment - does it act via serotonin?, Pharmacopsychiatry, 18:214-217.

Ortiz, J., Artigas, F., and Gelpi, E., 1988, Serotonergic status in human blood, Life Sci., 43:983-990.

Perez-Cruet, J., Tagliamonte, A., Tagliamonte, P., and Gessa, G.L., 1971, Stimulation of serotonin synthesis by lithium, J. Pharmacol. Exp. Ther., 178:325-330.

Poirier, M.F., Galzin, A.M., Pimoule, C., Schoemaker, H., Le Quan Bui, K.H., Meyer, P., Gay, C., Loo, H., and Langer, S.Z., 1988, Short-term lithium administration to healthy volunteers produces long-lasting pronounced changes in platelet serotonin uptake but not imipramine binding, Psychopharmacology, 94:521-526.

Raleigh, M.J., McGuire, M.T., Crammer, G.L., and Yuwiler, A., 1984, Social and environmental influences on blood serotonin concentrations in monkeys, Arch. Gen. Psych., 41:405-410.

Sarrias, M.J., Artigas, F., Martinez, E., and Gelpi, E., 1989, Seasonal changes of plasma serotonin and related parameters: correlations with environmental measures, Biol. Psych., 26:695-706.

Sarrias, M.J., Artigas, F., Martinez, E., Gelpi, E., Alvarez, E., Udina, C., and Casas, M., 1987, Decreased plasma serotonin in melancholic patients: a study with clomipramine, Biol. Psych., 22:1429-1438.

Schwörer, H., Racke, K., and Kilbinger, H., 1987, Cholinergic modulation of the release of 5-hydroxytryptamine from the guinea pig ileum, Naunyn-Schmiedeberg's Arch. Pharmacol., 336:127-132.

Vanhoutte, P.M., ed., 1985, "Serotonin and the Cardiovascular System", Raven Press, New York.

KYNURENINE AND MESANGIAL-PROLIFERATIVE GLOMERULONEPHRITIS

V. Rudzite, V. Gromm, and A. Martinsons

Latvian Research Institute of Cardiology
Riga
USSR

INTRODUCTION

The increased kynurenine and xanthurenic acid excretion in urine observed in calcium oxalate urolithiasis is related to the decreased activity
of kynureninase (Grimm et al., 1986), i.e. a pyridoxal-5-phosphate (P-5-P)-
dependent enzyme. The present work reports on relationships between kynurenine accumulation in blood serum and renal cell failure studied by analyzing
kidney biopsies and histological sections in cases with experimentally induced P-5-P deficiency in rats.

MATERIALS AND METHODS

The present work is divided into clinical and experimental studies. Serum kynurenine has been determined in fasting subjects before and after an
oral load of 30 mg L-tryptophan/kg body weight (Joseph and Risby, 1975). We
have investigated 7 healthy individuals, 16 patients with membranous-proliferative and 37 patients with mesangial-proliferative glomerulo-nephritis.
The diagnoses were verified by analyzing kidney biopsies.

In experimental studies, kynurenine accumulation in blood serum was induced in Wistar male rats by giving isoniazide (SERVA) in an oral dose of 20
mg/kg body weight once a day for 3 months. 20 isoniazide-treated and 20 control (intact) rats were used in these experiments. Kynurenine levels in
blood serum (Joseph and Risby, 1975), and P-5-P concentrations (Serfontein et
al., 1985) in blood serum and kidney tissues were determined in control and
experimental animals. Histological sections of kidney and electron microscope appearance of kidney cells obtained from control and experimental rats
were investigated.

The statistical significance of values obtained in normal and pathological conditions was calculated (Weber, 1957).

RESULTS AND DISCUSSION

Kidney biopsies revealed an increased amount of mesangial matrix in 37
renal patients. Therefore, mesangial-proliferative glomerulonephritis was
diagnosed in these patients. In the remaining 16 patients, the above men-

Table 1. Kynurenine concentration in blood serum (μmol/l) in
          healthy individuals and patients with mesangial-prolif-
          erative glomerulonephritis

Time of observation	Healthy	Mesangial-proliferative glomerulonephritis	
	(N – 7)	(N – 15)	(N – 22)
On an empty stomach	4.1 ± 0.1	3.3 ± 0.1	5.0 ± 0.2*
After tryptophan load			
3 h	14.5 ± 0.7	12.4 ± 0.2	27.2 ± 0.9*
24 h	4.3 ± 0.1	3.6 ± 0.2	7.1 ± 0.3*

Data are the mean ± quadratic deviation. *P < 0.001 (t-test).

tioned renal cell pathology was followed by thickening of capillary walls.
Therefore, membranous-proliferative glomerulonephritis was diagnosed.

Patients with membranous-proliferative glomerulonephritis had normal
kynurenine levels in blood serum both before and after tryptophan loading.
15 patients with mesangial-proliferative glomerulonephritis had normal kynurenine levels in blood serum, 22 patients had elevated levels (Table 1).

Light microscopical analysis did not reveal essential differences in the
change of structural elements of the kidney between the two groups of patients. Electron microscopic investigations revealed changes in the configuration of nephrocytes of renal tubules. The apices of the microvilli had
flasklike protuberances, and remnants of membranes were found in the apical
part of nephrocytes among separate microvilli (Fig. 1). On the free surface
of the proximal part of nephrocytes of the proximal and collecting tubules,
single cilia could be observed. Their possible function is still discussed.
Various and numerous vesicles, vacuoles, lysosomes, autophagosomes, fatty
elements, ribosomes, mitochondria with a matrix of low electronic density and
narrow, shortened, irregular and partly reduced cristae were seen in the
cytoplasm. There were one or two active nucleoli in the nucleus (Fig. 2).
Thus, mesangial-proliferative glomerulonephritis in patients with increased
kynurenine accumulation in blood serum is usually observed without specific
structural deviations in the tubular part of the kidney.

Kynurenine levels in serum of isoniazide-treated rats increased from 9.9
μmol/l (controls) to 15.0 μmol/l. The concentration of P-5-P decreased both
in blood serum (3.0 ± 0.3 mg/ml in controls and 1.9 ± 0.2 mg/ml in isoniazide-treated rats; P < 0.01) and kidney tissue (3.7 ± 0.2 mg/g dry weight in
controls and 2.8 ± 0.1 mg/g dry weight in isoniazide-treated rats; P < 0.05).

Structural changes were seen in endotheliocytes and nephrocytes. Light
and dark endotheliocytes could be observed. At the bottom and top of cytoplasm of light endotheliocytes were picnotic vesicles. In both kinds of
endotheliocytes, lysosomes were present. The configuration of nephrocytes

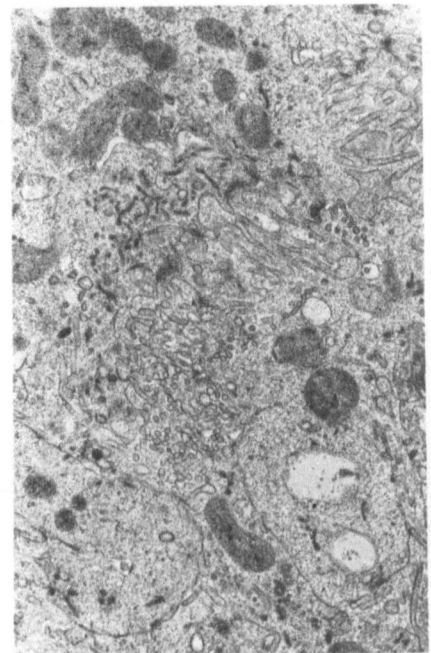

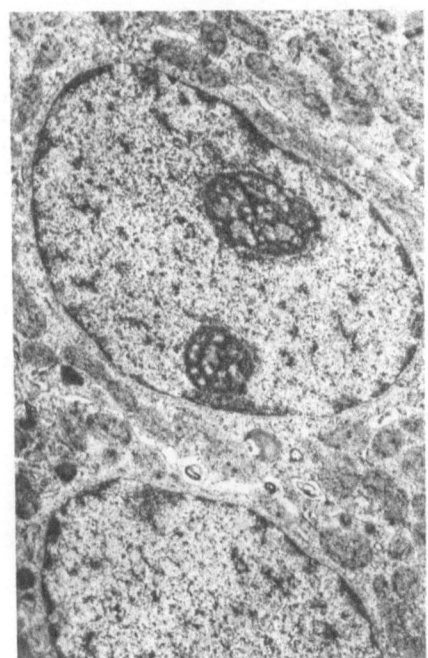

Fig. 1. Apical parts of some nephrocytes of the proximal part
of nephron. Protrusion of
the apical part of the cytoplasm is seen. There are
flashlike protuberances of
microvilli at their ends.
In the cytoplasm, there are
protuberances of micro-
villi at their ends. In
the cytoplasm, there are
small vesicles, vacuoles,
lysosomes, mitochondria and
remnants of membranes.
(x 9,000)

Fig. 2. Fragments of cytoplasm of
neighboring nephrocytes.
There are two active nuc-
leoli in the nucleus. In
the cytoplasm, there are
mitochondria, cysterns of
endoplasmatic reticulum
and ribosomes. In the
intracellular space, there
are free membranous out-
lines (x 9,000).

were changed, i.e. large mitochondric aggregates were seen in the cytoplasm.
Mitochondrial structures were polymorphous and did not contain cristae. A
concentration of nuclei and Schick-positive substances in the tubular lumen
was observed both in isoniazide-treated rats and in patients with mesangial-
proliferative glomerulonephritis.

CONCLUSION

These findings suggest that kynurenine accumulation in blood serum is
one of the pathogenic bases for mesangial-proliferative glomerulonephritis
due to endothelial and renal cell failure.

REFERENCES

Grimm, U., Steinhauser, I., Wulf, K., Knapp, A., and Zschiesche, M., 1987, Tryptophan metabolic studies in calcium oxalate urolithiasis, in: "Progress in Tryptophan and Serotonin Research 1986", Bender, D.A., Joseph, M.H., Kochen, W., and Steinhart, H., eds., de Gruyter, Berlin, pp. 295-300.

Joseph, M.H., and Risby, D., 1975, The determination of kynurenine in plasma, Clin. Chim. Acta, 63:197-204.

Serfontein, W.J., Ubbink, J.B., De Viller, L.S., Rapley, C.H., and Becker, P.J., 1985, Plasma pyridoxal-5-phosphate level as risk index for coronary artery disease, Atherosclerosis, 55:357-361.

Weber, E., 1957, "Grundriss der biologischen Statistik für Naturwissenschaftler, Landwirte und Mediziner", VEB Gustav Fischer Verlag, Jena.

IMPAIRMENT OF KYNURENINE METABOLISM IN CARDIOVASCULAR DISEASE

V. Rudzite, G. Sileniece, D. Liepina, A. Dalmane, and R. Zirne

Latvian Research Institute of Cardiology
Riga
USSR

INTRODUCTION

Our previous observations allowed us to suggest that kynurenine, a key metabolite of the tryptophan-nicotinic acid pathway, promotes vasoconstriction via noradrenaline release (Rudzite et al., 1987a) and forms the pathogenic bases for bradyarrhythmias and cardiomyopathies due to myocardial cell failure (Rudzite et al., 1987b). The present work reports further clinical studies of the relationship between kynurenine accumulation in blood serum and efficiency of antihypertensive drugs in cases of arterial hypertension as well as in experimental investigations of heart cell failure induced by kynurenine accumulation due to pyridoxal-5-phosphate (P-5-P) depletion in the organism.

MATERIALS AND METHODS

In clinical investigations, the effect of a 3-year antihypertensive treatment was studied in 24 patients with essential hypertension with different levels of kynurenine in blood serum after an oral load of 30 mg L-tryptophan/kg body weight. Therapeutic doses of clonidine hydrochloride (0.15-0.45 mg/day), propranolol (60-120 mg/day) and dihydrochlorthiazide (25 mg/day) were used in these investigations. The left ventricular myocardial mass (Teischolz et al., 1972; Troy et al., 1972), systolic and diastolic blood pressure and kynurenine levels in blood serum (Joseph and Risby, 1975) were measured before treatment and yearly during treatment. The statistical difference between the values obtained from patients at the beginning of the observation and during the treatment was calculated (Weber, 1957).

In experimental studies, kynurenine accumulation in blood serum was induced in Wistar male rats by giving isoniazide (SERVA) at an oral dose of 20 mg/kg once a day for 3 months. 20 isoniazide-treated and 10 intact (control) rats were used in these experiments. Kynurenine level were measured in blood serum (Joseph and Risby, 1975), and P-5-P was determined in blood serum, liver, kidney and heart tissue (Serfontein and Becker, 1985) of controls and experimental animals. The statistical difference between the values obtained in control and experimental animals was calculated (Weber, 1957). The electron microscopic appearance of cardiomyocites obtained from control and experimental rats was investigated as well.

Kynurenine and Serotonin Pathways
Edited by R. Schwarcz *et al.*, Plenum Press, New York, 1991

RESULTS AND DISCUSSION

At the beginning of clinical observation, elevated kynurenine accumulation in blood serum after tryptophan loading was noted in 13 patients and normal accumulation in 11 patients. Increased kynurenine accumulation after tryptophan loading is the most sensitive test for P-5-P deficiency in the organism (Brown, 1980). Therefore, all patients with increased kynurenine accumulation after tryptophan loading received supplementary treatment with vitamins B_6, B_2 and nicotinic acid (Rudzite et al., 1984). Patients with normal kynurenine accumulation in blood serum after tryptophan loading received only clonidine hydrochloride, propranolol and dichlorthiazide for treatment. Fig. 1 shows the results obtained at the beginning of the observation period and after a 3-year treatment. In 6 of 11 patients with normal kynurenine accumulation in blood serum before the investigation, 3-year treatment with the hypotensive drugs did not result in changes in serum kynurenine content after tryptophan loading (see first stripped bars on the upper part of Fig. 1). In 5 of 11 patients (see second stripped bars on the upper part of Fig. 1), treatment with hypotensive drugs was accompanied by an increased accumulation of kynurenine in blood serum after tryptophan loading. There were no changes in the left ventricular myocardial mass or in blood pressure in these patients after a 3-year antihypertensive treatment.

In 13 of 24 patients, elevated kynurenine accumulation after tryptophan loading was noted at the beginning of the investigation. Therefore, they received supplementary vitamin B_6, B_2 and nicotinic acid in addition to hypotensive drugs. In 8 of 13 patients, the treatment caused a decrease in kyn-

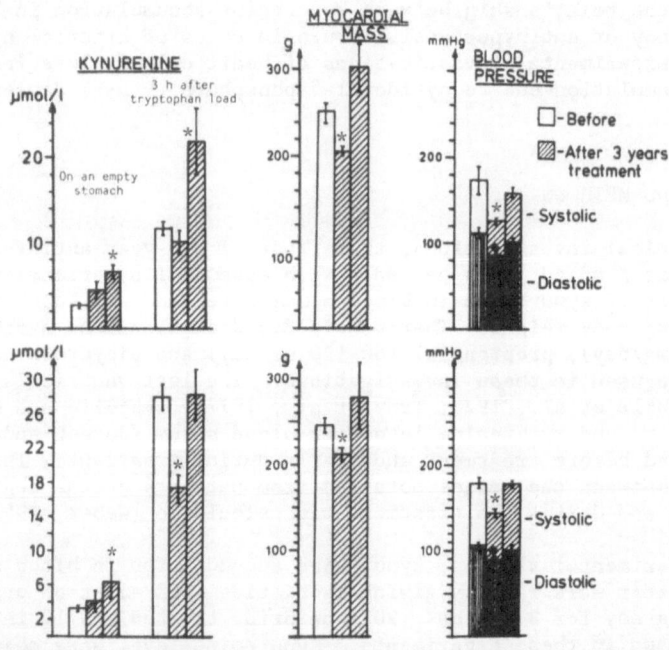

Fig. 1.   Results of a 3-year treatment of patients with essential hypertension and with different levels of kynurenine in blood serum after tryptophan loading at the beginning of the investigation. Top: patients with normal kynurenine accumulation; Bottom: patients with elevated kynurenine accumulation. *P < 0.05 to P < 0.0002 (t-test).

urenine accumulation after tryptophan loading and a decrease in the left
ventricular myocardial mass and arterial blood pressure (see first stripped
bars on the lower part of Fig. 1). In the remaining 5 patients (see second
stripped bars on the lower part of Fig. 1), the treatment was not accompanied by changes in kynurenine accumulation in blood serum or by changes in
the left ventricular myocardial mass and arterial blood pressure. These results suggest that kynurenine accumulation in blood serum prevented the positive effect of the above-mentioned antihypertensive drugs in patients with
essential hypertension and could be added to the negative signs of illness.

The enzyme systems splitting kynurenine and 3-hydroxykynurenine require
P-5-P for activation (Meister, 1965). Therefore, drugs which cause P-5-P
deficiency in the organism induce kynurenine accumulation in blood and urine
(Gorjachenkova, 1974). Administration of isoniazide, a direct antagonist of
P-5-P, for 3 months caused an elevation of kynurenine levels in blood serum
on an empty stomach (from 9.9 μmol/l to 15.0 μmol/l). The concentration of
P-5-P was decreased in isoniazide-treated rats both in blood serum and in the
liver, kidney and heart (Fig. 2). The isolated rat atria frequency in control rats was 249 ± 11/minute and in isoniazide-treated rats 146 ± 5/minute
(P < 0.01). We also observed a decrease in cardiac frequency in patients
with kynurenine accumulation in blood serum (Rudzite et al., 1987b).

Structural changes in cardiomyocytes, endocrine cardiomyocytes and endotheliocytes were seen in experimental animals. Light and dark endotheliocytes were observed as well. At the bottom and top of the cytoplasm of light
endotheliocytes were picnotic vesicles, and lysosomes were seen in both kinds
of endotheliocytes. Cardiomyocyte sarcolemma was folded at the level of the
Z-line, and the marginal hyaloplasm of the cell was light and edematous. Under the cytolemma in the cytoplasm were many picnotic vesicles. Moreover,
large mitochondrial aggregates were found with interconnected outer mitochondrial members between the myofibrils. Mitochondrial structures were polymor-
phus without cristae. Five to seven supercontracted sarcomers were also observed in myofibrils. Gradually, the Z-lines disappeared and chaotic proto-
fibrils, fatty inclusions, lysosomes and autophagosomes were observed in some
places. Endocrine cardiomyocytes myofibriles and mitochondria were changed
like those of the usual mitochondria and myofibrils in cardiomyocytes. The
great Golgi apparatus was found in cytoplasm around the nucleus and in the
middle of the cell. In and around vacuoles, many immature and mature secretion granules, differing in size and density, were seen. Granules were found
among myofibrils and the cell margin (Fig. 3). The morphology of the endocrine apparatus indicates active secretion of atrial natriuretic peptide.

The observed changes in cardiomyocytes are not specific, characteristic
only of P-5-P deficiency and kynurenine addition to the incubation medium for

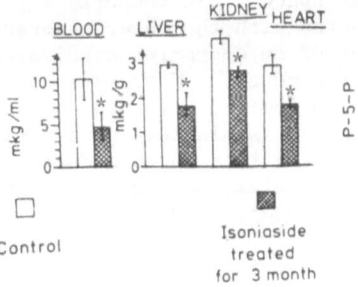

Fig. 2. Concentration of P-5-P in blood serum, liver,
kidney and heart tissue of control and isoniazide-treated rats. *P < 0.05 to P < 0.01
(t-test).

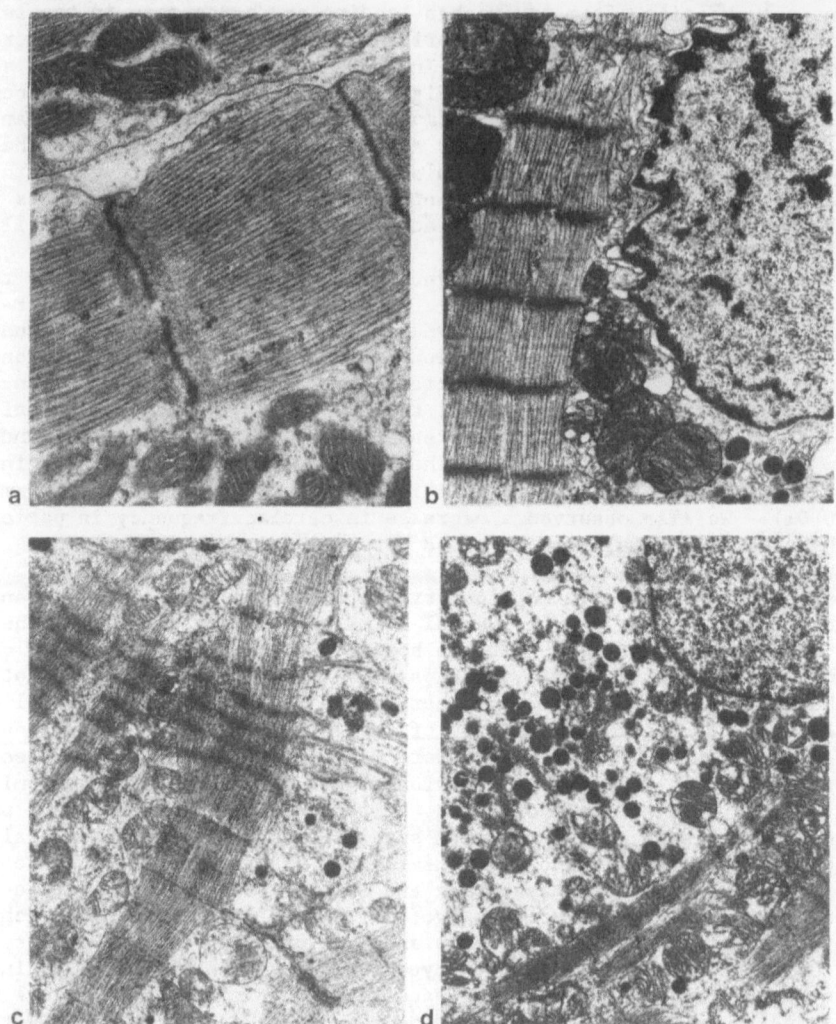

Fig. 3. Ultrastructure of cardiomyocytes from intact and isoniazide-treated rats. a: Cardiomyocytes from an intact rat. Note normal myofibrils and mitochondrial
structure (x 25,000); b: Fragment of endocrine cardiomyocyte from an isoniazide-treated rat. Note folded
perinuclear space. Mitochondric aggregates are seen
interconnected with the outer membranes (x 20,000);
c: Fragment of an endocrine cardiomyocyte from an iso-
oniazide-treated rat. Note supercontraction of myofibrils (x 20,000); d: Fragment of an endocrine cardiomyocyte from an isoniazide-treated rat. Note the folded
sarcolemma, with interspersed enlarged vesicles of endoplasmatic reticulum. Granules are of various size and
density (x 20,000).

the isolated heart. They are found in cells in which calcium has accumulated
because of the impairment of cytolemma. The supercontraction of myofibrils
is one of the morphological expressions of increased intracellular calcium
content. The disappearance of the mitochondrial membrane is also connected

666

with calcium accumulation in the cell (calcium activates phospholipase A_2, which dissolves membranes of mitochondria) as well as the disintegration of Z-lines (calcium activates the neutral proteasis, which disintegrates the Z-line of myofibrils). The active secretory action of endocrinic cardiomyocytes can be regarded as an adaptive compensatory reaction, because the atrial natriuretic peptide regulates electrolyte and water homeostasis.

CONCLUSION

We conclude that increased kynurenine accumulation in blood serum is a negative sign for cardiovascular diseases. It is a factor for inefficiency of antihypertensive treatment due to noradrenaline release. It impairs cytolemma, promotes calcium accumulation in the cell and calcium dependent cell dystrophy. We suggest that active secretory action of endocrine cardiomyocytes is the compensatory reaction for keeping the electrolyte and water homeostasis in impairment of the tryptophan-nicotinic acid pathway.

REFERENCES

Brown, R.R., 1980, Tryptophan metabolism in humans: perspectives and predications, in: "Biochemical and Medical Aspects of Tryptophan Metabolism", Hayaishi, O., Ishimura, Y., and Kido, R., eds., Elsevier North-Holland, Amsterdam, pp. 227-230.
Gorjachenkova, E.V., 1974, in: "Vitamins", Smirnov, E.J., ed., Medicine, Moscow, pp. 236-263.
Joseph, M.H., and Risby, D., 1975, The determination of kynurenine in plasma, Clin. Chim. Acta, 63:197-204.
Meister, A., 1965, "Biochemistry of the Amino Acids", Academic Press, New York.
Muharlamov, N.M., and Belenkov, P.N., 1981, "The Ultrasounds: Diagnostic in Cardiology", Medicine, Moscow.
Rudzite, V., Jurika, E., Arajs, J., and Andrejev, N., 1987a, The relationship between kynurenine and arterial blood pressure, in: "Progress in Tryptophan and Serotonin Research 1986", Bender, D.A., Joseph, M.H., Kochen, W., and Steinhart, H., eds., de Gruyter, Berlin, pp. 127-130.
Rudzite, V., Sileniece, G., and Jirgensons, J., 1984, Tryptophan-nicotinic acid pathway in cardiovascular diseases. in: "Progress in Tryptophan and Serotonin Research", Schlossberger, H.J., Kochen, W., Linzen, B., and Steinhart, H., eds., de Gruyter, Berlin, pp. 365-381.
Rudzite, V., Sileniece, G., Jirgensons, J., Skards, J., Zirne, R., and Dalmane, A., 1987b, Bradyarrhythmias and myocardial cell failure induced by kynurenine, in: "Progress in Tryptophan and Serotonin Research 1986", Bender, D.A., Joseph, M.H., Kochen, W., and Steinhart, H., eds., de Gruyter, Berlin, pp. 131-136.
Serfontein, W.J., Ubbink, J.B., De Viller, L.S., Rapley, C.H., and Becker, J.B., 1985, Plasma pyridoxal-5-phosphate level as risk index for coronary artery disease, Atherosclerosis, 55:357-361.
Teischolz, L., Kreulen, T., Herman, M., and Goirlin, R., 1972, Circulation, Suppl. 11, 465:75-81.
Troy, B., Pambo, J., and Rackley, C., 1972, Measurement of left ventricular wall thickness and mass by echocardiography, Circulation, 45:602-608.
Weber, E., 1957, "Grundriss der biologischen Statistik fur Naturwissenschaftler, Landwirte und Mediziner", VEB Gustav Fischer Verlag, Jena.

TRYPTOPHAN AND ITS METABOLITES IN A FAMILY WITH HARTNUP DISEASE

D.D. Milovanović, L. Milovanović[1], B. Stanković[2], and
D. Radulović[3]

Clinic for Neurology and Psychiatry for Children
and Young People
Belgrade

[1]Laboratory for Drug Control
Belgrade

[2]Institute of Analytical Chemistry
Faculty of Pharmacy
University of Belgrade

[3]Institute of Pharmaceutical Chemistry
Faculty of Pharmacy
University of Belgrade
Yugoslavia

INTRODUCTION

A study dedicated to research of tryptophan and its metabolites in urine of Hartnup disease was published 30 years ago (Milne et al., 1960). The present paper pertains to the investigation of L-tryptophan (TRY), kynurenine (KYN), xanthurenic acid (XA), 5-hydroxyindoleacetic acid (5-HIAA) and indoxyl sulfate (IS) in plasma and urine of a four member family, with one symptomatic and one asymptomatic patient with Hartnup disease, prior to and after a TRY load. In parallel, one healthy subject was studied.

MATERIAL AND METHODS

Subjects: The investigation was performed under uniform conditions with a hospital diet lacking bananas, walnuts and tomatoes.

Patient D.R., daughter, 16 years of age (50 kg body weight), with the clinical picture of Hartnup disease, was first described by Baron et al. (1956). Therapy was discontinued four days before investigation, except for Lincomycin (lincocin) which was stopped one day prior to the TRY load. Mother S.R., 41 years of age (60 kg body weight), had severe migraine headache before investigation. Father D.R., 39 years of age, (80 kg body weight) and son, V.R., 10 years of age (27 kg body weight), were asymptomatic. The control subject was a 25 year old healthy female (65 kg body weight). The oral TRY loading test (50 mg/kg body weight) was performed at 7:00 a.m. after overnight fasting. After oral loading of TRY, blood pressure was measured at specific time intervals. TRY was administered in tablets of 500 mg (Leopold Co., Graz, Austria).

Urine and blood samples: For determination of TRY, 2 ml blood samples were taken on an empty stomach into heparinized tubes every 2 hours. The first meal was taken after blood sampling at 1:00 p.m. Blood was prepared for analysis in accordance with the method of Lato et al. (1974). Urine was collected at several intervals and stored in a cooler.

Standard solutions: For analytical purposes, a freshly prepared aqueous solution containing 25 mg/50 ml of L-tryptophan (Fluka) was used. L-kynurenine sulfate, xanthurenic acid and indoxylsulfate (Serva) standard solutions were prepared at a concentration of 20 mg/100 ml in distilled water. 5-Hydroxy-3-indoleacetic acid (Fluka) was made up at 10 mg/50 ml.

Apparatus: A Pye Unicam SP-6-500 spectrophotometer (Cambridge, UK) provided with 10 mm quartz cells was used, and the test compounds were detected by UV absorption at 254-356 nm.

Chromatographic analyses: TRY was separated by unidimensional paper chromatography. Urine and plasma samples were run twice simultaneously with standard solution. The chromatograms were developed in freshly mixed isopropanol-ethyl acetate-acetone-methanol-ammonia-water (Lato et al., 1974). After drying for two hours in a stream of air, the chromatograms were developed with 0.1% ninhydrin in N-butanol-acetic acid-water (237.5:1.5:11).

Metabolites of the tryptophan-kynurenine pathway were separated by unidimensional paper chromatography with acetate buffer (pH = 5.4) as a solvent. The position of fluorescent spots was noted under UV light.

Quantitative analysis: The method of Mortreuil and Khouvine (1954) was used to determine TRY. A linear relationship between absorbance and concentration was established over the range of 2.5-25 μg. The regression equation was $y = 85.3x - 1.3$ and the correlation coefficient (r) was 0.999. The detection limit was 0.5 μg/ml. KYN and XA were measured by the method of Walsh (1965). Optical density plots from pure standards were linear in the range of 2-20 μg for KYN and XA. The regression equation was $y = 153.0x - 0.7$ and $r = 0.997$ for KYN. The equation for XA was $y = 42.9x - 0.3$ and $r = 0.997$. 5-HIAA was determined in urine samples immediately after urine collection using the spectrophotometric method of Udenfriend et al. (1955). The regression equation was $y = 155.2x - 1.7$ and $r = 0.999$. IS in urine was visualized by two-dimensional paper chromatography (Jepson, 1955). Color intensity and spot size were marked as - (negative), ± (slightly decreased), + (slightly increased), and ++ (increased).

RESULTS AND DISCUSSION

TRY concentrations in plasma in Hartnup disease after an oral load of TRY are reportedly unchanged or show minimal changes (Wong and Pillai, 1966; Shih et al., 1971).

Concentration of TRY in plasma after a load showed two peaks (Table 1), the first occurring after 2 hours and the second after 8 hours (after the first meal) except in the mother whose TRY concentration was lower. After an oral load, the concentration of TRY in the plasma of patient (D) was consistent with these findings in Hartnup disease (Wong and Pillai, 1966; Shih et al., 1971). TRY concentrations in plasma decreased in later samples. Nonsignificant increases of TRY after a load were also found in the son, who is heterozygous for Hartnup disease. However, TRY concentration in the plasma of patient (F) after a load were significantly higher.

Table 1. Plasma tryptophan in a family with Hartnup disease and the control subject

	Tryptophan (50 mg/kg body weight)				
Time of day	7:00[x]	9:00	11:00	13:00[i]	15:00

Tryptophan

D	69	78	11	19	27
M	61	186	132	144	90
F	44	395	169	144	236
S	115	153	136	115	211
C	117	414	295	81	115

[x]Before oral loading. [i]Meal. D: Daughter (patient with clinical picture of Hartnup disease); M: Mother; F: Father (asymptomatic patient with Hartnup disease); S: Son; C: Control subject. Plasma TRY levels are expressed in nmol/ml.

The peak levels of urinary TRY after a load were found between 7:00 and 10:00 a.m. and were different in each subject. Mild tryptophanuria in the patient and significant tryptophanuria in her father were found before and after an oral load (Table 2). The findings in patient (D) suggest impaired absorption of this amino acid in the intestine and impairment of reabsorption in proximal renal tubules. TRY levels in patient (F) suggest a greater renal

Table 2. Urinary tryptophan in a family with Hartnup disease and the control subject

	L-Tryptophan (50 mg/kg body weight)				
Time period	22:00-7:00[x]	7:00-10:00	10:00-13:00	13:00-19:00	19:00-7:00
D	13.5 (460)	14.3 (80)	11.3 (50)	9.2 (60)	6.0 (260)
M	2.4 (600)	7.1 (180)	15.5 (140)	4.4 (280)	4.3 (650)
F	26.4 (380)	122.8 (400)	97.9 (430)	87.5 (550)	34.6 (570)
S	4.0 (330)	7.3 (50)	5.6 (190)	7.4 (150)	2.0 (220)
C	6.8 (380)	11.1 (110)	8.3 (80)	6.1 (225)	8.8 (610)

[x]Before oral loading. All values are expressed as μmol TRY per urine volume (presented in parentheses in ml) per 1 hour.

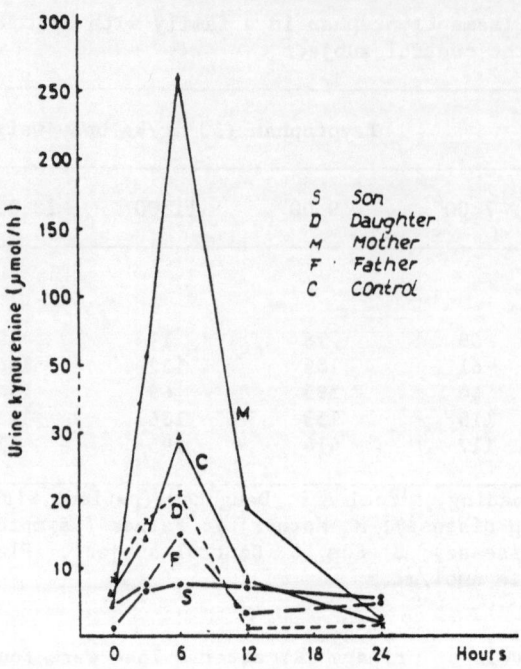

Fig. 1. Urinary excretion of kynurenine after oral loading.

loss of TRY, while the intestinal absorption was significantly less impaired. The son of patient (F) had relatively low levels of plasma TRY after an oral load. This indicates that he had some impairment of TRY absorption. The plasma concentration of TRY in Hartnup disease may, however, be reduced by poor absorption from the gut and by abnormal urinary loss (Milne et al., 1960).

Excretion of KYN in both patients (D,F) was significantly reduced six hours after the oral load. The characteristics of KYN excretion after a load

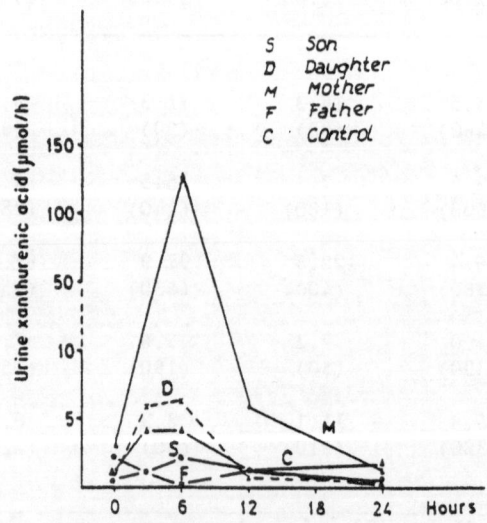

Fig. 2. Urinary excretion of xanthurenic acid after oral loading.

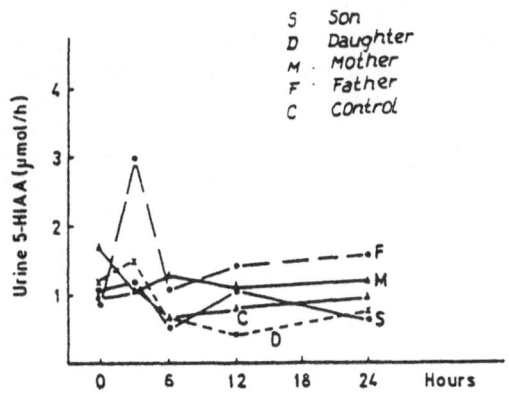

Fig. 3. Urinary excretion of 5-HIAA after oral loading.

in both patients and the son agree with published reports (Milne et al., 1960; Wong and Pillai, 1966).

Enormous excretion of KYN and XA was found in the mother 3 to 6 hours after an oral load. Paper chromatography results of the mother's urine revealed some abnormalities of Ehrlich reagent-positive compounds. The mother is heterozygote for Hartnup disease. One day prior to the load she had severe migraine headache. During the past two decades, a patient who excreted large amounts of KYN, 3-hydroxy-KYN and XA in the urine has been reported (Komrower et al., 1964). Increased excretion of these compounds is caused by a deficit of kynureninase. Recently, a high basal excretion of KYN in urine was described in one patient, which suggests a defect in the activity of kynureninase (Salih et al., 1985). However, the expected greater excretion of KYN in urine was not found after a TRY load. Basal KYN values in the mother were within normal limits (Fig. 1). In two families, gene carriers were found for two disorders, Hartnup disease and for methylmalonic aciduria (Shih et al., 1984). The question must therefore be posed if the mother has some other disease. In patient (F) and his son, excretion of XA was not increased in urine after an oral load (Fig. 2). In patient (D) and the control subject, the increased excretion after a load was decreased after six hours.

A continuous increase of 5-HIAA in urine was observed in the mother at all intervals before and after an oral load (Fig. 3). In patient (F), this increase was observed at all intervals after the load. These increases point out that metabolism of TRY via the 5-hydroxyindole route is increased. Since the biochemical findings in the family are unexpected and interesting, further work and research is necessary. Excretion of IS before and after oral TRY loading were not significantly increased. On the paper chromatogram of 21 urine samples, a slight decrease was seen, in 3 findings were negative, and 2 slightly increased. On paper chromatography of the mother's urine after loading, six orange and 2 blue spots were detected. In two urine samples of the mother, 3 and 6 hours after the load, the color became dark with a yellowish shade. Similar phenomena have been described in the patient with hydroxykynureninuria (Komrower et al., 1964).

Blood pressure, measured at 1,4,7,12 and 24 hours after loading, did not reveal significant changes in any subject. However, stabilizing of blood pressure after a load was observed in this group.

CONCLUSION

Determination of TRY in plasma and urine before and after a TRY load suggests different degrees of a TRY absorption defect in the intestine and its renal reabsorption in both the symptomatic (daughter) and the asymptomatic (father) patients. In the youngest member of this family, the son, the existence of a partial defect in absorption of TRY in the intestine can be assumed with certainty, such as may be found in a heterozygote. The characteristics of KYN excretion in Hartnup disease after a load were found in both patients (symptomatic and asymptomatic) and the son.

Enormous excretion of KYN and XA was found in the mother's urine after a load. The probable genetic nature of the condition is discussed.

REFERENCES

Baron, D.N., Dent, C.E., Harris, H., Hart, E.W., and Jepson, J.B., 1956, Hereditary pellegra-like skin rash with temporary cerebellar ataxia, constant renal aminoaciduria, and other bizarre biochemical feature, Lancet, 2:421-428.

Jepson, J.B., 1955, Paper chromatography of urinary indoles, Lancet, 2:1009-1011.

Komrower, G.M., Wilson, V., Clamp, J.R., and Westall, R.G., 1964, Hydroxykynureninuria, Arch. Dis. Childh., 39:250-256.

Lato, M., Rufini, S., Ghebergzabher, M., Cuiffini, G., and Mezzetti, T., 1974, A sensitive chromatographic technique for screening of amino acid metabolic defects in the newborn, Clin. Chim. Acta, 53:273-280.

Milne, M.D., Crawford, M.A., Girao, C.B., and Loughridge, L.M., 1960, The metabolic disorder in Hartnup disease, Quart. J. Med., 29:407-421.

Mortreuil, M., and Khouvine, Y., 1954, Dosage des acides aminés par les complexes colorés dicetohydrindylidene-hydrindamine-sels de cadmium, Bull. Ste. Chim. Biol., 36:425-428.

Salih, M.A.M., Bender, D.A., and McCreanor, G.M., 1985, Lethal familial pellagra-like skin lesion associated with neurologic and developmental impairment and the development of cataracts, Pediatrics, 76:787-793.

Shih, V.E., Bixby, E.M., Alpers, D.H., Bartsocas, C.S., and Thier, S.O., 1971, Studies of intestinal transport defect in Hartnup disease, Gastroenterology, 61:445-453.

Shih, V.E., Coulombe, J.T., Wadman, S.K., Duran, M., and Waelkens, J.J.J., 1984, Occurrences of methylmalonic aciduria and Hartnup disorder in the same family, Clin. Genet., 26:216-220.

Udenfriend, S., Titus, E., and Weissbach, H., 1955, The identification of 5-hydroxy-3-indoleacetic acid in normal urine and method for its assay, J. Biol. Chem., 216:499-505.

Walsh, M.P., 1965, Separation and estimation of tryptophan-nicotinic acid metabolites in urine by thin-layer chromatography, Clin. Chim. Acta, 11:263-267.

Wong, P.W.K., and Pillai, P.M., 1966, Clinical and biochemical observations in two cases of Hartnup disease, Arch. Dis. Childh., 41:383-388.

EFFECT OF HUNTINGTON'S AND ALZHEIMER'S DISEASES ON THE TRANSPORT OF NICOTINIC ACID OR NICOTINAMIDE ACROSS THE HUMAN BLOOD-BRAIN BARRIER

L. V. Hankes, H.H. Coenen, E. Rota, K.J. Langen,
H. Herzog, W. Wutz, G. Stoecklin, and L.E. Feinendegen

Institut für Medizin
Institut für Chemie I
Kernforschungsanlage Jülich GmbH
D-5170 Jülich
Germany

INTRODUCTION

Although Elvehjem et al. (1937) discovered that nicotinamide relieved the neurological symptoms of pellagra in dogs, and in spite of the discovery by Hankes and Elvehjem (1949) that nicotinamide or tryptophan reversed the neurological syndrome produced in rats by a diet high in phenylalanine, the importance of nicotinamide and related compounds in neurology remained dormant for almost two decades. The recent surge of interest in the effects of the tryptophan-kynurenine-nicotinamide pathway compounds in neurology (Stone and Connick, 1985) and the transport of these compounds through the blood-brain-barrier prompted a study of the transport of nicotinic acid or nicotinamide through the blood-brain barrier and a subsequent study of the rate of metabolism of these two compounds in the human brain in various neurological disease states.

COMPOUNDS AND SUBJECTS

Carboxy-^{11}C-nicotinic acid used in the study was synthesized by a carboxylation of ^{3}Li-pyridine with $^{11}CO_2$ produced by a minicyclotron. Carboxamide-^{11}C-nicotinamide was made by amidation of carboxyl-^{11}C-nicotinic acid. Both compounds were purified with a high-pressure liquid chromatography system and dissolved in pyrogen-free saline solution (Machulla et al., 1976). Table 1 shows that 1.8 to 15 mCi of a labeled compound was intravenously injected into normal volunteers and patients with Huntington's or Alzheimer's disease.

EXPERIMENTAL METHODS

The heads of the subjects were centered in a Scandtronix-type PC 4096-15 WB high resolution (5 mm) positron emission tomographic camera system containing eight rings. This orientation provided 15 sections of an organ at any one time. Positron emission tomography (PET) images were produced over a period of 40 to 60 minutes after the injection of a labeled compound. These images were used to construct time-activity curves of the activity concentra-

Table 1. Subjects, compounds and dosages (mCi)

	Code	Carboxyl-^{11}C-nicotinic acid	Carboxamide-^{11}C-nicotinamide
Control	1A	9.0	3.4
	1B		5.3
Huntington	2A	15.0	
	2B		2.5
Alzheimer	3A	8.0	1.8

tion in the cerebral tissue. Blood samples were drawn at specified intervals for the measurement of the activity present in whole blood and plasma. Pictures of the scans were printed out about 15 minutes after the administration of the compound (Rota et al., 1988).

RESULTS AND DISCUSSION

The K3 values in Table 2 (rate constants of flow of substrate into metabolism) show that nicotinic acid and nicotinamide passed through the blood-brain barrier and were slowly accumulated in cerebral tissue. As seen in Table 3, the cerebral uptake of nicotinamide was much higher than that for nicotinic acid. The values in Table 4 show that when the activity in the cerebrum was compared to the activity in plasma, the differences in uptake became smaller than when the uptake was compared to the activity in the injected dose. Apparently, more of the nicotinic acid entered other tissues such as red blood cells.

This is again illustrated by the data in Table 5, where the ratio of the activities in blood vs. plasma steadily increased when nicotinic acid was given to the subjects. The values of the ratio for the samples from patients receiving nicotinamide were much lower than those for the samples from patients given nicotinic acid. The data showed that after a slight drop at

Table 2. Rate constants K3 determined by Patlak plot (1/min)

	Code	Carboxyl-^{11}C-nicotinic acid	Carboxamide-^{11}C-nicotinamide
Control	1A	0.040	0.024
	1B		0.022
Huntington	2A	0.012	
	2B		0.022
Alzheimer	3A	0.020	0.012

K3: Rate constant of flow of substrate from compartment of free tracer in tissue to compartment of metabolized tracer in tissues. Patlak plot: Graphic evaluation of transfer to blood flow (Patlak and Blasberg, 1985).

Table 3. Cortical activity concentration per injected dose ($\mu Ci/ml/mCi$)

	Code	Carboxyl-^{11}C-nicotinic acid	Carboxamide-^{11}C-nicotinamide
Control	1A	3.6	17.7
	1B		17.9
Huntington	2A	2.7	
	2B		16.8
Alzheimer	3A	3.8	9.4

three minutes post injection, the values returned to a constant level, suggesting some sort of equilibrium status for this period of time. The large difference in the ratio for nicotinic acid compared to the ratio for nicotinamide again suggests that more of the nicotinamide went into the red cells. There were no significant differences in the observations on the human control subjects compared to the Huntington or Alzheimer patients.

SUMMARY

The cerebral uptake of ^{11}C-nicotinic acid (^{11}C-NAC) and ^{11}C-nicotinamide (^{11}C-NAM) was quantified by the use of PET. Based on the amount of activity injected, the PET images showed a low cerebral uptake of ^{11}C-NAC, while ^{11}C-NAM was clearly visualized in the cortical areas. This discrepancy was found to be the result of the binding of ^{11}C-NAC to the red blood cells by a factor of 5 to 20 above that for ^{11}C-NAM. ^{11}C-NAM was better extracted by the cerebrum than ^{11}C-NAC, as shown by the mean values of the cortical tissue/plasma ratio of 1.9 for ^{11}C-NAC and 5 for ^{11}C-NAM at 30 min. post-injection. An analysis of Patlak-Gjedde plot curves revealed a metabolic compartment for ^{11}C-NAC and ^{11}C-NAM with similar values of about 0.02 1/min for the accumulation constant K3. This was indicative of a slower transport rate for ^{11}C-NAC. A significant finding of the study was the increasing ratio of activity concentrations in red blood cells versus the concentrations in plasma (over time). There were no significant differences between the data from normal volunteers and patients with Huntington's or Alzheimer's disease.

Table 4. Ratio of cortical activity concentration/plasma activity concentration (mean between 20-40 min., p.i.) ($\mu Ci/\mu Ci$)

	Code	Carboxyl-^{11}C-nicotinic acid	Carboxamide-^{11}C-nicotinamide
Control	1A	2.0	3.5
	1B		4.5
Huntington	2A	2.0	
	2B		8.0
Alzheimer	3A	1.7	4.0

Table 5. Ratio of whole blood vs. plasma activity (nCi/ml)

Sample	Nicotinic acid		Nicotinamide			
Time (min)	2A	3A	2B	3A	1B	1A
0.25			0.74	1.02	1.24	0.73
0.5		2.5	0.70	0.80	1.18	0.62
0.75			2.26		1.10	0.93
1.0		4.9	1.63	2.47	1.14	1.20
1.5	14.2	5.5	1.25	2.29	1.07	1.09
2.0			1.18	2.11	1.02	1.01
3.0			1.28	1.95	1.05	0.98
4.0		10.8	1.64	2.39	1.13	1.00
6.0			2.91	3.31	1.20	1.10
10.0	25.9	13.7	3.04	2.85	1.45	1.18
15.0			3.31	3.64	1.64	1.27
20.0		16.7	3.34	3.74	1.79	1.47
25.0			3.33	3.32	1.91	1.46
30.0	40.3		3.74	3.61	1.99	1.35
40.0	47.1	20.8				
45.0			3.99	3,76	2.18	1.52
50.0	59.9					
60.0			3.31	3.25	2.12	1.53

The number-letter code is the same as in Table 1.

REFERENCES

Elvehjem, C.A., Madden, R.J., Strong, F.M., and Wollen, D.W., 1937, Relation of nicotinic acid and nicotinic acid amide to canine black tongue, J. Am. Chem. Soc., 59:1767-1768.

Hankes, L.V., and Elvehjem, C.A., 1949, A nervous syndrome produced with phenylalanine and methionine, Proc. Soc. Exp. Biol. Med., 72:349-351.

Machulla, H.J., Laufer, P., and Stoecklin, G., 1976, Radioanalytical quality control of ^{11}C, ^{18}F and ^{123}I labelled compounds and radiopharmaceuticals, Radioanal. Chem., 32:32:381-400.

Patlak, C.S., and Blasberg, R.G., 1985, Graphical evaluation of blood to brain transfer constants from multiple-time uptake data, J. Cereb. Blood Flow Metab., 5:584-590.

Rota, E., Herzog, H., Schmid, A., Holte, S., and Feinendegen, L.E., 1988, Performance of machine parameters and its influence on metabolic data using the PET-scanner Pc-4096-15WB, J. Nucl. Med., 29:877.

Stone, T.W., and Connick, J.H., 1985, Quinolinic acid and other kynurenines in the central nervous system, Neuroscience, 15:597-617.

EFFECTS OF PROFOUND INSULIN-INDUCED HYPOGLYCEMIA ON QUINOLINIC ACID IN

HIPPOCAMPUS AND PLASMA

M.P. Heyes[1,2], M. Papagapiou[3], C. Leonard[3],
S.P. Markey[2], and R.N. Auer[3]

[1]Section on Analytical Biochemistry
Laboratory of Clinical Science
National Institute of Mental Health
Bethesda, Maryland 20892

[2]Laboratory of Neurophysiology
National Institute of Mental Health
Bethesda, Maryland 20892
USA

[3]Neuroscience Research Group
University of Calgary
Calgary, Alberta T2N 4N1
Canada

INTRODUCTION

Profound insulin-induced hypoglycemia is associated with neurodegeneration which resembles lesions produced by administration of exogenous excitotoxins (Auer, 1986). Neuronal lesions in hypoglycemia begin within minutes
of the hypoglycemic insult and are attenuated by local application of
N-methyl-D-aspartate (NMDA) receptor antagonists (Wieloch 1985; Simon et al.,
1986).

Recent studies have indicated that the L-tryptophan metabolite quinolinic acid (QUIN) is an agonist of NMDA receptors (Perkins and Stone, 1983)
and an excitotoxin (Schwarcz et al., 1983), but the effects of hypoglycemia
on brain QUIN metabolism are unknown. To investigate whether QUIN concentrations are increased in brain during hypoglycemia, we measured QUIN (Heyes and
Markey, 1988a,b) in hippocampus and plasma after 40 minutes of electroenceph-
alogram (EEG) isoelectricity as well as after one to two hours of normoglycemic recovery. Because NMDA receptors are localized on the outer membranes
of neurons and are activated by compounds in the extracellular fluid, we also
quantified QUIN in the extracellular fluid of the hippocampus by in vivo
microdialysis.

METHODS

Male Wistar rats (230 - 243 g) were fasted overnight but had free access
to tap water. Brains from four groups of six rats were studied.

1) Control group: Killed immediately upon removal from the cage.

2) Sham-operated group: Exposed to all procedures except injection of insulin.

3) Hypoglycemic isoelectricity group: Exposed to 40 min of insulin-induced EEG isoelectricity.

4) Recovery group: 40 min of EEG isoelectricity followed by 60 min of normoglycemic recovery.

For the hypoglycemia groups, rats were given an intraperitoneal injection of 10 to 15 IU/kg insulin 1 to 2 h prior to operation. Sham-operated rats received an equivalent volume of saline. Rats were artificially ventilated, and blood gases, blood pressure and bipolar interhemispheric EEG monitored. Cerebral isoelectricity began 60 to 170 min after the injection of insulin. After 40 min of EEG isoelectricity, normoglycemic recovery was induced with an infusion of 1.5 ml of 25% glucose. The concentrations of L-TRP, 5-HT and 5-HIAA were qualified in brain tissue by high pressure liquid chromatography (Heyes and Markey, 1988b).

Microdialysis probes were inserted one day prior to induction of hypoglycemia. Dialysates were collected in one hour aliquots during control and recovery periods and for 40 min during isoelectricity. In two additional rats, 3-hydroxyanthranilic acid (100 nmol/l) was perfused through the microdialysis probe for 1 h beginning after a 3 h control period.

Values presented are the mean ± one SEM. Results were analyzed by one-way analysis of variance and Dunnett's 't'-test.

RESULTS

There were no significant differences in the hippocampal concentrations of QUIN, L-TRP, 5-HT or 5-HIAA between control and sham rats and the data were pooled for statistical comparisons: QUIN, 21.1 ± 3.4 fmol/mg; L-TRP, 41.3 ± 6.3 pmol/mg; 5-HT, 4.3 ± 0.6 pmol/mg and 5-HIAA, 3.0 ± 0.3 pmol/mg.

Plasma QUIN concentrations were 273.2 ± 32.1 pmol/ml in control rats. Plasma QUIN increased significantly by 2 h after injection of insulin (to 375% of control; $P < 0.01$) and increased further during hypoglycemic isoelectricity to 647% of control ($P < 0.001$). Plasma QUIN concentrations began to return to control values during the recovery period, but were still increased by 396% of control at 2 h ($P < 0.01$).

In the hippocampus, QUIN concentrations increased to 573% of control ($P < 0.005$) during hypoglycemia and increased further to 706% of control ($P < 0.005$) during recovery. However, QUIN recoveries from microdialysates did not change during either hypoglycemia or recovery. In contrast, 3-hydroxyanthranilic acid, added to the microdialysate, resulted in a 18- and 30-fold increase in QUIN in the two rats studied.

L-TRP concentrations did not change during hypoglycemia but were increased to 160% of control ($P < 0.01$) after 1 h of recovery. 5-HT concentrations decreased to 40% and 69% of control ($P < 0.01$) after both hypoglycemia and recovery, respectively. 5-HIAA concentrations increased to 261% and 325% of control ($P < 0.01$) after hypoglycemia and recovery, respectively.

DISCUSSION

The increase in hippocampus QUIN concentrations during 40 min of cerebral isoelectricity and normoglycemic recovery could be interpreted as being
consistent with the notion that QUIN, via agonist effects on NMDA receptors,
may be involved in the neuropathology of hypoglycemia. Neuronal damage produced by profound hypoglycemia is characteristic of excitotoxic damage including dendritic swelling, mitochondrial disruption, subsequent acidophilic
staining of neurons and necrosis (Auer, 1986). Neuronal damage is observed
within minutes of isoelectricity (Auer, 1986), which emphasizes the importance of early changes in the concentrations of potential neurotoxic candi-
dates. QUIN excitotoxic lesions are similar to hypoglycemic damage and result when, for example, 12 nmol of QUIN is injected into the striatum
(Schwarcz et al., 1983). Assuming a wet weight of 50 mg for a rat striatum,
12 nmol may be expected to produce peak QUIN concentrations of 240 micromolar. This resultant concentration of QUIN far exceeds the levels of QUIN
observed in brain tissue in hypoglycemia or calculated extracellular fluid
QUIN concentrations (During et al., 1989), which dilutes the potential for
QUIN to mediate hypoglycemic brain damage.

In view of the increased QUIN concentrations in hippocampus, in vivo
microdialysis was used to evaluate changes in QUIN in the extracellular fluid
during hypoglycemia (During et al., 1989) where QUIN would have access to
NMDA receptors. The hippocampus was chosen because of its susceptibility to
hypoglycemic damage. However, QUIN recoveries from the microdialysate probe
indicate that extracellular fluid QUIN concentrations were not increased during either hypoglycemia or the recovery period. The fact that 3-hydroxyan-
thranilic acid, a known precursor of QUIN in vivo (Heyes et al., 1988), added
to the microdialysate perfusion medium, increased QUIN synthesis indicates
that the procedure employed would have been able to detect changes in extracellular fluid QUIN concentrations. Therefore increases in brain tissue QUIN
during hypoglycemia may be localized to the intracellular compartment.

Plasma QUIN concentration increased during hypoglycemia. Assuming a
brain plasma volume of 0.9 μl (Heyes and Markey, 1988a), the contribution of
plasma QUIN to measured brain QUIN content in control and sham rats was approximately 4%, whereas in hypoglycemic rats the contribution was approximately 12%. This calculation does not include the small perivascular fluid
volume, which presumably contains QUIN. While it is possible that the increases in hippocampal QUIN concentration during hypoglycemia and recovery
may reflect the presence of QUIN in the blood of the tissue samples, it is of
note that hippocampus QUIN concentrations increased further after 1 h of recovery, whereas plasma QUIN levels began to decrease towards control values
during the recovery period.

Systemic L-TRP administration increases plasma QUIN concentration (Heyes
and Markey, 1988b) and reduces the rate of hepatic gluconeogenesis, most
likely because of increased hepatic QUIN concentrations, chelation of iron by
QUIN and concomitant inhibition of hepatic phosphopyruvate carboxylase (Snoke
et al., 1971). Gluconeogenesis is also reduced during hyperglycemia induced
by glucose administration and an increase in plasma insulin concentrations.
If insulin per se increases the synthesis of QUIN in liver, as indicated by
the plasma data, QUIN may have a physiological role in the inhibition of hepatic gluconeogenesis at the site of phosphopyruvate carboxylase during hyper-
glycemia-induced secretion of insulin.

During hypoglycemia, regional brain 5-HT concentration decreased at the
same time as 5-HIAA concentration increased. This increase in 5-HIAA may reflect release of 5-HT from storage granules due to reduced energy stores to
maintain the integrity of vesicular transport of the transmitter (Agardh et
al., 1979). The restoration of 5-HT concentrations during the recovery per-

iod may be related to increases in tryptophan hydroxylase activity (Agardh et al., 1979) as well as increases in the concentrations of L-TRP and conversion to 5-HT by a mass action ratio effect (Fernstrom, 1983).

We conclude that although insulin-induced hypoglycemia is associated with increased systemic synthesis of QUIN and, perhaps, an increase in QUIN concentration within the intracellular compartment of brain, the concentration of QUIN in the extracellular fluid space within brain are unchanged. Our data do not support a direct role for QUIN in the early excitotoxic lesions associated with profound insulin-induced hypoglycemia and cerebral isoelectricity.

REFERENCES

Agardh, C.-D., Carlsson, A., Lindqvist, M., and Siesjö, B.K., 1979, The effect of pronounced hypoglycemia on monoamine metabolism in rat brain, **Diabetes**, 28:804-809.

Auer, R.N., 1986, Progress review: hypoglycemic brain damage, **Stroke**, 17:699-708.

During, M.J., Heyes, M.P., Freese, A., Markey, S.P., Martin, J.B., and Roth, R.H., 1989, Quinolinic acid concentrations in striatal extracellular fluid reach potentially neurotoxic levels following systemic L-tryptophan loading, **Brain Res.**, 476:384-387.

Fernstrom, J.D., 1983, Role of precursor availability in control of monoamine biosynthesis in brain, **Physiol. Rev.**, 63:484-564.

Heyes, M.P., Hutto, B., and Markey, S.P., 1988, 4-chloro-3-hydroxyanthranilic acid inhibits 3-hydroxyanthranilic acid oxidase in brain, **Neurochem. Int.**, 13:405-409.

Heyes, M.P., and Markey, S.P., 1988a, Quantification of quinolinic acid in rat brain, whole blood and plasma by gas chromatography and negative chemical ionization mass spectrometry: Effects of systemic L-tryptophan administration on brain and blood quinolinic acid concentrations, **Anal. Biochem.**, 174:349-359.

Heyes, M.P., and Markey, S.P., 1988b, [^{18}O]-quinolinic acid: its esterification without back exchange for use as internal standard in the quantification of brain and CSF quinolinic acid, **Biomed. Environ. Mass Spectrom.**, 15:291-293.

Lapin, I.P., 1982, Convulsant action of intracerebroventricularly administered l-kynurenine sulphate, quinolinic acid and other derivatives of succinic acid, and effects of amino acids: structure-activity relationships, **Neuropharmacology**, 21:1227-1233.

Schwarcz, R., Whetsell, W.O., and Mangano, R.M., 1983, Quinolinic acid: an endogenous metabolite that produces axon-sparing lesions in rat brain, **Science**, 219:316-318.

Simon, R.P., Schmidley, J.W., Meldrum, B.S., Swan, J.H., and Chapman, A.G., 1986, Excitotoxic mechanisms in hypoglycemic hippocampal injury, **Neuropathol. Appl. Neurobiol.**, 12:567-576.

Snoke, R.E., Johnston, J.B., and Lardy, H.A., 1971, Response of phosphopyruvate carboxylase to tryptophan metabolites and metal ions, **Eur. J. Biochem.**, 24:342-346, 1971.

Wieloch, T., 1985, Hypoglycemia-induced neuronal damage prevented by an N-methyl-D-aspartate antagonist, **Science**, 230:681-683.

BRAIN AND CEREBROSPINAL FLUID QUINOLINIC ACID CONCENTRATIONS IN PATIENTS WITH

INTRACTABLE COMPLEX PARTIAL SEIZURES

M.P. Heyes[1], A.R. Wyler[2], O. Devinsky[3], J.A. Yergey[4],
S.P. Markey[1], and N.S. Nadi[5]

[1]Section on Analytical Biochemistry
Laboratory of Clinical Science
NIMH
Bethesda, Maryland 20892

[2]Department of Neurosurgery
University of Tennessee School of Medicine
Memphis, Tennessee 38103

[3]UMDNJ
Newark, New Jersey 07103

[4]Section of Analytical Chemistry
Laboratory of Clinical Studies
NIAAA
Bethesda, Maryland 20892

[5]Section of Neuronal Excitability
NINDS
Bethesda, Maryland 20892
USA

INTRODUCTION

Excitatory amino acids have been implicated in development and maintenance of seizures (Anderson et al., 1987; Meldrum 1987). In accordance with
this notion are the observations that antagonists of excitatory amino acid
receptors, including the NMDA receptor, are potent anticonvulsants (Meldrum
et al., 1988). In the widely used model of kindling, electrophysiologic
studies have demonstrated an increased sensitivity of NMDA receptors at
stages IV and V (Mody et al., 1988). In man, increases in the concentrations
of glutamate and aspartate, but not of GABA, have been reported in the focal
compared to non-focal regions of cerebral cortex (Nadi et al., 1987; Sherwin
et al., 1988). Additional studies have shown an increased number of NMDA receptors in the epileptic focus (Wyler et al., 1987). Studies in experimental
animals have shown that local application of quinolinic acid (QUIN) increases
neuronal activity (Perkins and Stone, 1983) and may cause seizures (Lapin,
1982). Conceivably, QUIN may therefore contribute to increased neuronal excitability in human seizure disorders. Direct measures of QUIN in human epilepsy have not been reported, although decreases in the degradative enzyme of
QUIN, quinolinic acid phosphoribosyltransferase (QPRT), have been published
(Feldblum et al., 1988), possibly indicating an impaired ability to catabolize QUIN.

Kynurenine and Serotonin Pathways
Edited by R. Schwarcz *et al.*, Plenum Press, New York, 1991

To investigate whether there is evidence of a disturbance in QUIN concentrations in human epileptics, the concentration of QUIN was measured in surgically removed focus and non-focus regions of temporal neocortex of 18 patients with complex partial seizures, intractable to medication. The concentration of QUIN was also measured in cerebrospinal fluid (CSF) from inter- and post-ictal samples from patients with complex partial seizures, intractable to medication, compared to CSF from neurologically normal volunteers.

METHODS

Specimen collection

Samples of focus and non-focus regions of brain were collected from epileptic patients admitted to the University of Tennessee Medical Center because of seizures intractable to drug therapy. Epileptic brain regions in the inferior mesial temporal cortex were first localized by chronic subdural electrode recording and were further delineated during surgery with the use of electrocorticography. The non-spiking region, which had to be resected in order to reach the spiking area, was also in the mesial temporal lobe.

CSF was collected from eleven neurologically normal volunteers and from 18 patients with complex partial seizures. Seizures were verified by telemetry as well as clinical observations. An 'inter-ictal' sample was defined as CSF collected in the absence of seizures for at least 24 h. Post-ictal CSF samples were defined as samples collected within 50 min following a seizure. QUIN was quantified in the seventh milliliter aliquot (Heyes and Markey, 1988a,b).

All epileptic patients studied were taking carbamazapine, phenytoin or valproate in monotherapy or polytherapy form. The blood levels of drugs were in the therapeutic range at the time of surgery and CSF collection. Control patients were not on any medication.

Results were analyzed by two-tailed 't'-tests and one way analysis of variance with Dunnett's 't'-test. Values presented are mean ± one SEM.

RESULTS

There was no significant difference in the concentration of QUIN in focus (217.7 ± 21.0 fmol/mg protein) versus non-focus (236.7 ± 20.3 fmol/mg protein) regions of cerebral cortex. However, the concentration of QUIN in CSF of the seizure patients was significantly lower in both the inter-ictal (15.17 ± 1.37 nmol/l; $P < 0.01$) and post-ictal samples (14.72 ± 0.61 nmol/l, $P < 0.01$) compared to controls (21.88 ± 2.85 nmol/l).

DISCUSSION

The convulsant effects of QUIN (Lapin, 1982; Perkins and Stone, 1983) and the fact that QUIN is present in brain raises the possibility that QUIN may have an etiologic role in the pathogenesis of human convulsant disorders. However, the present results show that brain QUIN concentration is similar in focus and non-focus regions of the cerebral cortex and that the CSF QUIN concentration is actually lower in seizure patients compared to normal controls.

QPRT activity has been reported to be decreased in focus regions obtained during surgery compared to samples obtained post mortem, which was predicted to result in a reduction in the rate of QUIN catabolism and thereby an increase in local QUIN concentrations (Feldblum et al., 1988). Our observations in brain tissue do not support this prediction; rather, the CSF data indicate that there may be a reduction in the rate of QUIN synthesis in brain or increased removal of QUIN from the brain. Because it is not possible to collect neurologically normal brain samples from appropriate controls under the same conditions as employed for the epileptic patients, we cannot say for certain whether actual brain concentrations are lower in seizure patients than in control subjects, as suggested by the CSF data. If CSF QUIN concentrations reflect brain tissue concentrations, then the CSF QUIN results may indicate a generalized disturbance in brain QUIN metabolism in this form of seizure disorder, perhaps secondary to antiepileptic medication. However, Young et al. (1983), based on their studies of kynurenine concentrations in the CSF of drug-treated epileptics versus untreated epileptics, have suggested that decreased CSF levels of kynurenine may be a function of the disease process per se rather than antiepileptic drugs, although it is of note that kynurenine in brain may originate predominantly from blood (Gál and Sherman, 1980). If QUIN in brain is sensitive to the availability of kynurenine in brain, our observed decrease in CSF QUIN may be considered consistent with the reduced CSF kynurenine concentrations observed by Young et al.

We conclude that chronic or local increases in brain QUIN concentration are not a feature of chronic intractable complex partial seizures in the patients we have studied. However, our results do not exclude the possibility that early increases in brain QUIN concentration are involved in the pathogenesis of the epileptic focus. QUIN has been postulated to be involved in seizures associated with fever (Heyes et al., 1988) and it is of note that a history of febrile convulsions is often found in patients with temporal epilepsy (cf. So et al., 1989).

ACKNOWLEDGEMENT

We appreciate the careful technical assistance of B.J. Quearry and M. Der.

REFERENCES

Anderson, W.W., Swartzwelder, H.S., and Wilson, W.A., 1987, The NMDA receptor antagonist 2-amino-5-phosphonovalerate blocks stimulus train-induced epileptogenesis but not epileptiform bursting in the rat hippocampal slice, J. Neurophysiol., 57:1-21.

Feldblum, S., Rougier, A., Loiseau, H., Loiseau, P., Cohadon, F., Morselli, P.L., and Lloyd, K.G., 1988, Quinolinic acid phosphoribosyltransferase activity is decreased in epileptic human brain tissue, Epilepsia, 29:523-529.

Gál, E.M., and Sherman, A.D., 1980, L-Kynurenine: its synthesis and possible regulatory function in brain, Neurochem. Res., 5:223-239.

Heyes, M.P., Kim, P., and Markey, S.P., 1988, Systemic lipopoly saccharide and pokeweek mitogen increase quinolinic acid content of mouse cerebral cortex, J. Neurochem., 51:1946-1948.

Heyes, M.P., and Markey, S.P., 1988a, [^{18}O]-Quinolinic acid: its esterification without back exchange for use as internal standard in the quantification of brain and CSF quinolinic acid, Biomed. Environ. Mass Spectrom., 15:291-293.

Heyes, M.P., and Markey S.P., 1988b, Quantification of quinolinic acid in rat brain, whole blood and plasma by gas chromatography and negative chemical ionization mass spectrometry: effects of systemic L-tryptophan administration on brain and blood quinolinic acid concentrations, Anal. Biochem., 174:349-359.

Lapin, I.P., 1982, Convulsant action of intracerebroventricularly administered L-kynurenine sulphate, quinolinic acid and other derivatives of succinic acid, and effects of amino acids: structure-activity relationships,
Neuropharmacology, 21:1227-1233.

Meldrum, B., 1987, Neurotransmitter amino acids in epilepsy, **Electroenceph-
alogr. Clin. Neurophysiol. Suppl.**, 39:191-199.

Meldrum, B., Millan, M., Patel, S., and de Sarro, G., 1988, Antiepileptic
effects of focal micro-injection of excitatory amino acid antagonists, J.
Neural Transm., 72:191-200.

Mody, I., Stanton, P.K., and Heinemann, U., 1988, Activation of N-methyl-D-
aspartate receptors parallels changes in cellular and synaptic properties
of dentate gyrus granule cells after kindling, J. **Neurophysiol.**, 59:1033-
1054.

Nadi, N.S., Wyler, A.R., and Porter, R.J., 1987, Amino acids and catecholamines in the epileptic focus from the human brain, **Neurology**, Suppl., 37:
106.

Perkins, M.N., and Stone, T.W., 1983, Pharmacology and regional variations of
quinolinic acid evoked excitations in rat central nervous system, J.
Pharmacol. Exp. Ther., 226:551-557.

So, N., Oliver, A., Andermann, F., Gloor, P., and Quesney, L.F., 1989, Results of surgical treatment in patients with bitemporal epileptiform abnormalities, **Ann. Neurol.**, 25:432-439.

Sherwin, A., Robitaille, Y., Quesney, F., Olivier, A., Villemure, J.,
Leblanc, R., Feindel, W., Andermann, E., Gotman, J., Andermann, F.,
Ethier, R., and Kish, S., 1988, Excitatory amino acids are elevated in
human epileptic cerebral cortex, **Neurology**, 38:920-923.

Wyler, A.R., Nadi, N.S., and Porter, R.J., 1987, Acetylcholine, GABA, benzodiazepine, glutamate receptors in the temporal lobe of epileptic patients,
Neurology, Suppl., 37:103.

Young, S.N., Joseph, M.H., and Gauthier, S., 1983, Studies of kynurenine in
human cerebrospinal fluid: lowered levels in epilepsy, J. **Neural Transm.**,
58:193-204.

CEREBROSPINAL FLUID QUINOLINIC ACID CONCENTRATIONS ARE INCREASED IN ACQUIRED

IMMUNE DEFICIENCY SYNDROME

M.P. Heyes[1], B. Brew[2], A. Martin[3], S.P. Markey[1], R.W. Price[2],
R.B. Bhalla, and A. Salazar[3].

[1]Section on Analytical Biochemistry
Laboratory of Clinical Science
NIMH
Bethesda, Maryland 20892

[2]Department of Neurology
Memorial Sloan-Kettering Memorial Cancer Center
New York, New York 10021

[3]Walter Reed Army Medical Center
Washington, District of Columbia 20307
USA

INTRODUCTION

Infection of man with the human immunodeficiency virus (HIV) is associated with a variety of pathological consequences including involvement of
the central nervous system. In addition to brain opportunistic conditions
that occur in patients with the acquired immune deficiency syndrome (AIDS),
a profile of neurologic dysfunctions has been described known as the AIDS dementia complex (ADC; Navia et al., 1986a,b; Price et al., 1988). At autopsy, diffuse pallor in the white matter, perivascular infiltrates of lymphocytes and macrophages, as well as loss of cortical neurons, have been described (Koenig et al., 1986; Navia et al., 1986a,b). Microglial nodules are
found predominantly in cerebral cortex and basal ganglia and may be associated with cytomegalovirus infection. Koenig et al. (1986) have identified
HIV in the brain in mononucleated and multinucleated macrophages.

While HIV has been shown to be present in severely demented individuals,
patients with less severe dementia may have little or no detectable evidence
of brain infection (Price et al., 1988). Consequently, indirect mechanisms,
including the production of toxins, may account for ADC. We have found that
the concentration of the L-tryptophan metabolite quinolinic acid (QUIN) is
increased in the cerebrospinal fluid of patients with AIDS (Heyes et al.,
1989) and is increased in mouse brain following systemic administration of
endotoxin, a model of infection (Heyes et al., 1989). QUIN is an agonist of
N-methyl-D-aspartate receptors (NMDA; Perkins and Stone, 1983) which are involved in the modulation of neuronal electrical activity, long term potentiation, memory and neuronal plasticity (cf. Morris et al., 1986). Changes
in the concentration of QUIN in the CSF of HIV-infected patients could
influence these functions of NMDA receptors.

To investigate the hypothesis that QUIN may act as a neuropathogen in ADC (Heyes et al., 1989), the concentration of QUIN was measured in the CSF of HIV-infected patients at different stages of the disease who presented for assessment of neurological or neuropsychological complaints, as well as 10 neurologically normal volunteers. The neurological status of the patients was evaluated at the time of sample collection.

METHODS

All HIV-infected patients were categorized based on both the Walter Reed Staging system (WR; Redfield et al., 1986) and a standardized functional ADC staging scheme (Price and Brew, 1988). Patients with an ADC score of ≥ 1 were classified as demented whereas patients who scored 0.5 or zero on the ADC scale were classified as not demented. The consequences of opportunistic conditions in AIDS patients were evaluated in 12 patients with either primary lymphoma, toxoplasmosis, cryptococcal meningitis, herpes zoster or progressive multifocal leukoencephalitis.

QUIN was quantified by electron capture negative chemical ionization gas chromatography/mass spectrometry (Heyes and Markey, 1988a,b), and L-TRP was quantified by high performance liquid chromatography.

RESULTS

Mean CSF QUIN concentration in control subjects was 21.1 ± 2.1 nmol/l. CSF QUIN concentrations were elevated in all HIV-positive groups studied. In WR1-2, CSF QUIN concentration in the first sample collected was 46.3 ± 6.6 nmol/l ($P < 0.0005$) and increased further to 97.3 ± 22.8 nmol/l ($P < 0.02$ compared to the first sample). Demented later-stage HIV-infected patients (WR3-6) had significantly higher CSF QUIN concentrations than did non-demented HIV-infected patients within the same WR stage group. As a group, non-demented patients had a mean CSF QUIN concentration of 87.9 ± 11.1 nmol/l while in demented patients CSF QUIN concentration was 1391 ± 457 nmol/l ($P < 0.0001$). While there was some overlap in CSF QUIN concentrations between HIV-infected patients who were not demented and those who were demented, all of the non-demented WR3-6 patients did have minor neurological signs and an ADC score of 0.5.

The highest three CSF QUIN concentrations exceeded 12,000 nmol/l. Importantly, in HIV-infected patients collectively, without the confounding influence of opportunistic infections or neoplasms, there was a correlation between the severity of ADC and the concentration of QUIN in CSF ($r = 0.427$, $P < 0.0002$). CSF QUIN concentrations were markedly elevated in 12 AIDS patients with opportunistic conditions to 1455 ± 394 nmol/l.

CSF beta-2 microglobulin concentrations were 1.5 ± 0.2 mg/l in neurological controls and were significantly increased in both non-demented patients (2.8 ± 1.0 mg/l, $P < 0.0001$) and in demented patients (4.1 ± 2.1 mg/l, $P < 0.0001$). There was a significant correlation between CSF beta-2 microglobulin and CSF QUIN concentrations ($r = 0.69$, $P < 0.00001$).

DISCUSSION

The strikingly large increases in the concentration of QUIN in the CSF of patients infected with HIV supports the hypothesis of a pathogenic role for QUIN in ADC (Heyes et al., 1989). The significance of our observations is highlighted by: 1) the increase in CSF QUIN in the early stages of HIV infection, and the increase in CSF QUIN concentrations with disease progression; 2) the correlation between the severity of ADC and the concentration of QUIN in the CSF; and 3) the increased concentration of QUIN in the CSF of patients with opportunistic conditions.

The metabolic origin of QUIN in the brain is unknown. In extrahepatic tissues, the first enzyme of the kynurenine pathway, indoleamine-2,3-dioxygenase (IDO) is induced by endotoxin and interferon (Takikawa et al., 1986; Yoshida et al., 1986). In the present study, the strong correlation between CSF QUIN concentration and CSF beta-2 microblobulin concentration (a surrogate marker of interferon activity (Brew et al., 1989) is consistent with activation of IDO in brain. It is of note that infiltration of the brain with macrophages is a prominent feature of HIV-infection (Koenig et al., 1986; Navia et al., 1986a). Macrophages in vitro respond to interferon-gamma by increasing IDO activity, and increased synthesis of metabolites of the kynurenine pathway, including 3-hydroxyanthranilic acid (Werner et al., 1987), a precursor of QUIN in vivo (Heyes et al., 1988).

The increased concentration of QUIN in the CSF of patients with AIDS suggests that the concentration of QUIN in the extracellular space of the brain is also increased. The localization of QUIN to the extracellular space is an important consideration because studies in experimental animals have established that at least some of the neuronal effects of QUIN are mediated via agonist effects on NMDA receptors localized on the outer membrane of neurons. Elevated QUIN may cause non-cytolytic neuronal dysfunction through its agonist properties on NMDA receptors (Perkins and Stone, 1983) and may disrupt the normal functioning of these receptors in neuronal plasticity, long term potentiation and memory (Morris et al., 1986). It is of note that basal ganglia structures and the hippocampus, areas rich in NMDA receptors, are a focus of demonstrable productive infection in ADC (Koenig et al., 1986; Navia et al., 1986a). Some of the neuropsychological deficits in the central nervous system in patients in the earlier stages of infection are consistent with involvement of the basal ganglia (Martin et al., in preparation). Studies of regional brain glucose consumption in patients with mild ADC have shown basal ganglia hypermetabolism followed by hypometabolism with progression of dementia (Rottenberg et al., 1987). In parallel, local infusion of QUIN into the striata of rats also acutely increases glucose consumption which subsequently decreases as neurodegeneration occurs (Heyes and Garnett, unpublished data).

REFERENCES

Brew, B.J., Bhalla, R.B., Fleisher, M., Paul, M., Khan, A., Schwartz, M.K., and Price, R.W., 1989, Cerebrospinal fluid B2 microglobulin in patients infected with human immunodeficiency virus, Neurology, 39:830-834.

Heyes, M.P., Hutto, B., and Markey, S.P., 1988a, 4-Chloro-3-hydroxyanthranilic acid inhibits 3-hydroxyanthranilic acid oxidase in brain, Neurochem. Int., 13:405-409.

Heyes, M.P., Kim, P., and Markey, S.P., 1988b, Systemic lipopolysaccharide and pokeweed mitogen increase quinolinic acid concentrations in mouse cerebral cortex, J. Neurochem., 51:1946-1948.

Heyes, M.P., and Markey, S.P., 1988a, Quantification of quinolinic acid in rat brain, whole blood and plasma by gas chromatography and negative chemical ionization mass spectrometry: effects of systemic L-tryptophan administration on brain and blood quinolinic acid concentrations, Anal. Biochem., 174:349-359.

Heyes, M.P., and Markey, S.P., 1988b, [^{18}O]-Quinolinic acid: its esterification without back exchange for use as internal standard in the quantification of brain and CSF quinolinic acid, Biomed. Environ. Mass Spectrom., 15:291-293.

Heyes, M.P., Rubinow, D., Lane, C., and Markey, S.P., 1989, Cerebrospinal fluid quinolinic acid concentrations are increased in acquired immune deficiency syndrome, Ann. Neurol., 26:275-277.

Koenig, S., Gendelman, E., Orenstein, J.M., Dal Canto, M.C., Pezeshkpour, G.H. Yungbluth, M., Janotta, F., Aksamit, A., Martin, M.A., and Fauci, A.S., 1986, Detection of AIDS virus in macrophages in brain tissue from AIDS patients with encephalopathy, Science, 233:1089-1093.

Morris, R.G.M., Anderson, E., Lynch, G.S., and Baudry, M., 1986, Selective impairment of learning and blockade of long-term potentiation by the N-methyl-D-aspartate receptor antagonist, AP5, Nature, 319:774-776.

Navia, B.A., Cho, E.S., Petito, C.K., and Price, R.W., 1986a, The AIDS dementia complex: II. Neuropathology, Ann. Neurol., 19:525-535.

Navia, B.A., Jordan, B.D., and Price, R.W., 1986b, The AIDS dementia complex: I. Clinical Features, Ann. Neurol., 19:517-524.

Perkins, M.N., and Stone, T.W., 1983, Pharmacology and regional variations of quinolinic acid-evoked excitations in rat central nervous system, J. Pharmacol. Exp. Ther., 226:551-557.

Price, R.W., and Brew, B.J., 1988, The AIDS dementia complex, J. Infect. Dis., 148:1079-1083.

Price, R.W., Brew, B.J., Sidtis, J., Rosenblum, M., Scheck, A.C., and Clearly, P., 1988, The brain in AIDS: central nervous system HIV-1 infection and the AIDS dementia complex, Science, 239:586-592.

Redfield, R.R., Wright, D.C., and Tramount, E.C., 1986, The Walter Reed staging classification for HTLV-III/LAV infection. New Engl. J. Med., 314:131-132.

Rottenberg, D.A., Moeller, J.R., Strother, S.C., Sidtis, J.J., Navia, B.A., Dhawan, V., Ginos, J.Z., and Price, R.W., 1987, The metabolic pathology of the AIDS dementia complex, Ann. Neurol., 22:700-706.

Schwarcz, R., Whetsell, W.O., and Mangano, R.M., 1983, Quinolinic acid: an endogenous metabolite that produces axon-sparing lesions in rat brain, Science, 219:316-318.

Takikawa, O., Yoshida, R., Kido, R., and Hayaishi, O., 1986, Tryptophan degradation in mice initiated by indoleamine-2,3-dioxygenase, J. Biol. Chem., 261:3648-3653.

Werner, E.R., Bitterlich, G., Fuchs, D., Hausen, A., Reibnegger, G., Szabo, G., Dierich, M.P., and Wachter, H., 1987, Human macrophages degrade tryptophan upon induction by interferon-gamma, Life Sci., 41:273-280.

Yoshida, R., Oku, T., Imanishi, J., Kishida, T., and Hayaishi, O., 1986, Interferon: a mediator of indoleamine-2,3-dioxygenase induction by lipopolysaccharide, Poly(I), Poly(C) and pokeweed mitogen in mouse lung, Arch. Biophys., 249:596-604.

PRODUCTION OF QUINOLINIC ACID AND KYNURENIC ACID BY HUMAN GLIOMA

A. Vezzani, J.B.P. Gramsbergen[1], C. Speciale[1], and R. Schwarcz[1]

Istituto di Richerche Farmacologiche "Mario Negri"
20157 Milan
Italy

[1]Maryland Psychiatric Research Center
Baltimore, Maryland 21228
USA

INTRODUCTION

Using biochemical and immunohistochemical techniques, the biosynthesis of the excitotoxin quinolinic acid (QUIN) and the anti-excitotoxin kynurenic acid (KYNA) in the rat brain has been demonstrated to take place preferentially in glial cells (see Schwarcz and Du, this volume, for review). Although a dysfunction of either of these two brain metabolites has been hypothetically associated with the occurrence of human neuro-psychiatric diseases and seizure disorders (Lapin, 1981; Schwarcz et al., 1984; Stone and Burton, 1988), very little is known about the cellular localization of QUIN and KYNA metabolism in the human brain. We have now used human gliomas obtained during neurosurgery to assess the activity of three enzymes involved in QUIN biosynthesis, and have studied, in separate experiments, the ability of glioma to produce KYNA from its bioprecursor L-kynurenine (KYN).

MATERIALS AND METHODS

Brain tumors (astrocytomas) were removed (Ospedale Nigmarda, Department of Neurosurgery, Milan, Italy) according to the standards of oncological brain surgery. All specimens were carefully isolated and inspected by a pathologist.

For the examination of enzyme activities in tissue homogenates, surgical samples were rapidly frozen on dry ice and stored at -80°C until analysis. On the day of the experiment, tissues were thawed, weighed, and sonicated (1:6 or 1:4, w/v) in 2[N-morpholino]ethanesulfonic acid (MES) buffer, pH 6.5, for incubation in the presence of KYN, 3-hydroxykynurenine (3HK) or 3-hydroxyanthranilic acid (3HANA). To evaluate QUIN production after exposure of tissue to its precursors KYN and 3HK, 170 μl of homogenate (equivalent to 28 mg of original tissue) were incubated for 2 hours at 37°C in the presence of 40 μM ferrous ammonium sulfate and 500 μM KYN or 3HK in a final volume of 295 μl. 100 μM NADPH was included in the incubation mixture containing KYN because of the NADPH-dependency of kynurenine hydroxylase, the enzyme converting KYN to 3HK (cf. Uemura and Hirai, this volume). The reaction was stopped by immersion of test tubes into a boiling water bath for 10 min. Subsequently, the samples were centrifuged in a Beckman 52B microfuge for 10 min. QUIN production was measured in the supernatant according to the radioenzymatic method of Foster et al. (1986a).

Production of QUIN from its immediate bioprecursor 3HANA was assessed by incubating the homogenate (5 mg of original tissue per tube) in the presence of 2.8 μM [carboxy-^{14}C]3-hydroxyanthranilic acid (6 mCi/mmol) for one hour at 37°C as described previously (Foster et al., 1986b).

An aliquot of each homogenate preparation was saved and subsequently assayed for protein content (Lowry et al., 1951).

KYNA production was studied in tissue samples obtained from seven different patients. For this purpose, 1 mm thick tissue slices were prepared within 1 hour after surgery and exposed to KYN in physiological buffer as described previously for experiments with rat brain slices (Turski et al., 1989). The slices were preincubated for 10 min at 37°C in culture wells containing 1 ml oxygenated Krebs-Ringer buffer with continuous supply of oxygen. Subsequently, the pre-incubation medium was removed, and the tissue incubated for an additional two hours under the same conditions in 1 ml fresh buffer containing 40, 200 or 1000 μM KYN. The number of experimental conditions was dictated by the amount of available tumor tissue. Thus, 2 or 3 wells were used for each data point. Following incubation, the wells were placed on ice, tissue was separated from the incubation buffer, and incubation media were frozen and stored at -70°C until analysis.

For KYNA determination, 200 μl 1 N HCl were added to each sample, and proteins were precipitated by boiling (10 min) and removed by centrifugation (8,730 x g, 10 min). The resulting supernatant was applied to a Dowex 50W (H$^+$ form; Sigma) cation-exchange column (0.5 x 2 cm). Subsequently, the column was washed with 1 ml 0.1 N HCl and 1 ml of distilled water, and the fraction containing KYNA was eluted with 2 ml of water, lyophilized and resuspended in 240 μl distilled water. 200 μl were subjected to high performance liquid chromatography (HPLC), and KYNA was detected by UV absorption at 340 nm as described previously (Turski et al., 1988 and 1989; sensitivity limit: 2 pmol). Tissue slices were used for protein determination according to the method of Lowry et al. (1951).

Table 1. Production of quinolinic acid by incubation of glioma tissue homogenates with bioprecursors

Patient	Kynurenine (500 μM)	3-Hydroxykynurenine (500 μM)	3-Hydroxyanthranilic acid (2.8 μM)
1	4.3	57.4	715
2	4.0	22.2	708
3	2.4	4.7	97
4	8.8	90.5	1143
5	3.6	13.8	630
6	2.5	12.4	787
Mean ± SEM	8.4 ± 2.1	67.1 ± 30.0	680 ± 151

Tissues were incubated and processed as described in the text. Data are the mean of duplicate samples and are expressed as pmoles QUIN formed per mg protein in 2 hours (from KYN and 3HK, respectively) or 1 hour (from 3HANA).

Table 2. KYNA synthesis by human astrocytomas

Patient	[Kynurenine]		
	40 μM	200 μM	1000 μM
7	n.d.	45.1	85.1
8	n.d.	90.1 ± 3.7	339.7
9	n.d.	21.7 ± 5.5	63.5 ± 8.9
10	n.d.	34.1	88.5
11	n.d.	27.4	91.2
12	6.6	n.d.	n.d.
13	6.2	n.d.	n.d.
Mean ± SEM	6.4	43.7 ± 12.2	133.6 ± 51.8

Tissue slices were incubated with kynurenine (40, 200 or 1000 μM) as described in the text. Data are expressed as pmol KYNA/2 h/mg protein and are the mean of 2 wells or the mean ± SEM of 3 wells per kynurenine concentration. n.d.: not done (tumor biopsies were too small to provide sufficient tissue for all kynurenine concentrations).

RESULTS

As shown in Table 1, astrocytoma homogenates were capable of converting KYN, 3HK and 3HANA into QUIN. In spite of the far lower concentration of 3HANA used, all six specimens produced substantially more QUIN from the immediate precursor than from KYN or 3HK. In average, 500 μM 3HK proved to be approximately eight times more effective as a QUIN bioprecursor than 500 μM KYN.

Exposure of tissue slices to various concentrations of KYN resulted in the dose-dependent production of KYNA, which could be recovered in the extracellular compartment (Table 2). In separate experiments, de novo synthesized KYNA was identified by several chromatographic procedures (Vezzani et al., 1990).

DISCUSSION

Taken together, the present data demonstrate the ability of human astrocytomas to synthesize both QUIN and KYNA from their respective bioprecursors. Unfortunately, limitations in tissue availability, i.e. the size of the surgically removed specimens, precluded a detailed examination of the relative ability of KYN to serve as the parent compound of QUIN and KYNA, respectively. It was also not possible to definitively delineate the relative activities of kynurenine 3-hydroxylase, kynureninase and 3-hydroxyanthranilic acid oxygenase (3HAO) in the tissue homogenates, since only one substrate concentration and assay condition could be investigated in each case. It appears safe to conclude, however, that, similar to the situation in the rat brain, KYN and 3HK are less readily incorporated into QUIN than 3HANA (cf. Schwarcz and Du, this volume). Notably, the activity of 3HAO measured in the gliomas was virtually identical to that found in homogenates of normal human brain, where the enzyme is likely to reside preferentially in astrocytes (Schwarcz et al., 1988). The activity of kynurenine 3-hydroxylase and kynureninase in human brain has not yet been examined, however, so that no comparisons with gliomas can be made at present.

Human astrocytoma slices produced substantially less KYNA from KYN than rat brain slices under identical experimental conditions (Turski et al., 1989). This is of particular interest in view of the much higher concentration of KYNA in the normal human (Turski et al., 1988) than in the rat brain (Carlà et al., 1988). The reason could lie in the pathological nature of the excised surgical tissue, which may contain only few kynurenine uptake sites (Speciale and Schwarcz, 1990) or low kynurenine aminotransferase (KAT) activity. Both the transport process and KYNA's biosynthetic enzyme, which are both preferentially associated with astrocytes in the rat brain (Schwarcz and Du, this volume), are therefore clearly worthy of further examination in astrocytomas.

In summary, we have demonstrated that pathological human brain tissue is capable of producing the neuroactive kynurenines QUIN and KYNA. Since glial cells can liberate both QUIN and KYNA (Speciale and Schwarcz, this volume; Gramsbergen et al., this volume), astrocytic tumors may release either of the two compounds into their microenvironment for interaction with excitatory amino acid receptors (Stone and Burton, 1988). It remains to be elaborated, how such glioma-derived kynurenines may play a role in pathophysiological conditions.

ACKNOWLEDGEMENTS

We gratefully acknowledge the assistance of neurosurgeon Dr. P. Versari, of anesthetist Dr. F. Procaccio, and of Drs. V. Monte, M.A. Stasi and Ms. L. Besozzi. This work was supported in part by USPHS grant NS 16102 (to R.S.) and Italian National Research Council Grant 87.00041.04.

REFERENCES

Carlà, V., Lombardi, G., Bensi, M., Russi, P., Moneti, G., and Moroni, F., 1988, Identification and measurement of kynurenic acid in the rat brain and other organs, Anal. Biochem., 371:267-277.
Foster, A.C., Okuno. E., Brougher, D.S., and Schwarcz, R., 1986a, A radio-enzymatic assay for quinolinic acid, Anal. Biochem., 158:98-103.
Foster, A.C., White, R.J., and Schwarcz, R., 1986b, Synthesis of quinolinic acid by 3-hydroxyanthranilic acid oxygenase in rat brain tissue in vitro, J. Neurochem., 47:23-30.
Lapin, I.L., 1981, Kynurenines and seizures, Epilepsia, 22:257-265.
Lowry, O.H., Rosebrough, N.J., Farr, A.L., and Randall, R.J. 1951, Protein measurement with the Folin phenol reagent, J. Biol. Chem., 193:265-275.
Schwarcz, R., Foster, A.C., French, E.D., Whetsell, Jr., W.O., and Köhler, C., Excitotoxic models for neurodegenerative disorders, Life Sci., 35:19-32.
Schwarcz, R., Okuno E., White, R.J., Bird, E.D., and Whetsell, W.O. Jr., 1988, 3-Hydroxyanthranilate oxygenase activity is increased in the brains of Huntington disease victims, Proc. Natl. Acad. Sci. USA, 85:4079-4081.
Speciale, C., and Schwarcz, R., 1990, Uptake of kynurenine into rat brain slices, J. Neurochem., 54:156-163.
Stone, T.W., and Burton, N.R., 1988, NMDA receptors and ligands in the vertebrate CNS, Prog. Neurobiol., 10:333-368.
Turski, W.A., Gramsbergen, J.B.P., Traitler, H., and Scharcz, R., 1989, Rat brain slices produce and liberate kynurenic acid upon exposure to L-kynurenine, J. Neurochem., 52:1629-1636.
Turski, W.A., Nakamura, M., Todd, W.P., Carpenter, B.K., Whetsell, W.O., and Schwarcz, R., 1988, Identification and quantification of kynurenic acid in human brain tissue, Brain Res., 454:164-169.

Vezzani, A., Gramsbergen, J.B.P., Versari, P., Stasi, M.A., Procaccio, F., and Schwarcz, R., 1990, kynurenic acid synthesis by human glioma, J. Neurol. Sci., 99:51-57.

CONTRIBUTORS

F.P. Abramson
Laboratory of Clinical Science
NIMH
Bethesda, Maryland 20892
USA

A. Adell
Department of Neurochemistry
Jorge Girona 18-26
08034 Barcelona
Spain

G.K. Aghajanian
Yale University School of Medicine
Dept. of Psychiatry and Pharmacology
34 Park Street
New Haven, Connecticut 06508
USA

G. Allegri
Istituto di Chimica Farmaceutica
Universita di Padova
Via Marzolo 5
I-35100 Padova
Italy

E. Alvarez
Dept. of Psychiatry
Hospital de S. Pau
Barcelona
Spain

G.M. Anderson
Yale Child Study Center
333 Cedar Street
New Haven, Connecticut 06510
USA

I. Anderson
Oxford University
Dept. Psychiatry and MRC
Clinical Pharmacology Unit
Littlemore Hospital Research Unit
Oxford OX4 4XN
UK

B.O. Anyanwu
Department of Biology
Texas Southern University
Houston, Texas 77004
USA

M. April
NIHAC, Lab. of Clinical Science
NIMH
Bethesda, Maryland 20892
USA

R. Arban
Department of Pharmaceutical Sciences
University of Padova
I-35131 Padova
Italy

F. Artigas
Department of Neurochemistry,
C.S.I.C.
Jorgi Girona 18-26
08034 Barcelona
Spain

R.N. Auer
Neuroscience Research Group
University of Calgary
Alberta, T2N 4N1
Canada

J.D. Bacher
Veterinary Resources Branch
NIH
Bethesda, Maryland 20892
USA

H. Baran
Maryland Psychiatric Research Center
P.O. Box 21247
Baltimore, Maryland 21228
USA

R.M. Beams
Medicinal Chemistry
The Wellcome Research Laboratories
Beckenham, Kent BR3 3BS
UK

D.A. Bender
Department of Biochemistry
University College
Gower Street
London WC1E 6BT
UK

A. Bettero
Dept. of Pharmaceutical Sciences
University of Padova
Via Marzolo 5
I-35100 Padova
Italy

J.L. Beverly
Department of Physiological Sciences and
Food Intake Laboratory
University of California, Davis
Davis, California 95616
USA

J. Beyer
III. Medical Clinic
Dept. of Endocrinology and Metabolism
University of Mainz
Germany

R.B. Bhalla
Section of Analytical Biochemistry
Lab. of Clinical Sci.
NIMH
Bethesda, Maryland 20892
USA

R.V. Bhat
Food and Drug Toxicology Research Centre
National Institute of Nutrition
Hyderabad 500 007
India

M. Biasiolo
Dept. of Pharmaceutical Sciences
University of Padova
Via Marzolo 5
35131 Padova
Italy

D. Blackmore
NIHAC, Laboratory of Clinical Science
NIMH
Bethesda, Maryland 20892
USA

J. Blanke
Dept. of Psychiatry
University of Göttingen
von-Siebold-Str. 5
3400 Göttingen
Germany

A. Blumhofer
Dept. of Psychiatry
Hospital of the University
Frankfurt
Germany

R.J. Boegman
Department of Pharmacology and Toxicology
Queen's University
Kingston, Ontario K7L 3N6
Canada

R. Boni
NIH
Building 10, Room 3D40
9000 Rockville Pike
Bethesda, Maryland 20892
USA

E.C. Borden
University of Wisconsin Clinical
Cancer Center
Madison, Wisconsin 53792
USA

C. Bradberry
Dept. Pharmacology
Yale School of Medicine
333 Cedar
New Haven, Connecticut 06510
USA

B. Brew
Memorial Sloane Kettering Cancer Center
New York, New York 10021
USA

R.R. Brown
Dept. of Human Oncology
Center for Health Sciences
University of Wisconsin
Madison, Wisconsin 53792
USA

M. Cairella
Dept. Human Biopathology
Serv. Dietologia
1st Terapia Medica Sistematica
Univ. "La Sapienza"
Rome
Italy

C. Cangiano
Lab. Clin. Nutrition
3rd Dept. Internal Med.,
Viale dell'Universita 37
00185 Rome
Italy

V. Carlà
Department of Pharmacology
University of Florence
Viale Morgagni 65
50134 Firenze
Italy

B.K. Carpenter
Cornell University
Department of Chemistry
Ithaca, New York 14853
USA

A. Cascino
Lab. Clin. Nutrition
3rd Dept. Internal Med.,
Viale dell'Universita 37
00185 Rome
Italy

F. Ceci
Dept. Human Biopathology
1st Terapia Medica Sistematica
Univ. "La Sapienza"
Rome
Italy

P. Celada
Dept. of Neurochemistry,
C.S.I.C.
Jorge Girona 18-26
08034 Barcelona
Spain

N.A. Clarkson
Biochemical Sciences
Wellcome Research Laboratories
Beckenham, Kent BR3 3BS
UK

H.H. Coenen
Institut fur Medizin und Institut
für Chemie I
Kernforschungsanlage Jülich GMBH
D 5170 Jülich
Germany

J.H. Connick
Dept. of Pharmacology
University of Glasgow
Glasgow G12 8QQ, Scotland
UK

C.V.L. Costa
Istituto di Chimica Farmaceutica
Universita di Padova
Via Marzolo 5
I-35100 Padova
Italy

P.J. Cowen
MRC Unit of Clinical Pharmacology and
University Department of Psychiatry
Littlemore Hospital
Littlemore, Oxford OX4 4XN
UK

M.A.E. Critchley
Dept. of Pharmacology
The Wellcome Research Laboratories
Beckenham, Kent BR3 3BS
UK

G. Curzon
Department of Neurochemistry
Institute of Neurology
1 Wakefield Street
London WC1N 2PJ
UK

A. Dalmane
Latvian Research Institute of Cardiology
Riga
USSR

P.S. Danielian
Department of Biochemistry
University of College London
Gower Street
London WC1E 6BT
UK

S.P. Datta
Departments of Human Oncology and Medicine
University of Wisconsin Medical School
Madison, Wisconsin 53792
USA

G. De Luca
Polifarma S.p.A.
Via Tor Sapienza 138
00155 Rome
Italy

M. Del Ben
Dept. Human Biopathology
Serv. Dietologia
1st Terapia Medica Sistematica
Univ. "La Sapienza"
Rome
Italy

L. Demisch
Zentrum für Psychiatrie
der JWG-Universität Frankfurt
Heinrich-Hoffmann-Str. 10
6000 Frankfurt/Main 70
Germany

O. Devinsky
UMDNJ
Newark, New Jersey 07103
USA

M. Diksic
McGill University
Montreal Neurological Institute
3801 University Street
Montreal, Quebec H3A 2B4,
Canada

G. Di Stazio
Polifarma Res. Centre
Via Tor Sapienza 138
Rome
Italy

M. Dragoslav
Ustanicka 174/52
11000 Beograd
Yugoslavia

F. Du
Maryland Psychiatric Research Center
P.O. Box 21247
Baltimore, Maryland 21228
USA

M.L. Dubocovich
Dept. Pharmacology
Northwestern University Medical School
303 E. Chicago Avenue
Chicago, Illinois 60611
USA

D.S. Duch
Wellcome Research Laboratories
Research Triangle Park,
North Carolina 27709
USA

M.J. During
Neuropsychopharmacology Research Unit
B254 SHM
Yale University School of Medicine
333 Cedar Street
New Haven, Connecticut 06510
USA

C.L. Eastman
Johns Hopkins School of Hygiene
615 N. Wolfe Street - Room 2001
Baltimore, Maryland 21205
USA

M. Ebadi
Dept. Pharmacology
Univ. Nebraska Medical Center
42nd & Dewey Avenue
Omaha, Nebraska 68105
USA

M.P. Edelstein
Wellcome Research Laboratories
Research Triangle Park,
North Carolina 27709
USA

A. Elderfield
8 Murray Park Road
Figtree NSW 2525
Australia

L.E. Feinendegen
Institut fur Medizin und Institut
für Chemie I
Kernforschungsanlage Jülich GMBH
D 5170 Jülich
Germany

J.D. Fernstrom
Univ. Pittsburgh
Western Psychiatric Institute and Clinic
3811 O'Hara Street
Pittsburgh, Pennsylvania 15213-2593
USA

A. Freese
Departments of Pharmacology and Psychiatry
Yale University School of Medicine
New Haven, Connecticut 06510
USA

M.J. Fregly
Department of Physiology
Box J274
College of Medicine
University of Florida
J. Hillis Miller Health Center
Gainesville, Florida 32610
USA

D. Fuchs
Institute for Medical Chemistry
and Biochemistry
University of Innsbruck
Fritz-Pregl-Str. 3
A-6020 Innsbruck
Austria

H. Fujimoto
Department of Biochemistry
Aichi Medical University
Nagakute, Aichi
Japan

P. Gebhardt
Dept. of Psychiatry
Hospital of the University
Frankfurt
Germany

M.D. Gershon
Dept. of Anatomy and Cell Biology
Columbia University
College of Physicians and Surgeons
630 West 168th Street (P&S 12-513)
New York, New York 10032
USA

D. Ghosh
1839 Harold
Houston, Texas 77098
USA

D. Gietzen
VM Physiological Sciences
University of California
Davis, California 95616
USA

K. Goda
Department of Nutrition
Kobe-Gakuin University
Ikawadani-cho, Nishi-ku
Kobe 673
Japan

A. Gorini
Polifarma Research Centre
Via Tor Sapienza 138
Rome
Italy

J.P.B. Gramsbergen
Maryland Psychiatric Research Center
P.O. Box 21247
Baltimore, Maryland 21228

M. Gravell
Infectious Disease Branch
Laboratory of Clinical Science
NIMH
Bethesda, Maryland 20892
USA

V. Gromm
Latvian Research Institute of Cardiology
Riga
USSR

T. Guilarte
Johns Hopkins School of Hygiene
615 N. Wolfe Street - Room 2001
Baltimore, Maryland 21205
USA

J. Guilford
College of Pharmacy and Health Sciences
Texas Southern University
Houston, Texas 77004
USA

A. Habara-Ohkubo
Department of Cell Biology
Osaka Bioscience Institute
Osaka 565
Japan

G. Hajak
Department of Psychiatry
Georg-August-Universitä
von-Siebold Str. 5
3400 Göttingen
Germany

J.E. Hall
Dept. Biol.
Texas A & M Univ.
College Station, Texas 77843
USA

V.A. Hammer
Department of Physiological Sciences and
Food Intake Laboratory
University of California, Davis
Davis, California 95616
USA

M. Haneda
Department of Biochemistry
Aichi Medical University
Nagakute
Aichi
Japan

L.V. Hankes
P.O. Box 1056
Setauket, New York 11733
USA

A. Hausen
Institute for Medical Chemistry
and Biochemistry
University of Innsbruck
Fritz-Pregl-Str. 3
A-6020 Innsbruck
Austria

T. Hayakawa
1-1-1 Sakuragaoka
Setagaya-ku
Tokyo 156
Japan

L.L. Henderson
Department of Biology
Texas Southern University
Houston, Texas 77004
USA

H. Herzog
Institut fur Medizin und Institut
für Chemie 1
Kernforschungsanlage Jülich GMBH
D-5170 Jülich
Germany

M.P. Heyes
Section of Analytical Biochemistry
Laboratory of Clinical Science
Bldg. 10, Room 3040
NIMH.
Bethesda, Maryland 20892
USA

A. Himeno
Section on Pharmacology
Lab. Clinical Sci.
NIMH,
Building 10, Room 2D45
9000 Rockville Pike
Bethesda, Maryland 20892
USA

T. Hino
Faculty of Pharmaceutical Sciences
Chiba University
1-33, Yayoi-cho
Chiba-shi 260
Japan

K. Hirai
Neurochemistry Dept.
Psychiatric Research Institute of Tokyo
2-1-8, Kamikitazawa
Setagaya-ku
Tokyo, 156
Japan

M. Hisaoka
Dept. of Nutrition
Kobe-Gakuin University
Kobe 673
Japan

H.F. Hodson
Dept. of Medicinal Chemistry
The Wellcome Research Laboratories
Beckenham, Kent BR3 3BS
UK

M. Ikeda
Suntory Institute for Bioorganic Research
Shimamoto-cho
Mishima-gun
Osaka 618
Japan

I. Ishiguro
Dept. of Biochemistry
School of Medicine
Fujita-Gakuen Health University
1-98 Kutsukake-cho
Toyoake, Aichi 470-11
Japan

T. Ishikawa
Dept. of Biochemistry
Wakayama Medical College
27 Kyuban-cho
Wakayama 640
Japan

K. Iwai
34-6, Santoji Otawa
Yamashina-Ku
Kyoto 607
Japan

R. Iver
Dept. of Medicinal Chemistry
The Wellcome Research Laboratories
Langley Court
Beckenham
Kent, BR3 3BS
England
UK

X.-d. Ji
Section on Analytical Biochemistry
Laboratory of Clinical Science
Bldg. 10, Room 3040
NIMH
Bethesda, Maryland 20892
USA

E. Jurika
Latvian Research Institute of Cardiology
Riga
USSR

P. Kaczmarczyk
Department of Psychiatry
Hospital of the University
Frankfurt a.M.
Germany

G. Kennett
Beecham Pharmac. Research
The Pinnacles
Harlow, Essex CM19 5AD,
England
UK

O.G. Kenunen
Laboratory of Psychopharmacology
Bekhterev Psychoneurological
Research Institute
Leningrad 193019
USSR

A. Khazai
Department of Psychiatry
Hospital of the University
Frankfurt
Germany

R. Kido
Dept. of Biochemistry
Wakayama Medical College
9 Kyubancho
Wakayama 640
Japan

R.G. Knowles
Biochemical Sciences
Wellcome Research Labs
Langley Court
Beckenham
Kent BR3 3BS
UK

V.L. Kozlovsky
Laboratory of Psychopharmacology
Bekhterev Psychoneurological
Research Institute
Leningrad 193019
USSR

K.J. Langen
Institut fur Medizin und Institut
für Chemie I
Kernforschungsanlage Jülich GMBH
D-5170 Jülich
Germany

I.P. Lapin
Bekhterev Psychoneurological
Research Institute
Leningrad 193019
USSR

M.V. Lavaggi
Polifarma Research Centre
Via Tor Sapienza 138
Rome
Italy

A. Laviano
Lab. Clin. Nutrition
3rd Dept. Internal Med.,
Viale Dell'Universita 37
00185 Rome
Italy

H. Lehnert
Abt. für Endokrinologie
11 Medizinische Klinik
Langenbeckstrasse 1
D-6500 Mainz
Germany

B. Lemmer
Dept. of Pharmacology
Hospital of the University
Frankfurt
Germany

C. Leonard
Neuroscience Research Group
University of Calgary
Calgary, Alberta T2N 4N1
Canada

D. Liepina
Latvian Research Institute of Cardiology
Riga
USSR

G. Lombardi
Dept. Pharmacology
University of Florence
Italy

W.T. London
Infectious Disease Branch
Laboratory of Clinical Science
NIMH
Bethesda, Maryland 20892
USA

H. Maass
Frankfurt University Hospital
D-6000 Frankfurt 70
Germany

D.J. Madge
Medicinal Chemistry
The Wellcome Research Laboratories
Beckenham, Kent, BR3 3BS
UK

T. Maesaki
Department of Biochemistry
Aichi Medical University
Nagakute, Aichi
Japan

D.J. Malone
Departments of Human Oncology and Medicine
University of Wisconsin Medical School
Madison, Wisconsin 53792
USA

A. Margonelli
Polifarma Research Centre
Via Tor Sapienza 138
Rome
Italy

S.P. Markey
NIH
Building 10, Room 3D40
Bethesda, Maryland 20892
USA

A. Martin
Walter Reed AMC
Washington, D.C. 20307
USA

J.B. Martin
Walter Reed AMC
Washington, D.C. 20307
USA

E. Martinez
Dept. of Neurochemistry
C.S.I.C.
Barcelona 08034
Spain

A. Martinsons
Latvin Research Institute of Cardiology
Riga
USSR

W.R. Matson
Department of Neurology
Mass. General Hospital
Boston, Massachusetts 02114
USA

I.S.L. Matsui
9 bancho 27
Wakayama City 640
Japan

M.T. McGuire
Department of Psychiatry
UCLA School of Medicine
Los Angeles, California 90024
USA

R. Metcalf
Department of Pharmacology & Toxicology
Queen's University
Kingston K7L 3N6
Canada

W. Miki
Suntory Institute for Bioorganic Research
Shimanoto-cho
Mischima-gun
Osaka 618
Japan

D.D. Milovanovic
Neuropsychiatric Clinic
Faculty of Medicine Beograd BOAL
Dr. Subotica 6
11000 Belgrad
Yugoslavia

L. Milovanovic
Neuropsychiatric Clinic
Faculty of Medicine Beograd BOAL
Dr. Subotica 6
11000 Belgrad
Yugoslavia

Y. Minatogawa
Dept. of Biochemistry
Wakayama Medical College
27 Kyu-bann-cho
Wakayama 640
Japan

J.R. Moffett
Department of Biology
Georgetown University
Washington, D.C. 20057
USA

F. Moroni
Dept. Pharmacology and Toxicology
Univ. Florence Med. School
Viale Morgagni 65
50134 Florence
Italy

K. Murata
608 Honyakushi
Nara City 630
Japan

M. Muscaritoli
Lab. Clin. Nutrition
3rd Dept. Internal Med.
Viale dell-Universita 37
00185 Rome
Italy

N.R. Myslinski
Dept. of Physiology
School of Dentistry
University of Maryland School of Medicine
666 W. Baltimore Street
Baltimore, Maryland 21201
USA

N.S. Nadi
Section of Neuronal Excitability
NINCDS
Bethesda, Maryland 20892
USA

Y. Nagamura
Department of Biochemistry
Fujita-Gakuen Health University
School of Medicine
Toyoake, Aichi 470-11
Japan

J. Naito
Fujita-Gakuen Health Univ.
School of Hygiene
Kutsukake-cho
Toyoake City 470-11
Japan

M. Nakagawa
Faculty of Pharmaceutical Sciences
Chiba University
1-33 Yayoi-cho
Chiba-shi, 260
Japan

703

M. Nakamura
Dept. of Biochemistry
Wakayama Medical College
27 Kyuban-cho
Wakayama 640
Japan

K. Nakanishi
Suntory Institute for Bioorganic Research
Shimanoto-cho
Mischima-gun
Osaka 618
Japan

M. Nakatsuka
Department of Biochemistry
Aichi Medical University
Nagakute
Aichi
Japan

M.A.A. Namboodiri
Department of Biology
Georgetown University
Washington, D.C. 20057
USA

Y. Naya
Suntory Institute for Bioorganic Research
Shimamoto-Cho, Mishima-Sun
Osaka 618
Japan

T. Nickelsen
Dept. of Endocrinology
University Hospital
Theodor-Stern-Kai 7
D-6000 Frankfurt 70
Germany

L.P. Niles
Dept. Biomed. Science
Division of Neuroscience
McMaster University
1200 Main Street West
Hamilton, Ontario L8N 3Z5
Canada

M. Nishimura
Sect. Analytical Biochemistry
Lab. Clinical Science
Bldg. 10, Room 3040,
NIMH
Bethesda, Maryland 20892
USA

Y. Nisimoto
Dept. of Biochemistry
Aichi Medical University
Nagakute
Aichi 480-11
Japan

E.N.M. Njagi
Dept. of Biochemistry
University College
Gower Street
London WC1E 6BT
UK

M. Nomura
Dept. of Physiology
School of Medicine
Fujita-Gakuen University
Toyoake
Aichi 470-11
Japan

H. Ogawa
Dept. of Biology
Fujita Gakuen University
Toyoake
Aichi 470-11
Japan

M. Ohnishi
Suntory Institute for Bioorganic Research
Shimanoto-cho
Mischima-gun
Osaka 618
Japan

T. Ohta
Department of Biochemistry
Aichi Medical University
Nagakute
Aichi
Japan

Y. Ohta
Department of Biochemistry
Fujita-Gakuen Health University
Toyoake, Aichi 470-11
Japan

E. Okuno
Dept. of Biochemistry
Wakayama Medical College
Wakayama 640
Japan

J. Ortiz
Dept. of Neurochemistry
C.S.I.C.
Jorge Girona 18-26
08034 Barcelona
Spain

Y. Ozaki
Univ. Wisconsin Clinical Cancer Center
K4/417 CSC
Madison, Wisconsin 53792
USA

M. Papagapiou
Neuroscience Research Group
University of Galgary
Alberta T2N 4N1
Canada

S.K. Patnaik
Enzyme Engineering Laboratory
Department of Pharmaceutics
Institute of Technology
Banaras Hindu University
Varanasi 221005
India

J.C. Peters
Procter & Gamble Co.
Miami Valley Labs
P.O. Box 39175
Cincinnati, Ohio 45239-8707
USA

C.I. Pogson
Head, Biochemical Sciences
Wellcome Research Laboratories
Langley Court
Beckenham, Kent BR3 3BS
UK

V. Politi
Polifarma S.p.A.
Via Tor Sapienza 138
00155 Rome
Italy

R.W. Price
Memorial Sloane Ketting Cancer Center
New York, New York 10021
USA

B.J. Quearry
Sect. Analytical Biochemistry
Lab. Clinical Science
Building 10, Room 3040
NIMH
Bethesda, Maryland 20892
USA

D. Radulović
Institute of Pharmaceutical Chemistry
Faculty of Pharmacy
University of Belgrade
Yugoslavia

M.J. Raleigh
Dept. Psychiatry
UCLA School of Medicine
Los Angeles, California 90024
USA

Y. Ramakrishna
Food and Drug Toxicology Reearch Centre
National Institute of Nutrition
Hyderabad-500 007
India

G. Reibnegger
Institute for Medical Chemistry
and Biochemistry
University of Innsbruck
Fritz-Pregl-Str.3
A-6020 Innsbruck
Austria

R.J. Reiter
Dept. Cellular and Structural Biology
Univ. Texas Health Science Center
7703 Floyd Curl Drive
San Antonio, Texas 78284
USA

A. Rodenbeck
Dept. of Psychiatry
University of Göttingen
von-Siebold-Str. 5
3400 Göttingen
Germany

Q. Rogers
Dept. of Physiology Science
School of Veterinary Medicine
University of California
Davis, California 95616
USA

H. Rommelspacher
Freie Universität Berlin
Inst. Neuropsychopharmakologie FU
Ulmenallee 30
D-1000 Berlin 19
Germany

F. Rossi-Fanelli
Clinica Medica 3
Universita degli Studi di Roma
Viale dell'Universita 37
I-00185 Rome
Italy

E. Rota
Institut fur Medizin und Institut
für Chemie I
Kernforschungsanlage Jülich GMBH
D-5170 Jülich
Germany

R.H. Roth
Depts. of Pharmacology and Psychiatry
Yale University School of Medicine
Connecticut Mental Health Center
New Haven, Connecticut 06510
USA

V. Rudzite
Latvian Research
Institute of Cardiology
13, Pilsonjustr.
226002 Riga
USSR

P. Russi
Department of Pharmacology
University of Florence
Viale Morgagni 65
50134 Firenze
Italy

E. Rüther
Psychiatrische Klinik
der Universität Göttingen
von-Siebold-Str. 5
3400 Göttingen
Germany

J.M. Saavedra
Section on Pharmacology
Lab. Clinical Sci.
NIMH
Building 10, Room 2D45
9000 Rockville Pike
Bethesda, Maryland 20892
USA

K. Saito
Department of Biochemistry
Fujita-Gakuen Health University
School of Medicine
1-98 Kutsukake-cho
Toyoake, Aichi 470-11
Japan

A. Sakata
Department of Biochemistry
Aichi Medical University
Nagakute, Aichi
Japan

A. Salazar
Walter Reed AMC
Washington, D.C. 20307
USA

M. Salter
Biochemical Sciences
Wellcome Research Labs
Langley Court
Beckenham, Kent, BR3 3BS
UK

A. Samel
Frankfurt University
D-6000 Frankfurt 70
Germany

G.S. Sarna
Dept. Neurochemistry
Institute of Neurology
1 Wakefield Street
London WC1N 1PJ
UK

M.J. Sarrias
Dept. of Neurochemistry,
C.S.I.C.
Jorge Girona 18-26
08034 Barcelona
Spain

E. Sasaki
Department of Biochemistry
Fujita-Gakuen Health University
School of Medicine
1-98 Kutsukake-cho
Toyoake, Aichi 470-11
Japan

B. Sashidhar Rao
Food and Drug Toxicology Res. Center
National Institute of Nutrition
Hyderabad-500 007
India

E.V. Savvateeva
Pavlov Institute of Physiology
199164 Leningrad
USSR

K. Schöffling
DRVLR Institute
D-5000 Köln 90
Germany

M. Schühle
Freie Universität Berlin
Inst. Neuropsychopharmakologie FU
Ulmenallee 30
D-1000 Berlin 19
Germany

R. Schwarcz
Maryland Psychiatric Research Center
P. O. Box 21247
Baltimore, Maryland 21228
USA

K. Shibata
Department of Food Science & Nutrition
Teikoku Women's University
6-173 Thoda-cho
Moriguchi
Osaka 570
Japan

Y. Shibata
Dept. of Biochemistry
Aichi Medical University
Nagakute, Aichi-gun
Aichi 480-11
Japan

R. Shinohara
Fujita-Gakuen Health University
Toyoake Aichi 470-11
Japan

T. Sielaff
Zentrum für Psychiatrie
der JWG-Universität Frankfurt
Heinrich-Hoffmann-Str. 10
6000 Frankfurt/Main 70
Germany

G. Sileniece
Latvia Research Institute of Cardiology
Riga
USSR

V. Simonneaux
Department of Pharmacology
University of Nebraska
College of Medicine
Omaha, Nebraska 68105
USA

R.V. Smalley
University of Wisconsin Clinical
Cancer Center
600 Highland Avenue
Madison, Wisconsin 53792
USA

P.M. Sondel
Departments of Human Oncology and Medicine
University of Wisconsin Medical School
Madison, Wisconsin 53792
USA

T.L. Sourkes
Dept. of Psychiatry
McGill University
1033 Pine Avenue West
Montreal, Quebec H3A 1AI
Canada

Carmela Speciale
Maryland Psychiatric Research Center
P.O. Box 21247
Baltimore, Maryland 21228
USA

J.S. Sprouse
Depts of Pharmacology and Psychiatry
Yale University School of Medicine
Connecticut Mental Health Center
New Haven, Connecticut 06510
USA

B. Stanković
Institute of Analytical Chemistry
Faculty of Pharmacy
University of Belgrade
Yugoslavia

H. Steinhart
Ordinariat für Lebensmittelchemie
der Universität Hamburg
Grindelallee 117
D-2000 Hamburg 13
Germany

G. Stoecklin
Institut für Medizin und Institut
für Chemie I
Kernforschungsanlage Jülich GMBH
D-5470 Jülich
Germany

T.W. Stone
Dept. Pharmacology
University of Glasgow
Glasgow G12 8QQ
UK

R. Susilo
Lab GmbH
Wegenerstraße 13
D-7910 Neu Ulm
Germany

T. Suzuki
Dept. Nutrition and Biochemistry
Institute of Public Health
6-1 Shirokanedai 4
Minato-ku
Tokyo 108
Japan

K.J. Swartz
Departments of Pharmacology and Psychiatry
Yale University School of Medicine
New Haven, Connecticut 06510
USA

O. Tagawa
Eisai Co., Ltd.
Kawashima Gifu 483
Japan

H. Taguchi
Mie University
Tsu, Mie 514
Japan

H. Takahashi
Dept. of Biochemistry
Toxicology Institute
Kumamoto University
Medical School
Kumamoto 860
Japan

F. Takeuchi
Dept. of Biochemistry
Aichi Medical University
Yazako, Nagakute-cho
Aichi-gun
Aichi 480-11
Japan

O. Takikawa
6-2-4 Furuedai
Suita
Osaka 565
Japan

T. Tamai
Department of Biochemistry
Aichi Medical University
Nagakute, Aichi
Japan

A. Taylor
Robens Institute
University of Surrey
UK

W.P. Todd
Cornell University
Department of Chemistry
Ithaca, New York 14853
USA

R. Truscott
Illawarra Area Biochemistry Service
The Wollongong Hospital
Wollongong
NWS 2500
Australia

R. Tsubouchi
Dept. of Biochemistry
Aichi Medical University
Nagakute
Aichi 480-11
Japan

M. Tsujimoto
27 Kyubancho
Wakayama 640
Japan

C. Udina
Dept. of Psychiatry
Hospital de S. Pau
Barcelona
Spain

T. Ueda
Dept. of Nutrition
Kobe-Gakuin University
Kobe 673
Japan

T. Ueda
Faculty of Nutrition
Kobe-Gakuin University
Arise
Ikawadeni-cho
Nishi-ku
Kobe 673
Japan

T. Uemura
Psychiatric Research Institute of Tokyo
2-1-8, Kamikitazawa
Setagaya-ku
Tokyo 156
Japan

Y. Umebachi
Department of Biology
Faculty of Science
Kanazawa University
Kanazawa 920
Japan

M. Valivullah
Department of Biology
Georgetown University
Washington, D.C. 20057
USA

M. Vejvoda
Frankfurt University Hospital
D-6000 Frankfurt 70
Germany

A. Vezzani
Istituto di Richerche Farmacologiche
"Mario Negri"
20157 Milan
Italy

H. Wachter
Institut für Medizinische
Chemie und Biochemie
Fritz-Pregel-Str. 5
A-6020 Innsbruck
Austria

K. Wainwright
Department of Pharmacology and Toxicology
Queen's University
Kingston, Ontario K7L 3N6
Canada

S.D. Wainwright
Department of Biochemistry
Dalhousie University
Tupper Building
Halifax, Nova Scotia B3H 4H7
Canada

J.L. Walsh
Maryland Psychiatric Research Center
P.O. Box 21247
Baltimore, Maryland 21228
USA

H. Wegmann
Frankfurt University Hospital
D-6000 Frankfurt 70
Germany

E.R. Werner
Institute Medical Chemistry & Biochemistry
University of Innsbruck
Fritz-Pregl-Str. 3
A-6020 Innsbruck
Austria

G. Werner-Felmayer
Institute Medical Chemistry & Biochemistry
University of Innsbruck
Fritz-Pregl-Str. 3
A-6020 Innsbruck
Austria

B. Witkop
3807 Montrose Driveway
Chevy Chase, Maryland 20815
USA

W. Wuttke
Clin. and Exp. Endocrinology
University of Göttingen
Robert-Koch-Str. 40
3400 Göttingen
Germany

W. Wutz
Institut fur Medizin und Institut
für Chemie I
Kernforschungsanlage Jülich GMBH
D 5170 Jülich
Germany

A.R. Wyler
Department of Neurosurgery
University of Tennessee
Memphis, Tennessee 38103
USA

J.A. Yergey
Laboratory of Clinical Studies, NIAAA
Bethesda, Maryland 20892
USA

T. Yokoyama
Department of Biology
Faculty of Science
Kanazawa University
Kanazawa, 920
Japan

R. Yoshida
Dept. of Medical Chemistry
Kyoto University of Medicine
Yoshida
Sakyo-ku
Kyoto 606
Japan

S.N. Young
Dept. of Psychiatry
McGill University
1033 Pine Avenue, West
Montreal H3A 1AI
Canada

S. Yuyama
Dept. Biochemistry and Nutrition
Institute of Public Health
6-1 Shirokanedai 4 chome
Minato-ku
Tokyo 108
Japan

R. Zirne
Latvian Research Institute of Cardiology
Riga
USSR

Gallamine, 595
Gelsemine, 3
Gepirone, 233, 236, 378, 379, 384
Glioma, 691-693
Glomerulonephritis, 659-661
Glucocorticoids, 167, 359, 366, 603, 604
Gluconeogenesis, 345, 352, 681
Glutamine phenylpyruvate aminotransferase, 203
Glutathione, 627, 628
Glycine, 186, 333, 334
Gonadal steroids, 154
Gramicidin A, 5, 6
Growth hormone, 246, 247, 412, 413, 607, 608
GTP-cyclohydrolase, 179-181

Hallucinogens, 107, 111, 112
Haloperidol, 338, 339, 380
Harmaline, 6
Hartnup disease, 669-674
Hepatic encephalopathy, 73, 194
Hepatocytes, 641-643
Herpes simplex virus, 556
Histidine decarboxylase, 116
Histidine-pyruvate aminotransferase, 202
Homovanillic acid, 75, 615-617
Honey bee, 320-327
Huntington's disease, 185, 194, 195, 334, 485, 675-677
Hydrocortisone, 352, 511-514
20-Hydroxyecdysone, 309-317
3-Hydroxyanthranilic acid, 163, 164, 177, 182, 186, 188, 194, 319, 347, 455, 457, 485-487, 556, 563-565, 579, 581-584, 626, 645, 680, 681, 689, 691-693
3-Hydroxyanthranilic acid oxygenase (3HAO), 188-191, 361, 485, 487, 531, 556, 557, 563-565, 579-585, 650, 693
8-Hydroxy-2(di-n-propyl-amino) tetralin (8-OH-DPAT), 71, 74, 224, 226, 232, 233, 236, 378-380, 384, 392-396, 401
6-Hydroxydopamine, 74, 121, 123, 225
5-Hydroxyindalpine, 223-226
6-Hydroxyindalpine, 223, 224
5-Hydroxyindoleacetic acid (5-HIAA)
  effect of antidepressants, 653-657
  effect of diet, 350, 390, 419, 422
  effect of endotoxin, 559-561
  effect of pyridoxine, 521
  effect of tryptophan, 245, 246, 250, 284, 286, 521, 615, 617, 621
  formation from indole-3-pyruvic acid, 305
  GC/MS assay, 445

5-Hydroxyindoleacetic acid (5-HIAA) (continued)
  in autism, 59
  in CSF, 54, 408, 422
  in Hartnup disease, 669, 673
  in hypoglycemia, 680, 681
  in microdialysate, 65, 67-77, 451, 536
  in plasma, 447-449
  in urine, 499, 520, 521, 592, 593, 673
  pineal, 139-144
3-Hydroxykynurenine, 33, 163, 170, 171, 174, 187, 191, 307-317, 320-327, 359-365, 467, 485-487, 497, 499, 502, 523-527, 513-533, 559-561, 625-628, 665, 673, 691-693
Hydroxykynureninuria, 673
Hydroxyl radical, 11
Hydroxyindole-O-methyltransferase (HIOMT), 141-155, 255, 256, 489
6-Hydroxymelatonin, 43, 44, 46, 257, 259
5-Hydroxytryptamine (5-HT, serotonin)
  bulbospinal pathway, 595-598
  degradation of, 127, 138, 139, 150, 256
  effect of diet, 369-374, 417-422
  effect of endotoxin, 559-561
  effect of tryptophan, 68-72, 347-351, 621, 622
  effect on dopamine release, 258
  fluorometric detection, 53-58
  in brain diseases, 11, 59, 107, 231-237
  in CSF, 54-56
  in hypoglycemia, 680, 681
  in microdialysate, 65-77, 81-87, 451-453, 535-538
  in plasma, 447-450, 653-657
  in urine, 499
  pineal, 139-146, 150
  platelet, 653-657
  receptors, see serotonin receptors
  release, 70, 81-87
  role in feeding, 389-400
  synthesis of, 93-103, 116, 119-121, 245, 305
  uptake, 59, 247, 292, 653, 656
5-Hydroxytryptophan (5-HTP), 5, 8, 32, 56, 93, 115, 120, 123, 137, 140, 143-146, 149, 231, 232, 245, 250, 286, 287, 291-296, 314, 347, 348, 352, 378, 383, 407, 408, 591-597, 622
5-hydroxytryptophan decarboxylase, 8, 9, 115-121
5-Hydroxytryptophol, 137-144
Hyperphagia, 378-385, 395
Hyperserotonemia, 57, 59